Porsche 911 Carrera

Service Manual

Coupe, Targa and Cabriolet

1984, 1985, 1986, 1987, 1988, 1989

Bentley Publishers®
.com

BENTLEY PUBLISHERS® | Automotive Reference™

Bentley Publishers, a division of Robert Bentley, Inc.
1734 Massachusetts Avenue
Cambridge, MA 02138 USA
800-423-4595 / 617-547-4170

Information that makes the difference®

BentleyPublishers®.com

Technical contact information

We welcome your feedback. Please submit corrections and additions to our Audi technical discussion forum at:

http://www.BentleyPublishers.com

Updates and corrections

We will evaluate submissions and post appropriate editorial changes online as updates or tech discussion. Appropriate updates and corrections will be added to the book in future printings. Check for updates and corrections for this book before beginning work on your vehicle. See the following web address for additional information:

http://www.BentleyPublishers.com/updates/

WARNING—Important Safety Notice

Do not use this manual for repairs unless you are familiar with basic automotive repair procedures and safe workshop practices. This manual illustrates the workshop procedures for some maintenance and service work. It is not a substitute for full and up-to-date information from the vehicle manufacturer or for proper training as an automotive technician. Note that it is not possible for us to anticipate all of the ways or conditions under which vehicles may be serviced or to provide cautions as to all of the possible hazards that may result.

We have endeavored to ensure the accuracy of the information in this manual. Please note, however, that considering the vast quantity and the complexity of the service information involved, we cannot warrant the accuracy or completeness of the information contained in this manual.

FOR THESE REASONS, NEITHER THE PUBLISHER NOR THE AUTHOR MAKES ANY WARRANTIES, EXPRESS OR IMPLIED, THAT THE INFORMATION IN THIS MANUAL IS FREE OF ERRORS OR OMISSIONS, AND WE EXPRESSLY DISCLAIM THE IMPLIED WARRANTIES OF MERCHANTABILITY AND OF FITNESS FOR A PARTICULAR PURPOSE, EVEN IF THE PUBLISHER OR AUTHOR HAVE BEEN ADVISED OF A PARTICULAR PURPOSE, AND EVEN IF A PARTICULAR PURPOSE IS INDICATED IN THE MANUAL. THE PUBLISHER AND AUTHOR ALSO DISCLAIM ALL LIABILITY FOR DIRECT, INDIRECT, INCIDENTAL OR CONSEQUENTIAL DAMAGES THAT RESULT FROM ANY USE OF THE EXAMPLES, INSTRUCTIONS OR OTHER INFORMATION IN THIS MANUAL. IN NO EVENT SHALL OUR LIABILITY WHETHER IN TORT, CONTRACT OR OTHERWISE EXCEED THE COST OF THIS MANUAL.

Before attempting any work on your Porsche, read the warnings and cautions on pages vii and viii, and any warning or caution that accompanies a procedure in the service manual. Review the warnings and cautions on pages vii and viii each time you prepare to work on your Porsche.

Your common sense and good judgment are crucial to safe and successful service work. Read procedures through before starting them. Think about whether the condition of your car, your level of mechanical skill, or your level of reading comprehension might result in or contribute in some way to an occurrence which might cause you injury, damage your car, or result in an unsafe repair. If you have doubts for these or other reasons about your ability to perform safe repair work on your car, have the work done at an authorized Volkswagen dealer or other qualified shop.

Part numbers listed in this manual are for identification purposes only, not for ordering. Always check with your authorized Porsche dealer to verify part numbers and availability before beginning service work that may require new parts.

Special tools required to perform certain service operations are identified in the manual and are recommended for use. Use of improper tools may be detrimental to the car's safe operation as well as the safety of the person servicing the car.

The vehicle manufacturer will continue to issue service information updates and parts retrofits after the editorial closing of this manual. Some of these updates and retrofits will apply to procedures and specifications in this manual. We regret that we cannot supply updates to purchasers of this manual.

This manual is prepared, published and distributed by Bentley Publishers, 1734 Massachusetts Avenue, Cambridge, Massachusetts 02138 USA. All information contained in this manual is based on the information available to the publisher at the time of editorial closing. Porsche has not reviewed and does not vouch for the accuracy or completeness of the technical specifications and work procedures described and given in this manual.

ISBN: 978-0-8376-1696-4 **Job code: P989-09** Editorial closing 03/2000

Library of Congress Cataloging-in-Publication Data for 2000 edition

Porsche 911 Carrera : service manual : Coupe, Targa, and Cabriolet, 1984, 1985, 1986, 1987, 1988, 1989.
p. cm.
Includes index.
ISBN 0-8376-0291-2 (pbk. : alk. paper)
1. Porsche 911 automobile--Maintenance and repair--Handbooks, manuals, etc. I. Robert Bentley, inc.

TL215.P75 P64 2000
629.29' 722--dc21 00-023006

The paper used in this publication is acid free and meets the requirements of the National Standard for Information Sciences-Permanence of Paper for Printed Library Materials. ∞

Manufactured in the United States of America
202607R001

Selected Books and Repair Information From Bentley Publishers

Engineering

The Hack Mechanic Guide to European Automotive Electrical Systems *Rob Siegel*

ISBN 978-0-8376-1751-0

Mechanical Ignition Handbook: The Hack Mechanic Guide to Vintage Ignition Systems *Rob Siegel*
ISBN 978-0-8376-1767-1

Physics for Gearheads
Randy Beikmann
ISBN 978-0-8376-1615-5

Bosch Automotive Handbook
Robert Bosch GmbH
ISBN 978-0-8376-1732-9

Bosch Fuel Injection and Engine Management *Charles O. Probst, SAE*
ISBN 978-0-8376-0300-1

Maximum Boost: Designing, Testing, and Installing Turbocharger Systems
Corky Bell ISBN 978-0-8376-0160-1

Motorsports

Alex Zanardi: My Sweetest Victory
Alex Zanardi and Gianluca Gasparini
ISBN 978-0-8376-1249-2

The Unfair Advantage *Mark Donohue and Paul van Valkenburgh*
ISBN 978-0-8376-0069-7

Equations of Motion - Adventure, Risk and Innovation
William F. Milliken
ISBN 978-0-8376-1570-7

Audi

Audi A4 Service Manual: 2002-2008, 1.8L Turbo, 2.0L Turbo, 3.0L, 3.2L including Avant and Cabriolet
Bentley Publishers
ISBN 978-0-8376-1574-5

Audi TT Service Manual: 2000-2006, 1.8L turbo, 3.2L, including Roadster and quattro *Bentley Publishers*
ISBN 978-0-8376-1625-4

Audi A6 (C5 platform) Service Manual: 1998-2004, includes A6, allroad quattro, S6, RS6 *Bentley Publishers*
ISBN 978-0-8376-1670-4

Audi A6 Electrical Wiring Manual
Bentley Publishers
ISBN 978-0-8376-0166-3

BMW

Memoirs of a Hack Mechanic
Rob Siegel ISBN 978-0-8376-1720-6

BMW 4 Series (F32, F33, F36) Service Manual: 2014-2016 *Bentley Publishers*
ISBN 978-0-8376-1765-7

BMW X3 (E83) Service Manual: 2004-2010 *Bentley Publishers*
ISBN 978-0-8376-1731-2

BMW X5 (53) Service Manual: 2000-2006 *Bentley Publishers*
ISBN 978-0-8376-1643-8

BMW 3 Series (F30, F31, F34) Service Manual: 2012-2015 *Bentley Publishers*
ISBN 978-0-8376-1752-7

BMW 3 Series (E90, E91, E92, E93) Service Manual: 2006-2011
Bentley Publishers
ISBN 978-0-8376-1723-7

BMW 3 Series (E46) Service Manual: 1999-2005 *Bentley Publishers*
ISBN 978-0-8376-1657-5

BMW Z3 (E36/7) Service Manual: 1996-2002 *Bentley Publishers*
ISBN 978-0-8376-1617-9

BMW 5 Series (E60, E61) Service Manual: 2004-2010 *Bentley Publishers*
ISBN 978-0-8376-1689-6

Corvette

Corvette: America's Star-Spangled Sports Car *Karl Ludvigsen*
ISBN 978-0-8376-1659-9

Corvette by the Numbers: The Essential Corvette Parts Reference 1955-1982 *Alan Covin*
ISBN 978-0-8376-0228-2

Zora Arkus-Duntov: The Legend Behind Corvette *Jerry Burton*
ISBN 978-08376-0858-7

Porsche

Porsche 911 (996) Service Manual: 1999-2005 *Bentley Publishers*
ISBN 978-0-8376-1710-7

Porsche 911 (993) Service Manual: 1995-1998 *Bentley Publishers*
ISBN 978-0-8376-1719-0

Porsche Boxster Service Manual: 1997-2004 *Bentley Publishers*
ISBN 978-0-8376-1645-2

Porsche 911 Carrera Service Manual: 1984-1989 *Bentley Publishers*
ISBN 978-0-8376-1696-4

Porsche 911 SC Service Manual: 1978-1983 *Bentley Publishers*
ISBN 978-0-8376-1705-3

Porsche: Excellence Was Expected
Karl Ludvigsen ISBN 978-0-8376-0235-6

Porsche — Origin of the Species
Karl Ludvigsen ISBN 978-0-8376-1331-4

Volkswagen

Volkswagen Rabbit, GTI Service Manual: 2006-2009 *Bentley Publishers*
ISBN 978-0-8376-1664-3

Volkswagen Jetta Service Manual: 2005-2010 *Bentley Publishers*
ISBN 978-0-8376-1616-2

Volkswagen Jetta, Golf, GTI Service Manual: 1999-2005 *Bentley Publishers*
ISBN 978-0-8376-1678-0

Volkswagen Jetta, Golf, GTI: 1993-1999, Cabrio: 1995-2002 Service Manual *Bentley Publishers*
ISBN 978-0-8376-1660-5

Volkswagen GTI, Golf, Jetta Service Manual: 1985-1992 *Bentley Publishers*
ISBN 978-0-8376-1637-7

Volkswagen Corrado Repair Manual: 1990-1994 *Bentley Publishers*
ISBN 978-0-8376-1699-5

Volkswagen Passat, Passat Wagon Service Manual: 1998-2005
Bentley Publishers
ISBN 978-0-8376-1669-8

MINI Repair Manuals

MINI Cooper Service Manual: 2007-2013 *Bentley Publishers*
ISBN 978-0-8376-1730-5

MINI Cooper Service Manual: 2002-2006 *Bentley Publishers*
ISBN 978-0-8376-1639-1

B BentleyPublishers®
Automotive Reference

Bentley Publishers has published service manuals and automobile books since 1950. For more information, please contact Bentley Publishers at 1734 Massachusetts Avenue, Cambridge, MA 02138 USA, or visit our web site at
BentleyPublishers.com

Foreword

This service manual covers 1984 through 1989 Porsche 911 Carrera models, and is specifically designed to cover only those models built for sale in the United States.

For the Porsche owner with basic mechanical skills, this manual gives detailed maintenance and repair information. In addition, the Porsche owner who has no intention of working on his or her own car will find that owning and reading this manual will make it possible to be better informed and to discuss repairs more intelligently with a professional technician. The aim throughout this manual has been clarity and understanding with practical descriptions, step-by-step procedures, and accurate specifications.

The Porsche owner intending to do maintenance and repair should have a set of tools including a set of metric wrenches and sockets, screwdrivers, a torque wrench, and feeler gauges, since these basic tools will be used to do the majority of the maintenance and repair procedures described in this manual. This manual includes detailed information on these basic tools and other tips for the beginner in the first section of the manual, entitled **010 General**. For some of the repairs described in this manual, Porsche technicians use special tools. The text will note when a repair requires these special tools and, where possible, will recommend practical alternatives.

We have endeavored to ensure the highest degree of accuracy possible. When the vast array of data presented in the manual is taken into account, however, no claim to infallibility can be made. We therefore cannot be responsible for the result of any errors that may have crept into the text. The publisher encourages comments from the readers of this manual with regard to any errors and, also, suggestions for improvement in the presentation of technical material. These communications have been and will be carefully considered in the preparation of future printings of this and other manuals. Please contact Bentley Publishers using the contact information on the copyright page at the beginning of this manual.

Porsche offers extensive warranties, especially on components of the fuel delivery and emissions control systems. Therefore, before deciding to repair a Porsche that may still be covered wholly or in part by any warranties issued by Porsche Cars North America Inc., consult your authorized Porsche dealer. You may find that he can make the repair either free or at minimum cost.

Regardless of its age and whether or not it is still protected by warranty, your Porsche is an easy car to get serviced. So if at any time a repair is needed that you feel is too difficult to do yourself, a trained Porsche technician is ready to do the job for you. Each authorized Porsche dealer service department has made a significant investment in service and diagnostic test equipment, specials tools, and Porsche original parts. It is also the best source of the most up-to-date repair and service techniques, which includes factory training and technical literature. Servicing your Porsche through an authorized Porsche dealer will insure that your investment will be protected while maintaining the highest degree of service standards.

Bentley Publishers

Please read these warnings and cautions before proceeding with maintenance and repair work.

WARNING—

- See also Cautions on page viii

- Some repairs may be beyond your capability. If you lack the skills, tools and equipment, or a suitable workplace for any procedure described in this manual, we suggest you leave such repairs to an authorized Porsche dealer service department or other qualified shop.

- Do not re-use any fasteners that are worn or deformed in normal use. Many fasteners are designed to be used only once and become unreliable and may fail when used a second time. This includes, but is not limited to, nuts, bolts, washers, self-locking nuts or bolts, circlips and cotter pins. Always replace these fasteners with new parts.

- Never work under a lifted car unless it is solidly supported on stands designed for the purpose. Do not support a car on cinder blocks, hollow tiles or other props that may crumble under continuous load. Never work under a car that is supported solely by a jack. Never work under the car while the engine is running.

- If you are going to work under a car on the ground, make sure that the ground is level. Block the wheels to keep the car from rolling. Disconnect the battery negative (–) terminal (ground strap) to prevent others from starting the car while you are under it.

- Never run the engine unless the work area is well ventilated. Carbon monoxide kills.

- Finger rings, bracelets and other jewelry should be removed so that they cannot cause electrical shorts, get caught in running machinery, or be crushed by heavy parts.

- Tie long hair behind your head. Do not wear a necktie, a scarf, loose clothing, or a necklace when you work near machine tools or running engines. If your hair, clothing, or jewelry were to get caught in the machinery, severe injury could result.

- Do not attempt to work on your car if you do not feel well. You increase the danger of injury to yourself and others if you are tired, upset or have taken medication or any other substance that may keep you from being fully alert.

- Illuminate your work area adequately but safely. Use a portable safety light for working inside or under the car. Make sure the bulb is enclosed by a wire cage. The hot filament of an accidentally broken bulb can ignite spilled fuel or oil.

- Catch draining fuel, oil, or brake fluid in suitable containers. Do not use food or beverage containers that might mislead someone into drinking from them. Store flammable fluids away from fire hazards. Wipe up spills at once, but do not store the oily rags, which can ignite and burn spontaneously.

- Always observe good workshop practices. Wear goggles when you operate machine tools or work with battery acid. Gloves or other protective clothing should be worn whenever the job requires working with harmful substances.

- Greases, lubricants and other automotive chemicals contain toxic substances, many of which are absorbed directly through the skin. Read the manufacturer's instructions and warnings carefully. Use hand and eye protection. Avoid direct skin contact

- Disconnect the battery negative (–) terminal (Ground strap) whenever you work on the fuel system or the electrical system. Do not smoke or work near heaters or other fire hazards. Keep an approved fire extinguisher handy.

- Friction materials (such as brake pads or shoes or clutch discs) contain asbestos fibers or other friction materials. Do not create dust by grinding, sanding, or by cleaning with compressed air. Avoid breathing dust. Breathing any friction material dust can lead to serious diseases and may result in death.

- Batteries give off explosive hydrogen gas during charging. Keep sparks, lighted matches and open flame away from the top of the battery. If hydrogen gas escaping from the cap vents is ignited, it will ignite gas trapped in the cells and cause the battery to explode.

- Connect and disconnect battery cables, jumper cables or a battery charger only with the ignition switched off, to prevent sparks. Do not disconnect the battery while the engine is running.

- Do not quick-charge the battery (for boost starting) for longer than one minute. Wait at least one minute before boosting the battery a second time.

- Do not allow battery charging voltage to exceed 16.5 volts. If the battery begins producing gas or boiling violently, reduce the charging rate. Boosting a sulfated battery at a high charging rate can cause an explosion.

- The air conditioning system is filled with chemical refrigerant, which is hazardous. The A/C system should be serviced only by trained technicians using approved refrigerant recovery/recycling equipment, trained in related safety precautions, and familiar with regulations governing the discharging and disposal of automotive chemical refrigerants.

- Do not expose any part of the A/C system to high temperatures such as open flame. Excessive heat will increase system pressure and may cause the system to burst.

- Some aerosol tire inflators are highly flammable. Be extremely cautious when repairing a tire that may have been inflated using an aerosol tire inflator. Keep sparks, open flame or other sources of ignition away from the tire repair area. Inflate and deflate the tire at least four times before breaking the bead from the rim. Completely remove the tire from the rim before attempting any repair.

- The ignition system produces high voltages that can be fatal. Avoid contact with exposed terminals and use extreme care when working on a car with the engine running or the ignition switched on.

- Place jack stands only at locations specified by the manufacturer. The vehicle lifting jack supplied with the vehicle is intended for tire changes only. A heavy duty floor jack should be used to lift vehicle before installing jack stands. See **010 General**.

- Battery acid (electrolyte) can cause severe burns. Flush contact area with water, seek medical attention.

- Aerosol cleaners and solvents may contain hazardous or deadly vapors and are highly flammable. Use only in a well ventilated area. Do not use on hot surfaces (engines, brakes, etc.).

continued on next page

Please read these warnings and cautions before proceeding with maintenance and repair work.

CAUTION—

- See also Warnings on page vii

- If you lack the skills, tools and equipment, or a suitable workshop for any procedure described in this manual, we suggest you leave such repairs to an authorized Porsche dealer or other qualified shop.

- Porsche is constantly improving its cars and sometimes these changes, both in parts and specifications, are made applicable to earlier models. Therefore, part numbers listed in this manual are for reference only. Always check with your authorized Porsche dealer parts department for the latest information.

- Before starting a job, make certain that you have all the necessary tools and parts on hand. Read all the instructions thoroughly, do not attempt shortcuts. Use tools appropriate to the work and use only replacement parts meeting Porsche specifications. Makeshift tools, parts and procedures will not make good repairs.

- Use pneumatic and electric tools only to loosen threaded parts and fasteners. Never use these tools to tighten fasteners, especially on light alloy parts. Always use a torque wrench to tighten fasteners to the tightening torque specification listed.

- Be mindful of the environment and ecology. Before you drain the crankcase, find out the proper way to dispose of the oil. Do not pour oil onto the ground, down a drain, or into a stream, pond or lake. Dispose of in accordance with Federal, State and Local laws.

- Always make sure ignition is off before disconnecting battery.

- Label battery cables before disconnecting. On some models, battery cables are not color coded.

- Disconnecting the battery may erase fault code(s) stored in control module memory. Using special Porsche diagnostic equipment, check for fault codes prior to disconnecting the battery cables.

- If a normal or rapid charger is used to charge battery, the battery must be disconnected and removed from the vehicle in order to avoid damaging paint and upholstery.

- Do not quick-charge the battery (for boost starting) for longer than one minute. Wait at least one minute before boosting the battery a second time.

- Connect and disconnect a battery charger only with the battery charger switched off.

- Sealed or "maintenance free" batteries should be slow-charged only, at an amperage rate that is approximately 10% of the battery's ampere-hour (Ah) rating.

- Do not allow battery charging voltage to exceed 16.5 volts. If the battery begins producing gas or boiling violently, reduce the charging rate. Boosting a sulfated battery at a high charging rate can cause an explosion.

010 General

TABLES

GENERAL

This section of the manual is intended to help the beginner get started. To begin with there is a discussion on **How to Use This Manual**. Tips on mechanic's skills and workshop practices that can be found under **Getting Started.** The basic tools needed to do most of the procedures in this manual are found under **Tools**. The section ends with a quick reference guide to **Emergencies**, including basic troubleshooting information.

> ***WARNING —***
> *Do not use this manual unless you are familiar with basic automotive repair procedures and safe workshop practices. This manual illustrates the workshop procedures required for most service work; it is not a substitute for full and up-to-date information from the vehicle manufacturer or for proper training as an automotive technician. Note that it is not possible for us to anticipate all of the ways or conditions under which vehicles may be serviced or to provide cautions as to all of the possible hazards that may result.*

> ***WARNING —***
> *Your common sense and good judgment are crucial to safe and successful service work. Read procedures through before starting them. Think about whether the condition of your car, your level of mechanical skill, or your level of reading comprehension might result in or contribute in some way to an occurrence that might cause you injury, damage your car, or result in an unsafe repair. If you have doubts for these or other reasons about your ability to perform safe repair work on your car, have the work done at an authorized Porsche dealer or other qualified shop.*

HOW TO USE THIS MANUAL

The manual is divided into ten sections:

0 GENERAL AND MAINTENANCE
1 ENGINE
2 ENGINE MANAGEMENT, EXHAUST AND ENGINE ELECTRICAL
3 TRANSMISSION
4 SUSPENSION, BRAKES AND STEERING
5 BODY–ASSEMBLY
6 BODY–COMPONENTS AND ACCESSORIES
7 BODY–TRIM, SEATS
8 HEATING, VENTILATION, AIR CONDITIONING
9 ELECTRICAL SYSTEM

0 General and Maintenance covers the recommended maintenance schedules and service procedures as well as a section covering basic instructions, tips and helpful hints for the do-it-yourselfer.

The next nine sections are repair based and are further broken down into three digit repair groups. Each major section begins with a **General** repair group, e.g. **100 Engine–General.** These "00" (double zero) groups are mostly descriptive in nature, covering topics such as theory of operation and troubleshooting. The remainder of the repair groups contain repair information. The last major section contains detailed electrical system information, wiring diagram schematics, and electrical component locations.

A master listing of the ten major sections and the corresponding individual repair groups can be found on the inside front cover.

Each repair group begins with a table of contents listing the major subject headings within the group. Page numbers throughout the manual are organized according to the repair group system. For example, you can expect to find repair information on steering (Repair Group **480**) beginning on page 480-1. A comprehensive index can be found at the back of the manual.

Warnings, cautions and notes

Throughout this manual are many passages with the headings **WARNING**, **CAUTION** or **NOTE**. These important messages have different meanings.

> ***WARNING —***
> *The text under this heading warns of unsafe practices that are very likely to cause injury, either by direct threat to the person(s) performing the work or by increased risk of accident or mechanical failure while driving.*

> ***CAUTION —***
> *A caution calls attention to important precautions to be observed during the repair work that will help prevent accidentally damaging the car or its parts.*

NOTE —
A note contains additional information or tips intended to clarify the repair procedure or aid in doing a better job or completing it more easily.

Always read each **WARNING**, **CAUTION** and **NOTE** at the front of the manual and when they appear in repair procedures. These messages are important. Read them before you begin any maintenance or repair job.

GETTING STARTED

Most of the necessary maintenance and minor repair that an automobile will need can be done with ordinary tools, even by owners with little or no experience in car repair. Below is some important information on how to work safely, a discussion of what tools will be needed and how to use them.

Safety

Although an automobile presents many hazards, common sense and good equipment can help ensure safety. Many accidents happen because of carelessness. Pay attention and stick to these few important safety rules.

WARNING —

- *Never run the engine in the work area unless it is well-ventilated. The exhaust should be vented to the outside. Carbon monoxide (CO) in the exhaust kills.*
- *Remove all neckties, scarfs, loose clothing, or jewelry when working near running engines or power tools. Tuck in shirts. Tie long hair and secure it under a cap. Severe injury can result from these things being caught in rotating parts.*
- *Remove rings, watches, and bracelets. Aside from the dangers of moving parts, metallic jewelry conducts electricity and may cause shorts, sparks, burns, or damage to the electrical system when accidentally contacting the battery or other electrical terminals.*
- *Disconnect the battery negative (–) cable whenever working on or near the fuel system or anything electrically powered.*
- *Never work under a lifted car unless it is solidly supported on jack stands. Do not support a car on cinder blocks, bricks, or other objects that may shift or crumble under continuous load. Never work under a car that is supported only by a jack.*
- *Keep sparks, lighted matches, and any open flame away from the top of the battery. Hydrogen gas emitted by the battery is highly flammable. Any nearby source of ignition may cause the battery to explode.*
- *Never lay tools or parts in the engine compartment or on top of the battery. They may fall into confined spaces and be difficult to retrieve, become caught in belts or other rotating parts when the engine is started, or cause electrical shorts and damage to the electrical system.*
- *Illuminate the work area adequately and safely. A fluorescent type light is best because it gives off less heat. If using a light with a normal incandescent bulb, use rough service bulbs to avoid breakage. The hot filament of an accidentally broken bulb can ignite spilled fuel or oil.*
- *The fuel system is designed to retain pressure even when the ignition is off. When working with the fuel system, loosen the fuel lines slowly to allow the residual pressure to dissipate gradually. Take precautions to avoid spraying fuel.*
- *Fuel is highly flammable. When working around fuel, do not smoke or work near heaters or other fire hazards. Keep an approved fire extinguisher handy.*

Lifting the car

For repairs that require raising the car, the specified jacking points should be used to raise the car safely and avoid damage. There are two side jacking points to temporarily lift the vehicle for tire changing. The jack supplied with the car by Porsche can only be used at the vehicle side points. When using a floor jack or automotive lift, the four jacking point closer to the wheels should be used.

> ***WARNING —***
> *Watch the jack closely. Make sure it stays stable and does not shift or tilt. As the car is raised, the car may roll slightly and the jack may shift.*

Raising car safely using vehicle jack

– Park car on flat, level surface.

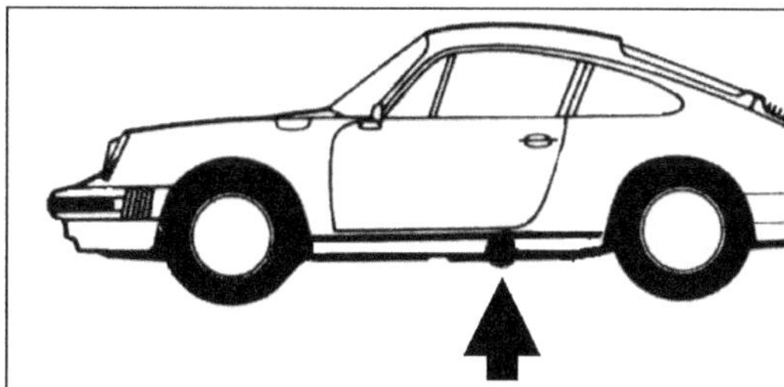

0019454

< Remove plastic plug from jack receptacle in rocker panel (**arrow**). Place jack fully into square receptacle in rocker panel. Make sure jack is resting on flat, solid ground. Use board or other support to provide a firm surface for jack, if necessary.

– Raise car slowly with jack while constantly checking position of jack and car.

– Once car is raised, block wheel on opposite side of vehicle. to prevent car from unexpectedly rolling.

> ***WARNING —***
> - *Do not rely on the transmission or the emergency brake to keep the car from rolling. They are not a substitute for positively blocking the opposite wheel.*
> - *Never work under a car that is supported only by a jack. Use jack stands that are properly designed to support the car. See* **Tools**.

Working under car safely

Lifting or raising the vehicle for repairs requires a hydraulic jack or lifting equipment. In addition to the instructions below, follow the manufacturer's directions for the use of the lifting equipment.

– Disconnect negative (–) battery cable.

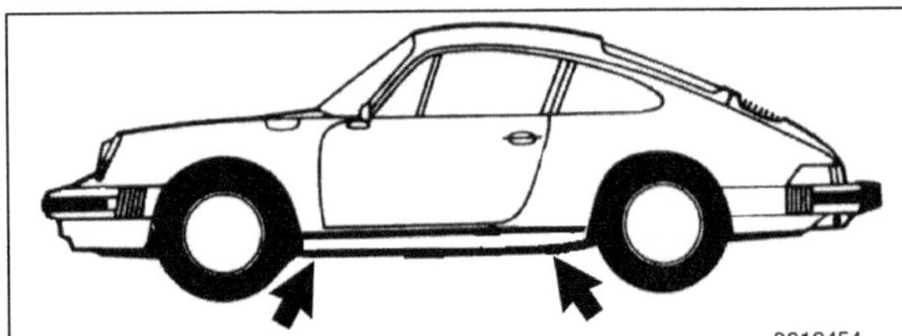

0019454

◀ Position jack or lift arms beneath specified lifting points (**arrows**). Make sure jack is resting on flat, solid ground.

> ***WARNING —***
> *When raising the car using a floor jack or a hydraulic lift, carefully position the jack pad to prevent damaging the car body. A suitable liner (wood, rubber, etc.) should be placed between the jack and the car to prevent body damage.*

– Raise car slowly while constantly checking position of lifting equipment and car.

– Use at least two jack stands to support the car. A jack is a temporary lifting device and should not be used alone to support the car while you are under it. Use jack stands designed for the purpose of supporting a car. For more information on jack stands, see **Tools** below.

> ***WARNING —***
> *Do not use wood, concrete blocks, or bricks to support a car. Wood may split. Blocks or bricks, while strong, are not designed for that kind of load, and may break or collapse.*

– Place jack stands on firm, solid surface. If necessary, use a flat board or similar solid object to provide a firm footing.

– Lower car slowly until its weight is fully supported by jack stands. Watch to make sure that the jack stands do not tip or lean as the car settles on them. Raise the jack again for additional support.

– Observe all jacking precautions when raising car to remove jack stands.

ADVICE FOR THE BEGINNER

The tips that follow are general advice to help the do-it-yourself Porsche owner perform repairs and maintenance tasks more easily and more professionally.

Planning ahead

Most of the repairs and maintenance tasks described in this manual can be successfully completed by anyone with basic tools and abilities. To prevent getting in too deep, know what the whole job requires before starting. Read the procedure thoroughly from beginning to end to know just what to expect and what parts will have to be replaced.

Cleanliness

Keeping things organized, neat, and clean is essential to doing a good job.

0019513

◄ When working in the engine compartment, a fender cover will protect the finish from scratches and other damage. Make sure the car is relatively clean so that dirt under the cover does not scratch the finish. Most custom covers have storage pockets for holding removed fasteners and parts and a special pad (**arrow**) for the A/C compressor. Removing the A/C compressor and laying it aside without disconnecting the refrigerant lines is an easy procedure, and is required for certain procedures.

Any repair job will be less troublesome if the parts are clean. For cleaning old parts, there are many solvents and parts cleaners commercially available.

For cleaning parts prior to assembly, commercially available aerosol parts cleaner or brake cleaner are handy to use, and the cleaner will evaporate completely.

> ***WARNING —***
> *Virtually all solvents used for cleaning parts are highly flammable and hazardous to health, especially in aerosol form. Use with extreme care. Do not smoke. Do not use these products indoors or near any source of heat, sparks or flame.*

Let any solvent or cleaning product dry completely. Low-pressure, dry compressed air is helpful if available. Also, use only lint-free rags for cleaning and drying.

> ***WARNING —***
> *Avoid getting tools or clothing near the battery. Battery electrolyte is a corrosive acid. Be careful with brake fluid, as it can damage the car's paint. Finally, keep rubber parts such as hoses and belts free from oil or gasoline, as they will cause the material to soften and fail prematurely.*

Non-reusable fasteners

Many fasteners used on the cars covered by this manual must be replaced with new ones once they are removed. These include but are not limited to: bolts, nuts (self-locking, nylock, etc.), cotter pins, studs, roll pins, clips and washers. Genuine Porsche parts should be the only replacement parts used for this purpose.

Some bolts are designed to stretch during assembly and are permanently altered rendering them unreliable once removed. These are known as torque-to-yield fasteners. Always replace fasteners where instructed to do so. Failure to replace these fasteners could cause vehicle damage and personal injury. See an authorized Porsche dealer for applications and ordering information.

Tightening fasteners

When tightening the bolts or nuts that attach a component, it is always good practice to tighten the bolts gradually and evenly to avoid misalignment or over stressing any one portion of the component. For components sealed with gaskets, this method helps to ensure that the gasket will seal properly and completely.

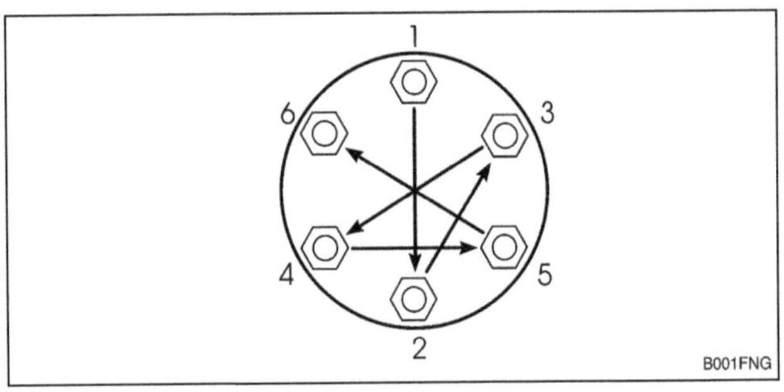

◄ Where there are several fasteners, tighten them in a sequence alternating between opposite sides of the component. Repeat the sequence in stages until all the bolts are evenly tightened to the proper specification.

For some repairs a specific tightening sequence is necessary, or a particular order of assembly is required. Such special conditions are noted in the text, and the necessary sequence is described or illustrated. Where no specific torque is listed, **Table a** can be used as a general guide for tightening fasteners.

> ***WARNING —***
> **Table a** *is a general reference only. The values listed in the table are not intended to be used as a substitute for torques specifically called out in the text throughout this manual.*

NOTE —

- *Metric bolt classes or grades are marked on the bolt head.*
- *Do not confuse wrench size with bolt diameter size. For a listing of the common wrenches used on various bolt diameters, see* **Basic tool requirements**.

Table a. General Bolt Tightening Torques in Nm (max. permissible)

Bolt diameter	Bolt Class (according to DIN 267)					
	5.6	**5.8**	**6.8**	**8.8**	**10.9**	**12.9**
M5	2.5	3.5	4.5	6	8	10
M6	4.5	6	7.5	10	14	17
M8	11	15	18	24	34	40
M10	23	30	36	47	66	79
M12	39	52	62	82	115	140
M14	62	82	98	130	180	220
M16	94	126	150	200	280	340
M18	130	174	210	280	390	470

Gaskets and seals

Gaskets should not be reused. Once a gasket has been used, it is no longer capable of making as good a seal as when new, and is much more likely to leak. For this reason, Always plan to use new gaskets for any reassembly. Some gaskets are directional. Make sure that these are installed correctly. This same logic applies to any part used for sealing, including rubber O-rings and copper sealing washers.

In places where a shaft must pass through a housing, flexible lip seals are used to keep the lubricating oil or grease from leaking out past the rotating shaft. Seals should never be reused once they have been removed. When removing a seal, be careful not to scratch or otherwise damage the metal surfaces.

When installing a new seal, it a good idea to coat the seal with oil to aid installation. Some seals are directional and special installation instructions apply. Make sure a seal is installed with the lip facing the correct way. Normally the lip faces the inside. Note the installation direction of the old seal before removing it.

Electrical testing

Many electrical problems can be understood and solved with only a little fundamental knowledge of how electrical circuits function.

Electric current only flows in a complete circuit. To operate, every electrical device in the car requires a complete circuit including a voltage source and a path to ground. The positive (+) side of the battery is the original voltage source, and ground is any return path to the negative (–) side of the battery, whether through the wiring harness or the car body. Except for portions of the charging system, all electrical current in the car is direct current (DC) and flows from positive (+) to negative (–).

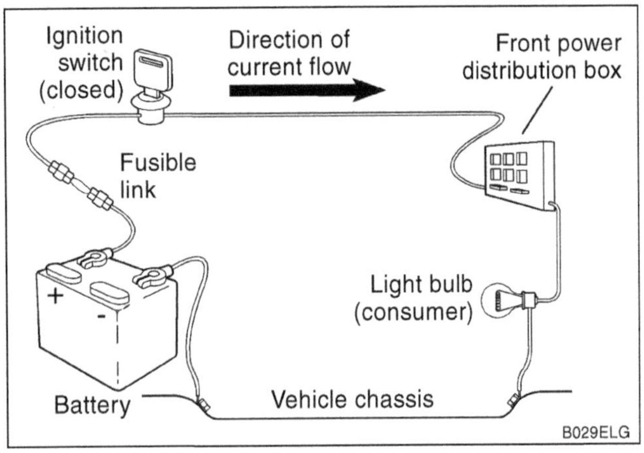

Switches are used to turn components on or off by completing or interrupting the circuit. A switch is "open" when the circuit is interrupted, and "closed" when the circuit is completed. See **900 Electrical System–General** for electrical troubleshooting.

Wire repairs

Repairs to a wiring harness require special care to make the repair permanent. The wire ends must be clean. If frayed or otherwise damaged, cut off the end. If the wire is too short, splice in a new piece of wire of the same size and make two connections.

Use connectors designed for the application. Crimped-on or soldered-on connectors are best. Crimp connectors and special crimping pliers are widely available. If soldering, use needlenose pliers to hold the wire near the solder joint and create a "heat dam". This keeps the heat and the solder from traveling up the wire. Always use a solder made specifically for electrical work (rosin core).

NOTE —
Twisting wires together to make a repair is not recommended. Corrosion and vibration will eventually spoil the connection and may lead to irreparable damage to sensitive electronic components.

Insulate the finished connection. Electronics stores can supply heat-shrinkable insulating tubing that can be placed onto the wire before connecting, slid over the finished joint, and shrunk to a tight fit with a heat gun or hair dryer. The next best alternative is electrical tape. Make sure the wire is clean and free of solder flux or other contamination. Wrap the joint tightly to seal out moisture. See **900 Electrical System–General** for more information.

BUYING PARTS

Many of the maintenance and repair tasks in this manual call for the installation of new parts, or the use of new gaskets and other materials when reinstalling parts. Most often, the parts that will be needed should be on hand before beginning the job. Read the introductory text and the complete procedure to determine which parts will be needed.

NOTE —
For some bigger jobs, partial disassembly and inspection are required to determine a complete parts list. Read the procedure carefully and, if necessary, make other arrangements to get the necessary parts while your car is disassembled.

Genuine Porsche parts

Genuine Porsche replacement parts from an authorized Porsche dealer are designed and manufactured to the same high standards as the original parts. They will be the correct material, manufactured to the same specifications, and guaranteed to fit and work as intended by the engineers who designed the car. Some genuine Porsche parts have a limited warranty.

Porsche is constantly updating and improving their cars, often making improvements during a given model year. Porsche may recommend a newer, improved part as a replacement, and your authorized dealer's parts department will know about it and provide it. The Porsche parts organization is best equipped to deal with any Porsche parts needs.

Non-returnable parts

Some parts cannot be returned for credit, even if they are the wrong parts for the car. The best example is electrical parts, which are almost universally considered non-returnable because they are so easily damaged internally.

Buy electrical parts carefully, and be as sure as possible that a replacement is needed, especially for expensive parts such as electronic control modules. It may be wise to let an authorized Porsche dealer or other qualified shop confirm your diagnosis before replacing an expensive part that cannot be returned.

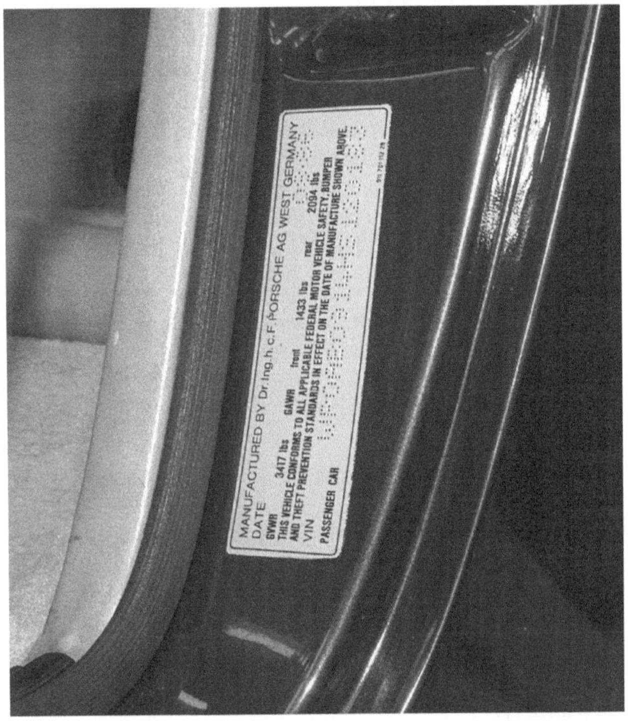

0019509

Information you need to know

Model. When ordering parts it is important that you know the correct model designation for your car. Models covered in this manual are 911 Carrera coupe, targa and cabriolet configurations.

Model Year. This is not necessarily the same as date of manufacture or date of sale. A 1986 model may have been manufactured in late 1985, and perhaps not sold until early 1986. It is still a 1986 model. Model years covered by this manual are 1984 to 1989.

Date of Manufacture. This information is helpful when ordering replacement parts or determining if any of the warranty recalls are applicable to your car. The label on the driver's door pillar (**arrow**) will specify the month and year that the car was built.

Vehicle Identification Number (VIN). This is a combination of letters and numbers that identify the particular car. The VIN appears on the state registration document, and on the car itself. One location is in on the left windshield pillar, another is on the luggage compartment lid (**arrow**). The sticker on the luggage compartment lid contains the following information:

1. Vehicle Identification Number
2. Vehicle code
3. Engine and transmission code
4. Paint and interior code
5. Option Codes

A duplicate of this label is also contained in the vehicle's original paper work (Warranty and Maintenance card).

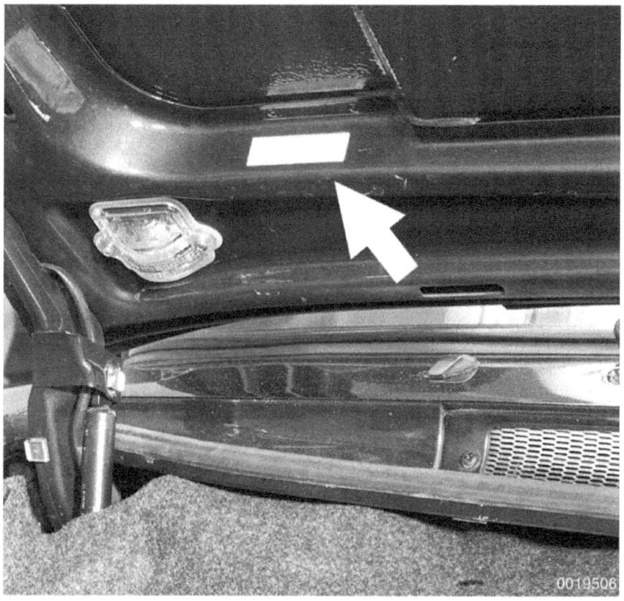
0019506

◀ As of 1987 model year, the National Highway Traffic Safety Administration (NHTSA) requires passenger cars with a high theft rate to have the VIN marked on specific parts of the car when manufactured. On Porsche cars, these parts are identified by an adhesive label bearing the VIN and Porsche script. Replacement parts have a similar label, bearing the letters DOT-R. These labels should not be removed as they will tear apart or painted over.

0019512

◀ **Engine number**. The engine number is stamped on the left crankcase half, below the fan housing (**arrow**). 911 cars covered in this manual are powered by a 6-cylinder engine. For information on engine codes and engine applications, see **100 Engine–General**.

Engine Numbers

- 1984-1986 (930/21) . . . . .200 hp (149 kW)
- 1987-1989 (930/25) . . . . .216 hp (160 kW)

0019510

◀ **Paint code**. The paint code is on a sticker on the right fender inside the luggage compartment (**arrow**).

Transmission. The transmission type with its identifying code may be important when buying clutch parts, seals, gaskets, and other transmission-related parts. Two transmissions were installed in the models covered by this manual, depending on model year.

Transmissions

- 1984-1986
  with cable-operated clutch . . . . 915, 5-speed manual
- 1987-1989
  with hydraulic clutch operation . . . . G50 5- speed manual

The transmission number is an eight digit number, i.e. **73 G 349533**

Digit 1	6-cyl. transmission (**7**)
2	5-speed (**3** or **4**)
3	Model year (**G**=1986)
Digits 4 - 8	Serial number (**349533**)

For more information, see **300 Transmission and Clutch–General**.

SERVICE

Porsche dealers are uniquely qualified to provide service for Porsche cars. Their authorized relationship with the large Porsche service organization means that they are constantly receiving new tools and equipment, together with the latest and most accurate repair information.

The Porsche dealer's service technicians are highly trained and very capable. Unlike most independent repair shops, authorized Porsche dealers are intensely committed to supporting the Porsche product. They share the owner's interest in Porsche value, performance, and reliability. On the other hand, there are many independent shops that specialize in Porsche service and are capable of doing high quality repair work. Checking with other Porsche owners for recommendations on service facilities is a good way to learn of reputable Porsche shops in your area.

Compression test

A test of compression pressures in the individual cylinders will tell a lot about the condition of the engine without the need for taking it apart. The test is relatively simple. It requires a compression tester, spark plug wrench, and a few hand tools.

To obtain accurate results, the battery and starter must be capable of turning the engine at normal cranking speed. The area around the spark plugs should be clean, to avoid getting debris inside the engine when the spark plugs are removed. Because engine temperature may affect compression, the most accurate results are obtained when the engine is at normal operating temperature.

Compression test

< Disable ignition system and fuel pump by removing harness connector from DME main/fuel pump relay (**2**), located beneath driver's seat.

> ***WARNING —***
> *The ignition system produces high voltages that can be fatal. Avoid contact with exposed terminals and use caution when working on a car with the ignition switched on or the engine running.*

– Remove heater blower elbow on left side of engine and air filter cover and filter element on right side.

– Remove and label spark plug wires.

– Remove and number spark plugs.

NOTE —
Used spark plugs should be installed in the same cylinder from which they were removed.

– Thread compression tester into cylinder no. 1 spark plug hole tight enough to seal hole.

– With parking brake set, transmission in neutral, and throttle held wide open, crank engine with starter. Crank engine for 4-5 full revolutions. Record value.

NOTE —
The gauge reading should increase with each engine revolution. The engine should be cranked an equal number of revolutions at each cylinder to obtain the most accurate readings.

– Remove gauge from cylinder no. 1 and release pressure from gauge.

– Repeat test for other cylinders. Record gauge reading for each cylinder.

– Compare readings from each cylinder.

NOTE —
Removing the spark plugs may cause built-up carbon on the plug to fall onto the exhaust valve seat. If this happens, the exhaust valve may not seat fully, resulting in a low compression reading at that cylinder. If you suspect a partially open valve, reinstall the plug and run the engine at 2,500 rpm for about 1 minute. Remove the plug and repeat the test.

– Reinstall spark plugs and spark plug wires in their original locations. The remainder of installation is reverse of removal.

Tightening Torque
- Spark plug to cylinder head . . . . 25 Nm (18 ft-lb)

Low compression is evidence of poorly sealed combustion chambers. Generally, compression pressures that are relatively even but low indicate worn piston rings and/or cylinder walls. Erratic values tend to indicate valve leakage or a broken cylinder head stud. There are other tests that can further isolate the problem.

Wet compression test

To further help analyze the source of poor compression, repeat the compression test, this time with about a teaspoon of oil squirted into each cylinder. The oil will temporarily help seal between the piston rings and the cylinder wall, practically eliminating leakage past the rings for a short time. If this test yields higher compression reading than the "dry" compression test, there is probably leakage between the piston rings and cylinder walls, due either to wear or to broken piston rings. little or no change in compression reading indicates other leakage, probably from the valves.

Leak-down test

The most conclusive diagnosis of low compression symptoms requires a leak-down test. Using a special tester and a supply of compressed air, each cylinder is pressurized. The rate at which the air leaks out of the cylinder, as well as where the air leaks out, can accurately pinpoint the magnitude and location of the leakage. Any engine compression diagnosis that will require major disassembly should first be confirmed by the more accurate leak-down test. Because the test requires special equipment and experience, it may be desirable to have it done by a Porsche dealer or other qualified repair shop.

TOOLS

Most maintenance can be accomplished with a small selection of the right tools. Tools range in quality from inexpensive junk, which may break at first use, to very expensive and well-made tools for the professional.

Many reputable tool manufacturers offer good quality, moderately priced tools with a lifetime guarantee. These are your best buy. They cost a little more, but they are good quality tools that will do what is expected of them.

Basic tool requirements

The basic hand tools described below can be used to accomplish most of the simple maintenance and repair tasks.

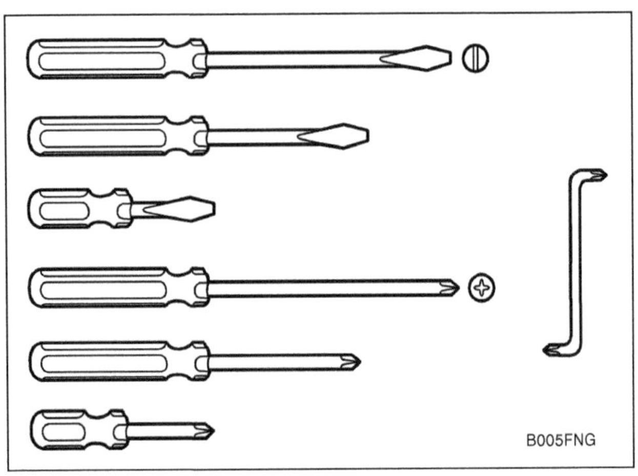

Screwdrivers. The common flat-blade type and the Phillips type will handle almost all screws used on Porsches. Two or three different sizes of each type will be required, since a screwdriver of the wrong size will damage the screw head.

A complete set of screwdrivers should also include Torx® type screwdrivers.

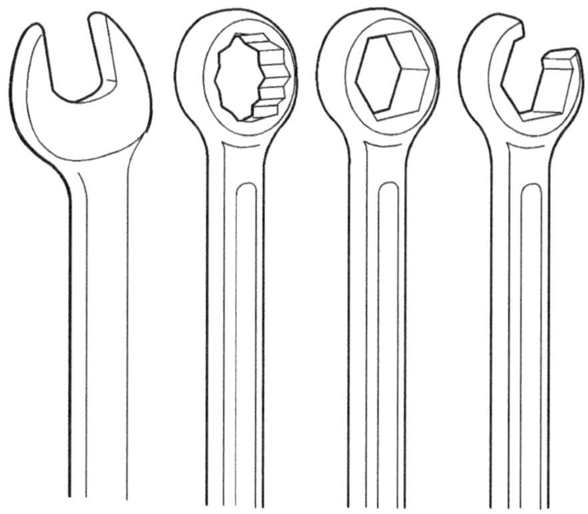

B002FNG

Wrenches. Wrenches come in different styles for different uses. The basic open-end wrench is the most widely used, but grips on only two sides. It can spread apart and slip off more easily. The box-end wrench has better grip, on all six sides of a nut or bolt.

A 12-point box-end can loosen a nut or bolt where there is less room for movement, while a 6-point box-end provides better grip. For hex fasteners on fluid lines, like brake lines and fuel lines, a flare-nut wrench offers the advantages of a box-end wrench with a slot that allows it to fit over the line.

The combination wrench is the most universal. It has one open-end and one box-end. 10 mm and 13 mm wrenches are the most common sizes needed. A more complete set of wrenches would include 6mm through 19 mm sizes.

Sockets. Sockets can be combined with extensions and universal joints (swivels) to reach fasteners more easily than hand wrenches. Sockets come with different drive size connections. The most common drive sizes are ¼ in., 3/8 in., and ½ in.

Sockets come in 6-point and 12-point styles. For use with a ratchet, the 6-point offers a better grip on tight nuts and bolts. 6 mm to 19 mm sockets are the most needed sizes. Below is a list of typical bolt diameters and the corresponding wrench sizes.

Fastener Diameters and Wrench Sizes

Fastener	Wrench size
• M5	8 mm
• M6	10 mm
• M8	12 mm, 13 mm
• M10	14 mm, 15 mm, 17 mm
• M12	19 mm
• M14	22 mm

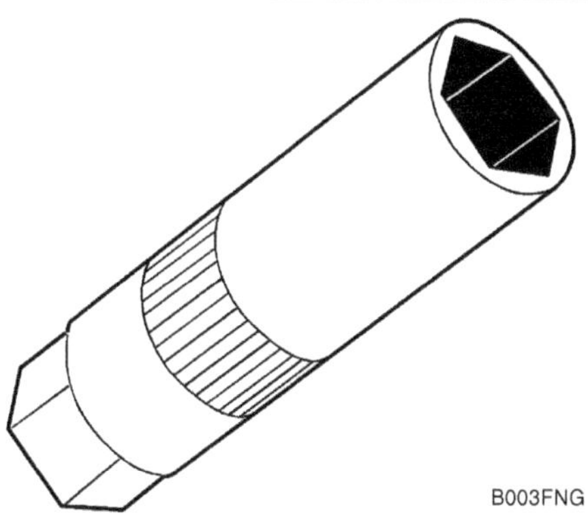

B003FNG

Spark Plug Socket. A special 13/16 in. socket for spark plugs is the correct size, is deep enough to accommodate a spark plug's length, and includes a rubber insert to both protect the spark plug from damage and grip it for easier removal.

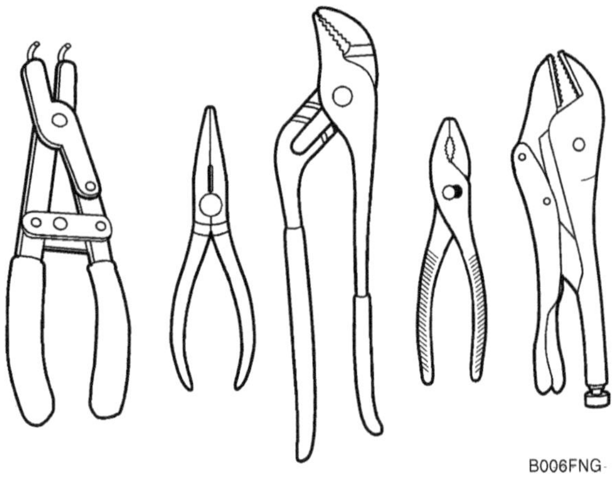

B006FNG

Pliers. Most are used for holding irregular objects, bending, or crimping. Some have special applications.

A needlenose plier is used for gripping small and poorly accessible objects, and is useful for wiring and other electrical work. A locking plier such as the Vise-Grip® is useful because of its tight grip.

Snap-ring and circlip pliers with special tipped jaws are used to remove and install snap-rings or circlips. A Channellock® or water pump plier has adjustable jaws that can be quickly changed to match the size of the object being held to give greater leverage.

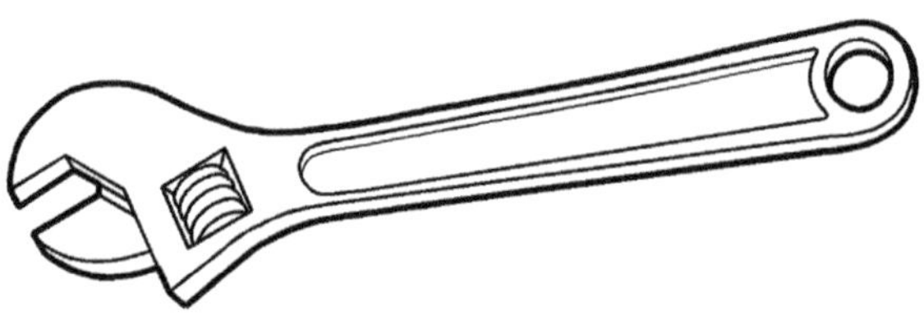

B008FNG

An adjustable wrench can be a useful addition to a small tool kit. It can substitute in a pinch, if two wrenches of the same size are needed to remove a nut and bolt. Use extra care with adjustable wrenches, as they tend to loosen, slip and damage fasteners.

Compared to a wrench of the correct size, an adjustable wrench is always second best. They should only be used when the correct size wrench is not available. Choose one of average size range, about 6 to 8 inches in length.

Jack stands

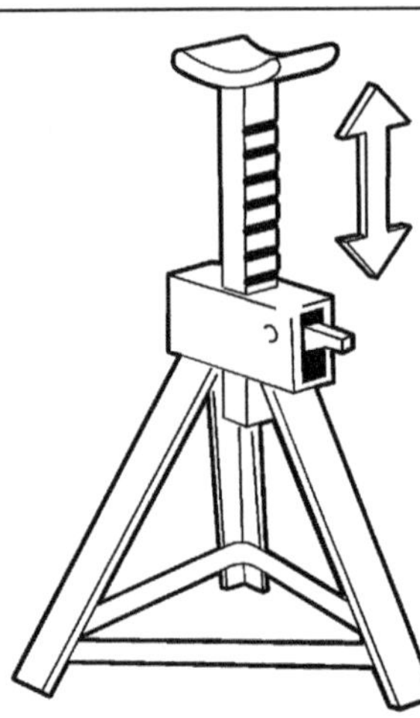

B021FNG

Strong jack stands are extremely important for any work that is done under the car. Use only jack stands that are designed for the purpose. Blocks of wood, concrete, bricks, etc. are not safe or suitable substitutes.

> ***WARNING —***
> *A jack should be used only to raise the vehicle and should not be used to support the car for a long period. Always use jack stands to support a raised vehicle.*

Jack stands are available in several styles. The best ones are made of heavy material for strength, have a wide base for stability, and are equipped to positively lock in their raised positions. Get the best ones available.

Oil change equipment

Changing engine oil requires a 15 mm socket or box-end wrench to loosen and tighten the drain plugs and a drain pan (at least 12 qt. capacity). A wide, low drain pan will fit more easily under the car. An oil filter removal wrench is also required. Use a funnel to pour the new oil into the engine.

Engine oil service is covered in **030 Maintenance**.

Torque wrench

A torque wrench is used to precisely tighten threaded fasteners to a predetermined value. Some of the repair procedures in this manual include Porsche-specified torque values in Newton-meters (Nm) and the equivalent values in pound-feet, more commonly written as foot-pounds (ft-lb).

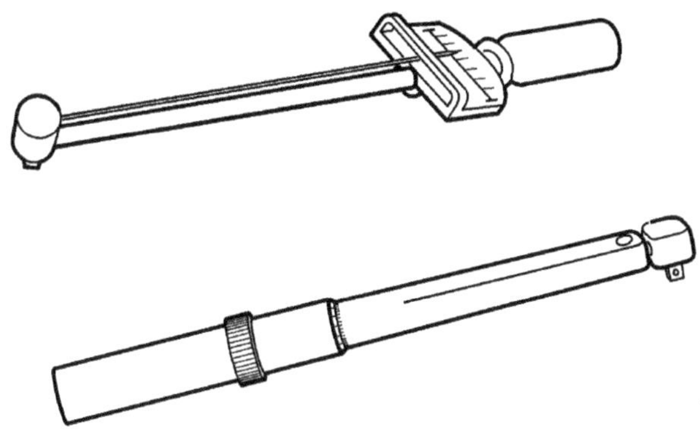

B018FNG

Several types of torque wrenches are available. They all do the same job, but offer different convenience features at different prices. The most convenient ones have a built-in ratchet, and can be preset to indicate when a specific torque value has been reached. Follow the wrench manufacturer's directions for use to achieve the greatest accuracy.

A torque wrench with a range up to about 250 Nm (185 ft-lb) has adequate capacity for most of the repairs covered in this manual. For recommended torque values of 10 Nm or below, the English system equivalent is given in inch-pounds (in-lb). These small values may be most easily reached using a torque wrench calibrated in inch-pounds. To convert inch-pounds to foot-pounds, divide by 12.

Feeler gauges

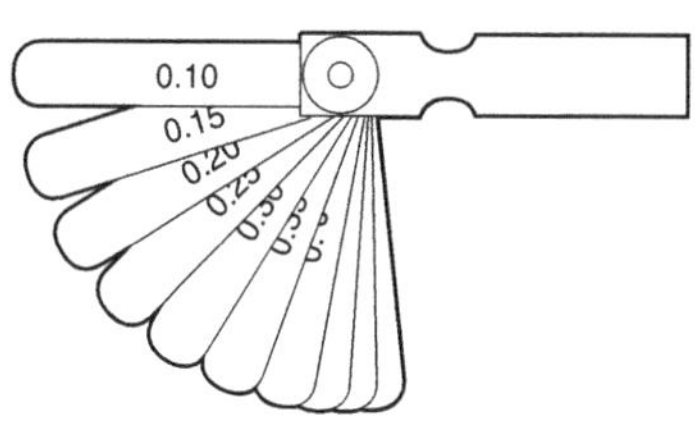

020FNG

Feeler gauges are thin metal strips of precise thickness, used to measure small clearances such as valve clearance. They are normally available as a set, covering a range of sizes.

Digital multimeter

0012223

Many of the electrical tests in this manual call for the measurement of resistance (ohms) or voltage values. For safe and accurate tests of ignition, fuel injection, and emission control systems, the multi-meter, should be digital, with high (at least 10,000 ohms) input impedance. Some meters have automotive functions such as dwell and pulse width that are useful for troubleshooting ignition and fuel injection problems.

CAUTION—
The Motronic (DME) fuel and ignition system, and other electronic systems, may be damaged by the high current draw of a test light with a normal incandescent bulb. Always use a high impedance digital multimeter or an LED test light for all electrical testing.

Porsche special tools

Some of the repairs covered in this manual call for the use of Porsche special tools. Many of the special tools are simply the best thing to use to do the job correctly. In these cases, the tool is identified with a part number and an ordering number, where available.

This, however, does not automatically mean that the job is too complicated or out of reach of the novice. For example an affordable and very useful special tool is a hose clamping tool. This tool is helpful to clamp off fluid lines for a neater job.

There are some jobs for which expensive special tools are essential, and not a cost-effective purchase for one-time repair by the do-it-yourself owner. An example is the rear wheel bearing press. This tool is more expensive than having the job done by a professional who already owns the tool. This manual includes these jobs for the benefit of those with the necessary experience and access to tools.

Most of the Porsche special tools can be purchased through an authorized Porsche dealer. As an alternative, some special tools mentioned may be purchased from the following tool manufacturers and/or distributors:

- **Baum Tools Unlimited. Inc.**
  P.O. Box 87, Longboat Key, FL 34228
  (800) 848-6657
  http://www.baumtools.com
- **Assenmacher Specialty Tools**
  6440 Odell Place, Boulder, CO 80301
  (303) 530-2424
  http://www.asttool.com
- **Schley Products Inc.**
  5350 E. Hunter Ave., Anaheim Hills, CA 92807
  (714) 693-7666
  http://www.sptool.com
- **Zelenda Automotive Inc.**
  66-02 Austin Street, Forest Hills, NY 11375
  (718) 896-2288
  http://www.zelenda.com

NOTE —
When ordering special tools from the dealer, have the tool ordering number on hand. This number normally has more digits in it than the tool number itself. When ordering tools directly from Baum Tools Unlimited, Inc., either the tool number or the ordering number will be sufficient.

EMERGENCIES

Changing a tire

Stop the car on as flat a surface as possible, in a place where you can be easily seen by other drivers. Turn on the emergency flashers, and set out flares or emergency markers well behind the car. Chock the wheel diagonally opposite to the one being changed. Firmly set the parking brake. Passengers should get out of the car and stand away from the road.

> ***WARNING —***
> *If a tire goes flat while driving, pull well off the road. Changing a tire on a busy street or highway is very dangerous. If necessary, drive a short distance on the flat tire to get to a safe place. It is much better to ruin a tire or rim than to risk being hit.*

- Take collapsible spare tire, jack and tool kit from luggage compartment. The jack and tool kit are located ahead of spare tire.

- Loosen wheel bolts while car is on ground, but leave them a little snug.

- Remove plastic plug from jack receptacle in rocker panel (**arrow**). Place jack fully into square receptacle in rocker panel. Make sure jack is resting on flat, solid ground. Use board or other support to provide a firm surface for jack, if necessary.

- Raise car just until wheel is fully off ground. Remove wheel nuts and wheel. Place removed road wheel in plastic bag provided and store in luggage compartment.

- Install spare wheel. Install wheel nuts and tighten them by hand.

- Remove electric air compressor from tool kit and connect it to cigarette lighter socket. Inflate tire to specified pressure. Check tire pressure using a tire gauge.

Tire Inflation Pressure

• Collapsible spare . . . . 2.2 bar (32 psi)

- Lower car to ground and fully tighten nuts in a crisscross pattern. Torque wheel nuts.

Tightening Torque
- Wheel to wheel hub . . . . 130 Nm (96 ft-lb)

WARNING —
Do not drive more than 50 mph with the collapsible spare tire. Adhere to a moderate a driving style. This tire is for emergency use and short distances only. Remount the full size road wheel as soon as possible.

NOTE —
When the air is removed to from the collapsible spare tire, it will return to its original shape after cooling down for several hours.

Car will not start

If the engine turns over slowly or not at all, especially on cold mornings, the battery may not be sufficiently charged. Jump-starting the battery from another car may help.

WARNING —
Push starting (or bump starting) a car is not recommended.

NOTE —
Be sure to read the cautions under **Jump starting car** *prior to boosting a low battery. Failure to follow the cautions may result in damage to the electronic components in the car.*

If the engine is turning over at normal speed with the starter motor, the battery and starter are fine. Check to make sure that there is fuel in the tank. Do not rely on the fuel gauge, it may be faulty. Instead, remove the gas filler cap and rock the car. If there is gas in the tank, you should hear a sloshing sound at the filler neck. If there is plenty of fuel in the tank, see **100 Engine–General** for in-depth diagnostics and troubleshooting procedures.

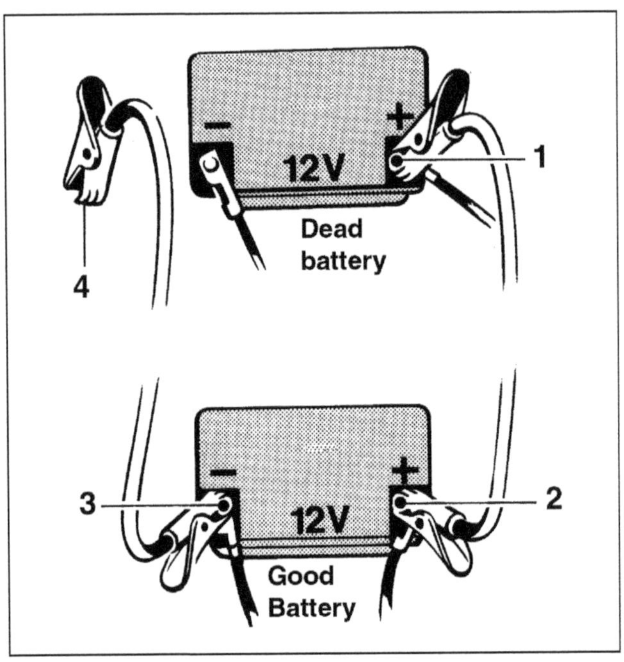

Jump starting car

Cars with discharged or dead batteries can be jump-started from another car. When jump-starting the engine, always note the following warnings.

– Place cars close together, but do not allow them to touch each other. Turn off engine of car with good battery.

◄ Open luggage compartment lid and locate battery (**arrow**) on left hand side of luggage compartment.

> ***WARNING —***
> - *Battery acid (electrolyte) can cause severe burns, and will damage the car and clothing. If electrolyte is spilled, wash the surface with large quantities of water. If it gets into eyes, flush them with water for several minutes and call a doctor.*
> - *Batteries produce explosive and noxious gasses. Keep sparks and flames away. Do not smoke near batteries.*
> - *Do not jump-start the engine if you suspect that the battery is frozen. Trapped gas may explode. Allow the battery to thaw first.*
> - *Do not quick-charge the battery (for boost starting) for longer than one minute, and do not exceed 15 volts at the battery with the boosting cables attached. Wait at least one minute before boosting the battery a second time.*

- Connect one end of positive (+) cable to positive (+) post of good battery (**1**). Connect other end of positive (+) cable to positive (+) post of dead battery (**2**).
- Connect one end of negative (–) cable to negative (-) battery post of good battery (**3**). Connect opposite end of negative cable (–) to engine block of car with dead battery (**4**).
- Start car with good battery and run engine at about 2,000 rpm, then start car with dead battery.
- With engine at idle, carefully disconnect jumper cables, starting with negative cable on engine block.

NOTE —
The engine should be run for at least an hour to recharge the battery.

Alternator warning light on

The alternator warning light monitors the alternator, the V-belt, and the engine cooling fan. If the light flickers on or comes on continuously, the alternator V-belt is most likely loose or broken. The belt should be tensioned or replaced as described **030 Maintenance**.

> ***CAUTION —***
> *Do not drive car with broken V-belt. Overheating of engine will result.*

Low oil pressure

If the red engine oil pressure warning light illuminates or the oil pressure gauge drops to below normal, immediately stop the car and turn off the engine. Have the vehicle flat-bedded or towed to an authorized Porsche service center and have the engine oil pressure checked.

CAUTION—
Do not drive car with oil pressure warning light on until the oil pressure system has been diagnosed and repaired as necessary.

NOTE—
Detailed information on checking the oil pressure is covered in **170 Engine Lubrication**.

Brake fluid level

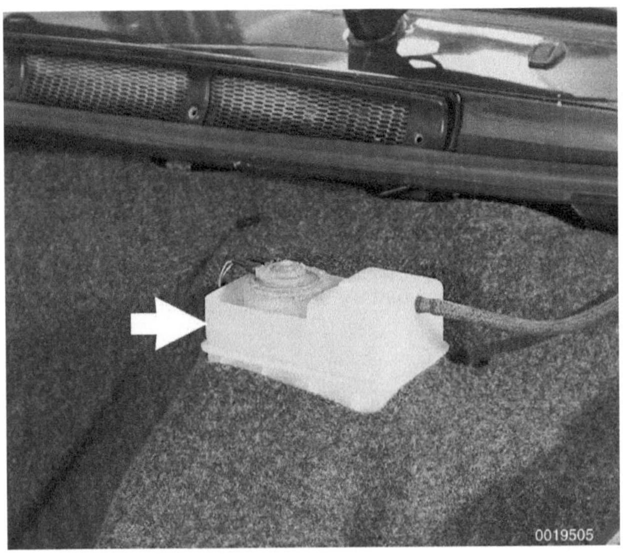

The brake warning light may be an indicator of brake fluid loss. Problems with the brake system should be checked and repaired immediately. Check the brake fluid level at the reservoir (**arrow**) in the luggage compartment. See **470 Brakes–Hydraulic** for more information.

Dim lights

Headlights and instrument panel illumination that are dim or gradually getting dimmer generally indicate a problem with the battery or charging system. The battery charge indicator light may come on as the lights are dimming. In either case, the engine and accessories are running off the battery, and will soon discharge it.

If possible, do not stop the engine unless you have the capability to jump start it. There may not be enough power in the starting system to restart the engine. Instead, turn off as many electrical consumers as possible. This will allow the car to be driven farther before you lose all battery power.

With the engine and ignition off, check for loose or corroded battery cables or wires at the battery or the alternator. Disconnecting, cleaning, and reinstalling corroded wires and connectors may solve the problem. Also check drive belt tension as described in **030 Maintenance**.

Towing

The cars covered by this manual should be towed with a tow truck using wheel lift or flat bed equipment. Do not tow the car on all four wheels except for very short distances to move it to a safe place.

Tow hooks are located under the front bumper.

When towing the car from the front with the rear wheels on the ground, a maximum distance of 20 miles is acceptable. If the car needs to be towed further, have the rear wheels placed on dollies.

> ***CAUTION —***
> *Do not tow with sling-type equipment. The front spoilers and bumper covers may sustain damage.*

Spare parts kit

Carrying a basic set of spare parts can prevent a minor breakdown from turning into a major annoyance. Many of the following items won't allow you to do major repair work on the car, but they will help in the event of the failure of something that can disable the car or compromise its safety.

Spare Parts Kit - Basic Contents

- DME main/fuel pump relay (relay is located under driver's seat.)
- Alternator and A/C drive belts (V-belts)
- Engine oil (one or two quarts)
- Fuse assortment (5A, 8A, 16A, 25A)

Spare Parts Kit - Additional Contents

- Exterior lighting bulbs (headlight, brakelight, turn signal, and taillight)
- Wiper blades
- Brake fluid (new unopened bottle, DOT 3 or higher specification)

030 Maintenance

TABLES

GENERAL

The information given in this repair group includes the routine checks and maintenance steps recommended by Porsche to ensure long and reliable vehicle operation.

NOTE—
Porsche is constantly updating their recommended maintenance procedures and requirements. The information contained here is as accurate as possible at the time of publication. If there is any doubt about what procedures apply to a specific model or model year, or what intervals should be followed, remember that an authorized Porsche dealer has the latest information on factory-recommended maintenance.

MAINTENANCE TABLES

Table a through **Table d** on the following pages list the routine maintenance tasks specified by Porsche. The intervals of these tasks are determined by elapsed vehicle mileage and/or time.

Except where noted, the maintenance items listed apply to all models and model years covered by this manual. The columns on the right side of each table give quick-reference information about the job. The bold text references the repair group where additional repair information can be found.

Table a. Oil Service – every 7,500 miles

Maintenance item	Tools required	New parts required	Warm engine required	Dealer service recommended
Engine compartment maintenance				
Change oil and oil filter	*	*	*	

Table b. Maintenance Schedule – every 15,000 miles

Maintenance item	Tools required	New parts required	Warm engine required	Dealer service recommended
Engine compartment maintenance				
Engine: Visually inspect for leaks.				
Crankcase ventilation: Inspect for leaks and check hose connections.				
Fuel system: Inspect for leaks and check hose connections.				
Intake air: Check hoses, lines and connections.				
Change oil and oil filter.	*	*	*	
Check and adjust valve clearance.	*			
Throttle cable, lubricating				
Under car maintenance				
Manual transmission: Inspect for leaks. Check oil level and add as necessary **340 Manual Transmission–Controls and Case**.	*	*	*	
Fuel system: Inspect for leaks and check hose connection.				
Fuel tank, fuel lines: Inspect all connections for leaks.				
Brake system: Check brake pedal play and correct if necessary. Visually inspect brake hoses and lines for leaks, visible damage, corrosion. Visually inspect brake pads and rotors for wear. Check brake fluid level and top up as necessary.				
Clutch: 1984 - 1986 models: check clutch cable for play. 1987 - 1989 models: check hydraulic clutch actuator for leaks.				
Control arms: Check control arm mounting bolts tightness.	*			
Exhaust system. Visually inspect for leaks and damage. Check exhaust system mounting bolts for tightness. **260 Exhaust System**				
Steering: Check steering rack and steering linkages. Check steering rack mounting bolts for tightness. Inspect bellows for damage. Check tie rod ends for play and inspect dust covers for damage.	*			
Drive axles: check boots for leaks or damage.				
Check parking brake operation, adjust cable if necessary. **460 Brakes–Mechanical**	*			
Front and rear suspension: Check for visible damage. Check ball joints for wear and inspect ball join t dust covers for damage. **401 Front Suspension**				*
Front wheel bearings: Check for play and, if necessary, correct.	*			
Inspect wheels and tires, including spare, check tire pressure and condition	*			

Continued on next page

Table b. Maintenance Schedule – every 15,000 miles (continued)

Maintenance item	Tools required	New parts required	Warm engine required	Dealer service recommended
Body and Interior Maintenance				
Hinges, latches: Lubricate door and hood hinges and hood latch. Check door stop function.				
Locks: Check central door lock operation. Check engine lid lock and luggage compartment lock and safety catch.				
Seals for door, hood and roof: inspect and replace as necessary				
Check headlight and fog light aiming and adjust as necessary.	*			*
Vehicle lighting: Check operation of headlights, parking lights, back-up lights, license plate lights, interior lights, glove compartment light, instrument panel lights, turn signals, emergency flashers, stop lights, horns, headlight flasher and dimmer switch				
Horn: Check operation. Optional electrical equipment: Check operation.				
Check function of air conditioning. Check refrigerant charge at sight glass, where applicable. Tighten A/C compressor mounting bolts. **870 Air Conditioner**.	*			
Check function and condition of seatbelts.				
Check windshield washer fluid level and concentration, add as necessary. Check operation of washer system. Check condition of wiper blades. **920 Windshield Wipers and Washers**		*		
Battery: Check fluid level and if necessary, top up				
Road Test				
Check braking, clutch operation, cruise control, steering, HVAC, and instruments.				

Table c. Maintenance Schedule – additional items every 30,000 miles

Maintenance item	Tools required	New parts required	Warm engine required	Dealer service recommended
Engine compartment maintenance				
Change oil and oil filter	*	*	*	
Check and adjust valve clearance.	*			
V-belt: Check condition and tension.				
Spark plugs: Replace (at least every two years)		*		
Air filter: Replace filter element		*		

Table d. Maintenance Schedule – Time/Mileage

Maintenance item	every 24 months	every 48 months	every 60,000 miles	New parts required	Dealer service recommended
Manual transmission: change gear oil.			*	*	
Fuel filter: Replace.			*	*	
Oxygen sensor: Replace.			*		
Brake fluid: Flush and replace brake fluid/clutch fluid. **340 Manual Transmission–Controls and Case**	*			*	

FLUID AND LUBRICANT SPECIFICATIONS

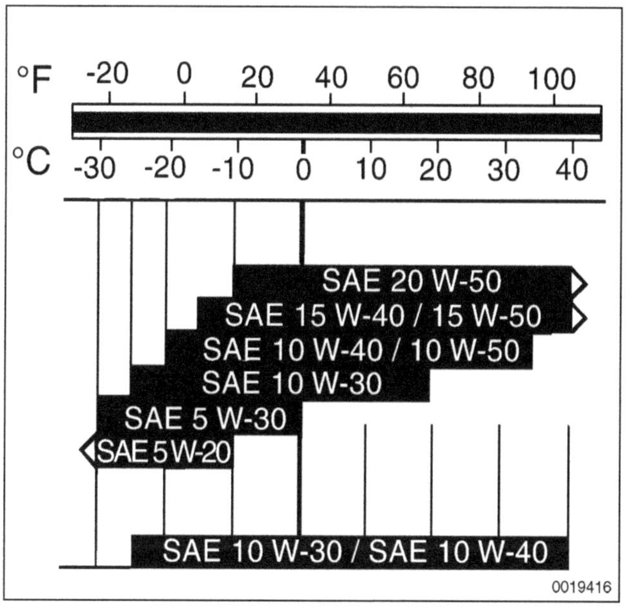

Engine oil viscosity (SAE grade) vs. operating temperature range shown for the Porsche engine covered in this manual.

The fluids and lubricants specified by Porsche are listed in **Table e**.

> ***WARNING —***
> *The use of fluids that do not meet Porsche's specifications may impair performance and reliability, and may void warranty coverage.*

Table e. Fluids and Lubricants

Fluid	Approximate capacity	Specification
Engine oil (with filter change)		
	10 L (10.5 US qt.)	API service SH or SJ
Manual transmission (drain and fill)		
915 G50	3.0 L (3.2 US qt.) 3.4 L (3.6 US qt.)	Hypoid gear oil SAE 90 (GL-5)
Brake fluid		
All models	—	SAE DOT 3 or DOT 4

NOTE —
Synthetic manual transmission fluid is not approved by the manufacturer, but its use may result in improved transmission performance and wear characteristics. Consult with an authorized Porsche Dealer for recommendations on synthetic transmission fluids.

ENGINE OIL SERVICE

If the car is used primarily for short trips in slow moving traffic, or routinely operated aggressively, the oil should be changed at shorter intervals (3,000 mile intervals).

Engine oil and filter, changing

- Run engine for a few minutes to warm oil. Turn engine off.
- With car on level ground, place a drain pan under oil drain plug (**arrow**) on left side of engine.

NOTE—
The car will not need to be raised if a shallow drain pan is used.

0019478

< Loosen crankcase oil drain plug. Remove plug and let oil drain into pan.

0018358

< Loosen supply tank oil drain plug (**arrow**). Remove plug and let oil drain into pan.

WARNING—
- *Pull the loose plug away from the hole quickly to avoid being burned by the hot oil. It will run out quickly when the plug is removed.*
- *Use gloves to protect your hands.*

0019501

< Remove oil filter. Lubricate oil filter seal and install new filter hand tight. Tighten filter an additional ½ turn.

CAUTION—
Overtightening the oil filter will make the next change much more difficult, and may deform the gasket, causing leaks.

- When oil flow has diminished to an occasional drip, reinstall drain plugs with new metal sealing washers and tighten plugs.

Tightening torque
- Engine oil drain plug to crankcase ..... 70 Nm (51 ft-lb)
- Engine oil drain plug to supply tank .... 42 Nm (31 ft-lb)

– Refill oil supply tank with oil. Oil specifications are found previously under **Fluid and Lubricant Specifications**.

– Start engine and allow to reach operating temperature. With engine running at idle, check engine oil level. Check for leaks at drain plug and oil filter.

> ***CAUTION—***
> *Dispose of used oil properly. Use tight-sealing containers and mark them clearly. Check with the place of purchase about disposal.*

ENGINE COMPARTMENT MAINTENANCE

Air filter element, replacing

The specified maintenance intervals for replacing the air filter element are based on normal use. If the car is operated primarily in dusty conditions, the air filter should be serviced more frequently. A dirty air filter starves the engine for air, reducing power output.

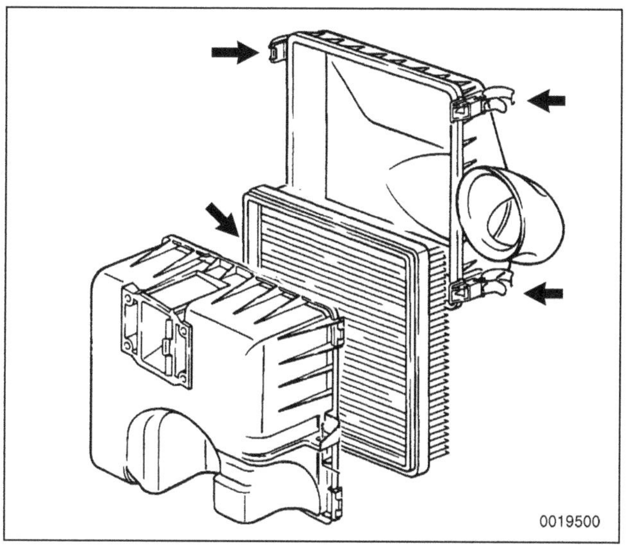

The two-piece air filter housing is fastened together with spring clips around outside edge (**arrows**).

- To replace air filter element, unfasten clips as necessary.

– Separate air filter housing and remove filter element.

– Wipe clean inside of air filter housing using a lint-free cloth.

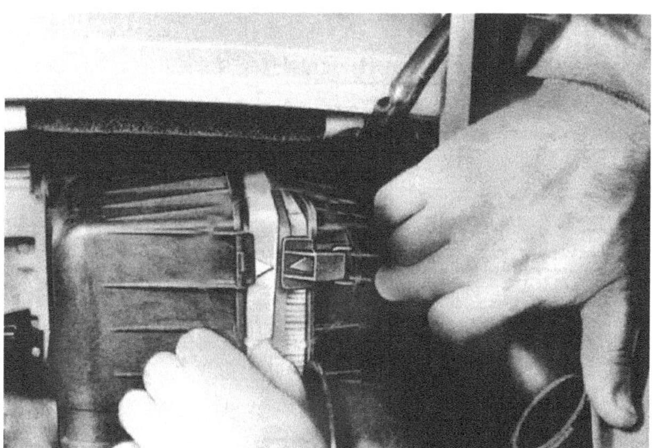

Install new filter, making sure that arrow on filter elements aligns with arrow on filter housing.

– Assemble air filter housing and refasten spring-clips.

> ***WARNING—***
> *When reinstalling a used air filter element, reinstall the filter in the same position that it was in before removal. Reversed installation will allow accumulated dirt to be drawn into the engine.*

Spark plugs, replacing

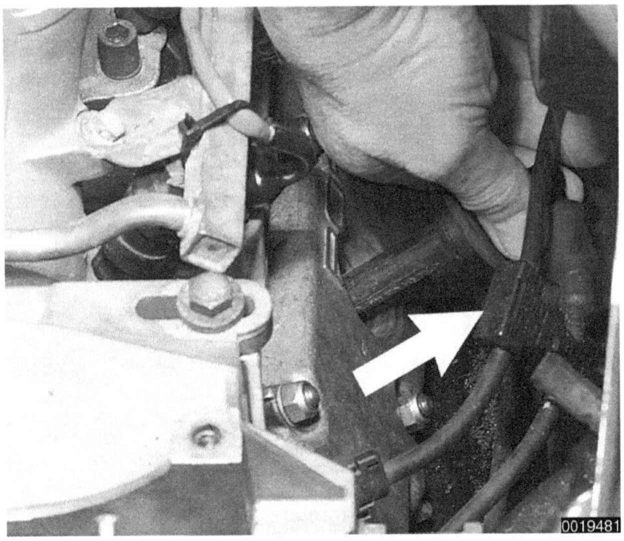

Carefully remove spark plug wire by pulling straight up on boot. Blow away any dirt from around the base of the plug to prevent it from entering the engine when the plug is removed.

> ***CAUTION—***
> *Spark plugs should be replaced one at a time so that the spark plug wires do not get mixed up. If all of the wires need to be removed together, label each wire so that they can be reinstalled on the correct spark plug.*

– Use 13/16 in. spark plug socket to remove and reinstall spark plugs.

- Use a spark plug gap gauge to check gap, if applicable. If necessary, bend outer electrode slightly to adjust gap to meet specification.
- Do not bend or file center electrode.
- Thread plugs into cylinder head by hand to prevent cross-threading.
- Torque spark plugs.

Tightening torque
- Spark plug to cylinder head .... 30 Nm (22 ft-lb)

NOTE—
Abnormal spark plug condition or engine damage could also result from spark plugs of the wrong specification. Check replacement plugs carefully and follow the spark plug manufacturer's recommendations.

Table f. Spark Plug Applications

Model	Spark plug (Bosch)	Electrode gap
911 Carrera	WR7DC (copper) WR7DP (platinum)	0.7-0.8 mm (0.028-0.032 in.)

Distributor cap/rotor/spark plug wires, replacing

The distributor cap, rotor, and spark plug wires deliver high-voltage spark to the spark plugs. They are subject to insulation breakdown, corrosion fouling, and electrode wear and damage. The components should be inspected and replaced as necessary at the intervals listed under **Table b** or **Table c** to ensure maximum engine efficiency. Guidelines for visual inspection and testing, and instructions for replacement are found in **280 Ignition System**.

Fuel filter, replacing

The fuel filter is located in the left rear of the engine compartment.

– Disconnect negative (–) battery cable.

◄ Loosen filter clamping bracket (**arrow**). Counterhold and loosen fitting at top and bottom of filter. Soak up spilled fuel.

- Install filter in correct direction and tighten fittings while counterholding fuel filter.

NOTE —

The direction of fuel flow is indicated on the filter.

> ***WARNING —***
>
> *Fuel will be spilled when the filter is removed. Do not smoke or work near heaters or other fire hazards. Keep a fire extinguisher handy.*

> ***CAUTION —***
>
> *Clean thoroughly around the filter connections before removing. Make sure no dirt gets into the fuel lines.*

Valve clearance, adjusting

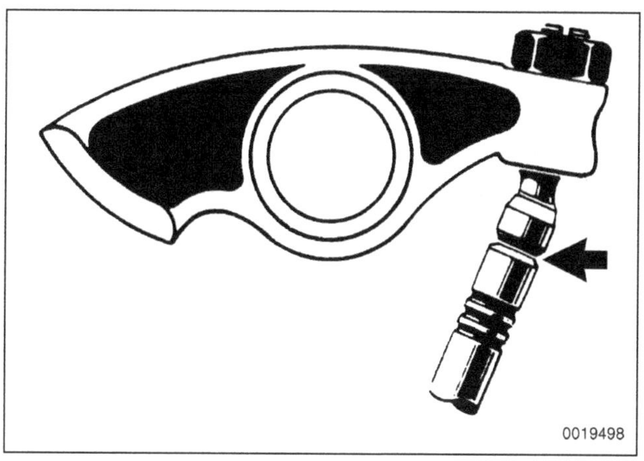

◄ The valve clearance is checked and adjusted with the engine cold (ambient temperature of approximately 20° C (68°F).

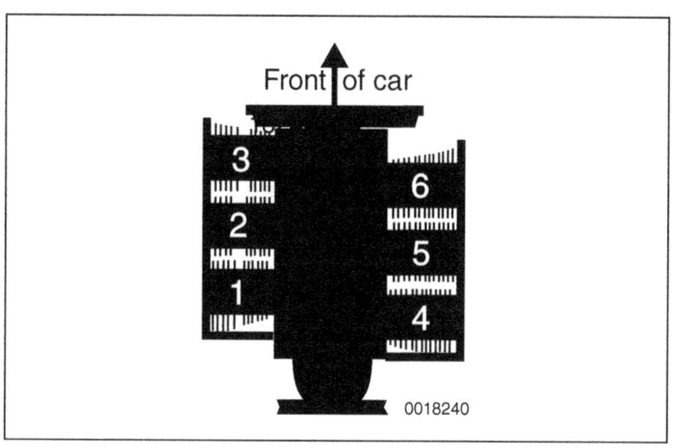

◄ The valve adjusting sequence is 1–6–2–4–3–5

– Drain oil from engine crankcase.

> ***WARNING —***
>
> *Do not remove the spark plugs to make the engine turn easier. Carbon deposits that have built up on the plug can fall between the valve and seat when the plug is removed. This can result in inaccurate valve adjustment.*

– Raise rear of car and support safely on jack stands.

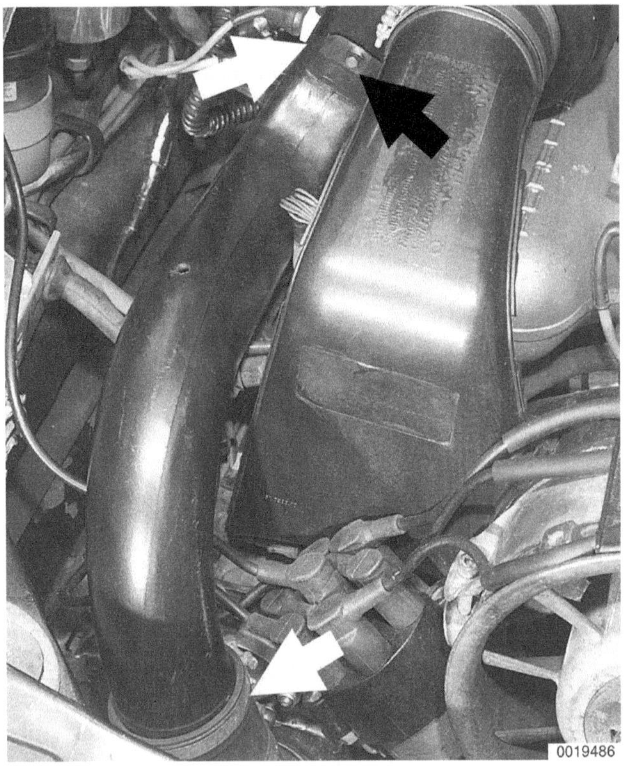

◀ Remove heater elbow and hose on left side of engine.

– Remove air filter cover and filter element.

◀ Remove A/C compressor from mount (**arrows**) and set aside without disconnecting A/C lines. Disconnect wire harness (**A**) before setting aside.

– Unplug all spark plug connectors and set aside.

– Remove distributor cap from distributor (do not remove spark plug wires from cap).

– Remove upper and lower valve covers.

- Using a 23 mm socket on fan pulley nut, rotate fan in a clockwise direction until distributor rotor is pointing to cylinder no. 1 mark (**arrow**) on distributor body.
  - **Z1** mark on crankshaft pulley must also align with mark on lower fan housing. If all marks align, engine is at top dead center (TDC).

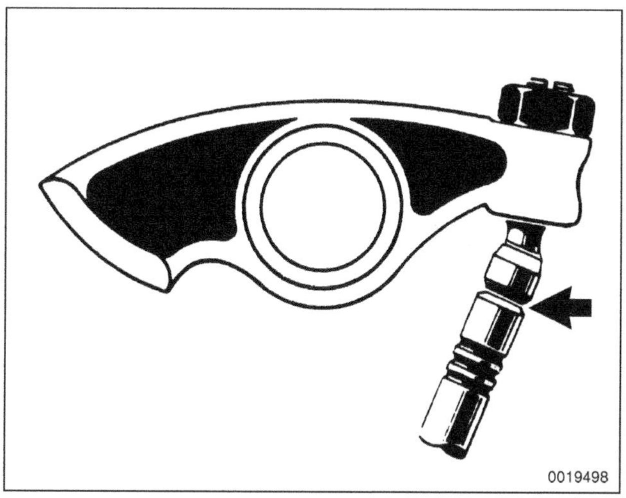

- Check clearance of cylinder no. 1 intake valve by sliding a feeler gauge between rocker arm adjusting screw and the valve stem (**arrow**).
  - If feeler gauge slides in easily, valve clearance is too big. If gauge will not slide in at all, clearance is too small.
  - Repeat procedure for exhaust valve.

- If necessary, loosen valve adjuster lock nut and turn adjuster with screwdriver until a light resistance is felt at feeler gauge.

- While holding adjuster in position, tighten adjuster locking nut.

- Recheck adjustment after tightening lock nut.

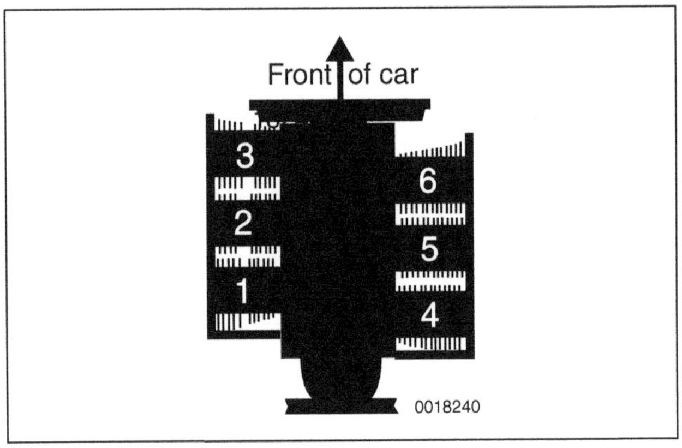

Rotate crankshaft pulley 120° (clockwise direction). Adjust valves for cylinder no. 6 using above procedure.

NOTE—
*The crankshaft is marked in 120 degree intervals starting with **Z1**. With the crankshaft at TDC, the **Z1** mark will be at 12 o-clock, the next mark will be at 4 o'clock and the last mark will be at 8 o'clock.*

- Turn crankshaft clockwise 120° and adjust valves for cylinder no. 2.
- Turn crankshaft clockwise 120° and adjust valves for cylinder no. 4.
- Turn crankshaft clockwise 120° and adjust valves for cylinder no. 3.
- Turn crankshaft clockwise 120° and adjust valves for cylinder no. 5.
- Clean valve cover surfaces on camshaft housing to remove any gasket material that may be stuck to the surface.
- Clean all gasket material and oil from valve covers.
- Install valve covers with new gaskets, washers and locknuts.

Tightening torques

• Valve cover nuts	8 Nm (6ft-lb)
• Crankcase drain plug	70 Nm (52 ft-lb)
• Oil tank drain plug	42 Nm (32 ft-lb)

- Replace engine oil filter and add new engine oil.

Battery maintenance

The battery is located in the luggage compartment, on the left side. When servicing the battery, always disconnect the negative (–) cable first, and connect it last.

- Simple maintenance of the battery and its terminal connections will ensure maximum starting performance, especially in colder temperatures. For a more detailed discussion of the battery and charging system, see **270 Battery, Starter, Alternator**.

NOTE—
A secure battery hold-down is important in order to prevent vibrations and road shock from damaging the battery internally.

Battery cable clamps should be tight. The terminals, the cable clamps, and the battery case should be clean and free of white deposits. Even a thin layer of dust containing conductive acid salts can cause battery discharge.

To remove battery corrosion, begin by disconnecting the cables. Disconnect the negative (–) cable first.

> ***CAUTION—***
> *Disconnecting the battery cables erases engine control module (ECM) adaptive memory. It may be necessary to drive the car for approximately 10 minutes after reconnecting the battery to reset ECM adaptive memory and restore normal engine performance.*

Clean the terminal posts and the cable clamps with a wire brush. Clean the main chassis ground terminal next to the battery.

Corrosion can be washed away with a baking soda and water solution that will neutralize the acid. Apply the solution carefully, though, since it will also neutralize the acid inside the battery. Lightly coat the outside of the terminals, hold down screw, and clamps with petroleum jelly, grease, or a commercial battery terminal corrosion inhibitor.

> ***WARNING—***
> - *Battery acid is caustic and dangerous. Take care to keep it from contacting eyes, skin, or clothing. Wear eye protection. Extinguish all smoking materials and do not work near any open flames.*
> - *Hydrogen gas given off by the battery during charging is explosive. Do not smoke. Keep open flames away from the top of the battery, and prevent electrical sparks by turning off the battery charger before connecting or disconnecting it.*

> ***CAUTION—***
> *To avoid causing battery acid damage to the interior, remove battery from car before using a battery charger.*

Battery electrolyte should be maintained at the correct level just above the battery plates and their separators. The correct level is approximately 5 mm (1/4 in.) above the top of battery plates or to the top of the indicator marks (if applicable). The battery plates and the indicator marks can be seen once the filler caps are removed. If the electrolyte level is low, replenish it by adding distilled water only.

Throttle housing and linkage, lubricating

The linkage at the throttle housing should be lubricated at the normal maintenance intervals. Use a general purpose oil on the joints and bearings of the linkage. Use a multipurpose lube on the bearing points of the throttle plate.

V-belts, checking and replacing

Replacement of V-belts every four years is recommended. When belts are replaced, store the old set in the luggage compartment for emergency use.

The V-belt is tensioned through a split pulley and shim arrangement. Six shims are installed at the factory and arranged on either side of the pulley depending on the adjustment required to correctly tension the belt. All six shims should remain on the alternator shaft.

Removing shims from the between the pulleys halves increases belt tension. Placing additional shims between the pulley halves decreases the tension.

A special spanner wrench is needed to hold the belt pulley stationary while loosening the large nut on the alternator shaft.

Inspect belts with the engine off. Twist the belt to inspect its sidewalls and bottom. Belt structural damage, glazed or shiny sidewalls caused by a loose belt, or layer separation caused by oil contamination are all reasons to replace a belt.

– Remove A/C compressor belt. See **870 Air Conditioner**.

While holding alternator belt pulley with spanner wrench (**arrow**), loosen and remove pulley mounting nut and separate pulley. Note arrangement of shims on either side of pulley.

– Remove belt from crankshaft pulley.

– Install new belt and reinstall shims and pulley. Install shims, washer, and nut. Tighten nut, making sure pulley seats properly.

Tightening torque
- Alternator pulley to alternator . . . . 40 Nm (30 ft-lb)

– Check belt tension by applying thumb pressure at belt midpoint. The belt should deflect approximately 5 mm (0.2 in.).
  - If adjustment is necessary, reposition shims. Run engine for a few minutes (at least 1,500 rpm), then recheck belt tension.

Oxygen sensor, replacing

The oxygen sensor monitors engine combustion efficiency by measuring the oxygen content of the exhaust gasses. The output signal from the sensor is used to precisely control fuel delivery. Any problems with the oxygen sensor will directly affect exhaust emissions and the way the engine runs.

Replacement of the oxygen sensor at the specified interval ensures that the engine and emission control system will continue to operate as designed.

To remove sensor (**arrow**) from catalyst:
- Detach oxygen sensor harness connector in engine compartment.
- Remove sensor from catalyst.
- Lightly coat the threads of new sensor with anti-seize compound before installing.

NOTE—
A special deep 22 mm socket for replacing the oxygen sensor is available from most automotive parts stores or from Baum Tools Unlimited. The socket has a groove cut down one side to allow the sensor to be installed without damaging the wire harnesss.

CHASSIS AND UNDER-CAR MAINTENANCE

Tires, checking inflation pressure

Correct tire inflation pressures are important to handling and stability, fuel economy, and tire wear. Tire pressures change with temperature. Pressures should be checked often during seasonal temperature changes. Correct tire pressure information can be found on an information label (**arrow**) on the rear engine compartment apron and in the owner's manual.

> ***WARNING—***
> *Do not inflate any tire to a higher pressure than the tire's maximum inflation pressure listed on the sidewall. Use care when adding air to warm tires. Warm tire pressures can increase as much as 4 psi (0.3 bar) over their cold pressures.*

Tires, rotating

The tires must not be rotated because of front-to-rear size difference.

Wheels, aligning

Porsche recommends checking front and rear wheel alignment once a year and whenever new tires are installed. See **440 Wheels, Tires, Alignment** for a more detailed discussion of alignment requirements and specifications.

Front wheel bearings, checking play

Front wheel bearing

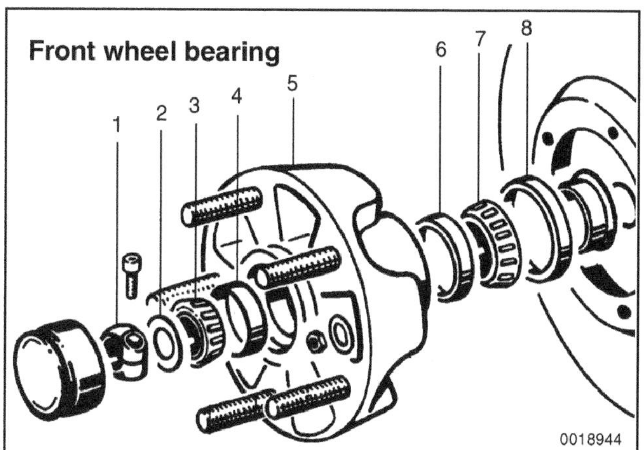

1. Bearing lock nut
2. Thrust washer
3. Outer bearing
4. Inner race, outer bearing
5. Bearing hub
6. Inner race, inner bearing
7. Inner bearing
8. Grease seal

Routine maintenance of the tapered front wheel bearings includes checking for play and adjusting as necessary.

- To check wheel bearing play, first loosen front wheel lug nuts and then raise front corner of car and support safely using jack stand(s).

Remove front wheel and pry off grease cap from wheel hub.

◀ Check that thrust washer can just barely be moved back and forth with a light push on the tip of a screwdriver (do not pry or twist screwdriver).

– If necessary, adjust wheel bearing play as described in **401 Front Suspension**.

– Install grease cap. Reinstall road wheel.

Tightening Torque
- Wheel to wheel hub . . . . 130 Nm (96 ft-lb)

Brake system, inspecting

Routine maintenance of the brake system includes maintaining an adequate level of brake fluid in the reservoir, checking brake pads for wear, checking hand brake function, and inspecting the system for fluid leaks or other damage.

Brake hoses and lines, inspecting

Check that all brake hoses are correctly routed to avoid chafing or kinking. Inspect unions and brake calipers for signs of fluid leaks. Inspect rigid lines for corrosion, dents, or other damage. Inspect flexible hoses for cracking. Replace faulty hoses or lines as described in **460 Brakes–Mechanical**.

CAUTION—
- *Incorrect installation or overtightening hoses, lines, and unions may cause chafing or leakage. This can lead to partial or complete brake system failure.*
- *Brake fluid is poisonous. Do not siphon brake fluid by mouth. Wear gloves when working with brake fluid or brake pads to prevent contamination of cuts.*

Brake pad/rotor wear, checking

Although the cars covered by this manual are equipped with a brake pad warning system, it is a good idea to check the pads and rotors anytime the wheels are off the car.

◄ Check brake pad thickness (**A**) through the opening in the caliper. See **460 Brakes–Mechanical** for brake pad replacement.

Brake Pad Minimum Thickness (Dimension A)	
• Front and rear disc brake pads	2.0 mm (0.08 in.)

– Check brake rotor for minimum thickness using a micrometer. Check rotor for deep scoring or uneven surface wear. Replace rotors in pairs if any faults are found. See **460 Brakes–Mechanical** for brake rotor replacement.

Brake Rotor Minimum Thickness (after machining)	
• Front/rear	22.6 mm (0.890 in.)

NOTE—
For parking brake lining wear, see **Parking brake, checking** *below.*

WARNING—
Friction materials such as brake linings may contain asbestos fibers. Do not create dust by grinding, sanding, or cleaning the pads with compressed air. Avoid breathing asbestos dust, as it may result in serious diseases and cancer, or in death.

Brake fluid, replacing

Porsche strictly recommends replacing the brake fluid every year. This will help protect against corrosion and the effects of moisture in the fluid. See **470 Brakes–Hydraulic** for brake fluid flushing procedures.

Parking brake, checking

The parking brake may require periodic adjustment depending on use. Adjust the parking brake if the brake lever can be pulled up more than 4-5 clicks on the ratchet mechanism. Check that the cable moves freely. A complete description of the parking brake and parking brake adjustment can be found in **460 Brakes–Mechanical**.

NOTE—
The parking brake may lose some of its effectiveness over time if it is not used frequently. This is due to corrosion build-up on the parking brake drum. To remove corrosion, apply the parking brake just until it begins to grip, then pull the lever up one more stop (click). Drive the car approximately 400 meters (1,300 ft.) and release the brake. To recheck the adjustment of the parking brake see **460 Brakes–Mechanical**.

Exhaust system, inspecting

Exhaust system life varies widely according to driving habits and environmental conditions. If short-distance driving predominates, the moisture and condensation in the system will not fully dry out. This will lead to early corrosion damage and more frequent replacement.

Check for restrictions due to dents or kinks. Check for weakness or perforation due to rust. Check that exhaust system fasteners are tight. Check to see that all support bands are in place and properly supporting the system. Exhaust system replacement is covered in **260 Exhaust System**.

Manual transmission service

Manual transmission service consists of inspecting for leaks, checking and changing the gear oil, checking the clutch and clutch cable (1984-1986) or hydraulic actuator (1987 and later). See **340 Manual Transmission–Controls and Case** and **301 Clutch**.

Manual transmission oil, checking and replacing

Porsche recommends checking the manual transmission oil level and changing the oil at specified intervals. Check and fill the transmission with the car on a level surface. Transmission oil level checking and oil replacement procedures are covered in **340 Manual Transmission–Controls and Case**.

Front suspension and steering linkages, inspecting

Inspection of the front suspension and steering includes a check of all moving parts for wear and excessive play. Also inspect the small rubber seals and boots for cracks or tears that could allow the entry of dirt, water, and other contaminants. See **401 Front Suspension**.

Rear suspension, inspecting

Rear suspension inspection consists of checking for damaged or worn suspension components, leaky shock absorbers, and checking the rear drive axle rubber boots for damage. See **420 Rear Suspension, Axle Shafts**.

Drive axle joint boots, inspecting

The rear drive axle inner and outer protective boots should be inspected for cracks and any other damage that will allow contaminants to get into the joint. If the rubber boots fail, the water and dirt that enters the joint will quickly damage it. Replacement of the CV joint boots and inspection of the joints are described in **420 Rear Suspension, Axle Shafts**.

Fuel tank and fuel lines, inspecting

– Inspect the fuel tank, fuel lines, and fuel system for damage or leaks.

- Check for fuel leaks in the engine compartment or fuel odors in the passenger compartment.
- Check for a damaged fuel tank.
- Check for any evaporative emissions hoses that may have become disconnected, checking carefully at the charcoal canister and evaporative emissions purge valve (where applicable).
- See **240 Fuel Injection** and **201 Fuel Supply** for component locations and additional information.

> ***WARNING —***
> *When checking for fuel leaks, the engine must be cold. A hot exhaust manifold or exhaust system could cause the fuel to ignite or explode causing serious personal injury. Ventilate the work area and clean up spilled fuel immediately.*

BODY AND INTERIOR MAINTENANCE

Windshield wiper blades, replacing

Normally, replacing the wiper blades twice a year (before and after the cold season) works well. See **920 Windshield Wipers and Washers** for more information about wiper system troubleshooting and wiper blade replacement.

NOTE —
When replacing the wiper blades, check the tightness of the wiper arm nuts. Loose wiper arms can lead to damaged wiper linkages. If found to be loose, align wiper arms and tighten the nuts to the torque listed in **920 Windshield Wipers and Washers**.

Body lubrication

The body and door hinges, the luggage compartment and engine lid latch, and the door check rods can be lubricated with a medium weight engine oil. Lubricate the seat runners with multipurpose grease.

Lubricate the sunroof guide rails with silicone spray. If door weatherstrips are sticking, lubricate them with silicone spray or talcum powder. Do not apply any oil to rubber parts.

The door locks and lock cylinders can be lubricated with an oil that contains graphite.

Seatbelts

Dirt and other abrasive particles will damage seatbelt webbing. If it is necessary to clean seatbelts, use a mild soap solution. Bleach and other strong cleaning agents may weaken the belt webbing.

> ***WARNING —***
> *Do not clean the seatbelt webbing using dry cleaning or other chemicals. Allow wet belts to dry before allowing them to retract.*

The condition of the seatbelt webbing and the function of the retractor mechanisms should be inspected.

100 Engine–General

TABLES

GENERAL

The 1984 - 1989 Porsche Carrera models are fitted with a 3.2 liter, air cooled, horizontally-opposed engine. General engine specifications are shown in **Table a**. Note that there were slight variations in the engine used in 1987 and later cars.

The 3.2 liter Carrera engine uses an advanced engine management system called Digital Motor Electronics (DME), sometimes referred to as the Motronic system. In the DME system, fuel injection, ignition, and other functions, are combined under the control of the engine control module (ECM). See **200 Engine Management–General** for more detailed information on system operation.

Table a. Engine Specifications

Model	Engine code	No. of cylinders	Displacement liters (cu. in.)	Compression ratio	Horsepower SAE net @ rpm	Torque lb-ft SAE net @ rpm
Carrera						
1984-1986	930/21	6	3164 (193.1)	9.5:1	200 @ 5,900	185 @ 4,800
1987-1989	930/25	6	3164 (193.1)	9.5:1	217 @ 5,900	195 @ 4,800

Full-power curves

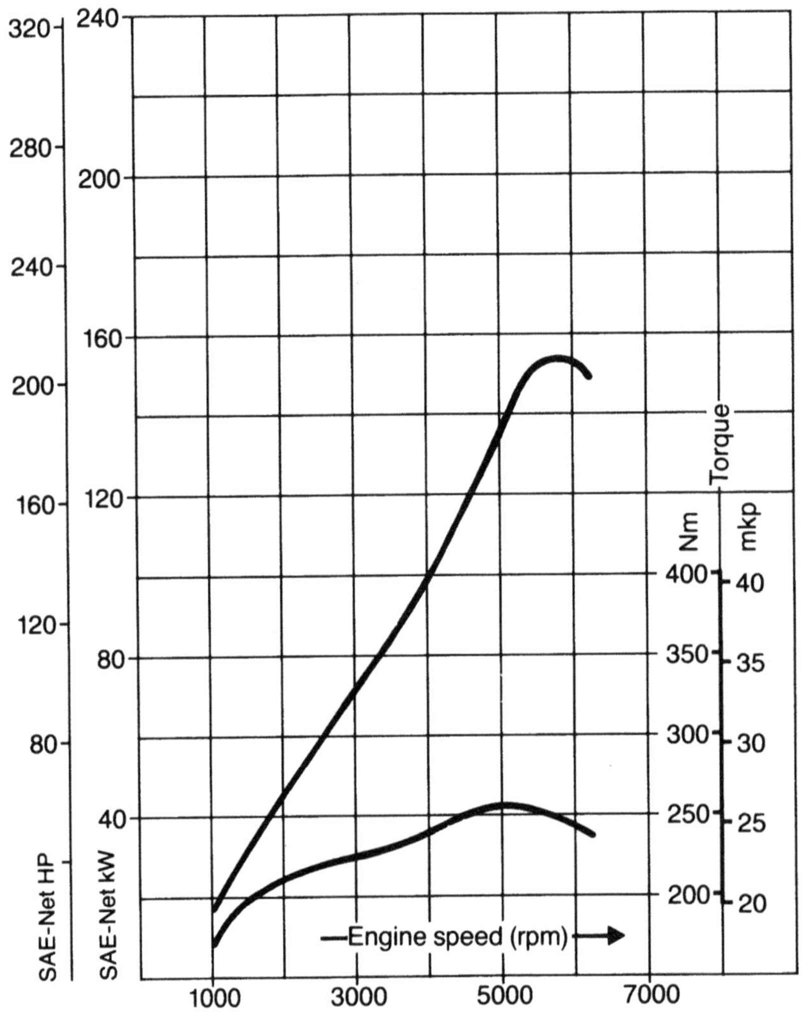

0019408

Acceleration curve

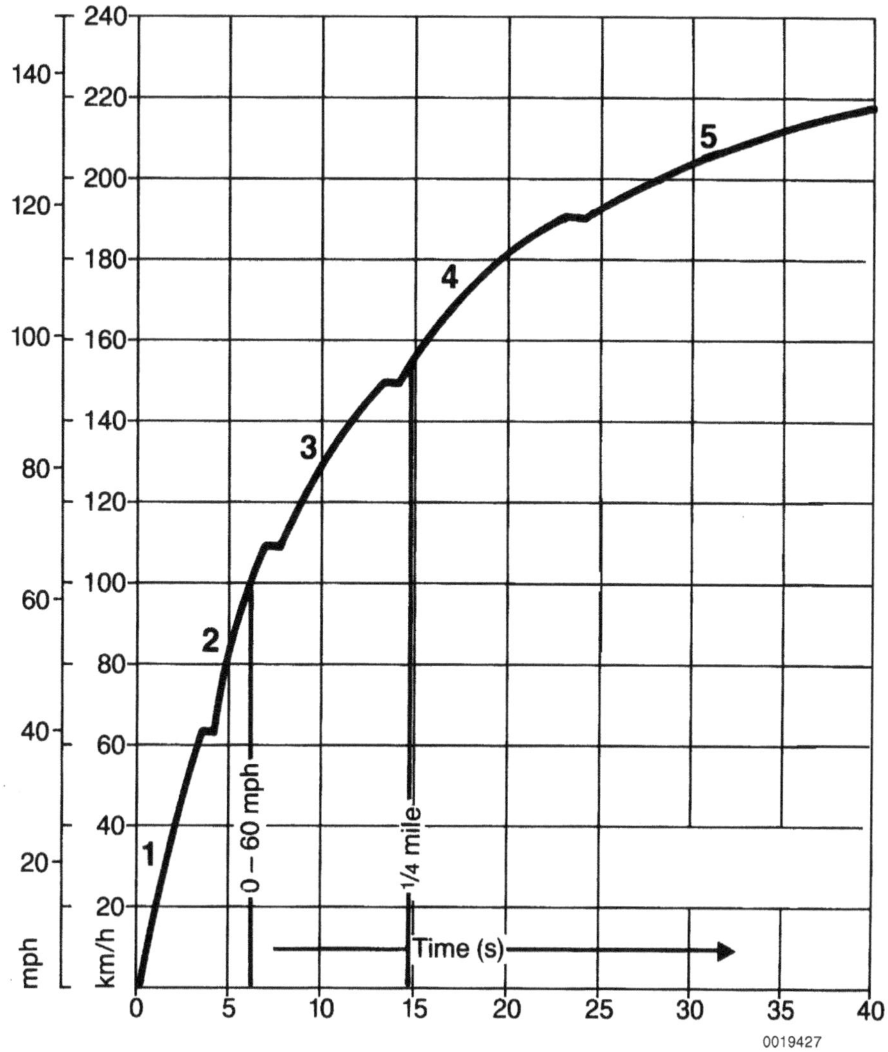

Cross section **Longitudinal section**

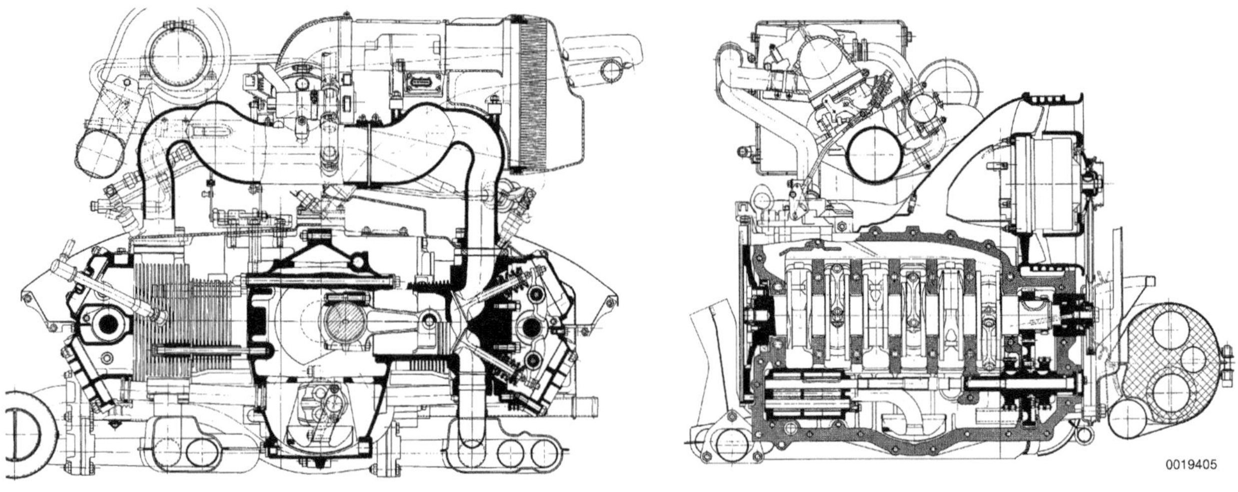

Engine case (crankcase)

The two-piece engine case is cast aluminum and is bolted together along the centerline of the engine around the crankshaft. The crankcase does not use an oil sump. Instead a dry-sump lubrication system is employed. The oil drain plug for the crankcase is located on the left (driver's) side of the case.

Crankcase and oil tank venting system

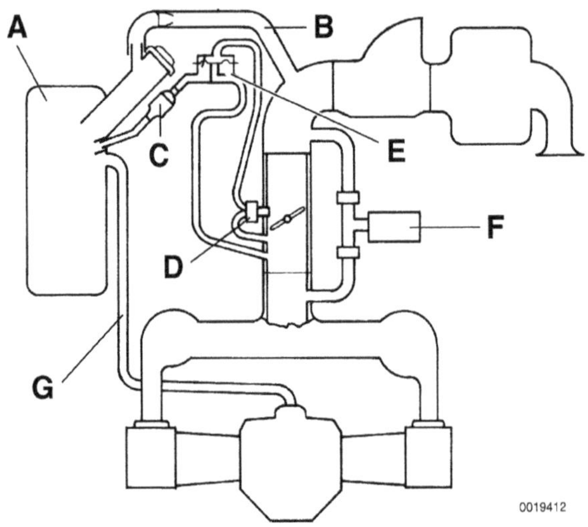

A Oil tank
B Main vent
C Vacuum control for auxiliary vent
D Temperature switch
E Vacuum switching valve
F Idle air control valve
G Crankcase vent

To prevent throttle body icing in the cold weather, an auxiliary crankcase venting system is used. The auxiliary vent is connected from the oil tank (**A**) to the intake system via a vacuum controlled switching valve (**E**). The switching valve is operated using engine vacuum when the temperature switch (**D**) is open. The temperature switch opens at temperatures below 10°C (50°F). At temperatures above 10°C (50°F), crankcase gasses are vented into the engine via the main vent.

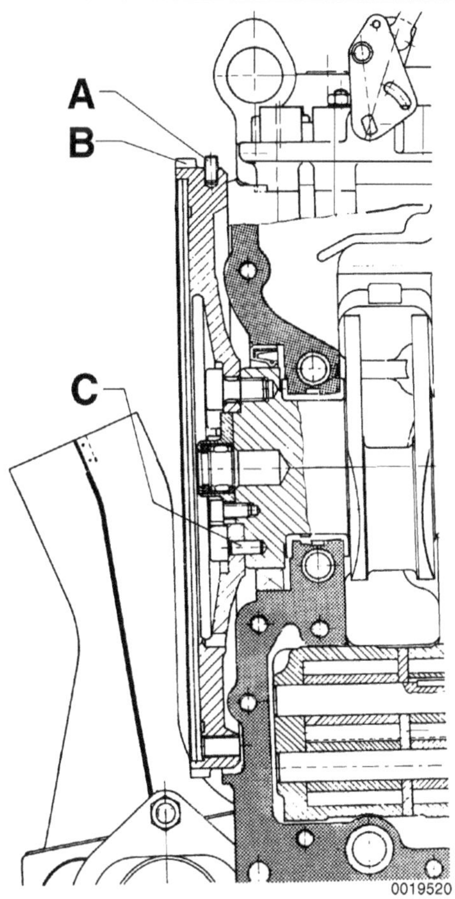

0019520

Crankshaft

The crankshaft rotates in replaceable split-shell main bearings. Oilways drilled into the crankshaft provide bearing lubrication. Flexible lip seals are installed at both ends of the crankshaft. The crankshaft has a stroke of 74.4 mm (2.93 in.).

The fully counterweighted crankshaft is the same crankshaft as used in the Turbo model, except that a different flywheel mounting arrangement is used owing to the DME engine management system. A threaded pin (**A**) for the TDC/reference signal, a toothed wheel (**B**) for the crankshaft speed /rpm signal, and a locating dowel pin (**C**) for timing the flywheel to the crankshaft are the main differences.

0019524

Pistons

The forged pistons are of the three-ring type with two upper compression rings and a lower one-piece oil scraper ring. The pistons have a nominal diameter of 95 mm (3.74 in.). Full-floating piston pins are retained with circlips.

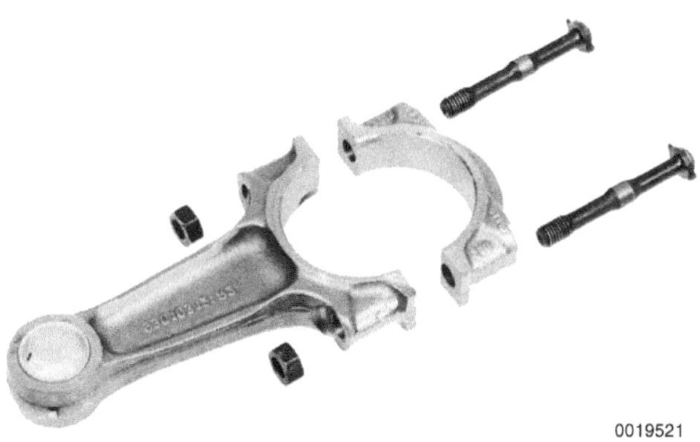

0019521

Connecting rods

The connecting rods use replaceable split-shell bearings at the crankshaft end and solid bushings at the piston pin end. The bolts are stretch type bolts and should always be replaced during engine assembly. The bolts are tightened using the torque angle tightening method for a more accurate torque.

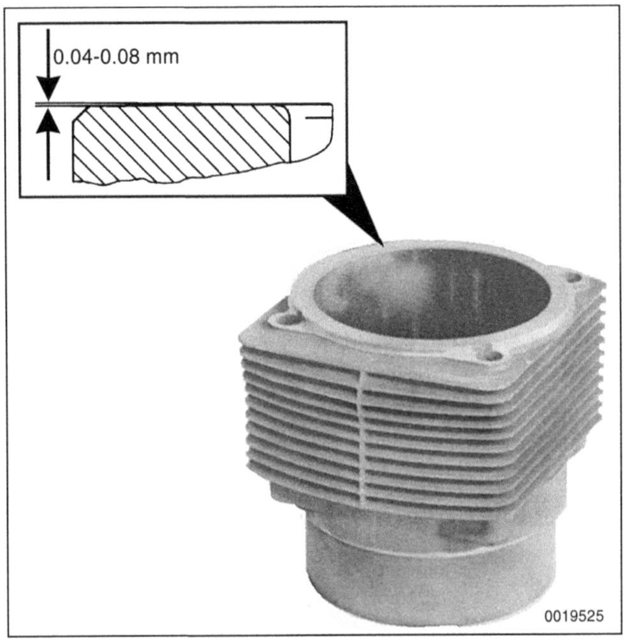

Cylinders

Individual aluminum cylinders are Alusil® or Nikasil® coated for long life. The cylinder sealing surfaces are machined with a small amount of runout towards the outer edge. The runout is approximately 0.04-0.08 mm (0.0016 - 0.0032 in.).

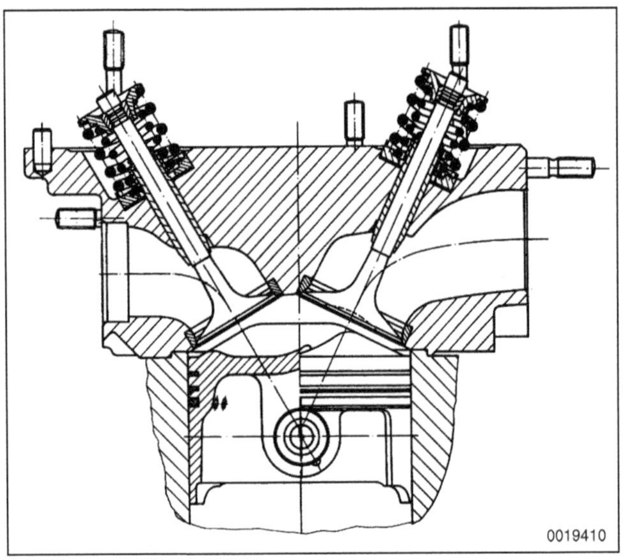

Cylinder head and valvetrain

The aluminum cylinder heads use chain-driven overhead camshafts and two valves per cylinder. The cylinder heads employ a crossflow design for greater power and efficiency. Intake air enters each combustion chamber from one side while exhaust gasses exit from the other. Oilways in the camshaft housings provide lubrication for the camshaft(s) and valvetrain. Oil is returned to the crankcase via six oil return tubes. The valve heads have a diameter of 49 mm (1.929 in.) on the intake side and 41.5 mm (1.634 in.) on the exhaust side.

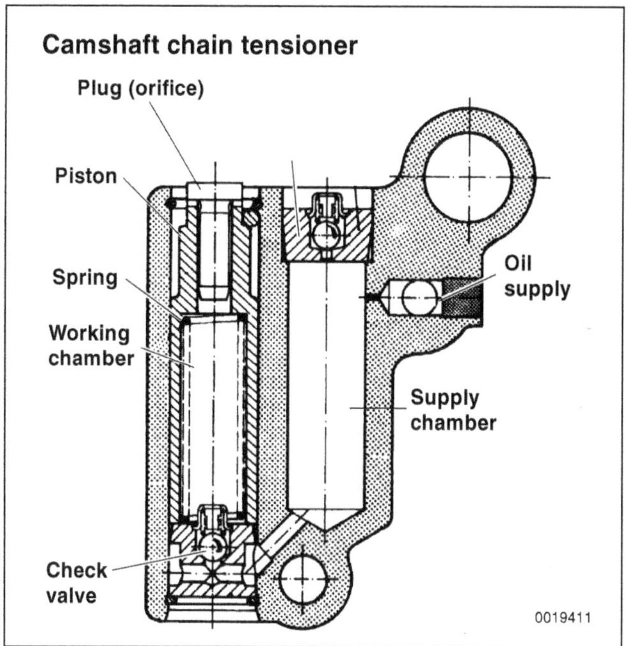

◀ The camshaft chains are tensioned via hydraulic tensioners. Oil feed lines at the front timing chain covers supply oil pressure directly to the tensioners.

Lubrication system

The unique 10.5-liter (11 quart) lubrication system is a dry sump design. A tandem oil pump (return/delivery) supplies pressurized oil to engine and also pumps the oil back to the oil tank. The pump is gear driven directly off the intermediate shaft.

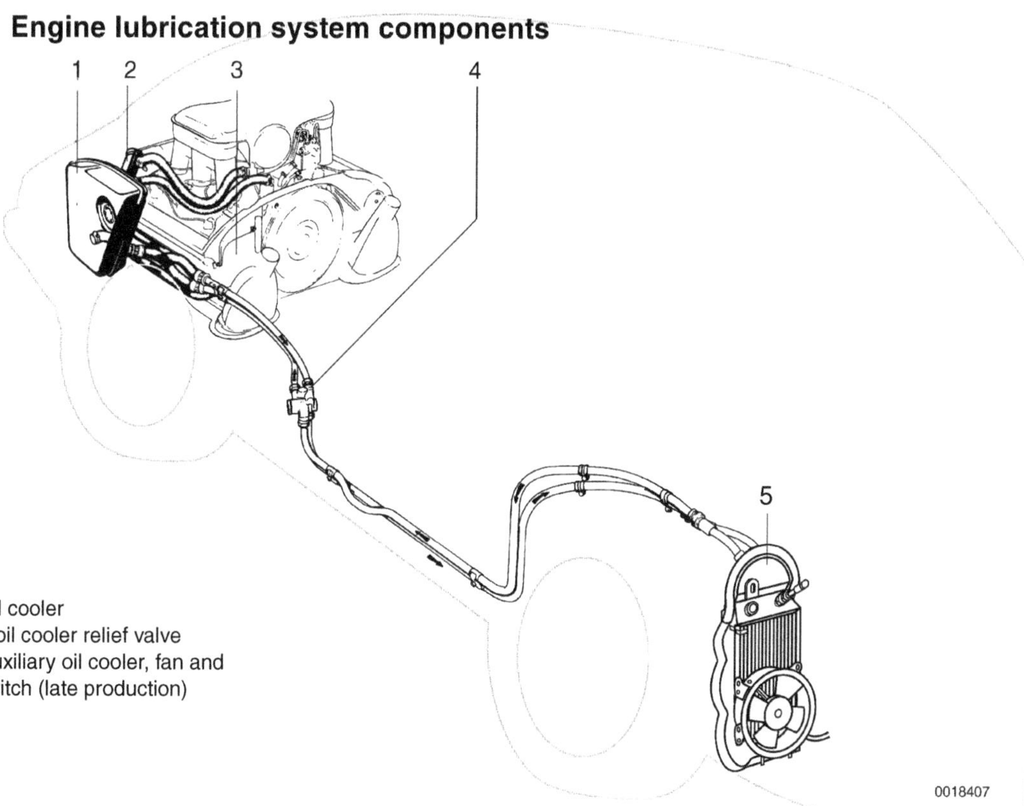

1. Oil tank
2. Oil filler
3. Engine oil cooler
4. Auxiliary oil cooler relief valve
5. Finned auxiliary oil cooler, fan and thermoswitch (late production)

The engine oil is cooled via two oil coolers. The primary oil cooler is mounted to the side of the crankcase. Flow through it is controlled via a thermostat that prevents oil from circulating until the oil has reached a specified temperature.

An auxiliary oil cooler is mounted ahead of the right front tire. An auxiliary oil cooler thermostat/pressure relief valve also controls flow into this cooler. On later models, a thermostatically-controlled fan is used with the auxiliary oil cooler. See **170 Lubrication System** for additional information.

Engine cooling

Engine cooling is provided by the alternator driven fan V-belt. Ducting provides control of the air over the engine. The fan aids in cooling the alternator via the slots in the fan hub (**arrow**). A 92 amp alternator is used.

The electric heater blower motor (**arrow**) also assists in engine cooling under the following conditions:

- Vehicle stationary
- Ignition switch on
- Engine compartment temperature of at least 128° C (260° F)

The heater blower fan will come on to cool the engine compartment. A temperature switch in the crankcase vent cover at the rear of the engine controls the heater blower motor operation.

MECHANICAL TROUBLESHOOTING

When troubleshooting an engine that fails to start or runs poorly, first check its mechanical condition—particularly if the engine has high mileage.

NOTE —
A cylinder compression test is given in **010 General**. *Also covered there is a brief description of a wet compression test and a cylinder leak down test.*

Table b lists the symptoms of common engine mechanical problems, their probable causes and the suggested corrective actions. The bold type indicates the repair groups where applicable test and repair procedures can be found.

Table b. Engine Mechanical Troubleshooting

Symptom	Probable cause and corrective action
1. Engine will not start or run. Starter cranks engine at normal speed.	a. No fuel reaching engine. Check for fuel in tank. b. Fuel pump not operating. Repair group **201** b. No voltage to spark plugs. Check ignition system. Repair group **280**
2. Blue-gray exhaust smoke and oily spark plugs. Indicates oil burning.	a. Contaminated or gasoline-diluted engine oil. Change engine oil and filter. Repair group **030** b. Faulty valve guide seals or valve guides. Replace valve guides oil seals or overhaul cylinder heads. Repair group **150** c. Worn piston rings. Overhaul or replace engine.
3. Pinging or rattling when under load, traveling uphill, or accelerating, especially from low speeds. Indicates pre-ignition or detonation.	a. Fuel octane level too low. b. Engine running too hot or overheating. Check engine oil cooler system. Repair group **170** c. Air/fuel mixture too lean. Repair group **240**
4. Light metallic tapping that varies directly with engine speed. Oil pressure warning light **not** illuminated.	a. Valve clearances excessive. Repair group **030** b. Low oil pressure or possibly defective warning light circuit. Check engine oil pressure and warning light circuit. Repair group **170**
5. Light metallic knock that varies directly with engine speed. Oil pressure warning light blinking or fully illuminated; may be most noticeable during hard stops or cornering.	a. Low oil level. Check and correct oil level. b. Contaminated or gasoline-diluted engine oil. Change engine oil and filter. Repair group **030** c. Low oil pressure. Worn or faulty oil pump. Oil pump pickup resticted. Check oil pressure. Repair group **170**
6. Growling or rumbling that varies with engine rpm. Indicates a bad bearing in an engine-driven accessory.	a. Remove A/C belt and fan/alternator belt. Check for play, bearing roughness, and loose mountings.

101 Engine Removal and Installation

Special Tools

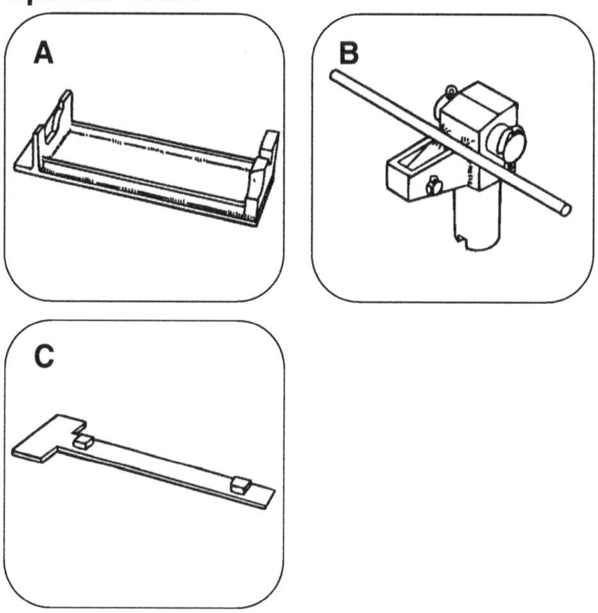

Special tool

A 9111 . . . . Engine fitting plate
(Porsche ordering #000 721 911 10)

B 9111/1 . . . . Adapter
(Porsche ordering #000 721 911 11)

C 9111/2 . . . . Support plate
(Porsche ordering #000 721 911 12)

GENERAL

This repair group covers engine removal and installation. The complete engine/transmission assembly is removed as a single unit through the bottom of the engine bay.

Removal and installation of the engine/transmission assembly is best accomplished by raising the car on an automotive lift. Support the engine/transmission from below using a heavy duty rolling cart. The complete engine with transmission weighs approximately 600 lbs.

Special tools

For an alternate removal procedure, some special jack adaptors are recommended to support the engine/transmission assembly. The jack adaptors allow the rear of the engine to pivot down and the engine/transmission assembly to be pulled out from the rear of the vehicle.

> ***WARNING —***
> *Due to risk of personal injury, be sure the engine is cold before beginning the removal procedure.*

> ***CAUTION —***
> - *Cover rear fender surfaces and painted areas surrounding engine bay before beginning removal procedure.*
> - *As an aid to installation, label all components, wires, and hoses before removing them.*
> - *Do not reuse gaskets, O-rings or seals during reassembly.*

ENGINE AND TRANSMISSION, REMOVAL AND INSTALLATION

Engine/transmission, removing

– Disconnect negative (–) battery cable and cover battery terminal to keep cable from accidentally contacting battery terminal.

> ***WARNING —***
> *Avoid getting tools or clothing near the battery. Battery electrolyte is a corrosive acid. Always wear eye protection when working on or near the battery.*

– Raise vehicle on lift to a comfortable working height to allow access to top of engine.

> ***WARNING —***
> *Be sure vehicle is stable and fully supported on the manufacturer's lift points. Removal of the engine/transmission assembly will upset the balance of the vehicle. Use tall jack stands to support the vehicle at all four corners.*

◄ Without loosening or disconnecting A/C lines, remove A/C compressor mounting bolts (**arrows**).

- Disconnect A/C harness connector **A**.
- Tilt compressor to disconnect drive belt from pulley.
- Swing compressor out of engine compartment onto quarter panel. Rest compressor on suitable pad.

> ***WARNING —***
> *A/C refrigerant hoses and components are under high pressure. Do not loosen, disconnect or damage A/C refrigerant lines or fittings.*

◄ Remove A/C mount fasteners (**arrows**). Remove mount.

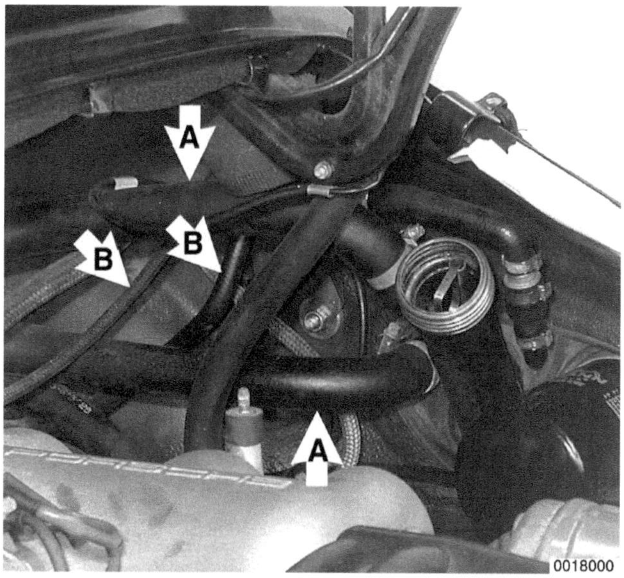

0018000

– Remove air filter cover and element from air filter housing.

◀ Disconnect emissions/breather hoses at right side of engine compartment.

- Oil tank breather hoses, top and bottom (**A**)
- Vacuum line and breather hose from emission control valve (**B**).

– Disconnect vent hose from rear of air filter housing.

NOTE —

- *A spring type clamp is used on one of this hose in place.*

0018002

◀ Remove large plastic elbow tube between heater blower fan on top of engine and heat exchanger feed tub

- Remove coil cover
- Remove bolts holding tube to blower housing, in front (**arrow**) and at rear of blower housing.
- Loosen hose clamp at heat exchanger end (bottom of elbow) and slide elbow out.

0018003

◀ Remove cover from electric panel in left side of engine compartment.

- Pull off coil secondary wire (**1**).
- Pull electrical harness sheath off electric panel bracket (**2**).
- Detach 14-pin harness connector from panel.

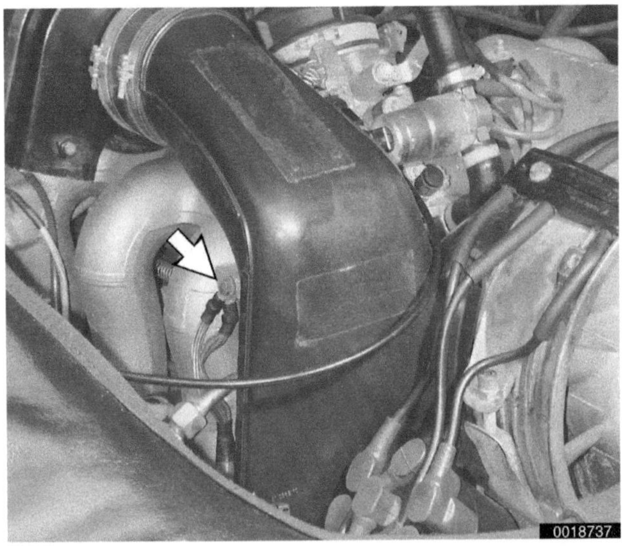

Remove ground wire mounting bolt (**arrow**) from intake runner 1.

Disconnect harness connectors at left of engine:

- Cylinder head temperature sensor (**1**)
- Engine speed sensor (**2**)
- Reference sensor (**3**)
- Oxygen sensor (**4**)

NOTE —

Later model connector shown in photo. Earlier models may have two connectors to oxygen sensor.

Disconnect hoses in left side of engine compartment. Counterhold fuel fittings when loosening.

- Disconnect fuel supply line at top of fuel filter (**A**).
- Disconnect fuel return line from rear of left side fuel rail (**B**).

WARNING —

Fuel will be discharged. Wrap a shop towel around fuel line fittings when loosening. Do not smoke or work around heaters or other fire hazards. Keep an approved fire extinguisher handy.

- Disconnect power brake vacuum hose (**C**).

◄ Disconnect engine wiring harness connectors:

- Multi-pin harness near left rear shock tower(**A**)
- Idle speed control valve (**B**)
- Throttle position switches (**C**)
- Air flow sensor (**D**).

◄ Disconnect cruise control cable from vacuum module at rear left of engine compartment:

- Remove screws holding cable end flange to vacuum diaphragm. Pull out end flange.
- Squeeze plastic retainer clip and push forward to release cable.

◄ Working inside passenger compartment, remove floor mats at rear of center tunnel.

NOTE —
Slide driver's seat forward to gain access to rear center tunnel.

- Remove access cover.
- Disconnect wires for speedometer sender (**arrows**). Push wires through bulkhead.
- On 1984 - 1986 models: Pull back rubber boot on shift linkage. Loosen 4 mm allen screw and pull shift rod out of shift coupler.

– Raise vehicle to gain access to bottom of engine.

◀ On 1987 - 1989 models: Working underneath car, push back rubber boot covering shift coupling.

- Remove hex bolt (**arrow**) and push coupling forward off transmission selector shaft.

◀ Disconnect accelerator pull rod ball-joint end (**arrow**) from throttle relay lever on left side of transmission.

◀ Remove clutch cable or clutch fluid supply hose retaining bracket (**arrow**).

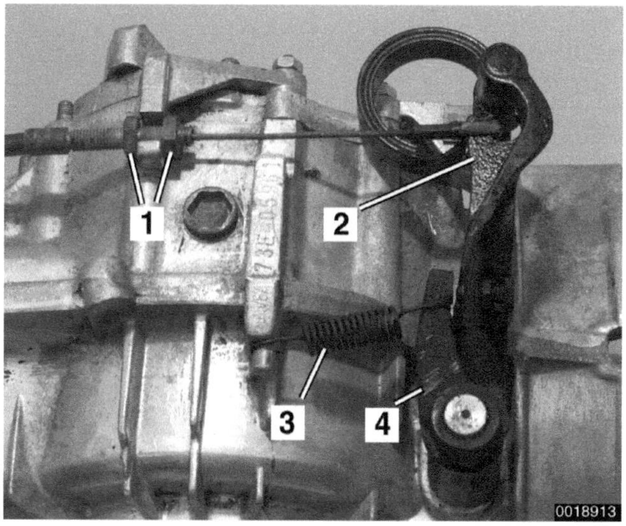

On 1984 - 1986 models: Remove clutch release lever:

- Loosen clutch cable mounting nuts and remove cable from boss (**1**) and then disconnect cable from release lever (**2**).
- Disconnect spring on positioning lever (**3**), remove circlip from lever (**4**), and slide lever off splined end of clutch cross-shaft.

On 1984 - 1986 models: Use large screwdriver to press clutch release lever forward. Remove release lever from clutch cross-shaft.

> ***WARNING —***
> *Once the clutch release lever passes a certain point, the auxiliary spring will force the release lever forward. Keep fingers and hands clear when pressing lever forward.*

On 1987 - 1989 models: Working above driver's side of transmission, remove clutch slave cylinder mounting nuts (**arrows**). Remove slave cylinder from transmission and support from body using stiff wire.

> ***CAUTION —***
> *Do not operate clutch pedal with clutch slave cylinder removed.*

– Drain crankcase oil.

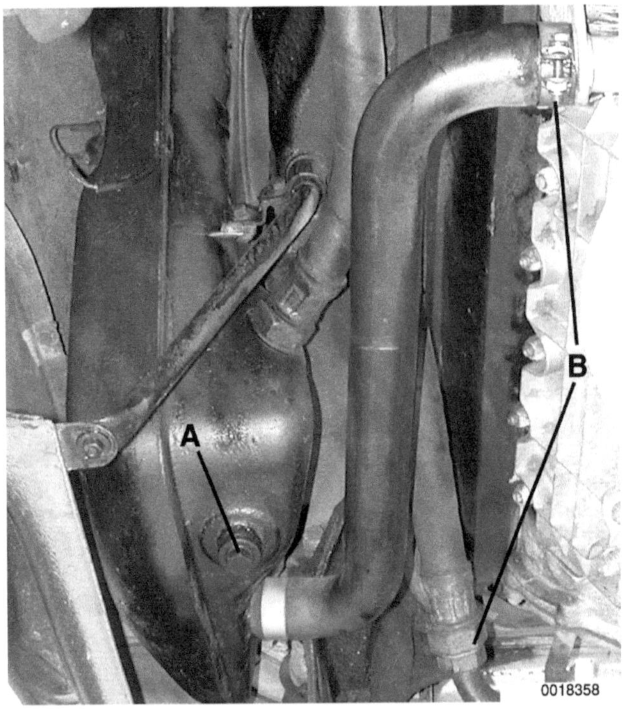

◄ Remove drain plug (**A**) and drain oil from oil tank. Disconnect oil lines (**B**). Counterhold line fitting when loosening.

◄ Remove rear stabilizer bar.

- Unbolt stabilizer bracket bolts (**arrows**) at left and right sides and lower bar out of car (left side shown).

– Disconnect wiring from starter motor.

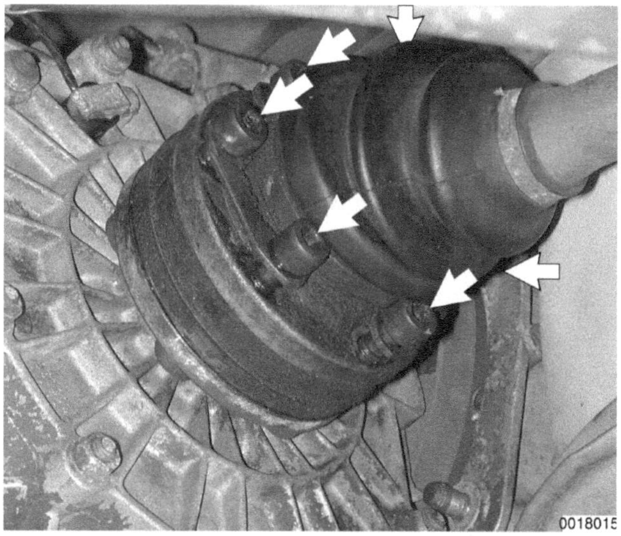

◄ Loosen inner constant velocity (CV) joint bolts (**arrows**) from differential drive flanges (left side shown). Suspend shafts from body with mechanic's wire.

NOTE —

On 1984 and early 1985 models cover CV joints with a plastic bag and tape bag to CV joint shaft to keep dirt out of the joints.

– Remove left and right heater hoses from heat exchanger connector.

– Lower vehicle until weight of engine/transmission assembly engine is supported by hydraulic support or equivalent.

NOTE —

If using a floor jack, raise it as much as possible to allow for maximum travel when it is lowered with the engine.

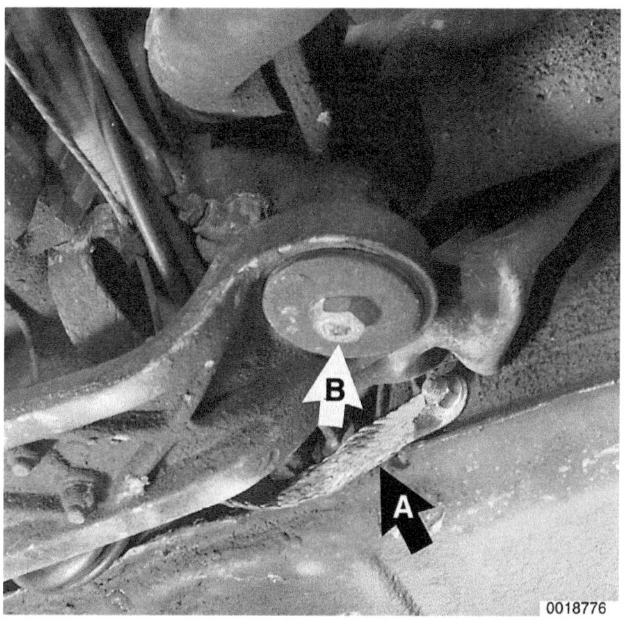

◄ Disconnect ground strap (**A**) between transmission and body.

- While supporting engine/transmission assembly, carefully remove two transmission mount bolts (**B**). Right side shown.

> ***WARNING —***
>
> *Check often to make sure that the car remains stable and well-supported while the engine is being removed.*

◄ Remove 2 engine support crossmember bolts (**arrow**) (right side shown).

0018153

Carefully lift car (or lower engine/transmission assembly) while allowing engine to tilt backward as assembly is removed from rear of car.

NOTE —

If separating the engine from the transmission, see **Engine/transmission, separating and rejoining.**

Engine/transmission, installing

Place engine/transmission assembly under engine/transmission bay. Lower car or raise engine/transmission.

- On 1984 - 1986 models: Tilt engine back and guide shift rod into center tunnel hole.

0018016

Raise engine/transmission assembly up to approximately 3 in. of mounting points in engine bay.

> ***CAUTION —***
>
> *Be very careful not to pinch any wiring harnesses or hoses between the body and the engine as it is raised up into the engine bay.*

- With engine 3 in. from its mounts connect heat exchanger hoses.

> ***CAUTION —***
>
> *Connecting heater hoses after engine is bolted in may result in damaged hoses.*

On 1987 - 1989 models: Working above driver's side of transmission, install clutch slave cylinder and torque bolts (**arrows**).

NOTE —
Be sure to align slave cylinder pushrod with clutch fork.

Tightening Torque
- Clutch slave cylinder to bell housing.... 25 Nm (18 ft-lb)

– Carefully raise engine/transmission assembly up to its installed position.

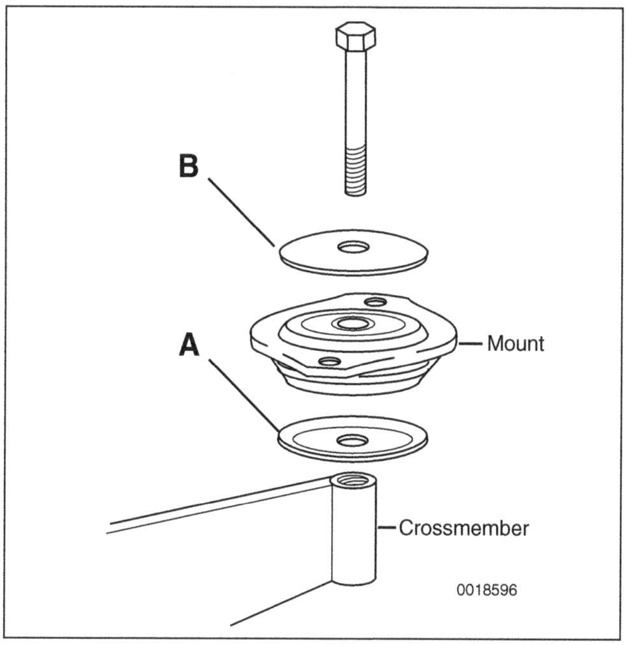

Place washer between engine mount and crossmember with washer concave facing up (**A**).

- Insert engine mounting bolt with concave washer facing down (**B**). Press bolt down through engine mount and washer into crossmember.
- Start engine mounting bolts into crossmember but do not tighten them.

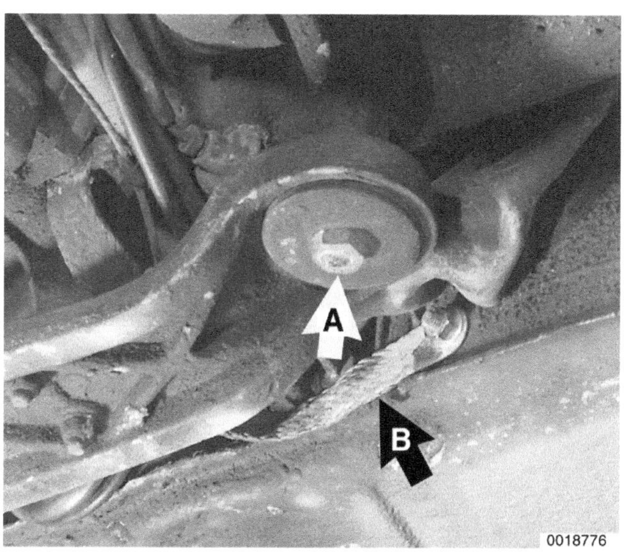

Install transmission mount bolts and washers (**A**). Arrange washers as per for engine mount (shown above).

– Reattach transmission to body ground strap (**B**).

– Tighten engine and transmission mounting bolts.

Tightening Torques
- Crossmember to transmission .... 25 Nm (18 ft-lb)
- Crossmember to crossmember mount .. 80 Nm (58 ft-lb)
- Ground strap nut to body stud .... 18 Nm (14 ft-lb)

0018913

◄ On 1984 - 1986 models: Install clutch release levers and clutch cable on bottom of transmission.

NOTE —
See **301 Clutch** *for detailed instructions on clutch adjustments on 1984 - 1986 models.*

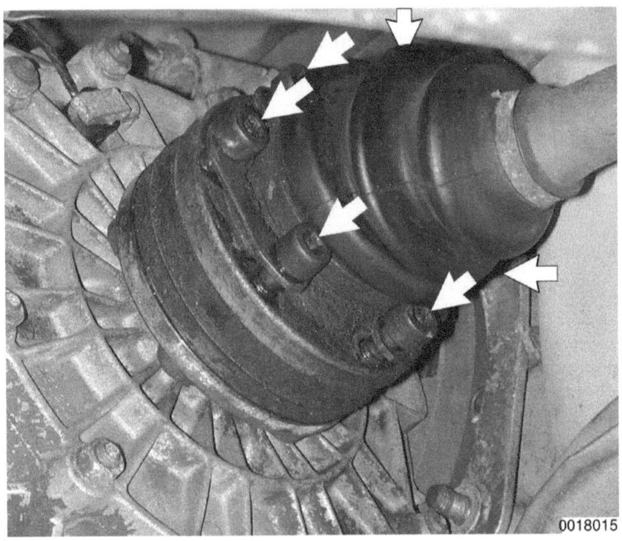
0018015

◄ Install right and left CV joints and tighten bolts (**arrows**).

Tightening Torque
- CV joint to final drive flange . . . . 42 Nm (30 ft-lb)

– Connect starter power feed cable.

0018026

◄ Reinstall stabilizer bar and tighten bolts (**arrows**).

Tightening Torques
- Stabilizer to body (M8) . . . . 25 Nm (18 ft-lb)
- Stabilizer to trailing arm link (M12) . . . . . 90 Nm (66 ft-lb)

– Install new sealing rings on both oil drain plugs and install.

Tightening Torques
- Drain plug to crankcase . . . . 70 Nm (52 ft-lb)
- Drain plug to oil tank. . . . . 42 Nm (31 ft-lb)

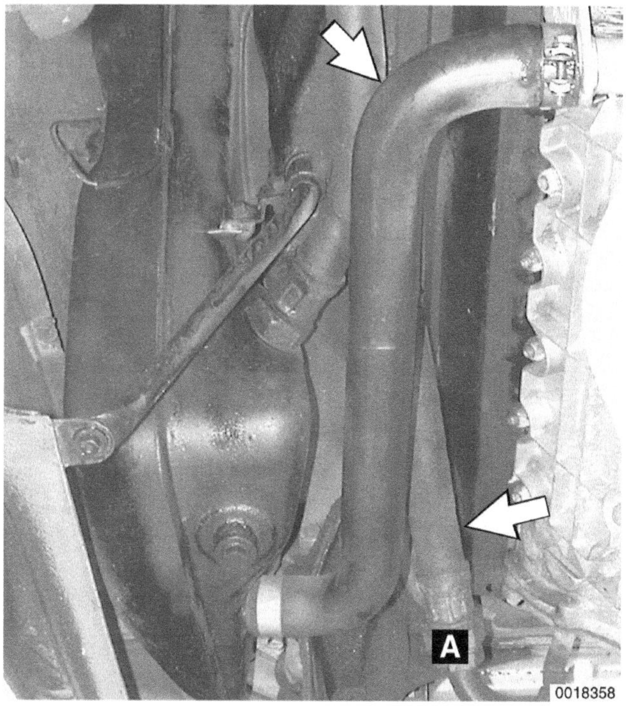

◀ Reconnect oil tank S-hose and return hose (**arrows**). Counterhold return hose fitting (**A**) while tightening.

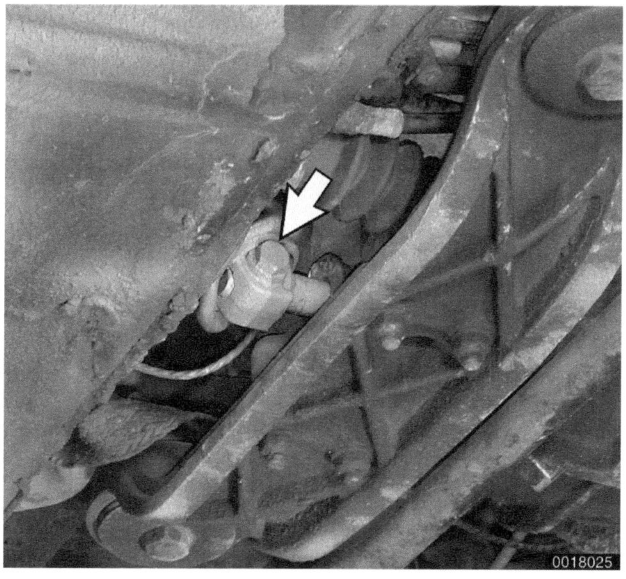

◀ On 1987 - 1989 models: Install shift rod bolt (**arrow**) and tighten.

Tightening Torque
- Shift rod bolt to coupler. . . . . 18 Nm (14 ft-lb)

– Connect throttle linkage to relay lever on side of transmission.

◀ Working inside passenger compartment, reconnect speedometer harness connectors (**arrows**) through hole in center tunnel.

– On 1984 - 1986 models: Connect shift rod to coupler and tighten.

- Install access cover and floor mats.

Tightening Torque
- Shift rod allen bolt to coupler . . . . 8 Nm (6 ft-lb)

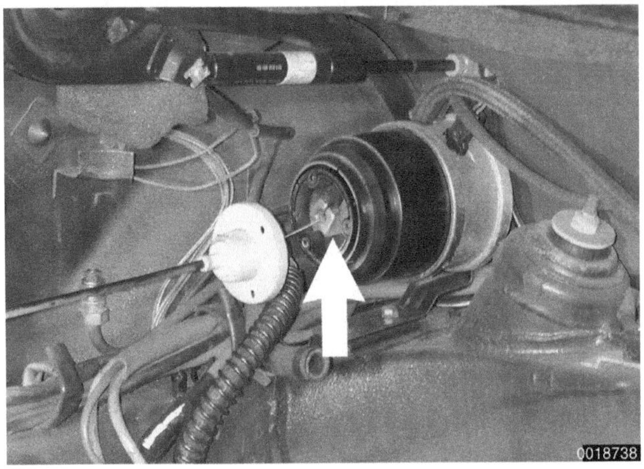

◀ Attach cruise control cable to diaphragm housing. Pull back on locking plastic clip until it snaps into metal retainer (**arrow**).

- Mount cruise control cable flange to diaphragm housing.

◀ Reconnect left side engine electrical harness connectors:

- Cylinder head temperature sensor (**1**)
- Engine speed sensor (**2**)
- Reference sensor (**3**)
- Oxygen sensor (**4**).

NOTE —

If the speed sensor and reference sensor harness connectors are reversed, the engine will not start.

– Reattach remaining electrical harness connectors:

- Air flow sensor
- Idle control valve
- Throttle switches
- Multi-pin connector beside left rear shock tower
- Multi pin connector in left rear fuse/relay panel
- Ignition coil secondary wire
- Ground connection to intake runner 1.

– Install rear fuse/relay panel cover.

– Remainder of installation is reverse of removal.

- Reinstall A/C compressor and adjust drive belt.
- Replace oil filter.
- Add 8 quarts of engine oil.
- Reconnect battery ground terminal.

Tightening Torque

- A/C compressor to mount . . . . 25 Nm (18 ft lb)

– Start engine with a helper checking for leaks or any other problems.

– With engine running and at normal operating temperature add 2.5 quarts of engine oil.

Engine/transmission, separating and rejoining

◄ Detach wiring harness connectors (**arrows**) from transmission.

◄ On 1987-1989 models: Remove clutch cross-shaft.

- Remove bolt (**arrow**) from cross-shaft retaining bracket.
- Pull out cross-shaft by attaching slide hammer with 6 mm bolt into end of cross-shaft.

◄ Remove 4 nuts holding transmission to engine and slide transmission off engine.

– Installation is reverse of removal, noting following.

- Cable-operated clutch: As transmission slides onto engine studs, turn clutch cross-shaft so that clutch release fork hooks into release bearing.
- Hydraulically operated clutch: Prior to installing transmission, use duct tape or equivalent to hold clutch release fork on release bearing. See **301 Clutch**. Clutch cross-shaft is installed after engine and transmission are joined.

Engine mount assembly Carrera Coupe and Targa

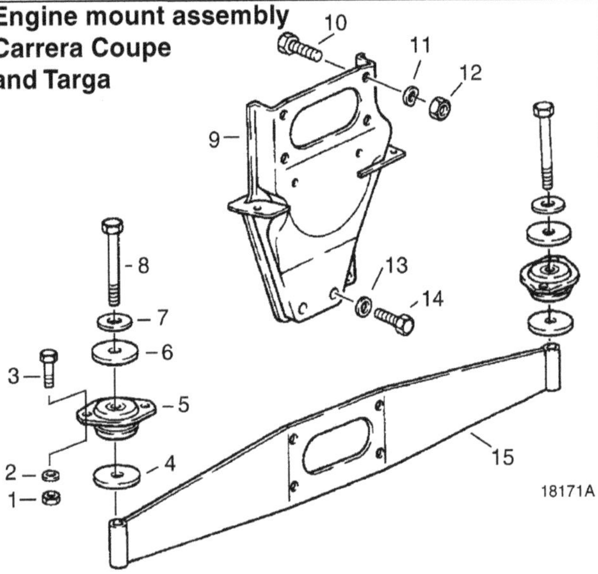

1. Hex nut M8
2. Lockwasher
3. Hex bolt M8x35
4. Washer
5. Bearing
6. Washer
7. Washer
8. Hex bolt M12x1.5x95
9. Engine mount carrier
10. Hex bolt M10x18
11. Spring washer
12. Hex nut M10
13. Washer
14. Hex bolt M10
15. Crossmember

Engine mount assembly Carrera Cabriolet

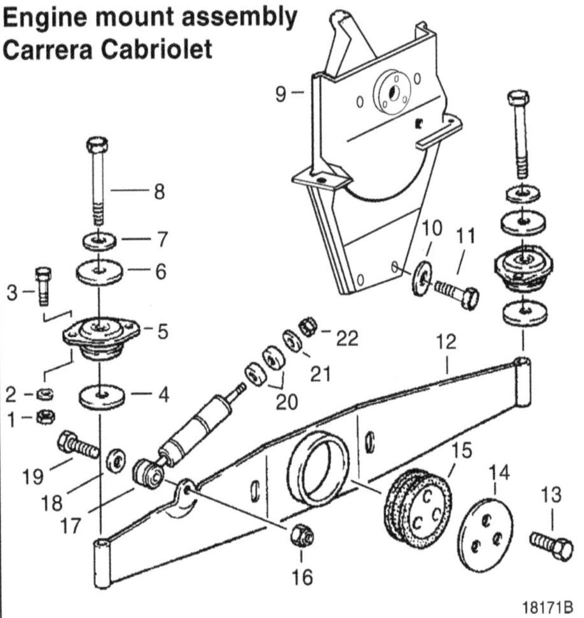

1. Hex nut M8
2. Lockwasher
3. Hex bolt M8x35
4. Washer
5. Bearing
6. Washer
7. Washer
8. Hex bolt M12x1.5x95
9. Engine mount carrier
10. Washer
11. Hex bolt M10
12. Crossmember
13. Hex bolt M10x35
14. End plate
15. Torsion bearing
16. Locknut M10
17. Vibration damper
18. Washer
19. Hex bolt M10x45
20. Spacer rings
21. Washer
22. Locknut M6

ENGINE/TRANSMISSION MOUNTS

The engine or transmission crossmember rubber mounts can be damaged or weakened due to age, oil contamination or accident. This damage can cause excessive noise or vibration in the passenger compartment.

Engine mount, removing and installing

Refer to illustrations for component locations.

Removal of the rear engine mounting crossmember and mount bracket is covered in **102 Engine Disassembly and Assembly**.

Tightening Torques

- Crossmember mount to body . . . . 25 Nm (18 ft-lb)
- Crossmember to crossmember mount . . 80 Nm (58 ft-lb)
- Engine mount carrier to crankcase . . . . . 40 Nm (30 ft-lb)
- Engine mount crossmember to carrier . . . . 40 Nm (30 ft-lb)

Transmission mount, replacing (915 transmission–1984-1986)

Transmission crossmember (1984-1986)

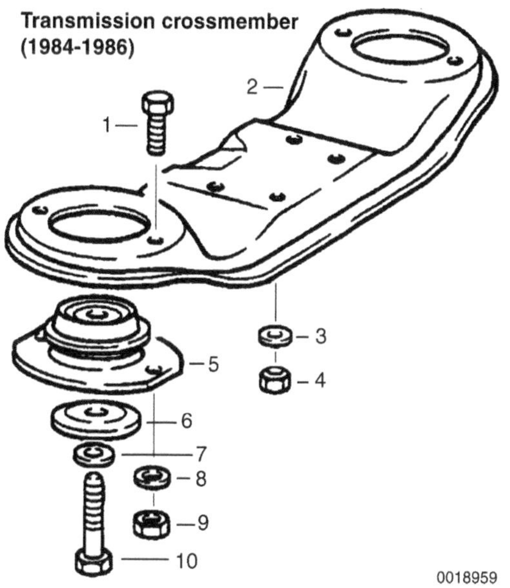

1. Hex bolt M8x16
2. Crossmember
3. Washer
4. Locknut
5. Mount
6. Concave washer
7. Lockwasher
8. Lockwasher
9. Hex nut M8
10. Hex bolt

Rubber transmission crossmember mounts in 1984-1986 models can be replaced without removing crossmember from transmission.

– Raise vehicle on lift and support safely.

> ***WARNING —***
> *Be sure vehicle is stable and fully supported on the manufacturer's lift points. Loosening of the engine/transmission assembly will upset the balance of the vehicle. Use tall jack stands to support the vehicle at all four corners.*

– Support transmission while removing transmission crossmember mount long bolts.

– Lower transmission just enough to reach over crossmember and remove bolts holding mount to crossmember.

– Installation is reverse of removal.

Tightening Torques

- Crossmember mount to body (M12) . . . . 80 Nm (58 ft-lb)
- Crossmember to crossmember mount (M8) . . . . 25 Nm (18 ft-lb)

Transmission mount, replacing (G50 transmission–1987-1989)

Transmission crossmember (1987-1989)

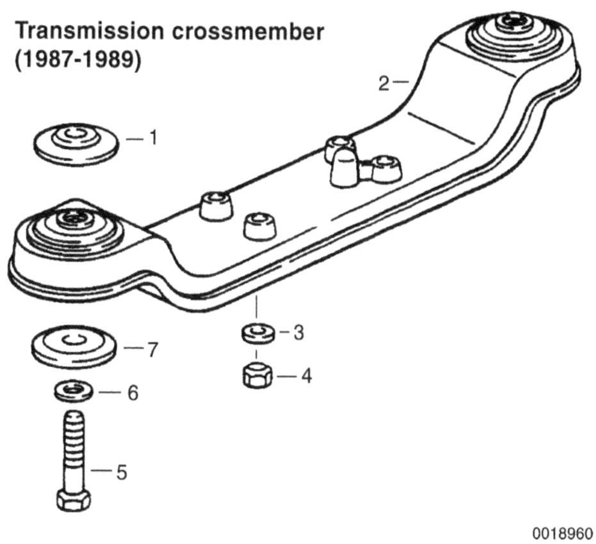

1. Concave washer
2. Crossmember
3. Washer
4. Locknut
5. Hex bolt M12
6. Washer
7. Concave washer

Rubber transmission crossmember mounts in 1987-1989 models cannot be replaced separately.

– Raise vehicle on lift and support safely.

> ***WARNING —***
> *Be sure vehicle is stable and fully supported on the manufacturer's lift points. Loosening of the engine/transmission assembly will upset the balance of the vehicle. Use tall jack stands to support the vehicle at all four corners.*

– Support transmission while removing transmission crossmember mount bolts from suspension crossmember.

– Remove crossmember mounting nuts from transmission studs.

– Installation is reverse of removal.

Tightening Torques

- Crossmember mount to body . . . . 80 Nm (58 ft-lb)
- Crossmember to transmission . . . . 25 Nm (18 ft-lb)

102 Engine Disassembly and Assembly

Special Tools

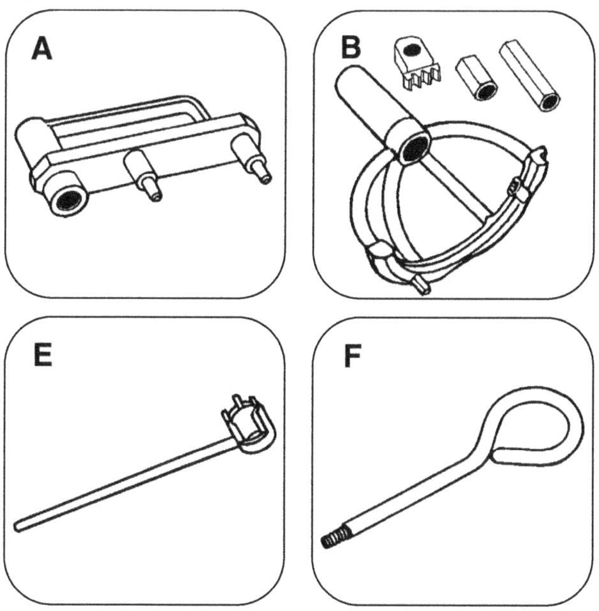

Special tool

A P 238 a. . . . . Flywheel locking tool (Porsche ordering #000 721 238 10)

B P 201/201 b . . . . Four-arm engine mount and adapters (Porsche ordering #000 721 201 00/20)

C 9191 . . . . Camshaft sprocket counterholding wrench (Porsche ordering #000 721 919 10)

D P 212 . . . . Camshaft sprocket lockpin tool (Porsche ordering #000 721 212 00)

GENERAL

This repair group covers engine disassembly and reassembly. Engine removal procedures, including separating the transmission from the engine, are covered in **101 Engine Removal and Installation**.

The repair information given here requires that the engine be removed from the car. The disassembly procedure should be carried out on an appropriate engine stand in a clean environment.

◄ Some of the procedures require special tools.

Reconditioning procedures and specifications are not covered here. Once the engine is disassembled to the component level, refer to the appropriate repair group for applicable reconditioning information.

- **130 Crankcase and Crankshaft**
- **131 Pistons and Cylinders**
- **150 Cylinder Heads**
- **151 Camshafts and Camshaft cases**
- **152 Camshaft Timing Chains**

Special Tools

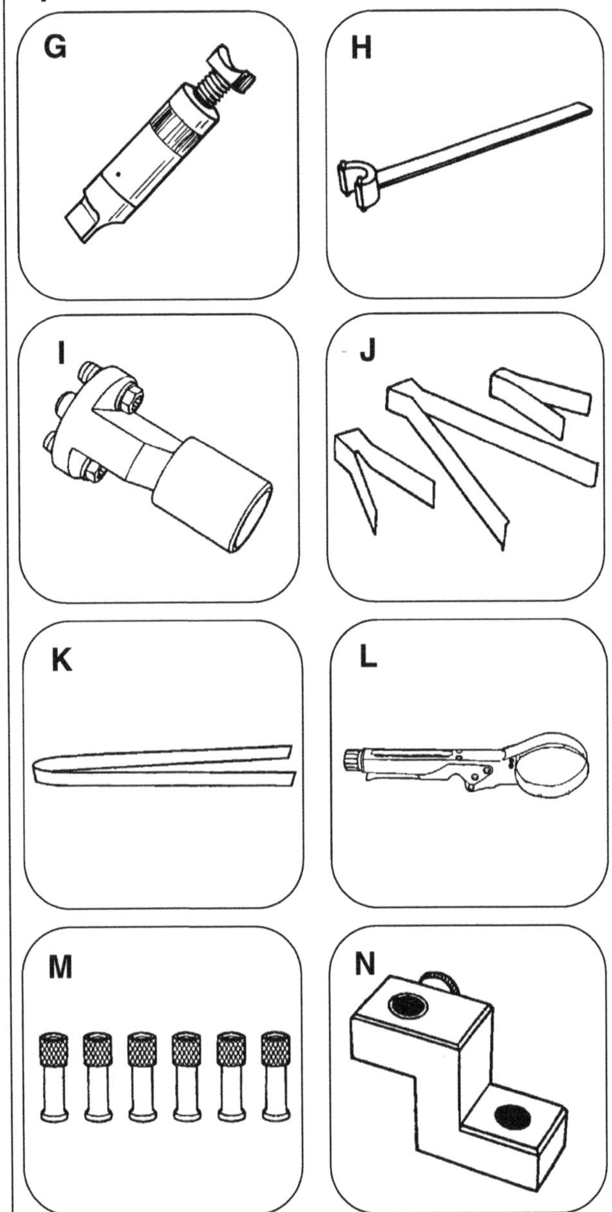

Special tool

E P 204 .... Timing chain idler pulley brace
F 9236 .... Pulley spanner wrench
(Porsche ordering #000 721 923 60)
G P 209 ....Crankshaft mounting bracket
H P 221 .. Left connecting rods support during engine assembly
(Porsche ordering #000 721 221 00)
I P 222 ..... Left timing chain support during engine assembly
(Porsche ordering #000 721 222 00)
J US 1008a .... Piston ring compressor
K P 140 ....Nuts for securing cylinders during engine assembly
(Porsche ordering #000 721 140 00)
L P 207 ....Dial gauge mounting bracket
(Porsche ordering #000 721 207 00)

CAUTION —
- *The pulley end of the engine is referred to as the **front**. The flywheel end of the engine is referred to as the **rear**.*
- *If the engine is disassembled because of bearing failure, a thorough flushing of the lubrication system must be carried out prior to reassembly. Refer to* **130 Crankcase and Crankshaft** *and* **170 Engine Lubrication**.

NOTE —
Some repair operations do not require engine removal and disassembly. Some of the more common engine-installed operations are as follows:

- *Camshaft timing adjustment and timing chain and tensioner replacement. See* **152 Camshaft Timing Chains**.
- *Camshafts and rocker arm replacement. See* **151 Camshafts and Camshaft Cases**.
- *Leaky engine oil return tube replacement. See* **170 Engine Lubrication**.

ENGINE, DISASSEMBLY

The disassembly procedure is divided into subsections as follows:

- **Engine peripherals, removing**
- **Camshaft timing chains and camshafts, removing**
- **Camshaft cases and cylinder heads, removing**
- **Pistons and cylinders, removing**
- **Crankcase, disassembling**
- **Crankshaft, disassembling**

Crankcase assembly

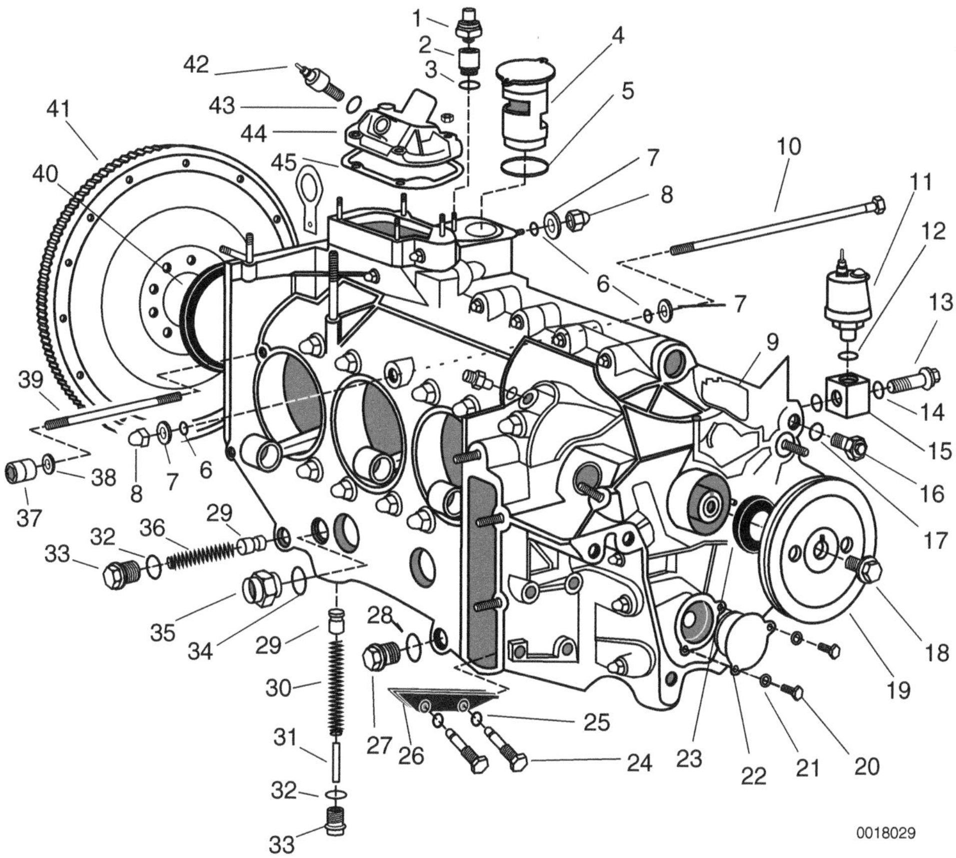

1. Oil pressure switch
   - torque to 20 Nm (15 ft-lb)
2. Switch adapter
   -torque to 35 Nm (26 ft-lb)
3. Sealing ring (18 x 24mm)
4. Oil thermostat
5. Thermostat sealing O-ring (39.2 x 3 mm)
6. Case bolt sealing ring
7. Case bolt beveled washer
8. Case bolt cap nut
   -torque to 35 Nm (26 ft-lb)
9. Crankcase
10. Case bolt
11. Oil pressure sensor
    -torque to 35 Nm (26 ft-lb)
12. Oil pressure sensor sealing ring (18 x 24 mm)
13. Oil pressure sensor adapter bolt,
    -torque to 35 Nm (26 ft-lb)
14. Adapter bolt sealing ring (12 x 16 mm)
15. Transmitter adapter housing
16. Oil temperature sensor
    -torque to 25 Nm (18 ft-lb)
17. Temperature sensor sealing ring
18. Front pulley bolt
    -torque to 80 Nm (59 ft-lb)
19. Front pulley
20. Intermediate shaft cover bolt
21. Intermediate shaft cover bolt washer
22. Intermediate shaft cover
23. Crankshaft oil seal, front
24. Chain guide rail bolt
25. Guide rail bolt sealing ring (10 x 16 mm)
26. Chain guide rail
27. Oil drain plug
    -torque to 70 Nm (52 ft-lb)
28. Drain plug sealing ring (22 x 27 mm)
29. Oil pressure relief valve piston
30. Oil pressure relief valve spring
31. Oil pressure relief valve sleeve
32. Sealing ring (18 x 24 mm)
33. Oil pressure relief valve plug
    -torque to 60 Nm (44 ft-lb)
34. Line adapter sealing ring, (22 x 27mm)
35. Oil line adapter
    -torque to 120 Nm (89 ft-lb)
36. Oil pressure safety valve spring
37. Cylinder head allen nut
38. Cylinder head washer
39. Cylinder head stud
40. Crankshaft oil seal, rear
41. Flywheel
42. Heater blower after-run temperature switch
    -torque to 25 Nm (18 ft-lb)
43. Temperature switch sealing ring
44. Breather cover
45. Breather cover gasket

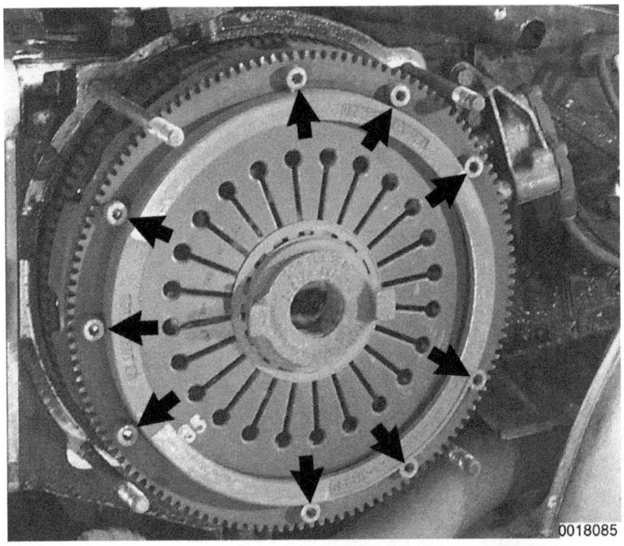

Engine peripherals, removing

< Remove clutch bolts (**arrows**). Remove clutch from flywheel.

< Hold flywheel stationary with Porsche special tool P 238 (remove inner pin from tool) or equivalent locking device and loosen flywheel bolts (**arrows**). Remove flywheel.

– Remove metal shrouding from flywheel end of engine.

< Remove TDC sensor bracket bolts from top of engine (**arrows**). Remove sensor.

– Mount engine on engine stand.

> ***WARNING —***
> *A 4-arm or 5-arm engine stand adapter (Porsche special tool P 201) should be used to mount the engine. 3-arm and "generic" engine stands are dangerous and may damage the engine block.*

– Remove metal shrouding from pulley end of engine.

– Remove engine support crossmember.

– Rotate engine on stand so exhaust system is on top. Lock engine stand in this position.

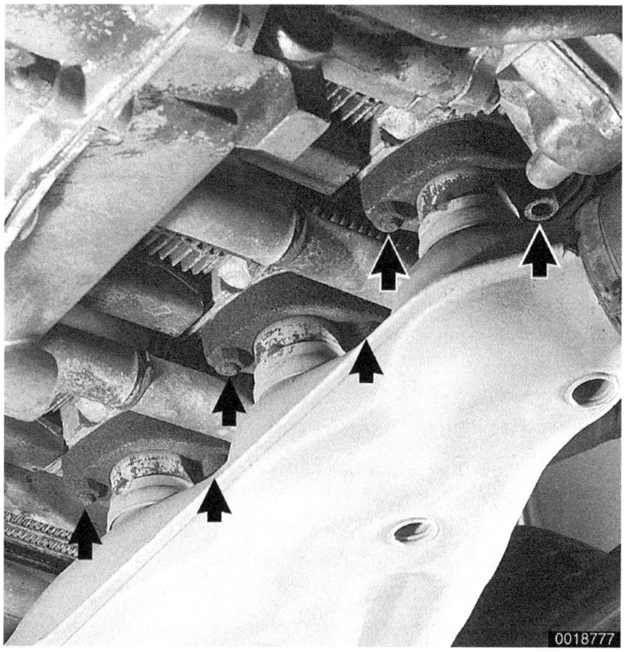

◀ Remove heat exchanger mounting nuts (**arrows**). Right side shown.

– Remove muffler retaining straps.

– Remove heat exchanger crossover tube.

– Remove exhaust system as a complete unit. See **260 Exhaust System**.

– Remove muffler support bracket.

– Rotate engine so that intake side is up. Lock engine stand in this position.

◀ Remove A/C compressor mounting bracket and engine support bracket bolts (**arrows**).

- Remove brackets.
- Note blind mounting nut on reverse side of engine support bracket.

NOTE —
*During disassembly, check for oil leakage at upper right mounting stud at rear engine carrier (**A**). If oil leakage is present, see **130 Crankcase and Crankshaft** for specific repair information.*

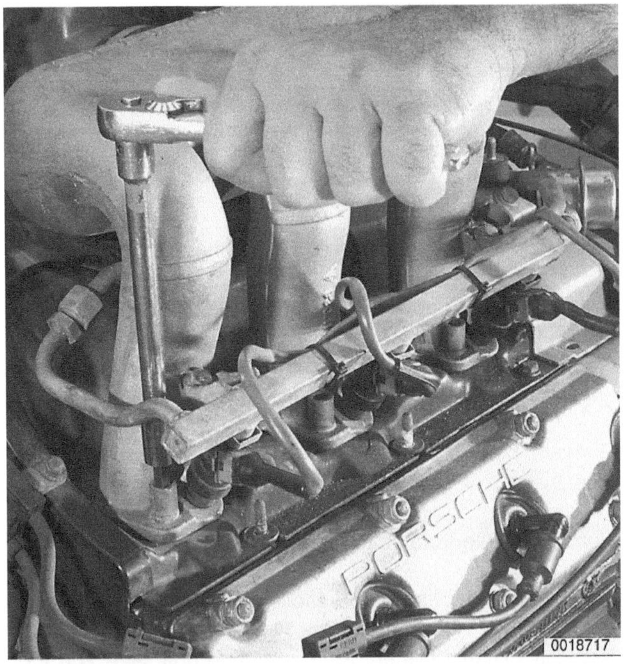

◀ Remove intake runner fasteners from cylinder heads.

◀ Disconnect electrical harness connectors on engine:

- Oil pressure gauge sender (**A**)
- Oil temperature sender (**B**)
- Oil pressure (warning light) switch (**C**)
- Heater blower after-run temperature sensor (**D**)

Remove cooling fan/alternator housing:

- Remove fan/alternator drive belt. See **030 Maintenance**.
- Remove allen bolt (**A**) from fan housing strap.
- Remove bolts (**B**) from engine cooling shroud.
- Remove spark plug wire supports (**C**).
- Pull alternator out partially and unbolt alternator ground strap from top of engine crankcase.

NOTE —
Once the drive belt is removed, reinstall alternator belt pulley and shims and tighten nut fingertight.

– Disconnect spark plug connectors. Unclip and remove distributor cap with wires.

– Remove left and right metal shrouding.

Pry off throttle linkage from bellcrank (**arrow 1**).

- Remove bell crank mounting nuts and remove bellcrank with spacer (**arrow 2**) from top of engine.

– Carefully lift off complete fuel system, intake manifolds and alternator with cooling shroud.

Remove oil cooler mounting nuts from top and bottom (**arrows**) of cooler.

– Slide oil cooler off mounting studs.

Check cast oil pipe webbing (**arrows**) in right crankcase half (lower oil cooler area) for oil leakage due to porous casting material.

- To repair, thoroughly clean crankcase in problem area using alcohol or acetone.
  - Fill area with Silastic® 732 RTV sealing compound or equivalent using light uniform pressure until depression is filled.
  - Allow 24 hours to harden.

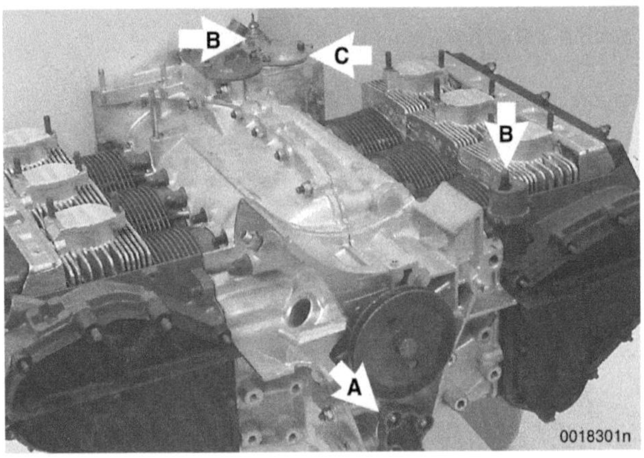

Remove following parts from engine:

- Intermediate shaft end cover (**A**).
- Oil pressure sensors (**B**).
- Oil thermostat (**C**).

Camshaft timing chains and camshafts, removing

Set crankshaft to **Z1** (TDC) cylinder 1 compression stroke (**arrow**).

- Remove intake and exhaust valve covers on both sides of engine.

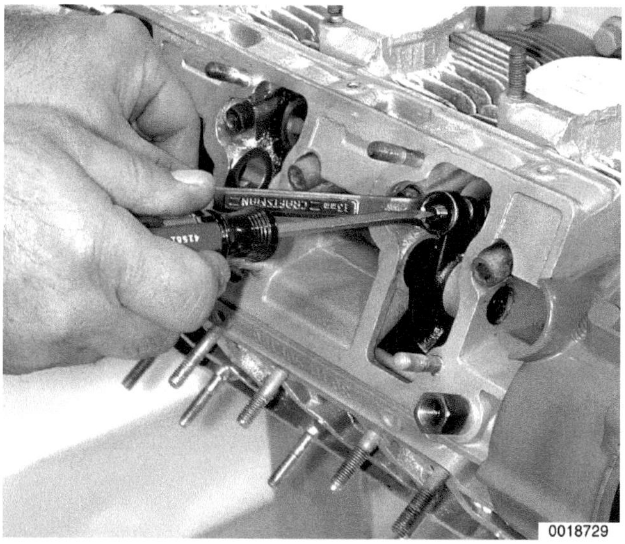

◀ Loosen all valve adjuster lock nuts and fully back off adjusting screws for all valves.

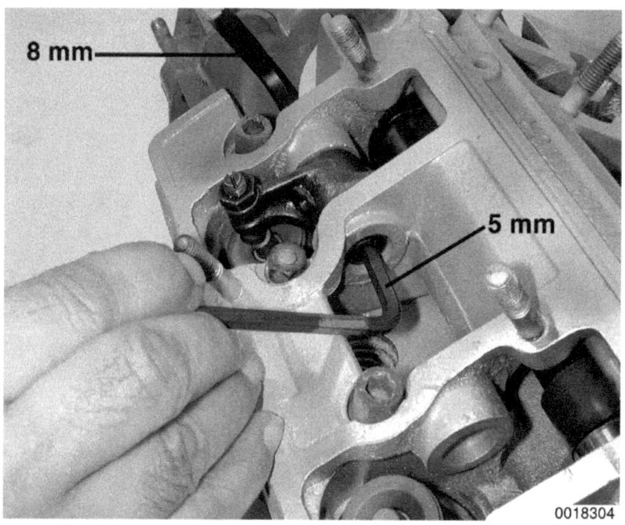

◀ Working at cylinder 1, counterhold rocker shaft nut with 8 mm allen wrench, then loosen opposite end with 5 mm allen wrench.

NOTE—

Mark location of each rocker arm. Do not interchange used parts.

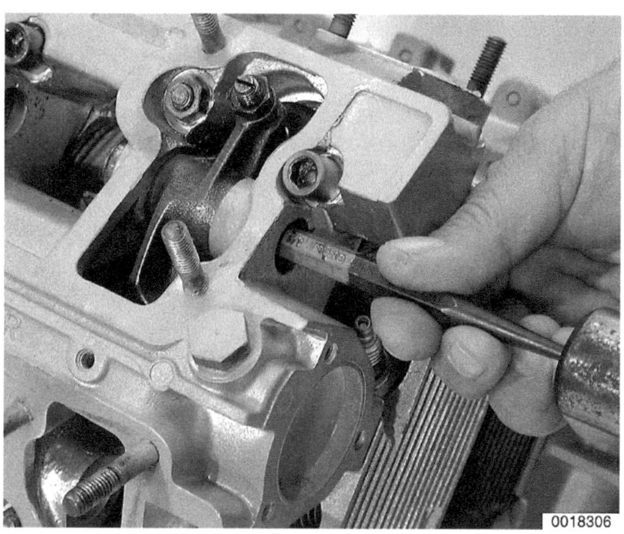

◀ Gently tap rocker shaft out and remove rocker arm.

- Turn crankshaft 120° in normal direction of rotation. Remove cylinder 6 rocker arms using above procedure.
- Following firing order, turn crankshaft 120° and remove rocker arms for next cylinder. Repeat until all rocker arms are removed.

NOTE—

The firing order is 1 - 6 - 2 - 4 - 3 - 5.

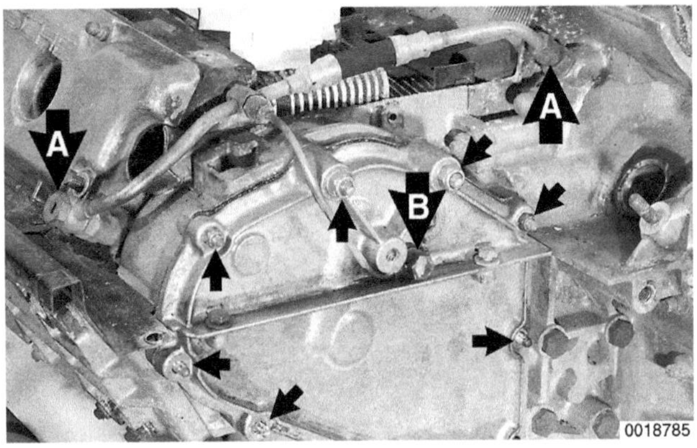

◀ Remove left camshaft case oil feed line pressure fittings (**A**).

- Remove left chain tensioner oil feed line pressure fitting (**B**).
- Remove left chain housing cover fasteners (**small arrows**), and remove cover.

◀ Loosen left camshaft sprocket mounting bolt.

NOTE —
Use Porsche special tool 9191 or equivalent to counterhold camshaft sprocket when loosening bolt.

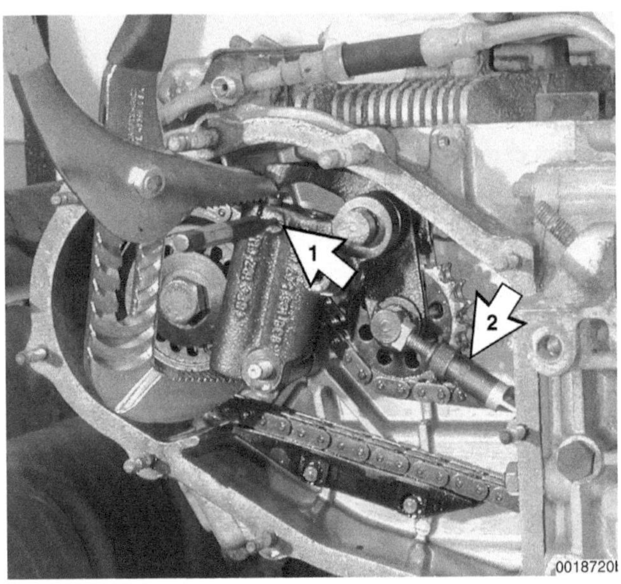

◀ Compress timing chain tensioner using large pliers.

- Lock tensioner down using 3 mm pin through hole in tensioner body (**1**).
- Special tool P 204 (**2**) can be used to brace idler sprocket before removing tensioner.

NOTE —
Check timing chain tensioner piston height before compressing. If piston is fully extended, carefully inspect timing chains and sprockets for wear. See **152 Camshaft Timing Chains**.

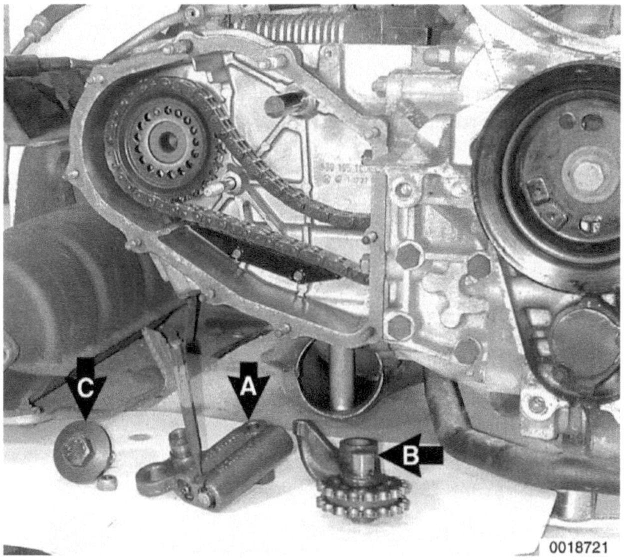

◀ Remove timing chain tensioner (**A**) and idler sprocket assembly (**B**) with special brace P 204 if installed.

– Remove camshaft sprocket bolt from left camshaft (**C**).

◀ Remove camshaft sprocket lockpin using Porsche special tool P 212 or equivalent.

– Remove camshaft sprocket and disengage from timing chain.

– Remove sprocket flange.

◀ Remove left camshaft:

- Remove woodruff key (**A**) from camshaft.
- Remove camshaft end cover retaining bolts (**arrows**) and remove camshaft end cover (**B**), shims and thrust spacer.
- Slide camshaft toward pulley end of engine to remove.

NOTE —

- *Wire tie end cover, shims and thrust spacer together and label.*
- *For identification, the left camshaft is marked "147-10" on the end.*

– Remove left timing chain housing.

– Repeat procedures on right side camshaft.

NOTE —

For identification, the right camshaft is marked "148-10" on the end.

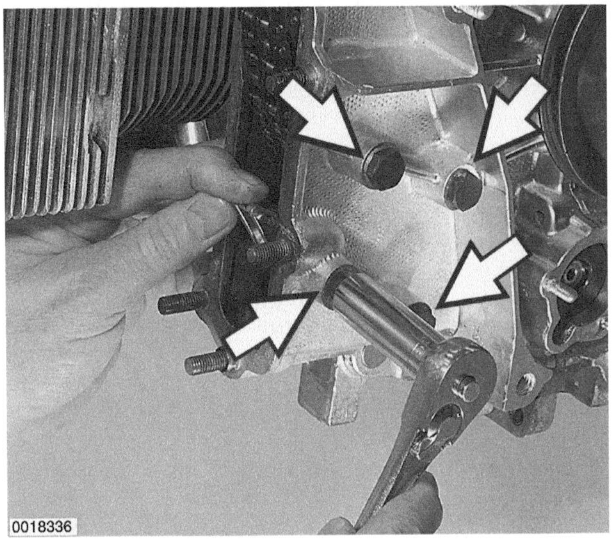
0018336

◀ Remove left side upper and lower timing chain guide rails by removing mounting bolts.

– Repeat procedures for right side chain guide rails. Note lower right guide rail is different from all other chain guide rails.

▶Camshaft cases and cylinder heads, removing

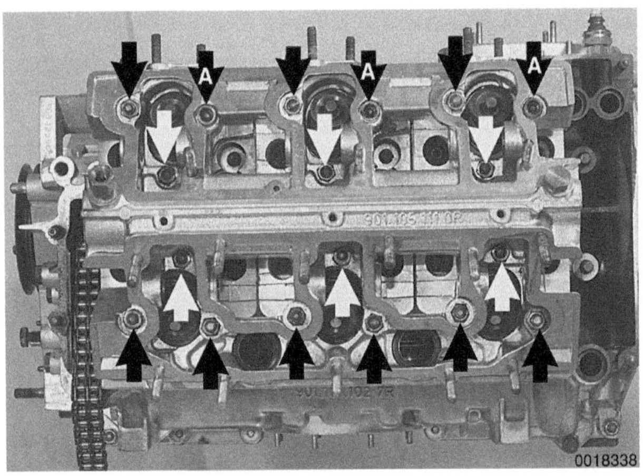

0018338

◀ Remove 18 nuts (**arrows**) from left camshaft case.

- Fifteen M8 (13 mm head) nuts.
- Three M8 (8 mm allen head) nuts (**A**).

– Remove camshaft case and mark for left side.

– Rotate engine on stand. Working on left side bottom of crankcase, remove oil pressure line and all brackets holding it to crankcase.

– Remove cylinder head nuts from each cylinder head using long 10 mm allen wrench.

– Remove cylinder heads, marking installed locations.

– Repeat procedures for right side.

NOTE —

- *Check for broken cylinder head studs when removing the cylinder heads. If any are found, check area around cylinder head and base of stud for signs of blow-by. If blow-by is present, mark that area on both cylinder and head for further inspection once the parts are thoroughly cleaned.*
- *Various head studs have been used on the Carrera 911 engines. Always check with an authorized Porsche parts department for the latest in parts information.*

0018711

◀ Disassemble cylinder heads using Porsche special tool P 200 or equivalent.

- Remove valve keepers, spring retainers, springs, valve stem seals, spacer shims and spring seats.
- Mark location of valves and remove from cylinder head.

Pistons and cylinders, removing

- Remove cylinder air shroud plates.
- Mark cylinder locations on each cylinder barrel.

NOTE —
Label all parts during disassembly. Used parts must be reinstalled in their exact location and position.

◄ Remove cylinders. If necessary, use a soft-faced hammer to tap cylinders off studs.

◄ Set crankshaft to **Z1** (TDC).

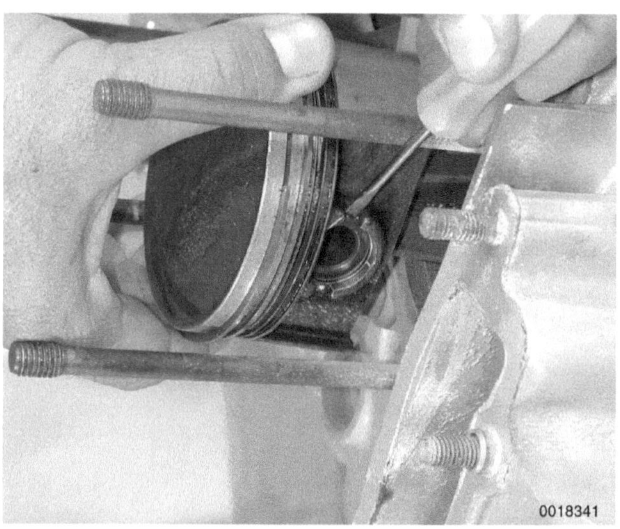

◄ Remove wrist pin circlips from piston 1.

0018345r

◀ Using a soft drift, tap wrist pin out of piston 1.

- Remove piston, marking its position.

– Rotate crankshaft 240°. Remove piston 2, marking its position.

> **CAUTION —**
> *When rotating crankshaft, manually guide pistons in and out of crankcase bores in order to avoid damaging the piston skirts or rings. Use shop towels around the pistons for protection.*

– Rotate crankshaft appropriate amount and remove remaining pistons, marking positions.

NOTE —
The firing order is 1 - 6 - 2 - 4 - 3 - 5.

▸Crankcase, disassembling

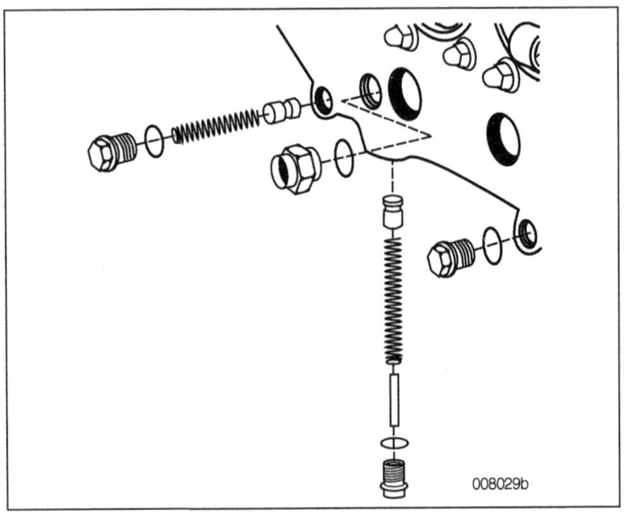

008029b

◀ Remove both oil pressure relief valve assemblies in front bottom of crankcase.

– Remove crankshaft pulley bolt using Porsche special tool 9236 or equivalent to counterhold pulley. Remove pulley.

0018333

◀ Remove M8 nuts (**arrows**) joining case halves together.

- 3 nuts (**A**) are in flywheel housing.
- 2 nuts (**B**) face right (bottom of case).

◀ Remove M10 case nuts (**arrows**) located in oil cooler housing.

◀ Remove M10 case nut (**arrow**) from left timing chain housing.

- Remove remaining M10 case nuts (quantity 11) on left side, counterholding case bolt heads on right side of crankcase.
- Carefully separate crankcase halves.

> ***CAUTION —***
> *Do not pry case apart at sealing surfaces. A damaged surface will result in an oil leak.*

- Lift left half of crankcase off and set aside.
- Lift crankshaft assembly out of crankcase and set aside.

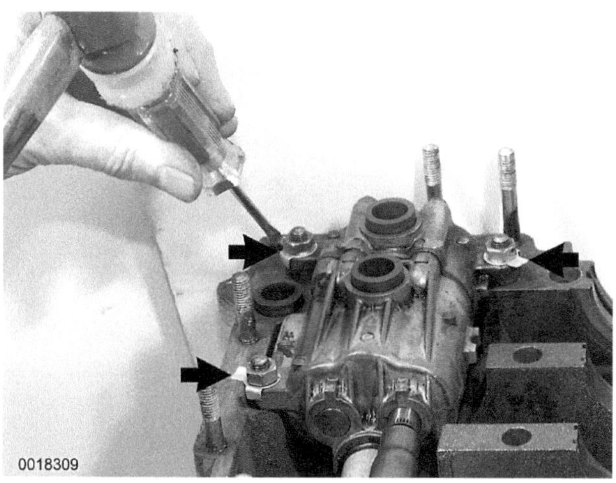

◀ Bend back locking tabs on oil pump mounting nuts (**arrows**). Remove nuts.

- Remove and disassemble intermediate shaft by removing timing chains, oil pump and oil pump drive shaft.

Crankshaft, disassembling

- Mount crankshaft to Porsche special tool P 209 or equivalent and clamp tool with crankshaft in shop vise.

NOTE —
An adequate substitute for special tool P 209 can be made from an old flywheel.

- Mark installed position of connecting rods and remove from crankshaft.
- Slide crankshaft nose bearing off crankshaft.
- Remove circlip holding gear assembly to crankshaft.

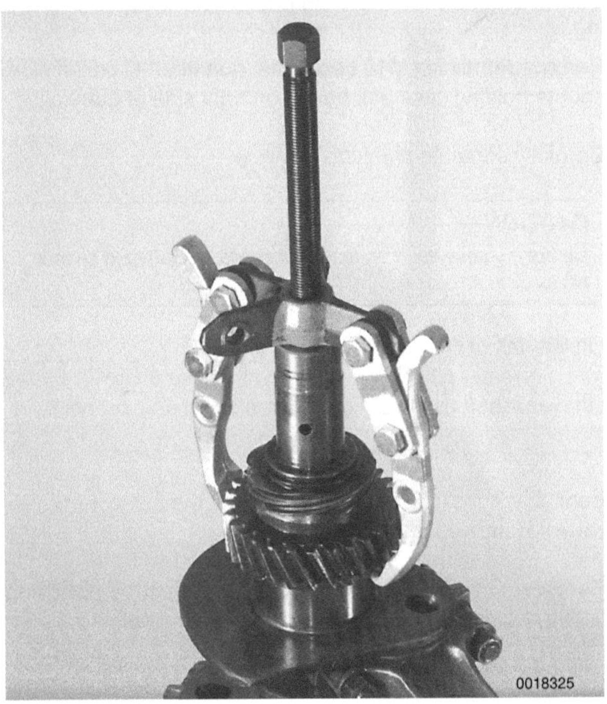

- Remove gear assembly on front of crankshaft using suitable gear puller.

ENGINE, ASSEMBLY

All parts must be clean and ready for reassembly at this point. All machine work should be completed and all new replacement parts should be on hand. Use suitable engine assembling lubricant during assembly.

> ***CAUTION —***
> - *If the engine failed mechanically, it is important to check the intake system for foreign objects.*
> - *If the engine is disassembled because of bearing failure, a thorough flushing of the lubrication system must be carried out prior to reassembly. Refer to* **130 Crankcase and Crankshaft** *and* **170 Engine Lubrication**.

NOTE —
- *Various head studs have been used on the Carrera 911 engines. Always check with an authorized Porsche parts department for the latest in parts information.*
- *If either the crankcase, the crankshaft drive gear or the intermediate shaft drive gear are replaced, special installation instructions apply. See* **130 Crankshaft and Crankcase**.

Cylinder numbering is shown in illustration. Cylinder position will be referred to during the assembly process.

The assembly procedure is divided into subsections, as follows:

- **Crankshaft, assembling**
- **Crankcase, assembling**
- **Pistons and cylinders, installing**
- **Cylinder heads, installing**
- **Camshaft cases, camshafts and camshaft timing chains, installing**
- **Camshaft timing, adjusting**
- **Engine peripherals, installing**

Crankshaft, assembling

Crankshaft reconditioning information is given in **130 Crankcase and Crankshaft**.

NOTE —
Check crankshaft drive gear for wear prior to installation. If gear is worn or damaged, replacement gear must be matched to crankcase and to gear on intermediate shaft. See **130 Crankcase and Crankshaft** *for checking and replacing.*

1. Crankshaft nose bearing
2. Circlip
3. Drive gear (for distributor)
4. Spacer
5. Drive gear (with mark **0** or **1**)

– Mount crankshaft to Porsche special tool P 209 or equivalent and clamp tool with crankshaft in shop vise.

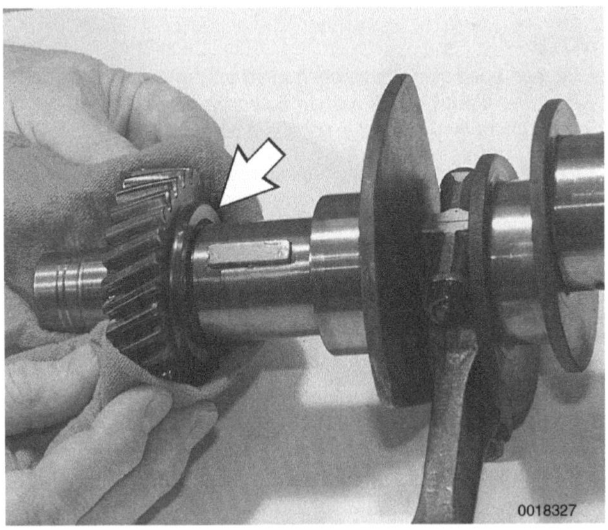

Install drive gear to crankshaft.

- Install woodruff key to crankshaft
- Heat gear to 150°C (300°F) in oil bath.
- With shoulder on gear facing in (**arrow**), slide drive gear fully onto crankshaft.

WARNING —
Use protective gloves or shop rag when handling the hot gear.

– Install spacer to crankshaft.

– Install distributor drive gear to crankshaft.

- Heat gear to 100°C (212°F) in water bath.

CAUTION —
Install distributor drive gear with marks on face of gear facing pulley end of crankshaft.

WARNING —
Use protective gloves or shop rag when handling the hot gear.

Install gear retaining circlip to crankshaft.

- Circlip used to hold gears in place on crankshaft comes in different thicknesses. See **Table a**.
- Select circlip to best obtain near zero clearance.

Table a. Crankshaft Gear Circlip Thickness

Porsche P/N	Thickness	Mark
901 102 148 00	2.4 mm	0
901 102 148 01	2.3 mm	1
901 102 148 02	2.2 mm	2
901 102 148 03	2.1 mm	3

– Install new connecting rod bolts in each connecting rod.

– Install connecting rod bearing shells in connecting rods.

◀ Install connecting rods on crankshaft in their original locations. Matching numbers on side of connecting rod and rod cap should all be facing same side of engine (i.e., all numbers face exhaust side of engine).

– Install new connecting rod nuts on all rod bolts.

Tightening Torque

- Connecting rod cap to connecting rod
  Stage 1 . . . . 20 Nm (14 ft-lb)
  Stage 2 . . . . 90° ± 2°

▶Crankcase, assembling

– Mount right side crankcase half on engine stand.

– Install and tighten oil drain plug using new sealing ring.

Tightening Torque

- Oil drain plug to crankcase . . . . 70 Nm (52 ft-lb)

◀ Install intermediate shaft bearing shells in crankcase halves.

NOTE —
Due to the high loads the timing chains put on the intermediate shaft, it is recommended that the intermediate shaft bearings be replaced during engine overhaul.

◀ Place oil pump sealing O-rings (**arrows**) in right crankcase half.

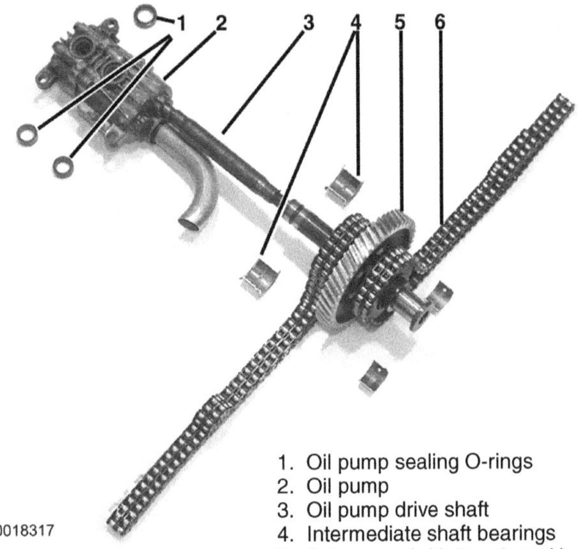

1. Oil pump sealing O-rings
2. Oil pump
3. Oil pump drive shaft
4. Intermediate shaft bearings
5. Drive gear (with **0** or **1** mark)
6. Timing chain

– Lubricate oil pump gears with engine assembling lubricant.

◄ Place both timing chains on intermediate shaft sprockets. Install oil pump drive shaft with oil pump to intermediate shaft.

NOTE —

- *Check intermediate shaft gear for wear prior to installation. If the intermediate shaft gear is worn or damaged, the replacement gear must be matched to the crankcase and crankshaft drive gear. See* **130 Crankcase and Crankshaft** *for checking and replacing.*
- *The intermediate shaft is hollow and should be cleaned out whenever the engine is being reconditioned and in case of engine bearing failure. See* **130 Crankcase and Crankshaft**.

◄ Lower complete intermediate shaft assembly into right crankcase half, guiding front timing chain (**arrow**) through opening in case.

◄ Place new locking tabs (**arrows**) on oil pump mounting studs and install and torque nuts. Bend lock tabs up into place.

Tightening Torque

- Oil pump to crankcase . . . . 25 Nm (18 ft-lb)

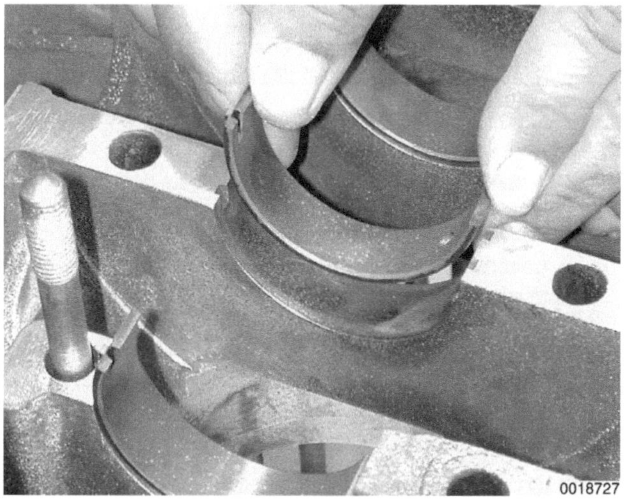

◀ Lubricate crankshaft main bearing shells with engine oil and install in crankcase halves.

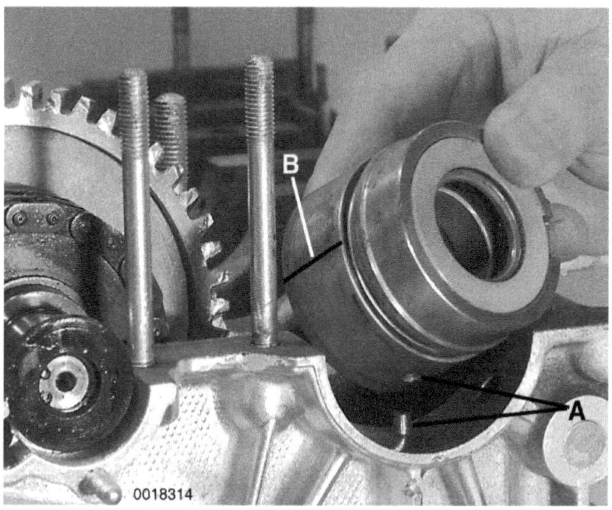

– Install crankshaft nose bearing locating pin in crankcase half.

◀ Without nose bearing outer sealing O-ring in place, lay crankshaft nose bearing in crankcase so that it locks into locating pin (**A**). Draw a line (**B**) on each side of bearing to use as a reference when installing crankshaft assembly.

– Remove nose bearing. Install nose bearing outer O-ring.

– Lubricate nose bearing and install on crankshaft.

◀ Place crankshaft assembly into right crankcase half.

- Make sure right side connecting rods for cylinders 4,5, and 6 are pointing down.
- Check that nose bearing (**A**) is correctly seated in case.
- Slowly turn crankshaft back and forth to check for binding or interference.

NOTE —

- *Porsche special tool P221 can be used to support the left side connecting rods during crankcase assembly.*
- *Porsche special tool P222 can be used to support the left timing chain during crankcase assembly.*

◀ Place two sealing O-rings (**A**) in oil pump.

- Check to see that oil passage sealing O-ring previously placed in right crankcase half (**B**) remains undisturbed.
- Lubricate all with engine oil.

NOTE —
Different size seals are used. Be sure to install the correct parts.

– Using Loctite® 574 sealant, apply even thin coat on both crankcase mating surfaces.

– Place left case-half in position over right case-half. Tap case-halves together lightly with a rubber mallet.

CAUTION —
Turn the crankshaft back and forth periodically to check for binding.

– Place sealing O-rings on two case studs in oil cooler housing.

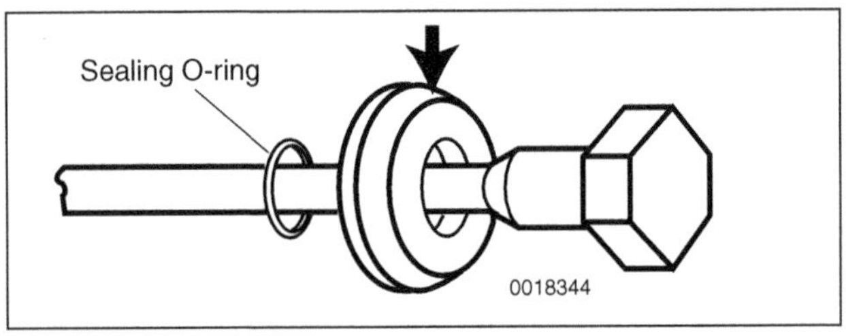

◀ When installing case washers, align bevel on washers (**arrow**) correctly. Bevels should face nut or bolt head.

◀ Install capped case nuts (**arrows**) in oil cooler housing.

Tightening Torque

- Crankcase capped nut .... 35 Nm (25 ft-lb)

Install washer and nut in timing chain cut-out on left side of engine.

Tightening Torque
- Timing chain cut-out case nut. . . . . 35 Nm (25 ft-lb)

CAUTION —
Check crankshaft for free movement by turning slowly back and forth.

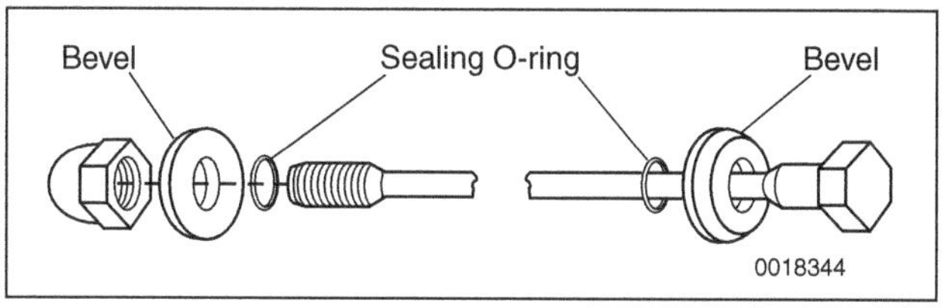

Insert M10 bolts through case.

- Make sure beveled washers and sealing O-rings are installed correctly.
- Install case nuts and tighten.

Tightening Torque
- Case bolt and nut (M10). . . . . 35 Nm (25 ft-lb)

– Install crankshaft seals on flywheel and pulley ends of engine. (special tools P 216 and 9126)

– Install crankshaft pulley, counterholding with Porsche special tool 9236 or equivalent when torquing bolt.

Tightening Torque
- Crankshaft pulley to crankshaft. . . . . 80 Nm (59 ft-lb)

NOTE —
If the crankshaft pulley locating hole or pin is damaged, valve timing may be affected. Replace a pulley with elongated/damaged hole.

Install 22 M8 nuts and bolts around outer edge of crankcase.

- Three nuts are in flywheel housing (**A**).
- Two nuts face right side (**B**).

Tightening Torque
- Crankcase perimeter fasteners (M8) . . . 25 Nm (18 ft-lb)

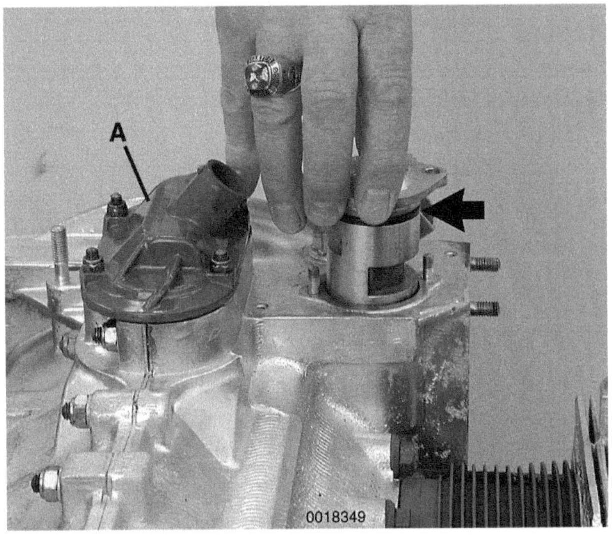

– Install oil pressure sending units.

Tightening Torques

- Oil warning light sending unit to crankcase . . . . 20 Nm (14 ft-lb)
- Oil gauge sending unit to crankcase. . . . 35 Nm (25 ft-lb)

◄ Install oil temperature thermostat with new sealing O-ring (**arrow**).

Tightening Torque

- Oil temperature thermostat to crankcase . . . . 8 Nm (6 ft-lb)

– Install breather cap (**A**) with a new gasket.

Tightening Torque

- Breather cap to crankcase . . . . 8 Nm (6 ft-lb)

– Install oil temperature sensor.

Tightening Torque

- Oil temperature sensor to crankcase . . . 25 Nm (18 ft-lb)

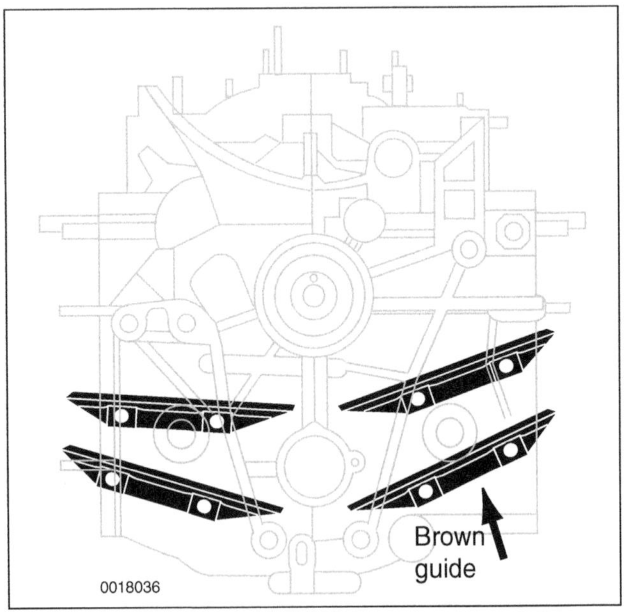

◄ Install chain guide rails inside crankcase chain housings.

- Note location of different guide rail in lower right slot.

◄ Use screwdriver to push chain guide rail toward pulley end of engine (**arrow**) while tightening retaining bolts.

Tightening Torque

- Timing chain guide retaining bolt to crankcase . . . . 25 Nm (18 ft-lb)

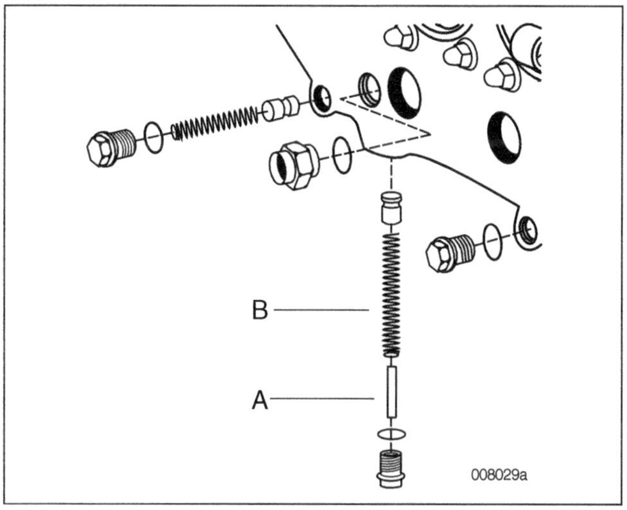

Install oil pressure relief valve assemblies in crankcase. Coat relief valves with assembly lube before installing.

Tightening Torque
- Oil pressure relief valve plug to crankcase .... 60 Nm (44 ft-lb)

NOTE —

*The main (vertical) pressure relief valve uses a sleeve (**A**) and longer spring (**B**). For further details, see* **130 Crankcase and Crankshaft**.

Pistons and cylinders, installing

Measurement and reconditioning of pistons and cylinders is covered in **131 Pistons and Cylinders**.

CAUTION —
- *Be sure to keep each piston and cylinder set clearly marked and together.*
- *Be sure used pistons and cylinders are reinstalled in the exact same locations and positions.*

Set crankshaft to **Z1** TDC.

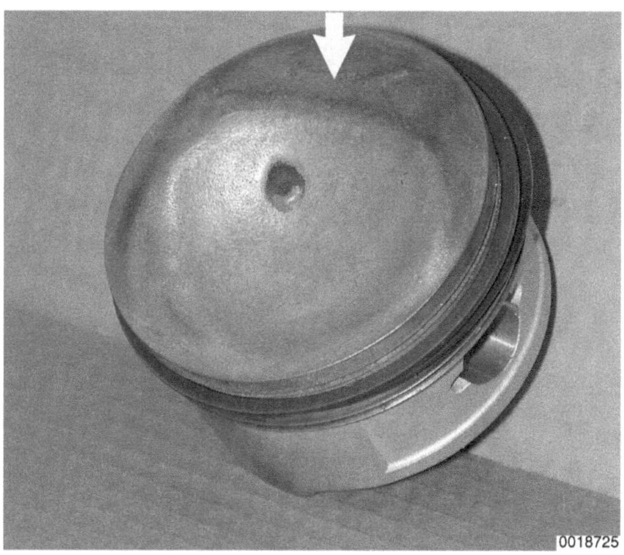

Install piston 1 with flat of intake valve relief (**arrow**) up.

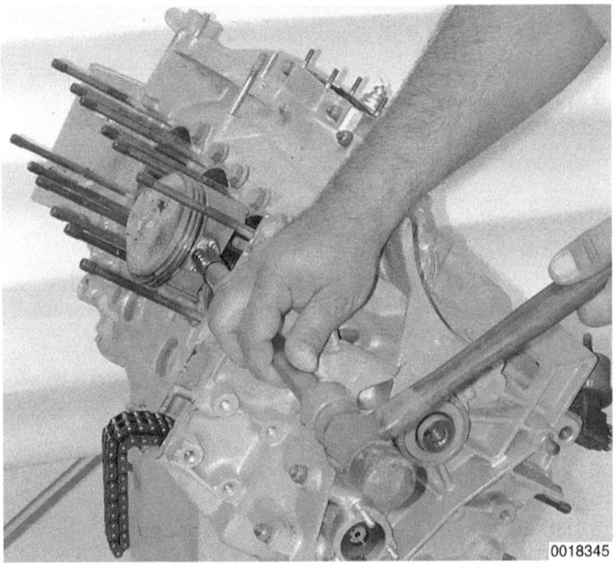

◀ Oil wrist pin and install. If necessary, tap pin in gently using soft punch. Install retaining circlips.

> ***CAUTION—***
> - *The wrist pin circlips installed on new pistons are for transportation purposes only. Replace these clips with ones provided separately with the pistons.*
> - *Always be sure that both piston pin clips are in place before installing the cylinder.*

– Rotate crankshaft and install pistons 2 and 3.

> ***CAUTION—***
> *When turning crankshaft, manually guide pistons in and out of crankcase bores in order to avoid damaging the piston skirts or rings. Use shop towels around the pistons for protection.*

– Reset crankshaft to **Z1** compression stroke.

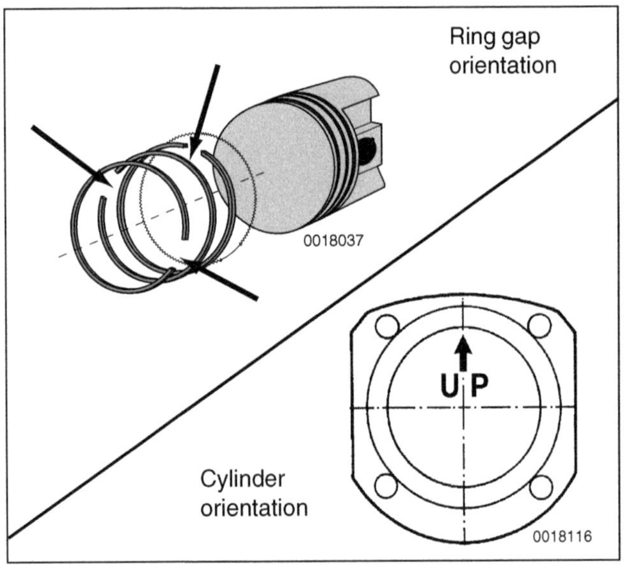

◀ Set piston ring gaps to clock positions 4, 10 and 2 (**arrows**).

– Place a new cylinder base gasket on cylinder 1 and lubricate cylinder wall.

> ***CAUTION—***
> *If cylinder heads have been machined, be sure to use 0.50 instead of 0.25 mm thick copper gaskets at cylinder bases. See* **150 Cylinder Heads**.

NOTE—
When installing cylinders, make sure the deeper cylinder cooling fins point down.

◀ Using a ring compressor such as Porsche special tool US 1008 a or equivalent, slide cylinder over piston and into side of case.

< Install new head gasket in groove at top of cylinder 1 (**arrow**).

< Temporarily hold cylinder in position using Porsche special tools P140 (**arrows**).

NOTE —

As a substitute for the special tools, slide two sockets over the cylinder head studs for each cylinder and install two cylinder head nuts finger tight.

– Rotate crankshaft and install cylinders 2 and 3 with base gaskets and head gaskets.

> ***CAUTION —***
>
> *When rotating crankshaft, manually guide pistons in and out of crankcase bores in order to avoid damaging the piston skirts or rings. Use shop towels around the pistons for protection.*

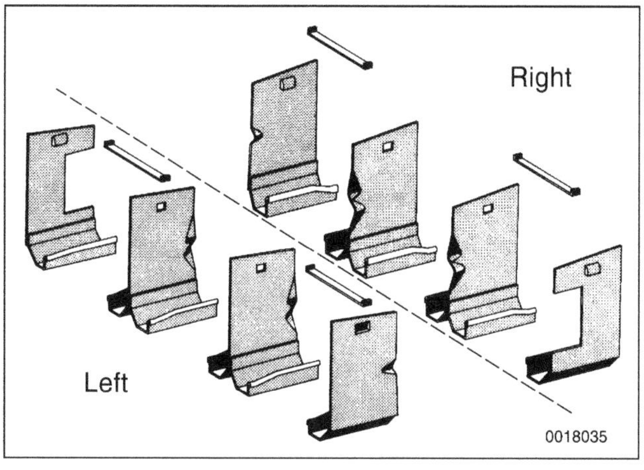

< Install left side cylinder air shroud plates.

– Install right side pistons and cylinders (4, 5, 6) with base gaskets and head gaskets.

– Install right side air guides.

Cylinder heads, installing

Repairs to cylinder heads and valves are described in **150 Cylinder Heads**.

- With engine on engine stand and pistons, cylinders, gaskets, and air guides installed, rotate engine on stand so that left side cylinders are up.

NOTE —
Prior to cylinder head installation be sure to remove temporary cylinder hold down nuts.

- Lubricate cylinder head studs, nuts and washers lightly with Optimoly®HT (Porsche part number 901 104 382 02) or equivalent. Starting with cylinder head 3, install left side cylinder heads.

NOTE —
Cylinder head 3 has a threaded boss for the cylinder head temperature sensor.

- Using screwdriver as guide, slide head washers over head studs.
- Install head nuts and hand tighten only. Temporarily fit camshaft case in position on heads. Alignment pins on cylinder heads must line up with alignment holes in camshaft case. Position cylinder heads as required until machined sealing surfaces meet. Temporarily install a few camshaft case mounting nuts (hand tight) to hold case in position.

NOTE —
Sealing surfaces on heads and camshaft cases must be clean, flat, and without burrs.

- Use cross pattern to torque cylinder head nuts.

Tightening Torques
- Cylinder head to crankcase
  - Step 1 .... 15 Nm (11 ft-lb)
  - Step 2 .... +90° (¼ turn)

- Remove camshaft case from cylinder heads 1 - 3. Rotate engine on stand so that right side cylinders 4 - 6 are up.
- Starting with cylinder 6, install right side cylinder heads.
- Slide head washers over head studs.
- Install head nuts and hand tighten only. Temporarily fit camshaft case in position on heads. Alignment pins on cylinder heads must line up with alignment holes in the camshaft case. Position cylinder heads as required until machined sealing surfaces meet. Temporarily install a few camshaft case mounting nuts (hand tight) to hold case in position.
- Use cross pattern to torque cylinder head nuts.

Tightening Torques
- Cylinder head to crankcase
  - Step 1 .... 15 Nm (11 ft-lb)
  - Step 2 .... +90° (¼ turn)

– Remove camshaft case from cylinder heads 4 - 6.

– Rotate engine on stand. Working on left side bottom of crankcase, install oil pressure line.

Camshaft cases, camshafts and camshaft timing chains, installing

Additional camshaft timing chain information can be found in **152 Camshaft Timing Chains**.

> ***CAUTION —***
> *Install camshaft cases and camshafts on left or right side according to marks made during disassembly.*

– Use Loctite®574 to coat machined sealing surfaces on camshaft cases.

– Place new sealing O-rings on oil return tubes.

– Holding oil return tubes in place, guide right camshaft case over its mounting studs.

NOTE —
Two piece oil return tube replacement is covered in **170 Engine Lubrication**.

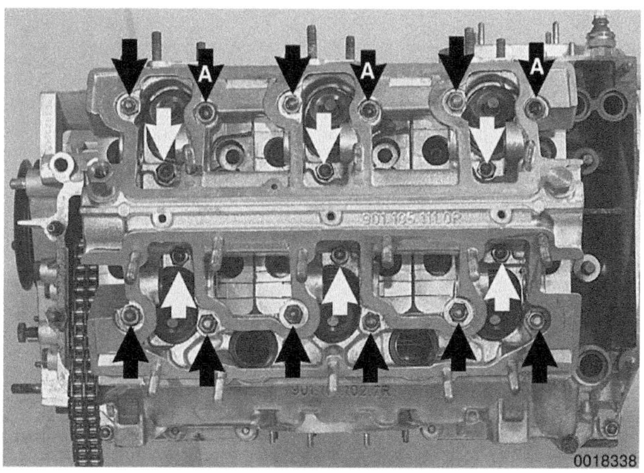

◄ Install 18 retaining nuts (**arrows**) and tighten after making sure oil return tubes are correctly positioned.

- There are fifteen M8 (13 mm head) nuts.
- There are three M8 (8 mm allen head) nuts (**A**).

Tightening Torque
- Camshaft case to cylinder head . . . . 25 Nm (18 ft-lb)

– Install right timing chain housing with new gasket.

Tightening Torque
- Timing chain housing to crankcase . . . . 25 Nm (18 ft-lb)

– Slide right camshaft into camshaft case.

– Install camshaft end cover with new O-ring and gasket. Install end cover retaining bolts and tighten.

– Reinstall right camshaft thrust spacer, sprocket shims, woodruff key and sprocket flange.

– Install right camshaft sprocket to timing chain. Slide sprocket over sprocket flange.

– Install sprocket washer and mounting bolt finger tight.

0018350

◄ Install right timing chain guide rail in housing, if previously removed.

– Install right timing chain tensioner and idler sprocket assembly using new O-ring. Tighten tensioner mounting nuts.

– Install left camshaft, camshaft case, timing chain guide rail, camshaft sprocket assembly and timing chain as described in previous steps.

NOTE —
Left and right camshaft sprockets are identical. Dished side of sprocket must face pulley side of engine on left camshaft (cylinders 1 - 3) and flywheel side of engine on right camshaft (cylinders 4 - 6)

Camshaft sprocket orientation

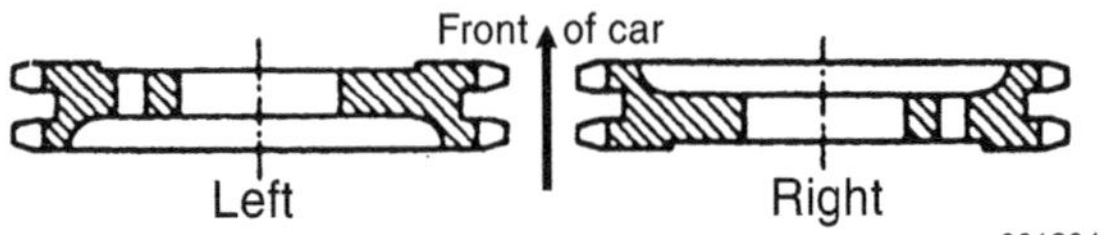

Camshaft sprocket assembly

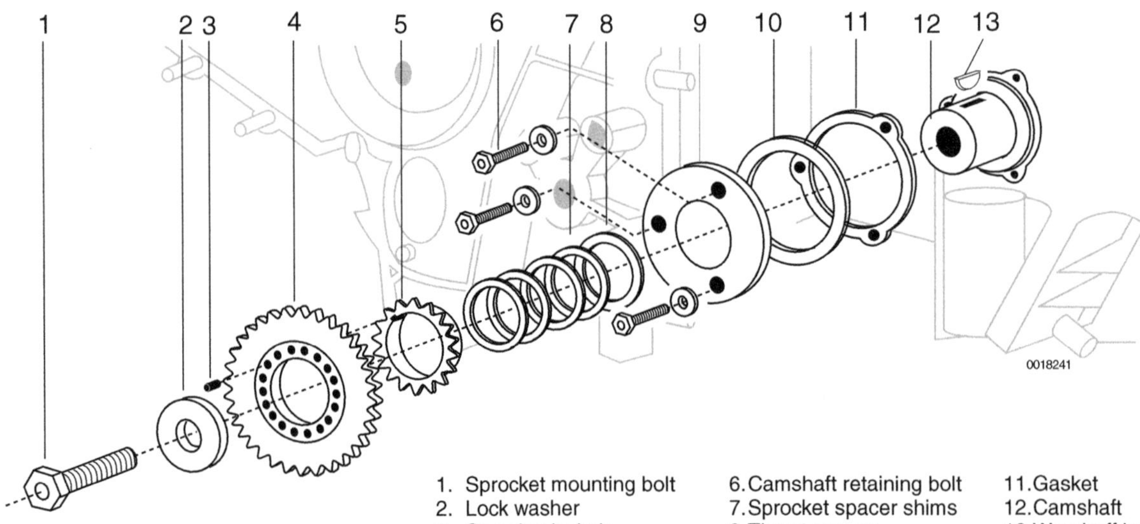

1. Sprocket mounting bolt
2. Lock washer
3. Sprocket lockpin
4. Camshaft sprocket
5. Sprocket flange
6. Camshaft retaining bolt
7. Sprocket spacer shims
8. Thrust spacer
9. Camshaft end cover
10. Sealing O-ring
11. Gasket
12. Camshaft
13. Woodruff key

– Check parallelism of intermediate shaft and camshaft sprockets by measuring distance from front of intermediate shaft to front edge of gear wheels. Use quality straight edge when making measurement.

NOTE —
- *Normally, if reinstalling camshafts and sprockets, 3 sprocket spacer shims will be necessary under the left sprocket and 4 shims under the right sprocket.*
- *If new camshafts or sprockets are being installed, camshaft sprocket spacing must be adjusted by adding or subtracting spacer shims so that each sprocket is parallel with the corresponding intermediate shaft sprocket. See* **152 Camshaft Timing Chains**.

Timing chain sprocket parallelism

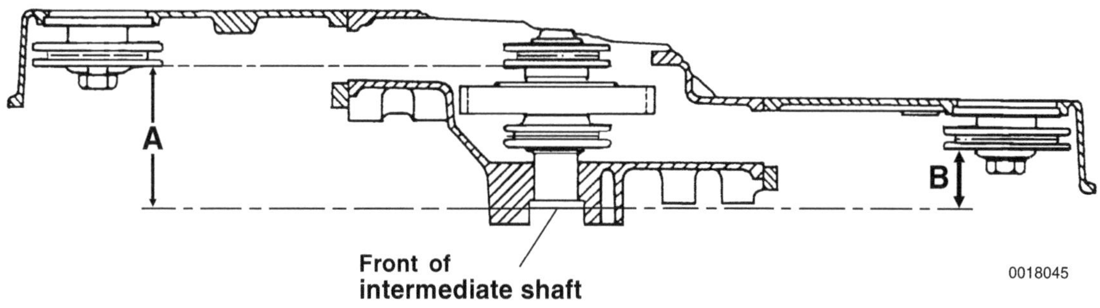

A Factory distance from end of intermediate shaft to face of cylinder 1 - 3 (left) timing chain sprocket = 98.07± 0.25 mm (3.86 ± 0.01 in.)
B Factory distance from end of intermediate shaft to face of cylinder 4 - 6 (right) timing chain sprocket = 43.27 ± 0.25 mm (1.70 ± 0.01 in.)

A preliminary setting of camshaft timing must be done as follows (refer to illustration on next page):

- Set front crankshaft pulley to **Z1** (TDC).
- Remove camshaft sprocket bolts.
- Rotate left camshaft until "**930**" points up.
- Install camshaft sprocket lockpin.
- Reinstall left camshaft sprocket bolt finger tight.
- Repeat for right camshaft.

Camshaft timing, adjusting

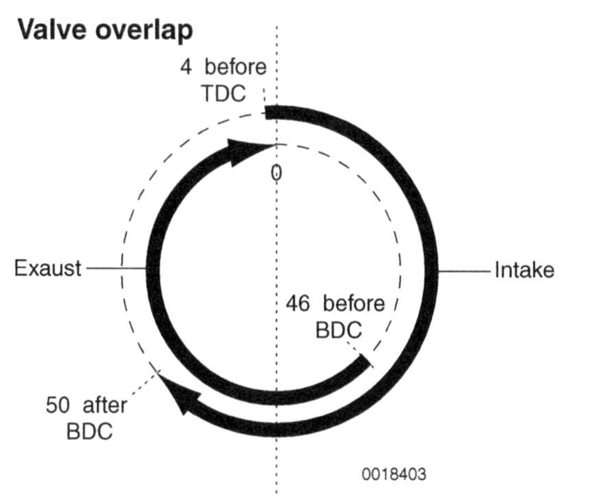

Camshaft timing is adjusted using a dial gauge setup. The dial gauge is used to precisely set the position of each camshaft in relation to partially open ("overlapped") valves.

NOTE —
Camshaft timing specification in degrees of rotation are given below.

Camshaft Timing Specifications

- Intake valve opens ....4° before TDC
- Intake valve closes ....50° after BDC
- Exhaust valve opens.................. 46° before BDC
- Exhaust valve closes .... At TDC

Rocker shaft and arm, installing

Set crankshaft to **Z1** cylinder 1 compression stroke.

- Check to see that punch marks or "**930**" on camshaft ends point *up*.

– Assemble rocker shaft.

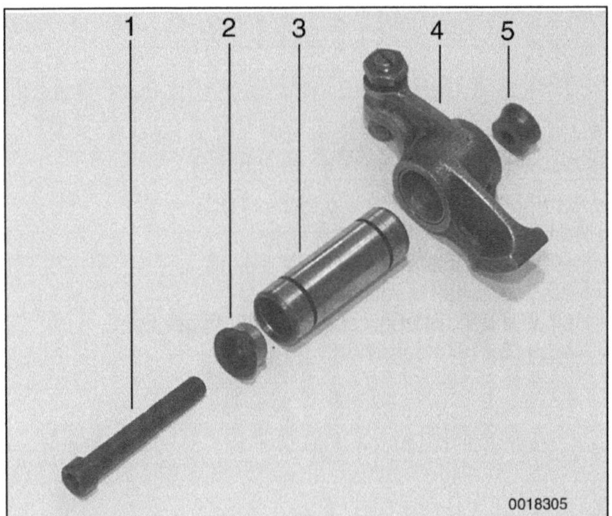

1. Bolt (5 mm allen)
2. Expansion sleeve
3. Rocker shaft
4. Rocker arm
5. Expansion nut (8mm allen)

Install cylinder 1 intake and exhaust rocker shafts with rocker arms:

- Install rocker shaft with 5 mm bolt facing cylinder 2.
- Tap rocker shaft into bore until shaft is flush with inner edge of support.
- Place rocker arm in position and tap rocker shaft in until shaft slides into far side support.

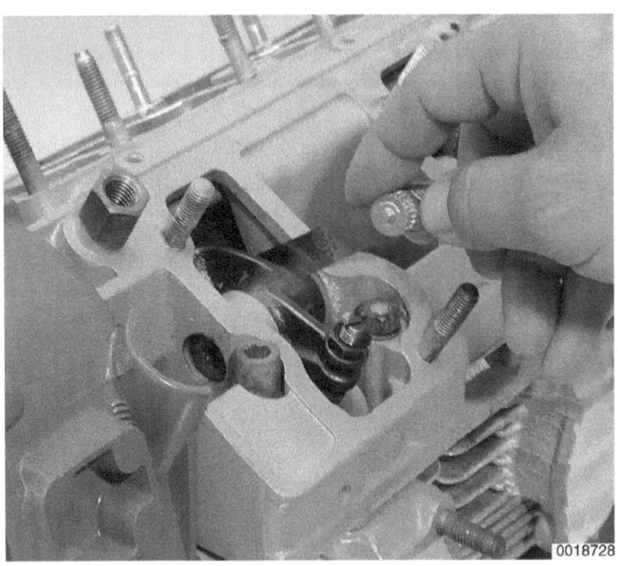

Place 0.5 mm (0.02 in.) feeler gauge between far side support and rocker arm.

- Continue to tap shaft until feeler gauge engages slot in rocker shaft.
- Remove feeler gauge and tap shaft in approximately 1.5 mm (0.06 in) further until shaft is approximately flush with outside edge of camshaft case (thinner bearing support side of casting).

Tightening Torque

- Rocker shaft pinch bolt . . . . 15 Nm (11 ft-lb)

Adjust cylinder 1 intake and exhaust valve clearance.

Valve Clearance
- Adjusting value . . . . 0.10 mm (0.004 in.)

Left (cylinders 1 - 3) camshaft timing

Set crankshaft to **Z1** (TDC) cylinder 1 compression stroke.

- Check to see that punch marks or "**930**" on camshaft ends point *up*.

Using Porsche special tool P 207, mount dial gauge on camshaft case stud next to cylinder 1 intake valve.

- Set dial gauge plunger against valve spring retainer on cylinder 1 intake valve.
- Zero dial gauge with approximately 10 mm (0.4 in.) preload.

– Starting at **Z1** (TDC) mark on pulley, turn crankshaft slowly in clockwise direction approximately 1 full turn while observing dial gauge. Turn crankshaft far enough to reach valve overlap adjusting value.

Valve Overlap Adjustment
- Adjusting value . . . . 1.25 mm (0.049 in.)

– Unscrew and remove left timing chain sprocket mounting bolt.

Pull out camshaft sprocket lockpin (**arrow**) using Porsche special tool P 212 or equivalent.

- Turn crankshaft (either clockwise or counterclockwise, as necessary) just until **Z1** mark on pulley is aligned accurately with joint of crankcase.
- Reinstall lockpin in hole with best alignment. Install camshaft sprocket bolt finger tight.
- Turn crankshaft clockwise two full turns. Recheck timing adjustment. Value must be within tolerance range.

Valve Overlap Adjustment
- Tolerance range . . . . 1.1 - 1.4 mm (0.043 - 0.055 in.)

- Torque camshaft sprocket bolt when settings are correct.

NOTE —
Use Porsche special tool 9191 or equivalent to counterhold camshaft sprocket when torquing bolt.

Tightening Torque
- Chain sprocket to camshaft . . . . 120 Nm (89 ft-lb)

Right (cylinders 4 - 6) camshaft timing

Set crankshaft to **Z1** (TDC) cylinder 4 compression stroke.
- Check to see that punch marks or "**930**" on camshaft ends point *down.*

- Install cylinder 4 rocker shafts and arms as described earlier.
  - Install 5 mm allen bolt with bolt head facing cylinder 5.

Tightening Torque
- Rocker shaft pinch bolt . . . . 18 Nm (16 ft-lb)

- Adjust cylinder 4 intake and exhaust valve clearance.

Valve Clearance
- Adjusting value . . . . 0.10 mm (0.004 in.)

◄ With crankshaft at cylinder 4 TDC compression stroke, use Porsche special tool P 207 to mount dial gauge on camshaft case stud next to cylinder 4 intake valve.

- Set dial gauge plunger against valve spring retainer on cylinder 4 intake valve.
- Zero dial gauge with approximately 10 mm (0.4 in.) preload.

— Starting at **Z1** (TDC) mark on pulley, turn crankshaft slowly in clockwise direction approximately 1 full turn while observing dial gauge. Turn crankshaft far enough to reach valve overlap adjusting value.

Valve Overlap Adjustment
- Adjusting value . . . . 1.25 mm (0.049 in.)

— Unscrew and remove right timing chain sprocket mounting bolt.

◄ Pull out camshaft sprocket lockpin using special tool P 212 (**arrow**).

— Turn crankshaft just until **Z1** mark on pulley is aligned accurately with joint of crankcase.

— Reinstall lockpin in hole with best alignment. Install camshaft sprocket bolt finger tight.

— Turn crankshaft clockwise two full turns. Recheck timing adjustment. Value must be within tolerance range.

Valve Overlap Adjustment
- Tolerance range . . . . 1.1 - 1.4 mm (0.043 - 0.055 in.)

— Torque camshaft sprocket bolt when settings are correct.

NOTE —

Use Porsche special tool 9191 or equivalent to counterhold camshaft sprocket when torquing bolt.

Tightening Torque
- Chain sprocket to camshaft . . . . 120 Nm (89 ft-lb)

Set crankshaft to **Z1** (TDC) cylinder 1 compression stroke.

NOTE —
Check to see that punch marks or "**930**" on camshaft ends point *up*.

NOTE —
The following instructions apply when installing rocker shafts.

- *Cylinder 2 and 5: 5 mm allen bolt head can be installed either way.*
- *Cylinder 3: install 5 mm allen bolt with bolt head facing cylinder 2.*
- *Cylinder 6: install 5 mm allen bolt with bolt head facing cylinder 5.*

– Turn crankshaft 120° in normal direction of rotation. Install cylinder 6 rocker shafts and arms as described earlier.

Tightening Torque
- Rocker shaft pinch bolt . . . . 18 Nm (16 ft-lb)

– Adjust cylinder 6 valve clearance.

Valve Clearance
- Adjusting value . . . . 0.10 mm (0.004 in.)

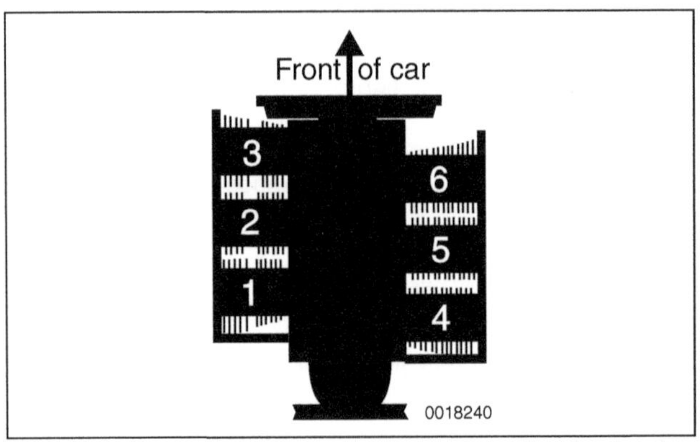

Following firing order, turn crankshaft 120° in normal direction of rotation and install rocker arms for next cylinder and adjust valves.

NOTE —
The firing order is 1 - 6 - 2 - 4 - 3 - 5.

– Repeat until all rocker arms are installed and valves adjusted.

– Install valve covers with new gaskets, sealing washers and lock nuts.

Tightening Torque
- Valve cover to camshaft case . . . . 8 Nm (71 in-lb)

– Install timing chain housing covers with new gaskets, sealing washers and lock nuts.

Tightening Torque
- Chain housing cover to housing . . . . 8 Nm (71 in-lb)

– Install right and left camshaft and timing chain tensioner oil feed lines with new sealing washers.

Engine peripherals, installing

◄ With crankshaft at TDC, install distributor so that mark on distributor rotor aligns with mark on distributor housing (**arrows**).

NOTE —
- *The marks line up approximately, not precisely.*
- *There is only one way to install the distributor housing.*

Tightening Torque
- Distributor base clamp (M8) . . . . 18 Nm (14 ft-lb)

– Install new spark plugs.

Tightening Torque
- Spark plug to cylinder head . . . . . 25-30 Nm (18-22 ft-lb)

– Install engine mount carrier and cross member.

– Install A/C compressor mounting brackets.

Tightening Torques
- Engine mount carrier to crankcase . . . . . 40 Nm (30 ft-lb)
- Engine mount crossmember to carrier . . . . 40 Nm (30 ft-lb)

◄ Install oil cooler from below. Install top and bottom (**arrows**) oil cooler mounting nuts.

CAUTION —
Always use new O-ring seals.

Tightening Torques
- Oil cooler to crankcase . . . . 23 Nm (17 ft-lb)

– Install all sheet metal shrouding around engine.

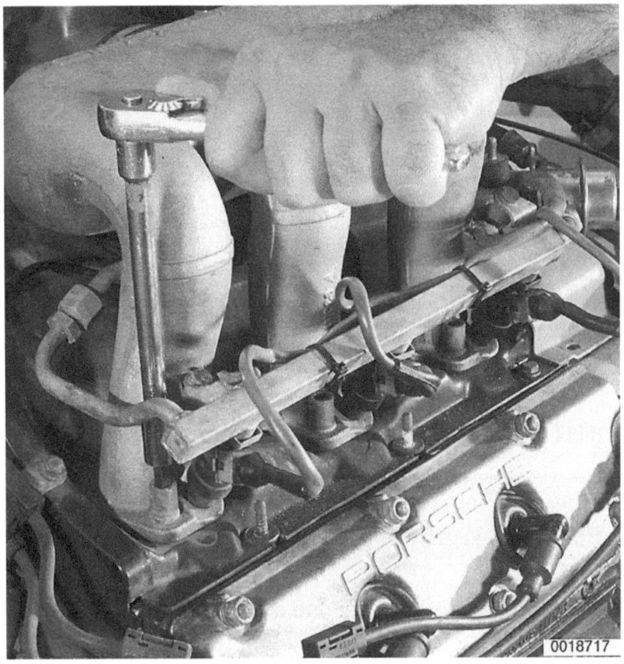

Set complete fuel system, intake manifold, alternator and cooling shroud in place.

- Reattach alternator ground to top of crankcase under shroud.
- Install intake runner washers and nuts.

Tightening Torque
- Intake manifold to cylinder heads . . . . . . 25 Nm (18 ft-lb)

– Connect all spark plug wires and install all wire holders.

– Rotate engine on stand and install heat exchangers, catalytic converter and muffler. See **260 Exhaust System**.

Tightening Torques
- Heat exchangers to cylinder heads. . . . . 25 Nm (18 ft-lb)

– Remove engine from engine stand.

– Install flywheel, clutch assembly and transmission as described in **101 Engine Removal and Installation**.

Tightening Torques
- Clutch to flywheel . . . . 25 Nm (18 ft-lb)
- Flywheel to crankshaft . . . . 90 Nm (66 ft-lb)
- Transmission to engine . . . . 45 Nm (33 ft-lb)

– Install engine. See **101 Engine Removal and Installation**.

130 Crankcase and Crankshaft

TABLES

GENERAL

This repair group provides reconditioning information for the crankcase, crankshaft and intermediate shaft. Engine disassembly is described in **102 Engine Disassembly and Assembly**.

> ***CAUTION —***
> *The information contained here is intended to be used as a reconditioning guide for the professional or experienced automotive technician.*

Special tools

Some of the procedures require special tools.

Special Tools

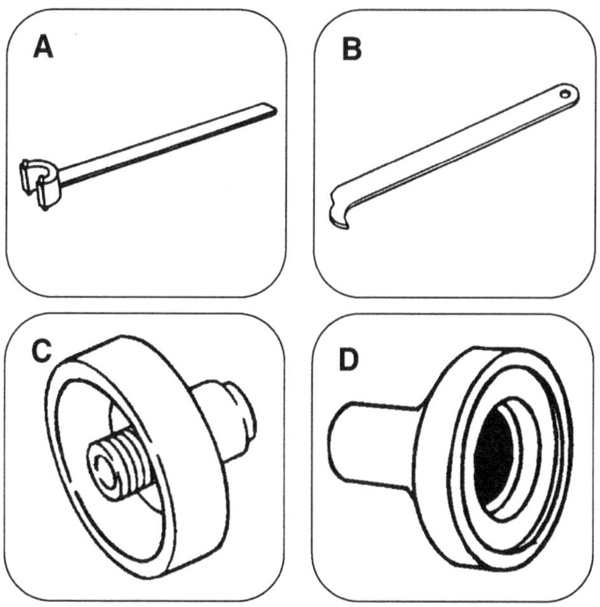

Special tools

A 9236 . . . . Pulley holding spanner wrench
(Porsche ordering #000 721 923 60)
B 681 . . . . Oil seal puller
(source: Baum Tools Unlimited)
C P 216 a. . . . . Pulley end oil seal installer
(Porsche ordering #000 721 216 10)
D 9126 . . . . Flywheel end oil seal installer
(Porsche ordering #000 721 912 60)

Special Tools

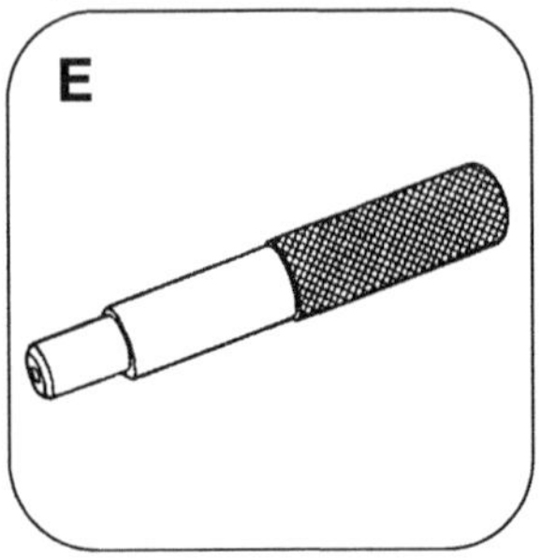

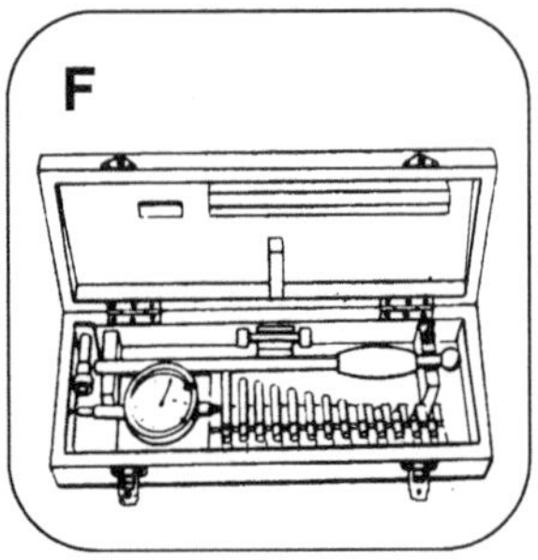

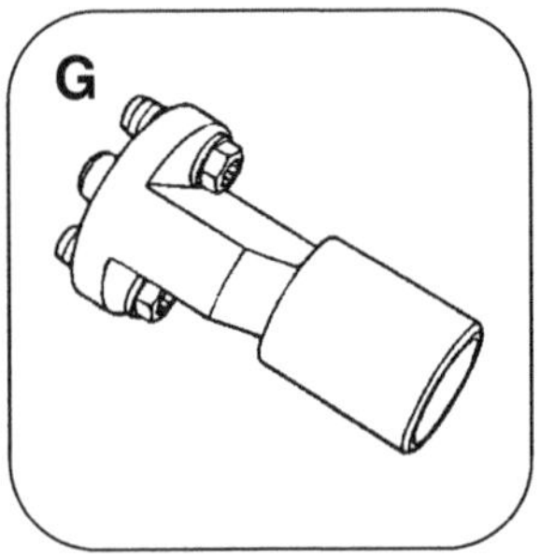

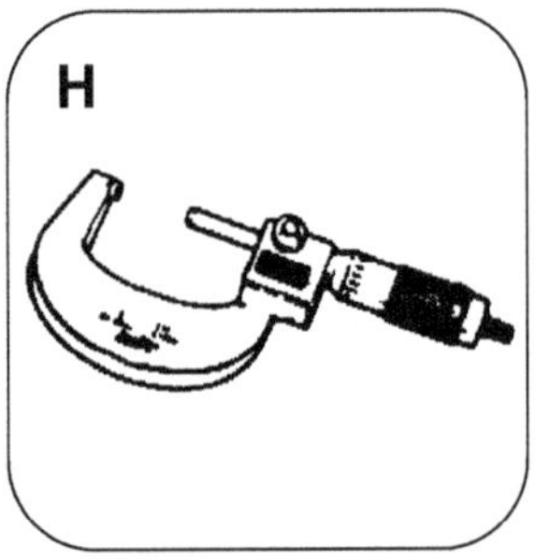

Special tool

E P 370 . . . . Clutch spline aligning tool (source: Baum Tools Unlimited)
F 17 . . . . Dial bore gauge 50-150 mm (source: Baum Tools Unlimited)
G P 209 a. . . . . Bracket for holding crankshaft in vise (Porsche ordering #000 721 209 10)
H 12 . . . . Micrometer (source: Baum Tools Unlimited)

CRANKSHAFT OIL SEALS

NOTE —

*The pulley end of the engine is referred to as the **front**. The flywheel end of the engine is referred to as the **rear**.*

Front crankshaft oil seal (pulley end), replacing

The front crankshaft oil seal can be replaced with the engine installed in the vehicle.

– Remove alternator and A/C compressor belts. See **270 Battery, Starter, Alternator**.

– Remove pulley mounting bolt and remove pulley.

NOTE —

Use Porsche special tool no. 9236 to counter hold pulley.

– Remove crankshaft oil seal with Porsche special tool no. 681 or equivalent.

◄ Using Porsche special tool P 216 or equivalent press new seal into position.

NOTE —
Lubricate inner lip of new seal with engine oil prior to installation.

– Install pulley and bolt. Counterhold pulley when tightening bolt.

Tightening Torque
- Pulley to crankshaft. . . . . 80 Nm (52-ft-lb)

– Install alternator and A/C compressor belts.

Rear crankshaft oil seal (flywheel end), replacing

Replacement of the rear crankshaft oil seal requires that the engine be removed from the car and the transmission be separated from the engine.

– Remove engine and transmission and separate transmission from engine as described in **101 Engine Removal and Installation**.

– Remove clutch mounting bolts and remove clutch pressure plate and disc.

– Remove flywheel bolts and remove flywheel.

– Remove crankshaft oil seal with Porsche special tool no. 681 or equivalent.

0018674

< Using Porsche special tool no. 9126 (**arrow**) or equivalent press seal into crankcase.

NOTE —
Lubricate inner lip of new seal with engine oil prior to installation.

– Reinstall flywheel and clutch using Porsche special tool no. P370 or equivalent to align clutch disc.

Tightening Torques
- Clutch to flywheel shaft. . . . . 25 Nm (18 ft-lb)
- Flywheel to crankshaft . . . . 90 Nm (66 ft-lb)

– Join transmission and engine and reinstall. See **101 Engine Removal and Installation**.

Tightening Torques
- Crankcase to transmission . . . . 48 Nm (36 ft-lb)

CRANKCASE

All disassembled parts should be thoroughly cleaned before inspection. If bearings are to be reused, they should always be reinstalled in their exact positions. Parts should never be interchanged between cylinders.

> ***WARNING —***
> - *When cleaning engine parts always wear eye protection.*
> - *Solvents used for cleaning parts are highly flammable, especially in aerosol form. Use with care. Do not smoke. Do not use these products near any source of sparks or flame.*

Crankcase assembly

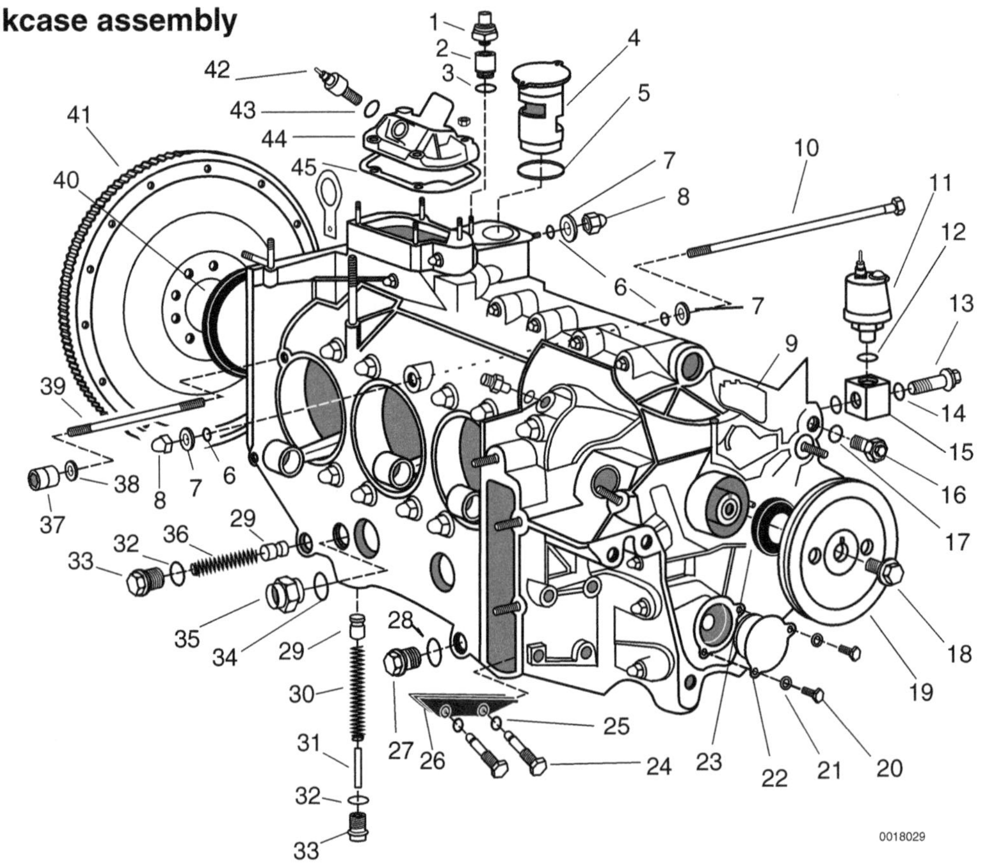

1. Oil pressure switch, 20 Nm (15 ft-lb)
2. Switch adapter, 35 Nm (26 ft-lb)
3. Sealing ring, (18 x 24mm)
4. Oil thermostat
5. Thermostat sealing O-ring, (39.2 x 3 mm)
6. Case bolt sealing ring
7. Case bolt beveled washer
8. Case bolt cap nut, 35 Nm (26 ft-lb)
9. Crankcase
10. Case bolt
11. Oil pressure sensor, 35 Nm (26 ft-lb)
12. Oil pressure sensor sealing ring, (18 x 24 mm)
13. Oil pressure sensor adapter bolt, 35 Nm (26 ft-lb)
14. Adapter bolt sealing ring, (12 x 16 mm)
15. Transmitter adapter housing
16. Oil temperature sensor, 25 Nm (18 ft-lb)
17. Temperature sensor sealing ring
18. Front pulley bolt, 80 Nm (59 ft-lb)
19. Front pulley
20. Intermediate shaft cover bolt
21. Intermediate shaft cover bolt washer
22. Intermediate shaft cover
23. Crankshaft oil seal, front
24. Chain guide rail bolt
25. Guide rail bolt sealing ring (10 x 16 mm)
26. Chain guide rail
27. Oil drain plug, 70 Nm (52 ft-lb)
28. Drain plug sealing ring, (22 x 27 mm)
29. Oil pressure relief valve piston
30. Oil pressure relief valve spring
31. Oil pressure relief valve sleeve
32. Sealing ring, (18 x 24 mm)
33. Plug, 60 Nm (44 ft-lb)
34. Line adapter sealing ring, (22 x 27mm)
35. Oil line adapter, 120 Nm (89 ft-lb)
36. Oil pressure safety valve spring
37. Cylinder head allen nut
38. Cylinder head washer
39. Cylinder head stud
40. Crankshaft oil seal, rear
41. Flywheel
42. Heater blower after-run temperature switch, 25 Nm (18 ft-lb)
43. Temperature switch sealing ring
44. Breather cover
45. Breather cover gasket

Crankcase, inspection

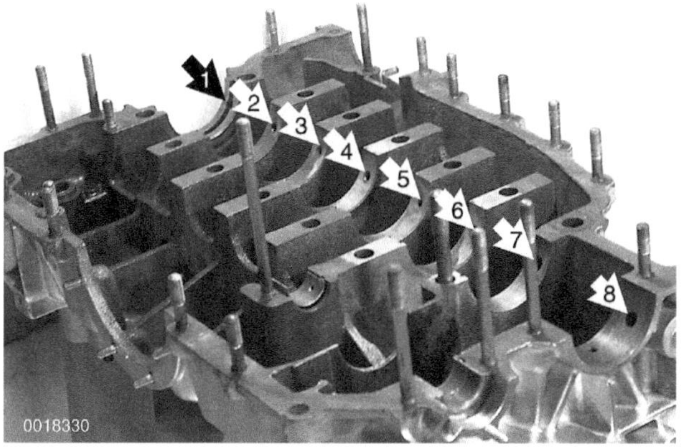

On engines that have had a mechanical failure, the complete lubrication circuit must be cleaned and flushed to remove foreign material before reassembling the engine. This includes cleaning all oil lines, oil pressure valves, crankcase oil passages (**arrows**) and oil reservoir. In addition to thorough cleaning, the engine oil coolers should be replaced. See **170 Engine Lubrication**.

Check cast oil pipe webbing in right crankcase half (lower oil cooler area) for oil leakage due to porous casting material.

- To repair, thoroughly clean crankcase in problem area using alcohol or acetone.
- Fill area with Silastic® 732 RTV sealing compound or equivalent using light uniform pressure until depression is filled.
- Allow 24 hours to harden.

Check for oil leakage at upper right mounting stud at rear engine carrier (arrow). If oil is present, metal between engine oil passages and stud hole threads may be porous. To correct this problem, replace M10 x 22 mm mounting stud with M10 x 15 mm stud, securing with Loctite®270.

When replacing cylinder head studs:

- Heat area around base of stud to loosen thread cement.
- Clean threads in crankcase thoroughly. Chase thread holes with tap.
- Install new studs to height of 135 mm (5.315 in.)

NOTE —

Various types of head studs have been used on the Carrera 911 engines. Always check with an authorized Porsche parts department for the latest in parts information.

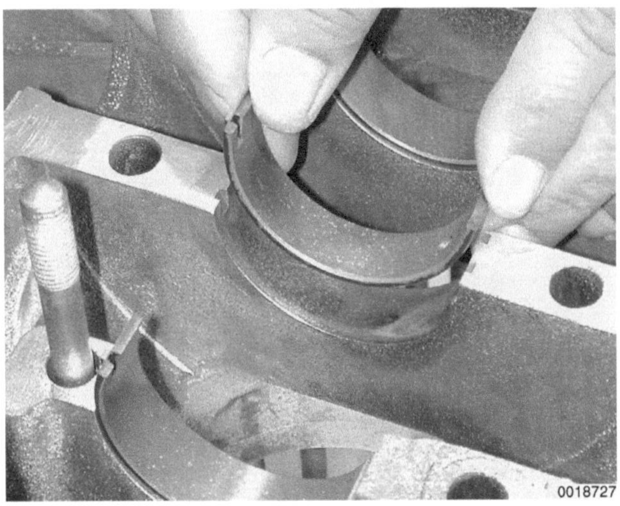

Remove and label main bearing shells from both case halves and check for wear, scoring or visible damage to case.

Remove and label intermediate shaft bearing shells from both case halves and check for wear, scoring or visible damage to case.

NOTE —

Always replace intermediate shaft bearing shells. DO NOT reuse intermediate shaft bearing shells.

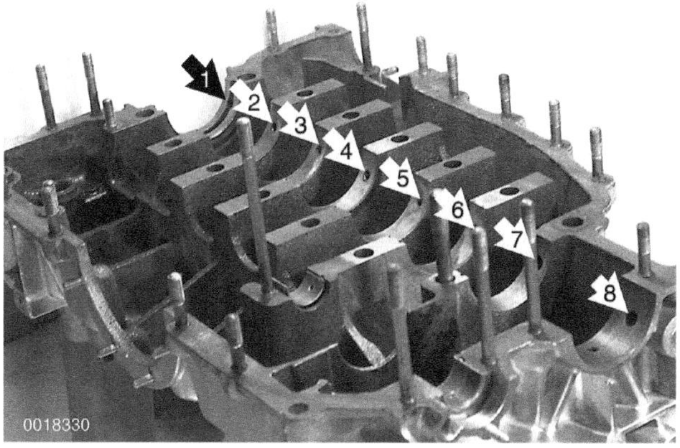

Blow out all engine crankcase oil feed bores (**arrows**) with compressed air.

– Install left half of crankcase and tap into place.

- Install all M10 crankcase bolts, washers and nuts.
- Install four M8 nuts on bearing 1 (flywheel end) and bearing 8 (pulley end).
- Finger tighten nuts.
- Align case sections with each other with light knocks from a plastic hammer. Bearing bore 8 joint must not be offset.
- Tighten all case bolts and nuts to specified torque.

Tightening Torques
- Crankcase bolts and nuts (M10)....... 35 Nm (26 ft-lb)
- No. 1 and 8 main journal nuts (M8)..... 25 Nm (18 ft-lb)

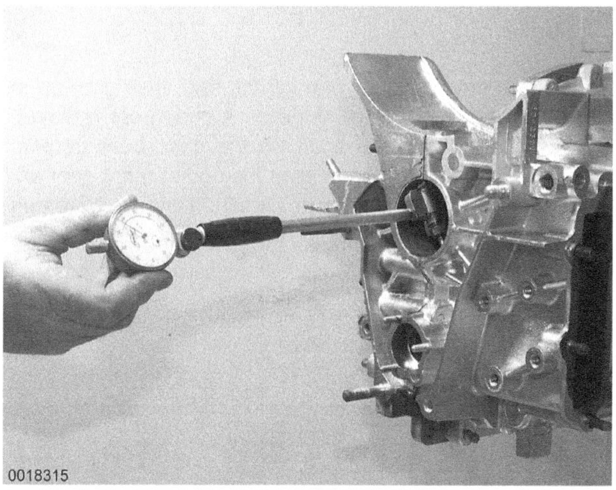

Measure bearing bores 1 to 8 with an internal micrometer. Check for wear, scoring or out of roundness.

Case Bore for Main Bearings (bearings removed)
- Bores 1 - 8
  standard .... 65.000-65.019 mm (2.5590-2.5598 in.)
  oversize........ 65.250-65.269 mm (2.5689-2.5696 in.)

Case Bore for Intermediate Shaft (bearings removed)
- Bearing 1 ...... 27.500-27.521 mm (1.0827-1.0835 in.)
- Bearing 2 ...... 26.500-26.521 mm (1.0433-1.0441 in.)

NOTE —
If bearing bore sizes are incorrect or out of round, the crankcase will need to be machined (line-bored) to accept oversized bearing shells. Check with an authorized Porsche dealer or professional machine shop.

Intermediate shaft, inspection

– Inspect intermediate shaft drive gear for wear or damage.

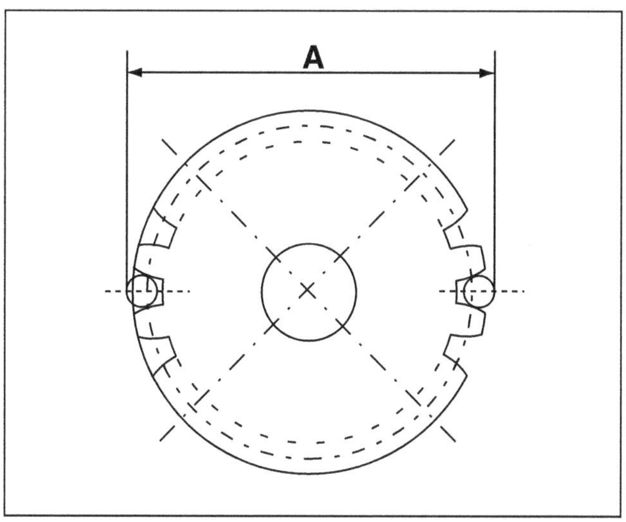

Measure gear by placing two 4.5 mm dowel pins (or drill bits) in gear teeth 180° apart.

- Replace intermediate shaft with drive gear if measurement **A** falls below specifications.

Intermediate Shaft Drive Gear Specifications
- Code 0 .... 136.50 mm (5.3740 in.)
- Code 1 .... 135.55 mm (5.3366 in.)

NOTE —
Intermediate shaft drive gear code is stamped on machined face of gear.

– Check shaft for wear or damage. See **Table a**.

Table a. Intermediate Shaft Clearances

Journal dia. bearing 1 bearing 2	 25.000 - 24.980 mm (0.9842 - 0.9835 in.) 23.980 - 23.967 mm (0.9441 - 0.9436 in.)
Intermediate shaft clearance (Plastigage®)	0.030 - 0.084 mm (0.00012 - 0.0033 in.)
Intermediate shaft axial clearance Wear limit	0.040 - 0.133 mm (0.0016 - 0.0052 in.) 0.16 mm (0.0063 in.)

NOTE —
The intermediate shaft and aluminum drive gear are one replacement part.

◄ Crankcase, drive gear on crankshaft, and drive gear on intermediate shaft are assembled as a coded set (**arrows**). This assembly code is based on the case bore centers. When replacing drive gears, note the code on the gear and use a gear with the same code. Additionally, the bore centers should be checked before final assembly as described below.

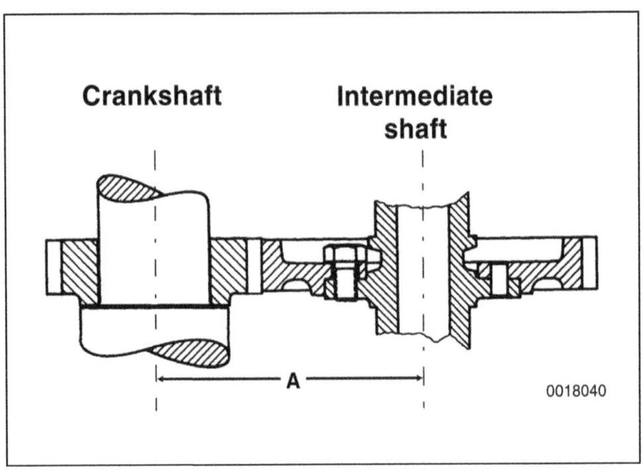

◄ Measure distance between centerline of crankshaft and intermediate shaft bores (**A**).

- Match gears and crankcase as shown in **Table b**.

– The intermediate shaft is hollow and should be cleaned out whenever the engine is being reconditioned and in case of engine bearing failure.

- Drill 6.4 mm (0.252 in.) diameter hole in aluminum plug located in end of intermediate shaft. Tap hole with M8 tap.
- Install M8 X 125 mm bolt into drilled hole and pull aluminum plug out of shaft.
- Flush residue from inside shaft and install new plug.

Table b. Crankcase/Gear Assembly Code

Distance A	Crankcase code	Crankshaft drive gear code	Intermediate shaft gear code	Backlash
103.975–103.990 mm (4.0935– 4.0941 in.)	0	0	0	0.029 - 0.049 mm (0.0011 - 0.0019 in.)
		installation permitted		
		1	0	0.016 - 0.042 mm (0.0006 - 0.0016 in.)
		0	1	0.017 - 0.043 mm (0.0006 - 0.0017 in.)
103.990–104.000 mm (4.0941–4.0945 in.)	1	1	1	0.012 - 0.041 mm (0.0004 - 0.0016 in.)
		installation permitted		
		0	1	0.025 - 0.049 mm (0.0009 - 0.0019 in.)
		1	0	0.025 - 0.048 mm (0.0009 - 0.0019 in.)

0018325

0018327

CRANKSHAFT AND CONNECTING RODS

Crankshaft gears, removing and installing

– Bolt crankshaft in special mounting tool (special tool P 209) and mount in a shop vise.

– Remove circlip holding gears on crankshaft.

◄ Remove complete gear assembly from crankshaft with a suitable puller. Check gear bore and shaft diameter (interference fit).

Crankshaft Drive Gear Fit

- Timing gear shaft dia. . . . . . . 42.002-42.013 mm (1.6536 - 1.6540 in.)
- Timing gear bore dia. . . . . . . 41.975-42.000 mm (1.6526 - 1.6535 in.)
- Fit tolerance -0.002 to -0.038 mm (-0.00008 to -0.0015 in.)

◄ Install drive gear to crankshaft.

- Install woodruff key to crankshaft.
- Heat gear to 150°C (300°F) in oil bath.
- With shoulder on gear facing in (**arrow**), slide drive gear fully onto crankshaft.

> ***WARNING —***
> *Use protective gloves or shop rag when handling the hot gear.*

Crankshaft assembly

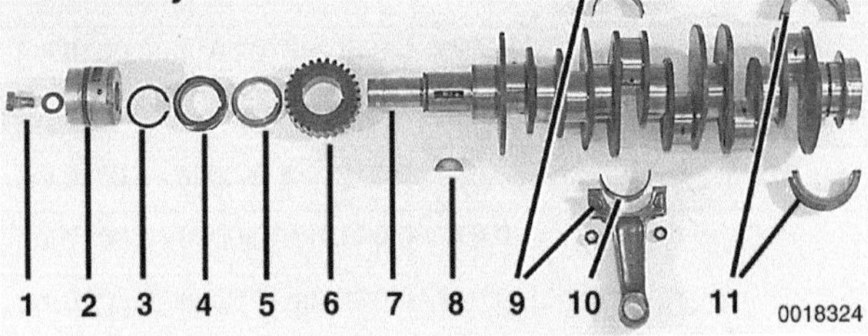

1. Pulley bolt
2. Nose bearing (no. 8)
3. Circlip
4. Distributor drive gear
5. Spacer
6. Intermediate shaft drive gear
7. Crankshaft
8. Woodruff key
9. Connecting rod
10. Connecting rod bearing
11. Main bearing

Install spacer to crankshaft.

– Install distributor drive gear to crankshaft.

- Heat gear to 100°C (212°F) in water bath. Install gear to crankshaft with gear mark facing out.

> ***WARNING —***
> *Use protective gloves or shop rag when handling the hot gear.*

NOTE —
Be sure marking on distributor drive gear faces V-belt drive pulley end of crankshaft. Incorrect installation will cause initial ignition timing to be off by approximately 13°.

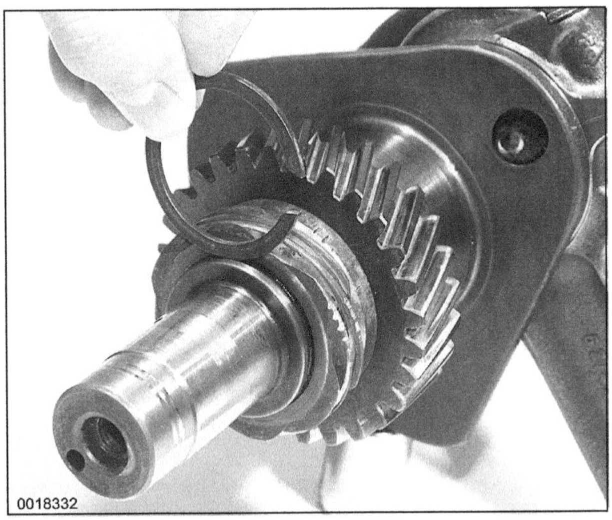

Install selective circlip to crankshaft so that zero clearance is best obtained.

NOTE —
Circlip used to hold gears in place on crankshaft comes in different thicknesses. See **Table c**. *Select circlip to best obtain near zero clearance.*

Table c. Crankshaft Gear Circlip Thickness

Porsche P/N	Thickness	Marking
901 102 148 00	2.4 mm	0
901 102 148 01	2.3 mm	1
901 102 148 02	2.2 mm	2
901 102 148 03	2.1 mm	3

Crankshaft main and rod bearings

Crankshaft main bearings are numbered from the rear of the engine (flywheel end) forward. Main bearing 8, (nose bearing) is at the pulley end.

Crankshaft main bearing shells, if they are to be reused, should only be installed in their original positions.

When servicing crankshaft, bolt crankshaft in special mounting tool (special tool P 209) and mount in a shop vise.

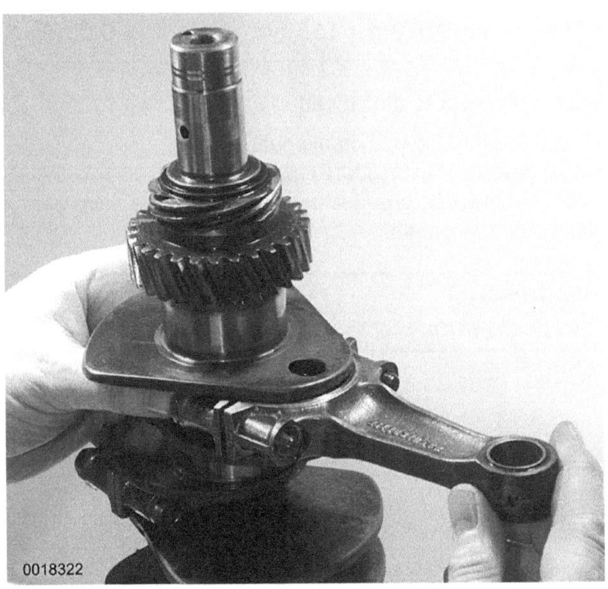

Mark installed position of connecting rods and remove them from crankshaft.

Crankshaft journal sizes are listed in **Table d**. Crankshaft clearance specifications are listed in **Table e**. Crankshaft bearings are available in standard size and four undersizes.

Crankshaft axial play (also called end play) is controlled by a two-piece thrust bearing. The thrust bearing is the no. 1 main bearing (flywheel end).

Table d. Crankshaft Journal Sizes

Size (mm)	**Main bearing 8 (nose bearing)**	**Main bearings 1 - 7**	Connecting rod bearing	Crankshaft drive gear	Crankshaft flywheel
Standard	30.980- 30.993 mm (1.2197- 1.2202 in.)	59.971- 59.990 mm (2.3611- 2.3618 in.)	54.971- 54.990 mm (2.1642 - 2.1650 in.)	42.002-42.013 mm (1.6536-1.6541 in.)	89.780-90.000 mm (3.5346-3.5433 in.)
Undersize 1 (- 0.25 mm)	30.730 - 30.743 mm (1.2098- 1.2103 in.)	59.721- 59.740 mm (2.3512- 2.3520 in.)	54.721 - 54.740 mm (2.1544 - 2.1551 in.)	–	–
Undersize 2 (- 0.50 mm)	30.480 - 30.493 mm (1.2000- 1.2005 in.)	59.471- 59.490 mm (2.3414- 2.3421 in.)	54.471 - 54.490 mm (2.1445 - 2.1453 in.)	–	–
Undersize 3 (- 0.75 mm)	30.230 - 30.243 mm (1.1901 -1.1907 in.)	59.221 - 59.240 mm (2.3315 - 2.3323 in.)	54.221 - 54.240 mm (2.1347 - 2.1354 in.)	–	–
Undersize 4 (-1.00 mm)	29.980 - 29.993 mm (1.1803 - 1.1808 in.)	58.971 - 58.990 mm 2.3217 - 2.3224 in.)	53.971 - 53.990 mm (2.1248 - 2.1256 in.)	–	–

Axial play should be measured at the center main bearing journal with the crankshaft supported on rollers on the outer main bearing journals.

Table e. Crankshaft Clearances

Crankshaft main bearing no. 1-7 radial clearance (Plastigage®)	0.010–0.072 mm (0.0003–0.0028 in.)
Crankshaft main bearing no. 8 radial clearance (Plastigage®)	0.048–0.104 mm (0.0019 –0.0041 in.)
Connecting rod bearing radial clearance (Plastigage®)	0.020–0.088 mm (0.0008–0.0035 in.)
Crankshaft axial clearance Wear limit	0.110–0.195 mm (0.0043–0.0078 in.) 0.20 mm (0.008 in.)
Maximum permissible crankshaft runout	0.15 mm (0.0060 in.)

Connecting rods

– Check weight of each connecting rod. Weigh a complete connecting rod, less the bearing shells. See **Table f**.

NOTE —
Replacement connecting rods are available in different weight classes. Each class can be identified by the last digit of the part number. The weight class is stamped on the shaft of the connecting rod.

CAUTION —
In a complete set of connecting rods installed in one engine no two must differ in weight by more than 9 grams.

– Hold connecting rod in soft-jawed vise and torque old rod bolts/nuts for measurement.

NOTE —
Do not use new fasteners when making this check. This check should be performed with used rod bolts.

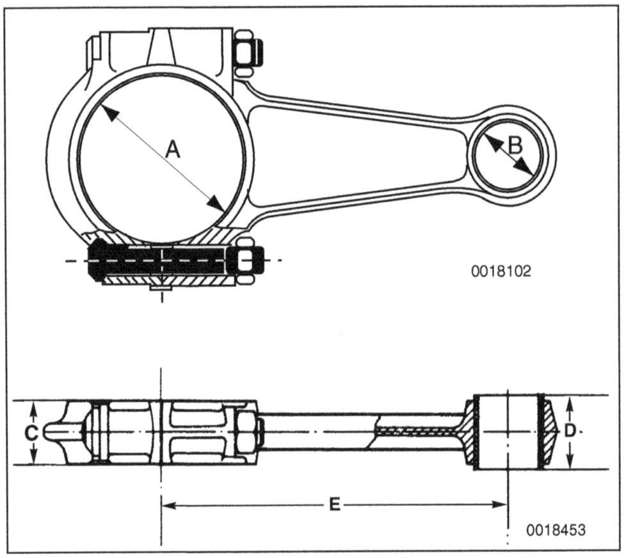

Checking Torque

- Connecting rod bolt/nut. . . . . 50-55 Nm (37-41 ft-lb)

Use micrometer and bore gauge to check connecting rod dimensions. See **Table g**.

Table f. Connecting Rod Weights

Weight in grams	Weight class	Porsche P/N
615 to 624	3	964 103 020 53
624 to 633	4	964 103 020 54
633 to 642	5	964 103 020 55
642 to 651	6	964 103 020 56
651 to 660	7	964 103 020 57
660 to 669	8	964 103 020 58
669 to 678	9	964 103 020 59
678 to 687	10	964 103 020 60
687 to 696	11	964 103 020 61

Table g. Connecting Rod Dimensions

Big end bore (**A**)	58.000 - 58.019 mm (2.2835-2.2842 in.)
Small end bore (**B**)	23.020 - 23.033 mm
Width at big end (**C**) 1984-1985 1986-1989	 21.70 - 21.80 mm (0.8543 - 0.8583 in.) 21.85 - 21.90 mm (0.8602 - 0.8622 in.)
Width at small end (**D**)	24.5 - 25.0 mm (0.9646 - 0.9842 in.)
Distance between centers (**E**)	126.95 - 127.00 mm (4.998 - 5.000 in.)
Connecting rod bushing/piston pin clearance Wear limit	0.020 - 0.037 mm (0.0008 - 0.0014 in.) 0.055 mm (0.0022 in.)

– Check piston pin fit to small end bushing. Pin should slide through bushing with light pressure.

– Check that numbers stamped on side of cap and rod are the same (cap and rod are a matched set).

– Insert bearing shells in connecting rod and corresponding rod cap.

– Lubricate bearing surfaces with suitable engine assembling lubricant.

0018321

During reinstallation, matching numbers on side of connecting rod and rod cap should all be facing same side of engine (for example, all marked numbers should face exhaust side of engine).

> ***CAUTION —***
> *Always use new connecting rod nuts and bolts during final assembly.*

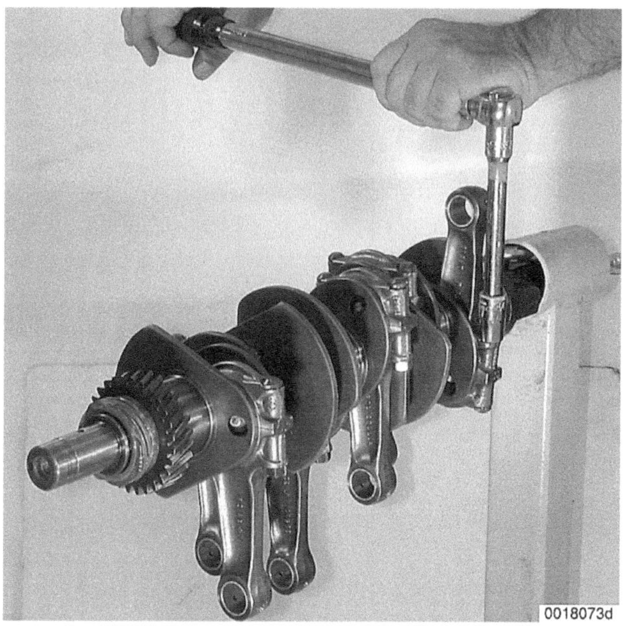

0018073d

With all connecting rods installed, torque each connecting rod nut.

Tightening Torques

- Connecting rod bolt/nut
  Stage 1 . . . . 20 Nm (14 ft-lb)
  Stage 2 . . . . 90°± 2°

NOTE —
Connecting rod must rotate easily under its own weight when the bearing is lubricated with engine oil.

131 Pistons and Cylinders

TABLES

GENERAL

This repair group provides reconditioning information for the pistons and cylinders. Engine disassembly is described in **102 Engine Disassembly and Assembly**.

> ***CAUTION —***
> *The information contained here is intended to be used as a reconditioning guide for the professional or experienced automotive technician.*

Special tools

Some of the procedures require special tools.

Special Tools

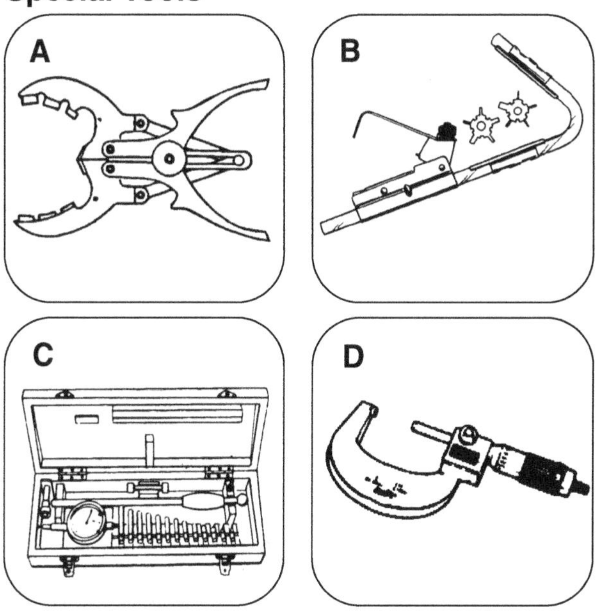

Special tool

A VW 121 b . . . . Piston ring expander
B 65. . . . . Piston ring groove cleaner
(source: Baum Tools Unlimited)
C 17. . . . . Inside bore gauge
(source: Baum Tools Unlimited)
D 12. . . . . Micrometer, 75 - 100 mm (3 - 4 in) range
(source: Baum Tools Unlimited)

PISTONS AND CYLINDERS

Pistons must be matched by weight. When replacing pistons and cylinders, always check with an authorized dealer for the latest in replacement parts information.

Mark all parts during engine disassembly. All used parts must be reinstalled in their exact installation position. Parts must not be interchanged between cylinders.

If engine is equipped with Nikasil® pistons and cylinders which meet all critical dimensions, hone cylinders lightly. Replace piston rings with Göetz® rings. If equipped with Alusil® pistons and cylinders, replace with Nikasil® pistons and cylinders.

Table a. Piston and Cylinder Identification

Manufacturer	Alloy	Cylinder	Piston	Oil control ring (groove 3)
Mahle®	Nikasil®	Coated aluminum, yellowish	Forged, leaded, dark grey	Bevelled edge, rubber lined spring, not chrome plated
KS®	Alusil®	Uncoated aluminum, light grey	Ferrocoat (cast iron coat), light grey	Double-bevelled, rubber lined spring, chrome plated

Piston weight, checking

Piston Weight Specifications

- Weight tolerance
  New piston: match to within 4 grams (0.14 oz.)
  Replacement piston: match to within 8 grams (0.28 oz.)

– Piston is weighed with wrist pin, wrist pin clips and rings installed.

Table b. Weight Classes for Mahle® Pistons

Weight class	Engine type 930/21	Code
Standard	613 – 617 g (21.46 – 21.60 oz.) 617 – 621 g (21.60 – 21.74 oz.)	– –
±4 grams max. permissible tolerance	621– 625 g (21.74 – 21.88 oz.) 625 – 629 g (21.88 – 22.02 oz.)	+ ++
±8 grams max. permissible tolerance	613–621 g (21.46– 21.74 oz.) 621– 629 g (21.74 – 22.02 oz.)	– – or – + or ++

Table c. Weight Classes for KS® Pistons

Weight class	Engine type 930/21 USA	Code
Standard	650 – 654 g (22.75 – 22.89 oz.) 654 – 658 g (22.89 – 23.03 oz.)	– – –
±4 grams max. permissible tolerance	658 – 662 g (23.03 – 23.17 oz.) 662 – 666 g (23.17 – 23.31 oz.)	+ ++
±8 grams max. permissible tolerance	650 – 658 g (22.75 – 23.03 oz.) 658 – 666 g (22.75 – 23.31 oz.)	– – or – + or ++

Cylinder height, checking

◄ Cylinder manufacturer identification and dimensions are indicated by marks on cylinder base.

Table d. Cylinder Height

Tolerance Group	Cylinder height
5	85.400 – 85.425 mm (3.3621 – 3.3631 in.)
6	85.425 – 85.450 mm (3.3631 – 3.3641 in.)

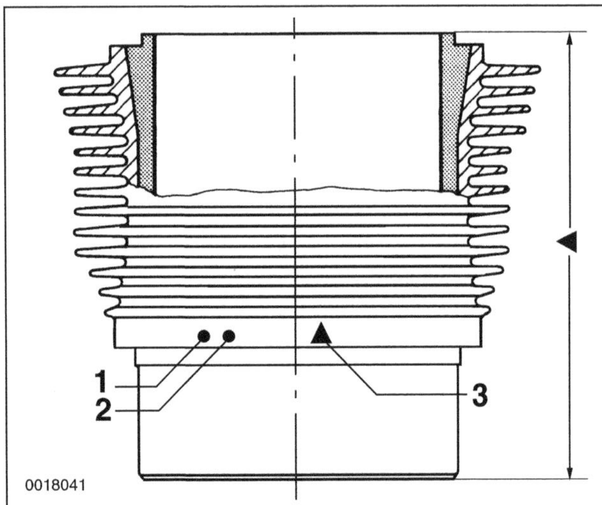

1. Manufacturer identification
2. Cylinder diameter tolerance group (see **Table e** or **Table f**).
3. Cylinder height tolerance group (5 or 6)

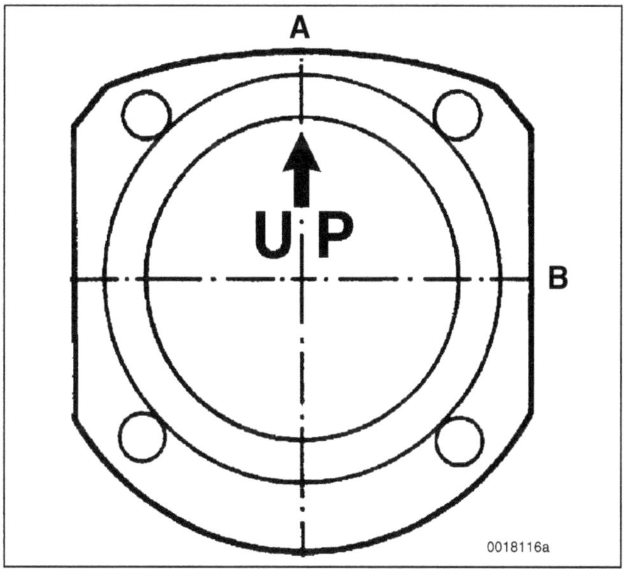

Cylinder roundness, checking

◄ Check cylinder roundness by measuring bore in two directions perpendicular to stud holes (**A** and **B**).

- Measure cylinder at a point 30 mm (1.181 in.) below upper edge.

Cylinder Roundness Wear Limit

- Difference between **A** and **B**. . . . . . 0.04 mm (0.0016 in.)

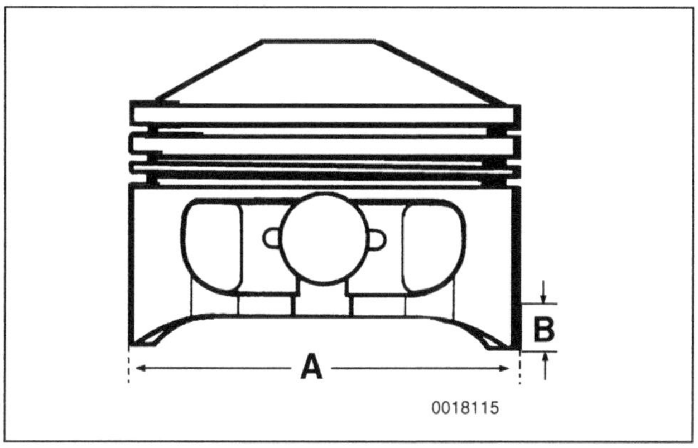

Piston and cylinder wear, checking

◄ Check piston for wear by measuring across piston skirt (**A**).

- Measure Mahle® pistons 18 mm (0.71 in.) from bottom (**B**).
- Measure KS® pistons 10 mm (0.39 in.) from bottom (**B**).

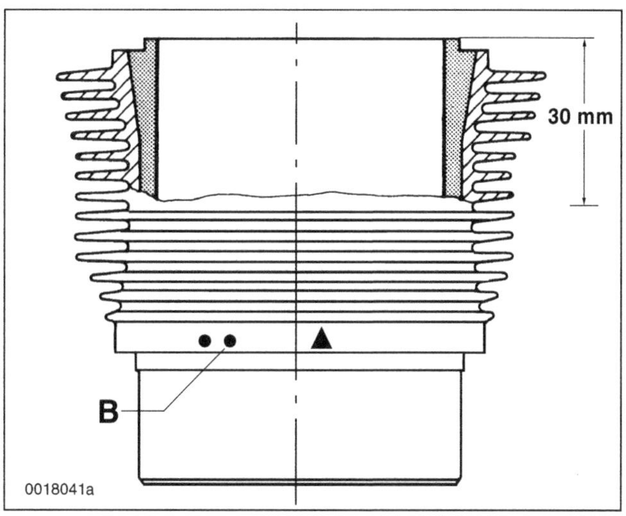

◄ Check cylinder for wear by measuring bore 30 mm (1.18 in.) below upper edge.

– Replace cylinders if measured diameter exceeds 0.08 mm (0.0031 in.) from installed size.

– Replace pistons and cylinders if running clearance exceeds 0.12 mm (0.005 in.).

Table e. Mahle® Piston and Cylinder Sizes

Code (B)	Cylinder diameter	Piston diameter
0	95.000 – 95.007 mm (3.7401 – 3.7404 in.)	94.965 – 94.975 mm (3.7387 – 3.7391 in.)
1	95.007 – 95.014 mm (3.7404 – 3.7407 in.)	94.972 – 94.982 mm (3.7390 – 3.7394 in.)
2	95.014 – 95.021 mm (3.7407 – 3.7409 in.)	94.979 – 94.989 mm (3.7393 – 3.7397 in.)
3	95.021 – 95.028 mm (3.7409 – 3.7413 in.)	94.986 – 94.996 mm (3.7395 – 3.7399 in.)

Table f. KS® Piston and Cylinder Sizes

Code (B)	Cylinder diameter	Piston diameter
0	95.000 – 95.005 mm (3.7401 – 3.7403 in.)	94.975 – 94.980 mm (3.7391 – 3.7393 in.)
1	95.005 – 95.010 mm (3.7403 – 3.7405 in.)	94.980 – 94.985 mm (3.7393 – 3.7395 in.)
2	95.010 – 95.015 mm (3.7405 – 3.7407 in.)	94.985 – 94.990 mm (3.7395 – 3.7397 in.)
3	95.015 – 95.020 mm (3.7407 – 3.7409 in.)	94.990 – 94.995 mm (3.7397 – 3.7399 in.)

Piston rings

– Check and adjust piston ring end gaps if reusing old pistons and cylinders.

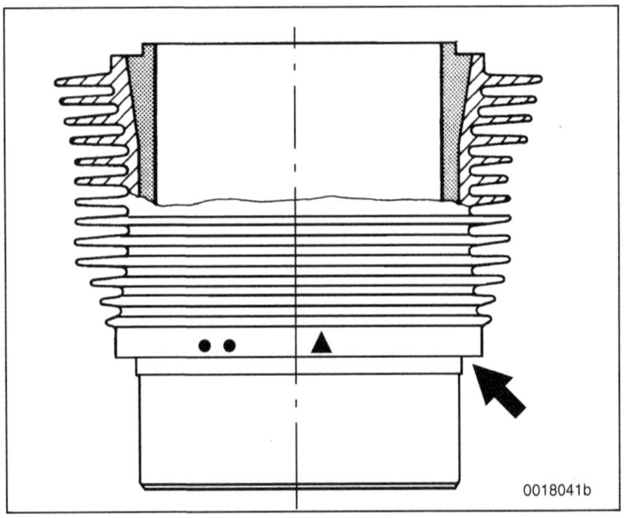

◄ Measure ring end clearance by pushing ring into bottom of cylinder to approximate height of cylinder base gasket surface (**arrow**).

Piston ring side and end clearance tolerances are given in **Table g** and **Table h**.

Table g. Piston Ring End Gap Clearance

Ring (groove)	Gap at installation	Wear limit
Compression (1)	0.2 – 0.4 mm (0.008 – 0.016 in.)	0.8 mm (0.031 in.)
Compression (2)	0.2 – 0.4 mm (0.008 – 0.016 in.)	1.0 mm (0.039 in.)
Oil control (3)	0.3 – 0.6 mm (0.012 – 0.024 in.)	2.0 mm 0.078 in.)

Table h. Piston Ring Side Clearance

Ring (groove)	Clearance	Wear limit
Compression (1)	0.070 – 0.102 mm (0.0027 – 0.0040 in.)	0.2 mm (0.0078 in.)
Compression (2)	0.040 – 0.072 mm (0.0016 – 0.0028 in.)	0.2 mm (0.0078 in.)
Oil control (3)	0.020 – 0.052 mm (0.0008 – 0.0020 in.)	0.1 mm (0.0039 in.)

150 Cylinder Heads

TABLES

GENERAL

This repair group provides reconditioning information for the cylinder heads, including valvetrain reconditioning specifications.

Cylinder head repair and reconditioning require that the engine be removed from the car and disassembled on an appropriate engine stand. Engine disassembly is described in **102 Engine Disassembly and Assembly**.

> ***CAUTION —***
> *The information contained in the repair group is intended to be used as a reconditioning guide for the professional or experienced automotive technician.*

Special tools

Some of the procedures require special tools.

Special Tools

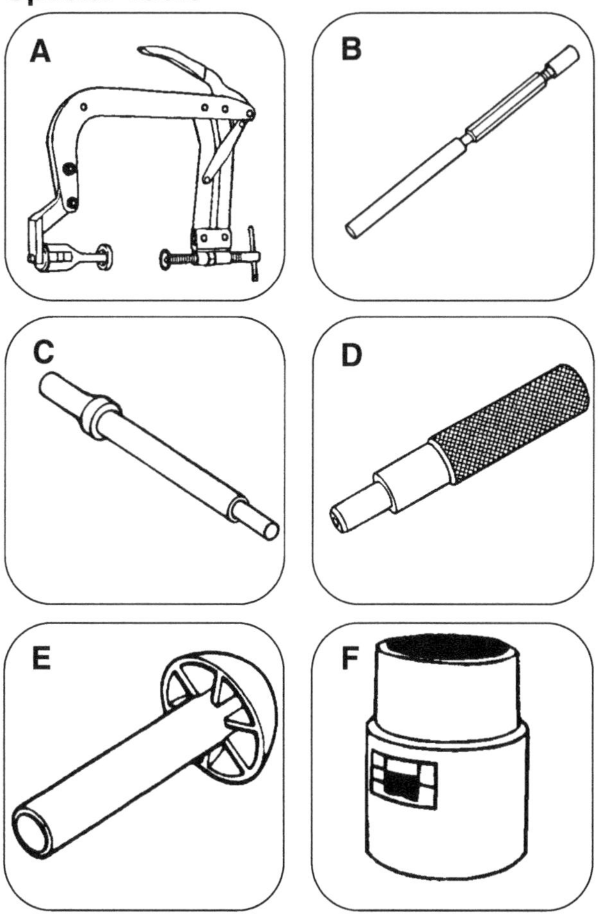

Special tool

A 1020 (source: Baum Tools Unlimited) Valve spring compressor
B P 206 . . . . Valve guide go/nogo gauge (Porsche ordering #000 721 206 00)
C 87a (source: Baum Tools Unlimited) . Valve guide removal drift
D 83 (source: Baum Tools Unlimited) Valve guide installation drift
E 10-204 (source: Baum Tools Unlimited)Valve stem seal installer
F P 10 c. . . . . Valve spring height gauge (Porsche ordering #000 721 010 30)

CYLINDER HEADS

Five of the cylinder heads are identical. Cylinder head 3 has a threaded boss for the temperature sensor.

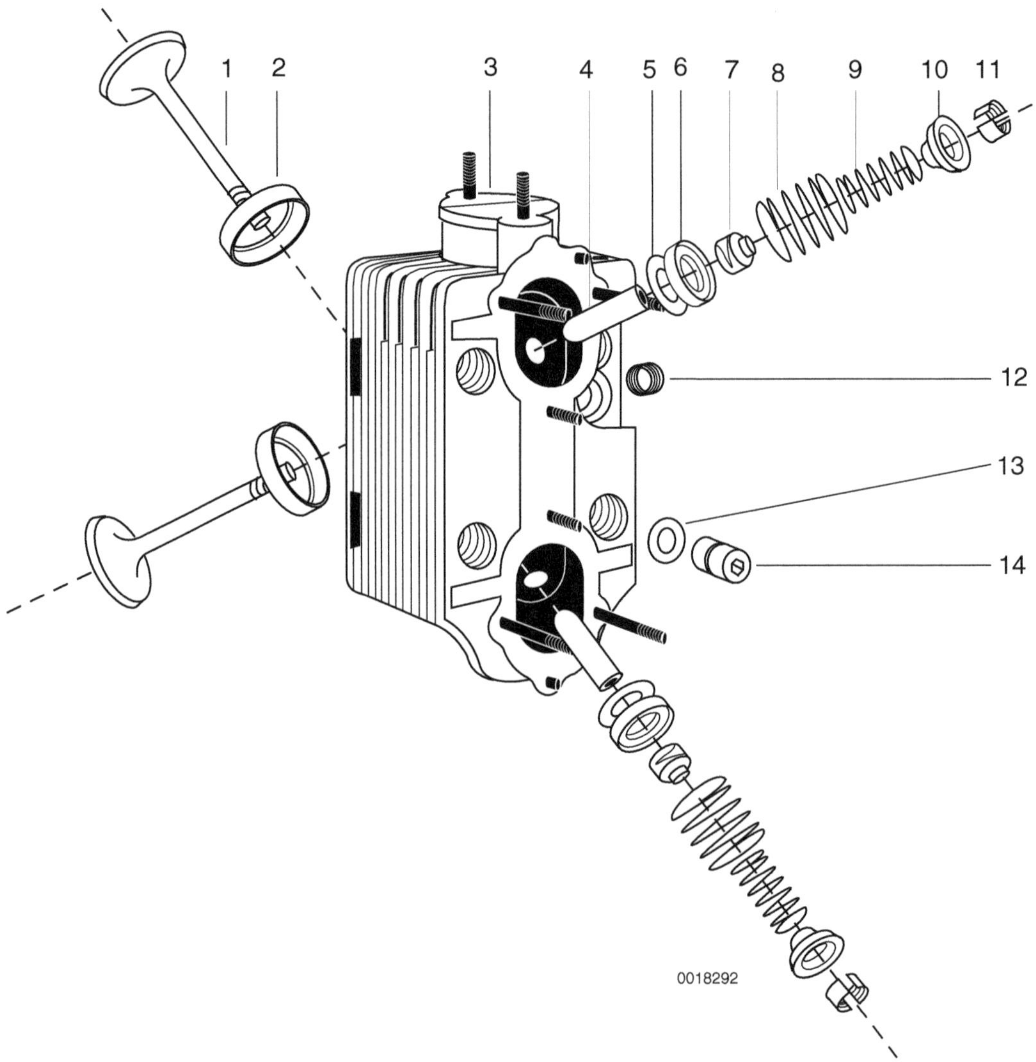

1. Valve
2. Valve seat
3. Cylinder head
4. Valve guide
5. Valve spring seat shim
6. Valve spring seat
7. Valve stem seal
8. Outer valve spring
9. Inner valve spring
10. Valve spring retainer
11. Valve keeper
12. Spark plug threaded insert
13. Cylinder head washer
14. Cylinder head nut (8 mm allen)

Cylinder head service

◀ Remove and install valve springs using Porsche special tool P 200 a or equivalent.

NOTE —
Keep parts for each valve assembly (keepers, springs, spring retainer etc.) together.

– On cylinder head no. 3: remove temperature sensor using Porsche special tool 9222/1 or equivalent.

– Wash cylinder head and clean thoroughly by bead blasting.

NOTE —
Check sealing surfaces on cylinder head prior to cleaning. If any head studs were found to be broken during disassembly, check carefully for damage to the castings caused by blow-by.

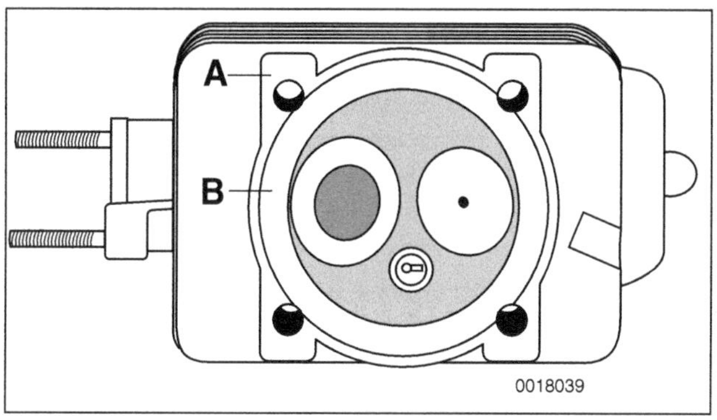

◀ Inspect cylinder head sealing surface. Cylinder heads can be machined, providing the following rules are adhered to:

- Mill surfaces **A** and **B** equal amounts.
- Mill all three heads on each side an equal amount.
- Bevel edges of machined surfaces slightly.
- Mark cylinder head with "–25", indicating that head was machined.

Cylinder Head Machining Limit
- Mill surfaces **A** and **B** . 0.25 ± 0.02 mm (0.01 ± 0.001 in.)

CAUTION —
- *Cylinder head surface may only be machined once.*
- *When assembling an engine with machined cylinder heads, be sure to use thicker 0.50 mm (0.02 in.) base gaskets. The 0.50 mm gaskets are available from an authorized Porsche parts department.*

VALVES

Valve guide, checking

Use Porsche special tool P 206 (valve guide go/nogo gauge) or bore gauge/micrometer to check guide for wear. Check against specification.

Valve Guide Wear Limit

- Intake valve guide inside dia. to intake valve stem clearance. . . . . . . 0.15 mm (0.006 in.)
- Exhaust valve guide inside dia. to exhaust valve stem clearance . . . . . 0.20 mm (0.008 in.)

Valve guide, replacing

- Working from camshaft side, machine away top of valve guide flush with top of cylinder head.
- Dislodge each valve guide with sharp blow from a hammer.
- Press guide out through combustion chamber using an appropriate drift.

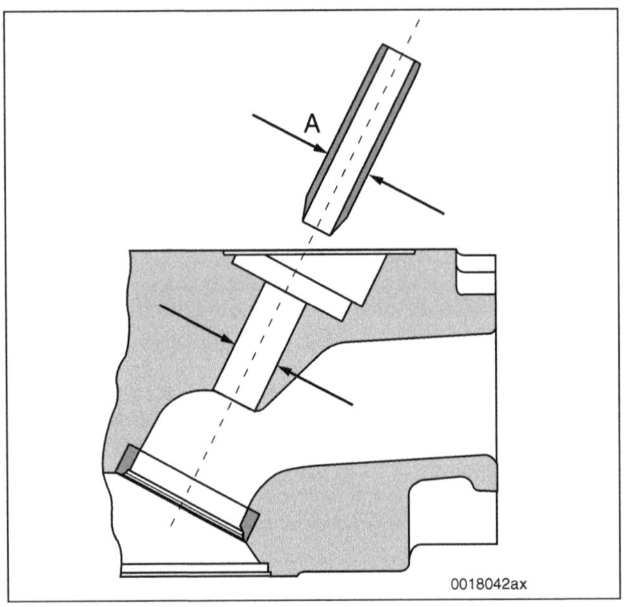

< Measure guide bores in cylinder head. Machine replacement valve guides so that valve guide diameter is 0.06–0.09 mm (0.0023–0.0035 in.) larger than valve guide hole.

NOTE —

When the valve guides are removed, the bores in the cylinder head will most likely be widened. Replacement valve guides from Porsche are oversized and must be machined (turned on a lathe) so that an interference fit is obtained.

Table a. Valve Guide Outside Diameters

Valve guide	Guide outside diameter (A)
Standard size	13.060 mm (0.514 in.)
Oversize 1	13.260 mm (0.522 in.)

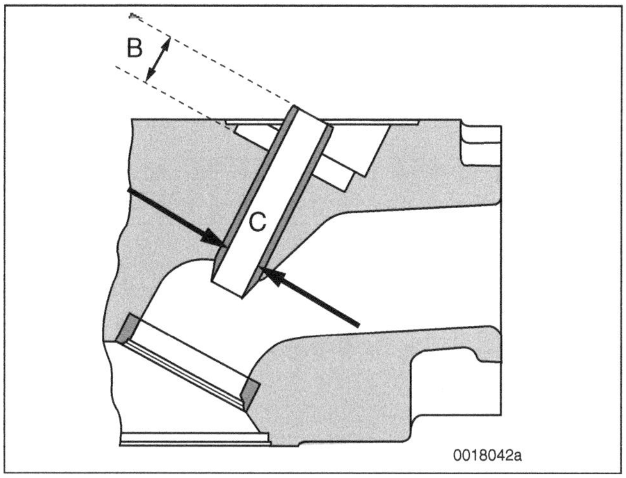

Apply lubricant (tallow-based) to valve guide and press new guides in to specified installed height using 9 mm stepped drift.

Valve Guide Installed Height

- Height above cylinder head (**B**) .... 13.2 mm (0.520 in.)
- Reamed diameter (**C**) 9.00 - 9.015 mm (0.354 - 0.355 in.)

– Ream new valve guides with special tool no. 78HSS (9 mm)

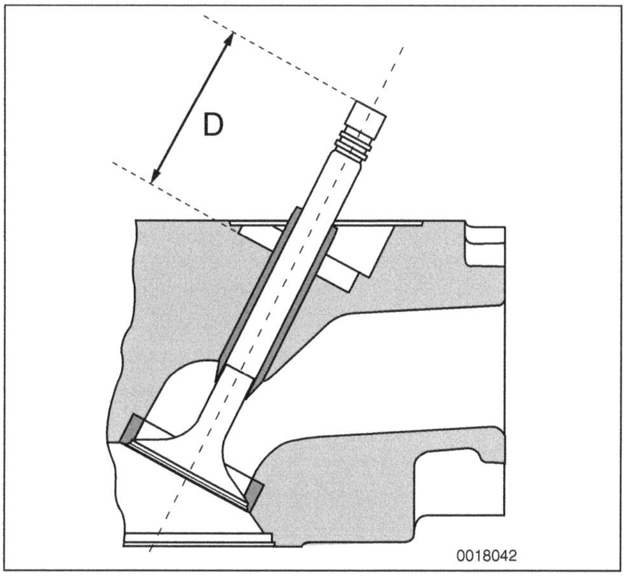

Valve seat, checking

Place valve in cylinder head and measure distance between top of valve stem and valve spring seat (shims removed).

Valve Seat Check

- Valve seat checking height **D** .... 46.0 ± 0.3 mm (1.811 ± 0.011 in.)

– If dimension **D** exceeds specifications, recheck with new valve. If measurement is still out of specifications, replace valve seat or cylinder head.

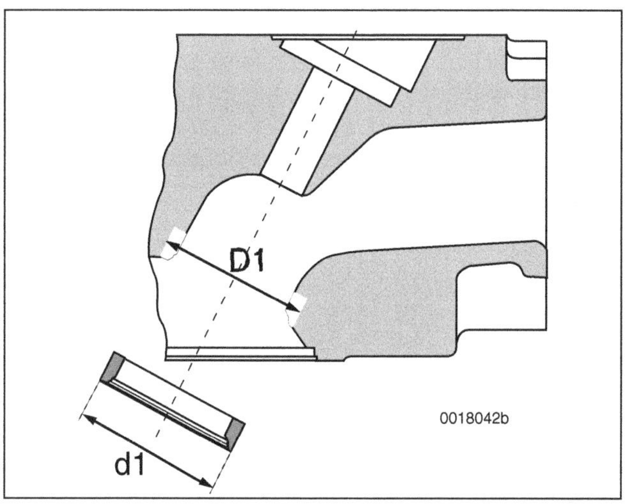

Valve seat, replacing

– Use rough cut stone to carefully grind away valve seat until seat becomes loose in cylinder head. Remove seat.

Measure valve seat outer diameter **d1**. Measure cylinder head bore diameter **D1**.

– Machine replacement valve seats so that seat diameter is larger than valve seat hole according to fit specifications. See **Table b**.

– Heat cylinder head slowly to 200°C (400° F).

– Install valve seat using appropriate drift.

– Allow cylinder head to slowly cool to room temperature.

– Reheat cylinder head to 200° C (400° F) and maintain temperature for 2 hours.

– Allow cylinder head to slowly cool to room temperature.

Table b. Valve Seat Dimensions

Valve Seat	Valve seat diameter (d1)	Cylinder head bore diameter (D1)
Standard size		
Intake seat	51.680 - 51.661mm (2.0346 - 2.0339 in.)	51.500 - 51.530 mm (2.0276 - 2.0287 in.)
Exhaust seat	44.200 - 44.184 mm (1.7402 - 1.7395 in.)	44.000 - 44.025 mm (1.7323 - 1.7333 in.)
Oversize 1		
Intake seat	52.000 - 51.981mm (2.0472 - 2.0465 in.)	51.820 - 51.850 mm (2.0401 - 2.0413 in.)
Exhaust seat	44.760 - 44.744 mm (1.7622 - 1.7616 in.)	44.560 - 44.585 mm (1.7543 - 1.7553 in.)
Valve seat to cylinder head clearance (interference fit)		
Intake valve	-0.16 to -0.20 mm (-0.0063 to 0.0078 in.)	
Exhaust valve	-0.14 to -0.18 mm (-0.0055 to 0.0071 in.)	

Valve seat, cutting

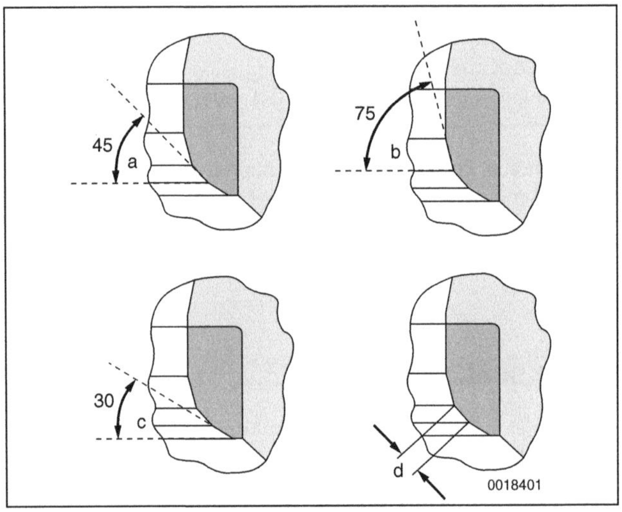

Use illustration and **Table c** as guide in cutting valve seat.

Table c. Valve Seat Cutting Guide

	Intake valve/exhaust valve
Valve seat angle (**a**)	45°
Valve seat lower edge (**b**)	75°
Valve seat upper edge (**c**)	30°
Valve seat width (**d**)	1.5 ± 0.1 mm (0.059 ± 0.004 in.)

Valves

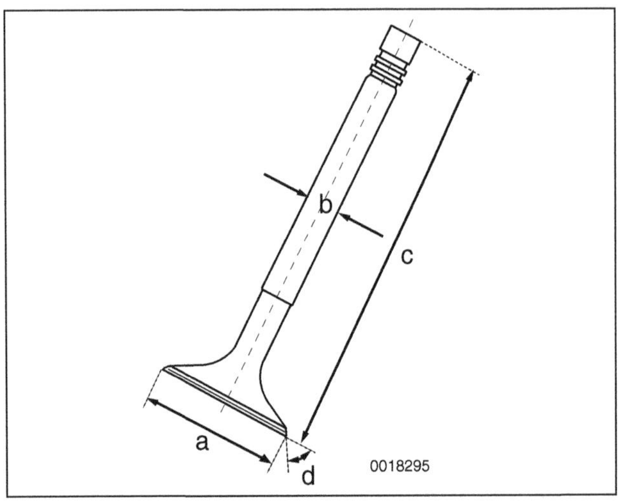

Use illustration and **Table d** for valve specifications and measurements.

Table d. Valve Specifications

Dimension	Intake	Exhaust
Valve head diameter (**a**)	49.00 ± 0.10 mm (1.929 ± 0.004 in.)	41.5 ± 0.10 mm (1.634 ± 0.004 in.)
Valve stem diameter (**b**)	8.97 – 0.012 mm (0.3531 – 0.0004 in.)	8.95 – 0.012 mm (0.3523 – 0.0004 in.)
Total valve length (**c**)	110.1 ± 0.25 mm (4.335 ± 0.010 in.)	108.4 ± 0.25 mm (4.268 ± 0.010 in.)
Valve seat angle (**d**)	45°	

Valve spring height, setting

Keep each set of valves, springs, keepers and spring retainers with the cylinder head that they came out of. Check both inner and outer springs for damage. The outer valve spring is progressively wound and should be installed with the tighter coils resting on the cylinder head. The inner spring can be installed either way.

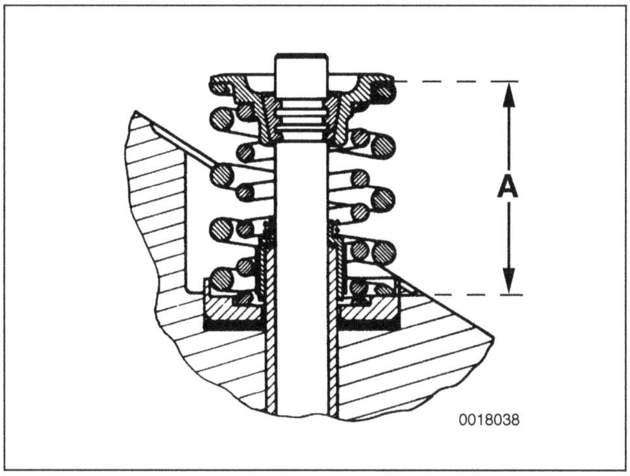

◀ Use Porsche special tool P 10 c or equivalent to measure distance **A**.

- Adjust valve spring height by adding or subtracting shims under valve spring.
- Correct valve spring height to specifications listed.

Valve Spring Height Specification

- Intake or exhaust **(A)** . . . 34.5–0.3 mm (1.358–0.012 in.)

---

151 Camshafts and Camshaft Cases

Special Tools

A

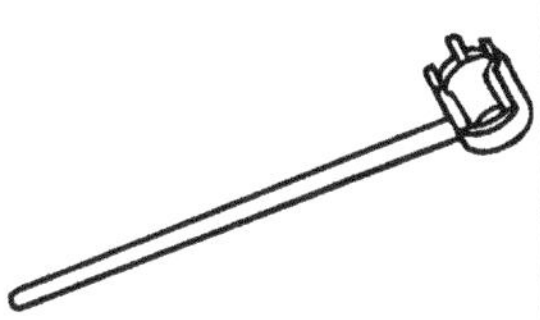

B

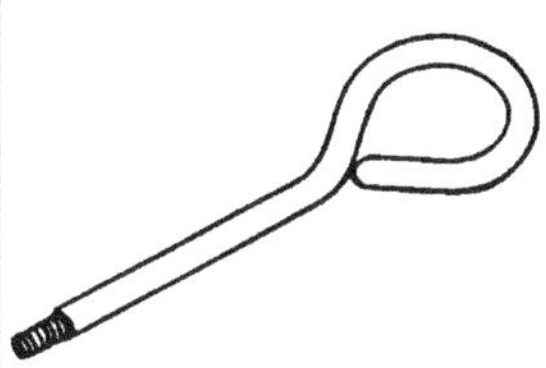

C

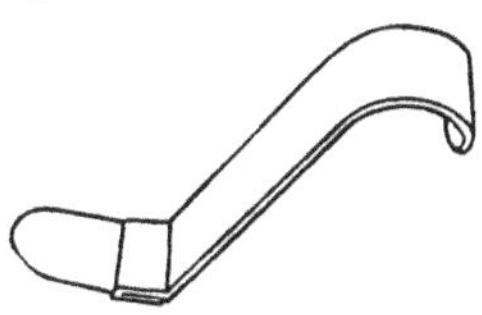

Special tool

A 9191 . . . . Cam sprocket counterholding wrench (Porsche ordering #000 721 919 10)
B P 212 . . . . Camshaft sprocket lockpin extractor (Porsche ordering #000 721 212 00)
C P 213 b. . . . . Feeler gauge (Porsche ordering #000 721 213 20)

GENERAL

This section covers information and repairs to the camshaft cases, camshafts, rocker shafts and rocker arms. Engine disassembly, including camshaft and camshaft case removal, is covered in **102 Engine Disassembly and Assembly**.

Timing chain replacement or adjustment is covered in **152 Camshaft Timing Chains**.

Special tools

Some of the procedures require special tools.

CAMSHAFTS

Camshafts and rocker shafts can be removed from the camshaft cases with the engine installed.

Camshafts, removing and installing

– Raise car and support in a safe manner.

> ***WARNING —***
> *Make sure the car is stable and well supported at all times. Use a professional automotive lift or jack stands designed for the purpose. A floor jack is not adequate support.*

– Disconnect negative (–) battery cable and cover battery terminal to keep cable from accidentally contacting terminal.

> ***WARNING —***
> *Battery electrolyte is a corrosive acid. Always wear eye protection when working on or near the battery.*

– Remove muffler as described in **260 Exhaust System**. Remove rear engine metal shroud.

◀ Set crankshaft to **Z1** cylinder 1 compression TDC (**arrow**).

NOTE —
At cylinder 1 TDC, the intake and exhaust valves for cylinder 1 will be closed.

– Remove intake and exhaust valve covers on both sides.

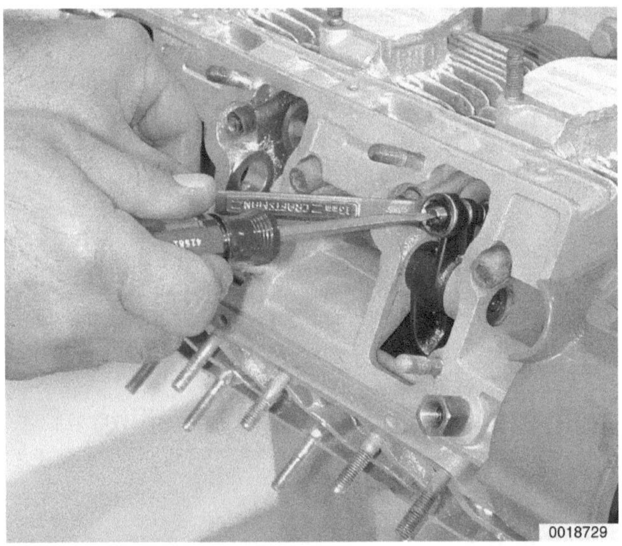

◀ Loosen cylinder 1 valve adjuster lock nuts and fully back out valve adjusting screws.

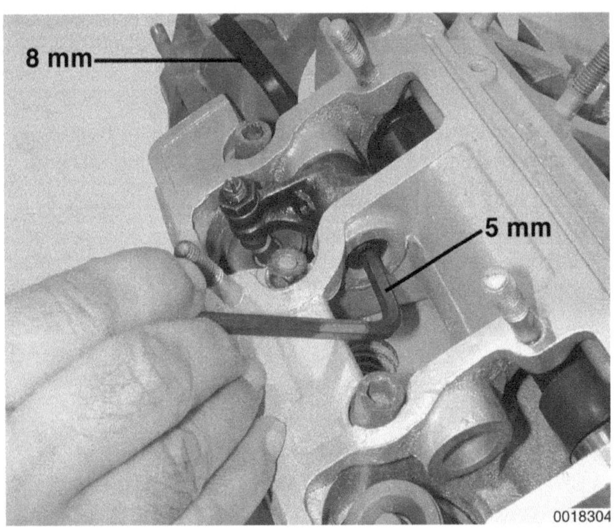

◀ Counterhold cylinder 1 intake rocker shaft nut with 8 mm allen wrench and loosen other end with 5 mm allen wrench.

NOTE —
Mark location of each rocker arm. Do not interchange used parts.

Tap rocker shaft out, removing rocker arm.

- Turn crankshaft 120° in normal direction of rotation. Remove cylinder 6 rocker shafts and arms.

- Following firing order, turn crankshaft 120° and remove rocker shafts and arms for next cylinder. Repeat until all rocker shafts and arms are removed.

NOTE —
The firing order is 1 - 6 - 2 - 4 - 3 - 5.

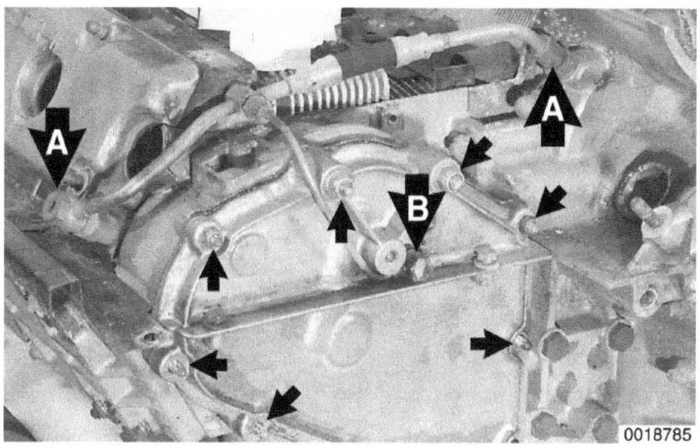

Remove left timing chain tensioner oil feed line (**A**).

- Remove left camshaft housing oil feed line (**B**).
- Remove left timing chain housing cover.

Loosen left camshaft sprocket mounting bolt.

NOTE —
Use Porsche special tool 9191 or equivalent to counterhold camshaft sprocket when loosening bolt.

Compress timing chain tensioner using large pliers.

- Install 3 mm (1/8 in.) pin through hole in upper tensioner body (**arrow**) to lock tensioner down.

NOTE —

Check position of timing chain tensioner piston in its installed position. If piston is fully extended, the timing chains and sprockets are most likely worn. See **152 Camshaft Timing Chains**.

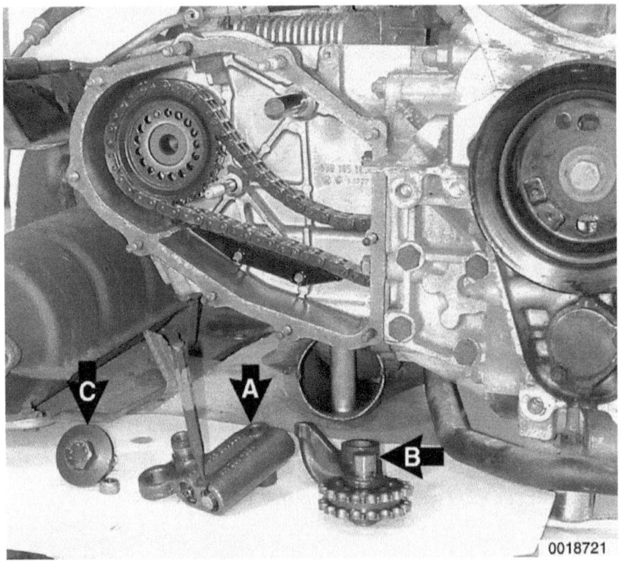

Remove timing chain tensioner (**A**) and idler sprocket assembly (**B**).

- Remove camshaft sprocket bolt from left camshaft (**C**).

Remove camshaft sprocket lockpin using Porsche special tool P 212 or equivalent.

– Remove camshaft sprocket and disengage from timing chain.

– Remove camshaft sprocket flange.

Camshaft sprocket assembly

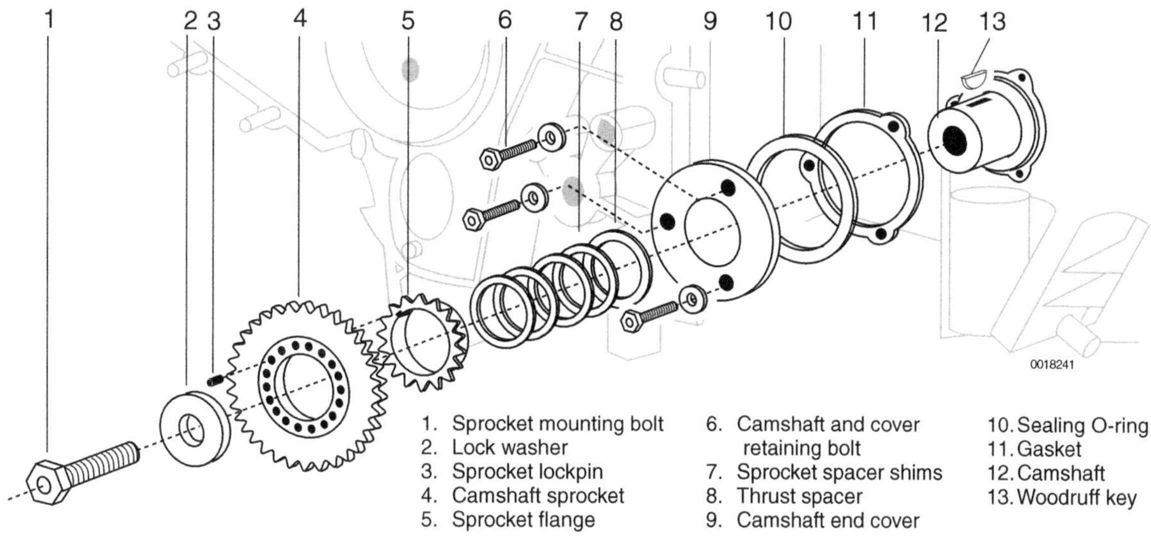

1. Sprocket mounting bolt
2. Lock washer
3. Sprocket lockpin
4. Camshaft sprocket
5. Sprocket flange
6. Camshaft and cover retaining bolt
7. Sprocket spacer shims
8. Thrust spacer
9. Camshaft end cover
10. Sealing O-ring
11. Gasket
12. Camshaft
13. Woodruff key

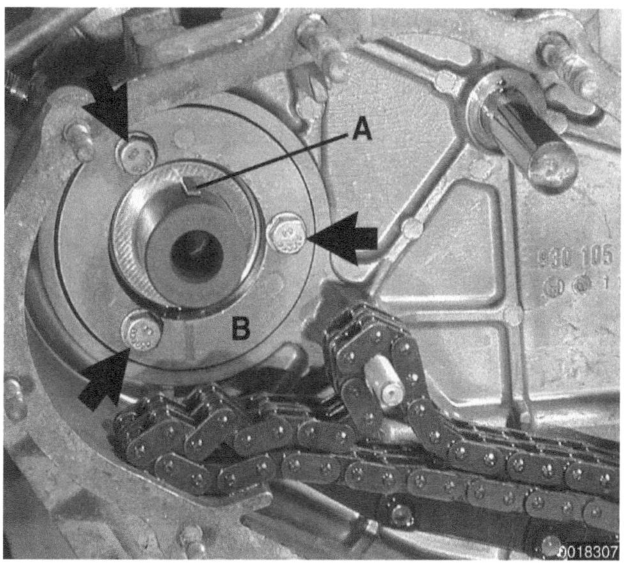

Remove left camshaft:

- Remove woodruff key (A) from camshaft.
- Remove camshaft end cover retaining bolts (**arrows**).
- Remove camshaft end cover (**B**), shims and thrust spacer.
- Slide camshaft toward pulley end of engine to remove.

NOTE—

- *Wire tie end cover, shims and thrust spacer together and label.*
- *For identification, the left camshaft is marked "147-10" on the end.*

– Repeat procedures on right side camshaft.

NOTE—

For identification, the right camshaft is marked "148-10" on the end.

– To reinstall, slide left camshaft into camshaft case.

- Install camshaft end cover with new O-ring and gasket.
- Install camshaft and cover retaining bolts and tighten.

– Reinstall left side cam sprocket spacer shims, woodruff key and camshaft sprocket flange.

– Put left timing chain over camshaft sprocket.

- Slide sprocket and lock washer over sprocket flange.
- Install sprocket mounting bolt finger tight.

– Install right camshaft, camshaft sprocket assembly and timing chain as described in previous steps.

Camshaft sprocket orientation

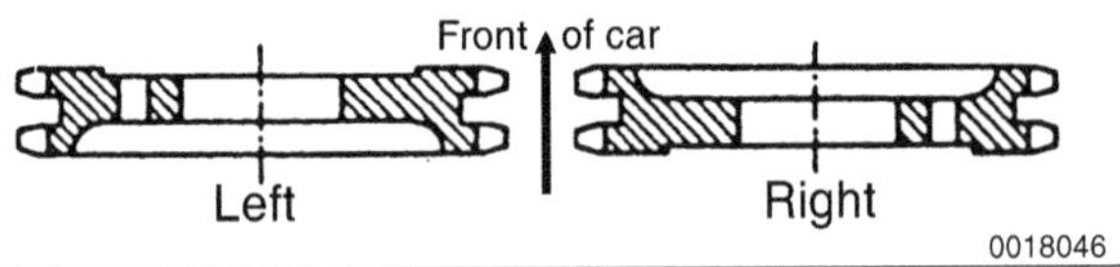

NOTE —

Left and right camshaft sprockets are identical. Dished side of sprocket must face pulley side of engine on left camshaft (cylinders 1 - 3) and flywheel side of engine on right camshaft (cylinders 4 - 6).

– Check that cam sprockets run parallel with intermediate shaft sprocket. See **152 Camshaft Timing Chains**.

NOTE —

- *Under most normal circumstances, if reinstalling previously used camshafts and sprockets, 3 sprocket spacer shims will be necessary under the left sprocket and 4 shims under the right sprocket.*
- *If new camshafts or sprockets are being installed, check to make sure that the sprockets are running parallel to each other to minimize chain wear. See* **152 Camshaft Timing Chains**.

– Install left timing chain tensioner and idler sprocket assembly.

- Remove locking cotter pin from chain tensioner, if applicable.

◄ Set crankshaft to **Z1**. Make sure it is at cylinder 1 compression TDC by aligning distributor rotor with scribe mark on distributor housing (**arrows**).

◄ Turn camshafts so that "**930**" marks point up.

– For initial timing adjustment:

- Remove left camshaft sprocket mounting bolt and lock washer.
- In this position one hole in camshaft sprocket will be aligned with one hole in sprocket flange. Insert camshaft sprocket lockpin.
- Reinstall sprocket retaining bolt finger tight.

– Repeat initial timing procedures for right camshaft.

– Turn crankshaft clockwise two full turns and recheck initial timing.

> ***CAUTION —***
> *Be sure to set fine camshaft timing as described in* **152 Camshaft Timing Chains**.

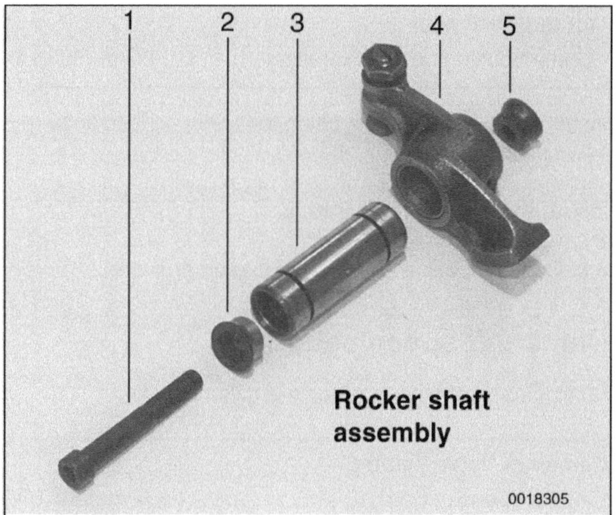

1. Pinch bolt (5 mm allen)
2. Expansion sleeve
3. Rocker shaft
4. Rocker arm
5. Expansion nut (8mm allen)

◄ Install rocker shafts and arms for each cylinder with camshafts set for TDC compression for that cylinder.

- Install cylinder 1 and 3 rocker shafts with 5 mm allen bolt facing cylinder 2.
- Install cylinder 4 and 6 rocker arms with 5 mm bolt facing cylinder 5.
- Cylinder 2 and 5 rocker shaft can be installed either way.

– Tap rocker shaft into bore until shaft is flush with inner edge of support.

– Place rocker arm in position and tap rocker shaft in until shaft slides into far side support.

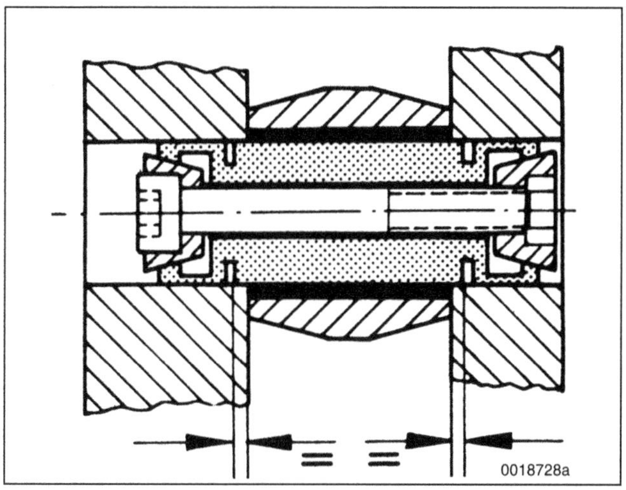

◄ Tap rocker arm shaft into position so that shaft is centered in camshaft housing.

- Tighten rocker arm shaft pinch bolt.

Tightening Torque
- Rocker shaft pinch bolt . . . . 15 Nm (11 ft-lb)

– Torque camshaft sprocket bolts.

NOTE —
Use Porsche special tool 9191 or equivalent to counterhold camshaft sprocket when torquing bolt.

Tightening Torque
- Chain sprocket to camshaft . . . . 120 Nm (89 ft-lb)

◄ Adjust all intake and exhaust valves to 0.1 mm (0.004 in.) as described in **030 Maintenance**.

– Install valve covers with new gaskets, sealing washers and new lock nuts.

Tightening Torque
- Valve cover to camshaft case (always replace) . . . . 8 Nm (71 in-lb)

– Install timing chain housing covers with new gaskets, sealing washers and lock nuts.

– Install right and left camshaft oil feed lines with new sealing washers.

Tightening Torque
- Chain housing cover to housing . . . . 8 Nm (89 in-lb)

– Install right and left timing chain tensioner oil feed lines with new sealing washers.

– Install rear engine metal shroud.

– Install muffler, using a new gasket and hardware.

Camshaft specifications

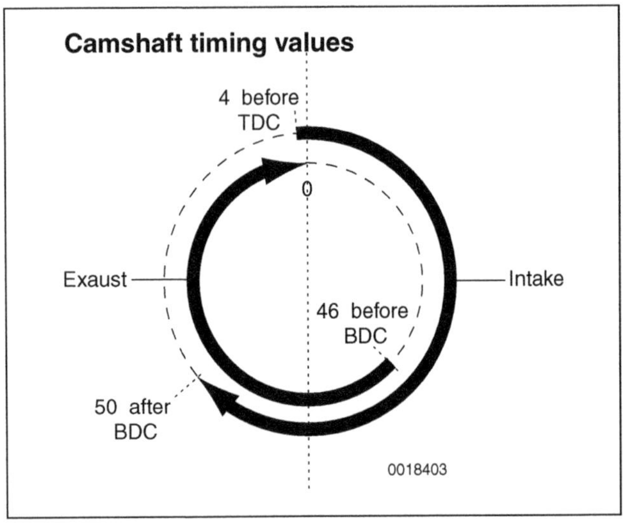

Stock camshaft timing values are given below.

Camshaft Valve Timing
- Intake opens . . . . .4° before TDC
- Intake closes. . . . . 50° after BDC
- Exhaust opens . . . . 46° before BDC
- Exhaust closes . . . . At TDC

Camshaft and valve train specifications are given in **Table a**.

Table a. Camshaft and Valvetrain Clearances

Measuring point	Tolerance	Wear limit
Camshaft bearing bore	48.967 - 48.992 mm (1.9278 - 1.9288 in.)	0.10 mm (0.004 in.)
Camshaft journal	48.926 - 48.942 mm (1.9262 - 1.9268 in.)	0.10 mm (0.004 in.)
Camshaft run-out	0.02 mm (0.0007 in.)	
Rocker arm shaft bearing bore	17.992 - 18.000 mm (0.7083 - 0.7086 in.)	
Rocker arm shaft	18.000 - 18.018 mm (0.7086 - 0.7093 in.)	
Rocker arm bore	18.000 - 18.018 mm (0.7086 - 0.7093 in.)	0.080 mm (0.003 in.)

CAMSHAFT CASES

The camshaft cases can only be removed with the engine out of the vehicle. See **102 Engine Disassembly and Assembly**.

Oil spray tube, replacing

The left and right camshaft cases each contain an oil spray tube for valvetrain lubrication. On high mileage engines and engines that have had catastrophic bearing failure, the tubes should be removed and cleaned prior to engine assembly.

0018709

With camshaft case removed and cleaned, blow compressed air in each camshaft tower oil feed line hole. Check each oil spray tube orifice (**arrows**) for air flow.

0018710

If a blocked hole is found, drill 5 mm (0.2 in.) hole approximately 9 mm (0.35 in.) deep in oil passage plug (**arrow**).

- Using an M6 tap, thread drilled hole.
- Pull out plug with 6 mm screw.
- Remove oil spray tube and clean thoroughly.
- Reinstall tube and replace end plug.

> ***CAUTION —***
> *Oil spray tubes must always be removed and cleaned thoroughly when engine bearings have failed.*

152 Camshaft Timing Chains

Special Tools

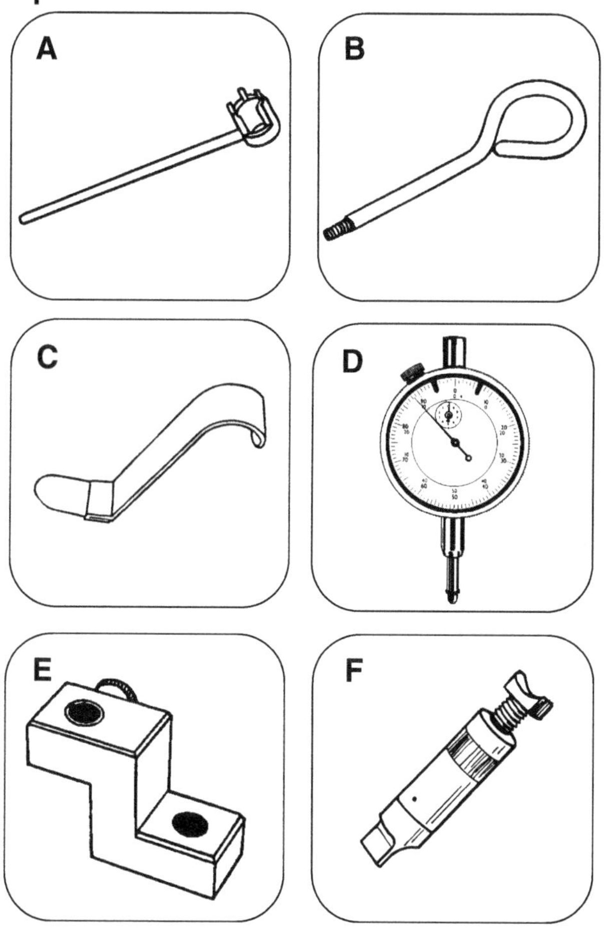

Special tool

- **A** 9191 . . . . Cam sprocket counterholding wrench (Porsche ordering #000 721 919 10) . . . .
- **B** P 212 . . . . Camshaft sprocket lockpin tool (Porsche ordering #000 721 212 10)
- **C** P 213 b. . . . . Valve adjusting feeler gauge (Porsche ordering #000 721 213 20)
- **D** 20A (source: Baum Tools Unlimited) . . . . Dial indicator
- **E** P 207 . . . . Dial indicator bracket (Porsche ordering #000 721 207 00)
- **F** P 204 . . . . Chain idler sprocket holder

GENERAL

This section covers repairs to the camshaft timing chain assembly. Timing chain replacement or adjustment can be carried out with the engine installed in the engine bay.

Special tools

Some of the procedures require special tools.

CAMSHAFT TIMING CHAINS

The original timing chains are continuous one-piece chains, (without master links). Replacing the one-piece chain with another one-piece requires that the engine be removed. As an alternative, replacement chains with master links are available from Porsche. The master link makes it possible to replace a timing chain with the engine installed.

Replacing the one piece timing chain with the engine removed is covered in **102 Engine Disassembly and Assembly**. When performing an engine overhaul, the continuous chain is the recommended replacement.

Timing chains, replacing (engine installed)

– Disconnect negative (–) battery cable and cover battery terminal to keep cable from accidentally contacting terminal.

> ***WARNING —***
> *Battery electrolyte is a corrosive acid. Always wear eye protection when working on or near the battery.*

– Raise car and support in a safe manner.

– Remove spark plugs so engine will turn easily.

> ***WARNING —***
> *Make sure the car is stable and well supported at all times. Use a professional automotive lift or jack stands designed for the purpose. A floor jack is not adequate support.*

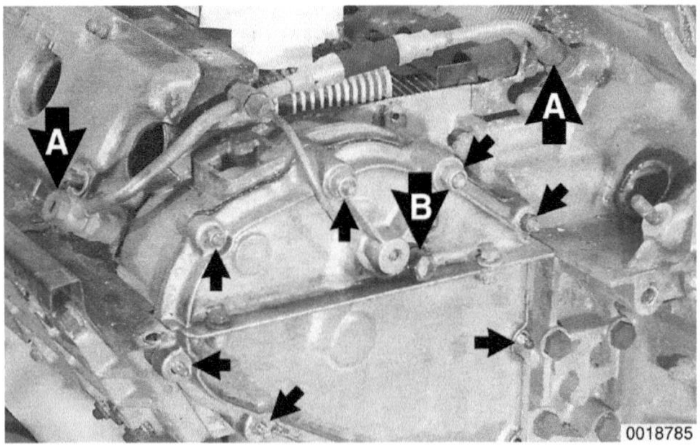

– Remove muffler as described in **260 Exhaust System**. Remove rear engine shroud.

– Remove spark plug wires to gain access to top of timing chain housing covers.

◄ Remove left timing chain tensioner oil feed line pressure fittings (**A**).

- Remove left camshaft case oil feed line pressure fitting (**B**).
- Remove left chain housing cover fasteners (**small arrows**), and remove cover.

– Repeat steps for right chain housing cover.

◄ Loosen and remove camshaft sprocket mounting bolts.

NOTE —
Use Porsche special tool 9191 or equivalent to counterhold camshaft sprocket when loosening bolt.

◄ Set crankshaft pulley to cylinder 1 compression TDC by lining up **Z1** mark with mark on fan housing.

- Check to see that punch marks or "**930**" on camshaft ends point *up*.

Rotate crankshaft counterclockwise until punch mark or "930" on end of left camshaft (**arrow**) points approximately 45° outward.

NOTE —

When the camshaft mark points 45° outwards, no valves are completely open and the crankshaft can be rotated safely without the pistons contacting the valves.

– Temporarily reinstall camshaft sprocket bolt finger tight.

Compress timing chain tensioner using large pliers.

- Install 3 mm (1/8 in.) pin through hole (**arrow**) in upper tensioner body to lock tensioner down.

NOTE —

- *Compressing the tensioner may take a little time. Oil flow out of the tensioner is restricted to a small oil relief hole.*
- *If the tensioner collapses easily it is most likely faulty.*

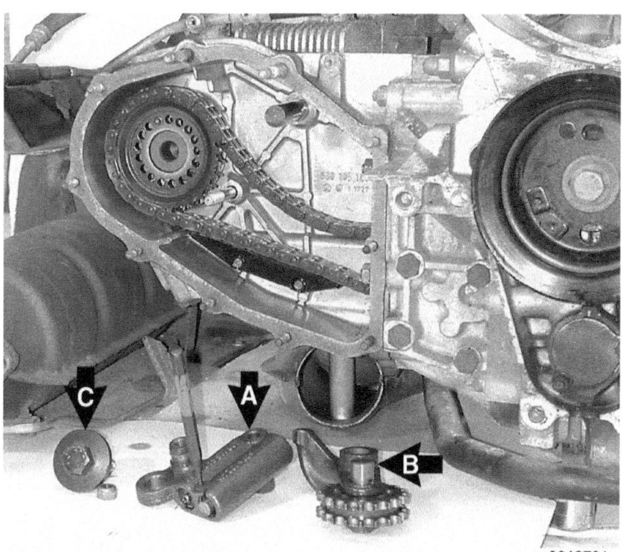

Remove timing chain tensioner (**A**) and idler sprocket assembly (**B**).

- Remove camshaft sprocket bolt from left camshaft (**C**).

◄ Remove camshaft sprocket lockpin using Porsche special tool P 212 or equivalent.

– Remove camshaft sprocket and remove sprocket from chain.

NOTE —
If necessary, the plastic chain rail can removed by prying with a large screwdriver.

– Grind rivet heads off one chain link and remove link.

> ***CAUTION —***
> *Place a shop towel in the crankcase opening to prevent metal filings from falling into crankcase.*

– Attach new chain with connecting link to end of old chain.

– Rotate crankshaft slowly in direction of rotation (clockwise as viewed from pulley end), keeping tension on chain at all times.

NOTE —
When rotating the crankshaft, make sure the valves do not contact the pistons. If any resistance is felt, reverse rotation direction of the crank slightly. Then rotate the camshaft enough to avoid contact and continue the operation.

– Rotate engine until master link can be installed on new chain. Remove old chain.

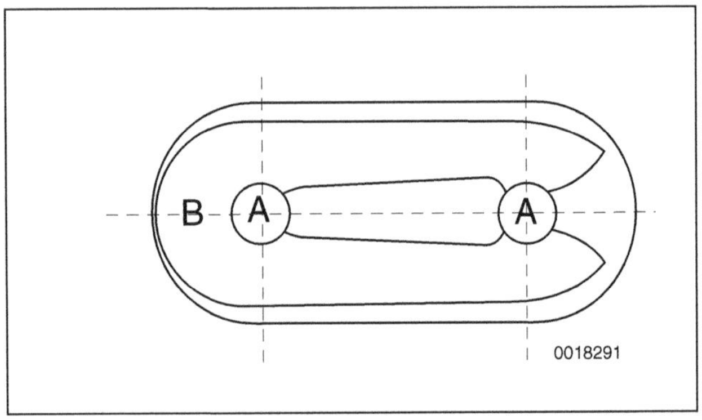

◄ Install master link. Connecting pins (**A**) should come in from back side of chain (flywheel side).

- Install spacers.
- Closed end of lock clip (**B**) must point in direction of rotation.

– Install guide rail, chain tensioner, sprocket carrier and camshaft sprocket.

◄ Set crankshaft pulley to cylinder 1 compression TDC by lining up **Z1** mark with mark on fan housing.

- Check to see that punch marks or "**930**" on camshaft ends point up.

> ***CAUTION —***
> *Turn camshaft and crankshaft slowly when setting up timing positions. If any interference is felt, stop and reposition camshaft before continuing operation.*

◄ Lock camshaft sprocket to camshaft hub by inserting dowel pin in hole with best alignment. Reinstall sprocket retaining bolt and washer finger tight.

– Rotate crankshaft forward (clockwise) until punch mark or "930" on end of right camshaft points approximately 45° outward.

NOTE —
When the camshaft mark points 45° outwards, no valves are completely open and the crankshaft can be rotated safely without the pistons contacting the valves.

– Repeat replacement steps for right timing chain.

– Adjust camshaft timing as described later.

– Engine reassembly is reverse of disassembly. Use new gaskets, seals and nuts. Replace sealing rings on oil feed lines.

Tightening Torque
- Cam sprocket to camshaft .... 120 Nm (89 ft-lb)
- Chain housing cover to chain housing .... 8 Nm (71 in-lb)

Timing chain sprockets, checking parallelism

The timing chain sprockets must be aligned parallel to each other to prevent premature wear of the timing chain and chain components.

Normally, when removing camshaft sprockets, 3 spacer shims will be found under the left sprocket hub and 4 shims under the right sprocket hub.

If new camshafts or sprockets are installed, camshaft sprocket alignment must be checked. Use vernier caliper and a quality straight edge to check sprocket depth in relation to the intermediate shaft. If adjustment is needed, shims can be added or subtracted.

Timing chain sprocket parallelism

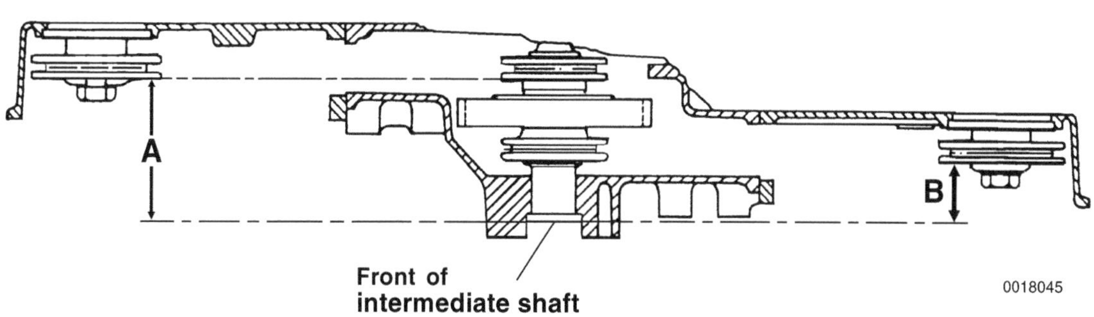

A Factory distance from end of intermediate shaft to face of cylinder 1 - 3 (left) timing chain sprocket = 98.07± 0.25 mm (3.86 ± 0.01 in.)
B Factory distance from end of intermediate shaft to face of cylinder 4 - 6 (right) timing chain sprocket = 43.27 ± 0.25 mm (1.70 ± 0.01 in.)

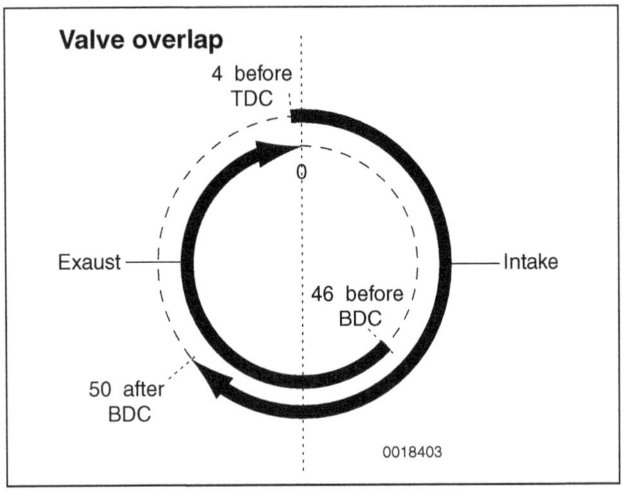

Camshaft timing, checking and adjusting

Camshaft timing is checked using a dial indicator setup. The cam timing is precisely set with the camshafts in the TDC overlap position. TDC overlap comes two strokes (one full crankshaft revolution) after the TDC compression stoke. Precise setting of cam timing results in smooth, vibration-free engine operation.

Valve overlap and camshaft timing values are shown in **Table a**.

Table a. Camshaft Timing

Intake opens	4° before TDC
Intake closes	50° after BDC
Exhaust opens	46° before BDC
Exhaust closes	At TDC

– Raise car and support in a safe manner.

> ***WARNING —***
> *Make sure the car is stable and well supported at all times. Use a professional automotive lift or jack stands designed for the purpose. A floor jack is not adequate support.*

– Remove spark plugs so engine will turn easily.

– Remove muffler as described in **260 Exhaust System**. Remove rear engine shroud.

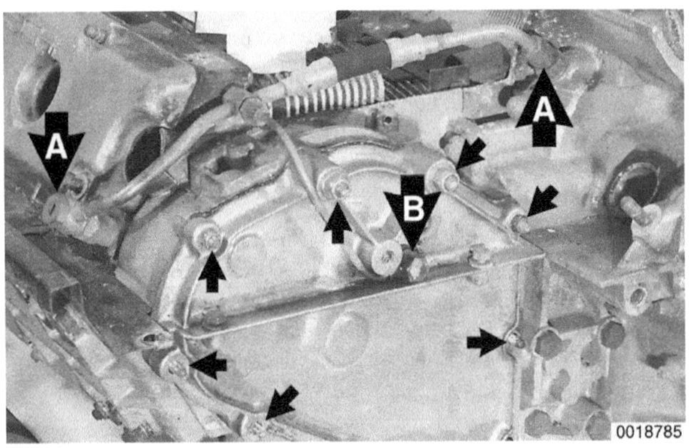

– Remove left camshaft case oil feed line pressure fittings (**A**).
- Remove left timing chain tensioner oil feed line pressure fitting (**B**).
- Remove left chain housing cover fasteners (**small arrows**), and remove cover.

– Repeat steps for right chain housing cover.

– Remove left and right valve covers.

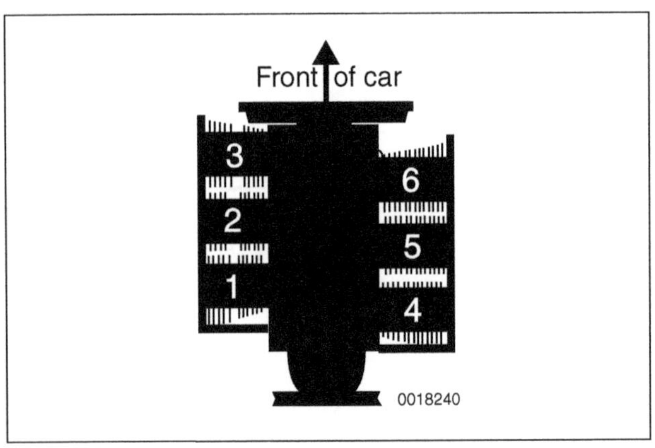

Note cylinder numbering.

- Set crankshaft pulley to cylinder 1 TDC by lining up **Z1** mark with mark on fan housing. (See figure above)

- While counterholding sprocket, remove left sprocket mounting bolt.

NOTE —
Use special tool no. 9191 to counterhold sprocket.

- Check to see that punch marks or "**930**" on camshaft ends point *up*.

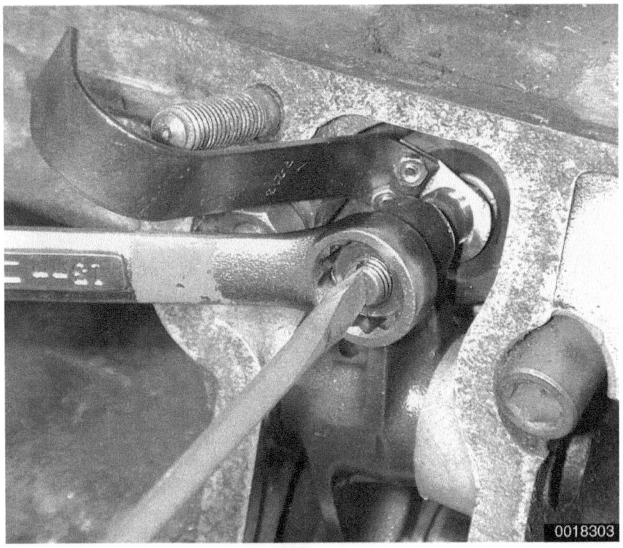

◄ Adjust cylinder 1 intake and exhaust valve clearance.

Valve Clearance
- Adjusting value . . . . 0.10 mm (0.004 in.)

◄ Using Porsche special tool P 207, mount dial gauge on camshaft case stud next to cylinder 1 intake valve.

- Set dial gauge plunger against valve spring retainer on cylinder 1 intake valve.
- Zero dial gauge with approximately 10 mm (0.4 in.) preload.

– Starting at **Z1** (TDC) mark on pulley, turn crankshaft slowly in clockwise direction approximately 1 full turn while observing dial gauge. Turn crankshaft far enough to reach valve overlap adjusting value.

Valve Overlap Adjustment
- Adjusting value . . . . 1.25 mm (0.049 in.)

– Remove left timing chain sprocket mounting bolt.

◄ Pull out camshaft locating dowel pin with Porsche special tool P 212 or equivalent.

– Turn crankshaft either clockwise or counterclockwise, as necessary, just until **Z1** mark on pulley is aligned accurately with mark on fan housing.

– Reinstall lockpin and install camshaft sprocket bolt finger tight.

– Turn crankshaft clockwise two full turns. Recheck timing adjustment. Value must be within tolerance range.

Valve Overlap Adjustment
- Tolerance range . . . . 1.1 - 1.4 mm (0.043 - 0.055 in.)

– Tighten camshaft sprocket bolt when settings are correct.

• Left camshaft is timed.

NOTE —
Use Porsche special tool 9191 or equivalent to counterhold camshaft sprocket when torquing bolt.

Tightening Torque
• Chain sprocket to camshaft . . . . 120 Nm (89 ft-lb)

◄ Set crankshaft pulley to cylinder 4 TDC by lining up **Z1** mark with mark on fan housing.

• Check to see that punch marks or "**930**" on camshaft ends point *down*.

– Adjust cylinder 4 intake and exhaust valve clearance.

Valve Clearance
• Adjusting value . . . . 0.10 mm (0.004 in.)

– Repeat camshaft timing adjusting procedures on cylinder 4 intake valve as described in steps for left camshaft timing adjustment.

– Reassemble engine using new gaskets, seals and nuts.

Tightening Torque
• Chain housing cover to chain housing . . . 8 Nm (71 in-lb)
• Chain sprocket to camshaft . . . . 120 Nm (89 ft-lb)

CAMSHAFT TIMING CHAIN TENSIONERS

The timing chain tensioner is spring loaded and hydraulically dampened. Engine oil under pressure fills a reservoir in the bottom of the tensioner.

When a timing chain tensioner fails (collapses), the engine will show a drop in power output. In addition, there will be a rattling noise coming from the timing housing area due to the loose chain.

Timing chain tensioner, checking and replacing

– Disconnect negative (–) battery cable and cover battery terminal to keep cable from accidentally contacting battery terminal.

WARNING —
Avoid getting tools or clothing near the battery. Battery electrolyte is a corrosive acid. Always wear eye protection when working on or near the battery.

– Raise car and support in a safe manner.

> ***WARNING —***
> *Make sure the car is stable and well supported at all times. Use a professional automotive lift or jack stands designed for the purpose. A floor jack is not adequate support.*

– Remove muffler. See **260 Exhaust System**. Remove rear engine shroud.

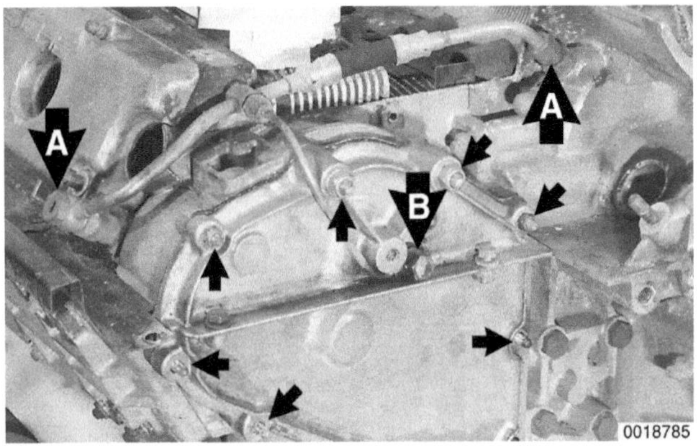

∢ Working at timing chain housing cover (left side shown):

- Remove timing chain tensioner oil feed line pressure fittings (**A**).
- Remove camshaft case oil feed line pressure fitting (**B**).
- Remove chain housing cover fasteners (**small arrows**), and remove cover.

∢ With a large screwdriver, check tensioner by prying down on idler sprocket bracket.

- If tensioner piston moves easily, it is faulty and should be replaced.

– Install new tensioner and adjust valve timing as described earlier.

– Reassembly is reverse of disassembly. Use new gaskets, seals and hardware.

Tightening Torques
- Chain housing cover to chain housing . . . 8 Nm (71 in-lb)

---

170 Engine Lubrication

Special Tools

A

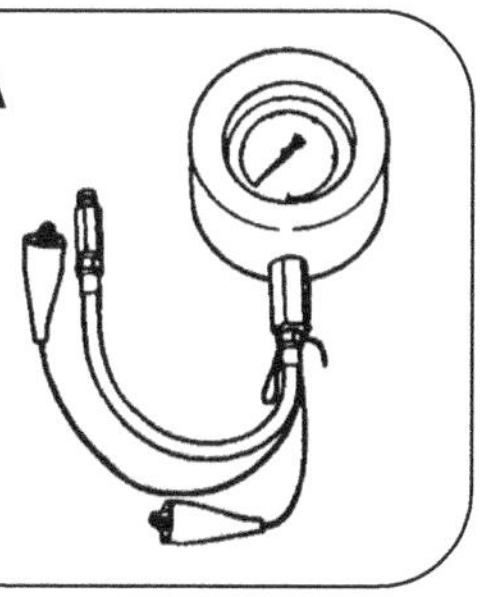

B

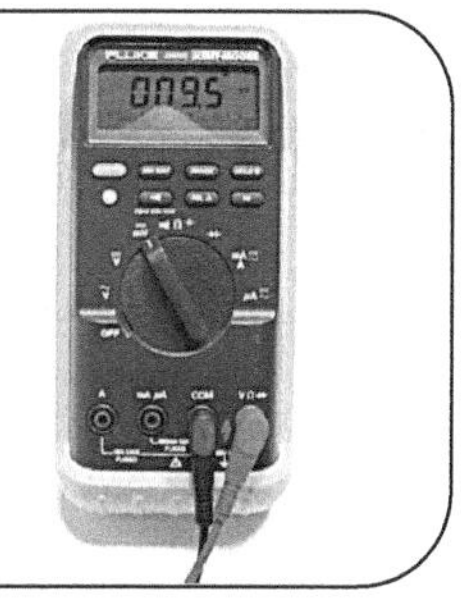

Special tool

A VW 1342 . . . . Oil pressure gauge
B Fluke 87 .Automotive digital multimeter

GENERAL

This section covers the engine lubrication system, including troubleshooting and component replacement.

If the engine is disassembled because of bearing failure, a thorough flushing of the lubrication system must be carried out prior to reassembly. That procedure is also covered here.

Some of the procedures require special tools.

NOTE —

- *Oil change procedure, including oil filter replacement, is covered in* **030 Maintenance**.
- *For information on the oil pressure gauge and its related circuit, see* **901 Instruments**.
- ***Front*** *of engine is the pulley side.* ***Rear*** *of engine is the flywheel side.*

Lubrication system

The dry sump lubrication system contains the tandem oil pump, oil supply tank, oil cooler, pressure safety system (including oil pressure gauge), filter, and connecting oil lines.

The tandem oil pump is in the base of the crankcase. As the name implies, the tandem oil pump contains two pumps. One pump draws oil from the oil tank and pumps it to the main oil circuit and main bearings. The second pump draws circulated oil from the crankcase and delivers it back to the oil tank via the oil filter. Oil pump removal requires that the crankcase halves be separated. This operation is described in **102 Engine Disassembly and Assembly**.

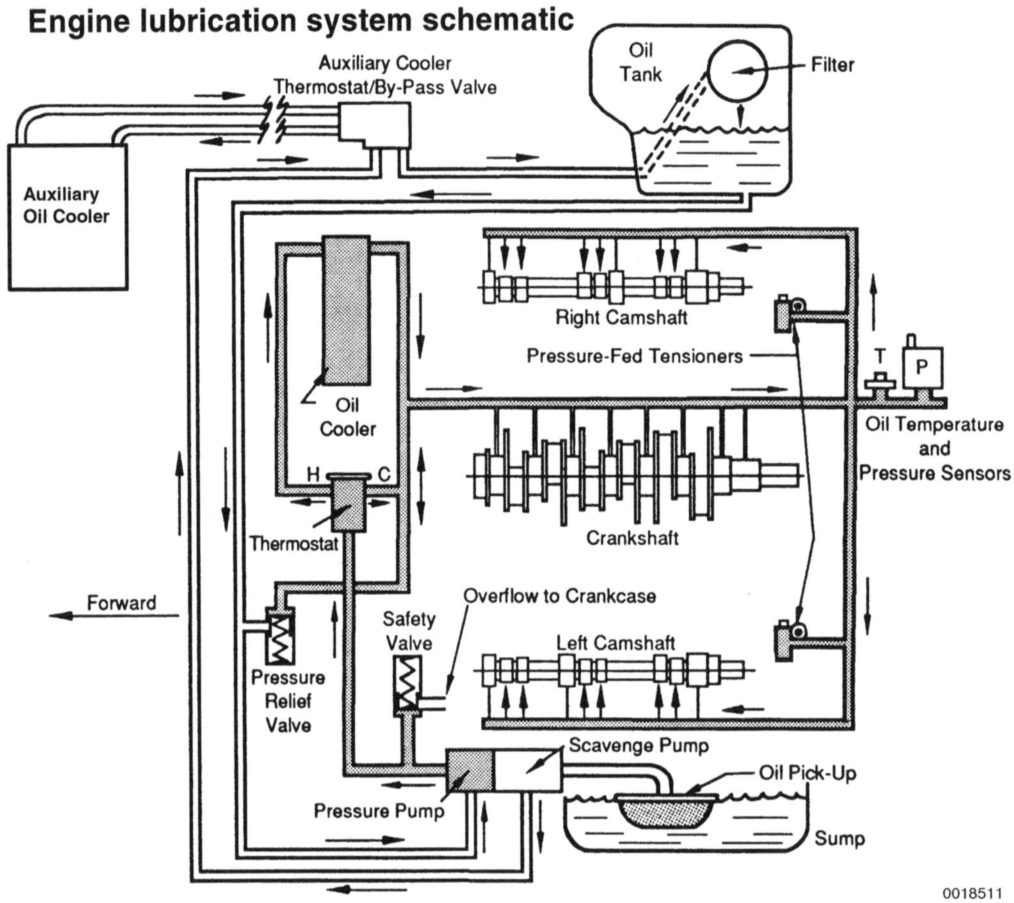

1. Gauge, oil temperature/pressure
2. Breather hose, crankcase to oil tank
3. Oil tank
4. Oil filter
5. Breather hose, oil tank to air filter
6. Oil filter neck
7. Sensor, oil pressure gauge
8. Oil thermostat (open at 80°C /176°F)
9. Sensor, oil temperature gauge
10. Oil cooler
11. Main oil pressure relief valve (opens at 6.2 ± 0.8 bar/88.2 ± 11.3 psi)
12. Tandem oil pump
13. Safety pressure relief valve (opens at approx. 8 bar/113.7 psi)
14. Crankcase oil strainer

The main bearings, rod bearings and camshaft bearings are supplied with pressurized oil. In addition, the undersides of the pistons are cooled using oil spray jets. Oilways in the camshaft cases provide splash lubrication for the rocker shafts and valve stems.

A thermostat in the main oil circuit regulates the flow of oil to the engine oil cooler. If the oil temperature is below 80°C (176°F), a thermostatically-operated flap is held closed to prevent oil from circulating through the cooler. When the flap is open, oil is circulated through the cooler before reaching the bearings.

An auxiliary oil cooler mounted behind the front right fender provides additional cooling.

Porsche 911 engines are equipped with an oil pressure warning system and an oil level warning system to help prevent engine damage. Other safety features include a filter bypass to guard against bursting the filter due to over pressure and three different oil pressure relief valves.

Engine lubrication system components

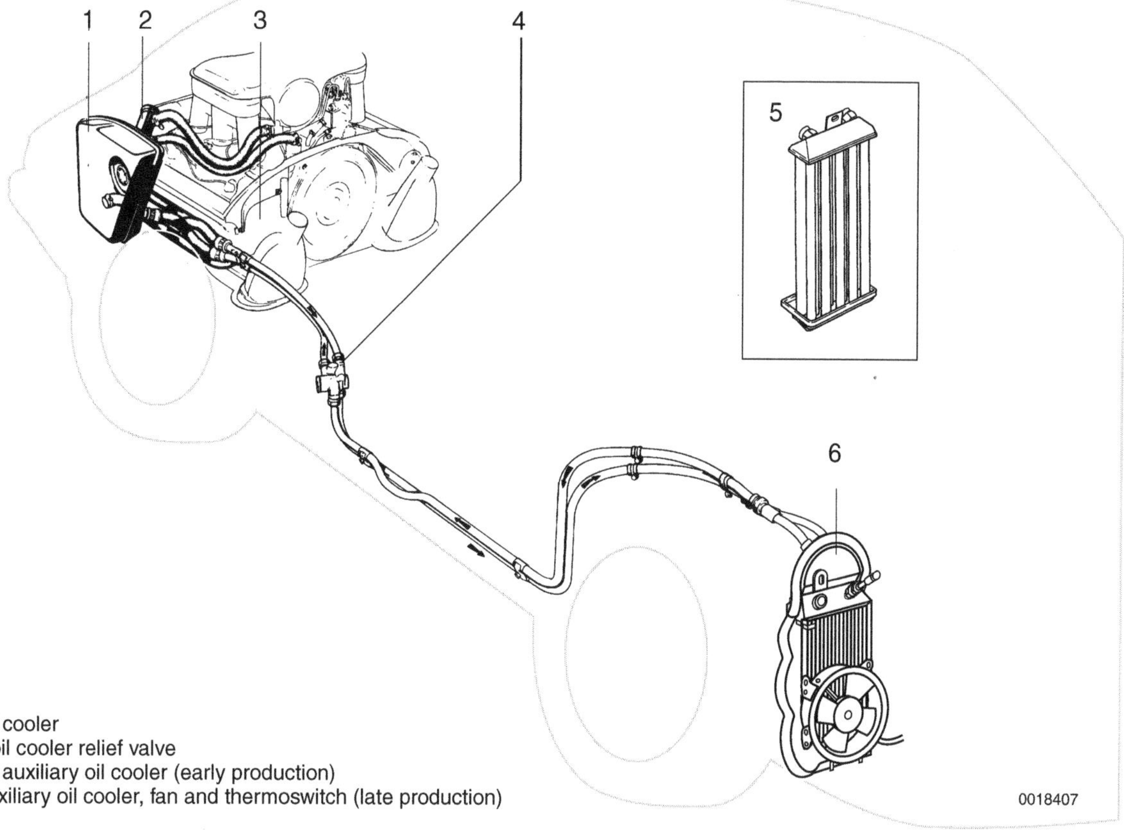

1. Oil tank
2. Oil filler
3. Engine oil cooler
4. Auxiliary oil cooler relief valve
5. Tube-type auxiliary oil cooler (early production)
6. Finned auxiliary oil cooler, fan and thermoswitch (late production)

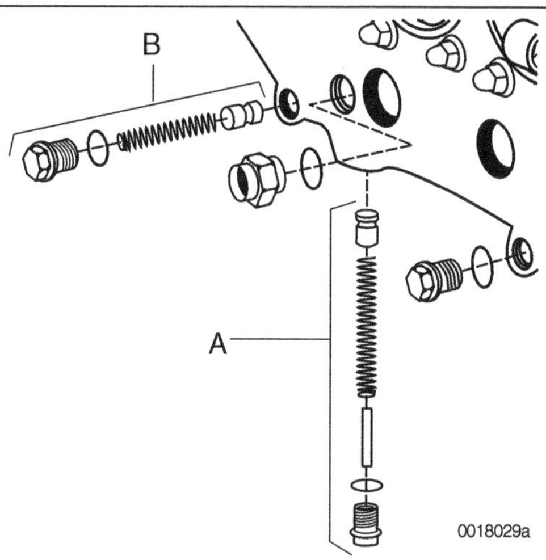

Oil pressure relief valves

The crankcase incorporates two different oil pressure relief valves. The main oil pressure relief valve (**A**) is mounted at the bottom of the crankcase. The safety relief valve (**B**) is mounted in the left side of the crankcase.

The oil pressure relief valve opens and allows oil to pass directly into the crankcase when oil pressure exceeds 6.2 ± 0.8 bar. (88.2 ± 11.3 psi). The oil pressure relief valve spring is 87 mm (3.43 in.) long.

In the event that the oil pressure relief valve doesn't open when oil pressure is excessive, the safety relief valve will open to prevent damage to the oil cooler, oil filter and oil lines. The oil pressure safety valve spring is 69 mm (2.71 in.) long. The safety valve is regulated to open at approximately 8.0 bar (113.7 psi).

The auxiliary oil cooler thermostat/pressure relief valve (**arrow**) controls oil flow to the auxiliary oil cooler at the front of the car. The thermostat/pressure relief valve is located under the car ahead of the right rear wheel. The thermostat/pressure relief valve contains two spring-loaded valves. One valve is thermostatically controlled and opens to allow pressurized oil to enter the front cooler when the temperature of the returning oil exceeds approximately 83°C (113°F). The other valve is pressure controlled and will open to bypass the cooler when the oil pressure is excessive.

TROUBLESHOOTING

Oil pressure, checking

> ***CAUTION —***
> *If the red oil pressure warning light comes on or flashes on while driving, always assume that the oil pressure is low.*

The oil pressure testing procedure given here works well using standard automotive oil pressure testing equipment.

Oil Pressure Sensor Location

- Gauge sensor . . . . on right top of engine

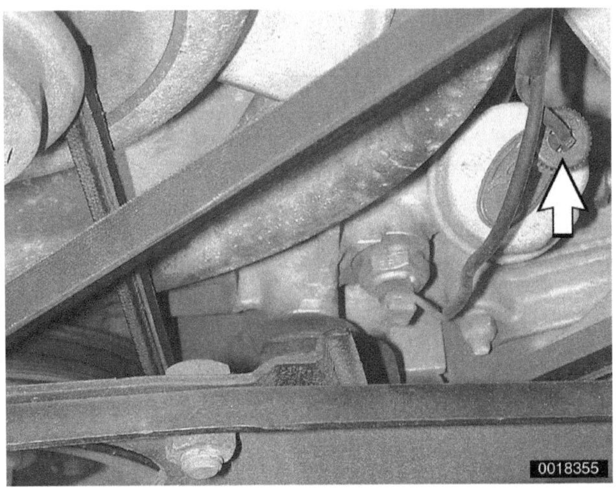

Disconnect harness connector (**arrow**) from oil pressure gauge sensor and remove sensor.

NOTE —
Thoroughly clean around the oil pressure sensor before removing it.

– Install pressure gauge in place of sensor.

NOTE —
The pressure gauge will need to be adapted to a 22 mm male thread.

– With gauge installed, start engine and allow to reach operating temperature.

NOTE —
- *For the most accurate test results, the engine oil (and filter) should be new and of the correct grade.*
- *As a general guide, oil pressure should rise approximately 10 psi for every 1,000 rpm.*

Oil Pressure Specification
- Regulated pressure (5,000 rpm @90°C/194°F) . . . . 4.5 bar (59 psi)

– Remove pressure gauge and reinstall sensor.
- Lubricate sensor threads prior to installation.

Tightening Torque
- Oil gauge sensor to crankcase . . . . 35 Nm (26 ft-lb)

If testing shows low oil pressure, one or more of the following conditions may be indicated:

- Worn or faulty oil pump assembly
- Worn or faulty engine bearings
- Severe engine wear

All of the conditions indicate the need for major repairs.

Lubrication warning system

The lubrication warning system consists of the following:

- Oil pressure warning light
- Oil pressure gauge
- Oil level gauge
- Oil temperature gauge

NOTE —
For troubleshooting and repair information on the oil pressure gauge and the oil temperature gauge, see **901 Instruments**.

Oil pressure warning light, testing

When the ignition is turned on, the oil pressure warning light comes on. When the engine is started and the oil pressure rises slightly, the oil pressure warning switch contacts open and the warning light goes out. Make sure the oil level is correct before making tests.

> ***CAUTION —***
> *If the oil warning light remains on while the engine is running, check the oil pressure as described earlier. Do not drive the car until the problem is corrected. The engine may be severely damaged.*

– Turn ignition switch on.
  - Warning light on instrument panel must light up.

0018718

◄ If warning light does not light when ignition is on, remove connector (**arrow**) from oil pressure warning switch and ground harness connector to a clean metal surface.
  - If warning light comes on, check switch as described later.
  - If warning light does not come on, wiring to instrument cluster or to light itself is faulty.

– Remove connector from oil pressure switch.
  - Warning light on instrument panel must go out.
  - If warning light does not go out, wiring to switch is most likely grounded somewhere between switch terminal and warning light. See **970 Electrical Wiring Diagrams**.

– To test switch, connect ohmmeter between switch terminal and ground.
  - With engine off, there should be continuity.
  - With engine running, oil pressure should open switch and there should be no continuity.

– Replace a faulty switch. Lubricate switch threads during installation.

Tightening Torque
- Oil warning switch to crankcase . . . . 20 Nm (15 ft-lb)

Oil level sensor, testing

The oil level sensor is similar in operation to the fuel level sensor in the fuel tank. Check that the oil level is correct before testing the sensor.

The oil level sensor is mounted in the oil tank in the right rear of the engine bay. It is accessed from below.

– Raise car and support in a safe manner.

> ***WARNING —***
> *Make sure the car is stable and well supported at all times. Use a professional automotive lift or jack stands designed for the purpose. A floor jack is not adequate support.*

– Remove right rear wheel.

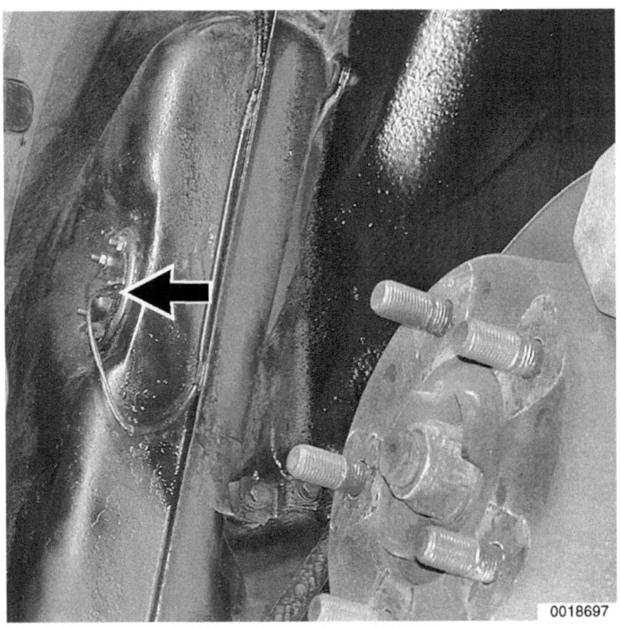

0018697

◄ Working behind right rear wheel housing, disconnect electrical connectors at oil level sensor (**arrow**).

– Check electrical resistance between sensor terminal and ground. If the sensor is electrically open, it is faulty and should be replaced.

NOTE —
The oil level sensor can be removed and installed with the oil tank in its installed position. It is not necessary to drain the engine oil to replace the oil level sensor.

– If no faults are found, ground disconnected wire. Turn ignition key on and check that gauge moves to full position.

- If gauge reacts correctly, sensor is faulty and should be replaced. If gauge does not respond, check electrical wiring to gauge. See **970 Electrical Wiring Diagrams**.

Lubrication system, flushing

After an engine bearing failure, the oil passages in the crankcase and crankshaft, cylinder heads and camshaft cases must be thoroughly flushed. See **130 Crankcase and Crankshaft**. In addition, the entire lubrication system must be flushed clean.

NOTE —
It is impossible to remove all metal partices from the engine oil cooler. It must be replaced after an engine bearing failure.

– Remove oil filter and oil tank as described earlier. Rinse oil tank with engine cleaner.

Auxiliary oil cooler pressure relief valve

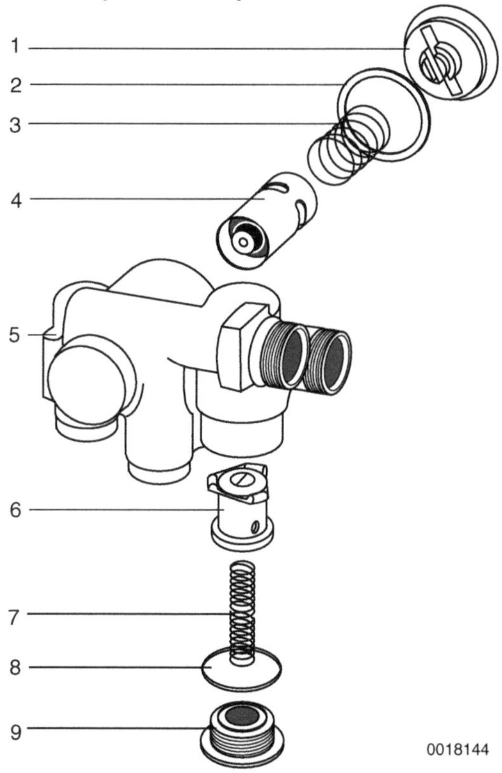

1. End plug
2. Gasket
3. Spring
4. Thermostatic plunger
5. Valve housing
6. Pressure relief plunger
7. Spring
8. Gasket
9. End plug

– Remove oil pressure relief valve for auxiliary oil cooler as described later.

- Disassemble and clean valve.

NOTE —
Apply heat to the pressure fitting nuts to avoid damaging the threads on the relief valve.

– Remove engine mounted oil cooler and replace it with a new cooler as described later.

– Remove auxiliary oil cooler as described later.

- Back-flush with engine parts cleaner.

– Flush oil lines that run front to rear with engine cleaner and carefully blow clean and dry with compressed air.

– Flush oil lines from oil tank to engine and from engine to oil cooler pressure regulator.

COMPONENT REPLACEMENT

NOTE —
Engine oil pump replacement is covered in **102 Engine Disassembly and Assembly***. Access to the oil pump requires that the engine/transmission be removed and separated. Engine can then be disassembled.*

Engine oil tank, removing and installing

The engine oil supply tank is located in the right rear fender well and can be accessed from below.

– Raise car and support in a safe manner.

> ***WARNING —***
> *Make sure the car is stable and well supported at all times. Use a professional automotive lift or jack stands designed for the purpose. A floor jack is not adequate support.*

– Remove right rear wheel.

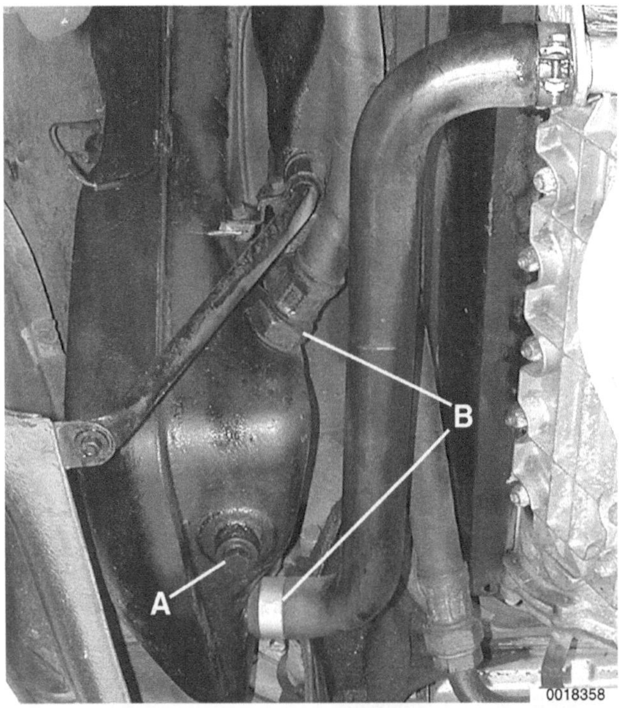

Remove oil tank drain plug (**A**) and drain oil.

- Disconnect both oil hoses (**B**) from tank.

NOTE —
Have a drain pan ready when disconnecting the hoses. The hoses will contain some residual oil. Do not let oil drip onto brake disc.

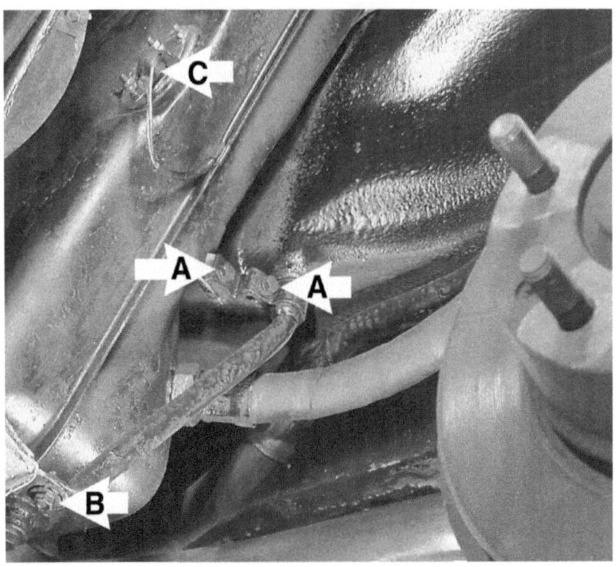

Remove bolts (**A**) from body panel support in right rear wheel housing and position bracket downward.

- Remove screw (**B**) holding support to quarter panel.
- Disconnect wires (**C**) from oil level sensor on oil tank.

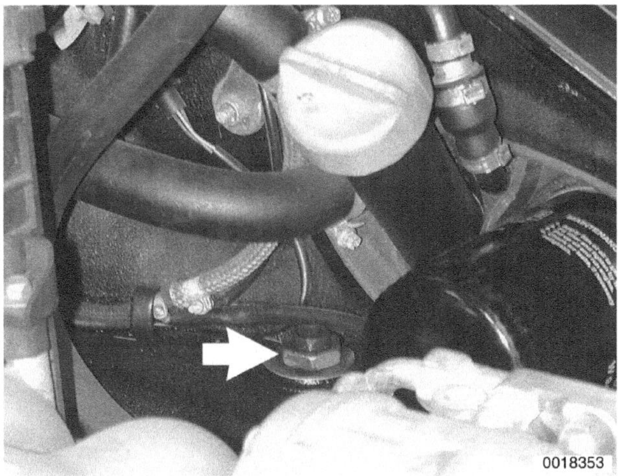

< Remove upper mounting nut (**arrow**) from support tube and remove tube from below.

- Support tube upper nut is locked in place. Bend back lock plate before loosening nut. Always replace lock plate.

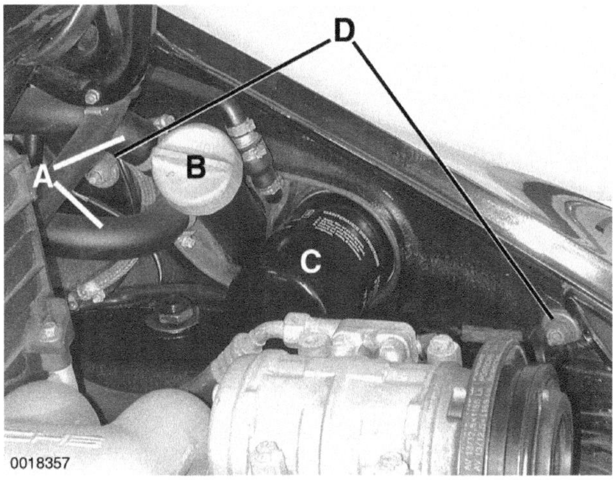

< Remove oil tank attachments.

- Detach two breather hoses in oil filter area (**A**).
- Remove oil filler cap (**B**).
- Remove oil filter (**C**).
- Remove oil tank mounting nuts (**D**).

– Remove oil tank by pulling it forward and down.

– Installation is reverse of removal.

- Be sure wheel housing liner gasket is correctly positioned on oil tank before installing tank.
- Top up engine and fill tank with oil.
- Run engine and check for leaks.
- Check oil level. See **030 Maintenance**.

Engine oil cooler, removing and installing

The engine oil cooler is mounted to the right side of the engine crankcase, next to the flywheel housing. Be sure to have new O-rings on hand before removing cooler.

– Raise car and support in a safe manner.

> ***WARNING —***
> *Make sure the car is stable and well supported at all times. Use a professional automotive lift or jack stands designed for the purpose. A floor jack is not adequate support.*

– Drain engine oil from crankcase and oil tank. See **030 Maintenance**.

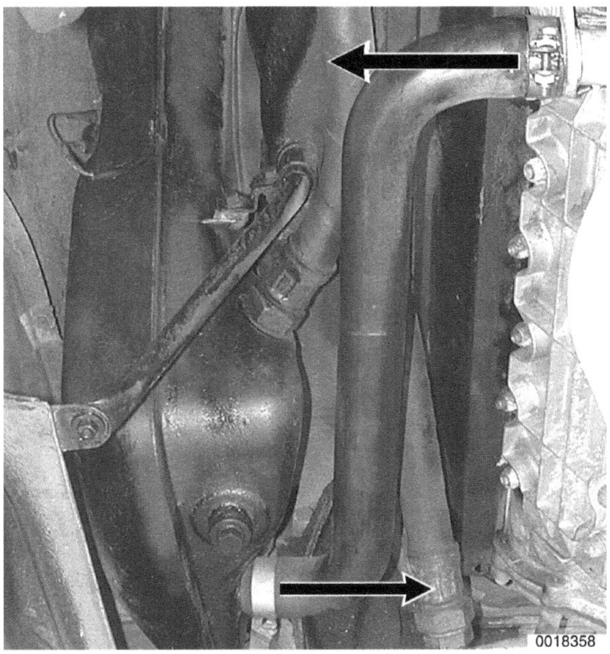

< Working underneath right side of engine, detach S-shaped oil return hose from oil cooler and oil tank (**arrows**).

NOTE —
Have a catch pan ready when disconnecting hose.

– Without disconnecting A/C pressure lines, remove A/C belt, A/C compressor and compressor bracket as described in **870 Air Conditioner**. Rest compressor on pad to the right of the engine compartment.

< Supporting engine securely from below, remove rear engine support crossmember bolts. Right side shown (**arrow**).

- Lower engine about 3 in. and support in that position.

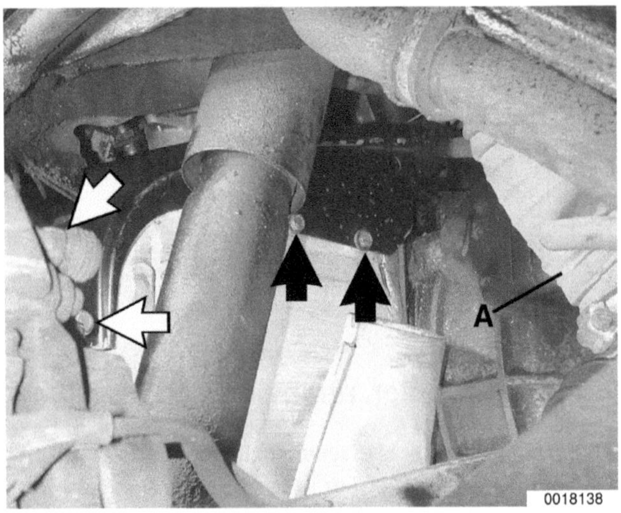

– Remove connector from spark plug 6.

< Working underneath right side of engine, remove right rear (flywheel-side) sheet metal shrouding after removing screws (**arrows**).

NOTE —

*To access shroud screws, detach right heat exchanger hose (**A**) and push forward.*

< Remove screws (**arrows**) holding plastic air shroud to top of oil cooler.

- Remove shroud.

< Remove upper oil cooler mounting nuts (**arrows**).

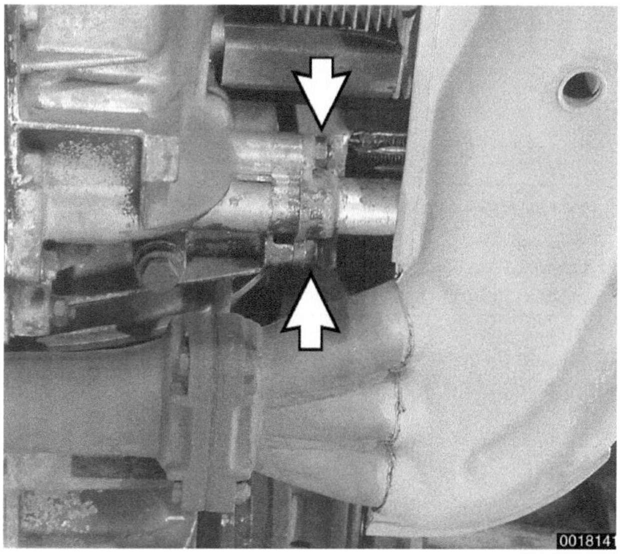

< Remove lower oil cooler mounting nuts (**arrows**).

NOTE —
For better access to oil cooler fasteners, remove right heat exchanger. See **260 Exhaust System**.

– Slide oil cooler out and down to remove.

– Installation is reverse of removal.
  - Use new O-rings at oil cooler.
  - Add engine oil and check oil level.

NOTE —
Always use new gaskets and seals.

Tightening Torques
- Engine oil drain plug to crankcase ..... 70 Nm (52 ft-lb)
- Engine oil drain plug to oil tank........ 42 Nm (30 ft-lb)
- Heat exchanger to crossover pipe ..... 25 Nm (18 ft-lb)
- Heat exchanger to cylinder head ...... 25 Nm (18 ft-lb)
- Oil cooler to crankcase.............. 25 Nm (18 ft-lb)
- Rear engine mount to crossover support.................. 80 Nm (58 ft-lb)

Auxiliary oil cooler pressure relief valve, removing and installing

The auxiliary oil cooler pressure relief valve is located under the car, ahead of the right rear wheel.

– Raise car and support in a safe manner. Remove right rear wheel.

WARNING —
Make sure the car is stable and well supported at all times. Use a professional automotive lift or jack stands designed for the purpose. A floor jack is not adequate support.

– Drain engine oil tank.

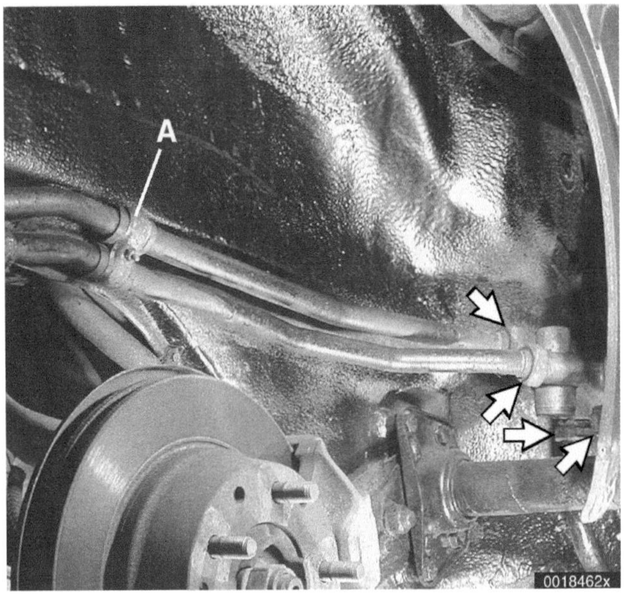

Loosen oil line mounting bracket (**A**). Detach oil lines (**arrows**) from relief valve.

NOTE —

Apply heat to the pressure fitting nuts to avoid damaging the threads on the relief valve.

– Installation is reverse of removal.

Tightening Torques

- Engine oil drain plug to oil tank. . . . . 42 Nm (30 ft-lb)

Auxiliary oil cooler, removing and installing

The auxiliary oil cooler is mounted in front of the right front wheel housing, just behind the front bumper. Oil flow to the cooler is controlled via the auxiliary oil cooler pressure relief valve.

– Raise front of car and support in a safe manner. Remove right front wheel.

> ***WARNING —***
>
> *Make sure the car is stable and well supported at all times. Use a professional automotive lift or jack stands designed for the purpose. A floor jack is not adequate support.*

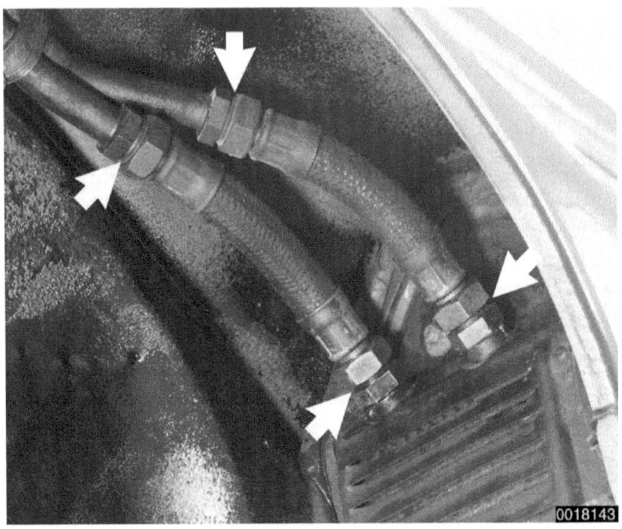

Separate oil cooler lines. Use a wrench to counterhold cooler and line fittings (**arrows**).

NOTE —

- *Apply heat to the pressure fitting nuts to avoid damaging threads.*
- *Plug all disconnected oil lines to keep dirt out of system.*
- *Have a drain pan ready to catch oil that will spill out.*

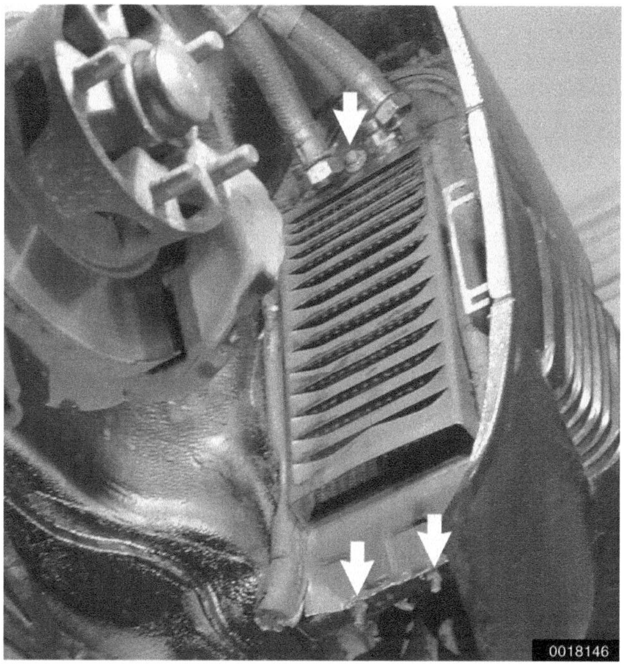

< Remove oil cooler mounting nuts (**arrows**).

– Remove luggage compartment bottom liner and spare tire.

< Detach auxiliary cooling fan harness connectors (**arrows**) in right front corner of luggage compartment.

- Push wiring harness together with rubber grommets to outside from inside trunk.

– Remove oil cooler by pulling up and out.

– Installation is reverse of removal. Clean threads on oil cooler fittings before reinstalling oil lines.

- Coat cooler line fittings with Optimoly® TA.

> ***CAUTION —***
> *Always check the auxiliary oil cooler and its connections for leaks after allowing the engine to reach operating temperature. The auxiliary oil cooler pressure relief valve must be open for oil to flow to the cooler.*

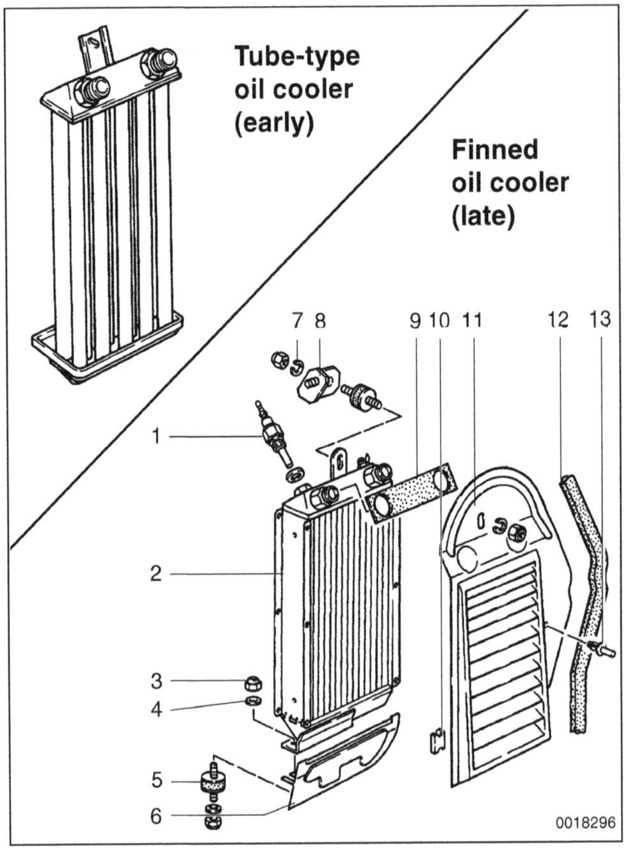

NOTE —
Early model 911 Carreras are fitted with a tube type front oil cooler. This can be upgraded to the new finned type oil cooler. A conversion parts list is given in **Table a**.

Table a. Parts List for Converting Tube Oil Cooler to Finned Oil Cooler

From diag.	New part req'd	Description	Qty.	Porsche part no.
1	no	Fan thermoswitch		
2	yes	Finned oil cooler	1	930.207.053.00
3	yes	Lock nut–8mm	6	900.910.022.02
4	yes	Washer	2	N.011.525.8
5	yes	Rubber/metal mount	3	930.207.239.00
6	yes	Air guide	1	930.207.121.00
7	yes	Washer	4	N.011.670.6
8	yes	Bracket	1	930.207.927.00
9	yes	Rubber Seal	1	930.207.361.00
10	yes	U-clamp	4	999.920.005.02
11	yes	Oil cooler cover	1	930.207.319.02
12	yes	Seal	1	930.207.353.00
13	yes	Rivet	2	999.591.429.40

Auxiliary oil cooling fan, replacing

If the car is equipped with an auxiliary oil cooling fan, the fan is thermostatically controlled. A thermoswitch (**arrow**) on the oil cooler turns the fan on when the oil temperature reaches 188°C (370°F). The fan operation can be quick-checked by applying battery voltage to the fan harness connector.

– Remove auxiliary oil cooler as described earlier.

– With oil cooler removed, cooling fan and temperature switch can both be changed.

- The cooling fan is mounted to oil cooler with four M6 fasteners.

Tightening Torque
- Auxiliary oil cooler thermoswitch to oil cooler . . . . 30 Nm (22 ft-lb)
- Oil cooler lines to auxiliary oil cooler . . . . 80-100 Nm (58-67 ft-lb)

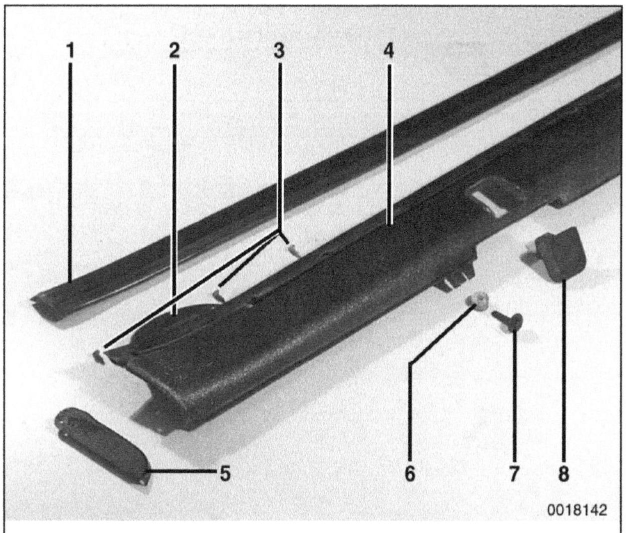

1. Rubber protector
2. Rear torsion bar end cover
3. Rocker panel mounting screws
4. Rocker panel
5. Rocker panel end cover
6. Plastic spacer
7. Rocker panel bottom mounting screw
8. Jack hole cover

Auxiliary oil cooler lines, removing and installing

– Raise car and support in a safe manner. Remove front and rear right side wheels.

> ***WARNING —***
> *Make sure the car is stable and well supported at all times. Use a professional automotive lift or jack stands designed for the purpose. A floor jack is not adequate support.*

◄ Remove right rocker panel.

- Pull off rubber rocker panel protector.
- Remove top and bottom mounting screws and remove rocker panel.

NOTE —
In early production cars the rear torsion bar end cover is one piece with the rocker cover. In later years it comes as a separate piece.

◄ Detach front oil lines (**arrows**) from auxiliary oil cooler relief valve.

NOTE —
Apply heat to the pressure fitting nuts to avoid damaging the threads on the relief valve.

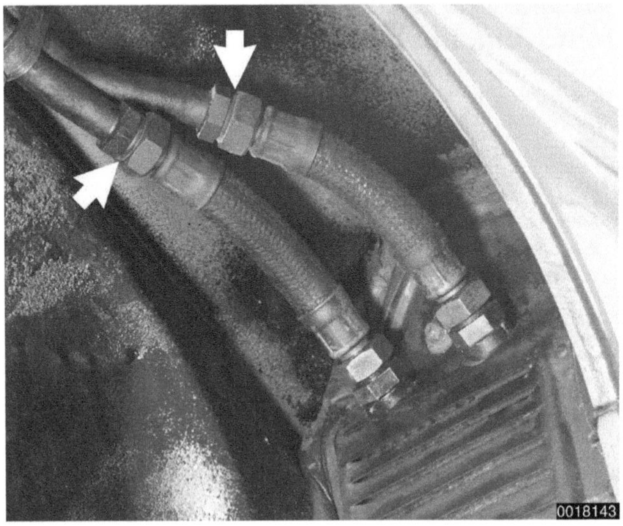

Disconnect oil lines from auxiliary oil cooler flex lines. Use a wrench to counterhold line fittings (**arrows**).

NOTE —
- *Apply heat to the pressure fitting nuts to avoid damaging the threads.*
- *Plug all disconnected oil lines to keep dirt out of system.*
- *Have a drain pan ready to catch oil that will spill out.*

– Installation is reverse of removal.

Oil return tube, replacing

The procedure below describes replacing the one-piece oil return tube with an expandable two-piece tube. This procedure can be done with the engine installed.

Oil return tubes can be replaced one at a time, although replacing in pairs is recommended.

If removing a left side tube, remove left heat exchanger. See **260 Exhaust System**.

- Disconnect heater hoses from front and rear of heat exchanger (**A**).
- Remove nuts holding heat exchanger to cylinder head (**B**).
- Remove catalytic converter mounting hardware (**C**).

NOTE —
If removing a right side tube, the right side heat exchanger can remain in place.

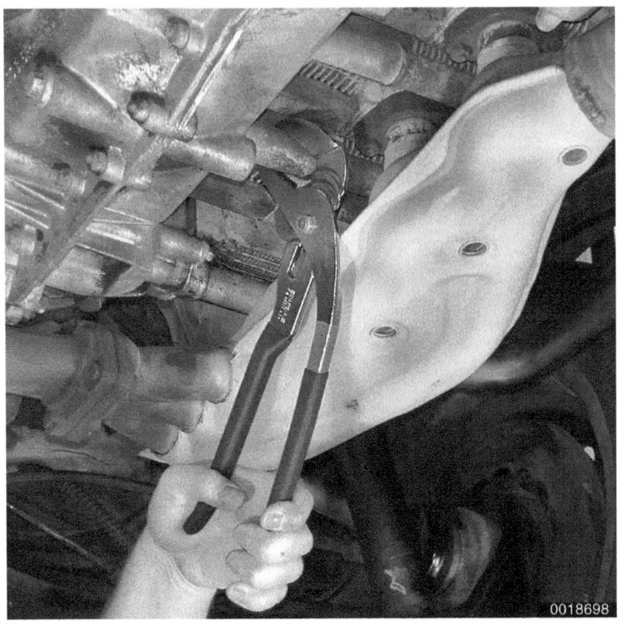

Crush center of oil return tube with large pliers and bend.

NOTE —
Have a drain pan ready to catch any engine oil that may spill.

– Remove destroyed return tube and clean sealing surfaces on crankcase and camshaft case.

Two-piece oil return tube

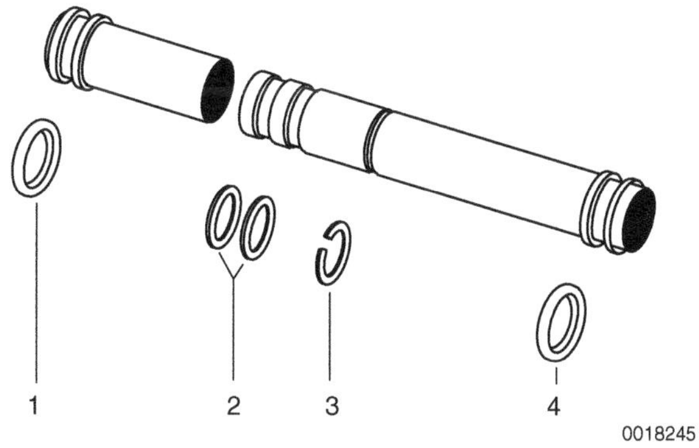

1. O-ring
2. O-rings
3. Locking ring
4. O-ring

◄ Mount O-ring seals to oil return tube sections and assemble.

– Push tube locking ring out of locking groove and slide approximately 40 mm (1.6 in.) away from locking groove.

– Apply a light film of tire mounting paste to seals and collapse sectioned oil return tube.

– Align telescoping end of tube with crankcase and camshaft housing holes. Expand tube so that it fits tightly into both holes.

– Slide locking ring into locking groove.

– Reinstall heat exchanger, if necessary.

Tightening Torques

- Heat exchanger to catalytic converter. . . 25 Nm (18 ft-lb)
- Heat exchanger to cylinder head . . . . . . 25 Nm (18 ft-lb)

– Run engine and check for oil leaks.

200 Engine Management–General

GENERAL

The Porsche 911 Carrera is equipped with a Bosch DME (Digital Motor Electronics or Motronic) engine management system. This section describes operation and troubleshooting of the DME system.

Refer to the following sections for repair information:

- **201 Fuel Supply.**
- **240 Fuel Injection.**
- **280 Ignition System.**

DME engine management system

The DME system uses electrical signals from the air flow sensor, the air and engine temperature sensors, the speed sensor, and an exhaust-mounted oxygen sensor as its primary inputs to electronically calculate fuel delivery and engine timing.

Ignition timing is electronically controlled and non-adjustable. The ignition distributor contains no mechanical timing-advance components. The only function the distributor serves is to distribute high voltage to the spark plugs. All ignition timing is determined by the engine (or DME) control module (ECM), interpreting signals from sensors that indicate engine crankshaft position and speed.

The ECM (**C**) is located under the driver's seat, next to the DME main/fuel pump relay (**A**) and altitude correction switch (**B**).

DME engine management system

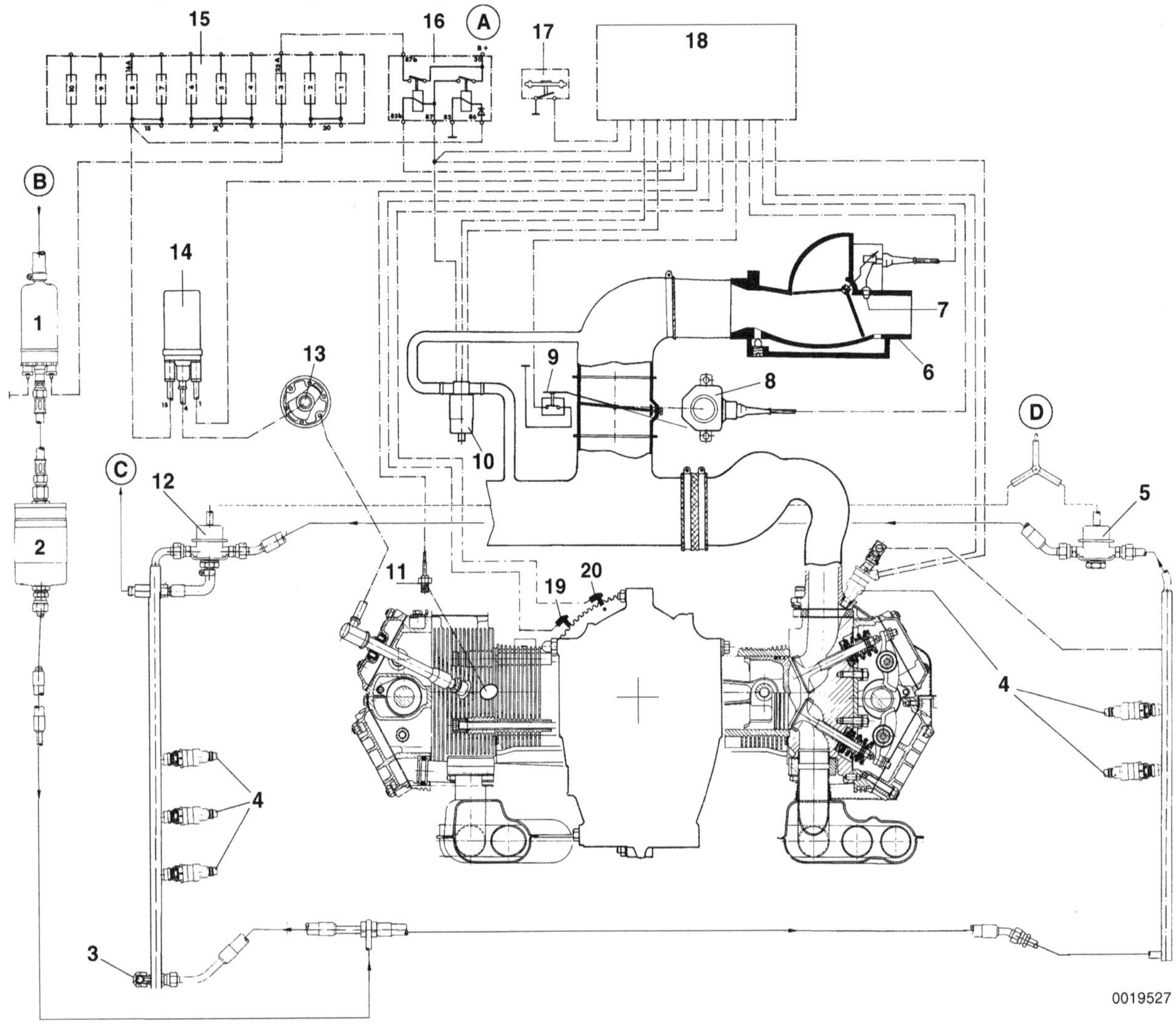

1. Fuel pump
2. Fuel tank
3. Fuel pressure test port
4. Fuel injectors
5. Fuel pressure damper
6. Air flow sensor
7. Intake air temperature (IAT)sensor
8. Full throttle position switch
9. Idle position switch
10. Idle air control valve
11. Cylinder head temperature sensor
12. Fuel pressure regulator
13. Ignition distributor
14. Ignition coil
15. Luggage compartment fusebox
16. Main/fuel pump relay
17. Altitude correction switch
18. DME engine control module (ECM)
19. Engine speed sensor
20. Reference (TDC) sensor

A Power (+) from battery
B Fuel from fuel tank
C Fuel return to fuel tank
D T-fitting to throttle housing

Fuel under pressure is injected via electronically controlled solenoid-type fuel injectors. The ECM electrically controls the opening and closing of the injectors.

The DME system features advanced adaptive control circuitry that actually changes its baseline values in the control module memory based on how that particular car is operating.

Principles of operation

The fuel injection is completely electronic in operation. The DME engine control module (ECM) controls engine fuel delivery and ignition timing using signals from the following sources:

- Air flow sensor with intake air temperature (IAT) sensor
- Cylinder head temperature sensor
- Reference position (TDC) sensor
- Crankshaft speed (RPM) sensor
- Throttle position switches
- Altitude correction sensor
- Oxygen sensor

Fuel supply and fuel delivery

Fuel from the tank is supplied to the engine compartment fuel system loop via an electric fuel pump, a fuel filter, and the connecting fuel lines. The fuel pump is controlled by the ECM via the DME main/fuel pump relay.

Engine compartment fuel system loop

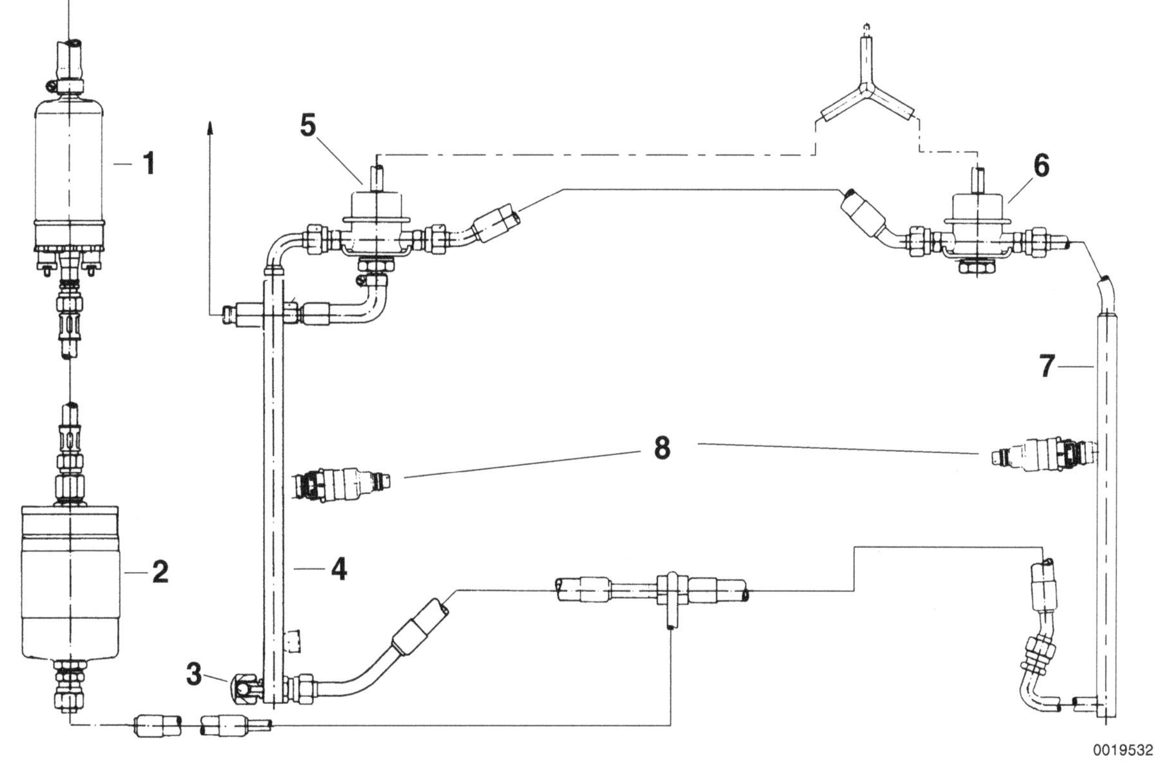

1. Fuel pump
2. Fuel filter
3. Fuel pressure test port
4. Left fuel rail
5. Fuel pressure regulator
6. Pressure damper
7. Right fuel rail
8. Fuel injector

Pulsed type fuel injector

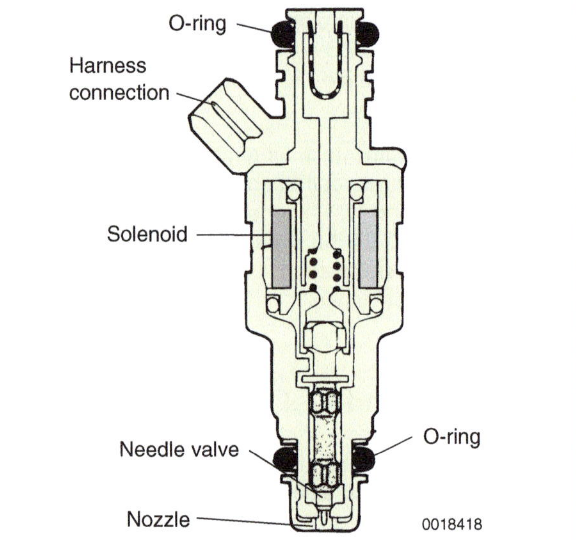

The DME system precisely meters fuel by changing the opening time (pulse width) of the fuel injectors. The injectors are switched on and off by the ECM at the ground side of the connector. Positive (+) battery voltage is always present at the connector when the car is running. The six injectors are mounted in two common fuel supply rails.

Fuel pressure regulator

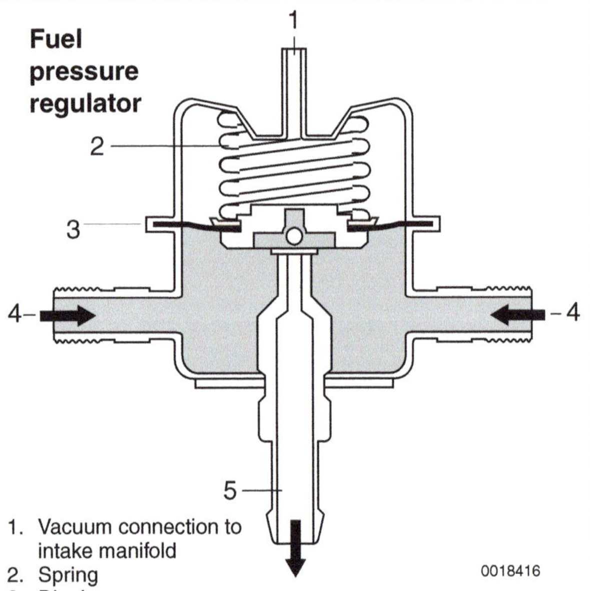

1. Vacuum connection to intake manifold
2. Spring
3. Diaphragm
4. From fuel loop
5. Return to fuel tank

To ensure that injector pulse width is the only factor that determines fuel metering, fuel pressure is maintained by the fuel pressure regulator. The pressure is controlled by returning surplus fuel back to the fuel tank. A change in fuel pressure results in a change in the amount of fuel (or fuel mixture) that is delivered to the engine.

Volume air flow sensor

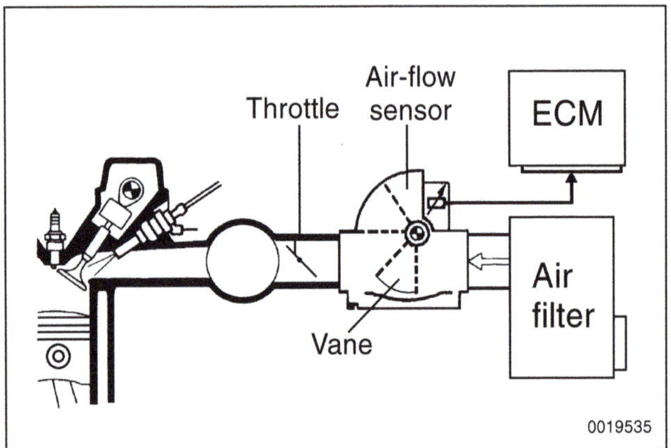

Air entering the engine passes through a pleated paper air filter in the air filter housing. Air flow is controlled by the throttle valve in the throttle housing. The throttle housing is connected to the volume air flow sensor by flexible rubber ducts. Air entering the engine is measured by the air flow sensor. An intake air temperature (IAT) sensor is integrated into the air flow sensor.

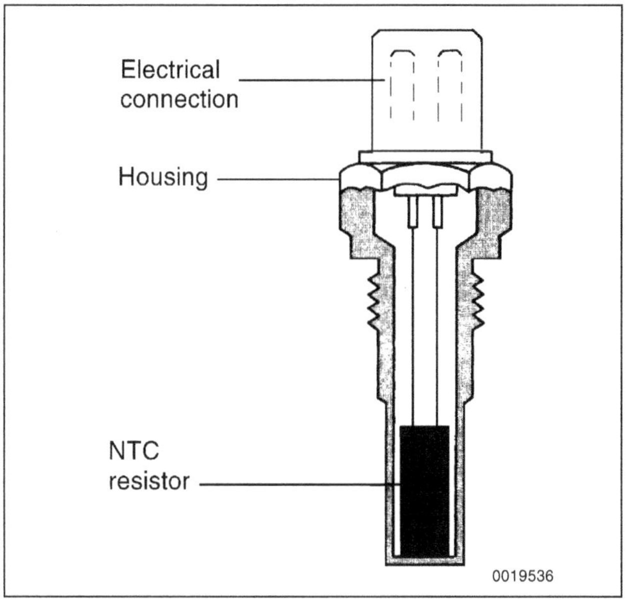

Cylinder head temperature sensor

The cylinder head temperature sensor is the main engine temperature input to the ECM. The temperature sensor sends continuous engine temperature information to the ECM. The temperature sensor is of the NTC (negative temperature coefficient) type; as temperature increases sensor resistance decreases.

- The sensor is mounted in cylinder no. 3 (left front cylinder).

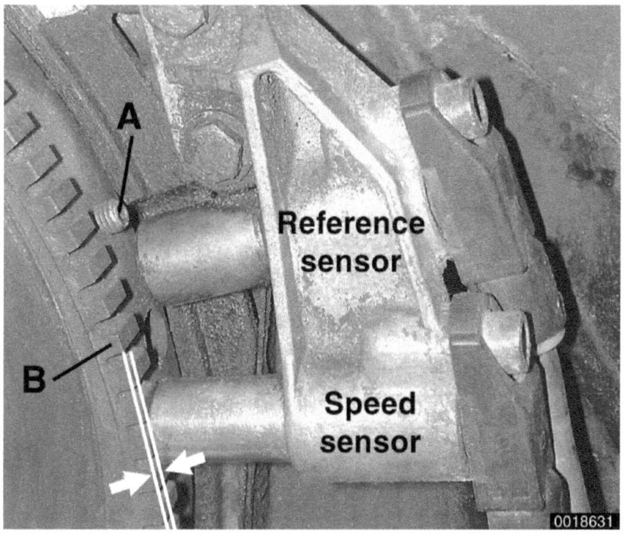

Reference and speed sensors

The reference (TDC) and engine speed sensors are mounted to the crankcase on the flywheel end of the engine. The ECM needs a signal from the reference position sensor for the engine to start.

The reference sensor responds to a set screw (**A**) in the flywheel during cranking. The engine speed sensor responds to a toothed wheel on the flywheel (**B**). Engine speed is determined by the rate at which the wheel's teeth pass the sensor. Two voltages pulses are generated for each tooth on the flywheel.

The sensor clearance (**arrows**) should be 0.8 ± 0.3 mm (0.03 ± 0.01 in.).

Throttle position switches

Two switches in the throttle housing are used by the ECM to determine throttle position.

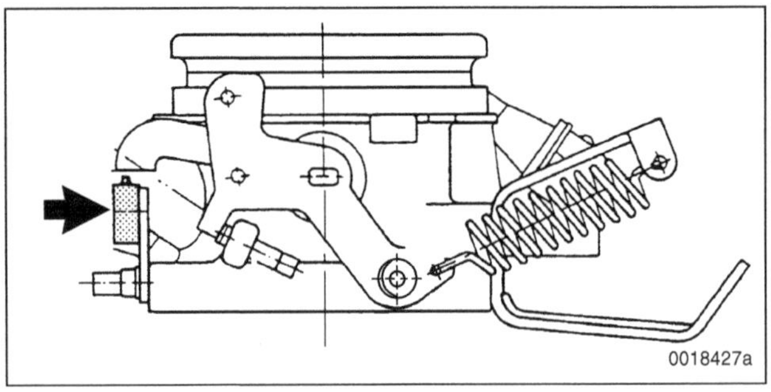

The idle position switch (**arrow**) provides a signal to the ECM when the throttle is in the rest position. When the idle switch is closed, the idle control valve is activated to regulate idle speed. The idle position signal is also used to for fuel cutoff during deceleration. The injectors are turned back on at 1,080 rpm when coasting from a higher speed.

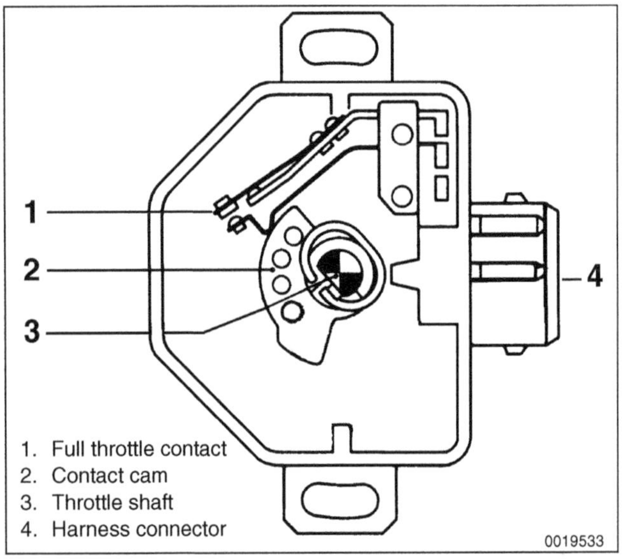

1. Full throttle contact
2. Contact cam
3. Throttle shaft
4. Harness connector

The full throttle switch provides a signal to the ECM when the throttle is in the rest position. When the full throttle switch is activated, the ECM substitutes an enrichment fuel delivery and ignition map for optimum power. The oxygen sensor signal is ignored under these conditions and the A/C compressor will be turned off if in use.

Altitude correction switch

Due to lower air pressures at high altitude, the air/fuel mixture will be richer when the car is operated at high altitudes. The altitude correction switch decreases the injection pulse width by 6% at 1,000 meter (3,300 ft.) and above.

Oxygen sensor

The oxygen sensor adapts the air-fuel mixture by sending a varying voltage signal to the engine ECM. The sensor is positioned in the exhaust stream and measures the amount of oxygen in the exhaust gas so that the fuel injection system can correctly adjust the air-fuel mixture.

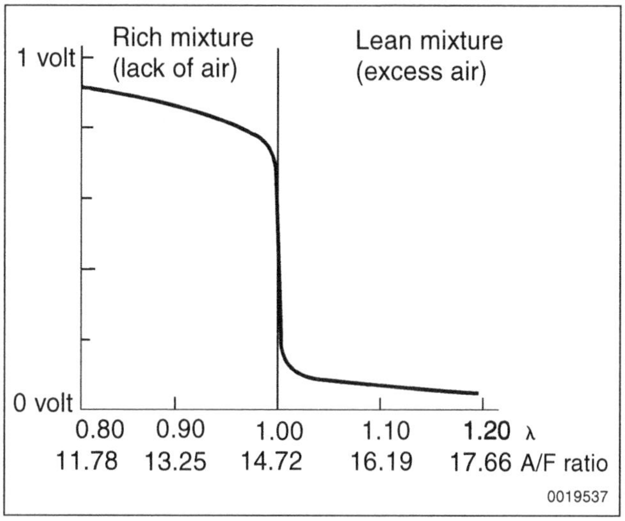

A high concentration of oxygen in the exhaust gas indicates a lean mixture and a low content indicates a rich mixture. The signal from the oxygen sensor is used by the ECM to influence engine performance and driveability.

NOTE —

The signal from the oxygen sensor is ignored by the engine ECM until the engine reaches a specified temperature. Therefore, when troubleshooting cold engine driveability problems, the oxygen sensor can be ruled out as a possible cause.

Idle speed control

Idle speed is electronically controlled via the idle speed control valve, which maintains idle speed by bypassing varying amounts of air around the closed throttle valve.

When the throttle is in the rest position, a programmed nominal speed (rpm) in the ECM is continuously compared to the engine's actual speed. If the actual speed varies from the programmed speed, the duty cycle signal to the valve is changed to maintain the desired engine speed. When the engine is warm, the regulated nominal engine speed is 800 ±20 rpm.

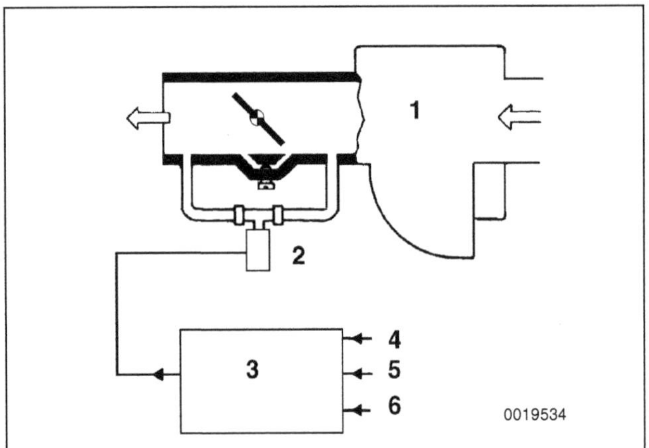

1. Air flow sensor
2. Idle speed control valve
3. DME control module (ECM)
4. Engine speed sensor
5. Engine temperature sensor
6. Idle contact

DME main/fuel pump relay

The main/fuel pump relay is energized via the DME engine ECM and supplies power to many of the DME components and subsystems, including the fuel pump.

The fuel pump relay is energized from terminal 20 of the ECM. The fuel pump only runs when the engine is running. If this relay is faulty, the engine will not start.

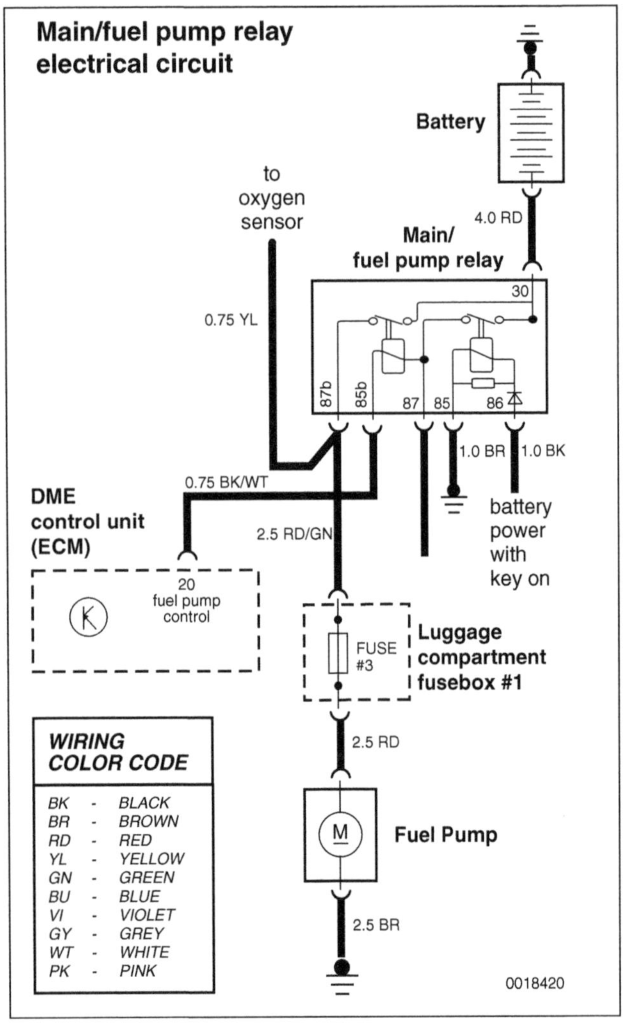

Ignition timing

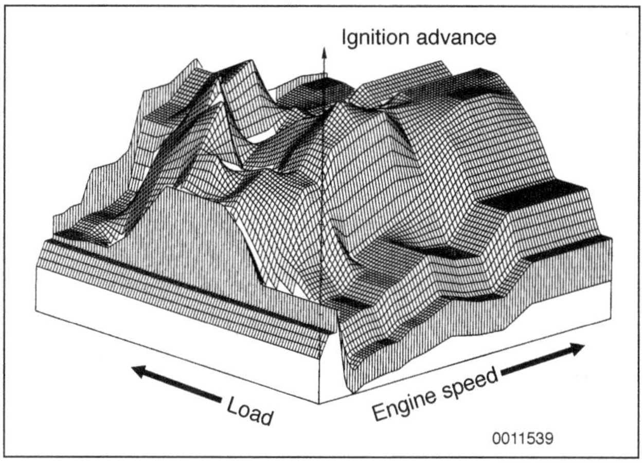

The ECM uses engine load, engine speed, coolant temperature, and intake air temperature as the basic inputs for timing control. The DME ignition characteristic map represents the optimum ignition point for varying engine speed and engine loads. A map similar to the one shown is digitally stored in the ECM.

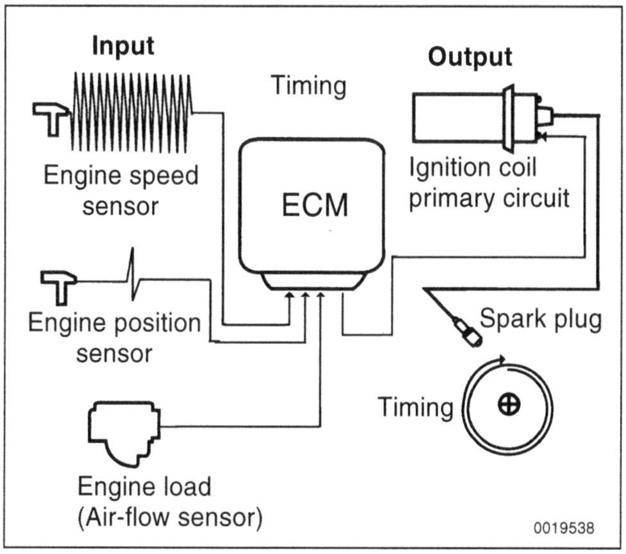

The initial ignition point is determined by the reference (TDC) sensor during cranking. Once the engine is running, timing is continually adjusted based on operating conditions.

> ***WARNING —***
> *The ignition system is a high-energy system operating in a dangerous voltage range that could prove to be fatal if exposed terminals or live parts are contacted. Use extreme caution when working on a car with the ignition on or the engine running.*

Emission controls

The emission control systems used in the 911 Carrera include the oxygen sensor system, the evaporative emission system, and the crankcase/oil tank ventilation system.

Oxygen sensor. This system provides the fuel injection system with feedback information about combustion efficiency.

Evaporative emission controls. This system provides venting for the fuel tank while at the same time trapping the fuel vapors that would otherwise be vented into the atmosphere. The evaporative emission system collects fuel vapors from the fuel tank in a charcoal canister. During certain engine operating conditions, the canister is purged, drawing the fuel vapors into the engine to be burned.

Crankcase and oil tank ventilation. This system provides venting for the crankcase and oil tank while at the same time trapping the fuel vapors that would otherwise be vented into the atmosphere. The crankcase and oil tank ventilation system also includes a secondary venting system to prevent throttle icing in cold weather. See **100 Engine–General** for more information.

DRIVEABILITY TROUBLESHOOTING

Driveability—the overall performance of the car, its ability to start quickly, run and accelerate smoothly, and deliver fuel economy as well as power—can only be achieved when all of the engine's major systems are working properly.

Poor driveability or faulty running may have a variety of causes. The fault may lie with the fuel system, the ignition system or a combination of both. Because of the interrelated functions of these systems it is often difficult to know where to begin looking for problems. For this reason, effective troubleshooting must always consider these system as one major system.

This section offers basic troubleshooting information that addresses symptoms of poor driveability. It is a guide to problem solving, intended to be used in conjunction with the other sections in this manual.

Observe the following cautions and warnings when performing any service or repair on the engine management system.

WARNING —
- *The fuel system is designed to retain pressure even when the ignition is off. When working with the fuel system, loosen the fuel lines very slowly to allow residual fuel pressure to dissipate gradually. Avoid spraying fuel.*
- *Fuel is highly flammable. When working around fuel, do not disconnect any wires that could cause electrical sparks. Do not smoke or work near heaters or other fire hazards. Keep an approved fire extinguisher handy.*
- *There are lethal voltages present at the ignition system when the engine is running or the key is on.*

CAUTION —
- *Before making any electrical tests with the ignition turned on, remove the harness connector from the DME main/fuel pump relay. The relay is located under the driver's seat.*
- *To prevent damage to the ignition system or the electronic fuel system components, including the ECM, always connect and disconnect wires and test equipment with the ignition off.*
- *Cleanliness is essential when working with the fuel system. Thoroughly clean the fuel line unions before disconnecting any of the lines.*
- *When replacing parts, install only new, clean components.Always replace seals and O-rings.*
- *Avoid nearby use of compressed air, and do not move the car while the fuel system is open.*
- *Do not use compressed air above 40 psi to blow out lines or components. Internal damage may result to components.*

Mechanical condition

Before troubleshooting a poorly running engine or an engine that will not start, determine the general condition of the engine, especially if it has high mileage. If the engine is severely worn or has mechanical problems, the only remedy is overhaul or repair. If a tune-up or scheduled maintenance is due, it should be done before proceeding to other repairs.

General engine condition can be assessed by checking the compression at each cylinder. Make sure that the valves are correctly adjusted before making the test. The compression check is described in **100 Engine–General**.

NOTE —
A compression test requires special test equipment. If the equipment is not available, most automotive repair shops can do these tests quickly and at a reasonable cost.

B7088

Carbon deposits on the fuel injectors and the intake valves will affect the way the engine idles and runs. Even a 10% decrease in the amount of fuel that the injectors deliver can cause driveability problems. A car that is predominantly driven on short trips around town or in city traffic seems to increase the likelihood of deposits forming.

Light carbon can usually be removed using a fuel additive. Specialized cleaning equipment is required to remove heavier carbon deposits. Check with an authorized Porsche dealer for information on carbon deposits and the best methods used to remove them.

CAUTION —
Always follow the manufacturer's directions when using fuel additives designed to remove carbon deposits and clean injectors. The excess amount of cleaner in the fuel can dilute engine oil and accelerate engine wear. Check with an authorized Porsche dealer before using fuel additives together with high detergent fuel.

Preventive maintenance

The condition of the tune-up and emission control components can affect engine performance and driveability. Extending maintenance schedules beyond those recommended by the manufacturer can adversely affect the way the engine runs. When experiencing driveability problems, a good starting point is to perform the manufacturer-specified maintenance. Many driveability problems are eliminated simply by replacing worn out or dirty spark plugs.

Replacement schedules for the spark plugs, spark plug wires, distributor cap, ignition rotor, fuel filter, air filter, oxygen sensor, and oil and oil filter are given in **030 Maintenance**.

Valve adjustment

Correctly adjusted valves are necessary for efficient engine operation. If the valve clearances are too small, the valves may not close all the way, resulting in low compression and a loss in power. If the valve clearances are too large, the valves may not fully open causing a reduction in engine efficiency. Procedures for checking and adjusting valve clearances are described in **030 Maintenance**.

Basic settings

In addition to tune-up component replacement, it is important that all of the basic setting and adjustments are within allowable limits. Check throttle plate adjustment, idle speed, idle mixture (%CO), and ignition timing to be sure they are all as specified. All of the models covered by this manual are equipped with a non-adjustable or self-correcting, ignition timing, idle mixture, and idle speed. Basic adjustment are outlined in **030 Maintenance**.

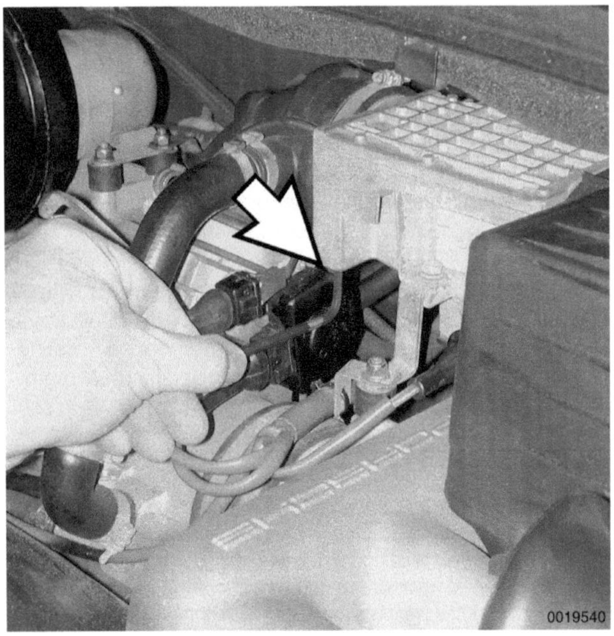

The idle mixture (%CO) is adaptive via the oxygen sensor feedback system. A CO correction screw is located in the air flow sensor, but is capped off by a non-reusable cap (**arrow**). If the idle mixture (%CO) is outside the allowable limits and the oxygen sensor has been replaced, the CO level may be corrected at the adjustment screw. This adjustment should be made by an authorized Porsche dealer or other qualified shop with the appropriate test equipment (gas analyzer).

NOTE —

- *The signal from the oxygen sensor is ignored by the engine ECM until the engine reaches a specified temperature. Therefore, when troubleshooting cold engine driveability problems, the oxygen sensor can be ruled out as a possible cause.*
- *As the oxygen sensor ages, it loses its ability to react quickly and it may eventually cease to produce any signal at all. When this happens, fluctuations in idle speed and increased fuel consumption may be noticed. The oxygen sensor should be replaced at the specified mileage interval as described in* **030 Maintenance**.

Air flow measurement and vacuum leaks

To control fuel injection the engine management system uses an air flow sensor to precisely measure incoming air. The sensor sends an electrical signal proportional to the measured air flow to the ECM, which uses this signal to determine the amount of fuel the engine needs.

Because proper fuel metering depends on accurately measuring the intake air, any unmeasured air entering the system will cause a lean fuel mixture and poor running.

There are many possible places for unmeasured air to enter the engine. Carefully inspect all hoses, fittings, duct work, seals and gaskets. Check the oil filler cap seal on the oil tank. For a thorough inspection, it may be necessary to remove hoses and ducts that cannot be completely checked in their installed positions.

Electrical system

All the cars covered by this manual use engine management systems that rely on precise electrical signals for proper operation. Some of the DME circuits operate on very low current and are sensitive to increased resistance due to faulty or corroded wiring or connectors. If any of these signals are distorted, incorrect, or missing, the car can develop major driveability problems.

NOTE —

In most cases, a faulty ECM is not the cause of driveability problems. These modules are extremely durable and reliable. Actual failures are not that common. Driveability problems are more often caused by missing or incorrect signals to the ECM, or by other faulty components.

First check to make sure all ground connections and harness connectors are tight, free of corrosion and damage-free. In particular, clean battery terminals and ground straps. For a complete listing of ground points, see **971 Electrical Component Locations**.

Make sure the battery is fully charged and in good condition. The DME system requires a specified operating voltage to function correctly. Battery voltage can be measured across its terminals with all cables attached. For the battery to maintain its proper voltage level, the charging system must be functioning correctly. See **270 Battery, Starter, Alternator**.

NOTE —
A digital voltmeter should be used to accurately measure battery voltage. A fully charged battery will measure 12.6 volts, or more, while a battery only 25% charged will measure 12.15 volts. Using an analog meter may result in inaccurate results.

The electrical system is subject to corrosion, vibration, roadway elements and general wear. Because of this, the integrity and freedom from corrosion in the connections, wires, and switches, including all ground connections, are important conditions for trouble-free operation of the engine management systems.

Loose or damaged connectors can cause intermittent problems, especially at the small terminals in each ECM connector. In most cases, a visual inspection will detect any faults. If a connector shows no visible faults but is still suspect, perform a voltage drop test at the connector. Even a small amount of corrosion in a connector can cause a large voltage drop to the circuit's load. See **900 Electrical System–General** for more troubleshooting information.

Poor ground connections are a major sources of driveability problems. There are only a few main ground connections or points for the engine management systems. These ground points are a grouping of many wires crimped or welded into an eyelet that is then bolted to the car's chassis or metal parts. If any of these ground points are faulty, the voltage to the circuit will be reduced or even eliminated.

To thoroughly check a circuit ground, check the voltage drop between the connector and a good ground source. Large voltage drops indicate too much resistance: The connection is corroded, dirty or otherwise damaged. Clean or repair the connection and retest. Also check both battery terminals and all ground straps between the engine and the body for voltage drops.

NOTE —
- *For a listing of the DME-related ground points, see* **240 Fuel Injection**. *A complete listing of the ground connections is given in* ***971 Electrical Component Locations***.
- *For voltage drop tests and other general electrical troubleshooting information, see* **900 Electrical System–General**.

Fuel delivery troubleshooting

Fuel from the tank is supplied to the engine via an electric fuel pump, a fuel filter, and the connecting fuel lines. If either the filter or a fuel line is restricted, the engine may not run properly. To check for a clogged fuel filter, a restricted fuel line, or a weak/worn fuel pump, perform a fuel delivery rate test as described **201 Fuel Supply**.

Power to run the fuel pump is controlled by the DME main/fuel pump relay. If the relay is not working, the car will not start or run. DME main/fuel pump testing is covered in **201 Fuel Supply**.

Idle speed

Erratic idle speed is one of the most common driveability problems encountered on the cars covered by this manual.

Be sure the idle position switch is correctly functioning as described in **240 Fuel Injection**. If no vacuum leaks can be found, the idle speed control valve may be worn out. The best way to check for a faulty valve it to substitute with a known good part.

Table a lists additional probable causes and corrective actions for driveability problems. The boldface type indicates the section of the manual where the applicable test and repair procedures are found.

Table a. Engine Driveability Troubleshooting

Symptoms or Faults
- a. Engine does not start or starts hard
- b. Erratic idle
- c. Poor acceleration
- d. Engine misfiring
- e High fuel consumption
- f. Low engine power output

a	b	c	d	e	f	Probable Causes of Fault	Corrective Action
X	X	X	X		X	Ground or plug connection	Check ground connections at intake runner 1, behind fuel filter. Check connectors at fuel injectors, other fuel injection components. **240 Fuel Injection.**
X			X			No power to ECM or fuel pump	Test DME main/fuel pump relay. **201 Fuel Supply**
	X	X	X	X	X	Speed sensor	Test sensor with oscilloscope for 2.5 V sine
X						Reference (TDC) sensor faulty	Test sensor with oscilloscope for 2.0 V sine
X		X	X		X	Ignition system fault	Test coil. **280 Ignition System.**
X	X	X	X	X	X	Air flow sensor faulty	Test air flow sensor. **240 Fuel Injection.**
X	X	X		X	X	Fuel pressure incorrect	Check fuel pressure/pressure regulator. **240 Fuel Injection.**
	X	X				Idle air control valve faulty or no control signal to valve.	Check for 12V at center terminal of valve with ignition key on. Check idle position switch. **240 Fuel Injection.**
		X			X	Full throttle position switch faulty	Check for continuity across switch terminals with throttle in wide open position.
X	X	X	X	X	X	DME cylinder head temperature sensor (Temperature sensor II) faulty	Test resistance value at room temperature = 2.2 to 2.8 k-ohms
X	X	X	X	X	X	Fuel injector(s) not working	Check for at least 10 V with key on at injector connector. Check for pulsed ground signal with engine running/cranking. **240 Fuel Injection.**
X	X	X				CO level and idle speed adjustments incorrect	Check for vacuum leaks, replace oxygen sensor, check air flow sensor. **240 Fuel Injection.**
	X	X	X		X	Intake air leak	Check for loose intake air ducting connections, splits or cracks in air boot, other sources for vacuum leaks.
		X	X		X	High altitude switch faulty	Bridge connectors at switch connector with engine running. CO should drop slightly (1% drop)
	X		X			Low system voltage.	Check battery voltage with engine running. Test alternator output 13.2 to 14.2 volts. **270 Battery, Starter, Alternator**.

201 Fuel Supply

TABLES

Special Tools

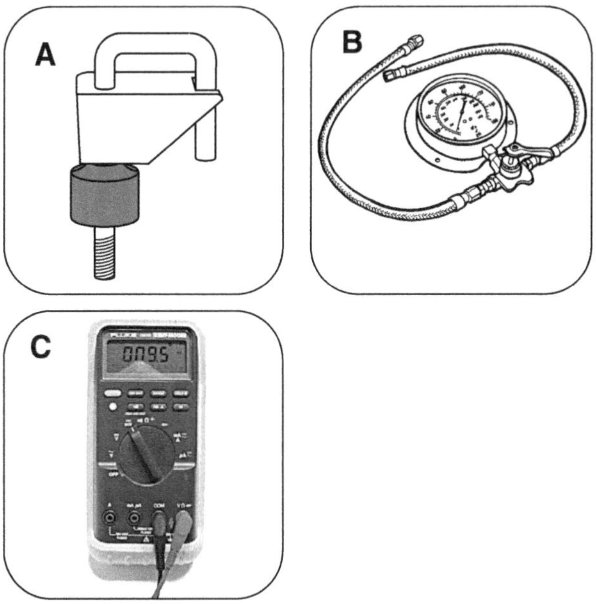

Special tool

A Baum 647b . . . . Fuel hose pinch clamp (source: Baum Tools Unlimited)
B Baum 1318 . . . . Fuel pressure gauge (source: Baum Tools Unlimited)
C Fluke 87 . . . . Automotive digital multimeter

GENERAL

This repair group covers repairs to the fuel tank, fuel pump, fuel level sender and connecting lines. The Motronic (DME) fuel injection and emission control systems are covered in **240 Fuel Injection**.

NOTE —
Fuel filter replacement is covered in **030 Maintenance***.*

Test equipment

> ***CAUTION —***
> *A high impedance digital multimeter should be used for all voltage and resistance tests.*

Some of the procedures require special tools.

Warnings and cautions

Please observe the following cautions and warnings when performing any service or repair on the fuel system.

WARNING —

- *When working on an open fuel system, wear suitable hand protection, as prolonged contact with fuel can cause illnesses and skin disorders.*
- *Gasoline is highly flammable and its vapors are explosive. Do not smoke or work on a car near heaters or other fire hazards when diagnosing and repairing fuel system problems. Have a fire extinguisher available in case of an emergency.*

CAUTION —

- *Disconnecting power from the engine control module (ECM) erases its adaptive memory. It may be necessary to drive the car for approximately 10 minutes after reconnecting the battery to re-set ECM adaptive memory and restore normal performance.*
- *Always unscrew the fuel tank cap to release pressure in the tank before working on the tank or lines.*
- *Before making any electrical tests with the ignition turned on, disable the ignition system as described in* **280 Ignition System**.
- *To prevent damage to the ignition system or the electronic fuel system components, including the ECM, always connect and disconnect wires and test equipment with the ignition off.*
- *Fuse and relay locations are subject to change. Use care when troubleshooting the electrical system at the fuse/relay panels. To resolve problems in identifying a relay, consult an authorized Porsche dealer for the latest information*
- *Absolute cleanliness is essential when working with the fuel system. Even a minute particle of dirt can cause trouble if it reaches an injector. Thoroughly clean the fuel line unions before disconnecting any of the lines.*
- *Use only clean tools. Keep removed components clean and sealed or covered with a clean, lint-free shop towel, especially if completion of the repair is delayed.*
- *When replacing parts, install only new, clean components.*
- *Avoid nearby use of compressed air, and do not move the car while the fuel system is open.*
- *Do not use compressed air above 40 psi to blow out lines or components. Internal damage may result to components.*
- *Always replace seals and O-rings.*

FUEL TANK AND LINES

Fuel Tank Capacity

• 1984 - 1989 Porsche Carrera . . . 22.5 US gals. (85 liters)

The fuel tank is located at the front of the car beneath the luggage compartment. The fuel level sending unit is mounted through the top of the tank. Connecting lines for the evaporative emission control system and fuel supply and return lines are also attached to the tank.

Fuel tank assembly

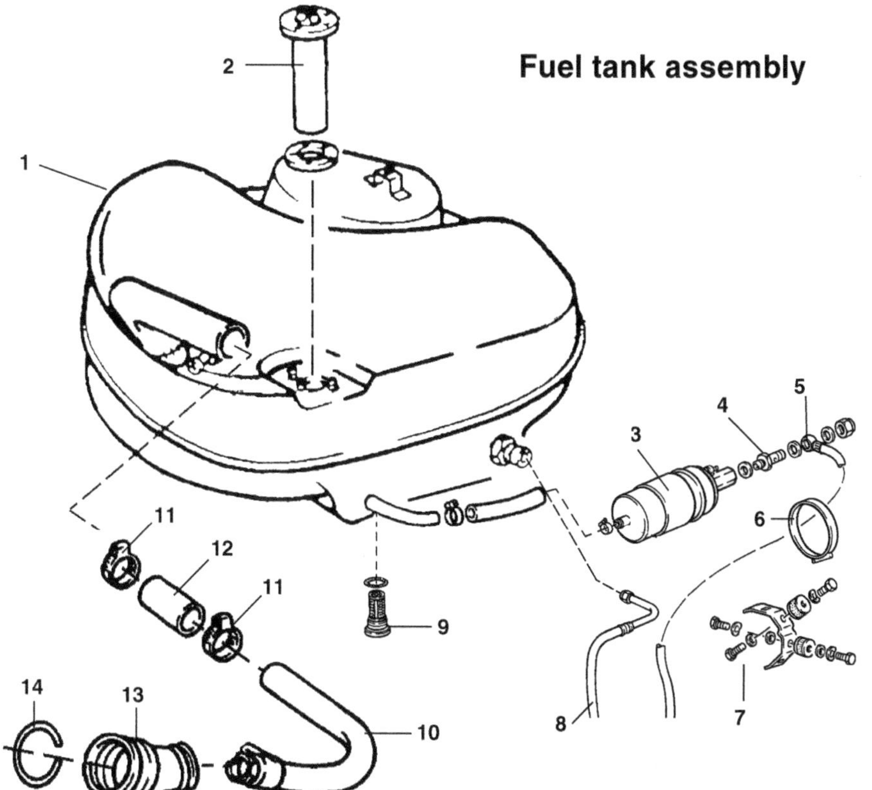

1. Fuel tank
2. Fuel level sending unit
3. Fuel pump
4. Fuel pump check valve
5. Fuel supply line (to fuel injection system)
6. Clamp
7. Fuel pump mounting assembly
8. Return line
9. Drain plug/screen filter
10. Connecting hose
11. Clamp
12. Connecting hose
13. Filler neck bellows
14. Circlip

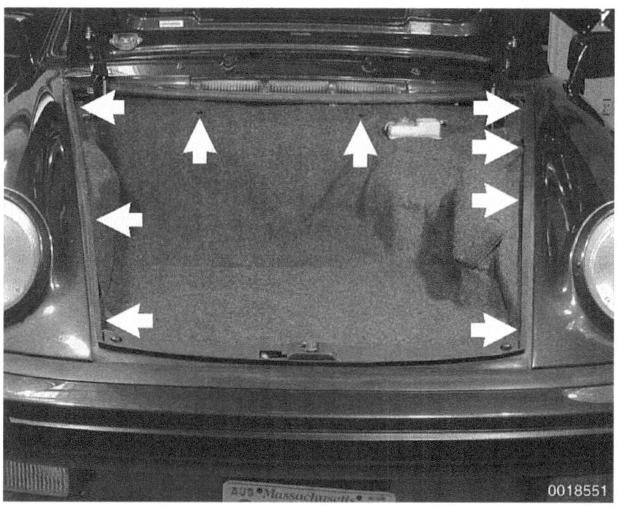

Fuel tank, draining

< Unsnap luggage compartment floor carpet (**arrows**) and lift off. Remove spare tire.

– Disconnect negative (–) battery cable and cover battery terminal to keep cable from accidentally contacting terminal.

> ***CAUTION —***
> *Avoid getting tools or clothing near the battery. Battery electrolyte is a corrosive acid. Always wear eye protection when working on or near the battery.*

– Remove fuel level sending unit as described later.

– Insert hose into tank opening and use an approved transfer pump and container to pump out fuel.

> ***CAUTION —***
> *Be sure the container is large enough to hold the volume of fuel that is to be pumped out of the tank.*

NOTE —
As an alternative, the drain plug (allen head) at the bottom of the fuel tank can be removed to drain tank.
– Use a large funnel and drum to catch draining fuel.
– Clean out filter element before reinstalling.

Tightening torques
- Fuel tank drain plug to tank . . . . 25 Nm (18 ft-lb)

Fuel tank, removing and installing

The fuel tank must be completely drained as described earlier before it is removed from the car.

– Disconnect battery negative (–) cable.

> ***CAUTION —***
> *Prior to disconnecting the battery, read the battery disconnection cautions given at the front of this manual on page viii.*

– Raise car and support in a safe manner.

> ***CAUTION —***
> *Make sure the car is stable and well supported at all times. Use a professional automotive lift or jack stands designed for the purpose. A floor jack is not adequate support.*

0018547

◄ Working underneath front of car, remove stone guard mounting bolts (**arrow**). Remove stone guard.

0018548

◄ Pinch off fuel feed line from tank (**arrow**) with a fuel line clamp.

– Lower car to ground.

◄ Working inside luggage compartment, detach hose from tank vent (**arrow**).

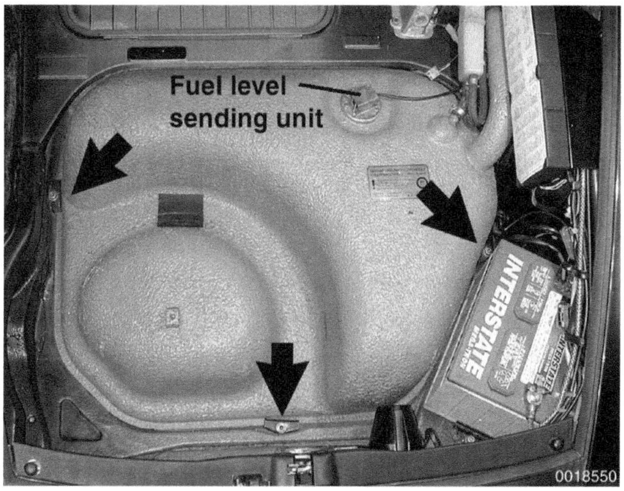

◄ Remove tank mounting bolts (**arrows**).

– Loosen filler neck clamps.

– Remove fuel tank by lifting it up and out through luggage compartment.

– Installation is reverse of removal.

- Always use new tank mounting gasket and hose clamps.
- Inspect all rubber parts including hoses and replace any that are worn, chafed, or cracked.

Tightening torques

- Fuel tank to body . . . . 20 Nm (15 ft-lb)
- Fuel drain plug . . . . 25 Nm (18 ft-lb)
- Stone guard to body
  M10. . . . . 45 Nm (33 ft-lb)
  M8. . . . . 25 Nm (18 ft-lb)

Fuel level sending unit

The fuel level sending unit is mounted in the top of the fuel tank. The unit has a float connected to a variable resistance for fuel level, and a switch that closes to turn on the low fuel light. When replacing the sending unit always replace hose clamps, gaskets and O-rings.

Fuel level sending unit, removing and installing

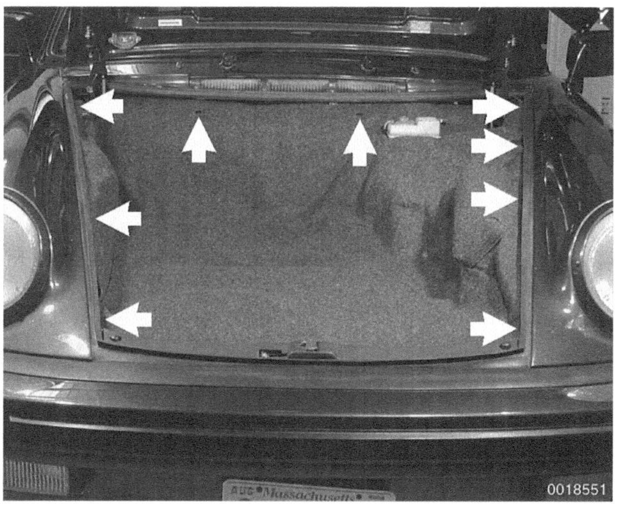

◄ Unsnap luggage compartment floor carpet (**arrows**) and remove carpet.

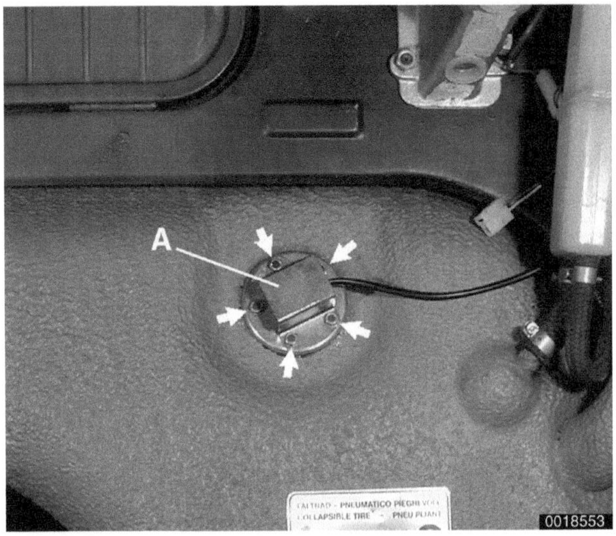

- Disconnect negative (–) battery cable and cover battery terminal to keep cable from accidentally contacting terminal.

> ***CAUTION —***
> *Avoid getting tools or clothing near the battery. Battery electrolyte is a corrosive acid. Always wear eye protection when working on or near the battery.*

- Remove nuts and washers (**arrows**) mounting sending unit to tank. Remove harness connector cover (**A**).

- Disconnect sending unit harness connector.

- Slowly withdraw sending unit from fuel tank, allowing it to drain as it is being removed.

- Test unit as described below.

- When reinstalling use a new gasket.

NOTE —
New replacement fuel level sending units from Porsche may be fitted with a guard for protection during shipping. This guard must be removed prior to installing the unit.

Tightening Torques
- Fuel level sending unit to fuel tank . . . . 2 Nm (18 in-lb)

- Reconnect harness connector. Fill fuel tank and check for leaks.

Fuel level sending unit, testing

- Test sending unit using an ohmmeter connected across terminals. Compare results to specification in **Table a**.

Table a. Fuel Gauge Sending Unit Specifications

Connector terminals	Float position	Test results in Ω
G and **31**	Empty	71.7 ± 2.3
G and **31**	Full	3.2 ± 0.7
G and **31**	Slowly tilting removed sending unit to move internal float from empty to full	Resistance continuously changing
W and **31**	Empty low fuel light on	Continuity

FUEL PRESSURE CHECKS

The fuel pump delivers fuel at high pressure to the fuel injection system. A pressure regulator maintains system pressure. The quantity of fuel supplied exceeds demand, so excess fuel returns to the fuel tank via a return line.

There are three significant fuel pressure and delivery checks:

- **System pressure** is created by the fuel pump and maintained by the pressure regulator.
- **Fuel delivery volume** is created by the fuel pump and affected by restrictions, such as clogged fuel filter.
- **Residual pressure** is maintained in the closed system after the engine and fuel pump are shut off.

NOTE —
Testing for residual pressure is covered in **240 Fuel Injection**.

Relieving fuel pressure and connecting fuel pressure gauge

To prevent fuel from spraying on a hot engine, system pressure should be relieved before disconnecting fuel lines. One method is to tightly wrap a shop towel around a fuel line fitting and loosen or disconnect the fitting.

Connect fuel gauge (Porsche special tool P 378 or equivalent) to fuel rail test port (**arrow**) and tighten to 12 Nm (9 ft-lb).

CAUTION —
When removing test port cap, note small sealing ball under cap. Be careful not to drop sealing ball.

NOTE —
The fuel pressure gauge should have a range of 0 to 5 bar (0 to 75 psi).

System pressure, testing

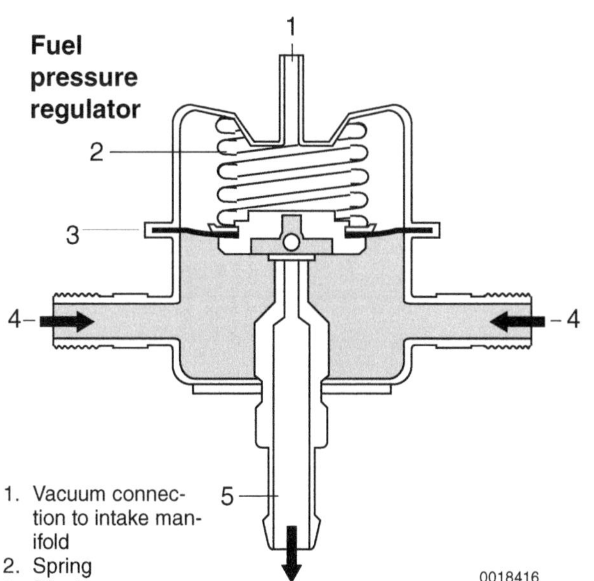

1. Vacuum connection to intake manifold
2. Spring
3. Diaphragm
4. From fuel loop

Fuel system pressure is created by the fuel pump and regulated by the fuel pressure regulator. System pressure is not adjustable.

Relieve fuel pressure and connect pressure gauge to fuel rail (**arrow**) as described earlier.

> ***CAUTION —***
> *The fuel pump is capable of developing a higher pressure than that regulated by the pressure regulator. If the fuel pump check-valve malfunctions (stuck closed), make sure the fuel pressure does not rise above 6.0 bar (87 psi). Damage to the fuel lines or fuel system components could result.*

– Operate fuel pump without running engine using a fuel pump relay jumper, as described later under **Operating fuel pump for tests**.

- Check that fuel pressure corresponds to specifications listed in **Table b**.
- If system pressure is low, repeat test while gradually pinching off return hose. Pressure should rise rapidly. If not, fuel pump is most likely faulty. If pressure does rise, fuel pressure regulator is faulty. See **240 Fuel Injection** for replacement information.
- If system pressure is too high, check return line from pressure regulator to tank. Check for kinks in hose. Blow compressed air through line to check for blockages.

> ***CAUTION —***
> *Do not use compressed air above 40 psi to blow out lines or components. Internal damage to components may result.*

– Remove fuel pump relay jumper and reattach relay to harness. Start engine and recheck fuel pressure.

- Remove vacuum hose from fuel pressure regulator. Recheck fuel pressure. Compare to specifications in **Table b**.
- If any faults are found, either fuel pressure regulator or vacuum hose is faulty and should be replaced.

When finished testing, disconnect pressure gauge and reinstall fuel pressure test port cap with sealing ball.

Tightening Torque
- Fuel pressure test port cap . . . . 12 Nm (9 ft-lb)

Table b. Fuel Pump Pressure Specifications

Condition	Fuel pressure
Engine off	2.5 bar (36 psi)
Engine running Engine running, vacuum hose removed from pressure regulator	2.0 bar (29 psi) 2.3 - 2.7 bar (33 - 39 psi)

Fuel delivery volume, testing

◄ Disconnect return line from fuel rail on rear left side of engine (**arrow**).

> ***WARNING —***
> *To prevent fuel from spraying on a hot engine, system pressure should be relieved before disconnecting fuel lines. One method is to tightly wrap a shop towel around a fuel line fitting and loosen or disconnect the fitting.*

– Connect a length of hose with threaded female fitting (14mm x 1.5) to fuel rail. Place open end of hose in a suitable container for catching fuel (2-quart capacity).

– Run fuel pump for exactly 30 seconds as described earlier under **Operating fuel pump for tests** and measure fuel collected. Compare to specification in **Table c**.

– When finished testing, reconnect fuel line. Tighten all hose clamps.

Table c. Fuel Pump Delivery Volume

Delivery condition	Delivery rate
30 seconds @ 12V	0.850 liter (0.90 qt.)

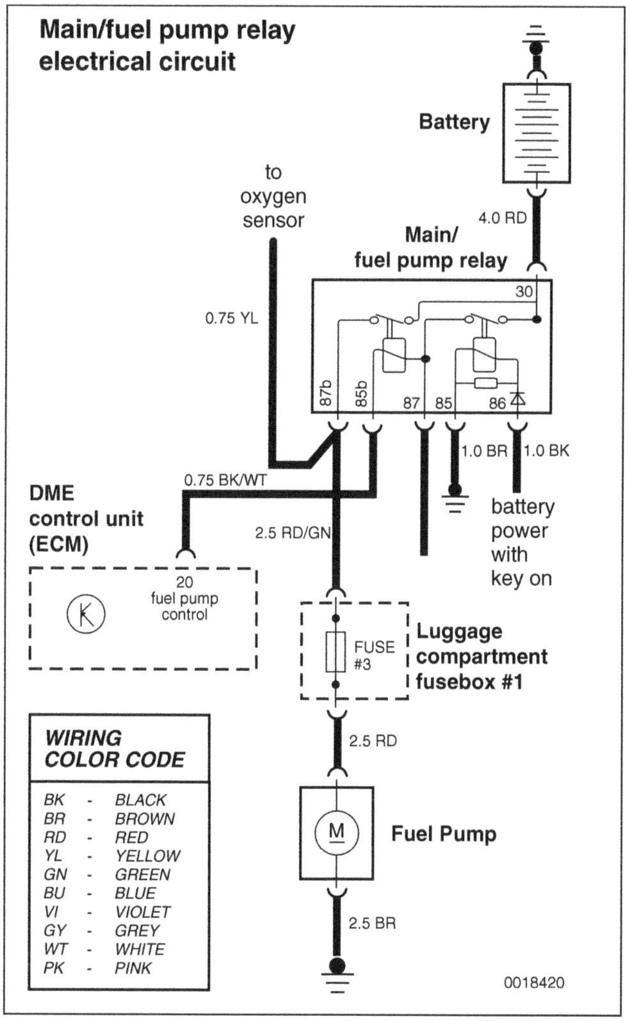

Main/fuel pump relay electrical circuit

FUEL PUMP ELECTRICAL CHECKS

Fuel pump, troubleshooting

During cranking and engine running, the DME main/fuel pump relay is energized and the pump runs. If an electrical system fault interrupts power to the fuel pump, the engine will not run.

– To quickly check if fuel pump is coming on, actuate starter while an assistant listens for a running pump.

– If pump is not running, check for faulty fuse or faulty DME main/fuel pump relay.

- Check fuel pump fuses. See **971 Electrical Component Locations**.
- Test DME main/fuel pump relay and circuit as described later.
- Check if voltage is reaching pump. See **Fuel pump wiring, checking**.

Operating fuel pump for tests

To operate fuel pump for testing purposes without having to run engine, the DME main/fuel pump relay can be bypassed to power pump directly.

– Disconnect DME main/fuel pump relay. Relay is located under driver's seat.

> ***CAUTION —***
> *Disconnecting power from the engine control module (ECM) erases its adaptive memory. It may be necessary to drive the car for approximately 10 minutes after reconnecting the battery to reset ECM adaptive memory and restore normal performance.*

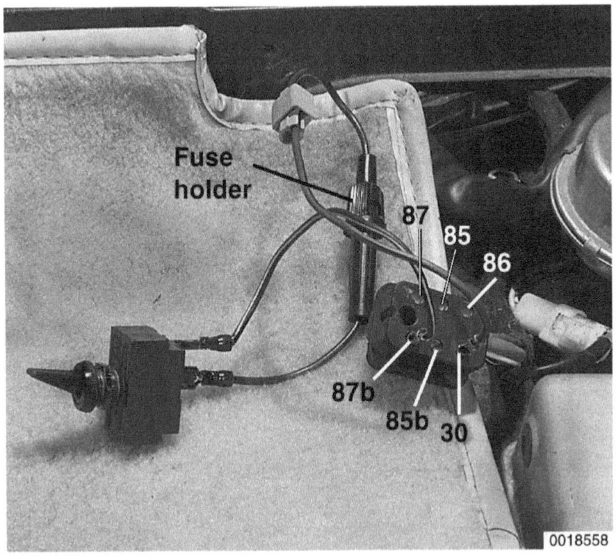

◄ Connect socket for relay terminal 30 to socket for relay terminal 87b with fused jumper wire to run pump. Ignition key does not need to be turned on.

– After completing tests, remove jumper wire.

> ***CAUTION —***
> *The main/fuel pump relay has a 4 mm² red wire at terminal 30 in the relay socket. Terminal 87b has a 2.5 mm² red/green wire. See* **970 Electrical Wiring Diagrams** *for additional wiring information.*

NOTE —
The jumper wire should be 1.5 mm² (14 gauge) and include an in-line fuse holder with a 15 amp fuse. To avoid harness connector damage from repeated connecting and disconnecting, also include a toggle switch.

DME main/fuel pump relay, testing

◄ The DME main/fuel pump relay (**arrow**) is located under the driver's seat next to the engine control module (ECM).

The DME main/fuel pump relay contains two relays within a single housing. One relay is the DME main relay and the other is the fuel pump relay.

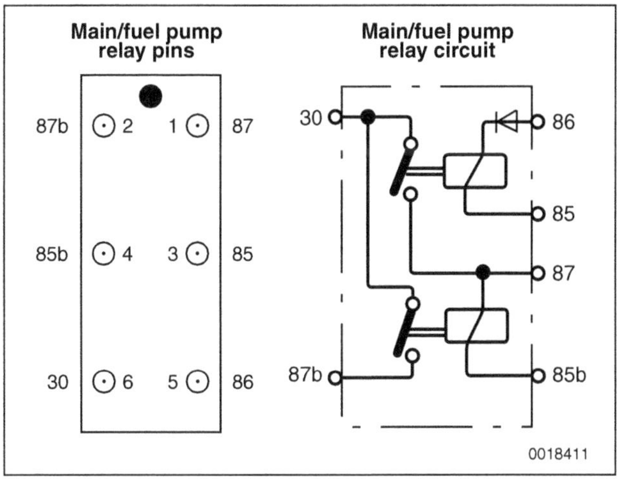

The Motronic DME main relay is energized when the key is turned on. The closed main relay supplies positive battery voltage (via terminal 87) to the fuel injectors, the engine control module (ECM), the idle speed control valve, and the fuel pump relay. The fuel pump relay is then energized (closed) by the ECM. The switched ground for this relay (terminal 85b) is supplied only if certain parameters are met, such as the presence of an ignition pulse and a starter input signal. This is a safety feature to prevent the pump from operating if the engine is not running.

> ***CAUTION —***
> *On cars with a factory alarm (alarm key on door edge), the alarm system interrupts power to terminal 86 of the main/fuel pump relay when the system is armed. When diagnosing a no-start condition be sure to check for key-on power at terminal 86.*

– Remove driver's seat. See **720 Seats**.

– With ignition switch in OFF position, remove DME main/fuel pump relay harness connector.

– With key off, check for battery voltage at terminal 30 in harness connector.

  • If battery voltage is not present, check for breaks in large red wire in relay harness. This wire is a direct (unfused) connection to battery. See **970 Electrical Wiring Diagrams**.

– With ignition switch in ON position check for battery voltage at terminal 86.

  • If battery voltage is not present, check for break in black wire in relay harness. See **970 Electrical Wiring Diagrams**.

> ***CAUTION —***
> *The vehicle alarm system interrupts power to terminal 86. When diagnosing a no-start condition be sure to test key-on power to terminal 86.*

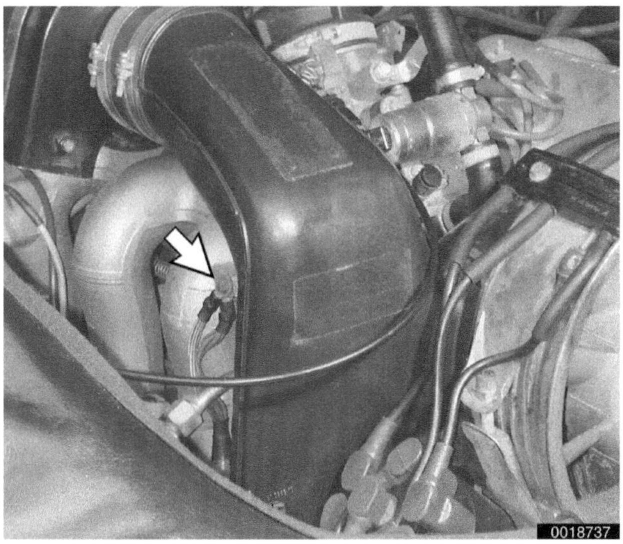

Check for ground at terminal 85.

  • If ground is present continue testing.
  • If ground is not present, check connections at ground point on intake runner 1 (**arrow**).

– Turn off ignition and reattach relay harness connector.

– With ignition switch in ON position and relay connected, check for battery voltage at terminal 87.

  • If battery voltage is present, first relay has energized and is functioning correctly.
  • If battery voltage is not present and all earlier tests are OK, replace DME main/fuel pump relay.

- If DME main relay has energized but fuel pump still won't run, test fuel pump relay ground and output. Relay must be plugged into harness for test.
  - With engine cranking, test for ground at terminal 85b.
  - With engine cranking check for battery voltage at terminal 87b.
  - If battery voltage is not present at terminal 87b but is present at terminal 87, replace DME main/fuel pump relay.

◄ If battery voltage is present but fuel pump doesn't run, check fuel pump fuse (**arrow**) in luggage compartment fuse box. See **971 Electrical Component Locations**.

Fuel pump wiring, checking

If fuel pump fuse and DME main/fuel pump relay are OK, but fuel pump will not run, check for voltage at pump.

- Raise car and support in a safe manner.

> ***CAUTION —***
> *Make sure the car is stable and well supported at all times. Use a professional automotive lift or jack stands designed for the purpose. A floor jack is not adequate support.*

◄ Working underneath front of car, remove stone guard mounting bolts (**arrow**). Remove stone guard.

- Working at fuel pump, pull back two rubber boots that protect fuel pump electrical connectors.

◄ With fuel pump relay bypassed as described above, check for voltage between fuel pump connectors.
  - If battery voltage is present and connectors are not loose or corroded, then fuel pump is faulty.
  - If there is no voltage, look for a faulty wire or connector.

- Also check fuel pump ground wire (brown wire) for continuity to ground.

Fuel pump current draw, checking

– Raise car and support in a safe manner.

> ***CAUTION —***
> *Make sure the car is stable and well supported at all times. Use a professional automotive lift or jack stands designed for the purpose. A floor jack is not adequate support.*

0018547

◄ Working underneath front of car, remove stone guard mounting bolts (**arrow**). Remove stone guard.

– Pull back two protective rubber boots from pump connectors. Disconnect 2.5 mm^2 red wire from pump.

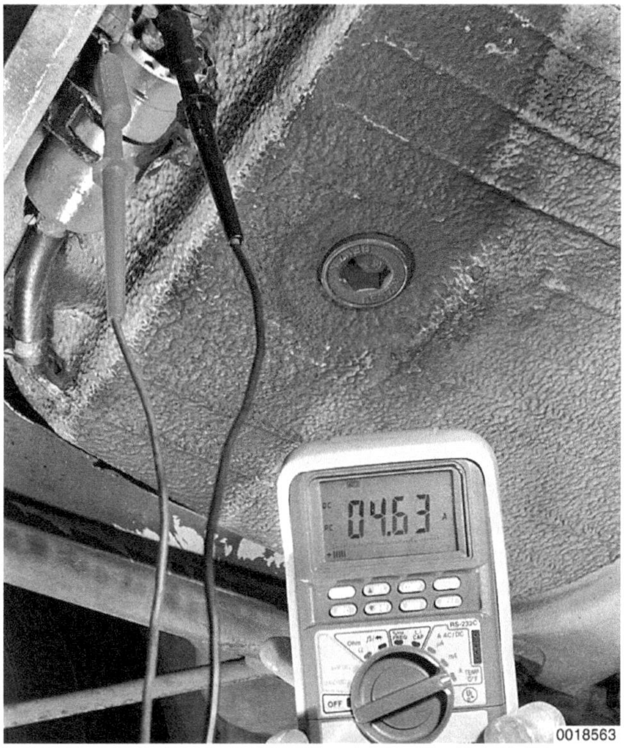

0018563

◄ Connect ammeter between disconnected red wire and fuel pump terminal as shown. The other wire remains connected to pump.

> ***CAUTION —***
> *Do not allow the test leads to short to ground.*

– Bypass fuel pump relay as described earlier. Fuel pump current should be in the range of 4 to 6 amps.

NOTE —
A higher than normal power consumption usually indicates a worn fuel pump, which may cause intermittent fuel starvation due to pump overheating and seizure. The only remedy is to replace the pump.

FUEL PUMP REPLACEMENT

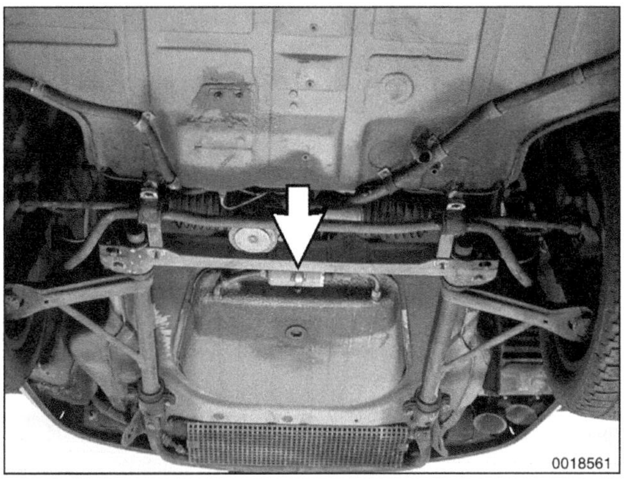
0018561

◄ The fuel pump is mounted beneath the car, behind fuel tank (**arrow**).

Fuel pump, replacing

– Disconnect negative (–) battery cable and cover battery terminal to keep cable from accidentally contacting terminal.

> ***CAUTION —***
> *Avoid getting tools or clothing near the battery. Battery electrolyte is a corrosive acid. Always wear eye protection when working on or near the battery.*

– Raise car and support in a safe manner.

> ***CAUTION —***
> *Make sure the car is stable and well supported at all times. Use a professional automotive lift or jack stands designed for the purpose. A floor jack is not adequate support.*

◄ Working underneath front of car, remove stone guard mounting bolts (**arrow**). Remove stone guard.

◄ Pinch off fuel feed line (**arrow**) from tank with a fuel line clamp.

– Disconnect two electrical wires from pump.

– Remove pressure side fuel line cap nut while counter holding pump fitting with wrench.

– Loosen pump mounting clamp and remove pump.

– Installation is reverse of removal.

NOTE —
Always replace sealing washers.

Tightening Torques

- Electrical connector nuts . . . . .1.5 Nm (13 in-lbs)
- Fuel pump clamp bolt . . . . 7 Nm (62 in-lbs)
- Hose clamps. . . . . 2 Nm (18 in-lbs)
- Stone guard to body (M10 bolt) . . . . 47 Nm (35 ft-lb)
- Stone guard to subframe (M10 nut) . . . . 28 Nm (21 ft-lb)

Fuel pump check valve, replacing

– Access fuel pump and pinch off fuel feed line as described earlier.

– Remove pressure side fuel line cap while counterholding pump fitting with wrench.

◄ Remove check valve (**arrow**) from end of pump.

- Replace copper sealing rings during reassembly.
- Start engine and check for leaks after tightening fittings, before reinstalling stone guard.

240 Fuel Injection

TABLES

Special Tools

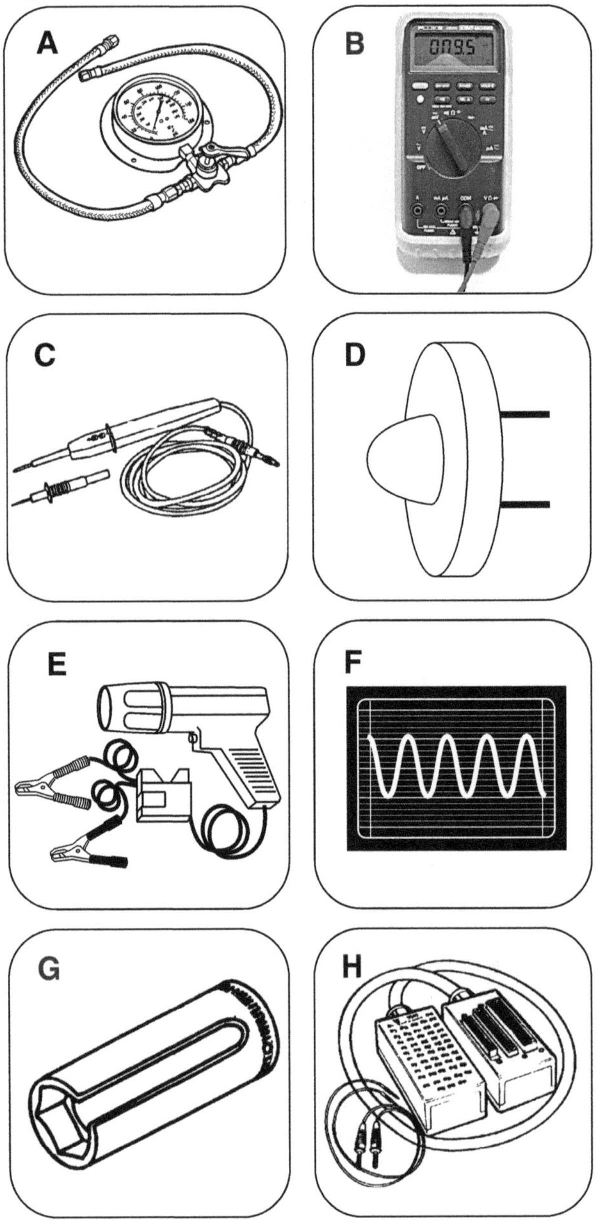

Special tool

- **A** 1318 .... Fuel pressure gauge (source: Baum Tools Unlimited)
- **B** Fluke 87 .... Automotive digital multimeter
- **C** 1115 .... LED tester with thin spade probes (source: Baum Tools Unlimited)
- **D** 1115a .... Low current test light ("noid") (source: Baum Tools Unlimited)
- **E** .... Strobe timing light
- **F** .... Oscilloscope
- **G** 9222 .... Deep slotted socket (Porsche ordering #000 721 922 21)
- **H** 9543 .... Pin out test harness (Porsche ordering #000 721 954 30)

GENERAL

This repair group covers specifically the fuel injection component of the DME (Motronic) system. DME ignition information, including testing of the reference and speed sensors, is covered in **280 Ignition System**.

NOTE —
- *For fuel pump and related fuel supply component testing and repair see* **201 Fuel Supply.**
- *Wiring diagrams for the DME engine management system can be found in* **970 Electrical Wiring Diagrams.**

Test equipment

CAUTION —
A high impedance digital multimeter should be used for all voltage and resistance tests. An LED test light or a "noid" should be used in place of an incandescent-type test lamp.

Special equipment is necessary for some of the procedures given in this repair group. If you do not have the equipment required to do the job, it is recommended that these repairs be left to an authorized Porsche dealer.

Warnings and cautions

Please observe the following cautions and warnings when performing any service or repair on the engine.

WARNING —
- *The fuel system is designed to retain pressure even when the ignition is off. When working with the fuel system, loosen the fuel lines very slowly to allow residual fuel pressure to dissipate gradually. Avoid spraying fuel.*
- *Fuel is highly flammable. When working around fuel, do not disconnect any wires that could cause electrical sparks. Do not smoke or work near heaters or other fire hazards. Keep an approved fire extinguisher handy.*
- *The ignition system produces high voltages that can be fatal. Avoid contact with exposed terminals. Use extreme caution when working on a car with the ignition switched on or the engine running.*
- *Do not touch or disconnect any high voltage cables from the coil, distributor, or spark plugs while the engine is running or being cranked by the starter.*

CAUTION —
- *Connect and disconnect the DME system wiring and test equipment leads only when the ignition is switched off.*
- *Before making any electrical tests that require the engine to be cranked using the starter, disable the ignition system as described in* **280 Ignition System**.
- *Wire colors may vary from those given in text and schematics. Always check component terminals to verify wire colors.*

Motronic engine management system

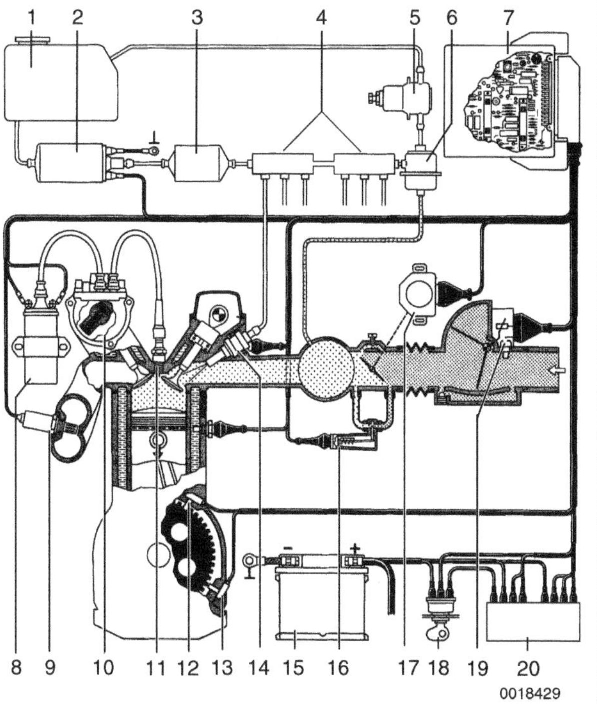

1. Fuel tank
2. Fuel pump
3. Fuel filter
4. Fuel rails
5. Fuel pressure regulator
6. Fuel damper
7. Engine control module (ECM)
8. Ignition coil
9. Oxygen sensor
10. Ignition distributor
11. Spark plug
12. Reference sensor
13. Crankshaft speed sensor
14. Fuel injector
15. Battery
16. Idle speed control valve
17. Throttle position switch
18. Ignition switch
19. Volume air mass sensor
20. Main/fuel pump dual relay

DME engine management, general

The 911 Carrera models covered in this manual are equipped with an electronic engine management system, controlling fuel delivery and ignition through single engine control module (ECM). This system is known as the Motronic system or Digital Motor Electronics (DME).

NOTE —

For an overview of the DME system, see ***200 Engine Management–General****.*

The fuel injection is completely electronic in operation. The DME engine control module (ECM) controls engine fuel delivery and ignition timing using signals from the following sources:

- Air flow sensor
- Intake air temperature (IAT) sensor
- Cylinder head temperature sensor
- Reference position (TDC) sensor
- Crankshaft speed (RPM) sensor
- Throttle position switches
- Oxygen sensor

Emission controls

The emission control system used in the 911 Carrera includes the oxygen sensor system and the evaporative emission system.

Oxygen sensor. This system provides the fuel injection system with feedback information about combustion efficiency. See **Oxygen sensor system**, given later.

Evaporative emission controls. This system provides venting for the fuel tank while at the same time trapping the fuel vapors that would otherwise be vented into the atmosphere. The evaporative emission system collects fuel vapors from the fuel tank in a charcoal canister. During certain engine operating conditions, the canister is purged, drawing the fuel vapors into the engine to be burned.

Basic engine settings

The DME system compensates automatically for changes in the engine due to age, minor wear or small vacuum leaks. As a result, idle speed and mixture do not need adjustment as part of routine maintenance. See **Table a**.

Table a. Idle and Emissions Specifications

Idle speed (1984 – 1986 models)	800 rpm	
Idle speed (1987 and later)	880 rpm	
%CO	0.6 – 1.0%	Measured at test port on front of catalytic converter, oxygen sensor disconnected
HC	<300 ppm	

NOTE —
Poor initial driveability may be encountered when the battery is disconnected and reconnected. When the battery is disconnected, the DME system adaptive memory is erased. The system will readapt after about ten minutes of driving.

Ignition timing

Ignition timing is electronically controlled and is not adjustable on DME systems. The initial ignition point is determined by crankshaft position during starting. This is signalled by the reference sensor. Once the engine is running, the ignition point is continually changed based on the various inputs to the ECM. Engine speed is signalled by the crankshaft speed sensor.

◀ If necessary, check ignition timing at the crankshaft pulley for test purposes using a strobe type timing light.

Ignition Timing Specification
- 800 rpm. . . . . 3.5° before TDC

0018751

Throttle plate basic adjustment

The factory-set throttle adjusting screw is not used to adjust idle speed. Its only function is to provide a mechanical stop for the linkage to prevent the throttle plate from contacting the inside of the throttle housing and causing wear. It should be adjusted only if the factory setting has been tampered with.

To correct faulty throttle adjustment:

- Remove tamper proof cap from throttle lever stop screw.
- Turn screw (**arrow**) counterclockwise until plate contacts throttle housing.
- Then turn screw clockwise just until throttle plate no longer contacts housing. Set clearance to specifications shown below using feeler gauge
- Apply paint or lacquer to screw threads. Install tamper-proof cap.

Throttle Plate Clearance

- Throttle plate to throttle housing . . . 0.04 mm (0.0015 in.)

FUEL DELIVERY TESTS

Checking fuel delivery is a fundamental part of troubleshooting and diagnosing the engine management system. Fuel pressure directly influences fuel delivery. An accurate fuel pressure gauge will be needed to make the tests.

Three significant fuel delivery values can be measured:

- **System pressure** is created by the fuel pump and maintained by the pressure regulator.
- **Fuel delivery volume** is created by the fuel pump and affected by restrictions, such as clogged fuel filter.
- **Residual pressure** is maintained in the closed system after the engine and fuel pump are shut off.

WARNING —
- *When working on an open fuel system, wear suitable hand protection, as prolonged contact with fuel can cause illnesses and skin disorders.*
- *Gasoline is highly flammable and its vapors are explosive. Do not smoke or work on a car near heaters or other fire hazards when diagnosing and repairing fuel system problems. Have a fire extinguisher available in case of an emergency.*

CAUTION —
- *Cleanliness is essential when working with fuel circuit components. Thoroughly clean the unions before disconnecting fuel lines.*
- *Renew fuel system hoses, clamps and O-rings any time they are removed.*

NOTE —
Fuel pump replacement is covered in **201 Fuel Supply**.

Relieving fuel pressure and connecting fuel pressure gauge

To prevent fuel from spraying on a hot engine, system pressure should be relieved before disconnecting fuel lines. One method is to tightly wrap a shop towel around a fuel line fitting and loosen or disconnect the fitting.

- Remove fuel pressure test cap (**arrow**) from end of left fuel rail.

> ***CAUTION —***
> *When removing cap be careful not to drop small sealing ball inside cap.*

- Connect fuel gauge (special tool 1318, available from Baum Tools Unlimited, or equivalent) to fuel rail and tighten to 12 Nm (9 ft-lb).

NOTE —
The fuel pressure gauge should have a range of 0 to 5 bar (0 to 75 psi).

Operating fuel pump for tests

- To operate fuel pump for testing purposes without having to run engine, main/fuel pump relay, located under driver seat, can be bypassed to power pump directly.

- Working below driver's seat, disconnect harness connector from main/fuel pump relay (**arrow**).

> ***CAUTION —***
> *Disconnecting power from the engine control module (ECM) erases its adaptive memory. It may be necessary to drive the car for approximately 10 minutes after reconnecting the battery to reset ECM adaptive memory and restore normal performance.*

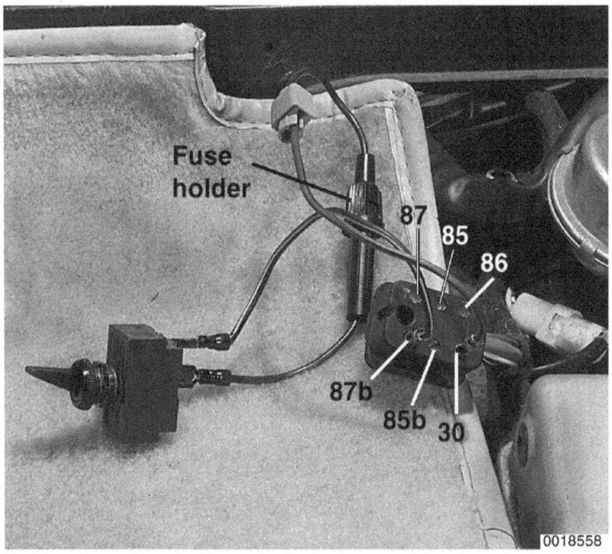

- Connect socket for relay terminal **30** to the socket for relay terminal **87b** with fused jumper wire.

- After completing tests, remove jumper wire.

> ***CAUTION —***
> *The main/fuel pump relay has a 4 mm² red wire at terminal 30 in the relay socket. Terminal 87b has a 2.5 mm² red/green wire. See* **970 Electrical Wiring Diagrams** *for additional wiring information.*

NOTE —
The jumper wire should be 1.5 mm² (14 gauge.) and include an in-line fuse holder with a 15 amp fuse. To avoid harness connector damage from repeated connecting and disconnecting, also include a toggle switch.

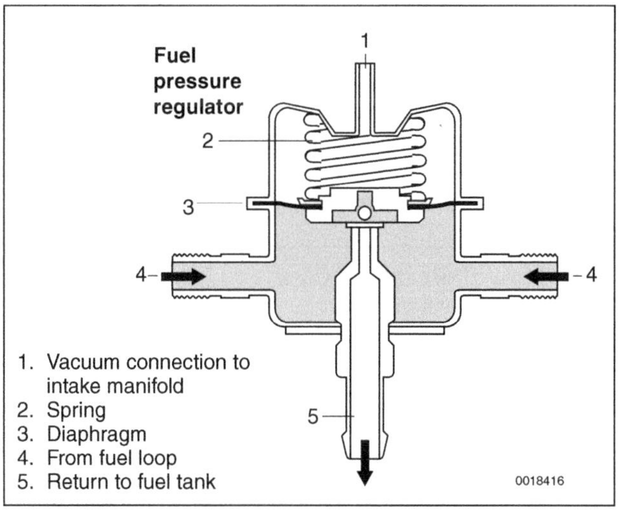

System pressure, testing

Fuel system pressure is created by the fuel pump and regulated by the pressure regulator. System pressure is not adjustable.

Relieve fuel pressure and connect pressure gauge to test port (**arrow**) as described earlier.

> ***CAUTION —***
> *The fuel pump is capable of developing a higher pressure than that regulated by the pressure regulator. If the fuel pump check valve malfunctions (stuck closed), make sure the fuel pressure does not rise above 6.0 bar (87 psi). Damage to the fuel lines or fuel system components could result.*

– Operate fuel pump without running engine. See **Operating fuel pump for tests**. Check that fuel pressure corresponds to specifications listed in **Table b**.

- If system pressure is low, repeat test while gradually pinching off return hose at rear of left fuel rail. Pressure should rise rapidly. If not, fuel pump is most likely faulty. See **201 Fuel Supply**.
- If system pressure is too high, check return line from pressure regulator to tank. Check for kinks in hose. Blow compressed air through line to check for blockages. If no faults can be found, replace pressure regulator as described later.

Table b. Fuel Pump Pressure Specifications

	Fuel pressure
Engine off	2.5 bar (36 psi)
Engine running Engine running, vacuum hose removed from pressure regulator	2.0 bar (29 psi) 2.3 - 2.7 bar (33 - 39 psi)

> ***CAUTION —***
> *Do not use compressed air above 40 psi to blow out lines or components. Internal damage to components may result.*

– Remove fuel pump relay jumper and reattach relay to harness connector. Start engine and recheck fuel pressure.
  - Remove vacuum hose from fuel pressure regulator. Recheck fuel pressure. See **Table b**.
  - If any faults are found, either fuel pressure regulator or hose (vacuum supply) is faulty and should be replaced.

◀ When finished testing, disconnect pressure gauge and reinstall fuel pressure test port cap with sealing ball.

Tightening Torque
- Fuel pressure test port cap. . . . . 12 Nm (9 ft-lb)

Fuel delivery volume, testing

◀ Disconnect return line from fuel rail in engine compartment.

> ***WARNING —***
> *To prevent fuel from spraying on a hot engine, system pressure should be relieved before disconnecting fuel lines. One method is to tightly wrap a shop towel around a fuel line fitting and loosen or disconnect the fitting*

– Connect a length of hose with female fitting (M14 x 1.5) to fitting on fuel rail and place open end of hose in a suitable container for catching fuel (2-quart capacity).

– Run fuel pump for exactly 30 seconds as described earlier under **Operating fuel pump for tests** and measure fuel collected. Compare to specification in **Table c**.

NOTE —
Delivery volume will vary depending on battery voltage. be sure the battery is fully charged for the most accurate test results.

– When finished testing, reconnect fuel line. Tighten all hose clamps.

Table c. Fuel Pump Delivery Volume

Delivery condition	Delivery rate
30 seconds @ 12V	0.850 liter (0.90 qt.)

Fuel pump replacement is covered in **201 Fuel Supply**.

Residual fuel pressure, testing

◀ For quick restarts and to avoid vapor lock, the fuel system is designed to retain fuel pressure after the engine has been turned off. This residual pressure is primarily maintained by a check valve (**arrow**) at the fuel pump outlet and by the fuel pressure regulator.

◀ Relieve fuel pressure and connect pressure gauge to test port (**arrow**) as described earlier.

– Operate fuel pump for approximately one minute as described under **Operating fuel pump for tests**.

– With pump off, observe fuel pressure gauge after 20 minutes. Compare to specification listed in **Table d**.

– If fuel system does not maintain specified pressure, check visually for leaks in fuel lines or at unions. Leaks can also be due to a leaking injector or a faulty fuel pump check valve.

Check fuel pump check valve by pinching off fuel return line (**arrow**) at end of fuel rail immediately after turning off pump.

- If pressure is now maintained, fault is most likely at fuel pump check valve.

– When finished testing, disconnect pressure gauge and reinstall fuel pressure test port cap with sealing ball.

Tightening Torque
- Fuel pressure test port cap . . . . 12 Nm (9 ft-lb)

Table d. Fuel Pump Residual Pressure

Fuel pump at rest	Minimum pressure
20 minutes	1 bar (14.5 psi)

NOTE —
- *The pump check valve is available as a replacement part. Fuel pump check valve replacement is covered in* **201 Fuel Supply**.
- *Fuel pressure regulator replacement is covered later in this repair group.*

ELECTRICAL CHECKS AND COMPONENT TESTING

CAUTION —
- *Use a high impedance digital multimeter for all voltage and resistance tests.*
- *An LED test light should be used in place of an incandescent-type test lamp.*
- *Avoid damaging harness connectors or relay panel sockets. Jumper wires should be fabricated with flat-blade end connectors that are the same size as the connector or relay terminal being tested.*
- *Always switch the test meter to the appropriate function and range before making test connections. Connect and disconnect the test equipment only with the ignition switched off.*
- *Wire colors may vary from those given in text and schematics. Always check component terminals to verify wire colors.*

Fuel injectors, testing

Pulsed type fuel injector

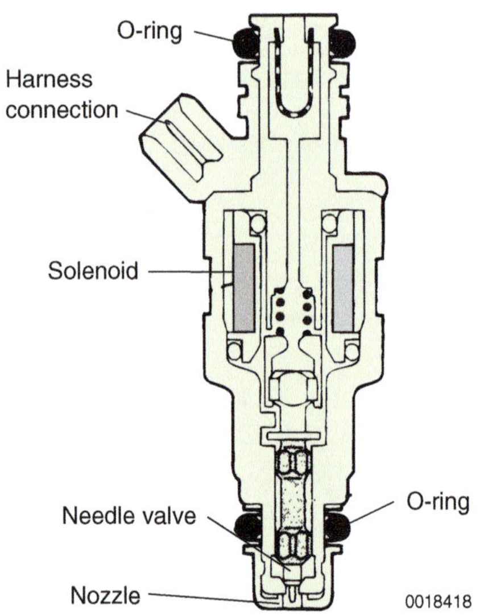

The injectors are turned on and off by the DME engine control module (ECM). The quantity of fuel delivered to the engine is determined by the amount of time the injectors remain open.

Injectors tests include electrical checks for resistance, power and ground signal from the ECM.

NOTE—

Injector flow tests require specialized equipment that opens each injector for a precise amount of time to measure fuel delivery. Most authorized Porsche dealers can perform an injector balance test to locate clogged or faulty injectors.

To quickly check if an injector is electrically functioning, start engine and place screwdriver or finger on injector body. A slight vibration or buzzing sound indicates that injector is functioning electrically.

With engine running, remove injector harness connectors one at a time.

- Engine RPM should drop each time one injector is disconnected.
- If disconnecting one injector does not affect engine RPM, continue testing that injector and harness connector.

– If any faults are found, turn engine off and disconnect electrical connector from suspected injector.

- Using an ohmmeter, check injector resistance. Compare to specification listed in **Table e**.

Table e. Fuel Injector Resistance Specification

Injector terminals	Resistance at 20°C (68°F)
Harness connector disconnected	2 to 3 Ω

– If injector resistance is within specifications, check that ECM signal is present at injector connector.

- Connect LED injector tester or back probe injector harness connector using digital voltmeter (select ms-pulse width range if available).
- Operate starter or run engine and check for pulsed voltage signal.
- Repeat for each injector.

– If no signal is present, check for power to injectors.

- There should be battery voltage (+) at red/white wire of each injector connector with ignition key on.
- If voltage is not present, check main/fuel pump relay output at terminal 87.

Motronic engine control module (ECM) pin assignments

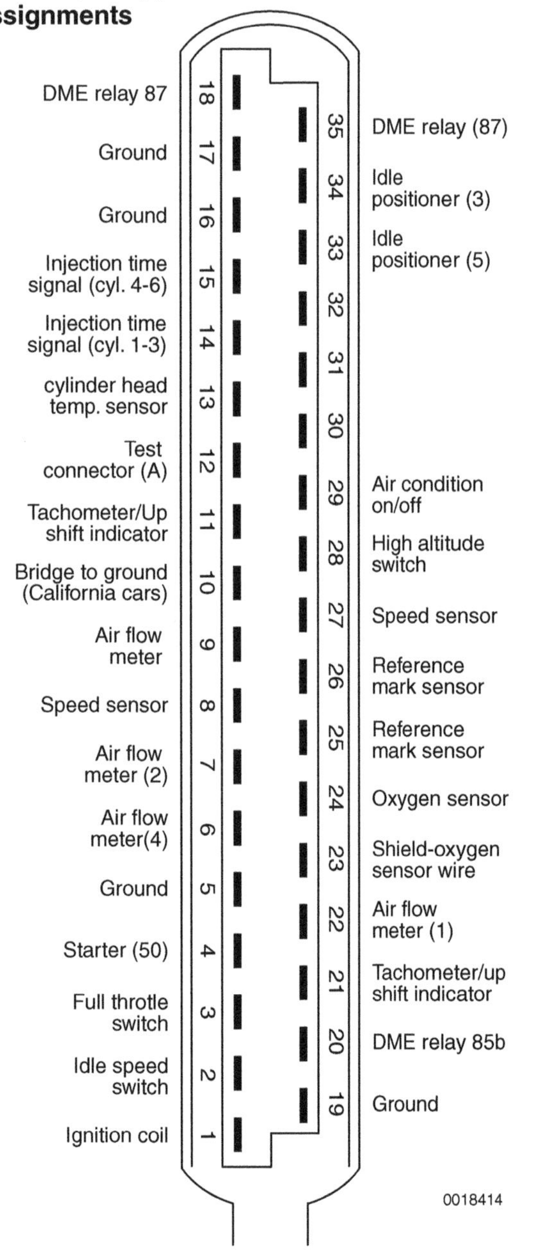

NOTE —
If there is positive (+) battery voltage at the injectors but no pulsed response at the voltmeter, check the wire(s) from the ECM to the injectors. If no wiring faults can be found, the pulsed ground signal from the control module may be missing. Check the outputs from the ECM.

If no wiring faults can be found after the above checks, the ECM may be faulty.

ECM harness electrical tests, as well as fuel injector replacement, is covered later in this repair group.

Engine control module (ECM)

When troubleshooting engine management problems, check the wiring and confirm that all input signals are correct before assuming the ECM is faulty.

ECM harness electrical tests are given in **Table f**, and ECM pin assignments are given in **Table g**. This information can be helpful when diagnosing faults to or from the ECM. If all inputs and wiring are OK but operational problems still exist, the ECM itself may be faulty.

Test results with incorrect values do not necessarily mean that a component is faulty. Check for loose, broken or corroded connections and wiring before replacing components. If the results are still incorrect, test the component itself.

ECM pin numbering is shown in illustration. When making checks at the ECM, a pinout test harness (sometimes referred to as a "breakout box") should be used to allow tests to be made with the connector attached to the ECM. As an alternative, the harness connector housing can be separated so that electrical tests can be made from the back of the connector with the connector still plugged in.

0018559a

Engine control module (ECM), accessing for tests

The DME engine control module (**C**) is located under the driver seat.

– Disconnect negative (–) battery cable. Wait approximately 1 minute.

> ***CAUTION —***
> *Prior to disconnecting the battery, read the battery disconnection cautions given at the front of this manual on page vii.*

– Remove driver's seat. See **720 Seats**.

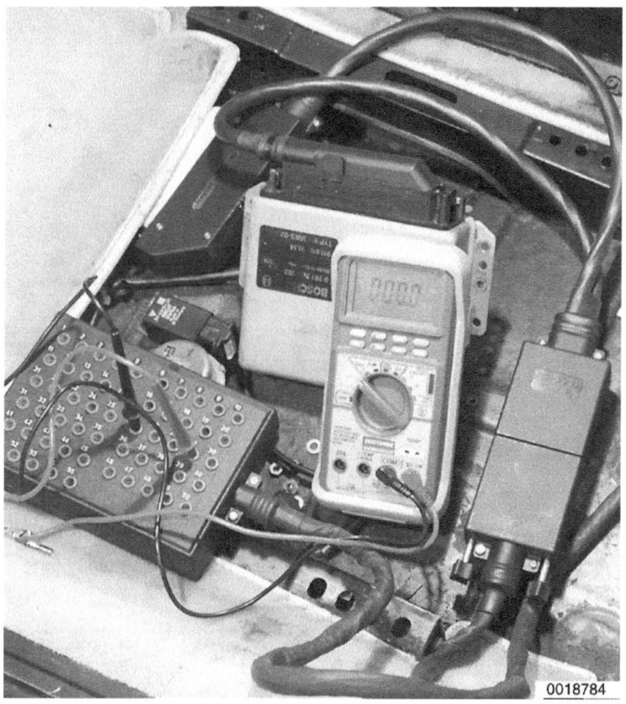
0018784

Disconnect ECM harness connector and connect pinout test harness to ECM for electrical tests of DME components.

Table f. DME Engine Control Module (ECM) Harness Electrical Tests

Component or circuit	Test terminals	Test conditions	Correct test value
Voltage supply to ECM	18 and ground 35 and ground	Ignition ON	Battery voltage (approximately 12 volts)
Main grounds	5 and ground 10 and ground 16 and ground 17 and ground 19 and ground	Ignition OFF	Continuity
Fuel pump relay control	20 and ground	Ignition ON	Fuel pump runs (audibly) when terminal 20 is touched to ground
Ignition coil	1 and ground	Key in start position	Pulse voltage signal
Starter input (terminal 50)	4 and ground	Actuate starter	8 volts minimum
Full throttle switch	3 and ground 3 and ground	Throttle closed or opened partially Throttle fully open	No continuity Continuity
Idle switch	2 and ground	Throttle closed Open throttle 1 mm	Continuity No continuity
Air temperature sensor	6 and 22	ignition OFF	Resistance varies with temp. 20°C (68°F) 2000 - 3000 ohms 80°C (176°F) 250 - 400 ohms
Idle speed control valve	34 and 33	Ignition OFF	40 Ω
Tachometer/up-shift light	11 and 21	Ignition OFF	Continuity
Crankshaft speed (rpm) sensor	8 and 27	Ignition OFF	0.6 to 1.6 kΩ
Reference (TDC) sensor	25 and 26	Ignition OFF	0.6 to 1.6 kΩ
Cylinder head temperature sensor	13 and ground	Ignition OFF	Resistance varies with temp.: 0°C (32°F): 4.4 - 6.8 kΩ 15° - 30°C (60° - 85°F): 1.4 - 3.6 kΩ 40°C (105°F): 1.0 - 1.3 kΩ 80°C (175°F): 250 - 390 Ω 100°C (212°F): 160 - 210 Ω 130°C (265°F): 90 Ω
Fuel injector control (injectors 1, 2, and 3	14 and ground	Ignition ON	Fuel injectors must click when terminal 14 is touched to ground
Fuel injector control (injectors 4, 5, and 6	15 and ground	Ignition ON	Fuel injectors must click when terminal 15 is touched to ground
Volume air flow sensor	7 and 9	Move sensor vane by hand or actuate starter	Resistance must fluctuate
Volume air flow sensor	22 and ground	Ignition ON	Variable voltage signal (0 - 5 volts)
High altitude switch	28 and ground	Ignition OFF	Continuity
Oxygen sensor	24 to ground	Connect green oxygen sensor wire to ground	Continuity
Air conditioner ON signal	29 and ground	Ignition ON A/C switch ON	Battery voltage

Table g. DME Engine Control Module (ECM) Pin Assignments

Pin	Signal	Component/function	Signal
1	output	Ignition coil	Primary signal, ignition coil
2	input	Idle switch	Ground with throttle closed
3	input	Full throttle switch	Ground with throttle fully open
4	input	Starter terminal 50	Battery voltage (+) during cranking
5		Ground	Shielding for reference sensor
6	input	Volume air flow sensor	Intake air temperature (IAT) sensor signal
7	input	Volume air flow sensor	Air flow sensor potentiometer
8	input	Crankshaft speed (rpm) sensor	Crankshaft speed sensor (-)
9	output	Volume air flow sensor	Reference signal (5 volt s) to volume air flow sensor potentiometer
10		Bridge to ground (California cars only)	DME ground to intake manifold runner
11		Tachometer/up-shift indicator	Terminal 1 up-shift indicator light
12		Test connector A	Terminal A in test connector
13	output	Cylinder head temperature sensor	Reference signal (5 volts) to cylinder head temperature sensor
14	output	Injection time signal	Ground pulse with engine running (injectors 1, 2, and 3)
15	output	Injection time signal	Ground pulse with engine running (injectors 4, 5, and 6)
16		Ground	DME ground to intake manifold runner
17		Ground	DME ground to intake manifold runner
18	input	DME main/fuel pump relay terminal 87	Battery voltage (+) from main/fuel pump relay to ECM
19		Ground	DME ground connection on # 1 intake manifold runner
20	output	DME relay terminal 85b	Ground signal to terminal 85b of main/fuel pump relay
21	output	Tachometer/up-shift indicator	Up-shift indicator light (term. 2), where applicable
22	output	Volume air flow sensor (terminal 1)	Variable voltage signal (0-5 volts) when vane door moves
23		Oxygen sensor wire shield	Shielded ground for oxygen sensor and speed sensor
24	input	Oxygen sensor	Variable voltage signal with engine running (0-1 volt)
25	output	Reference sensor	Reference sensor (-)
26	input	Reference sensor	Reference sensor (+), voltage signal (variable) with engine cranking
27	input	Crankshaft speed (rpm) sensor	Crankshaft speed sensor (+), voltage variable with engine cranking
28	input	High altitude switch	Ground signal from altitude switch when switch is closed
29	input	Air conditioner on/off	A/C temperature switch
30		Vacant	Open
31		Vacant	Open
32		Vacant	Open
33	output	Idle speed control valve	Idle speed control valve (open signal)
34	output	idle speed control valve	Idle speed control valve (closed signal)
35	input	Main/fuel pump relay terminal 87	Battery voltage (+) from terminal 87 main/fuel pump relay

Main/fuel pump relay

The main/fuel pump relay (**arrow**) is mounted under the driver's seat next to the engine control module (ECM). If the main/fuel pump relay is not functioning correctly, the engine will not run. Complete testing of the main/fuel pump circuit is covered in **201 Fuel Supply**.

Volume air flow sensor, checking

The amount of air entering the engine is electrically measured by the volume air flow (VAF) sensor (**arrow**). When the engine is running, intake air is drawn past the sensor vane and opens the spring-tensioned vane.

The vane is mechanically connected to a potentiometer inside the sensor housing. A reference voltage is sent through the air flow sensor potentiometer and converted into an air volume signal based on the position of the vane.

The VAF sensor also contains an integral intake air temperature (IAT) sensor. The IAT sensor is used to modify/correct the intake air signal from the air flow sensor based on the temperature of the intake air.

– Disconnect harness connector (**arrow**) from air flow sensor.

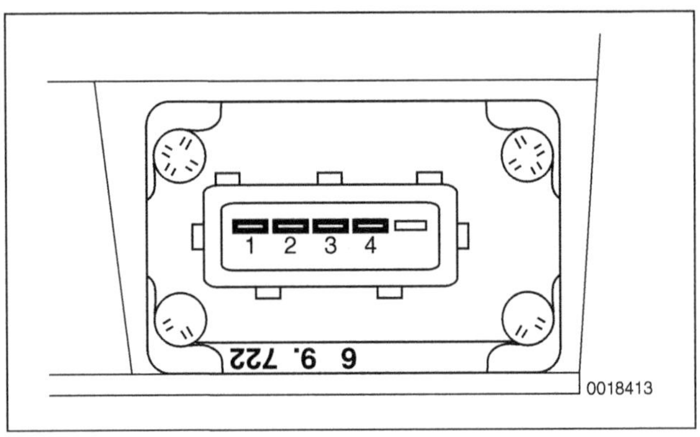

With ignition on, use a digital voltmeter to test for voltage at terminal **3** in connector. Compare against specification in **Table h**.

– Turn ignition off and remove intake air boot. Connect a digital ohmmeter across sensor terminals **2** and **3**.

- Move air flow sensor vane through its range of travel.
- Compare readings to specifications in **Table h**.
- Replacement procedure is covered later in this group.

Table h. Volume Air Flow Sensor Checks

Test	Test value
Terminal **3** and ground	approx. 5 volts
Terminals **2** and **3**, move vane through full range of travel	Resistance should change without interruption

NOTE —
The sensor's operation depends upon the free movement of the vane inside the main air passage. Check the vane by moving it through its full range of travel. Check that the vane returns to the closed position smoothly. If the vane binds at any point, remove the sensor as described later to check for problems. If the vane cannot be made to move freely, it should be replaced.

0018770

Intake air temperature (IAT) sensor, testing

– Disconnect harness connector from air flow sensor.

< Connect ohmmeter across terminals **1** and **4** of volume air flow sensor connector. Compare readings to specifications in **Table i**.

Table i. Intake Air Temperature (IAT) Sensor

Air temperature	Resistance in kΩ
0°C (32°F)	4.4 – 6.8
15° – 30°C (60 – 85°F)	1.4 – 3.6
40°C (105°F)	1.0 – 1.3

NOTE —
The IAT sensor is integrated into the air flow sensor housing and is not a separately available replacement part.

0018760

Cylinder head temperature sensor, testing

< The cylinder head temperature sensor (**arrow**) is installed in cylinder head 3. The cylinder head temperature sensor signal is the main temperature signal to the ECM.

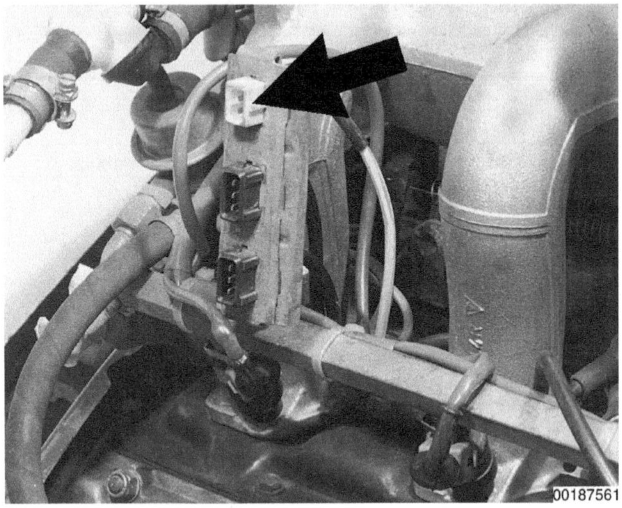

To test sensor:

- Disconnect harness plug from white connector on left side of engine (**arrow**).
- Connect ohmmeter across white connector terminals.
- Resistance should correspond to specifications listed in **Table j**.
- If any faults are found, replace sensor. Replacement procedure is covered later in this group.

NOTE —
An open (break) in the temperature sensor will cause the DME system to run rich. A short circuit in the temperature sensor causes leaner mixtures.

Table j. Cylinder Head Temperature Sensor Resistance Specifications

Test temperature	Resistance
0°C (32°F)	4.4 – 6.8 kΩ
15 – 30 °C (60 – 85°F)	1.4 – 3.6 kΩ
40°C (105°F)	1.0 – 1.3 kΩ
80°C (175°F)	250 – 390 Ω
100°C (212°F)	160 – 210 Ω
130°C (265°F)	90 Ω

Throttle position switches

The full throttle switch is used for full throttle enrichment. The switch signals the ECM when the throttle is fully opened.

A separate idle switch is used for the closed throttle signal. The closed throttle signal is also used for idle speed control and deceleration fuel cutoff.

Both throttle position switches are mounted on the throttle housing and can be tested with the throttle housing installed.

– Before testing full throttle or idle switch, check that each is receiving reference voltage signal (5 VDC) from ECM.

- Disconnect harness plugs. With ignition on, there should be 5 VDC between plug terminals.
- If voltage is not present, test ECM harness output as described earlier.

Full throttle switch, testing and adjusting

– With engine off, remove harness connector from switch.

< Connect ohmmeter leads to switch terminals **3** and **18**.

- Open throttle to full position. There should be continuity with throttle fully open and no continuity in all other positions.

NOTE —
*Terminal **2** is inactive.*

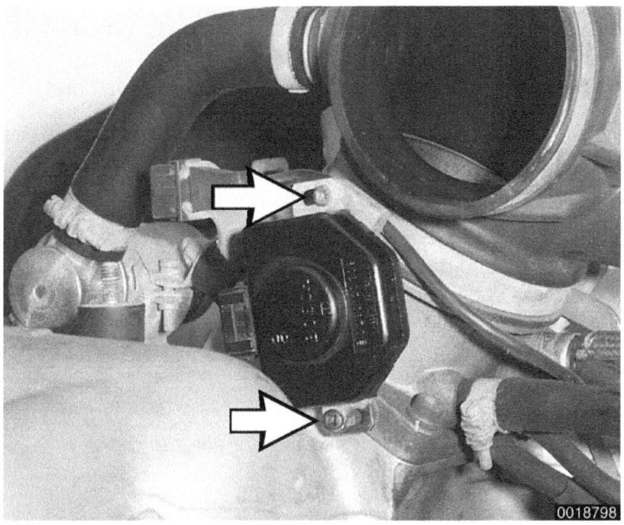

< If test results are incorrect, try adjusting switch.

- Loosen switch mounting screws (**arrows**).
- Connect ohmmeter leads to switch terminals. Hold throttle fully open.
- Twist switch body clockwise until ohmmeter shows continuity.
- Tighten screws and seal with paint or lacquer.

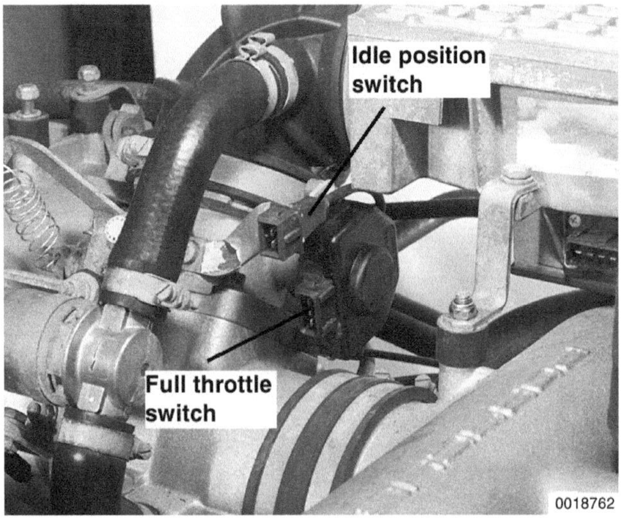

Idle position switch, testing

Prior to testing or adjusting idle switch, make sure that the throttle position is set according to instructions in **Throttle plate basic adjustment**, given earlier.

> ***NOTE —***
> *A defective idle switch (open circuit) will cause high idle speed (approximately 1,200 rpm) and advanced ignition timing (12° before TDC).*

– To test idle switch mechanical setting:

- Attach strobe type timing light according to manufacturer's instructions.
- Start engine and allow to idle.
- Open throttle plate approximately 1 mm while watching idle microswitch contact lever.
- Throttle switch should open before throttle plate begins to open.
- If switch is working idle speed will increase approximately 500 rpm and ignition timing will advance to 12° before TDC.

To test idle switch electrically:

- Engine should be stopped. Remove harness connector from switch.
- Connect ohmmeter leads to switch terminals (**42** and **44**).
- Open throttle approximately 1 mm.
- There should be continuity with throttle fully closed and no continuity in all other positions.

– To test idle switch harness and connector:

- Start engine and allow to idle.
- Disconnect idle switch harness connector.
- Engine speed must increase approximately 500 rpm.
- If engine speed does not increase check wiring to ECM.
- If no problems are found replace ECM.

– To test decel fuel shut-off function:

- Start engine and allow to idle.
- Disconnect idle switch harness connector and bridge harness terminals.
- Raise engine rpm to 1,300.
- Engine should start surging, indicating injectors are turning off. If not, check wiring between switch and ECM.
- If no problem is found with switch or wiring, ECM may be faulty.

Idle position switch, adjusting

The throttle linkage moves a drag arm approximately 1 mm before the throttle plate starts to open. When the drag arm has been moved approximately 1 mm, a click will be heard from the idle microswitch. This indicates that the switch has opened, interrupting power to the idle speed control valve.

Although the idle position switch itself needs no adjustment, the throttle valve drag arm clearance must be set in order to not damage the idle switch upon rapid closing of the throttle.

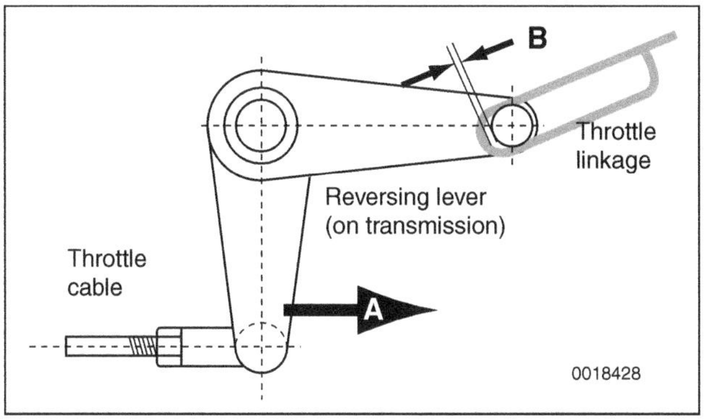

To adjust switch, push throttle cable reversing lever (on transmission housing) in direction **A**.

- Throttle linkage to reversing lever clearance (**B**) should be 1 – 2 mm (0.04 – 0.08 in.)

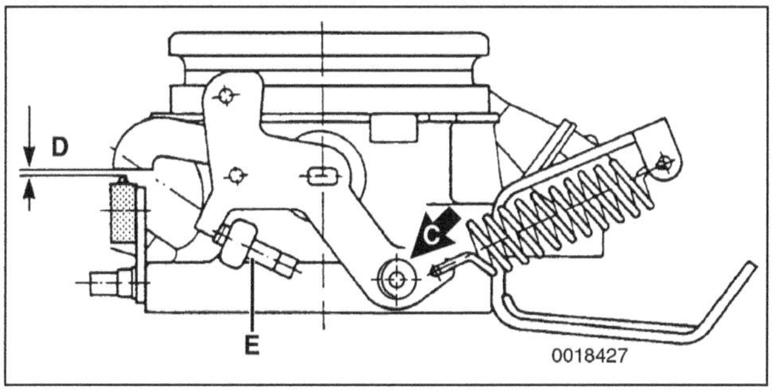

Working at throttle housing:

- Push drag arm in direction **C**.
- Use set screw (**E**) to adjust drag arm clearance (**D**) to idle switch housing.

Throttle Linkage Clearance Specifications

- Throttle linkage to reversing lever (**B**). . . . . 1 – 2 mm (0.04 – 0.08 in.)
- Drag arm to idle switch housing (**D**) . . 0.2 mm (0.008 in.)

Idle speed control valve, testing

The idle speed control valve controls the amount of air allowed to bypass the throttle to increase or decrease idle speed. Whenever the idle position switch is closed, the valve is in operation.

– To quick-check valve:
  - Start engine.
  - Listen and feel for valve vibration or humming.
  - If valve is not operating, check that idle switch is operating correctly as described above.
  - If valve is functioning but idle speed is erratic, check valve as described below.

Remove two hoses (**A**) and electrical harness connector (**C**) from valve. Remove mounting nuts (**B**). Remove idle speed control valve.

- Check valve piston for free movement by moving/shaking removed valve abruptly. Piston should slide freely.
- Test resistance values in valve electrical components. Compare to specifications below.
- If no faults are found, reinstall valve and reconnect wiring harness connector.

Idle Speed Control Valve Resistance

- Between outer terminals . . . . 40 Ω
- Between center terminal and each outer terminal . . . . 20 Ω

– Start engine and allow it to reach operating temperature. Then turn off engine and all electrical accessories.
  - Remove top hose from valve so that piston can be seen.
  - Turn ignition on. Valve piston should move approximately halfway across valve opening.

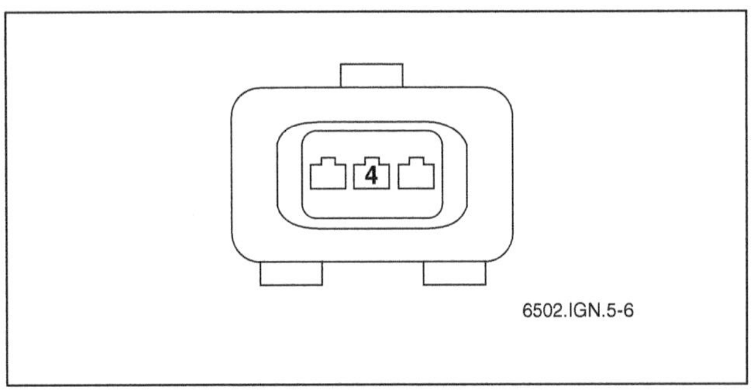

If idle speed is not within specifications and no faults are detected with idle speed control valve, or if valve piston did not move correct distance in previous test, test signal from ECM.

- With ignition off, disconnect electrical harness connector from idle valve.
- With ignition on, there should be battery voltage at terminal **4** (center terminal).
- If voltage is not present, test ECM harness inputs/outputs as described later.

Altitude correction switch

Due to lower air pressure, the engine mixture will be richer at higher altitudes. The function of the altitude switch is to regulate fuel mixture at higher elevations. The altitude switch closes at 1,000 meters (3,280 feet) above sea level. This leans the fuel mixture by 0.5% to 1.0%.

The altitude correction switch (**B**) is located under the driver seat.

- To test altitude switch, remove driver seat as described in **720 Seats**.
- Connect ohmmeter across switch terminals. There should be no continuity. If testing above 1,000 meters (3,280 ft.), there should be continuity.
- With engine running at operating temperature, check CO level.
- Remove harness plug from altitude correction box. CO value should not change. Bridge harness terminals. CO should drop 0.5% - 1.0%.

Evaporative emission control system

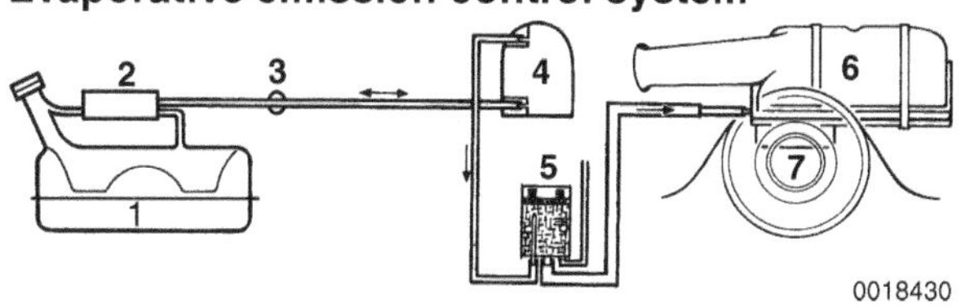

1. Fuel tank
2. Expansion chamber
3. Restrictor
4. Evaporation chamber
5. Carbon canister
6. Air filter housing
7. Engine fan

Evaporative emission control system

The charcoal canister collects and stores fuel vapors from the tank when the engine is idling or stopped. During normal driving, the fuel vapors are drawn into the engine and burned.

A faulty evaporative emission control system can cause hard warm starting, erratic idle, or poor acceleration if the fuel vapors are drawn into the engine at the wrong time. A restrictor prevents rapid purging of the charcoal canister.

NOTE —

A faulty evaporative emission control system usually affects driveability only when the engine is warm and/or outside temperatures are high, such as during prolonged idling on a hot day. The evaporative emission system is not normally the cause of cold running problems.

Zirconia oxygen sensor

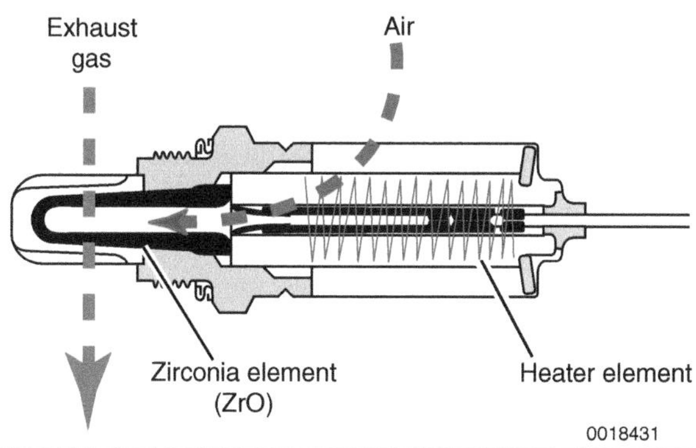

Oxygen sensor system

The oxygen sensor system provides the ECM with feedback information about combustion efficiency. The signal is used to continuously control the air-fuel mixture.

One surface of the exhaust-mounted oxygen sensor is exposed to the exhaust gas, while the other surface is exposed to the atmosphere. The difference in oxygen content between the two surfaces causes a chemical reaction that generates a low voltage electrical signal (100 - 900 mV).

The low voltage signal of the oxygen sensor is not reliably produced until the sensor reaches a temperature of approximately 300°C (600°F). For this reason, a separate heater circuit is used to bring the oxygen sensor up to operating temperature rapidly after a cold start.

The heated three-wire oxygen sensor (**arrow**) is mounted in the inlet end of the catalytic converter. The heater circuit gets power from terminal 87b of the main/fuel pump relay and is unfused.

> ***CAUTION —***
> - *Lubricating sprays containing silicone can contaminate the oxygen sensor.*
> - *Handle the oxygen sensor with care. Hitting or dropping will damage it.*

Oxygen sensor, checking

Basic checking of Oxygen sensor efficiency using a pinout test harness (Porsche special tool 9543 or equivalent) is described below.

> ***WARNING —***
> *Exhaust system parts can be hot enough to cause serious burns. When working near hot pipes or mufflers, use heavy gloves and other appropriate protection.*

NOTE —
For the most accurate test results, the engine must be fully warmed up, the exhaust system must be free of leaks, and all electrical consumers must be off.

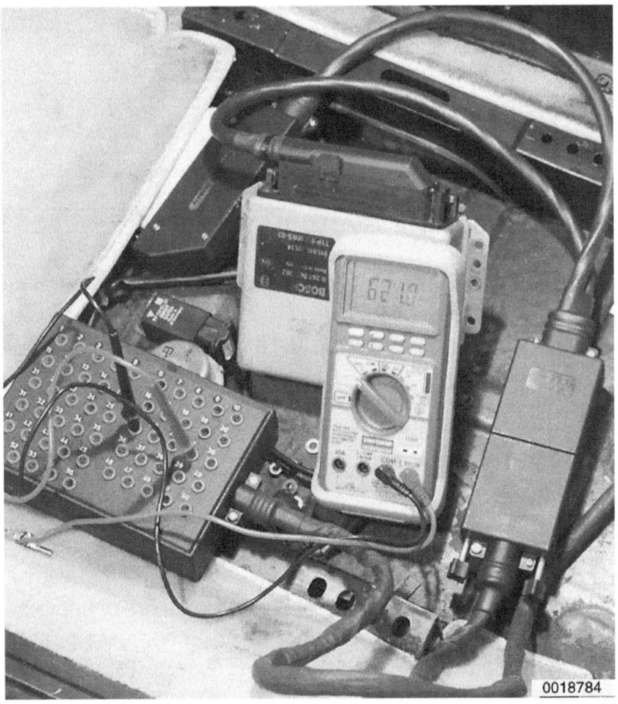

Disconnect ECM harness connector and connect pinout test harness to ECM under driver's seat.

- Connect a digital voltmeter (millivolt scale) to terminals 24 (Oxygen sensor signal) and 16 or 17 (ground).

– Start engine and let it idle. After a maximum of two minutes oxygen sensor output voltage should begin fluctuating.

Oxygen Sensor Voltage Output
- Engine warm. . . . . 0.1–0.9 VDC fluctuating

– With engine running, loosen oil filler cap to admit unmeasured intake air and simulate a lean running condition.

- Oxygen sensor output voltage should drop.

If voltage signal does not fluctuate or if there is no change when oil filler cap is removed, check for a faulty sensor wire in harness or connector (**arrow**) in engine compartment. If no wiring faults can be found, sensor is probably faulty and should be replaced.

NOTE —

Oxygen sensor replacement is a maintenance procedure scheduled at a specified time or mileage interval set by Porsche. See **030 Maintenance.**

Oxygen sensor heater circuit, testing

Check oxygen sensor heater circuit at harness side of oxygen sensor connector (**inset**).

- Attach digital multimeter set on volts to harness terminals **2** and **3**.

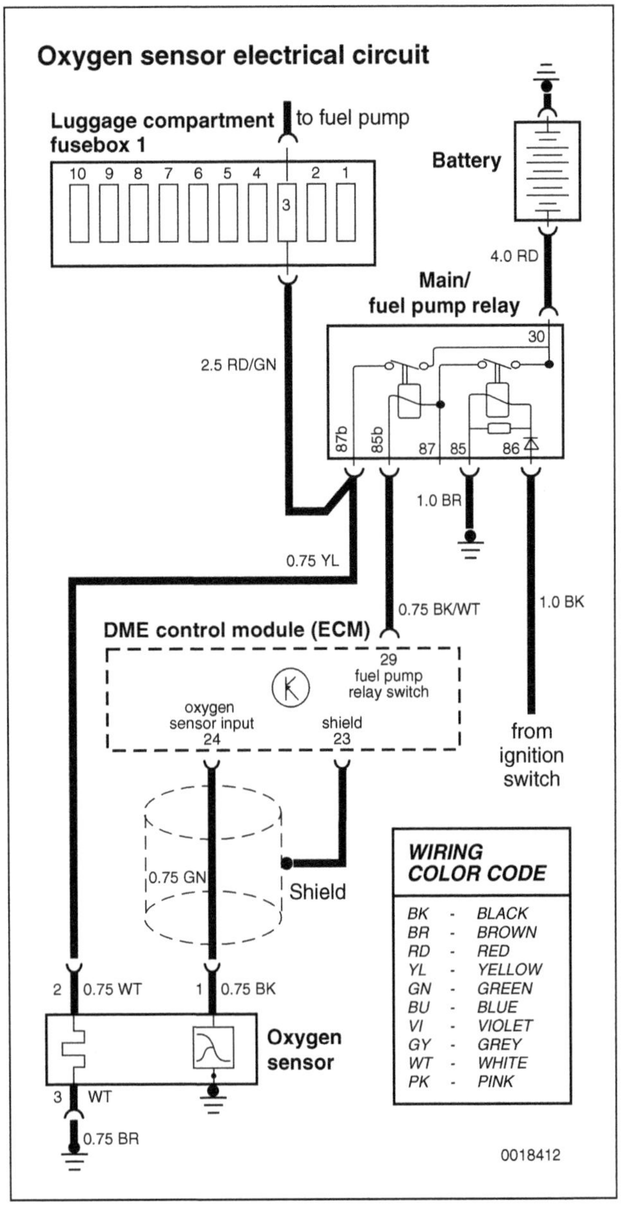

◄ With engine running there should be battery voltage (12 volts) to sensor heater.

- If not, check output at terminal **87b** of main/fuel pump relay.
- Check oxygen sensor harness ground wire for continuity to ground.

NOTE —

Wire colors leading to the sensor plug are different from wire colors leading to the sensor itself.

Table k. Oxygen Sensor Wire Colors

Function	Harness	Sensor
Signal	Green	Black
Heater	Yellow	White
Ground	Brown	White

COMPONENT REPLACEMENT

Main/fuel pump relay, replacing

- Remove driver seat as described in **720 Seats**.
- Detach wiring harness to main relay (**arrow**).
- Remove one M6 mounting nut and remove relay.
- Installation is reverse of removal.

Volume air flow sensor, replacing

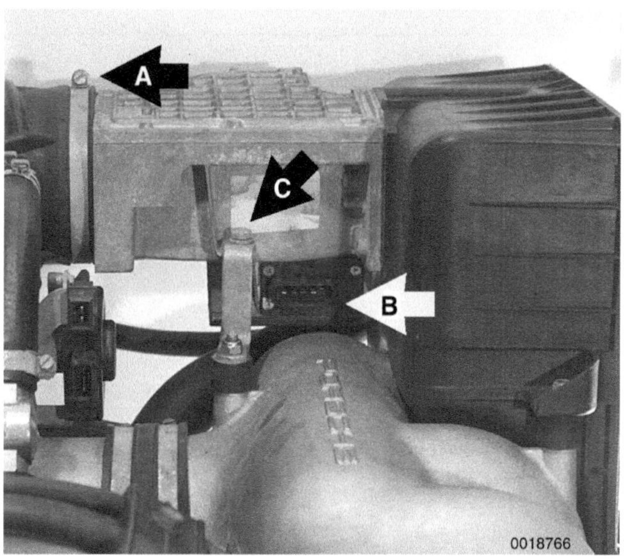

- Loosen large hose clamp (**A**) holding intake rubber boot to volume air flow sensor.
- Disconnect wiring harness (**B**) from air flow sensor.
- Remove mounting bolt (**C**) from air flow sensor body.

Unclip (**arrows**) air filter housing cover. Remove air filter cover and element.

Remove air flow sensor mounting bolts (**arrows**) from inside air filter housing.

– Slide volume air flow sensor out.

– Installation is reverse of removal.

Tightening Torques

- Volume air flow sensor to intake manifold . . . . 8 Nm (6 ft-lb)
- Air filter housing to air flow sensor . . . . 8 Nm (6 ft-lb)

Cylinder head temperature sensor, replacing

- Raise car and support in a safe manner.

> ***WARNING —***
> *Make sure the car is stable and well supported at all times. Use a professional automotive lift or jack stands designed for the purpose. A floor jack is not adequate support.*

0018769

◄ Pull out connector locking plate at sensor connectors in left front of engine compartment.

- Disconnect cylinder head temperature sensor connector (**white connector**).

- Working underneath engine remove rubber grommet at left rear (flywheel end) engine shroud.

> ***WARNING —***
> *Allow the engine to cool fully before working near the exhaust system or at the cylinder heads.*

0018760

◄ Using Porsche special tool 9222 or equivalent (a deep socket with a slit cut in the side), remove temperature sensor (**arrow**) from cylinder head 3.

- Pull wire with temperature sensor attached through rubber grommet opening.

- Installation is reverse of removal.

NOTE —
Before installing, apply a light coat of Molykote® HTP to the threads.

Tightening Torque

- Cylinder head temperature sensor to cylinder head . . . . 15 Nm (11 ft-lb)

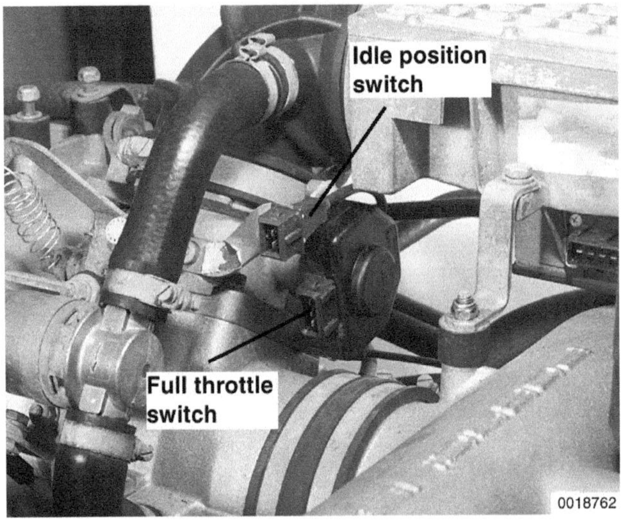

Full throttle switch, replacing

The full throttle and the idle switch are both attached to the throttle housing.

◀ Disconnect throttle switch wiring harness connector from throttle switch.

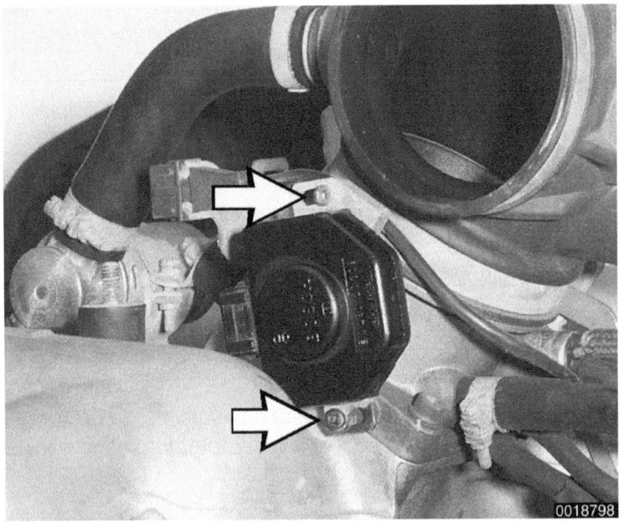

◀ Remove switch mounting screws (**arrows**).

– Pull switch off side of throttle body.

– Installation is reverse of removal.

NOTE —
After replacing idle switch, check idle switch adjustment as described earlier in **Full throttle switch, testing and adjusting**.

Idle speed control valve, replacing

◀ Remove two hoses (**A**) and electrical harness connector from valve (**C**). Remove mounting nuts (**B**).

– Remove idle air control valve.

– Installation is reverse of removal.

Fuel injectors, replacing

> ***WARNING —***
> *Gasoline is highly flammable and its vapors are explosive. Do not smoke or work on a car near heaters or other fire hazards when diagnosing and repairing fuel system problems. Have a fire extinguisher available in case of an emergency.*

NOTE —
Always use new injector sealing kits when servicing injectors.

Left side injectors, replacing

Remove large plastic elbow tube between heater blower fan on top of engine and heat exchanger feed tube.

- Remove bolts holding tube to blower housing, in front and at rear of blower housing (**arrows**).
- Loosen hose clamp at heat exchanger end (bottom of elbow) and slide elbow out.

Remove left fuel rail and injectors:

- Disconnect left fuel rail supply and return lines (**A**).

> ***CAUTION —***
> *Have a shop towel ready to soak up spilled fuel.*

- Cut wire ties and disconnect injector harness connectors 1, 2 and 3 (**B**).
- Remove fuel rail mounting bolts (**C**) from intake runners 1 and 3.
- Carefully pull up on fuel rail and remove with injectors.

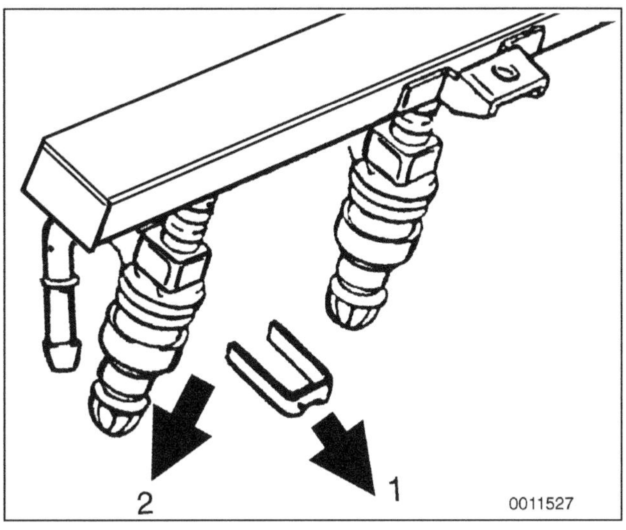

To remove an injector:

- Remove locking clip holding injector to fuel rail (**1**).
- Pull injector from rail (**2**).

– Installation is reverse of removal.

Tightening Torque
- Fuel rail to intake runners . . . . 7.5 Nm (5.5 ft-lb)

Right side injectors, replacing

- Unclip (**arrows**) air filter housing cover. Remove air filter cover and element.

- Remove right fuel rail and injectors:
  - Disconnect both fuel lines, one at rear of fuel rail, one at front (**A**).

> ***WARNING —***
> *Have a shop towel ready to soak up spilled fuel.*

  - Cut wire ties and disconnect injector harness connectors 4, 5 and 6 (**B**).
  - Remove fuel rail mounting bolts (**C**) from intake runners 4 and 6.
  - Carefully pull up on injector rail and remove with injectors.

– Remove locking clips holding injectors to fuel rail and remove injectors from rail.

– Installation is reverse of removal.

Tightening Torque
- Fuel rail to intake runners . . . . 7.5 Nm (5.5 ft-lb)

Fuel pressure regulator, replacing

The fuel pressure regulator is located at the rear (flywheel end) left of the engine, at the end of the left fuel rail.

> ***WARNING —***
> *Gasoline is highly flammable and its vapors are explosive. Do not smoke or work on a car near heaters or other fire hazards when diagnosing and repairing fuel system problems. Have a fire extinguisher available in case of an emergency.*

◄ Remove fuel pressure regulator vacuum line (**A**).

- Remove both fuel lines (**B**).

> ***WARNING —***
> *Have shop towels ready to soak up fuel.*

- Loosen fuel line clamp (**C**) at bottom of regulator and remove line.
- Remove 24 mm nut (**D**) at base of regulator housing and remove regulator.

– Installation is reverse of removal.

Fuel pressure damper, replacing

– The fuel pressure damper is located at the rear (flywheel end) right of the engine, at the end of the right fuel rail.

◄ Remove fuel pressure damper vacuum line (**A**).

- Remove both fuel lines (**B**).

> ***WARNING —***
> *Have shop towels ready to soak up fuel.*

- Remove M6 mounting bolt (**C**) and remove regulator.

– Installation is reverse of removal.

FUEL SYSTEM ELECTRICAL CIRCUITS

The electrical information and schematics provided here can be used as guide in diagnosing fuel system electrical problems.

The entire electrical system of the car is covered in **970 Electrical Wiring Diagrams** and **971 Electrical Component Locations**.

Ground locations

Tight and corrosion free ground connections, including the negative terminal of the battery and the ground straps, are essential to trouble-free operation.

Ground locations for the DME engine management system are shown in **Table I**.

Battery ground location **G104** (**arrow**).

Engine-body-transmission ground strap **G302** (**arrow**).

Left rear engine compartment ground location **G402** (**arrow**) in front of fuel filter. (Fuel filter removed for photo.)

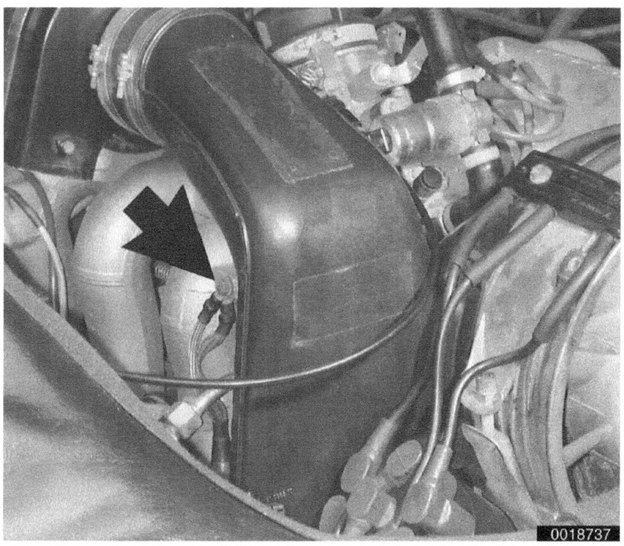

Intake runner 1 ground location **G407** (**arrow**).

Table I. DME System Ground Locations

Ground	Location	Connections
G104	In left front of luggage compartment	Battery ground Oil cooler fan Windshield washer system Headlight washer system Dual horns
G302	Engine-body-transmission ground strap	Engine Body Transmission
G402	In left rear of engine compartment	Rear window wiper motor Rear window defogger Heater system Oil tank level sender
G407	On intake runner 1	Throttle valve switch

Wiring schematics

The schematic diagrams on the following pages cover the DME fuel injection electrical system. DME ignition system wiring is shown in **280 Ignition System**.

NOTE —
The DME wiring diagrams shown are based on model year 1987. Wiring colors may vary slightly for other model year vehicles.

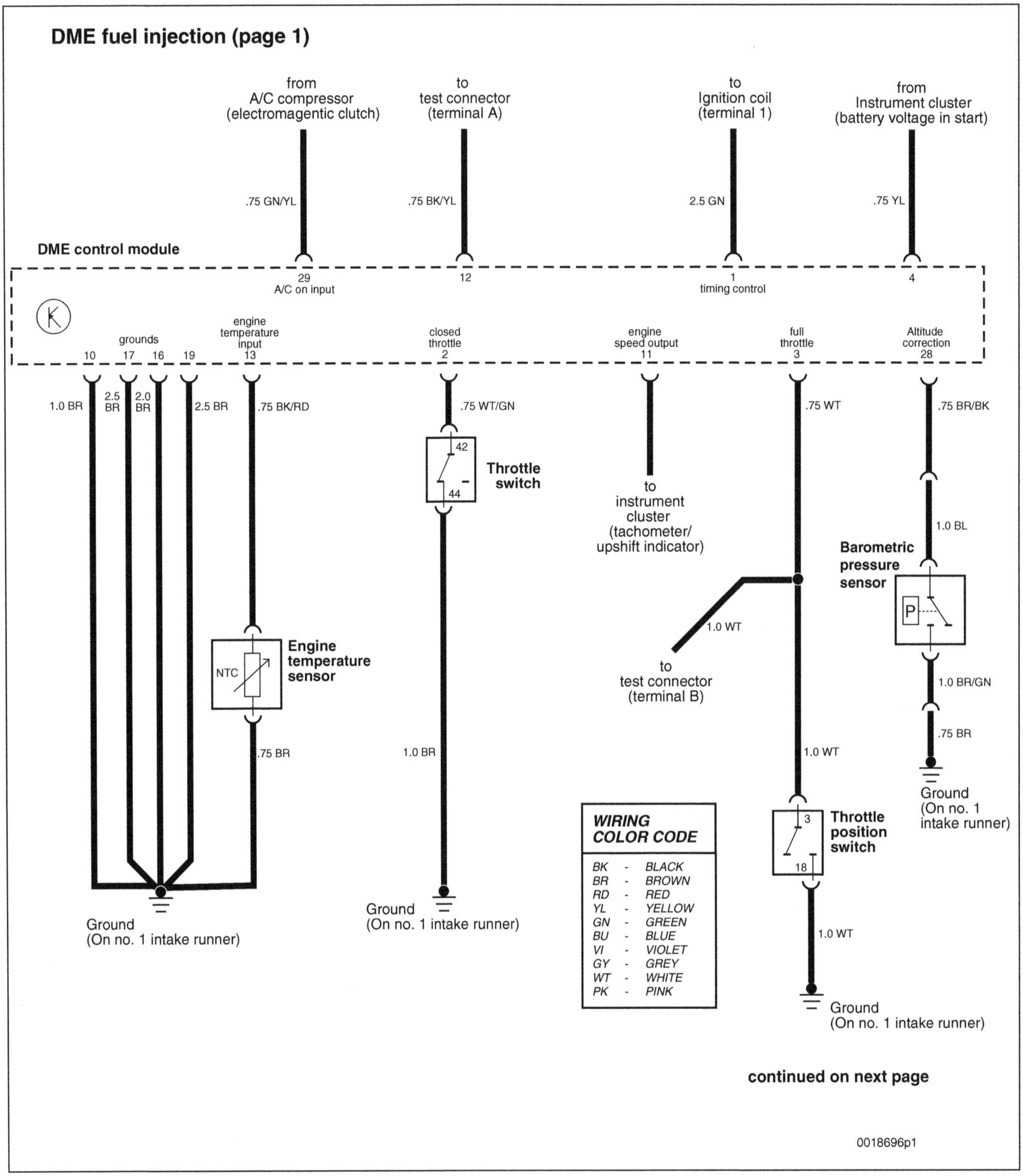
DME fuel injection (page 1)
from A/C compressor (electromagentic clutch)
to test connector (terminal A)
to Ignition coil (terminal 1)
from Instrument cluster (battery voltage in start)
.75 GN/YL
.75 BK/YL
2.5 GN
.75 YL
DME control module
29 A/C on input
12
1 timing control
4
grounds 10 17 16 19
engine temperature input 13
closed throttle 2
engine speed output 11
full throttle 3
Altitude correction 28
1.0 BR
2.5 BR
2.0 BR
2.5 BR
.75 BK/RD
.75 WT/GN
.75 WT
.75 BR/BK
Throttle switch
42
44
to instrument cluster (tachometer/ upshift indicator)
1.0 BL
Barometric pressure sensor
P
1.0 WT
to test connector (terminal B)
Engine temperature sensor
NTC
1.0 BR/GN
.75 BR
.75 BR
1.0 BR
1.0 WT
Ground (On no. 1 intake runner)
Throttle position switch
3
18
1.0 WT
Ground (On no. 1 intake runner)
Ground (On no. 1 intake runner)
Ground (On no. 1 intake runner)
WIRING COLOR CODE
BK - BLACK
BR - BROWN
RD - RED
YL - YELLOW
GN - GREEN
BU - BLUE
VI - VIOLET
GY - GREY
WT - WHITE
PK - PINK
continued on next page
0018696p1

DME fuel injection (page 2)

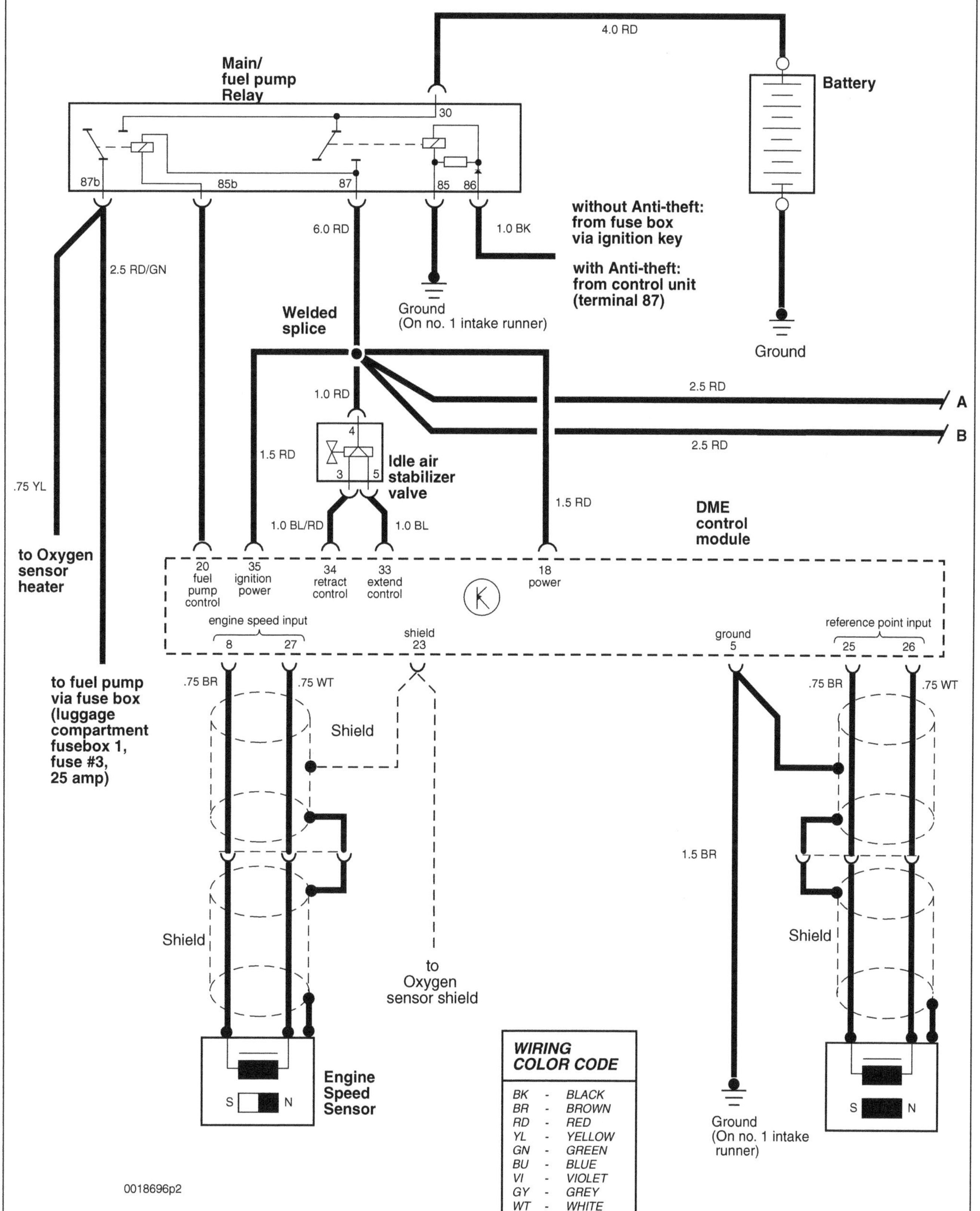

continued on next page

DME fuel injection (page 3)

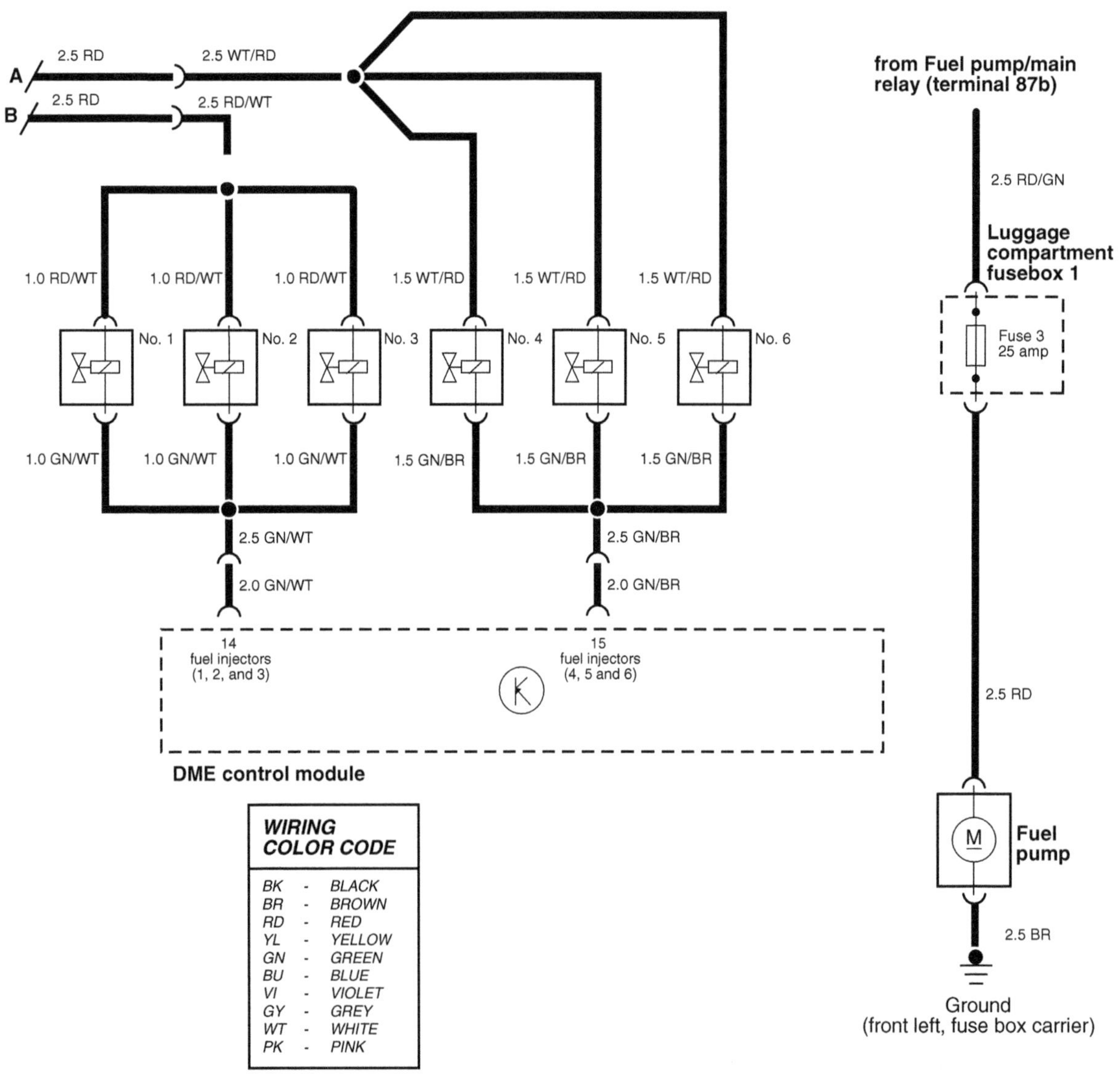

WIRING COLOR CODE		
BK	-	BLACK
BR	-	BROWN
RD	-	RED
YL	-	YELLOW
GN	-	GREEN
BU	-	BLUE
VI	-	VIOLET
GY	-	GREY
WT	-	WHITE
PK	-	PINK

continued on next page

0018696p3

DME fuel injection (page 4)

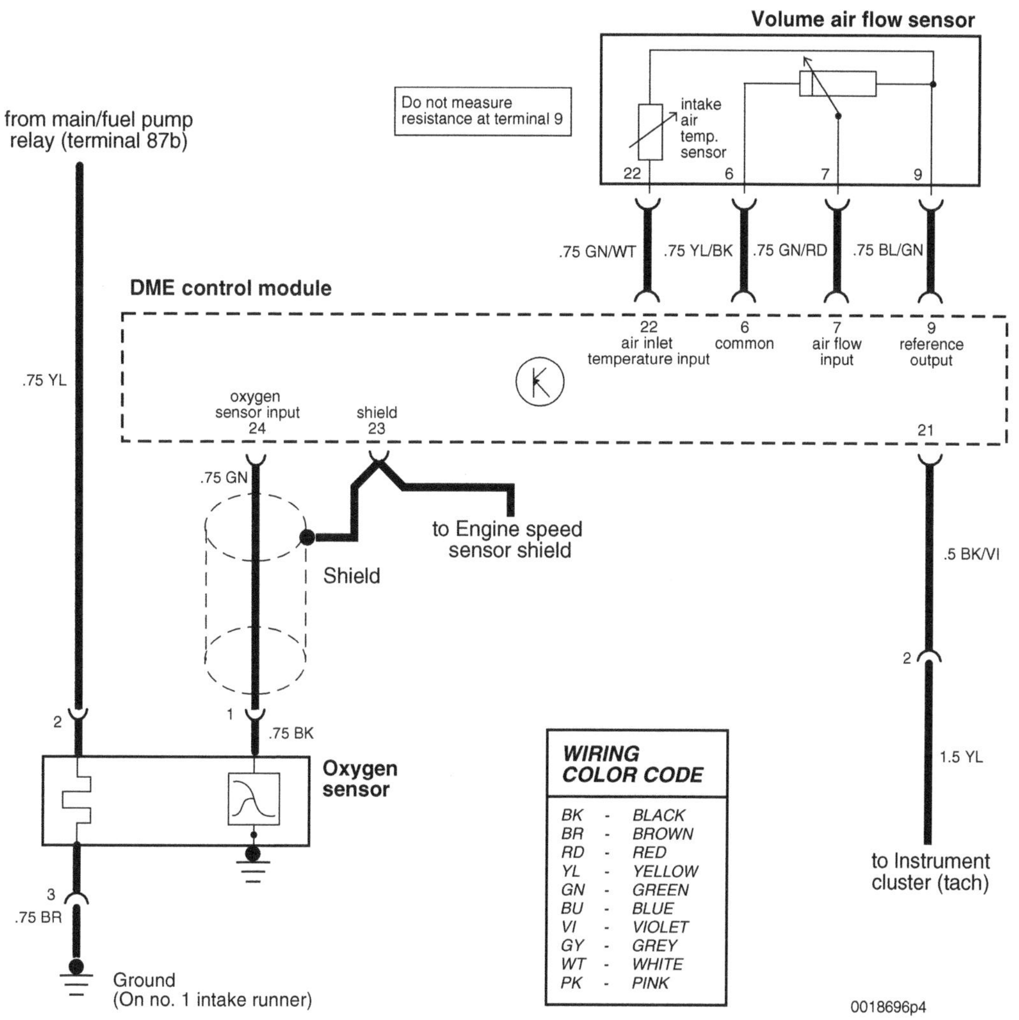

WIRING COLOR CODE		
BK	-	BLACK
BR	-	BROWN
RD	-	RED
YL	-	YELLOW
GN	-	GREEN
BU	-	BLUE
VI	-	VIOLET
GY	-	GREY
WT	-	WHITE
PK	-	PINK

260 Exhaust System

GENERAL

The exhaust system is designed to be maintenance free, but regular inspection is warranted due to harsh operating condition. The only scheduled maintenance is replacement of the oxygen sensor at specified time and mileage intervals.

This section covers maintenance, troubleshooting, and repair of the exhaust system. For information on testing the oxygen sensor, see **240 Fuel Injection.**

> ***WARNING —***
> - *To avoid injury, wear eye protection and heavy gloves when working on rusty parts of an old exhaust system.*
> - *Do not use flammable workshop chemicals near a hot catalytic converter. They may ignite.*

Exhaust system

The basic exhaust system components are the exhaust headers that are encased within the heat exchangers, a cross-over pipe, a catalytic converter and a muffler.

Separate exhaust headers/heat exchangers are mounted to each bank of three cylinders. Exhaust gases are channelled from each exhaust port to a single collector for each cylinder bank. The exhaust headers from each bank of cylinders are housed in a heat exchanger that supplies heated air for car interior heating and windshield defrosting.

A cross-over (connector) pipe brings the right side exhaust gases across to the left side. Both the cross-over pipe and left exhaust header connect to the catalytic converter. The muffler connects directly to the rear of the catalytic converter to complete the exhaust system.

Exhaust system components

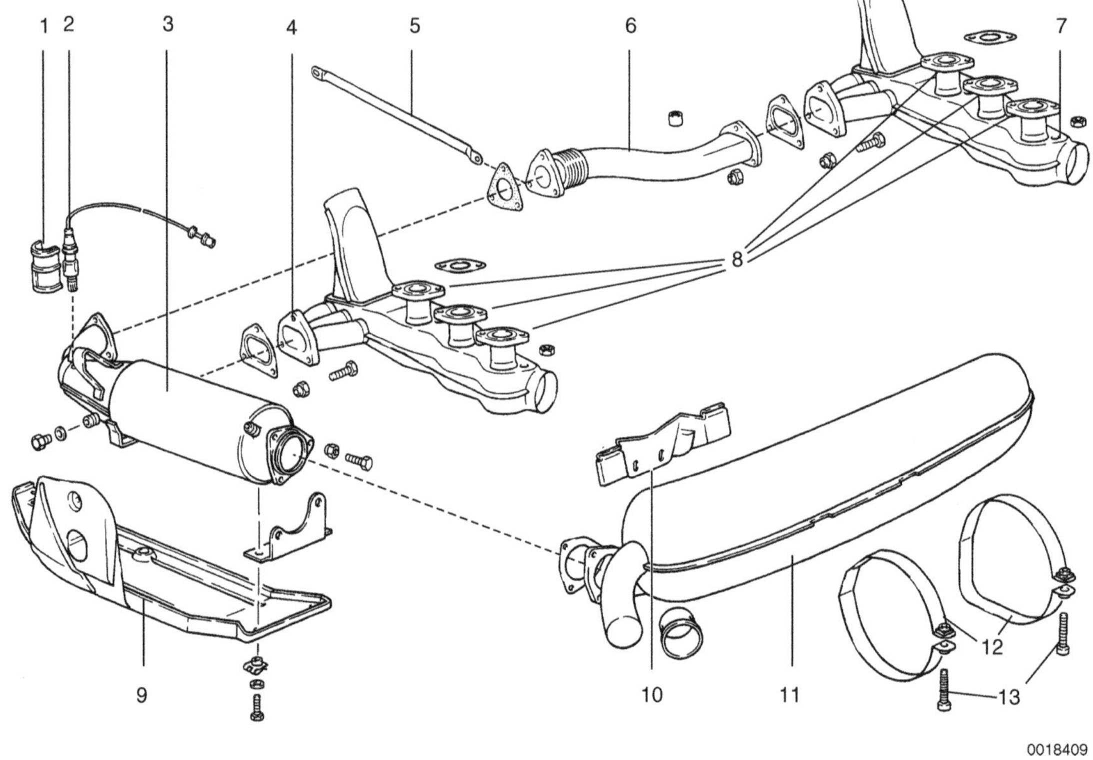

1. Oxygen sensor heat shield
2. Oxygen sensor
3. Catalytic converter
4. Left heat exchanger
5. Support bracket (optional)
6. Crossover pipe
7. Right heat exchanger
8. Exhaust headers
9. Catalytic converter heat shield
10. Muffler support bracket
11. Muffler
12. Muffler support straps
13. Muffler strap bolts

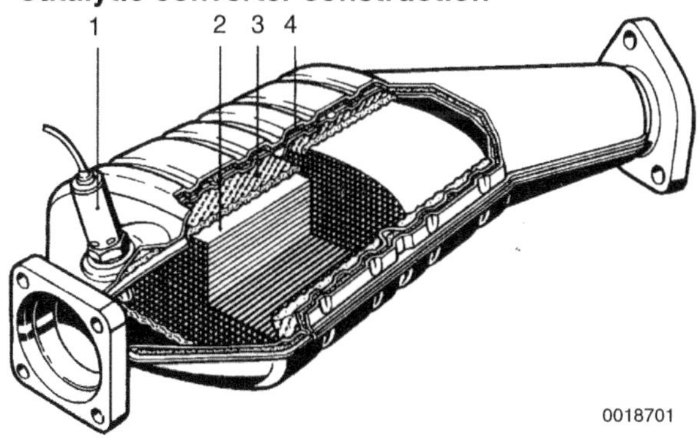

1. Oxygen sensor
2. Ceramic core
3. Wire screen
4. Heat resistant casing

Catalytic converter

All models covered in this manual are fitted with a three-way catalytic converter.

The catalytic converter is similar in appearance to a small muffler. Its honeycombed ceramic core contains hundreds of tiny passages whose surfaces are coated with a thin layer of precious metal(s). The internal construction of the catalytic converter is designed to maximize the surface area of the catalyst-coated ceramic and promote the chemical reactions that convert harmful compounds into nitrogen, carbon dioxide, water, and heat.

Under normal conditions the catalytic converter does not require replacement.

Oxygen sensor

The oxygen sensor is mounted in the inlet side of the catalytic converter. The oxygen sensor monitors the exhaust gas and provides feedback about combustion efficiency to the engine control module (ECM). The catalytic converter reduces emissions most efficiently when the percentage of oxygen in the exhaust falls within a narrow range.

Oxygen sensor diagnosis is covered in **240 Fuel Injection**. Oxygen sensor replacement is covered in **030 Maintenance.**

CAUTION—
Some carburetor and fuel injection additives or sprays can damage the oxygen sensor or catalytic converter. Check with an authorized Porsche dealer for approved fuel additives.

TROUBLESHOOTING

NOTE—
A quick-check for detecting exhaust leaks is to hold a wet shop towel over the tailpipe outlet with the engine running. If any hissing is heard under the car, the system has leaks.

WARNING—
Toxic exhaust gases are colorless and odorless. Do not run the engine in a non-ventilated area. Repair any structural damage to the body or exhaust system immediately.

Catalytic converter, checking

– Possible indications of faulty catalytic converter:
  - Reduced power
  - Stalling at idle
  - Rattles in the exhaust system
  - High exhaust emissions measured at the tailpipe.

– To visually check catalytic converter, remove muffler and check for evidence of physical damage inside converter.

– Alternatively, tap converter with a block of wood.
  - If rattling noise is heard from inside, replace converter.

– Test efficiency of catalytic converter:
  - Run engine until it and converter are at operating temperature.
  - Using infrared thermometer, measure temperature differential at inlet and outlet of converter.
  - If temperature difference is not more than approximately 55°C (100°F) replace converter.

> ***WARNING —***
> *The catalytic converter operates at a very high temperature and can cause severe burns.*

Tightening Torques

- Muffler to catalytic converter (M8) . . . . . 25 Nm (18 ft-lb)
- Muffler strap bolt (M10). . . . . 40 Nm (30 ft-lb)

COMPONENT REPLACEMENT

> ***WARNING —***
> - *To avoid injury, wear eye protection and heavy gloves when working on rusty parts of an old exhaust system.*
> - *Do not use flammable workshop chemicals near a hot catalytic converter. They may ignite.*

Use of non-original exhaust system components can change system backpressure and may affect fuel mixture. It may be necessary to check and adjust the fuel injection system after installing a non-original muffler.

NOTE —
- *Use new gaskets and mounting hardware whenever removing and installing exhaust system parts.*
- *When mounting exhaust system components, tighten the mounting nuts in two stages.*

Muffler, replacing

– Raise car and support in a safe manner.

> ***WARNING —***
> *Make sure the car is stable and well supported at all times. Use a professional automotive lift or jack stands designed for the purpose. A floor jack is not adequate support.*

Muffler and muffler support

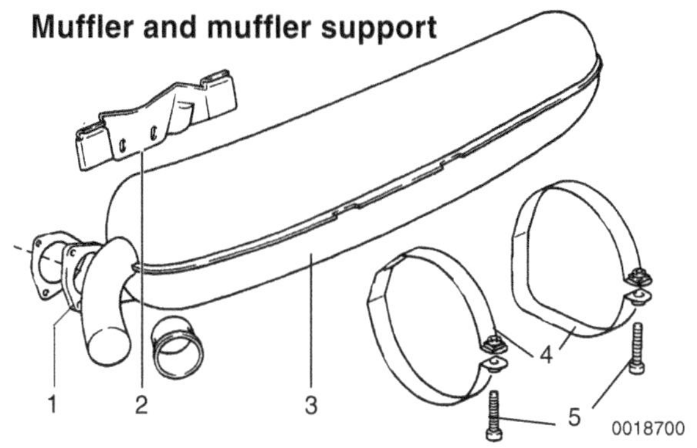

1. Flange to catalytic converter
2. Muffler support
3. Muffler
4. Muffler support straps
5. Muffler strap allen bolts

◄ Remove three M8 bolts and nuts that secure muffler to catalytic converter flange.

– Lubricate threads of two M10 allen bolts in muffler straps. Loosen both muffler strap bolts and slide each strap off muffler support while supporting muffler.

– Remove muffler.

– Installation is reverse of removal.
  - Use new gaskets and hardware.

Tightening Torques

- Muffler to catalytic converter (M8) ..... 25 Nm (18 ft-lb)
- Strap bolt (M10) .... 40 Nm (30 ft-lb)

Heat exchanger, removing and installing

– Raise car and support in a safe manner.

> ***WARNING —***
> *Make sure the car is stable and well supported at all times. Use a professional automotive lift or jack stands designed for the purpose. A floor jack is not adequate support.*

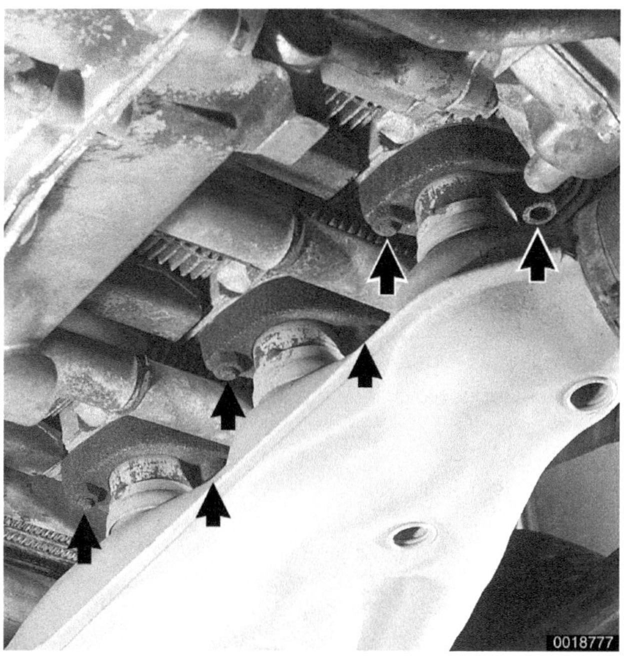

◄ Lubricate exhaust header to cylinder head fasteners (**arrows**).

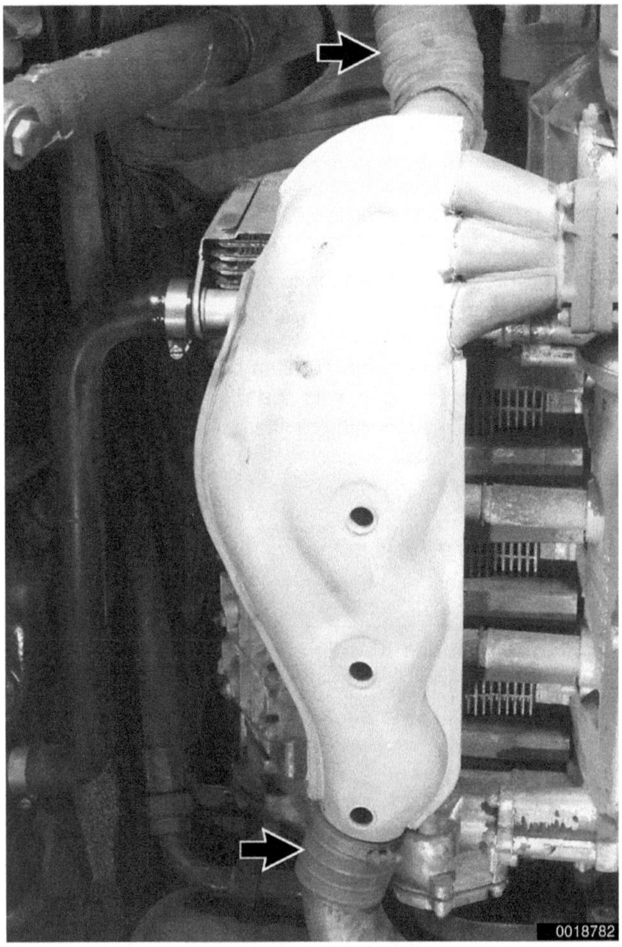

◄ Detach heater hoses (**arrows**) from heat exchanger.

- Loosen clamps and slide rubber boot back on heater blower cross-over tube.
- Loosen clamp and slide off heater hose that runs between body and heat exchanger.

– Left side: Remove three M8 fasteners connecting heat exchanger to catalytic converter.

– Right side: Remove three M8 fasteners connecting heat exchanger to crossover pipe.

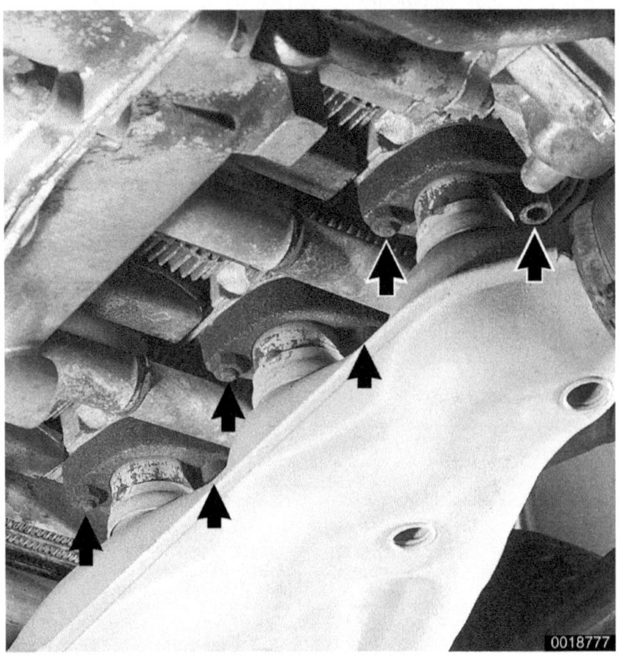

◄ Remove exhaust header to cylinder head mounting nuts (**arrows**).

- Three fasteners are locking nuts.
- Three are allen-head nuts.

> ***CAUTION —***
> *Support heat exchanger while removing the fasteners.*

– Lower heat exchanger down and out.

– Installation is reverse of removal.

- Use new gaskets and hardware.

Tightening Torques

- Exhaust runner to cylinder head....... 25 Nm (18 ft-lb)

Catalytic converter assembly

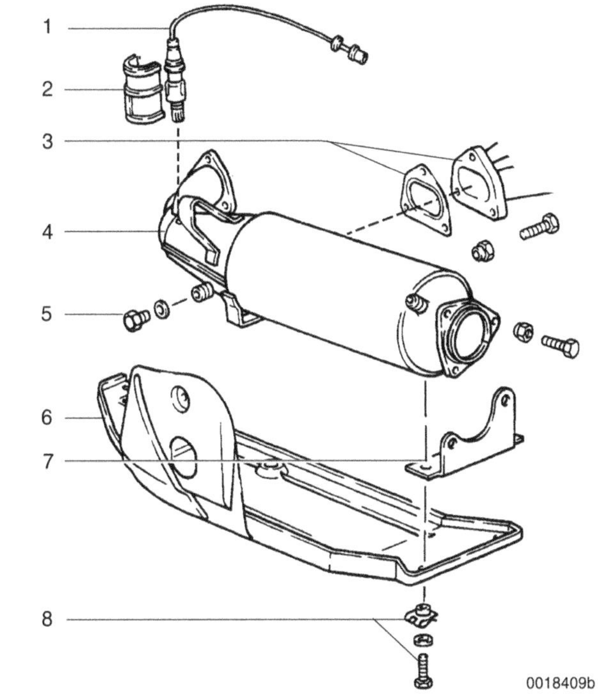

1. Oxygen sensor
2. Oxygen sensor heat shield
3. Heat exchanger flange and gasket
4. Catalytic converter
5. CO test port
6. Catalytic converter heat shield
7. Heat shield support bracket
8. Heat shield mounting hardware

Catalytic converter, removing and installing

– Raise car and support in a safe manner.

> ***WARNING —***
> *Make sure the car is stable and well supported at all times. Use a professional automotive lift or jack stands designed for the purpose. A floor jack is not adequate support.*

– Remove muffler as described earlier.

◄ Remove catalytic converter heat shield mounting bolts. Remove heat shield.

◄ Open engine compartment and locate oxygen sensor wiring.

- Detach connector (**arrow**).
- Push wire with rubber grommet through engine compartment bulkhead.

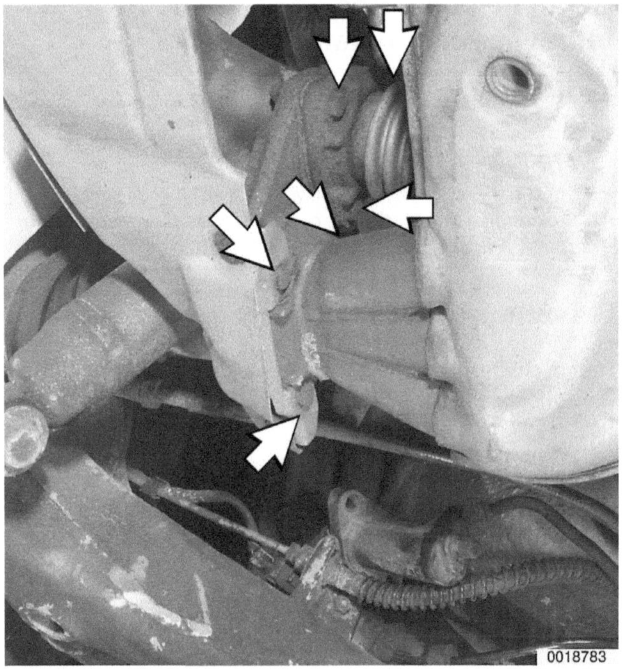
0018783

Remove converter flange fasteners (**arrows**) at left heat exchanger and right crossover pipe.

> ***CAUTION —***
> *Support catalytic converter while removing the fasteners.*

– Remove catalytic converter.

– Installation is reverse of removal.

- Use new gaskets and hardware.

Tightening Torques

- Catalytic converter to muffler (M8) . . . . . 25 Nm (18 ft-lb)

270 Battery, Starter, Alternator

TABLES

GENERAL

This section covers the battery, starter and alternator components of the electrical system. For additional electrical troubleshooting information that may apply to these components, see **900 Electrical System–General**.

The battery, starter and alternator are the three major component of the electrical system. The battery supplies electrical power when the engine is not running. The starter is an electric motor that rotates the engine via a flywheel when the key is in the start position.The alternator converts some of the engine's mechanical energy into electric energy whenever the engine is running.

The starter and alternator are not protected by fuses. To prevent accidental shorts that might blow a fuse or damage wires and electrical components, the negative (–) battery cable should always be disconnected before working on the electrical system.

Special tools

Some of the procedures require special tools.

Special Tools

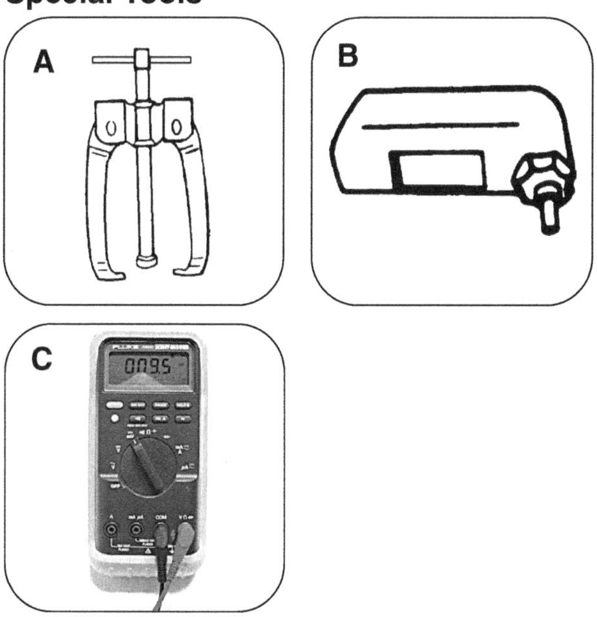

Special tool

A 236P (source: Baum Tools Unlimited) . . Battery terminal puller
B 1080 (source: Baum Tools Unlimited) . Battery electrolyte filler
C Fluke 87 . . . . . Automotive digital multimeter

Voltage and polarity

The cars covered by this manual have a 12-volt, direct current (DC), negative ground electrical system. The voltage regulator maintains the voltage in the system at approximately the 12-volt rating of the battery. All circuits are grounded by direct or indirect connection to the negative (–) terminal of the battery.

Battery

The battery is located in the front luggage compartment behind the left headlight.

The battery stores electricity chemically. Lead plates inside the battery interact with a mixture of sulfuric acid and water, called electrolyte, to create a flow of current.

Starter

The starter and its attached solenoid are mounted on the right side of the transmission just above the right drive axle.

Charging system

The charging system consists of a belt-driven alternator and voltage regulator. The voltage regulator, is mounted on the back side of the alternator and also serves as the alternator brush holder.

Maintenance

The following electrical system maintenance is described in **030 Maintenance**.

- Checking battery level.
- Checking alternator belt tension.
- Changing alternator V-belt.

Troubleshooting

The cause of a dead battery or an inoperative starter are often related. Before troubleshooting any battery-related problems, check that all ground connections are tight and free from corrosion.

Table a lists symptoms of trouble with the battery, starter and charging system, their probable causes, and suggested corrective actions. Most corrective actions are described later in this group.

Table a. Battery, Starter and Charging System Troubleshooting

Sympton	Probable Cause	Corrective Action
Engine turns slowly or not at all, solenoid clicks when starter is operated	a. Battery cables loose, dirty, or corroded b. Battery discharged c. Body ground strap loose, dirty, or corroded d. Poor connection or no voltage at starter motor terminal 30 e. Starter motor or solenoid faulty	a. Clean or replace cables. See **030 Maintenance**. b. Charge battery, test and replace if necessary. c. Inspect ground strap, clean, tighten and replace if necessary. d. Check connection, test for voltage at starter. e. Test starter.
Battery will not stay charged (more than a few days)	a. Short circuit draining battery b. Alternator V-belt loose or damaged c. Battery faulty d. Battery cables loose, dirty, or corroded e. Alternator or voltage regulator faulty	a. Test for static current draw. b. Inspect alternator V-belt. See **030 Maintenance**. c. Load test battery and replace if necessary. d. Clean or replace cables. See **030 Maintenance**. e. Test alternator and regulator.
Battery losing water	a. Battery being overcharged	a. Test voltage regulator.
Lights dim, light intensity varies with engine speed	a. Altenator V-belt loose or damaged b. Alternator or voltage regulator faulty c. Body ground strap loose, dirty, or corroded	a. Inspect alternator V-belt. See **030 Maintenance**. b. Test alternator and voltage regulator. Replace as required. c. Inspect ground strap, clean, tighten, and replace as required.

0018600

The battery to chassis ground cable connection (**arrow**) in left front of luggage compartment is shown.

0018776

Engine/transmission to body ground strap connection is shown (**arrow**).

BATTERY

> ***WARNING —***
> - *Electrolyte contains sulfuric acid and can cause severe burns.*
>   *–Wear goggles, rubber gloves, and a rubber apron when working with battery electrolyte.*
>   *–If electrolyte is spilled on your skin or clothing, flush the area at once with large quantities of water.*
>   *–If electrolyte gets into your eyes, bathe them with large quantities of clean water for several minutes and call a physician.*
> - *Batteries generate explosive gasses. Keep sparks and open flame away. Do not smoke.*

> ***CAUTION —***
> *Replace a battery if the case is cracked or leaking. Electrolyte can damage the car. If electrolyte is spilled, clean the area with a solution of baking soda and water.*

Low-maintenance batteries have been supplied as original equipment in the cars covered by this manual. These batteries have vented removable filler caps. The battery requires distilled water added periodically to maintain electrolyte level.

The six-cell, 12-volt lead-acid battery capacity is rated by ampere/hours (Ah) and cold cranking amps (CCA). The Ah rating is determined by the average amount of current the battery can deliver over time without dropping below a specified voltage. The CCA is determined by the battery's ability to deliver starting current at 0°F (–18°C).

Temperature affects the efficiency of the battery. The current-producing capacity of a battery at 5°F (–15°C) is only half its capacity at 68°F (20°C), and partly-discharged batteries can freeze due to the higher proportion of water in the electrolyte. A frozen battery produces no current, but can usually be restored when thawed.

For more information on battery maintenance, see **030 Maintenance**.

Recommended Battery Type and Rating
- Battery type. 12 V negative ground
- Ampere/hours (Ah) 88

Battery testing

Battery testing determines the state of battery charge. The most common methods are open-circuit and load voltage testing. Batteries that have filler caps can also be tested by checking the specific gravity of the electrolyte. The specific gravity test checks the amount of acid in the electrolyte as an indication of battery charge. Inexpensive specific gravity testers are available at most auto supply stores.

Open-circuit voltage test

An open-circuit voltage test checks battery voltage by connecting an accurate digital voltmeter to the battery posts. Before making any open circuit voltage test on the battery, first load the battery with 15 amps for one minute, for example by turning on the headlights without the engine running. Open-circuit voltage levels and their corresponding percentages of charge are shown in **Table b.**

CAUTION —
- *Never operate the engine with the battery disconnected.*
- *Prior to disconnecting the battery, read the battery disconnection cautions given on page viii.*

Table b. Open Circuit Voltage and Battery Charge

Open-circuit voltage	State of charge
12.6 volts or more	Fully charged
12.4 volts	75% charged
12.2 volts	50% charged
12.0 volts	25% charged
11.7 volts	Fully discharged

The battery is in satisfactory condition if the open-circuit voltage is at least 12.4 volts. If the open-circuit voltage is at this level or above, but the battery still lacks power for starting, make a load voltage test to determine the battery's service condition. If the open-circuit voltage is below 12.4 volts, recharge the battery. If the battery cannot be recharged to at least 75% it should be replaced.

Battery load voltage test

A battery load tester is required for a load voltage test. The test is made by applying a high resistive load to the battery terminals and then measuring battery voltage. The battery should be fully charged for the most accurate results. The battery cables must be disconnected before making the test. If the voltage is below that listed in **Table c**, the battery should be replaced.

WARNING —
Always wear protective goggles and clothing when performing a load test.

Table c. Battery Load Test–Minimum Voltage (apply 200 amp load for 15 seconds)

Ambient temperature	Voltage
80°F (27°C)	9.6 V
60°F (16°C)	9.5 V
40°F (4°C)	9.3 V
20°F (–7°C)	8.9 V
0°F (-18°C)	8.5 V

Static current draw test

If the battery discharges over time, there may be a constant drain or current draw on the battery. A small static drain on the battery is normal, but a large drain will cause the battery to quickly discharge. Make a static current draw test as the first step when experiencing battery discharge.

- Make sure ignition and all electrical accessories are switched off.
- Disconnect battery negative (–) cable.

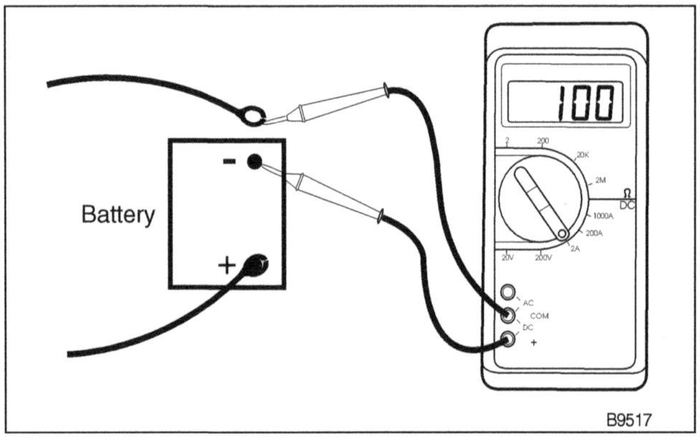

Connect a digital ammeter between negative battery post and negative battery cable to measure current. Wait at least one minute to get an accurate reading.

A range of about 0 to 100 milliampere is normal, depending on the number of accessories that need constant power. A current of 400 milliampere (0.4 amp) or more may indicate a problem.

To determine the circuit or component causing the problem, remove one fuse at a time until the current drops to a normal range.

NOTE —
Large current drains can be detected by using a test light connected in series with the negative cable and negative battery post.

Battery charging

Discharged batteries can be recharged using a battery charger, but a battery can never be charged to a voltage in excess of that which it is capable of producing electro-chemically. Prolonged charging causes gassing that will evaporate the electrolyte to a level that can damage the battery.

Always read and follow the instructions provided by the battery charger's manufactures. Do not use a charger if the instructions are not available. **Table a** lists charging rates and times that should be followed when charging a battery.

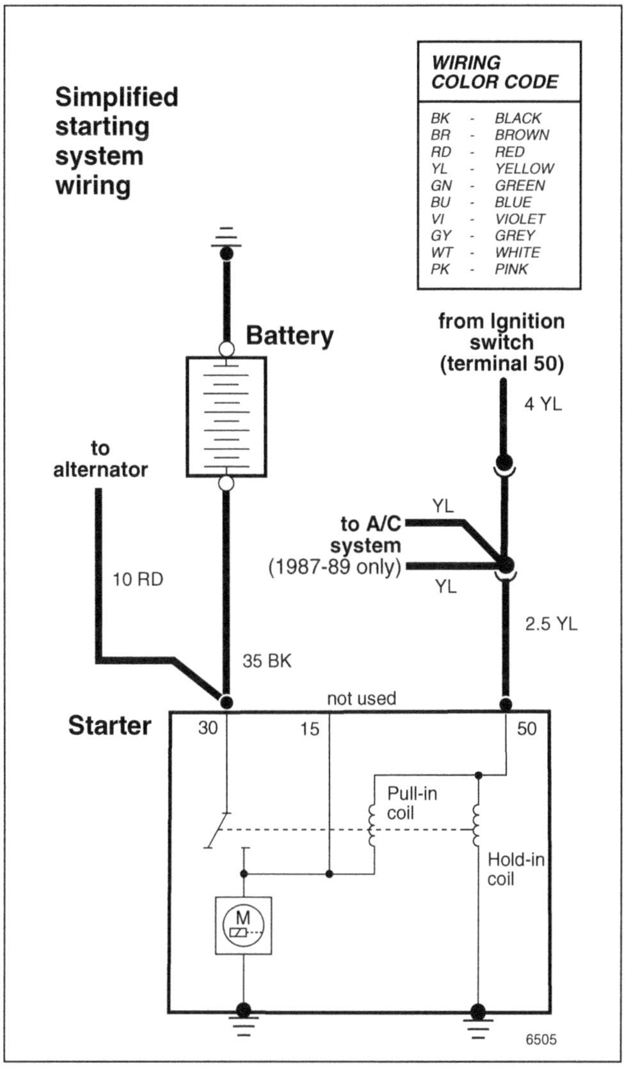

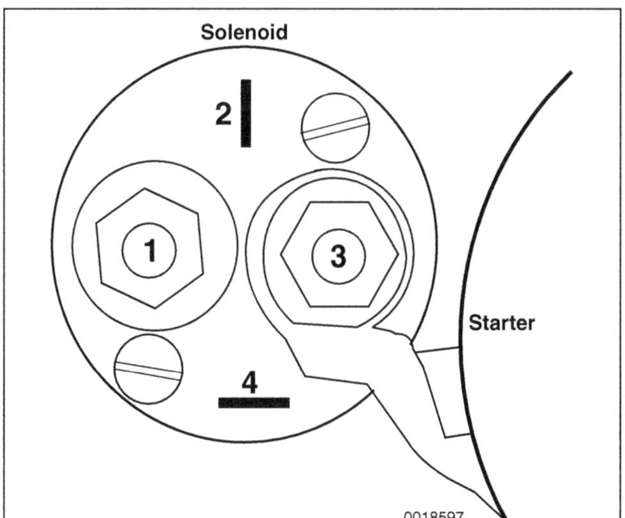

1. Terminal 30 from battery
2. Terminal 50 from ignition switch
3. To starter field winding
4. Terminal 15 (not used)

Table d. Battery Charging Specifications

Charging Rate Low-Maintenance Battery	Specific Gravity	Approximate Charging Time
Fast Charge (at 80% to 90% of battery's capacity. Example: 44 to 50 amperes for a 55 ampere-hour battery	1.150 or less 1.150 to 1.175 1.175 to 1.200 1.200 to 1.225	1 hour ¾ hour ½ hour ¼ hour
Slow charge at 10% of battery's capacity. Example: 5.5 ampere for a 55 ampere-hour battery	Above 1.225	Slow charge only to a specific gravity of 1.250 to 1.265

STARTING SYSTEM

When the ignition key is in the **start** position, battery power energizes a solenoid switch on the starter motor. The solenoid then engages the starter drive pinion with the ring gear on the engine flywheel/clutch assembly. To minimize wear and stress on the drive pinion and ring gear, the solenoid does not switch current to the starter until the drive pinion is fully engaged. Also the drive pinion has an overrunning clutch to prevent the starter from being driven by the engine.

Starting system troubleshooting

The battery and its cables should be in good condition when troubleshooting the starter. If in doubt about the battery, test as described earlier. Starter efficiency is affected by engine oil viscosity. This is especially true in cold weather.

Table e lists symptoms of trouble with the starting system, their probable causes, and suggested corrective actions, most described later in this group.

See **970 Electrical Wiring Diagrams** for a wiring schematic of the starter system that can be used as a guide for troubleshooting.

When troubleshooting starting system, refer to terminals in figure.

Table e. Starting System Troubleshooting

Symptom	Probable Cause	Corrective Action
Starter does not operate when ignition switch is turned to **start**	a. Ignition switch or wire leading from ignition switch to solenoid faulty (less than 7.5 volts to solenoid switch) b. Solenoid switch faulty (less than 7.5 volts to starter motor) c. Starter motor fault	a. Test for voltage at terminal 50 of solenoid switch with ignition switch at **start**. If not at least 7.5 volts, test for voltage at terminal 50 of ignition switch with switch at **start**. Replace ignition switch (See **480 Steering**) or eliminate open circuit between ignition switch and solenoid switch. b. Test for voltage at field-winding connector with ignition at **start**. If not at least 7.5 volts, replace solenoid. c. Test for voltage at field-winding connector with ignition at **start**. If at least 7.5 volts, repair or replace starter motor.
Starter turns slowly or fails to turn engine	a. Dirty, loose, or corroded starter connections b. Dirty, loose, or corroded ground strap between engine/transmission and body c. Starter worn or faulty	a. Remove, clean, and tighten connections. b. Remove and clean or replace ground strap. c. Repair or replace starter.
Starter makes unusual noises, turns erratically, or fails to turn	a. Drive pinion defective b. Starter ring gear damaged	a. Repair or replace starter motor. b. Replace starter ring gear. See **301 Clutch**.
Starter operates, but does not turn engine	a. Starter drive pinion or armature shaft faulty b. Solenoid switch mechanism faulty	a. Repair or replace starter. b. Replace starter solenoid switch.

Starter, removing and installing (1984 to 1986 models)

– Disconnect negative (–) battery cable and cover battery terminal to keep cable from accidentally contacting terminal.

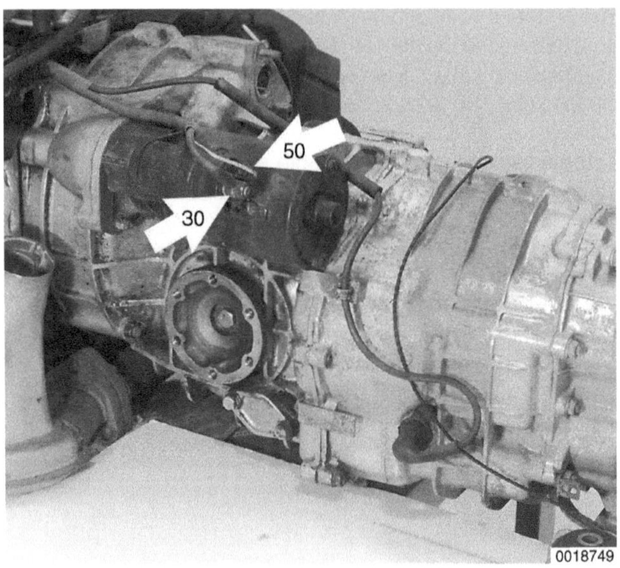

◄ Remove wires from starter terminals 30 and 50.

– Using 10 mm allen wrench, remove starter mounting nuts.

– Slide starter toward front of car to remove.

– Installation is reverse of removal.

Tightening Torques
- Positive cable to starter . . . . 10 Nm (7.5 ft lb)
- Starter to transmission or engine . . . . . . .45 Nm (33 ft lb)

Starter, removing and installing (1987 to 1989 models)

– Disconnect negative (–) battery cable and cover battery terminal to keep cable from accidentally contacting terminal.

– Raise vehicle to gain access to bottom of engine.

> ***WARNING —***
> *Make sure the car is stable and well supported at all times. Use a professional automotive lift or jack stands designed for the purpose. A floor jack is not adequate support.*

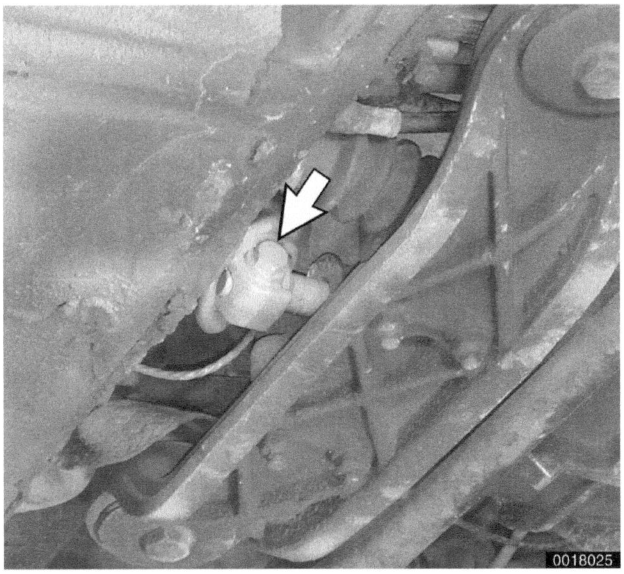

< Working underneath car, push back rubber boot covering shift linkage.

- Loosen hex bolt (**arrow**) and pull shift rod out of shift coupler.

< Remove rear stabilizer bar.

- Unbolt stabilizer bracket bolts (**arrows**) at left and right sides and lower bar out of car (left side shown).

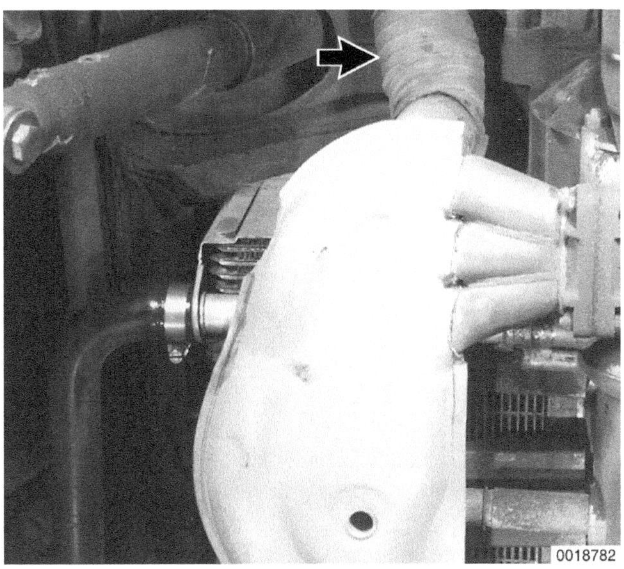

< Disconnect heater hoses from heat exchangers. Right side shown (**arrow**).

– Remove accelerator cable end from bellcrank on side of transmission.

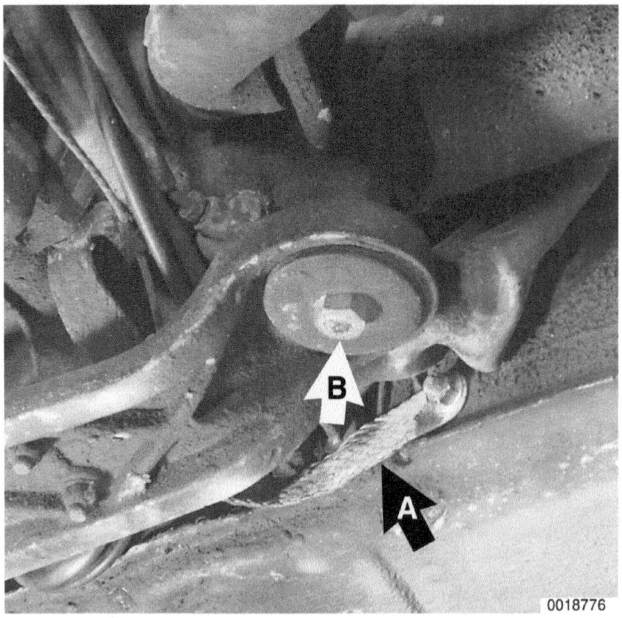

– Support transmission with transmission jack.

◄ Disconnect ground strap (**A**) between body and transmission.

- Remove transmission crossmember bolts. Right side shown (**B**).

– Carefully lower transmission about 15 cm (6 in.).

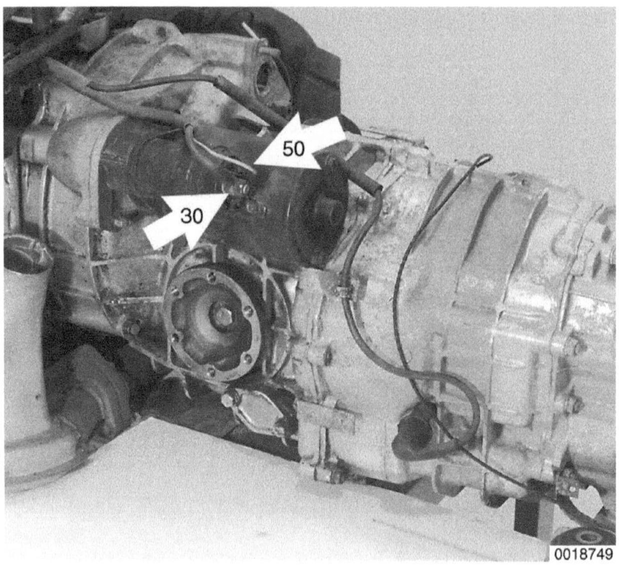

◄ Remove wires from starter terminals 30 and 50.

– Using 10 mm allen wrench, remove starter mounting nuts.

– Slide starter toward front of car to remove.

– Installation is reverse of removal.

Tightening Torques

- Positive cable to starter .... 10 Nm (7.5 ft lb)
- Shift rod linkage to shift rod ....25 Nm (18 ft lb)
- Stabilizer bar to body (bolts)............25 Nm (18 ft lb)
- Stabilizer bar to body (nuts) ....85 Nm (63 ft lb)
- Starter to transmission ....45 Nm (33 ft lb)
- Transmission crossmember to body ....80 Nm (58 ft lb)

CHARGING SYSTEM

WARNING —
Disconnect the battery negative (–) cable before disconnecting any wires from the rear of the alternator. Battery voltage is wired directly to the alternator without fuse protection. Reconnect the battery cable after all wires have been safely disconnected and insulated.

CAUTION —
- *Never operate the engine with the battery disconnected.*
- *Never operate the alternator with its output terminal (B+ or 30) disconnected and the other terminals connected.*
- *Never short, bridge, or ground any terminals of the charging system except as specifically described in* **Charging system troubleshooting**.

The charging system provides the current necessary to keep the battery charged and to operate the car's electrical accessories. The system includes an alternator driven from the engine crankshaft by a V-belt to generate the charging current, and a voltage regulator to control the charging current.

To prevent damage to the alternator or regulator when making tests or repairs, make all connections with negative (–) to negative, and positive (+) to positive unless directed otherwise. Even momentary contact with a conductor of the wrong polarity can damage the alternator's diodes. Make certain that the battery negative (–) cable is securely fastened, and that the cable to terminal 30 on the starter is connected to the battery's positive (+) terminal.

NOTE —
Rated output of the alternator when measured at the battery with the engine running can be 10 to 15 amps less than specified in the specifications listed below. The difference is accounted for by electrical consumers and is a normal condition.

Alternator Output
- 1984-1989 911 Carrera .... 90 amp

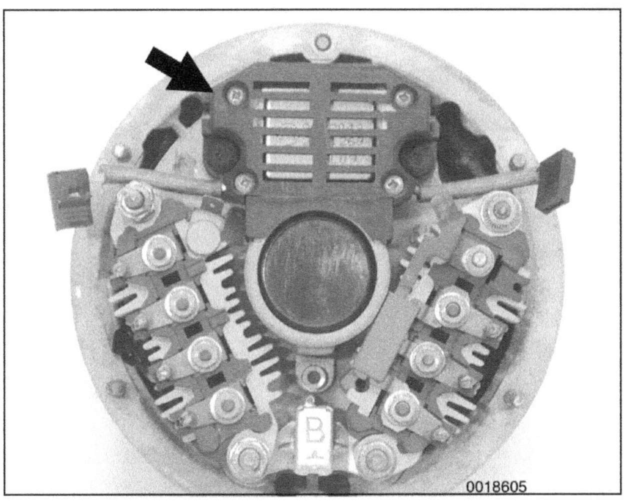
0018605

The voltage regulator (**arrow**) is mounted to the back of the alternator housing. Voltage regulators and alternator are available as replacement parts from an authorized Porsche dealer. Robert Bosch remanufactured alternators are also available from many auto parts stores.

Simplified charging system circuit

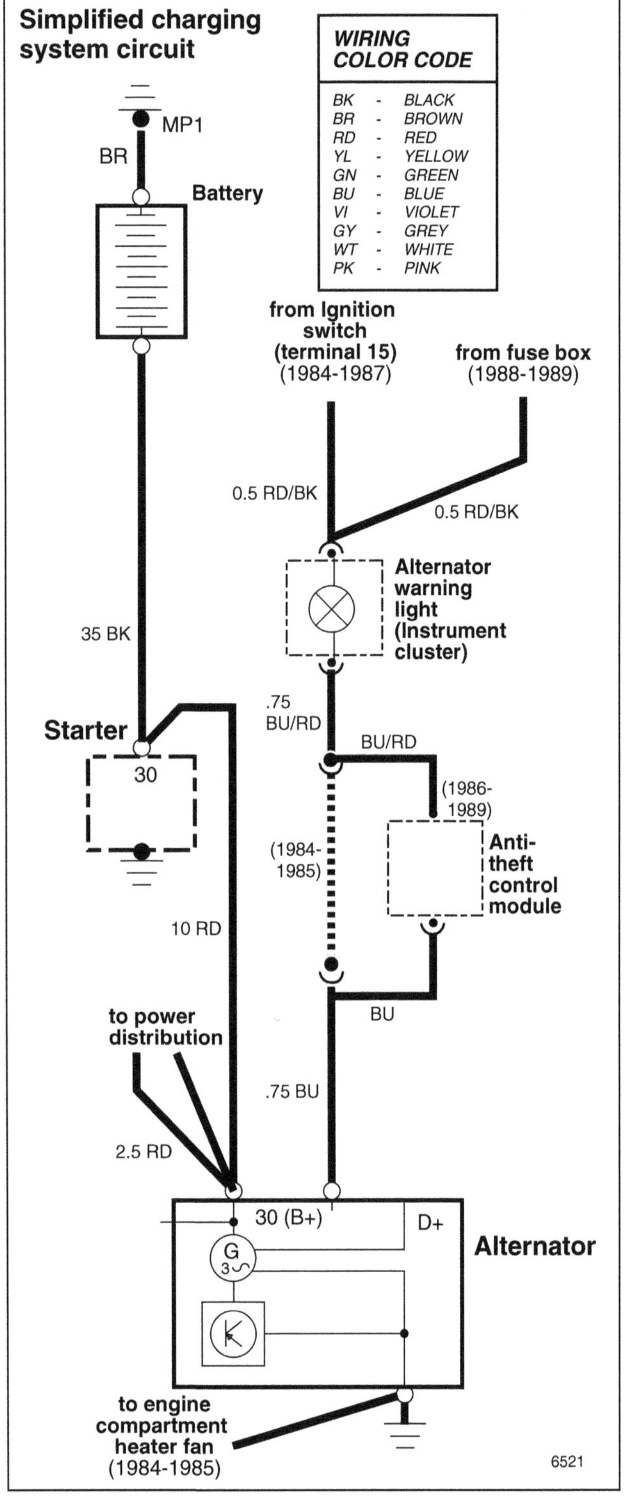

Charging system troubleshooting

Charging system trouble is indicated by an illuminated alternator warning light on the instrument panel, or by an under- or overcharged battery.

The alternator generates electrical current by electrical induction. That is, a magnetic field is placed in motion to induce a current in a stationary coil. When the engine is running and the alternator is spinning, part of the current it produces energizes its electromagnetic field. When starting, some other current must be provided to initially energize the field and begin the current generating process. This current is provided by the battery through the alternator warning light in the instrument panel.

NOTE —

On some models of Porsche cars a burned out instrument cluster alternator warning light will prevent the alternator from charging.

As soon as the alternator output equals battery voltage, the warning light goes out. Normally, the warning light should be off when the ignition is off and the engine is stopped. The light should only come on when the ignition is turned on (current to the alternator rotor) and go out again when the engine is started and the battery is being charged.

Table f lists symptoms of trouble indicated by the alternator warning light, their probable causes, and suggested corrective actions, most described later in this group.

An undercharged battery is usually associated with starting trouble. Again, make sure that the battery is in good condition and capable of accepting a full charge before blaming the charging system.

Causes of an undercharged battery are:

- The simultaneous use of many electrical accessories for long periods of time.
- Leaving accessories or lights in operation with the engine stopped.
- Frequent long periods of starter usage.
- Frequent short-trip driving.
- Improper alternator V-belt condition or tension.
- Broken or frayed charging system wiring.
- Corroded connectors at the D+ and B+ terminals of the alternator.
- Worn, corroded, or loose battery cable connections at the chassis.

Table f. Warning Light Troubleshooting

Symptom	Test and probable cause	Corrective action
Ignition off, engine off, warning light on	*Test:* Disconnect blue wire (D+) from alternator. a. Light goes out: Faulty alternator diodes. b. Light does not go out: short to ground in wiring harness or wiring connector.	a. Repair or replace alternator. b. Repair or replace faulty wiring.
Ignition on, engine off, warning light off	a. Battery fully discharged. b. Bulb burned out. *Test:* Disconnect blue wire (D+) from alternator. With battery connected and ignition on, touch blue wire to ground. c. Light does not come on: Faulty bulb socket, open circuit between socket and terminal 15 of ignition switch, or open circuit between blue wire (D+) on alternator and instrument cluster. d. Light comes on: Loose connection between regulator and alternator or loose connection between brushes and regulator. e. Light comes on, no faults with regulator: Internal alternator faults or faulty regulator.	a. Charge battery. b. Remove and test bulb. Replace faulty bulb. c. Replace instrument cluster bulb holder. Repair wiring to alternator. d. Inspect brushes. Correct loose connections. e. Repair or replace alternator or voltage regulator.
Engine running at any speed, warning light stays on	a. Loose or broken alternator V-belt. *Test:* Disconnect red wire (B+) from alternator. Do not short wire. Using a voltmeter, test between red wire and ground with battery terminal connected. b. No voltage to alternator: Open circuit between red wire on alternator and starter or between starter and positive (+) pole. c. Exciter diodes burned out. d. Faulty regulator or faulty alternator windings. e. High voltage drop between red alternator positive (+) wire and starter due to broken, loose, or corroded wires.	a. Replace or adjust V-belt. b. Repair wire or connections between alternator and starter. c. Repair or replace alternator. d. Test charging system and replace faulty components as needed. e. Repair wires or connectors.

Charging system, testing

The tests described here will help determine the cause of charging system trouble. The battery should be fully charged and the alternator V-belt correctly tensioned.

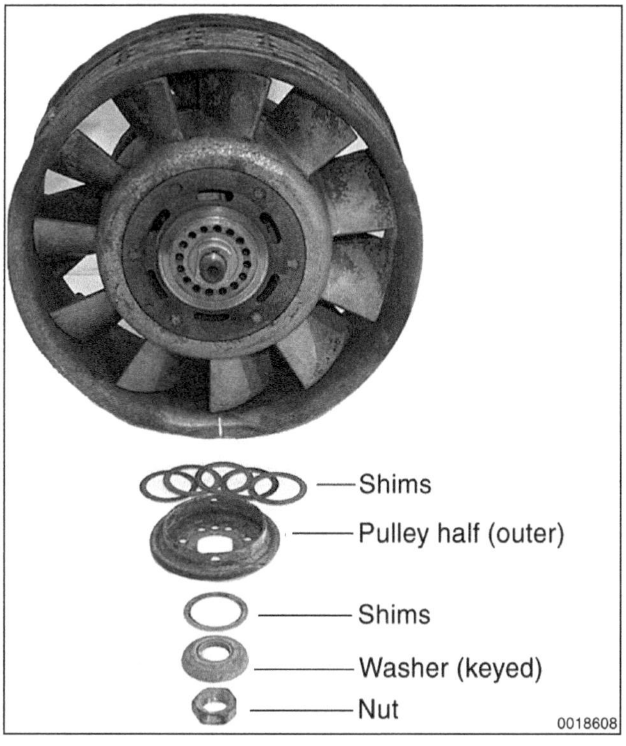

V-belt adjusting shims shown in illustration. See **030 Maintenance**.

The charging system warning light should be operating correctly. All electrical connections should be clean and tight. Replace wires that are hard and cracked.

If tests show that the alternator and regulator are operating correctly, but the battery still continually runs down, there may be a short in the electrical system causing a continuous current drain. Remember that the clock and some other components require constant current.

See **Static current draw test**, earlier in this group.

A general test of charging system output can be made with an accurate digital voltmeter. Such a test is given below. The most accurate testing is done using an alternator and regulator tester that applies a high-current load to the alternator. Conclusive tests using this equipment can be made inexpensively by an authorized Porsche dealer or other qualified shop.

Alternator and regulator, checking

0018606

Attach digital voltmeter to negative (–) and positive (+) battery terminals.

- Make sure tester is connected to clean areas of terminals.
- Set voltmeter to DCV scale.

– Start engine and run it at about 1500 rpm with all electrical accessories turned off. Measure voltage at battery.

– A reading much higher than 14.2 volts most likely indicates a faulty voltage regulator.

– A reading below 13.5 volts means that the battery is not adequately charged. This may be due to:

- Faulty V-belt
- Faulty voltage regulator
- Faulty alternator.

– To determine which component is faulty:

- Visually check V-belt. See **030 Maintenance**.
- Keep engine running at about 1500 rpm and turn on all electrical accessories.
- Check voltage across battery terminals.
- A reading that is the same as the first most likely indicates a faulty voltage regulator.
- A reading that is lower most likely indicates a faulty alternator.

Noisy alternator

0018607

Alternator mechanical noises are usually the result of misalignment between the V-belt and the pulley, a loose or broken pulley, worn bearings, or a bent rotor shaft. Check for bad bearings by removing the V-belt as described below and rotating the alternator pulley by hand. If the shaft grinds, the alternator should be replaced.

Alternator, replacing

– Disconnect negative (–) battery cable and cover battery terminal to keep cable from accidentally contacting terminal.

> ***CAUTION —***
> *Be sure to read the battery disconnection cautions given on page viii before disconnecting the battery.*

– Remove A/C drive belt as described in **870 Air Conditioner**.

– Remove V-belt. Reinstall pulley and nut fingertight.

Remove allen bolt (**A**) from cooling housing strap.

- Remove bolts (**B**) from engine cooling shroud.
- Remove spark plug wire supports (**C**).

– Pull alternator and fan out (toward rear of car) partially. Also lift engine shroud slightly and unbolt alternator ground strap from top of engine case.

Remove nuts (**arrows**) holding alternator air guide and slide guide back over wiring harness.

Lift off insulating rubber boots from alternator terminals.

- Detach electrical connections (**arrows**) in back of alternator.
- Remove alternator/cooling fan assembly.
- Working at bench, remove alternator to fan housing mounting nuts (**A**).
- Push alternator out of fan housing.

– Pull fan off alternator shaft using a puller.

– Installation is reverse of removal.

Tightening Torques

- Fan housing mounting strap allen bolt . . .25 Nm (18 ft lb)
- Fan/alternator pulley to alternator shaft . .55 Nm (40 ft lb)
- Terminal B+ . . . . 6 Nm (53 in lb)
- Terminal D+ . . . . 3 Nm (27 in lb)

Voltage regulator, replacing (with alternator removed)

The brushes contact the alternator's slip rings to supply the current that magnetizes the alternator rotor. The voltage regulator maintains a norminal voltage in the electrical system by feeding excess output from the alternator back to ground. In addition, it regulates the amount of current supplied to the alternator rotor, turning it on and off as needed. If either component is faulty, the result will be an over or undercharged battery. The brushes wear under normal use, and will eventually need to be replaced.

The regulator and brush holder are mounted on the rear of the alternator. To remove the regulator and brushes, remove alternator as described earlier.

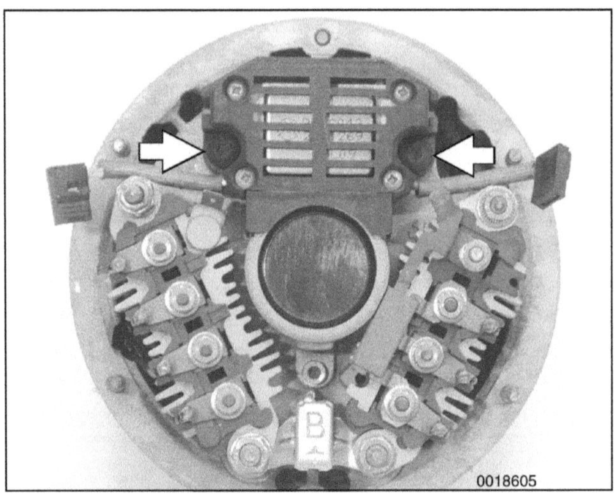

0018605

◄ With alternator on a bench remove two 5 mm Torx® (T20) screws (**arrows**) holding regulator to alternator housing.

– Slide regulator away from shaft and lift out of alternator housing.

– Installation is reverse of removal.

280 Ignition System

Special Tools

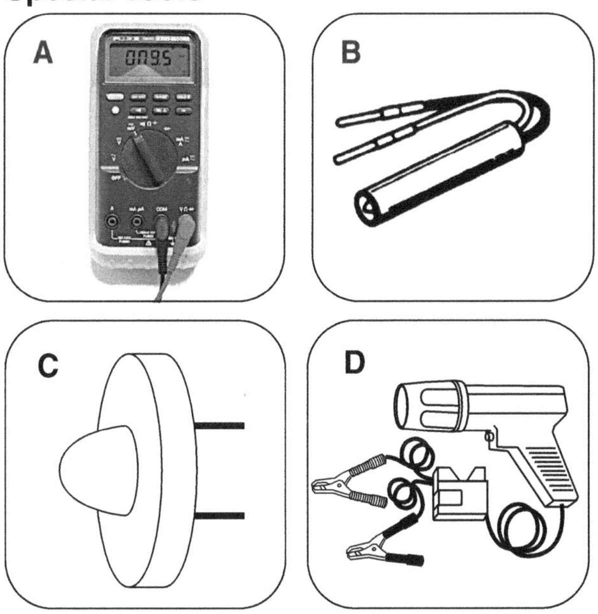

Special tool

- **A** Fluke 87 . . . . . Automotive digital multimeter
- **B** 1115 . . . . . LED tester with thin spade probes (source: Baum Tools Unlimited)
- **C** 1115a . . . . . Low current test light ("noid") (source: Baum Tools Unlimited)
- **D** . . . . . Strobe timing light

GENERAL

The ignition system installed in the cars covered by this manual incorporates many components that maximize engine performance and reliability in the face of increasing restrictions on engine exhaust emissions. The ignition components are part of the engine management system known as digital engine electronics (DME) or Motronic. For more information on the DME system, see **240 Fuel Injection.**

Test equipment

Some of the procedures require special tools.

Many tests in this section require checking for voltage, continuity or resistance at harness connector terminals. The blunt tips of a multimeter's probes can spread open the small connector terminals and cause poor connections. To prevent damage, use flat male connectors to probe the harness connector terminals.

> ***CAUTION —***
> *A high impedance digital multimeter should be used for all voltage and resistance tests. An LED test light or a "noid" should be used in place of an incandescent-type test lamp.*

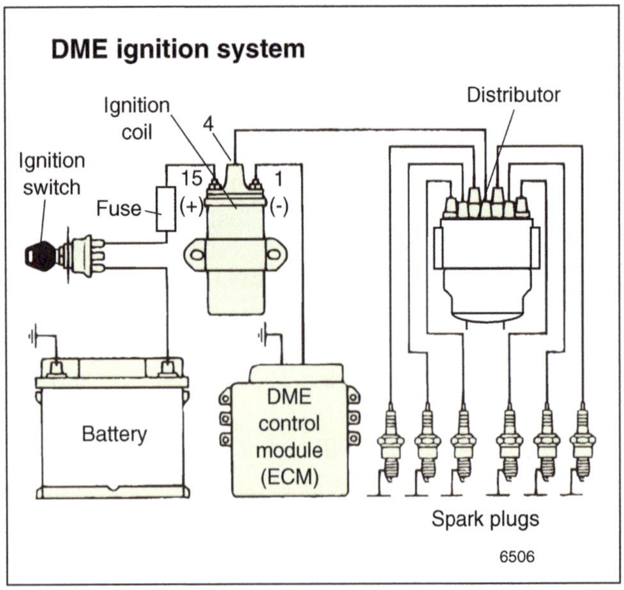

DME ignition system

In the DME system, all ignition and fuel injection control functions are controlled by a single engine control module (ECM). Ignition timing and fuel control are based on inputs the control module receives for engine load, engine speed, ignition quality, engine temperature, intake air temperature, and altitude sensor (barometric pressure). The only function that the distributor serves is to distribute the high voltage to the individual spark plugs.

The DME ignition system provides each spark plug with a precisely timed high-voltage charge to ignite the air fuel mixture in the combustion chamber. The system also makes adjustments to the ignition timing in response to changes in engine speed and load.

The high-voltage charge is created by the ignition coil. In the primary circuit, battery voltage is applied to the ignition coil to charge it. When the primary circuit is opened, the coil discharges its high voltage. The secondary ignition circuit—ignition rotor, distributor cap, spark plug wires and spark plugs—distributes the high voltage to the cylinders to ignite the air fuel mixture. Ignition timing refers to the position of the piston in the cylinder when the coil discharges.

Maintenance

Porsche specifies the maintenance steps below to be carried out at particular time or mileage intervals for proper maintenance of the ignition system.

- Replace spark plugs.
- Inspect distributor cap, ignition rotor and spark plug wires.

Information on other ignition system maintenance and on the prescribed maintenance intervals can be found in **030 Maintenance.**

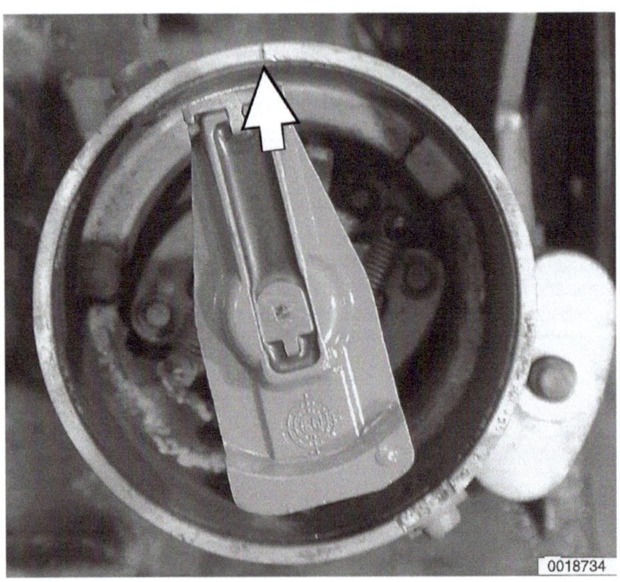

Firing order

To determine start of firing order, inspect lip of ignition distributor body for cylinder 1 scribe mark (**arrow**).

– The ignition rotor turns in a clockwise direction.

Spark plug wires must be installed so that the spark plugs fire in the correct order.

Ignition Firing Order

- Porsche 911 Carrera. 1-6-2-4-3-5

– Cylinder 1 is at left rear when viewing engine from rear of engine compartment.

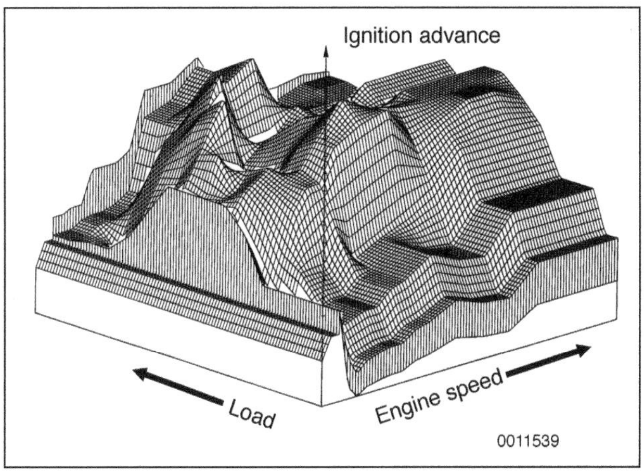

Ignition timing

Ignition timing is electronically controlled and is not adjustable on DME systems. The initial (baseline) ignition point is determined by crankshaft position during starting. This is signalled by a crankshaft reference sensor. Once the engine is running, the ignition point is continually changed based on the various inputs to the control module. Engine speed is signalled by a crankshaft speed sensor.

A DME ignition characteristic map illustrating all the possible ignition points is shown. A map similar to the one shown is digitally stored in the ECM.

– If It is necessary to check ignition timing for testing purposes, use a strobe type timing light. Power to the light should come from 110V wall plug, but it is possible to get 12V power from the small fuse box in the left side of the engine compartment. Attach signal probe to spark plug wire 1 (left rear engine compartment).

Disabling ignition system

The ignition system should be disabled when making certain tests such as compression test or starter current draw test.

Detach electrical harness connector to DME main relay (**arrow**) located under driver seat to disable ignition system. See **971 Electrical Component Locations** for more information.

> ***WARNING —***
>
> *Do not touch or disconnect any of the high tension cables from the coil, distributor cap or spark plugs while the engine is running or being cranked by the starter motor. Fatal voltage are present.*

> ***CAUTION —***
> - *Connect or disconnect ignition system wires, multiple connectors, and ignition test equipment leads only while the ignition is off.*
> - *A high impedance digital multimeter should be used for all voltage and resistance tests. An LED test light should be used in place of an incandescent-type test lamp.*
> - *Switch multimeter functions or measurement ranges only with the test probes disconnected.*
> - *Do not disconnect terminal 4 (center terminal) from the coil or remove the distributor cap to disable the ignition.*
> - *Do not disconnect battery while engine is running.*
> - *Do not connect a test instrument with a 12-volt supply to terminal 15 (+) of the ignition coil. The voltage backflow can damage the ECM. In general, make test connections only as specified by Porsche, this manual or the instrument manufacturer.*
> - *Before operating the starter without starting the engine (e.g., doing a compression test), always disable the ignition, as described above.*
> - *Do not quick charge the battery (for boost starting) for longer than one minute, and do not exceed 16.5 volts at the battery with the boosting cables attached. Wait at least one minute before boosting the battery a second time.*
> - *Do not wash engine while it is running, or any time the ignition is switched on.*
> - *Disconnect the battery when doing any electric welding on the vehicle or when charging the battery.*
> - *Do not start the engine if the car has been heated above 176°F (80°C) (e.g., in a paint drying booth). Wait for the car to cool to normal temperature.*
> - *Do not connect terminal 1 of the coil to ground as a means of preventing the engine from starting.*

TROUBLESHOOTING

The DME ignition system contains very sensitive electronic components. To protect the system, and for general safety, the cautions on this page should be observed during any ignition system troubleshooting, maintenance or repair work.

Poor driveability may have a variety of causes. The fault may lie with the ignition system, the fuel system, parts of the emission control system, or a combination of the three. Because of the interrelated functions of these systems and their effects on each other, it is often difficult to know where to begin looking for problems. For this reason, effective troubleshooting should always consider these systems in unison, as one major system.

This troubleshooting section applies to starting and running problems caused specifically by faults in the ignition system, including the coil, the distributor cap, and rotor, and the spark plug wires. A complete failure of the ignition system to produce spark at the spark plugs is self-evident. For other problems such as rough idle, misfiring, or poor starting, however, the cause is not so clear. For troubleshooting engine management and the way the car runs, see **240 Fuel Injection.**

NOTE —

If the harness connectors to the crankshaft reference and speed sensors have been reversed, the engine will not start. This topic is covered later in this group.

Basic troubleshooting principles

The function of the ignition system is to provide a properly timed high-voltage spark to the combustion chamber.

On DME ignition systems, the crankshaft reference sensor (in the transmission bellhousing) determines the initial crankshaft position and gives the electronic system its baseline ignition timing. Ignition timing is then controlled by the ECM based on engine load, engine speed (using the crankshaft speed sensor), engine temperature, and throttle position.

An engine that starts and runs indicates that the ignition system is fundamentally working—delivering voltage to the spark plugs. A hard starting or poor-running engine, however may indicate a problem with how well the spark is delivered. A faulty coil, cracked or deteriorated spark plug wires, worn or cracked distributor cap or rotor, and worn or fouled spark plugs are all causes of reduced spark intensity and inefficient combustion.

WARNING —

Inefficient combustion can cause serious problems in the catalytic converter. The poorly burned mixture can overload the converter with raw fuel, leading to converter overheating or plugging. An overheated catalytic converter can be a fire hazard.

An engine that has good cranking speed but will not even begin to start may indicate a complete failure of the system to produce spark. Inspect the ignition system visually. Make sure the spark plug wires have not been interchanged. The firing order is given earlier.

If no faults are located, make a basic check to see if spark is being produced as described below under Ignition system quick-check. This will be the most important first troubleshooting step. If a strong spark is observed, then the failure to start is due to another cause, such as no fuel being delivered to the engine.

Ignition system quick-check

If the engine does not start, the most fundamental step in troubleshooting the ignition system is to determine whether or not the system is making a spark at the spark plug. If no spark is present, then more detailed testing of the ignition system is necessary.

To check for spark, turn ignition off and remove connector from a spark plug.

- Connect to known good spark plug.
- Attach jumper wire with alligator ends to spark plug threads and a good engine compartment ground.

> ***WARNING —***
> *Do not hold the spark plug or its connector, even if using insulated pliers. The ignition system is a high-energy system operating in a dangerous voltage range which could prove to be fatal if exposed terminals or live parts are contacted.*

> ***CAUTION —***
> *Any test set-up other than the one described above may lead to damage or inconclusive results.*

– While a helper actuates starter, look and listen for spark in spark plug gap.

- Bright blue spark indicates healthy ignition system.
- Yellow-orange spark is weaker and indicates that, while spark is present and system is functioning, it is not operating at peak efficiency.
- Check condition of ignition system components as described below under **Ignition system visual inspection**.

> ***CAUTION —***
> - *If ignition system failure is not the problem, the engine may start during this test. Be prepared to turn off the ignition immediately. Running the engine with a spark plug wire disconnected may damage the catalytic converter.*
> - *Before checking the ignition system when there is no spark or a weak spark, make sure that the battery is fully charged.*

– If there is no spark, test for primary voltage at ignition coil.

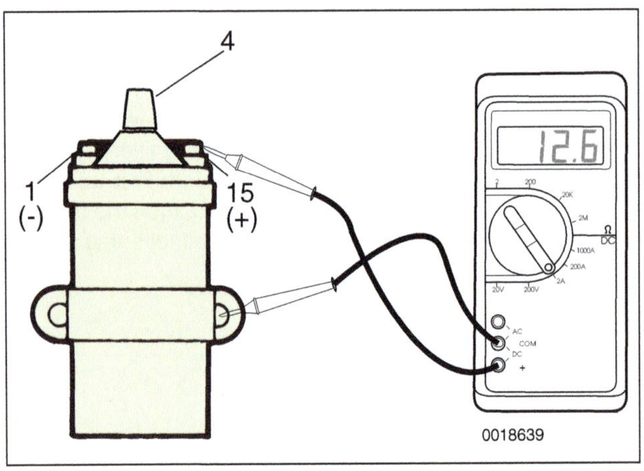

Connect digital voltmeter between terminal 15 (+) of ignition coil and ground (a clean, bare metal part).

- When ignition is turned on, there should be battery voltage at terminal.

- If battery voltage is not present, there is either a fault in wire between terminal 15 and ignition switch, in ignition switch itself, or in wiring from battery to ignition switch.
  - Check ignition coil fuse (**arrow**) in luggage compartment fusebox.
  - See **970 Electrical Wiring Diagrams** for engine management schematics.

– If no faults have been detected up to this point but there is still no spark or a weak spark, refer to **Table a** for more troubleshooting information.

– If coil is receiving voltage, or if a strong spark is observed but engine still will not start, refer to **240 Fuel Injection** for more troubleshooting information.

Ignition system harness and ground connections are listed below.

Ignition System Harness and Ground Locations

- Temperature, speed sensor, reference sensor harness connectors . . . Intake runner 3
- G407 DME ground . . . . Intake runner 1

Table a. Ignition System Troubleshooting

Symptom	Probable cause	Corrective action
No spark or weak spark observed during spark test	Wet or damp distributor cap and/or spark plug wires	Remove cap and wires. Dry and reinstall.
	Faulty wires or connectors (primary circuit)	Inspect and repair wiring as needed
	Weak or faulty coil	Test and replace as needed.
	Defective spark plug wires	Test and replace as needed.
	Worn or fouled spark plugs	Replace spark plugs. See **030 Maintenance**.
	Faulty reference or speed sensor	Test and replace as needed.
	Faulty ECM	Test and replace as needed. See **240 Fuel Injection**.

Ignition system visual inspection

The spark plug wires, ignition coil, distributor cap and ignition rotor are subject to wear and electrical breakdown which will impair their ability to deliver a crisply timed and powerful spark. Many of these conditions are most easily detected by a thorough visual inspection. Dirt and moisture on any of these components are also potential causes of poor spark at the spark plugs.

To inspect cap and rotor, use flat-tipped screwdriver to press down and turn cap hold down pins ¼ turn each.

- Lift off cap.
- Inspect contacts inside cap and at tip of the rotor for corrosion, wear, or pitting.
- Parts with corroded contacts can be cleaned and reused, but if there is wear, pitting, or heavy corrosion, replacements is recommended.
- Center black carbon brush inside the cap should spring back when compressed.

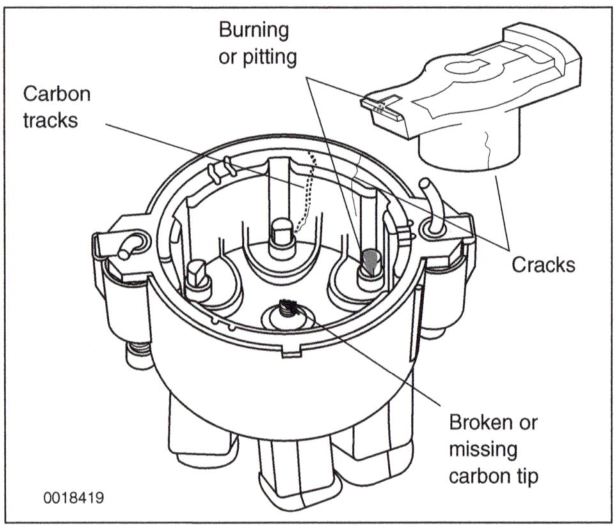

Cracks or carbon tracks in distributor cap may cause shorts to ground.

- Cracks may be fine and difficult to see. Check carefully, especially around the contacts. Carbon tracks are faint lines, usually running between two contacts or to ground, left over from high-voltage arcing.
- A distributor cap that shows any sign of cracks or carbon tracking should be replaced.

– Inspect black dust shield mounted under rotor.

- Excessive oil residue on shield, rotor or cap can lead to engine misfire and poor running under load.
- If any signs of oil residue are found on shield, carefully inspect distributor shaft seal.

– For a quick-check of distributor cap and spark plug wire condition, listen for sound of voltage arcing or watch while engine runs at night.

- In the darkness, the arc of high voltage to ground because of a crack in the cap or poorly insulated wires may be visible as a blue spark.

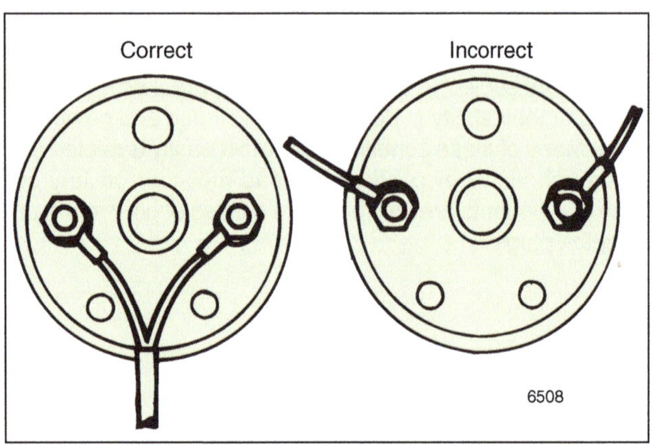

The coil should be closely examined for cracks, burns, carbon tracks, or any leaking fluid.

- Coil tower, terminal 4, should be clean and dry.
- If necessary, remove coil for cleaning and close examination.
- Check that wiring at top of coil is routed as shown. Loosen nuts and reposition wires if necessary.

– Inspect all primary wires and connections for any corrosion or damage.

- Clean or repair any faults found. Corroded or loose connections may interfere with ignition function.

NOTE —

A failed engine control module (ECM) can sometimes damage the ignition coil. Be sure to check the ignition coil anytime a faulty ECM is replaced and a no-start condition still exists.

– Inspect reference and speed sensors, connectors and wiring. See **Reference sensor and speed sensor**.

Ignition components, testing

Ignition coil, testing

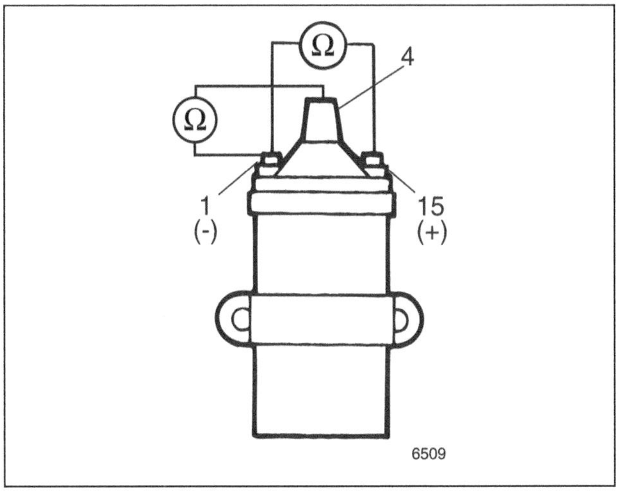

Primary coil resistance is measured with an ohmmeter between terminals **1** and **15**.

- Measurement of secondary resistance is between terminals **1** and **4**.
- Replace a coil which does not meet specifications. Correct values are in table below.

Ignition coil resistance
- Primary: Terminals 1, 15. . . . . 0.4 - 0.7Ω
- Secondary: Terminals 1, 4 . . . . 5.0 - 8.7kΩ

Spark plug wires, testing

– Check resistance of each spark plug end connector as well as wire itself.

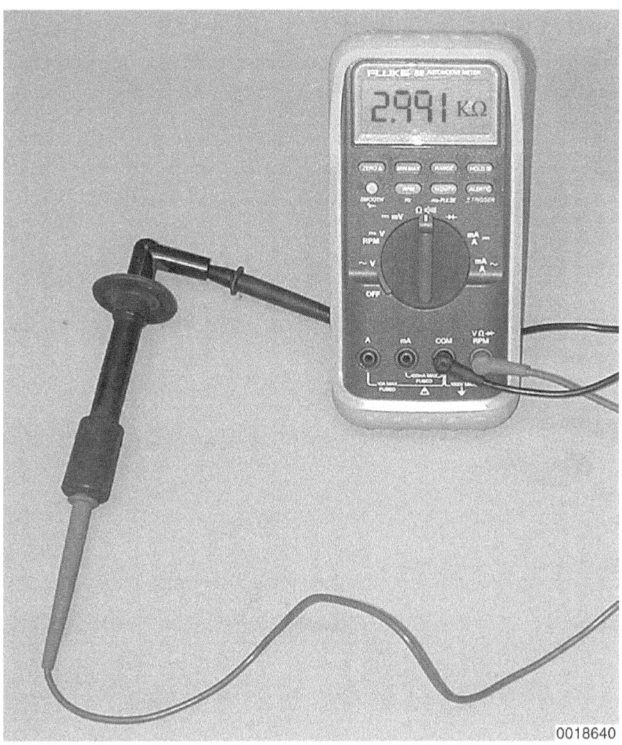

To test connector, use ohmmeter probe on both ends.

– Test shielded ends of coil and spark plug wires at distributor cap and coil using same method.

> ***CAUTION —***
> *To avoid damaging the distributor cap, do not wiggle the connectors when removing the spark plug wires. If necessary, twist to loosen. Then, pull straight out from the cap.*

- Check for corrosion at connectors.
- If measured resistance is too high, wire assembly should be replaced.

Ignition cable resistance
- Plug connector to distributor cap end . . . . 3 kΩ

NOTE —
Individual spark plug wire end connectors are available as replacement parts from an authorized Porsche dealer.

0018633

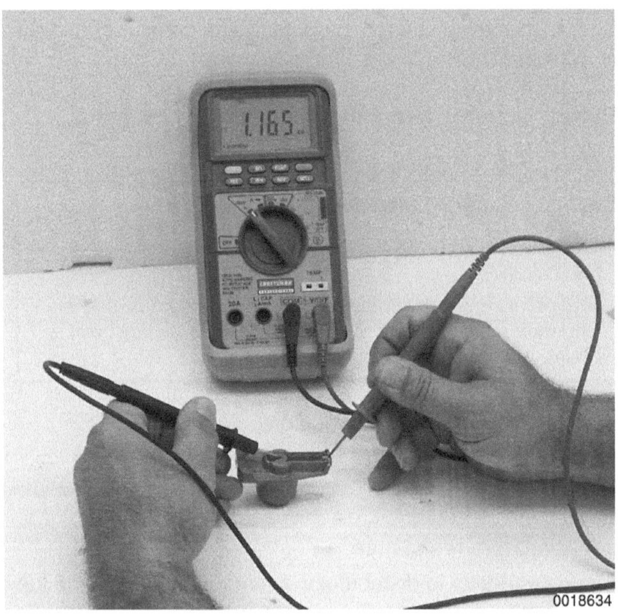
0018634

Distributor cap, testing

< To test cap and rotor, use flat-tipped screwdriver to press down and turn cap hold down pins ¼ turn each.

- Lift off cap.
- Detach spark plug wires and coil wire.

> ***CAUTION —***
> *To avoid damaging the distributor cap, do not wiggle the connectors when removing the spark plug wires. If necessary, twist to loosen. Then, pull straight out from the cap.*

NOTE —
Be sure to label the spark plug wires and their locations when disconnecting them from the cap.

- Check resistance of cap between tower and its matching contact inside cap. Resistance should be nearly zero ohms.
- If any resistance is detected, replace cap.

Ignition rotor, testing

– After removing cap, pull rotor straight up off drive shaft.

< Check rotor resistance with ohmmeter.

- If measured resistance is too high, replace rotor.

Ignition Rotor Resistance

- Center to tip of rotor . . . . 1 kΩ

Reference sensor and speed sensor

Two separate sensors, mounted on the driver side of the transmission bellhousing, supply the engine control module (ECM) with engine speed and crankshaft reference (position) signals.

- The speed sensor uses the flywheel teeth to determine engine speed.
- The crankshaft reference sensor signals the crankshaft position, or angle, when a raised pin on the flywheel passes the sensor.

– If the ECM does not receive a crankshaft reference signal from the sensor, the engine will not start. If the engine consistently misses at constant speeds, test the speed sensor. DME electrical tests other than those described below can be found in **240 Fuel Injection.**

NOTE —
A top dead center (TDC) position sensor is mounted on the top right of the flywheel end of the engine. This is for use with the Porsche service-test unit and is not linked to the DME system or the engine in any way and does not affect timing or the way the engine runs.

Reference and speed sensors, testing

The crankshaft reference and speed sensor wiring harnesses are routed around the left side of the engine and lead to connectors above the left fuel rail. The wires are marked with tape collars.

Sensor Wire Markings
- Speed sensor . . . . DG
- Reference sensor . . . . BG

When testing the sensors, the temperature should be approximately 77°F (25°C) to obtain the most accurate results.

Detach sensor electrical connectors located on left side of engine.

- Check resistance between sensor terminals.
- Correct resistance values are given in **Table b**.

NOTE —
The two sensors are electrically identical.

Table b. Reference or Speed Sensor Terminal Resistance

Terminals	Resistance
1 and 2	960 ± 96 Ω
1 and 3	> 100,000 Ω
2 and 3	> 100,000 Ω

– Remove sensors, as described later.
  - Check for dents or cracks in sensors, or chafed spots in wiring.

NOTE —
Foreign material in the bellhousing can destroy a sensor.

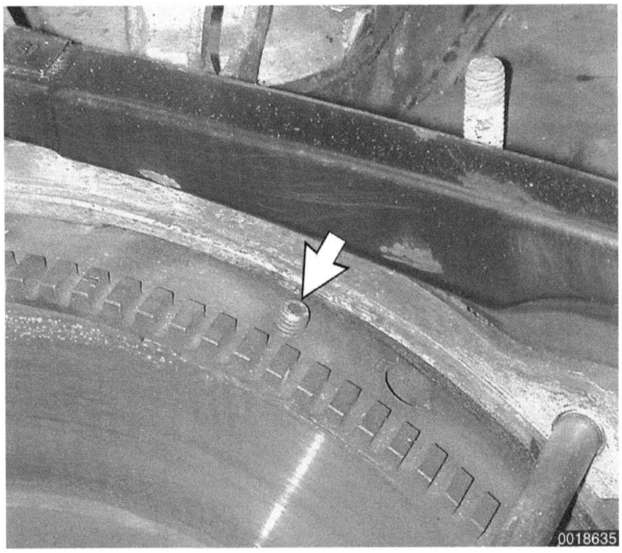

◀ If no faults are found with reference sensor or speed sensor, inspect raised pin on flywheel (**arrow**).
  - Use wrench on front crankshaft pulley to hand turn engine clockwise until pin is visible through hole at bottom of bell-housing.
  - If pin is damaged or missing, it should be replaced with an original pin available from a Porsche dealer.

COMPONENT REPLACEMENT

The cap, rotor and spark plug wires all carry high voltage to the spark plugs. Proper engine performance depends on getting the best possible spark at the spark plug. Worn and corroded contacts or poor insulation which allows the spark to short to ground are the primary reasons for replacement of these components.

Distributor cap and ignition rotor, removing and installing

NOTE —
It is recommended that the cap and rotor be replaced as a set during a major maintenance.

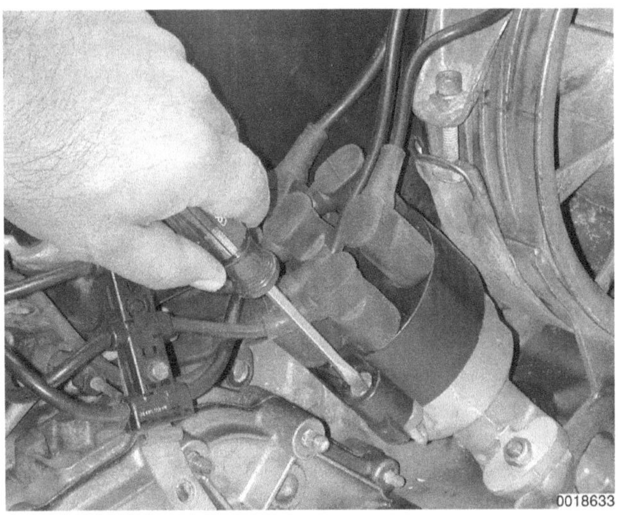

◀ To replace cap and rotor, use flat-tipped screwdriver to press down and turn cap hold down pins ¼ turn each.
  - Lift off cap.

– If replacing cap, temporarily leave spark plug wires in old cap.
  - Mount new cap on distributor and lock both pins.
  - Change over one spark plug wire at a time from old cap to new cap.

> ***CAUTION —***
> *To avoid damaging the distributor cap, do not wiggle the connectors when removing the spark plug wires. If necessary, twist to loosen. Then, pull straight out from the cap.*

NOTE —
The cap can only be mounted in one position. Make sure the cap is properly seated before locking the mounting pins.

– To remove rotor, first remove distributor cap. Next pull straight up on rotor to remove it from distributor shaft.

NOTE —
The rotor can only be mounted in one position. Align tab in the rotor shaft with the notch in the distributor shaft and push the rotor into place.

Spark plug wires, replacing

– When removing spark plug wires, label their positions so that they can be reinstalled in the proper places. If the wires get mixed up, see **Firing order** above.

Crankshaft sensors, removing and installing

◄ Pull out connector locking plate at sensor connectors in left side of engine compartment.

• Detach one or both black connectors.

Remove reference or speed sensor mounting bolt using a 5 mm allen wrench.

NOTE —

- *If the sensor is seized in the mounting bracket, remove bracket mounting bolts (**arrows**). Remove bracket with both sensors to work bench and replace sensor(s) as needed.*
- *Once bracket has been removed, sensor clearance gap must be reset as described below.*

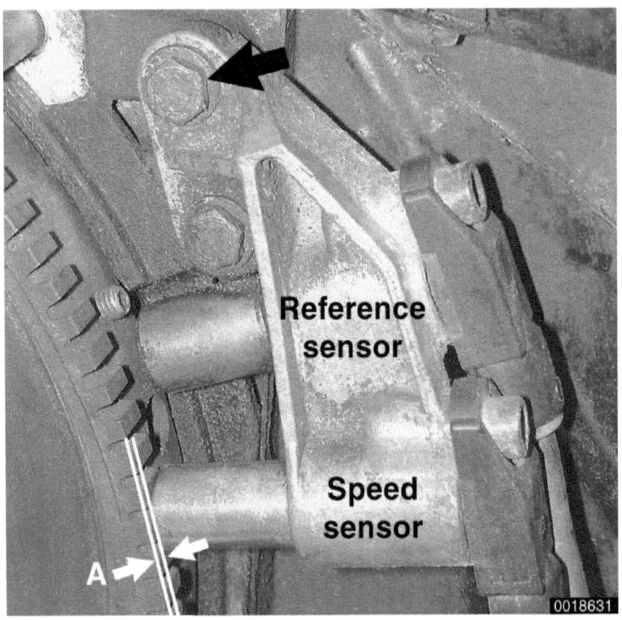

When installing sensor mounting bracket, position sensor tip correct distance (**A**) from toothed wheel using slotted bracket mounting bolt hole (**arrow**).

Crankshaft Sensor Clearance Specifications
- Sensor tip to toothed wheel (**A**) .... 0.08 mm (0.031 in.)

- Tighten sensor mounting bolt.
- Be sure wiring is correctly routed into engine compartment.

NOTE —

- *A discarded sensor can be used to set sensor gap. Glue an 0.8 mm washer to tip of used sensor and install temporarily in order to adjust gap.*
- *When replacing sensors make sure the connectors are not interchanged, and that each sensor is placed in the correct position in the bracket. The bellhousing is marked with a B for reference sensor and a D for speed sensor.*
- *If the sensors or sensor harness connectors are reversed, the engine will not start.*
- *Do not overtighten sensor bolt.*

Tightening Torque
- Sensor to bracket .... 8 Nm (6 ft-lb)

300 Transmission & Clutch–General

TABLES

GENERAL

This section covers general information about the transmission and clutch. Transmission removal is covered in **101 Engine Removal and Installation**. Refer to the following repair groups for repair information:

- **301 Clutch**
- **340 Manual Transmission–Controls and Case**
- **350 Manual Transmission–Gears and Shafts**
- **390 Differential**

TRANSMISSION

The models covered by this manual are equipped with either the 915 or the G50 transmission. The 915 transmission can be identified by its cable operated clutch system. The G50 can be identified by the hydraulic operated clutch.

Cross-section of 915 transmission

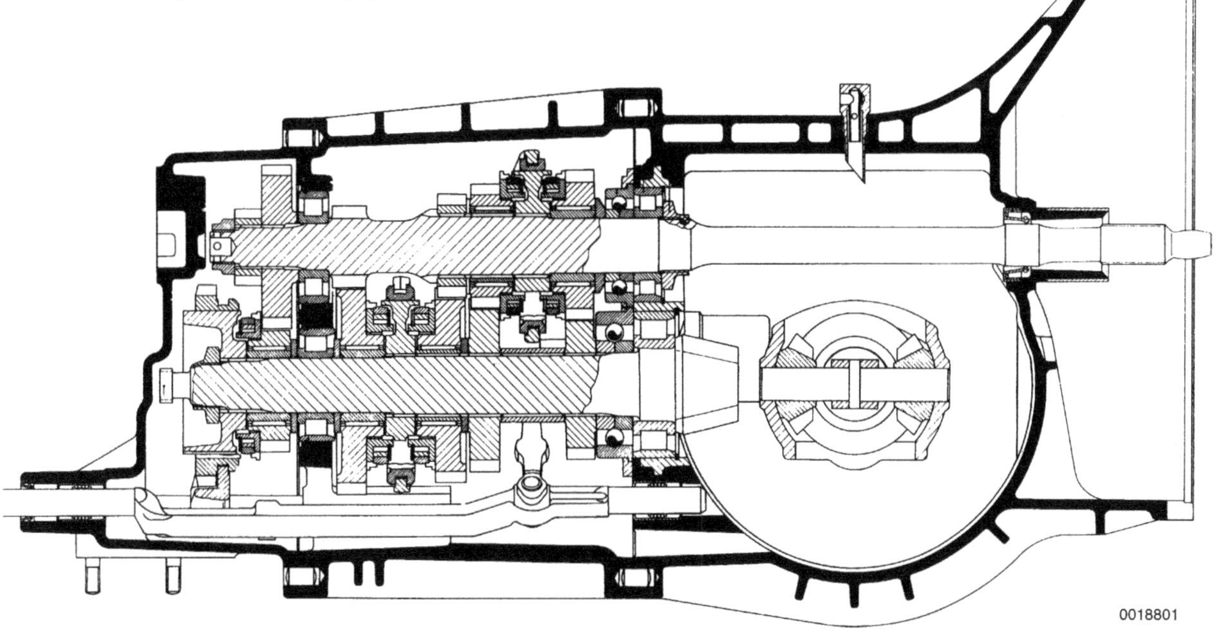

915 transmission speed in gear

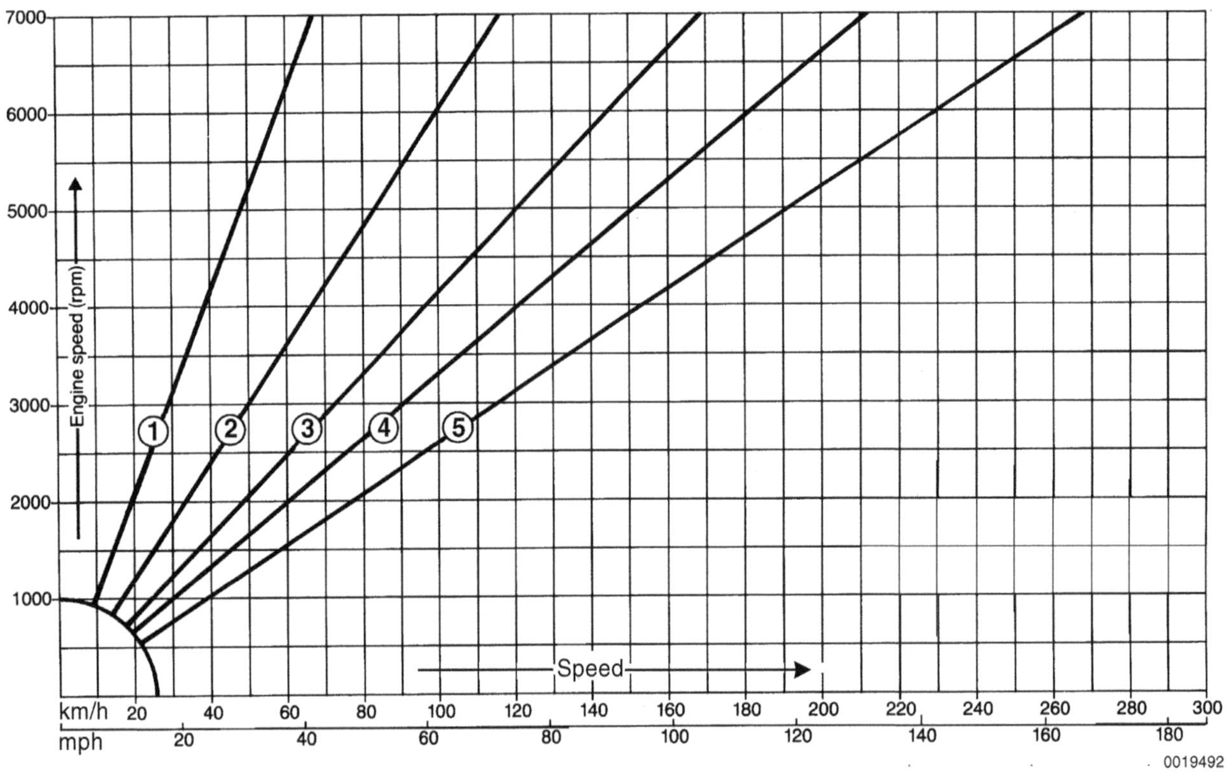

G50 transmission speed in gear

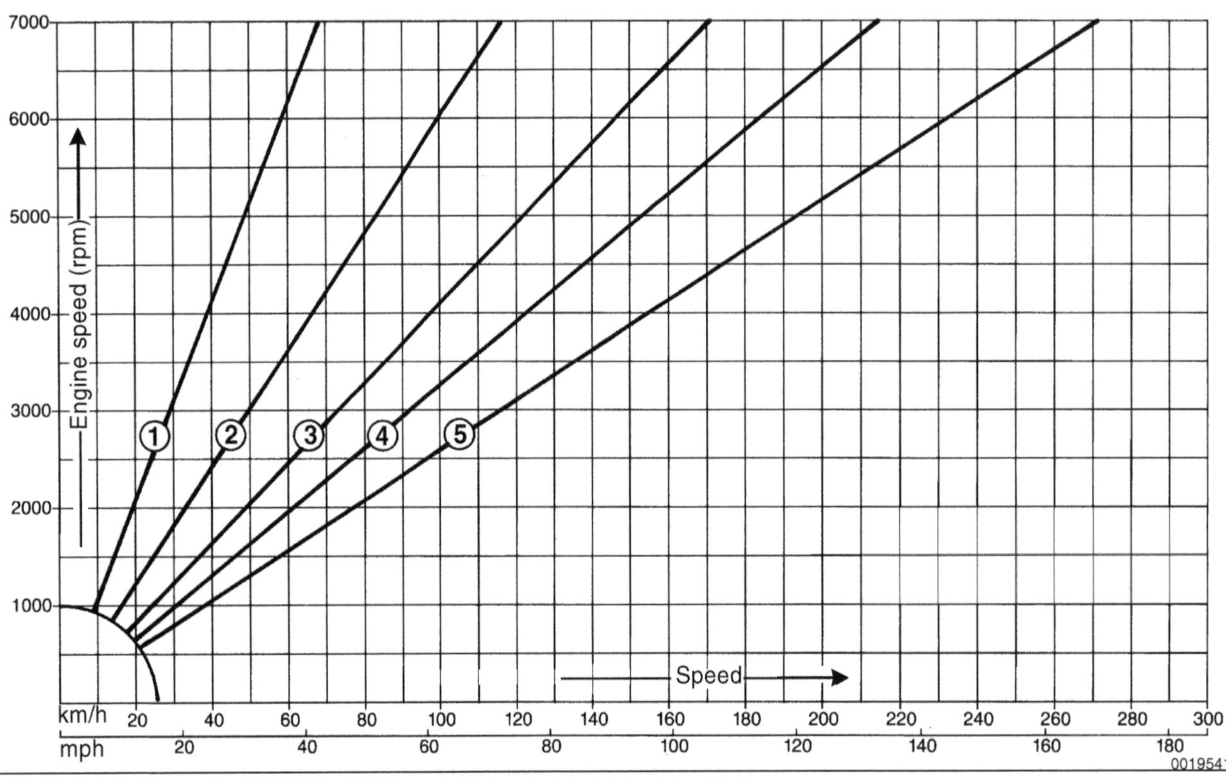

Transmission applications are listed in **Table a**. (R.o.W. in table is an abbreviation for "rest of world.") Transmission gear ratios are in **Table b**.

Table a. Transmission Applications

Year, type	Feature	Market	Serial numbers
1984			
915/67	oil cooler	Europe, R.o.W.	73 E 00001 - 10000
915/68	w/o oil cooler	USA, Japan	74 E 00001 - 10000
915/69	oil cooler	Europe, R.o.W.	73 E 10001 - 11000
915/70	w/o oil cooler	USA, Japan	74 E 10001 - 11000
1985-86			
915/72	oil cooler	Europe, R.o.W.	73 G 00001 - 10000
915/73	w/o oil cooler	USA, Japan	74 G 00001 - 10000
1987			
G50.00		Europe, R.o.W.	73 H 00001 - 20000
G50.01		USA, Japan	74 H 00001 - 20000
1988			
G50.01		USA, Canada	2J10001 - 2J20000
G50.01	40% limited slip		2J10001 - 2J20000
1989			
G50.01		USA, Canada	1K00001 - 1K10000
G50.01	40% limited slip		1K00001 - 2K20000

Table b. Transmission Gear Ratios

	915/67 915/69 916/72	915/68 915/70 915/73	G50.00	G50.01	G50.02
Gear ratios					
1st	3.1818	3.1818	3.500	3.500	3.154
2nd	1.8333	1.7777	2.059	2.059	1.895
3rd	1.2608	1.2608	1.409	1.409	1.333
4th	0.9655	1.0000	1.074	1.125	1.036
5th	0.7631	0.7895	0.861	0.889	0.861
reverse	3.3250	3.3250	2.857	2.857	2.857
Final drive	3.8750:1		3.444:1		

The transmission case, which contains the transmission gear train, is a lightweight alloy die casting. The transmission gears are of the constant-mesh type . Each forward gear (1st through 5th) is actually a mating pair of gears, and constant-mesh simply means the mating gears are always meshing.

The 915 transmission uses Porsche-designed gear type synchronizers. The G50 transmission uses more traditional balk ring synchronizers.

Gear selection is made using the gearshift lever in the passenger compartment, which is linked to the transmission by the gear shift linkage. When shifting, the synchronizer matches the speed of the selected gear to the speed of the output shaft. The synchronizer hub then locks the selected gear into rotation with the output shaft. This eases shifting and minimizes gear wear.

Transmission repairs

Repairs fall into two categories: those that can be performed externally, and those that require disassembly of the transmission. Problems that at first appear to be caused by internal faults, such as gear shifting difficulty or noisy operation, can often be corrected externally.

Repairs to the internal parts of the transmission require special tools, knowledge and experience. If you lack the skills and tools, or a suitable workplace for servicing the transmission, we suggest you leave these repairs to an authorized Porsche dealer or other qualified shop. Completely reconditioned transmissions are available from the authorized Porsche dealer or the aftermarket.

Though you may not have the skills and knowledge for doing the actual internal repairs, it is possible to save some of the expense of professional repair by removing and installing the transmission yourself, using the procedures described in **101 Engine Removal and Installation**. It is important to realize, however, that a partially disassembled transmission may be a problem for a mechanic. We strongly advise against taking the transmission apart to begin any repairs that cannot be properly finished.

CLUTCH

The 915 transmission models are equipped with a cable operated clutch release system with a spring dampening clutch disc. The G50 transmission models are equipped with a hydraulic clutch release system with a rubber dampening clutch disc.

MAINTENANCE

Information on clutch and transmission maintenance and on the prescribed maintenance intervals can be found in **030 Maintenance**.

Fluid capacities

- 915 transmission 1984 - 1986 . . . . 3.0 L (3.2 qts.).
- G50 transmission 1987 - 1989 . . . . 3.4 L (3.6 qts.).

301 Clutch

Special Tools

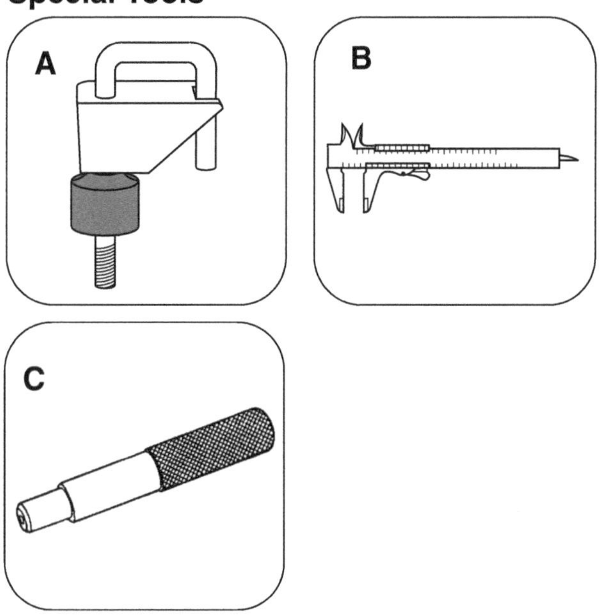

Special tool

A 647 b (source: Baum Tools Unlimited) . . Flexible line clamp tool
B 14 (source: Baum Tools Unlimited) . . . . Vernier tool
C P 370 . . . . Clutch disc alignment tool (source: Baum Tools Unlimited)

GENERAL

There are two clutch systems used on the 1984 through 1989 911 models. See **Table a**.

Table a. Clutch Applications

Model Year	Trans. Code	Operation	Clutch Disc Diameter	Disc design
1984 - 1986	915	Cable	225 mm (8.86 in.)	Spring damper center
1987 - 1989	G50	Hydraulic	240 mm (9.45 in.)	Rubber damper center

Special tools

Some of the procedures in this group require special tools.

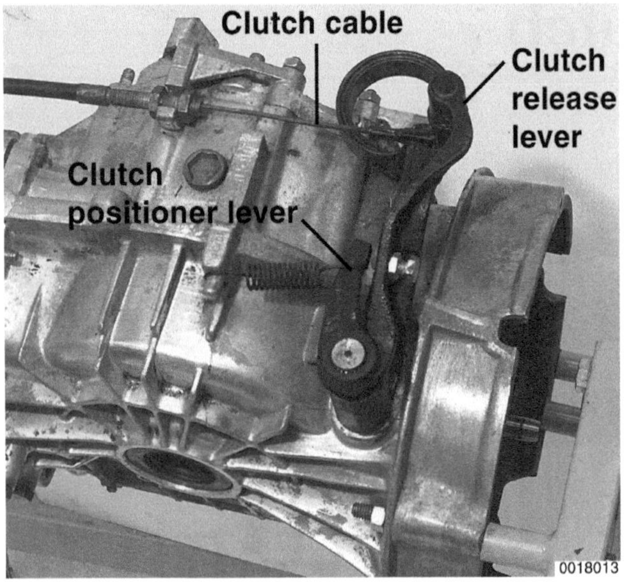

CLUTCH CABLE AND PEDAL CLUSTER (1984-1986 MODELS)

When the clutch pedal is depressed, the clutch cable actuates the clutch release lever via the clutch positioner lever on the transmission.

The positioner lever is splined to the clutch cross-shaft. The clutch release fork is also splined to the cross-shaft in the transmission bellhousing. Turning the cross-shaft pulls the release fork forward and disengages the clutch.

Clutch cable, adjusting

Clutch pedal and cable free play is adjusted at the clutch positioning lever on the transmission.

– Raise car and support in a safe manner.

> ***WARNING —***
> *Make sure the car is stable and well supported at all times. Use a professional automotive lift or jack stands designed for the purpose. A floor jack is not adequate support.*

Working under rear of car, loosen cable housing lock nuts (**arrow**) and disconnect cable from clutch release lever.

Set basic adjustment:

- With cable end disconnected, loosen lock nut and adjust stop bolt until specified gap is obtained between positioning lever and stop bolt.
- Tighten lock nut.

Clutch Cable Basic Adjustment (cable disconnected)

- Gap between clutch positioning lever and stop bolt. . . . . 1.2 mm (0.047 in.)

Set cable preload:

- Reattach clutch cable to release lever and cable support on transmission.
- Using cable housing locknuts, remove all slack in cable.
- Tension cable further (at housing locknuts) until specified gap between positioning lever and stop bolt is obtained.

Clutch Cable Preload (cable connected)

- Gap between clutch positioning lever and stop bolt. . . . . 1.0 mm (0.04 in.)

- Tighten cable housing lock nuts. Recheck gap.

NOTE —

If clutch cable housing does not provide the adjustment needed, the clevis at the front of the cable may be adjusted as described under **Clutch cable, replacing**. *Note that the clevis must threaded onto the cable within a range of 17 - 22 mm (0.67 - 0.87 in.).*

Clutch release distance, checking and adjusting (cable operated clutch)

Clutch release distance is limited by an adjustable stop on the floor cover above the pedal cluster. This adjustment should be checked only after adjusting the clutch cable as described earlier. See **Clutch cable, adjusting**.

– Raise car and support in a safe manner.

WARNING —

Make sure the car is stable and well supported at all times. Use a professional automotive lift or jack stands designed for the purpose. A floor jack is not adequate support.

Working under transmission, use vernier calipers to measure distance between clutch cable lock nut and end of cable.

– Have helper push clutch pedal to floor stop. Measure distance again.

– Compare difference between two measurements.

Clutch Release Specification (cable-operated clutch)

- Clutch release distance (measured at transmission) 25 ± 0.5 mm (0.98 ± 0.02 in.)

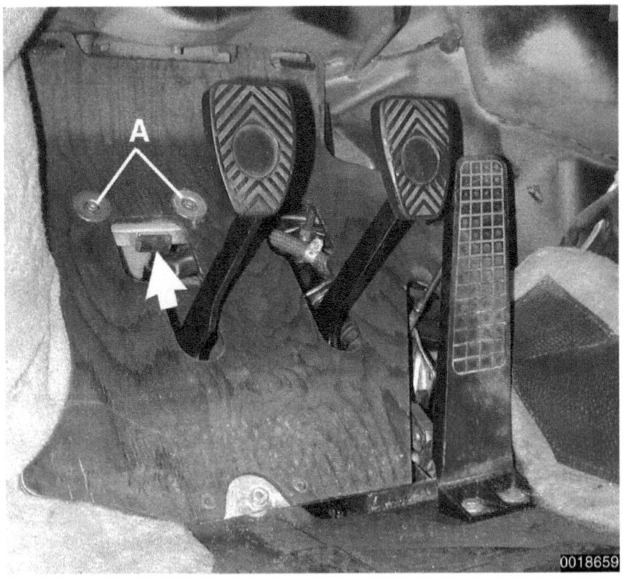

◄ If clutch release distance is out of tolerance, adjust clutch pedal stop (**arrow**) by loosening allen bolts (**A**).

Clutch cable, replacing

– Raise car and support in a safe manner.

> ***WARNING —***
> *Make sure the car is stable and well supported at all times. Use a professional automotive lift or jack stands designed for the purpose. A floor jack is not adequate support.*

◄ Working under rear of car, loosen clutch cable housing locknuts (**A**).

- Detach cable from clutch release lever.

– Working inside car, remove driver's side floor mat.

– Pull accelerator pedal back and disconnect accelerator push rod from pedal.

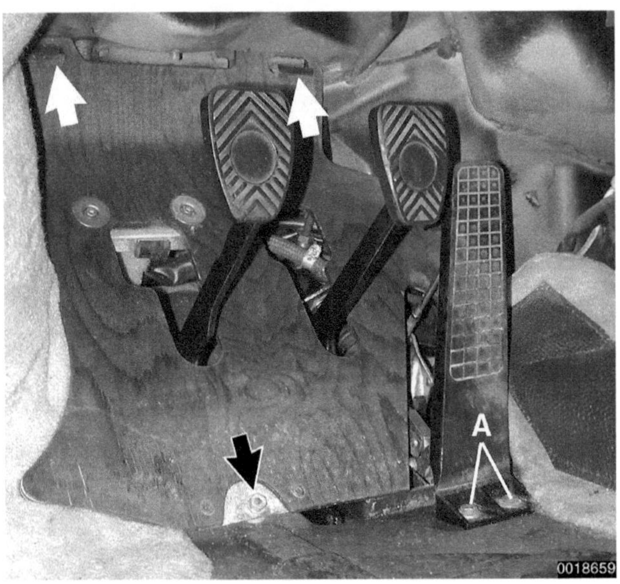

◄ Remove accelerator pedal mounting screws (**A**) and remove pedal.

– Remove pedal cluster cover by removing fastener at bottom and unhooking from above (**arrows**).

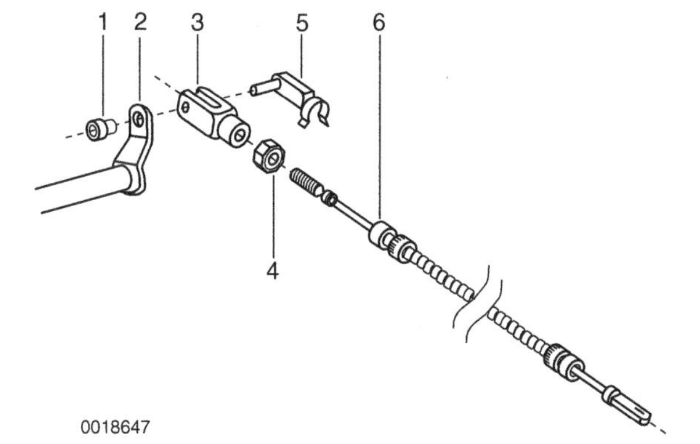

1. Bushing
2. Clutch pedal lever
3. Clevis
4. Clevis lock-nut
5. Clevis clevis pin/locking clip
6. Clutch cable

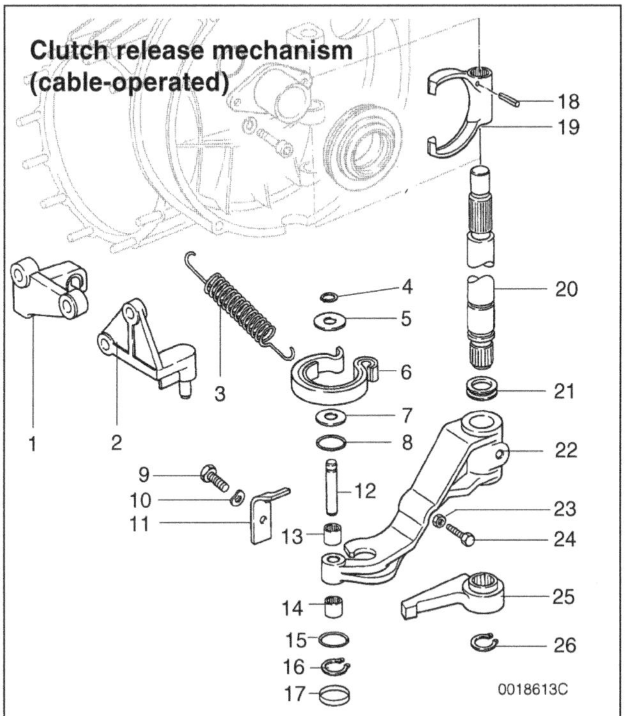

1. Clutch cable anchor
2. Auxiliary spring anchor
3. Positioning lever spring
4. Circlip
5. Washer
6. Auxiliary return spring
7. Washer
8. Seal
9. Bolt
10. Washer
11. Spring anchor
12. Pin
13. Auxiliary spring bushing
14. Auxiliary spring bushing
15. Washer
16. Circlip
17. Plug
18. Dowel pin
19. Clutch release fork
20. Clutch release cross-shaft
21. Seal
22. Clutch release lever
23. Lock-nut
24. Adjusting bolt
25. Clutch positioning lever
26. Circlip

< Locate clutch cable connecting clevis in center tunnel.

- Swivel locking clip upward and pull pin out of clevis.
- Loosen nut behind clevis and remove both clevis and nut.

– Working under rear of car, pull cable out of guide hole in body.

– Install new cable from rear.

– Working at pedal cluster area, install nut and clevis on new cable.

- Adjust clevis so that 17 - 22 mm (0.67 - 0.87 in.) of threaded cable end is screwed into clevis.
- Tighten clevis lock nut.

– Place clevis on pedal lever. Install and lock clevis pin.

– Reinstall pedal cluster cover, accelerator pedal and floor mat.

– Working under rear of car, install clutch cable to transmission.

– Check clutch cable adjustment. See **Clutch cable, adjusting**.

Clutch release lever, lubricating (cable operated clutch)

If the clutch pedal is sticking or not returning all the way, the problem may be in the clutch release lever or auxiliary spring bushings.

> ***CAUTION —***
> *Do not apply lubricant to the pedal bushings to alleviate stiff pedal operation. Most lubricants will cause the bushings to swell.*

– Raise car and support in a safe manner.

> ***WARNING —***
> *Make sure the car is stable and well supported at all times. Use a professional automotive lift or jack stands designed for the purpose. A floor jack is not adequate support.*

< Working under car, disconnect clutch cable from clutch release lever (**22**).

– Disconnect positioning lever spring (**3**).

– Remove circlip holding clutch positioning lever to clutch cross-shaft (**26**).

– Remove positioning lever (**25**).

– Pry release lever (**22**) forward with a large screwdriver.

– Pull release lever (**22**) off clutch cross-shaft (**20**).

> *CAUTION —*
> *Use care when prying release lever forward. The release lever will forcefully snap forward at a certain point.*

NOTE —
If the release lever will not slide off the clutch cross-shaft, the pedal stiffness problem may have been found. Use heat and back and forth motion to get the lever off.

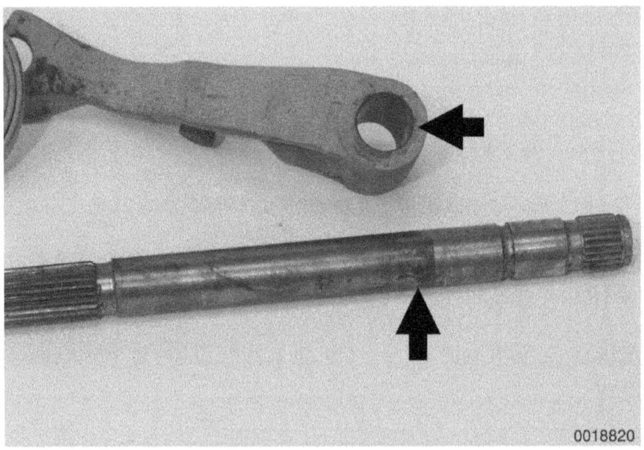

Check cross-shaft and cross-shaft bore (**arrows**) in release lever for rust or corrosion.

- Clean bore and cross-shaft.
- Lubricate cross-shaft and bore, release lever splines, auxiliary spring bushings and clutch cable mount on release lever, using multipurpose grease.

– Replace auxiliary return spring bushings if necessary.

NOTE —
If clutch cross-shaft bore or needle bearings (in transmission bellhousing) are damaged, repairs can only be carried out with the transmission removed form the car. See **Clutch cross-shaft, removing and installing (cable-operated clutch)** *later in this section.*

– Reassembly is reverse of removal.

– Check and adjust clutch cable and pedal throw (clutch release distance). See **Clutch cable, adjusting** and **Clutch release distance, checking and adjusting (cable-operated clutch)**.

Pedal cluster, removing and installing (cable operated clutch)

If the pedals malfunction and routine cleaning and maintenance does not relieve the problems, remove and overhaul the pedal cluster as described below.

> *CAUTION —*
> *Do not apply lubricant to the pedal bushings to alleviate stiff pedal operation. Most lubricants will cause the bushings to swell.*

NOTE —
Road dirt and grime will collect around the pedal cluster area and can eventually cause corrosion and stiff pedal operation. A yearly cleaning in the pedal cluster area will keep it working trouble-free.

– Raise car and support in a safe manner.

> *WARNING —*
> *Make sure the car is stable and well supported at all times. Use a professional automotive lift or jack stands designed for the purpose. A floor jack is not adequate support.*

◀ Working under car at transmission, loosen clutch cable mounting nuts (**A**).

- Detach clutch cable from clutch release lever.

– Working inside car, remove driver's floor mat.

– Pull accelerator pedal back to release accelerator push rod.

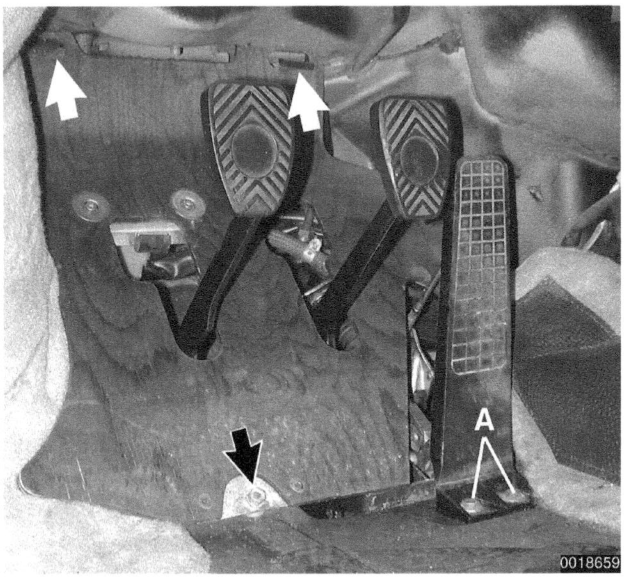

◀ Remove accelerator pedal mounting screws (**A**) and remove pedal.

– Remove pedal cluster cover after removing carpet securing straps at top and fastener at bottom (**arrows**).

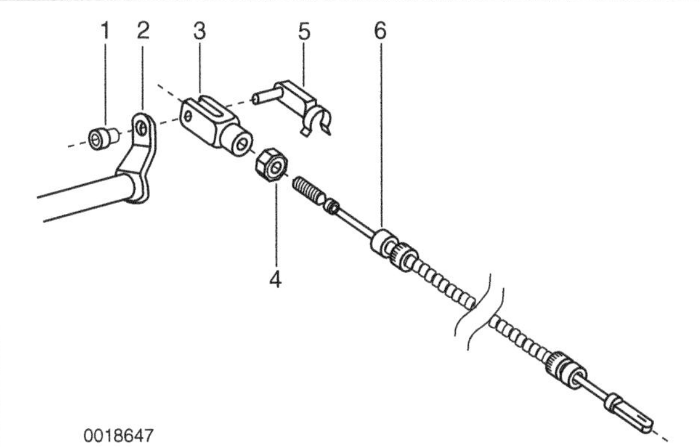

1. Bushing
2. Clutch pedal lever
3. Clevis
4. Clevis lock-nut
5. Clevis clevis pin and clip
6. Clutch cable

◀ Locate clutch cable connecting clevis (**arrow**) in center tunnel.

- Remove locking clip and pull pin out of clevis.
- Push clutch cable forward.

Disconnect/remove the following from pedals and/or pedal cluster:

- Vertical brace at left side of pedal cluster (**1**).
- Brake master cylinder pushrod (**2**).
- Return spring at clutch pedal (**3**). Push clutch pedal forward.
- Accelerator cable (**4**) from accelerator bell-crank.
- Pedal cluster mounting nuts (**5**).

Working underneath front of car, remove stone guard mounting bolts (**arrow**). Remove stone guard.

Remove two M8 (8 mm) pedal cluster mounting nuts (**arrows**) on left side of front bulkhead.

– Installation is reverse of removal.

Tightening Torque

- Pedal cluster to floor (M8). . . . . 17 Nm (12 ft-lb)

Pedal cluster assembly (cable-operated clutch)

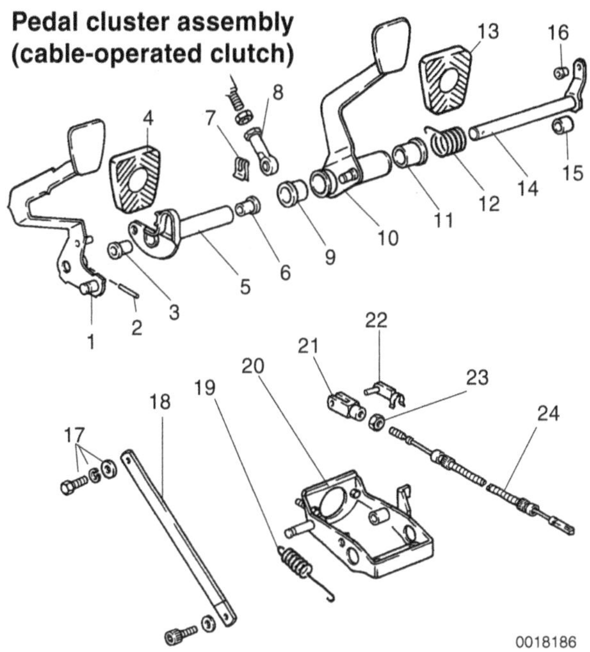

1. Clutch pedal
2. Roll pin
3. Bushing
4. Pedal pad
5. Brake pedal bearing tube
6. Bushing
7. Locking clip
8. Brake master cylinder pushrod
9. Bushing
10. Brake pedal
11. Bushing
12. Brake pedal return spring
13. Pedal pad
14. Clutch pedal cross-shaft
15. Bushing
16. Bushing
17. Brace mounting hardware
18. Vertical brace
19. Clutch pedal return spring
20. Pedal cluster housing
21. Clutch cable clevis
22. Clutch cable clevis pin
23. Clevis lock-nut
24. Clutch cable

Pedal cluster bushings, replacing (cable operated clutch)

– Remove and clean pedal cluster assembly before starting to disassemble.

– Support pedal cluster in soft jawed vise.

– Drive out dowel pin that locks clutch pedal to cross-shaft. Support shaft to prevent deformation of pedal cluster housing.

NOTE —
Use 4 mm pin punch (Automotion special tool no. TL-4600) to drive out dowel pin.

– Remove clutch pedal and cross-shaft.

– Remove bearing tube from brake pedal base.

– Remove brake pedal and spring.

< Remove spring clip from throttle bellcrank mounting shaft and slide bellcrank off shaft.

– Remove all bushings and clean all parts. Sand off rust and paint metal parts.

– Install new bushings.

NOTE —
- *Nylon replacement bushings are available from the Porsche dealer parts department.*
- *More durable brass bushings are available from aftermarket sources.*

– Reassembly is reverse of disassembly.

NOTE —
- *Check and replace parts that show any wear.*
- *When assembling, use universal grease.*

CLUTCH HYDRAULICS AND PEDAL CLUSTER (1987-1989 MODELS)

The hydraulic clutch is not adjustable. The clutch slave cylinder automatically compensates for normal wear of the clutch disc. The clutch shares a common fluid reservoir with the brake system. The fluid reservoir is located in the luggage compartment.

The same problems that typically affect brake operation—air in the lines and moisture in the fluid—can also affect clutch operation. Periodic clutch and brake fluid changes are recommended to minimize moisture absorption and corrosion in the hydraulic components.

Failure of the clutch to fully disengage (leading difficult gear shifting) may be caused by one or more of the following:

- Leaky or faulty clutch slave cylinder or master cylinder.
- Inadequate travel or misalignment of the slave cylinder pushrod.
- Faulty bent master cylinder pushrod.
- Binding clutch disc.
- Seized pilot bearing.

NOTE —
Although seal rebuild kits are available for clutch hydraulic components, replacing leaking components with new parts is recommended to insure reliability. In addition, rebuilding these components successfully depends on their internal condition, which can only be determined after the unit is removed and fully disassembled.

Bleeding clutch hydraulic system

Top off brake master cylinder fluid reservoir with brake fluid and remove screen.

- Clamp off overflow hose coming from reservoir using special tool 647 b or equivalent.
- Attach brake pressure bleeder to reservoir.

– Raise car and support in a safe manner.

WARNING —
Make sure the car is stable and well supported at all times. Use a professional automotive lift or jack stands designed for the purpose. A floor jack is not adequate support.

Attach bleeder hose and catch bottle to clutch slave cylinder.

- Open bleeder screw (**arrow**) on clutch slave cylinder and allow system to bleed until no air bubbles escape.

– With air bleed screw open slowly depress clutch pedal fully several times.

– With pedal fully depressed close bleeder screw.

– Remove pressure bleeder and top off brake fluid. Release clamp from overflow hose.

– Install screen and reservoir cap.

Brake Fluid Specification

• Brake/clutch fluid. . . . . .DOT 4

CAUTION —
Do not operate the clutch pedal with the slave cylinder removed from the transmission.

Clutch slave cylinder, replacing

< Remove brake fluid reservoir cap. Using a clean syringe, remove brake fluid until the level is below the feed hose connection (**arrow**) for the clutch master cylinder.

NOTE —
An alternate technique is to pinch off the clutch master cylinder feed hose, as shown.

WARNING —
Brake fluid is poisonous. Do not siphon brake fluid with your mouth. Wear safety glasses when working with brake fluid and rubber gloves to prevent brake fluid from entering the bloodstream through cuts or scratches. Always dispose of old fluid carefully.

– Raise car and support in a safe manner.

< Working on top of transmission on driver's side:

- Loosen clutch slave cylinder fluid line fitting (**A**).
- Remove slave cylinder mounting nuts (**B**).
- Pull slave cylinder out of transmission and disconnect fluid line.

Inspect contact surfaces of slave cylinder push rod and release bearing lever for wear.

NOTE —
- *The push rod should only be worn on the tip.*
- *The release lever should only show wear in the pushrod recess.*

Inspect clutch slave cylinder for fluid leaks. Lift up on dust book to check for leaks (**arrow**).

- Replace slave cylinder if fluid is present.

– Lightly coat slave cylinder pushrod tip with molybdenum disulfide grease (Molykote® Longterm 2 or equivalent) and make sure pushrod tip engages recess in clutch release lever.

– Install slave cylinder mounting nuts and washers.

– Install hydraulic line and tighten fitting.

– Refill brake fluid reservoir and bleed clutch system as described earlier.

Tightening Torques
- Slave cylinder to transmission . . . . 23 Nm (16 ft-lb)
- Hydraulic line fitting to slave cylinder . . . 28 Nm (20 ft-lb)

Clutch master cylinder, replacing

– Working inside car, remove driver's floor mat.

– Pull accelerator pedal back to release accelerator push rod.

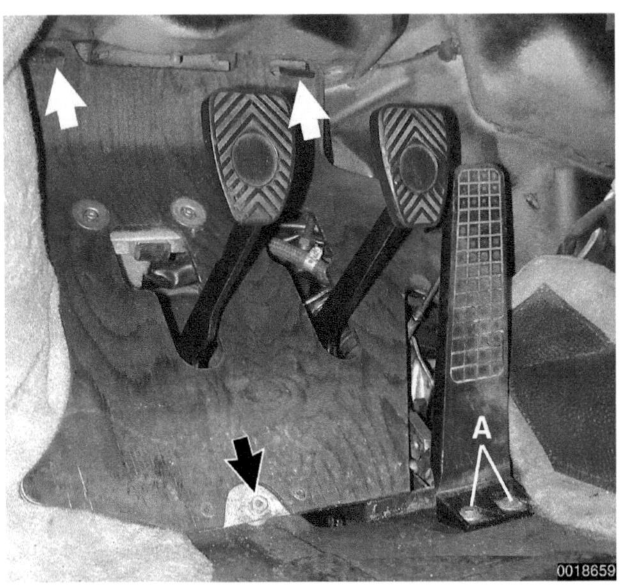

Remove accelerator pedal mounting screws (**A**) and remove pedal.

– Remove pedal cluster cover after removing fastener at bottom and unhooking from above (**arrows**).

– Disconnect brake push rod from brake pedal. Disconnect return spring from brake pedal.

– Disconnect clutch master cylinder push rod from clutch pedal.

Clamp off master cylinder supply line at reservoir (special tool 647b or equivalent).

- Working at pedal cluster, loosen and remove hydraulic line fitting from clutch master cylinder.
- Remove clutch master cylinder mounting nuts.
- Slide cylinder forward and out of pedal bracket.
- Installation is reverse of removal.
- Adjust master cylinder pushrod as described later, if necessary.
- Refill brake fluid reservoir and bleed clutch system as described earlier.

Tightening torques
- Master cylinder mounting nuts . . . . 23 Nm (16 ft-lb)
- Hydraulic line fitting to master cylinder . . 28 Nm (20 ft-lb)

NOTE —

Incorrect installation of either the master cylinder pushrod or the slave cylinder pushrod may cause the clutch release mechanism to bind. This will prevent the clutch pedal from returning or cause the release action to feel jerky. It may even prevent the clutch pedal from being depressed.

Clutch master cylinder push rod, basic setting

Check and adjust the clutch master cylinder pushrod basic setting if:

- the master cylinder has been replaced.
- the master cylinder push rod has been removed.
- the master cylinder push rod has been adjusted incorrectly.

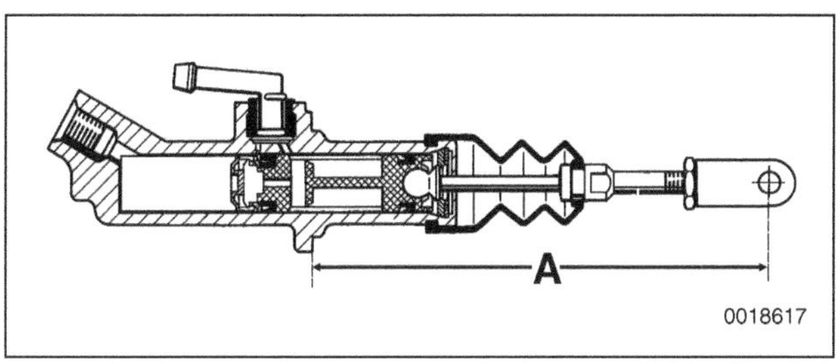

Check basic setting (**A**) and adjust threaded clevis if necessary. The clutch pedal pushrod basic setting must not be adjusted.

Clutch Pedal Pushrod Basic Setting
- Master cylinder flange to center of clevis pin hole (**A**) . . . . . . 158 ± 1 mm (6.221 ± 0.039 in.)

Clutch pedal travel, adjusting (hydraulic operated clutch)

Clutch pedal travel is limited by an adjustable stop (**arrow**) on pedal cluster cover.

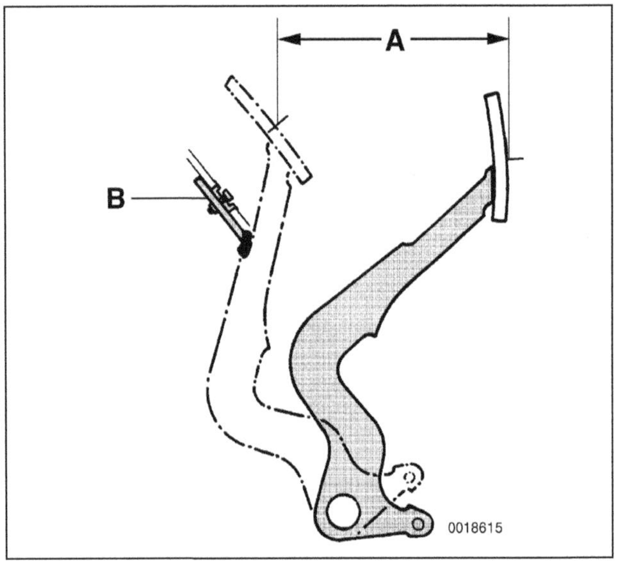

Measure clutch pedal travel distance (**A**) and compare to specification.

- Measure at center of pedal pad.
- Adjust pedal stop (**B**) if necessary.

Clutch Pedal Travel Specification (hydraulic clutch)

- Pedal at top rest position to pedal at bottom of travel .... 150 - 10 mm (5.91 - 0.39 in.)

Clutch pedal servo spring tension, adjusting (hydraulic operated clutch)

A servo spring, installed on the clutch pedal bracket, serves two functions:

- To reduce the amount of force needed to operate the clutch.
- To help return the clutch pedal.

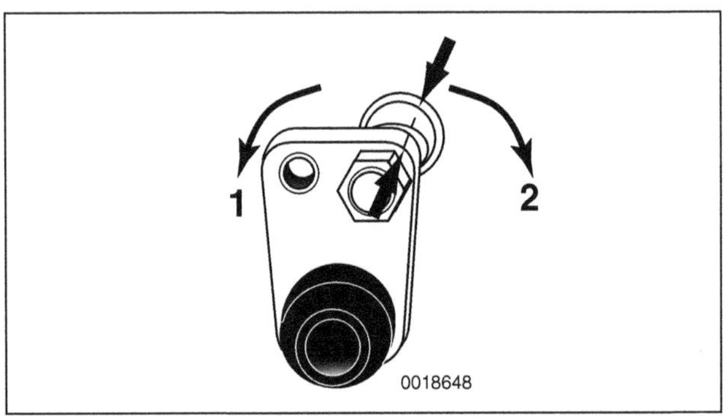

To adjust servo spring tension:

- Basic setting: Align marks (**large arrows**) on adjusting cam and tensioning lever.
- Slow clutch pedal return: Turn adjusting cam approximately 1 - 2 mm (0.039 - 0.078 in.) toward driver (**1**).
- Too much force required to disengage clutch: Turn adjusting cam approximately 1 mm (0.039 in.) away from driver (**2**).

NOTE —
Some earlier models did not have a mark on the cam or lever. If this is the case, scribe a mark on the forward radius of the lever. Then set the highest spot on the cam to the scribed mark and mark the cam.

Tightening Torque

- Servo spring cam to tensioning lever . . . 17 Nm (12 ft-lb)

Pedal cluster, removing and installing (hydraulic clutch)

If the pedals malfunction and routine cleaning and maintenance does not relieve the problems, remove and overhaul the pedal cluster as described below.

– Raise car and support in a safe manner.

> ***WARNING —***
> *Make sure the car is stable and well supported at all times. Use a professional automotive lift or jack stands designed for the purpose. A floor jack is not adequate support.*

– Working inside car, remove driver's floor mat.

– Pull accelerator pedal back to release accelerator push rod.

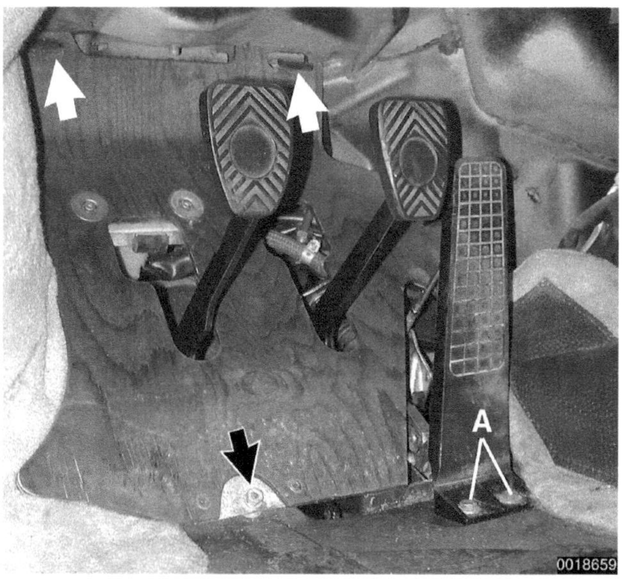

◄ Remove accelerator pedal mounting screws (**A**) and remove pedal.

– Remove pedal cluster cover after removing fastener at bottom and unhooking from above (**arrows**).

◄ Disconnect vertical brace (**1**) from left side of pedal cluster.

– Disconnect brake master cylinder pushrod (**2**) from brake pedal. Snap return spring off brake pedal. Push brake pedal forward.

– Disconnect clutch master cylinder push rod (**3**) from clutch pedal. Remove clutch master cylinder as described earlier. Pull pedal backward to relieve servo spring tension. Unhook spring from cam.

– Disconnect accelerator cable ball joint (**4**) from accelerator bell-crank (**5**).

– Remove nuts (**6**) mounting pedal cluster to floor.

Working underneath front of car, remove stone guard mounting bolts (**arrows**). Remove stone guard.

Remove two pedal cluster mounting nuts (**arrows**) on left side of bulkhead.

– Installation is reverse of removal.

Tightening Torque

- Pedal cluster to floor (M8). . . . . 17 Nm (12 ft-lb)

Pedal cluster, disassembling and reassembling (hydraulic operated clutch)

Remove and clean pedal cluster assembly before starting to disassemble.

– Support pedal cluster in soft jawed vise.

Drive out shaft roll pins (**arrows**). Support shaft to prevent deformation of pedal cluster housing.

NOTE —

Use 4 mm pin punch (Automotion special tool no. TL-4600) to drive out dowel pins.

– Remove clutch pedal and cross-shaft.

Remove bearing tube from brake pedal base. Remove brake pedal and brake pedal return spring.

– Remove servo spring shaft.

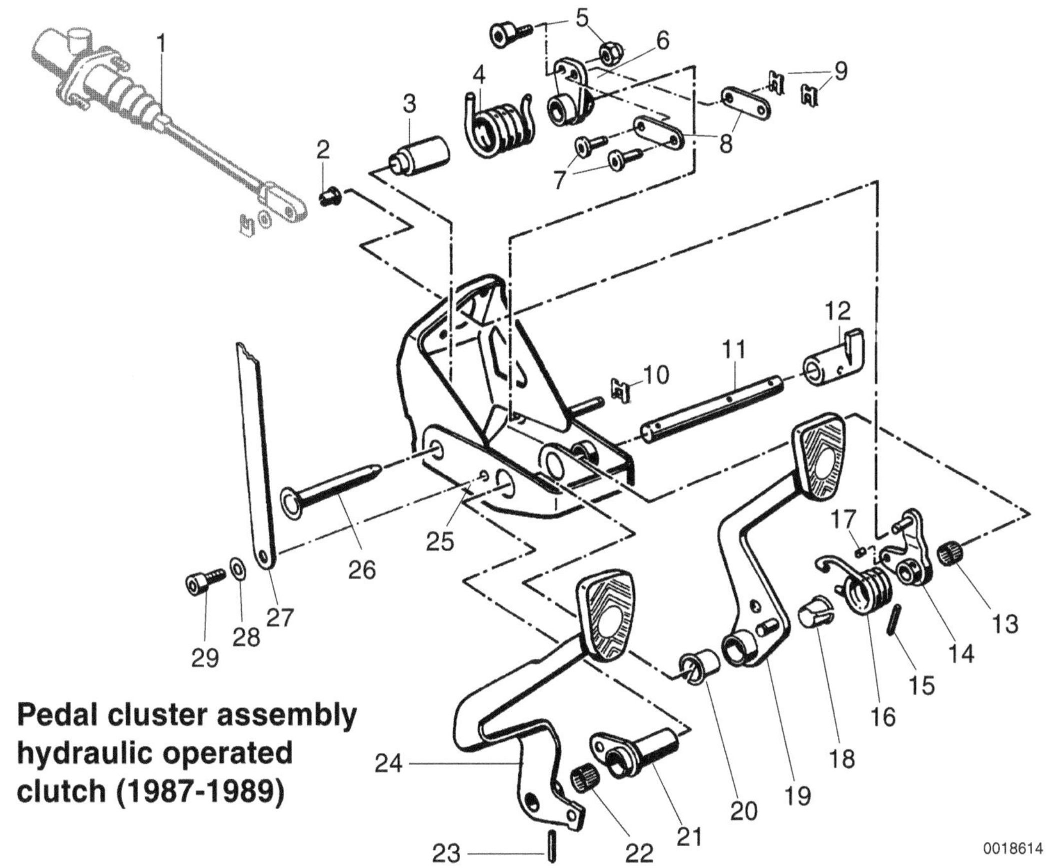

Pedal cluster assembly hydraulic operated clutch (1987-1989)

1. Clutch master cylinder
2. Bushing
3. Clutch servo spring shaft
4. Clutch servo spring
5. Eccentric cam and nut
6. Servo spring lever
7. Pins
8. Clutch servo spring links
9. Locking clips
10. Locking clip
11. Clutch pedal cross-shaft
12. Clutch cross-shaft inner end
13. Bushing
14. Clutch master cylinder bellcrank
15. Dowel pin
16. Brake pedal return spring
17. Bushing
18. Bushing
19. Brake pedal
20. Bushing
21. Brake pedal support tube
22. Bushing
23. Dowel pin
24. Clutch pedal
25. Pedal cluster housing
26. Servo spring support pin
27. Pedal cluster side brace
28. Washer
29. Bolt

◄ Replace clutch cross-shaft support bushings (**arrows**).

NOTE —
- *Note orientation of bushings in photo.*
- *Nylon replacement bushings are available from the Porsche dealer parts department. More durable brass bushings may be available from aftermarket sources.*

Install servo spring (**arrow**) with lever to pedal bracket.

Tightening Torque
- Servo spring cam to lever . . . . 17 Nm (12 ft-lb)

Install brake pedal and support tube along with brake pedal spring (**arrows**) into bracket.

– Reassemble clutch pedal assembly:
  - Hold clutch master cylinder bellcrank in installed position and insert clutch pedal and cross-shaft into pedal cluster housing and through bellcrank. Install dowel pin.
  - Install cross-shaft inner end and install locking pin.

NOTE —
Take care to install clutch pedal assembly parts in the correct orientation.

Attach clutch servo spring links to clutch master cylinder bellcrank with pins and locking clips (**arrows**).

– Install cluster in car.

Tightening Torque
- Pedal cluster bracket to body . . . . 17 Nm (12 ft-lb)

CLUTCH ASSEMBLY

Clutch cross-shaft, removing and installing (cable-operated clutch)

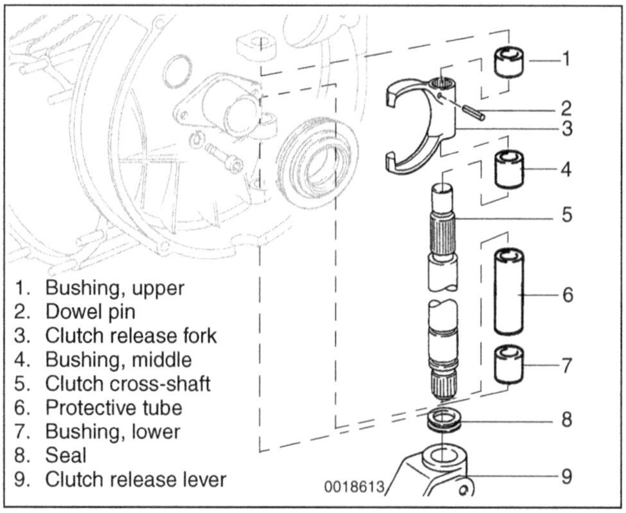

◀ Clutch cross-shaft repairs and upgrades are recommended during a clutch replacement job.

– Remove engine/transmission from car. Separate engine from transmission. See **101 Engine Removal and Installation**.

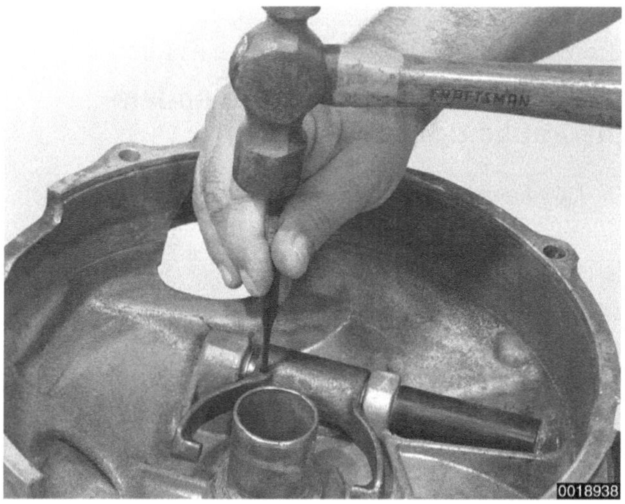

◀ Drive dowel pin out of clutch fork.

– Pull cross-shaft out of transmission, disengaging from fork.

NOTE —

If the clutch release lever is frozen to the clutch cross-shaft, remove them together by sliding the cross-shaft out of the bottom of the transmission bellhousing.

– Remove release lever from cross-shaft. Clean shaft and release lever.

◀ Check for wear or damage in clutch release fork. Look for cracks in casting at stress points (**arrows**).

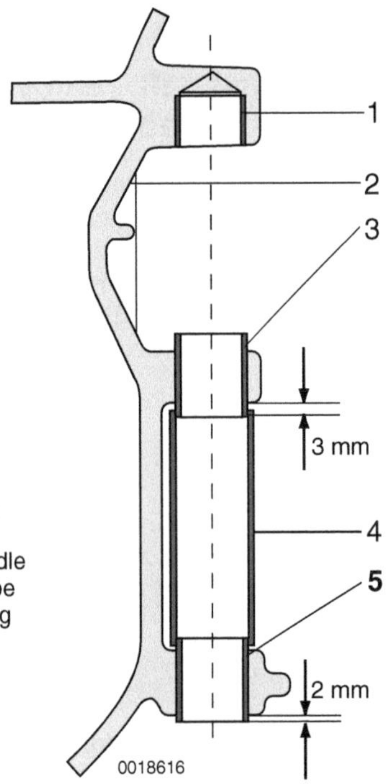

1984-1986 clutch cross-shaft bushings

1. Top bushing
2. Transmission bellhousing
3. Bushing, middle
4. Protective tube
5. Lower bushing

- Using a chisel, carefully remove old cross-shaft bushings from transmission housing.

< Using suitable drift, tap new top bushing into place.

- Install center and lower bushings, making sure protective tube is in place between them.

- Install clutch fork between top and center shaft supports.

- Install clutch cross-shaft through bushings and clutch fork.

NOTE —
Use a multipurpose grease on cross-shaft bushings and on clutch release lever.

- Tap roll pin through fork to lock it in place on cross-shaft.

- Join transmission to engine and reinstall engine/transmission unit in car. See **101 Engine Removal and Installation**.

- Adjust clutch as described earlier in this section.

Clutch cross-shaft, bearing update (hydraulic operated clutch)

On G50 transmission built before April 6,1989, the clutch cross-shaft should be converted to the updated cross-shaft and release fork assembly. The cross-shaft update should be made whenever transmission or clutch repairs are being performed.

NOTE —
On transmissions built from April 6, 1989, this update was part of the factory production.

Clutch Cross-shaft Update Required
- Standard differential .... up to 950/01 1K 03381
- Limited-slip differential .... up to 950/01 2K 01159

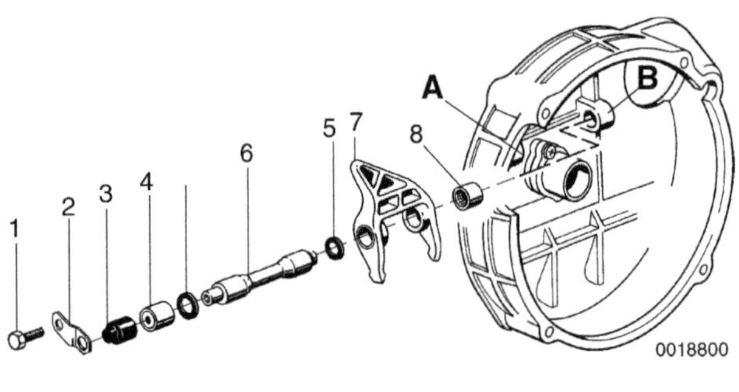

1. Bolt
2. Bracket
3. Bearing cover
4. Needle bearing
5. O-ring seal
6. Clutch cross-shaft
7. Clutch release fork
8. Needle bearing

< The cross-shaft update consists of cutting the bearing bores using a special cutter and installing needle bearings into the case. This update requires the following replacement parts:

Parts Required for Cross-shaft Update

Qty.	Description (no. in illustration)	Number
1	Bolt (1)	N 010 287 1
1	Bracket (2)	950 116 713 02
1	Cover (3)	950 116 725 00
1	Bearing (4)	999 201 339 00
1	Release shaft (6)	950 116 710 04
2	Seals (5)	999 113 418 40
1	Release fork (7)	950 116 086 05
1	Bearing (8)	999 201 365 00

– Special tools are required to update cross-shaft assembly:

Tools Required for Cross-shaft Update

- Special cutter kit #9292 (Porsche ordering #000 721 929 20)
- ATF lubricant
- M6 bolt approx. 80 mm (3 in.) long
- M6 nut
- Long life bearing grease (Olista® Longtime 3 EP) (Porsche ordering #000 043 024 00)
- Electric drill (variable speed)

– Remove engine and transmission, and separate transmission from engine as described in **101 Engine Removal and Installation**.

– Remove clutch release fork from release bearing. Discard release fork and cross-shaft.

– Cut/drill cross-shaft bores in transmission case to accept new bearings:

- Assemble 9292 cutter.
- Use ATF to lubricate cutter and cross-shaft bores in transmission case.
- Set electric drill speed to approximately 500 rpm.
- Slide 9292-3 guide pin into shaft bores in case and cut outer shaft bearing bore (**A**).
- Change guide pin to 9292-2 and cut inner bearing bore (**B**). Stop when cutter reaches bottom (increased resistance is noticed).
- Remove burrs from bearing bores. Clean clutch bell-housing.

– Lubricate new cross-shaft bearings with a long life grease. Place inner bearing on cross-shaft and install new bearing into inner bore in case.

0018910

◀ Install clutch fork on release bearing and hold in place with adhesive tape.

– Assemble transmission and engine.

– Install seal on each end of new cross-shaft. Lubricate outer cross-shaft bearing and install on cross-shaft.

NOTE —
Check seal direction. Refer to instructions with tool kit.

– Thread 6 mm nut approx. 15 mm onto M6X80 bolt.

– Screw 6 mm bolt into threaded end of cross-shaft until nut seats against bearing.

– Insert cross-shaft through bore in transmission case and clutch release fork.

NOTE —
It may be necessary to lightly tap the end of the bolt in order to start the cross-shaft into the release fork.

– Remove 6 mm bolt. Remove tape holding release fork in place.

– Lubricate outer cross-shaft bearing and install in transmission case.

– Install rubber bearing cover.

– Place cross-shaft bracket in position and install 6 mm bolt.

Tightening torques
- Cross-shaft bracket to transmission case . . . . 9 Nm (80 in-lb)

Clutch, replacing

The engine and transmission must be removed as a unit and than separated in order to service the clutch. It is normally recommended that the pressure plate and release bearing be replaced when a new clutch disc is installed.

The flywheel and the transmission pilot shaft bearing should be carefully inspected when replacing the clutch. Replacing the clutch disc without replacing other worn components may accelerate clutch disc wear.

– Remove engine/transmission from car. Separate engine from transmission. See **101 Engine Removal and Installation).**

– Insert clutch alignment tool into center of clutch.

< Loosen 9 bolts (**arrows**) holding clutch assembly to flywheel.

> ***CAUTION —***
> *To avoid warping the pressure plate, loosen bolts evenly, one turn at a time, until clutch pressure is relieved.*

– Once bolts have been loosened, remove all 9 bolts. Remove starter ring gear and clutch.

> ***WARNING —***
> *The clutch disc contains asbestos fibers. Breathing asbestos dust can cause asbestosis. Always wear an approved respirator and protective clothing when handling components containing asbestos. Do not use compressed air, do not grind, heat, weld or sand on or near any asbestos materials. Only approved cleaning equipment should be used to service the clutch disc or areas containing asbestos dust or asbestos fibers.*

– Check components for wear and damage. Repair or replace as needed.
- Check starter ring-gear teeth.
- Check release bearing guide tube.
- Check clutch fork.

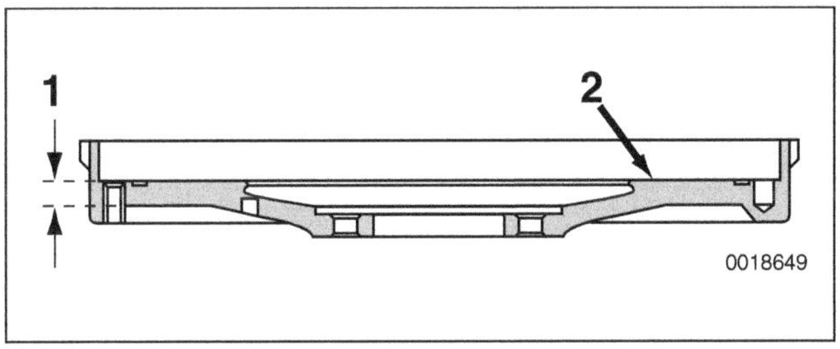

◀ Check flywheel face for damage.

- On 1984-1986 models: If damaged or scored, machine to specifications.
- On 1987-1989 models: If damaged or scored, replace.

1984-1986 Flywheel Service Specifications

- Wear limit (**1**). 8.5 mm (0.335 in.)
- Machining limit . . . . minimum needed for smooth surface
- Max runout (**2**) . . . . 0.1 mm (0.004 in.)

– Check for signs of oil at bottom of transmission or in bell-housing. If found, inspect engine and transmission oil seals.

- To replace transmission input shaft seal, see **340 Manual Transmission–Controls and Case**.
- To replace engine rear main seal (also referred to as the rear crankshaft oil seal), see **130 Crankcase and Crankshaft**.

◀ Check transmission pilot shaft bearing:

- On 1984-1986 models: Pilot bearing is lightly pressed into center of flywheel.
- On 1987-1989 models: Pilot bearing is held by three bolts to holes (**arrows**) at rear end of crankshaft.
- Replace bearing if dry or worn out.
- Lubricate bearing with a multipurpose grease.

– Place new release bearing on pressure plate as described later.

◀ Using special clutch disc alignment tool P 370, mount clutch assembly to flywheel. Tighten clutch bolts uniformly one turn at a time.

Tightening Torque

- Clutch to flywheel . . . . 25 Nm (18 ft-lb)

– Prior to installing transmission on engine, lubricate transmission input shaft splines with Optimoly® HT or equivalent. Lubricate release bearing guide tube with multipurpose grease containing molybdenum disulfide (MoS_2).

– On 1984-1986 models: As transmission slides onto the engine studs, turn clutch cross-shaft so that clutch fork hooks into release bearing.

On 1987-1989 models: Prior to installing transmission, use duct tape or equivalent to hold clutch release fork on release bearing (**arrow**).

- Install engine to transmission nuts and torque.

Tightening Torque
- Engine to transmission . . . . 45 Nm (32 ft-lb)

On 1987-1989 models: Install clutch cross-shaft (**arrow**), making sure it engages clutch fork inside bellhousing.

NOTE —
Remove the adhesive tape used to keep the clutch release fork in place through the hole in the side of the bellhousing.

- Install engine/transmission as described in **101 Engine Removal and Installation**.

Clutch release bearing, replacing

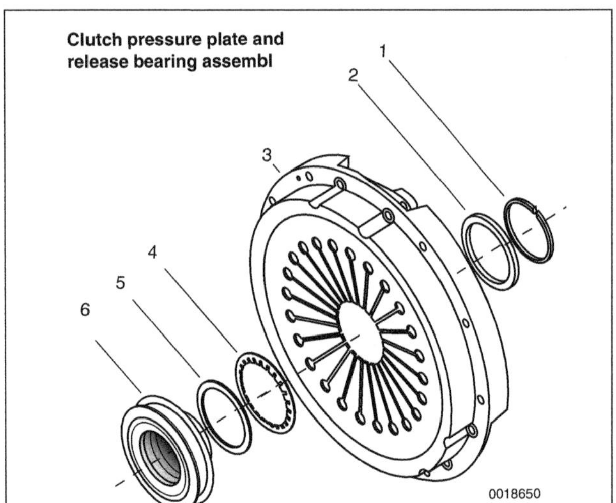

1. Locking ring
2. Thrust washer
3. Pressure plate
4. Spring washer
5. Shim(s)
6. Release bearing

- Lay clutch pressure plate with release bearing facing down in a press.
- Lightly press down on pressure plate, using suitable spacers and a steel bar.

Remove release bearing locking ring. (Early pressure plate shown. Later models are similar.)

- Remove from press and remove release bearing from pressure plate.

NOTE —
Release bearings are filled with a special grease and consequently must not be washed.

- When reinstalling release bearing, place shims and spring washer on first.

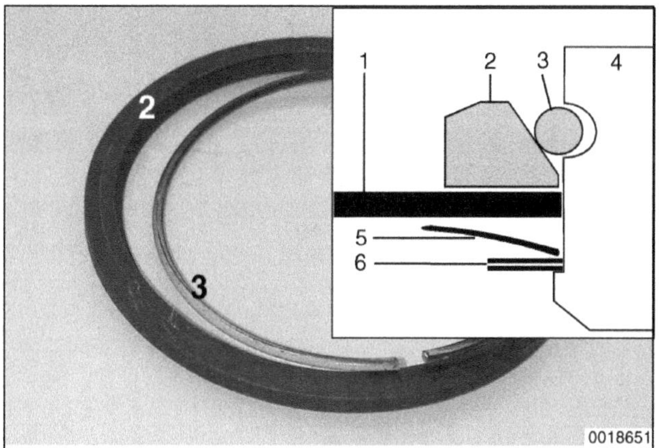

1. Clutch pressure plate
2. Thrust washer
3. Locking ring
4. Release bearing
5. Spring washer
6. Shims

– Use press to lightly press down on pressure plate, using suitable spacers and a steel bar.

– Install release bearing thrust washer (**2**) with beveled inner edge up (toward clutch disc).
  - Install locking ring (**3**) inside thrust washer bevel.
  - Be sure locking ring snaps into release bearing groove.

– Release pressure on clutch and remove from press.

– Check movement of release bearing in pressure plate. It should be tight but still moveable.

340 Manual Transmission–Controls and Case

TABLES

GENERAL

This repair group covers repairs to the shifter mechanism, transmission fluid and seal service, as well as disassembly and assembly of the transmission down to the input and output shaft assemblies.

Other repair groups dealing with transmission removal and repairs are:

- **101 Engine Removal and Installation**
- **301 Clutch**
- **350 Manual Transmission–Gears and Shafts**
- **390 Differential**

NOTE—
Transmission removal is accomplished by first removing the complete engine and transmission assembly and then separating the transmission from the engine. See **101 Engine Removal and Installation**.

Special tools

Some of the procedures in this group require special tools. If you lack the skills and tools, or a suitable workplace for servicing the transmission, we suggest you leave these repairs to an authorized Porsche dealer or other qualified shop. Completely reconditioned transmissions are available from qualified Porsche repair facilities.

Special Tools

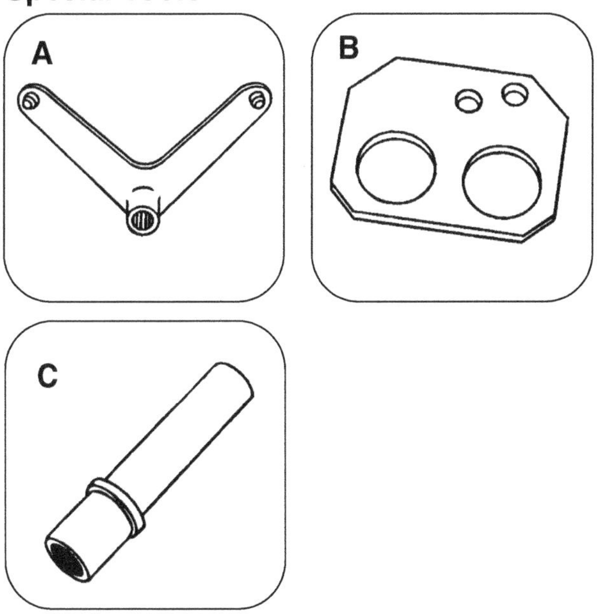

Special tool
A P 37 a. ... Input shaft lock
(Porsche ordering #000 721 037 10)
B P 260 a. ... Shift rod holder
(Porsche ordering #000 721 260 10)
C P 381 ... Input shaft installing sleeve
(Porsche ordering #000 721 381 00)

Transmission identification

The models covered by this manual are equipped with one of two five-speed manual transmissions, the 915 or the G50 (sometimes referred to as 950). The transmission can be identified by a code stamped into the case near the transmission drain plug.

Transmission applications are shown in **Table a**. Gear ratios are shown in **Table b**.

NOTE —
R.o.W. is an abbreviation for 'rest of world' vehicle production.

Table a. Transmission Applications

Type, year	Feature	Market	Serial numbers
1984			
915/67	oil cooler	Europe, R.o.W.	73 E 00001 - 10000
915/68	w/o oil cooler	USA, Japan	74 E 00001 - 10000
915/69	oil cooler	Europe, R.o.W.	73 E 10001 - 11000
915/70	w/o oil cooler	USA, Japan	74 E 10001 - 11000
1985-86			
915/72	oil cooler	Europe, R.o.W.	73 G 00001 - 10000
915/73	w/o oil cooler	USA, Japan	74 G 00001 - 10000
1987			
G50.00		Europe, R.o.W.	73 H 00001 - 20000
G50.01		USA, Japan	74 H 00001 - 20000
1988			
G50.01		USA, Canada	2J10001 - 2J20000
G50.01	40% limited slip		2J10001 - 2J20000
1989			
G50.01		USA, Canada	1K00001 - 1K10000
G50.01	40% limited slip		1K00001 - 2K20000

Table b. Transmission Gear Ratios

Gear	915/67 915/69 916/72	915/68 915/70 915/73	G50.00	G50.01	G50.02
1st	3.1818	3.1818	3.500	3.500	3.154
2nd	1.8333	1.7777	2.059	2.059	1.895
3rd	1.2608	1.2608	1.409	1.409	1.333
4th	0.9655	1.0000	1.074	1.125	1.036
5th	0.7631	0.7895	0.861	0.889	0.861
reverse	3.3250	3.3250	2.857	2.857	2.857
Final drive	3.8750:1		3.444:1		

TRANSMISSION FLUID SERVICE

Transmission fluid level, checking

Checking the transmission fluid level involves removing the side drain plug and inserting a finger into the drain plug hole. If the fluid level is up to the bottom of the drain hole (finger wetted by transmission fluid), the level is correct.

Transmission fluid, replacing

– Drive vehicle for a few miles to warm transmission.

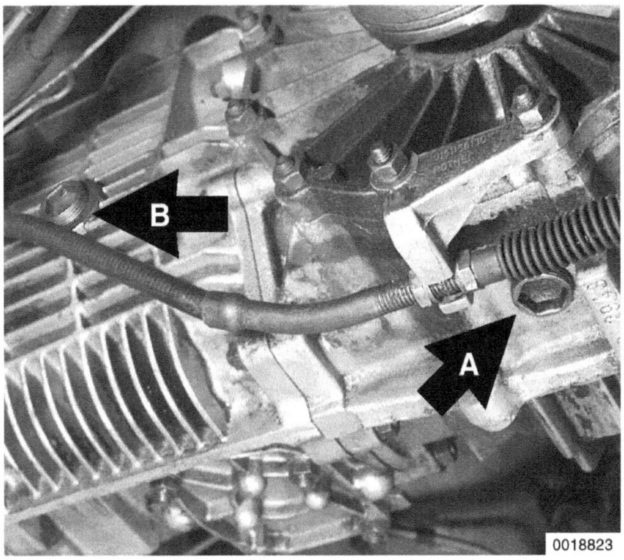

< Raise and safely support vehicle to access drain (**A**) and fill (**B**) plugs.

> ***WARNING —***
> *Make sure the car is stable and well supported at all times. Use a professional automotive lift or jack stands designed for the purpose. A floor jack is not adequate support.*

– Place drain pan under transmission and remove drain plug at bottom of transmission. Allow fluid to drain fully.

– Install and torque drain plug.

Tightening Torque
- Drain plug to transmission housing
  915 transmission. . . . . 24 Nm (18 ft-lb)
  G50 transmission . . . . 30 Nm (22 ft-lb)

– Remove fill plug from side of transmission. Slowly fill transmission until fluid overflows fill hole.

Transmission Fluid
- Type . . . . SAE 75W90
- API classification. GL5
- Alternate type MIL-L2105B

Transmission Fluid Capacity
- Without oil cooler . . . . 3.0 liter (6.34 pint)
- With oil cooler. . . . . 3.1 liter (6.55 pint)

– Install and torque fluid fill plug.

Tightening Torque
- Fill plug to transmission housing
  915 transmission. . . . . 24 Nm (18 ft-lb)
  G50 transmission . . . . 30 Nm (22 ft-lb)

SHIFTER MECHANISM (1984-1986 MODELS)

Shifter housing, removing and installing (1984-1986)

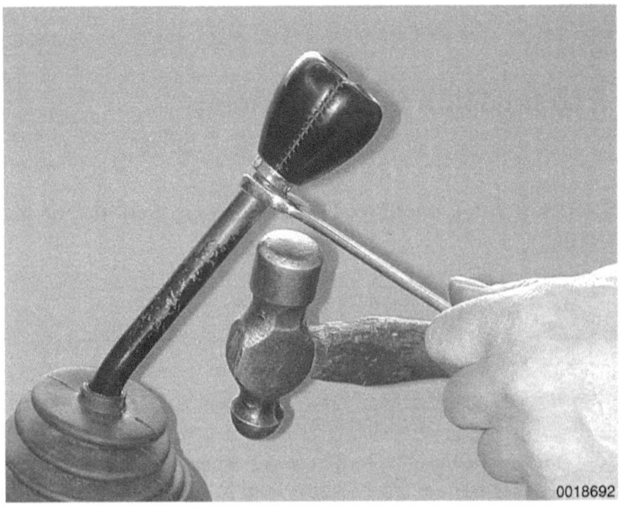

◀ Remove shifter knob by tapping up from below.

NOTE —
A special tool for forcing the shifter knob off the lever can be formed out of thick steel plate cut into a Y-shape. Alternatively, a 15 mm open-end wrench can be used.

– Pull up and remove shift lever boot.

– Remove center console:

- Remove 4 retaining screws on left and right sides of console front cover. Slide cover off center console towards front of car.
- Remove 3 center console retaining screws (2 in front, 1 in rear of console).

– Remove 5 screws at base of shifter housing.

– Remove shifter and housing by pulling straight up and off.

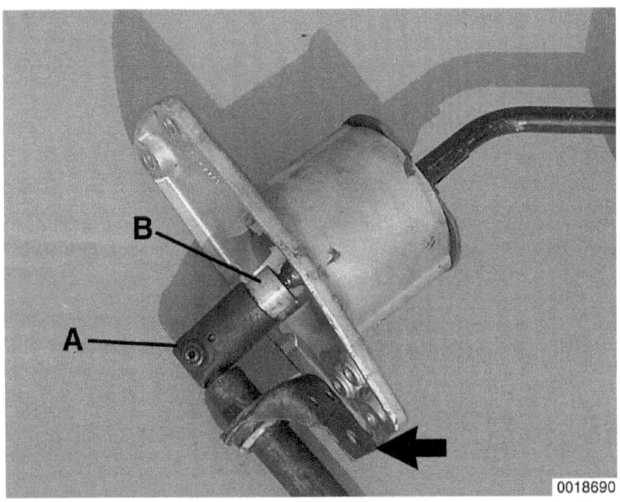

◀ During reinstallation:

- Line up shift rod coupling (**A**) with shift lever ball socket (**B**).
- Position shift rod bushing bracket (**arrow**) in place and start bracket mounting screws before fully lowering shifter housing into tunnel.

Tightening Torques
- Shift rod bushing bracket to housing. . . . .10 Nm (89 in-lb
- Shifter housing to tunnel. . . . . 25 Nm (18 ft-lb)

Shifter mechanism, disassembling (1984-1986 models)

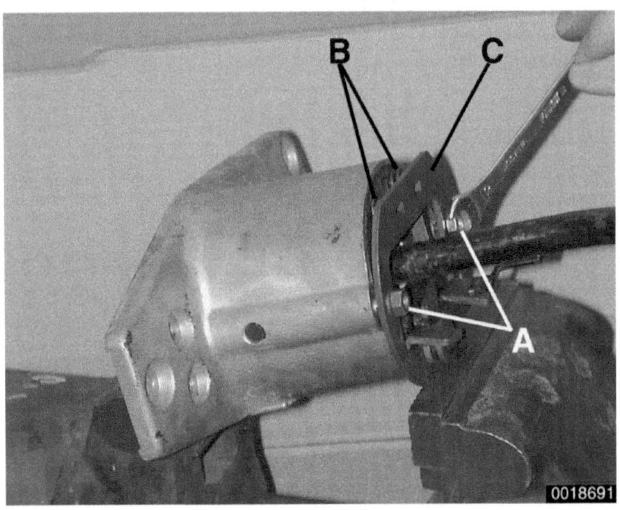

– Place removed shift mechanism on its side in a vise.

◀ Remove lock pawl plate retaining nuts (**A**).

– Slowly open vise, releasing pawl plate springs (**B**). Remove pawl plate (**C**) and springs.

> ***WARNING —***
> *Be careful when opening vise. The springs under the pawl plate are strong.*

– Remove shifter guide plate through top of housing.

Shifter mechanism (1984-1986 models)

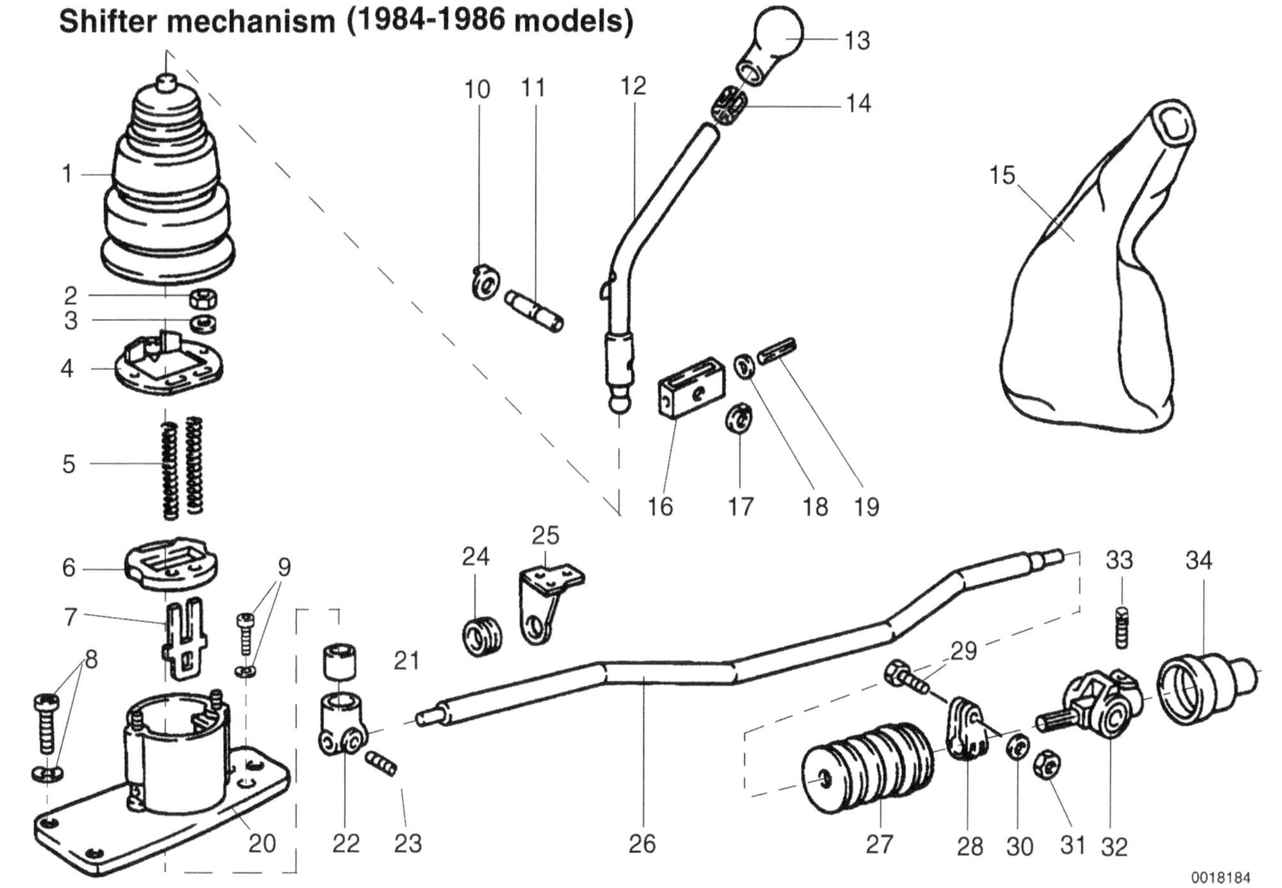

1. Dust boot
2. Nut (M6)
3. Lock washer (M6)
4. Lock pawl plate
5. Springs
6. Stop plate
7. Shifter guide plate
8. M8 bolt and washer (qty. 3)
9. M6 bolt and washer (qty. 2)
10. Locking clip
11. Shifter pivot pin
12. Shift lever
13. Shifter knob
14. Shift knob retaining ring
15. Boot
16. Shift pivot housing
17. Locking clip
18. Spacers
19. Roll pin
20. Shifter housing
21. Ball socket
22. Shift lever coupling
23. Set screw
24. Shift rod bushing
25. Shift rod bushing bracket
26. Shift rod
27. Boot
28. Coupling clamp
29. Clamp bolt (M8)
30. Lock washer (M8)
31. Nut (M8)
32. Shift rod coupling
33. Set screw
34. Transmission selector shaft boot

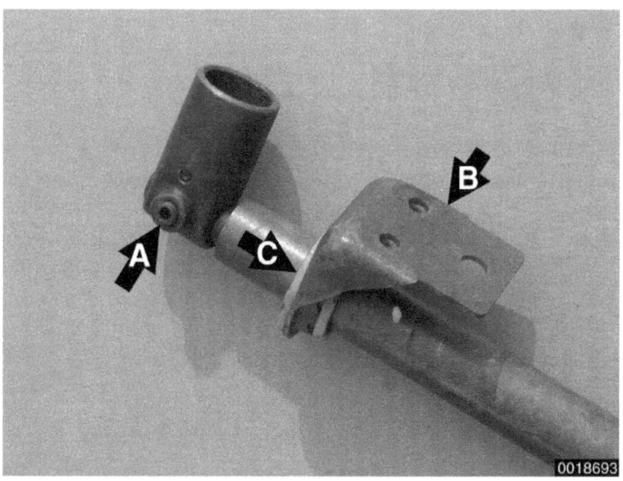

– Remove locking clips from shifter pivot pin and pull pin out of housing.

– Remove gear shift lever from shifter housing.

– Reassembly is reverse of disassembly.

Shift rod bushing, replacing (1984-1986 models)

– Remove shifter mechanism as described earlier.

◁ Remove set screw (**A**) from shift lever coupling and remove coupling from shift rod.

– Slide shift rod bushing bracket (**B**) off shift rod.

– Snap out old bushing (**C**) and replace with new.

– Installation is reverse of removal.

Shift rod coupling, replacing (1984-1986 models)

◄ Pull up center floor mat behind front seats. Remove 4 screws (**arrows**) from access panel and remove panel.

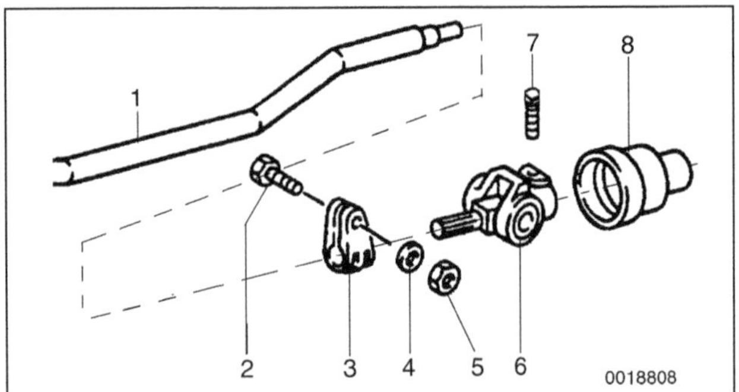

1. Shift rod
2. Pinch bolt
3. Clamp
4. Lock washer
5. Nut
6. Coupling
7. Set screw
8. Selector rod boot

◄ Remove set screw (**7**) from shift rod coupling to disengage transmission selector shaft.

– Loosen shift rod clamp pinch bolt (**2**).

– Slide coupling forward to remove from transmission selector shaft, then pull back to remove from shift rod.

– Installation is reverse of removal. Adjust as described below.

Tightening Torques	
• Coupling clamp to shift rod	23 Nm (17 ft-lb)
• Set screw to transmission selector shaft.	23 Nm (17 ft-lb)

Shift rod coupling, adjusting (1984-1986 models)

– Access shift rod coupling in rear part of center tunnel as described above.

◄ Loosen shift rod clamping nut (**A**) on coupling.

– Turn transmission selector shaft counterclockwise as facing rear of car (**arrow 1**).

– With shift lever in neutral, pull shift lever towards driver's seat so that shift lever contacts stop. This will preload the shift rod in a counterclockwise direction (**arrow 2**) as facing rear of car.

– Temporarily tighten shift rod coupling clamp, then test for correct adjustment:

- Shift transmission into all gears including 5th and reverse. All gears must engage with approximately the same effort and throw.
- Shift transmission into 5th gear. Push shift coupling boot back. Check transmission selector shaft and coupling for play. Some play must be felt.

– Tighten coupling clamping nut.

Tightening Torques

- Coupling clamp to shift rod . . . . 23 Nm (17 ft-lb)

– Install access cover and floor mat.

SHIFTER MECHANISM, (1987-1989 MODELS)

Shifter housing, removing and installing (1987-1989 models)

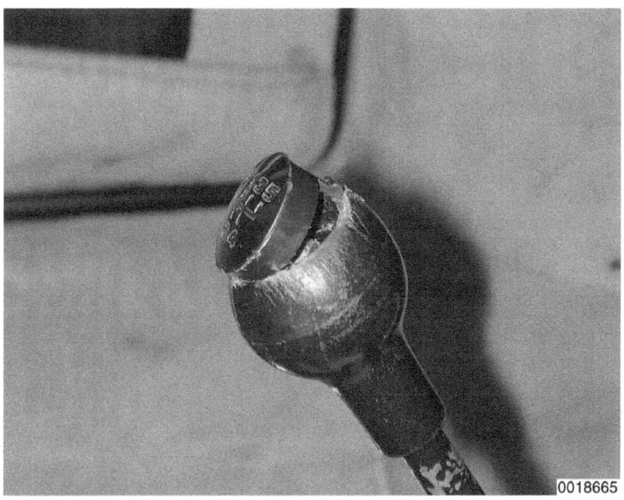

0018665

Gently pry cap from top of shift knob.

NOTE —
Cap locking tabs are at 4, 8 and 12 o'clock.

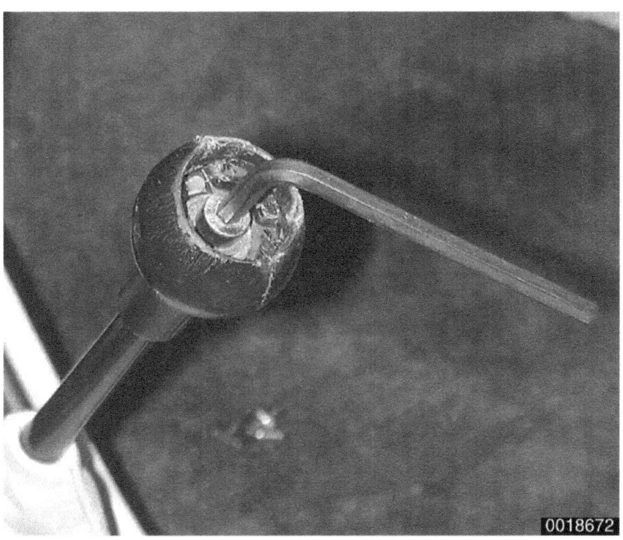

0018672

Remove allen bolt holding knob in place. Remove knob.

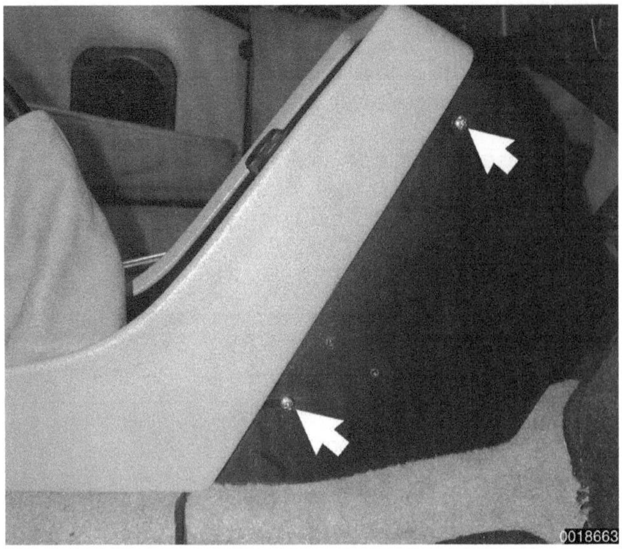

Remove left and right side screws securing front cover to center console. Slide cover forward of center console.

Remove three screws holding center console to floor.

- Screw on left shown (**A**). Right side is similar.
- Pull back rug slightly to reveal screw in middle of rear (**B**).
- Without disconnecting wiring harnesses, lift center console over shifter and lay to one side.

Detach leather shifter boot from foam rubber cover (**arrow**).

◄ Pull shifter boot up to expose rubber o-ring (**arrow**) securing boot to shift lever. Slide O-ring down off boot and remove boot.

– Remove inner boot and foam rubber cover from shift assembly.

– Slide driver's seat forward. Pull up center floor carpet from between rear seats.

◄ Remove 4 screw (**arrows**) holding center tunnel access panel in place. Remove panel.

◄ Remove two nuts holding shifter rod to coupling. Separate rod from coupling.

– Place shift lever in 4th gear position.

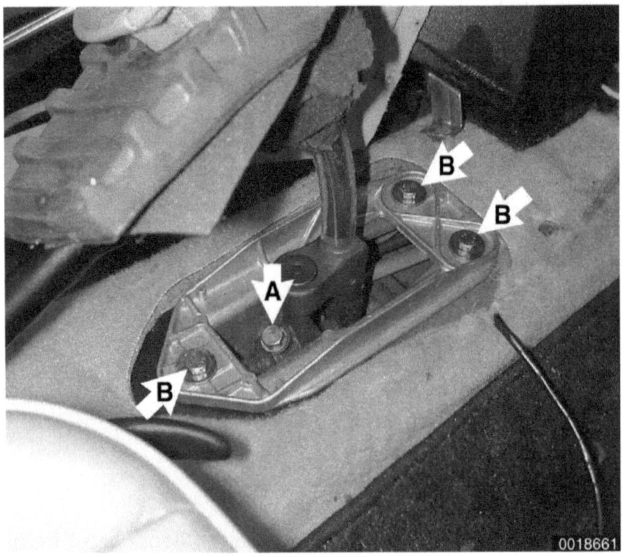

Remove front selector rod bolt (**A**) and 3 bolts securing shifter housing to floor (**B**).

NOTE—
Mark location of bolts securing shift housing to floor.

- Remove shift lever assembly.
- Installation is reverse of removal.

Tightening Torques

- Front selector coupling to shift rod ..... 23 Nm (17 ft-lb)
- Shift rod to shift rod coupling .... 23 Nm (17 ft-lb)
- Shifter housing to center tunnel .... 23 Nm (17 ft-lb)
- Shifter knob to shifter .... 10 Nm (7 ft-lb)

Shifter mechanism, disassembling (1987-1989 models)

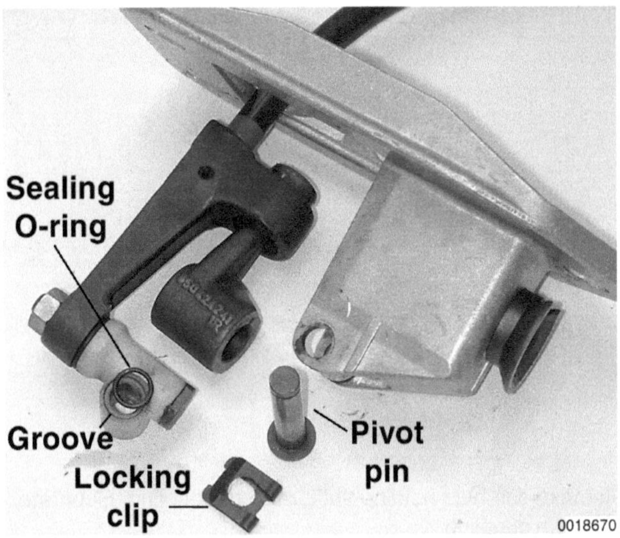

Remove locking clip from shift lever pivot pin and remove shifter.

- Remove shift lever from shifter housing.
- When reassembling shift lever pivot pin be sure to install sealing O-ring on side of shifter pivot with groove for O-ring.
- Reassembly is reverse of disassembly.

Shifter mechanism (1987 - 1989 models)

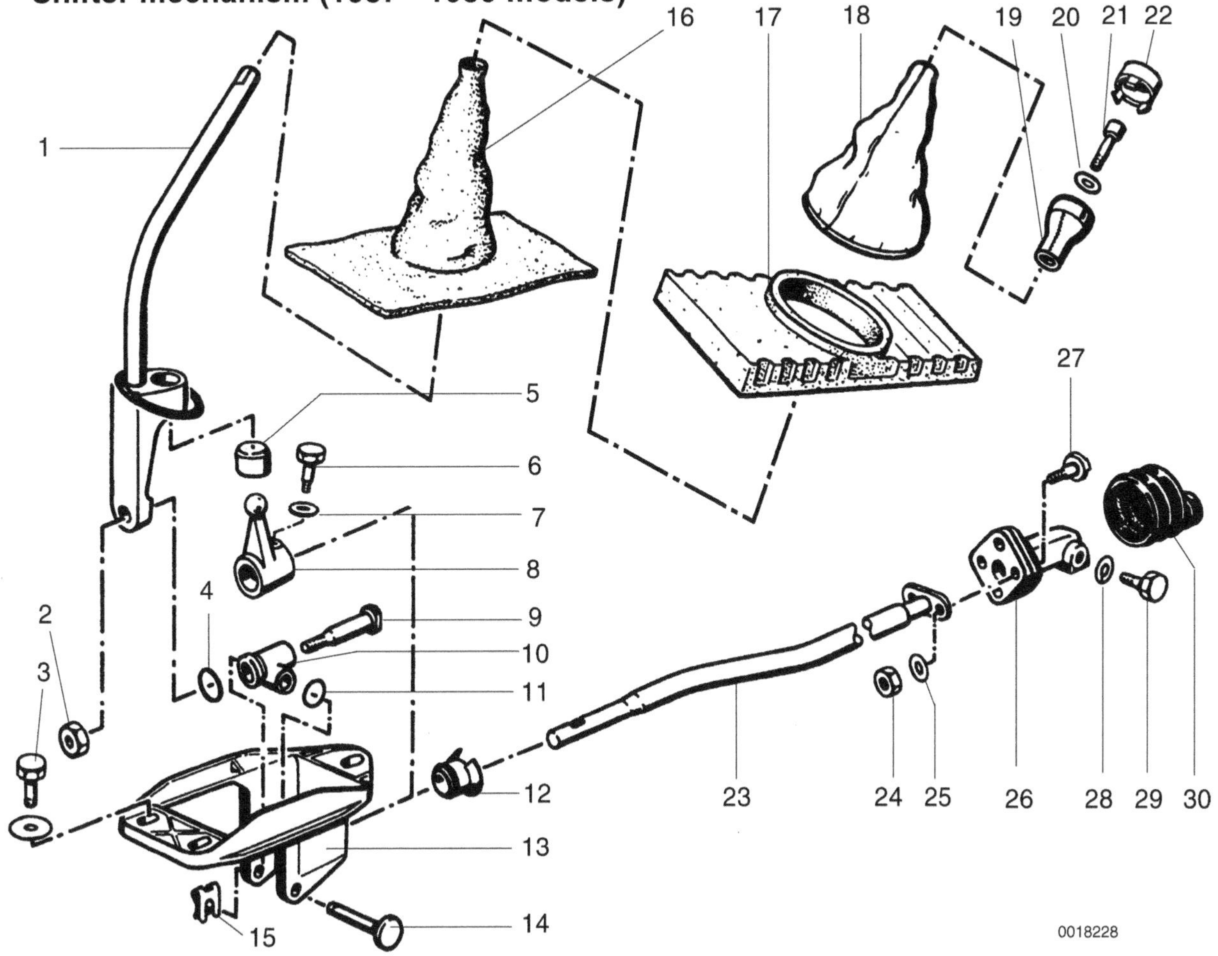

1. Shift lever
2. Nut
3. Bolt and washer
4. O-ring
5. Ball socket
6. Pinch bolt
7. Washer
8. Front shift selector
9. Bolt
10. Shift lever mount
11. O-ring
12. Bushing
13. Shifter housing
14. Pivot pin
15. Locking clip
16. Inner shift boot
17. Foam rubber cover
18. Leather shift boot
19. Shift knob
20. Washer
21. Allen bolt
22. Shift symbol cap
23. Shift rod
24. Nut
25. Washer
26. Shifter coupling
27. Bolt
28. Washer
29. Bolt
30. Selector shaft boot

TRANSMISSION SERVICE

The transmission case, which contains the transmission and differential gears, is a lightweight alloy die casting. The 915 transmission uses constant-mesh type gears, with Porsche's own design synchronizing system. Each forward gear(1st through 5th) is a mating pair of gears, and constant-mesh simply means the mating gears are always meshing.

This section covers disassembly of the 915 transmission used on 1984 through 1986 models. Disassembly of the G50 transmission is not covered here.

NOTE —
For information on removing and installing the transmission, see **101 Engine Removal and Installation**.

Back-up light switch, replacing

– Raise and safely support vehicle.

> ***WARNING —***
> *Make sure car is stable and well supported at all times. Use a professional automotive lift or jack stands designed for the purpose. A floor jack is not adequate support.*

– Place oil drain pan under transmission.

< Remove reverse light switch (**arrow**). Install new switch with new sealing washer.

> ***WARNING —***
> *Be sure transmission is cool during this operation. Hot oil can scald.*

Reverse Light Switch Location
- 915 transmission (1984-1986) .... Transmission end
- G50 transmission (1987-1989) .... Right side

Tightening Torque
- Reverse light switch to transmission .... 35 Nm (25 ft-lb)

– Top up transmission oil as necessary.

Selector shaft seal, replacing

– Raise and safely support vehicle.

> ***WARNING —***
> *Make sure the car is stable and well supported at all times. Use a professional automotive lift or jack stands designed for the purpose. A floor jack is not adequate support.*

< On 1984-1986 models: Pull up center floor mat behind front seats. Remove 4 screws (**arrows**) from access panel and remove panel.

– Push back shift coupling protective boot and remove selector shaft set screw.

< On 1987-1989 models: working underneath rear of car, push back rubber boot covering shift coupling.

- Loosen bolt (**arrow**) and push shift coupling forward off selector shaft.

– Place oil drain pan under transmission.

– Pry seal out of selector shaft housing and install new seal.

> ***WARNING —***
> *Be sure transmission is cool during this operation. Hot oil can scald.*

– Top up transmission fluid as necessary.

– Reassemble shift coupling.

Tightening Torques
- Set screw to selector shaft (1984-1986) . 23 Nm (17 ft-lb)
- Shift coupling to selector shaft (1987-1989) . . . . 23 Nm (17 ft-lb)

Input shaft seal, replacing

– Remove transmission and separate from engine as described in **101 Engine Removal and Installation**.

< Remove clutch release bearing guide tube screws (**arrows**).

Remove guide tube using a shop made tool and pry bar.

NOTE —

- *A tool to remove the release bearing guide tube can be made using 3/16" piece of steel rod. Bend hooks on both ends of wire to engage holes in guide tube. Lever the guide tube off using a pry bar.*
- *For information on clutch cross-shaft repairs, see* **301 Clutch**.

– Press input shaft seal out through front of guide tube using Porsche special tool P 381 or equivalent.

– Press new seal in through front of guide tube.

– Replace O-ring seal on outside of guide tube.

– Reinstall guide tube and torque screws.

Tightening Torque
- Guide tube to differential housing .... 10 Nm (7 ft-lb)

– Reinstall transmission.

Transmission, disassembling (915 transmission)

Mount transmission on engine stand.

– Drain oil from transmission. Replace drain plug.

Tightening Torque
- Transmission drain plug .... 24 Nm (18 ft-lb)

Remove back-up light switch and remove switch operating pin from opening in transmission.

915 transmission components

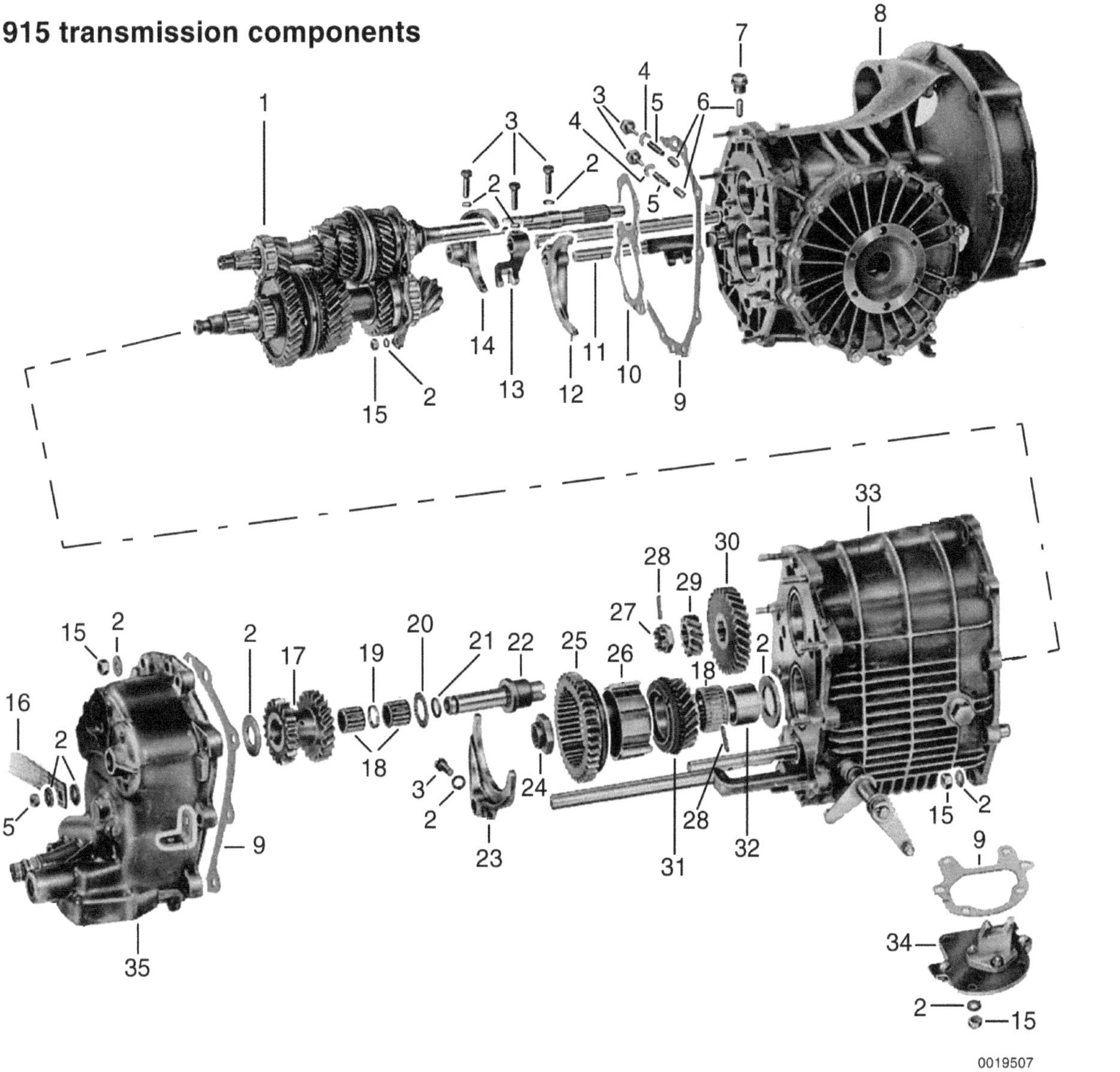

1. Input shaft
2. Washer
3. Bolt
4. Seal
5. Spring
6. Short detent
7. Plug
8. Differential housing
9. Gasket
10. Shim
11. Selector fork rod, 1st/2nd
12. Selector fork, 1st/2nd
13. Shift guide
14. Selector fork, 3rd/4th
15. Nut
16. Ground strap
17. Idler gear, 5th/reverse
18. Needle bearing cage
19. Intermediate ring
20. Thrust needle bearing cage
21. O-ring
22. Shaft, 5th/reverse idler gear
23. Shift fork, 5th/reverse
24. Flange nut
25. Sliding gear, 5th/reverse
26. Guide sleeve
27. Castellated nut
28. Roll pin
29. Gear, reverse
30. Gear, 5th
31. Gear, 5th
32. Bushing
33. Gear housing
34. Cover with guide fork
35. Front transmission cover

Remove end cover mounting nuts and washers (arrows).

- Slide cover off selector shaft.

Lock transmission input shaft with special tool P 37 a.

Push 5th - reverse operating sleeve in to engage 5th gear.

NOTE —
With input shaft locking tool installed and transmission in 5th gear, both shafts should be locked in place.

< Remove roll pin from castle nut on input shaft.

- Remove nuts (**arrows**) from input and output shafts.

– Remove input shaft locking tool (special tool P 37 a).

– Remove reverse light switch operating pin from 5th - reverse gear selector rod.

– Loosen and remove 5th - reverse gear shift fork pinch bolt.

NOTE —
Use a chisel to spread the opening at the shift fork pinch clamp slightly so that the fork moves on the selector rod without binding.

– Side 5th gear operating sleeve and shift fork off selector rod together.

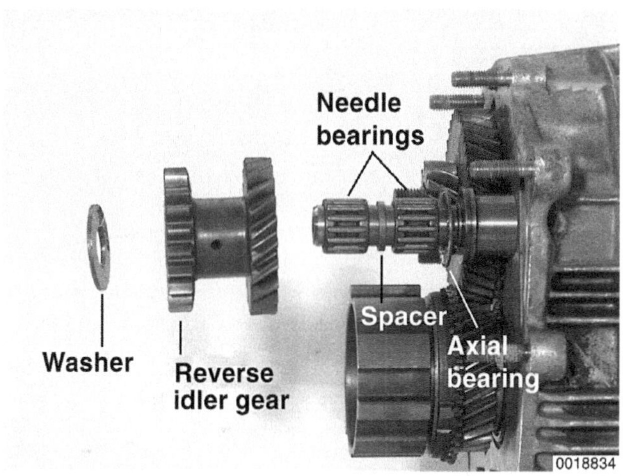

< Remove O-ring from end of reverse idler shaft. Remove washer, reverse idler gear (with needle bearings and spacer), and axial bearing.

NOTE —
Keep bearing, spacer, washer and 5th gear together as a set.

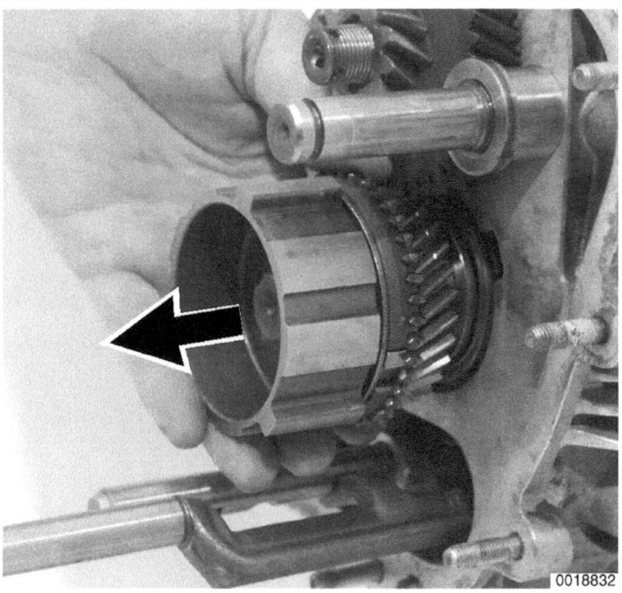

< Slide 5th - reverse gear operating hub together with 5th gear assembly off output shaft.

0018871

Label and remove needle bearing from output shaft. Carefully pry from behind and remove sleeve and washer from output shaft.

0018872

Remove reverse gear from input shaft. Pull off reverse gear idler shaft together with 5th gear.

0018836

Turn transmission over and remove nuts (**arrows**) holding shift cover in place. Carefully pry cover from transmission case to remove.

0018837

◄ Remove nuts (qty. 10) holding gear housing to differential housing.

– Place 5th gear selector rod in neutral.

– Slide gear housing off with selector shaft.

0018838

◄ Unscrew upper (3rd - 4th gear) shift rod detent plug (**arrow**) from side of differential housing.

– Remove spring and detent from behind plug.

NOTE —
A small magnet can be used to remove the spring and detent.

0018839

◄ Remove pinch bolt (**arrow**) from 1st - 2nd gear shift fork.

NOTE —
Use a chisel to spread the opening at the shift fork pinch clamp slightly so that the fork moves on the selector rod without binding.

◄ Remove nuts (**arrows**) from bearing flanges on input and output shafts. Left side nuts shown. Remove nuts and lock washers from both sides.

◄ Slide shafts with gears and forks out of differential housing, leaving 1st - 2nd gear selector rod in transmission case.

NOTE —

*Check the number and thickness of shims (**arrow**) between differential housing and bearing flange. This shimmed surface controls the pinion shaft depth. If no repairs are being carried out on the differential bearing or the differential, the exact shim pack should be reinstalled.*

0018847

Transmission, reassembling (915 transmission)

– Before assembly clean and check all parts to be sure they are in good operating condition. Lay parts out in order of assembly on a work bench.

◀ Install shims (**arrow**) on differential housing studs.

NOTE—
Unless the pinion bearing is being replaced or repairs have been made to the differential, install the same shims that were removed during disassembly. If the differential bearing is replaced or repairs have been made to the differential, special differential setup procedures must be carried out. Differential rebuild and setup is not covered in this manual.

0018848

◀ Insert 1st - 2nd selector rod into lower hole in housing, making sure locking detent is in correct position.

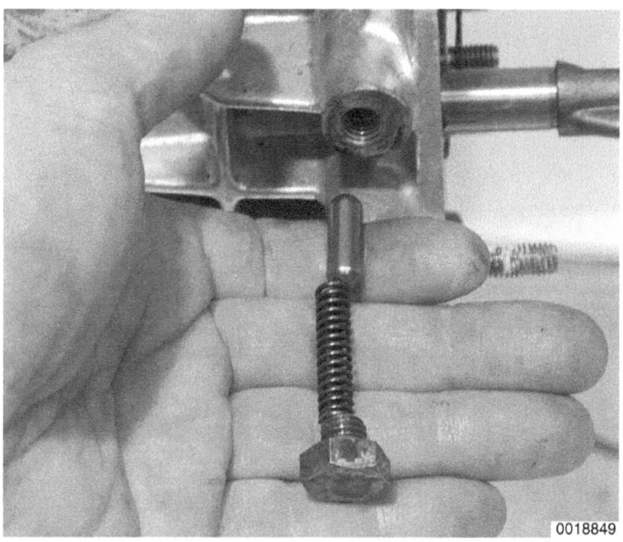

0018849

◀ Install 1st - 2nd gear shift rod detent, spring and plug,

Tightening Torque
- Detent plug to transmission case . . . . 17 Nm (12ft-lb)

◀ Place 1st - 2nd gear selector fork on operating sleeve. Insert output shaft into case so that bearing is supported on case.

◀ Mesh input shaft to output shaft and push both shafts into place so that bearing flanges are fully seated.

– Install bearing flange nuts with lock washers and tighten.

Tightening Torque

- Bearing flange to differential housing . . . 25 Nm (18 ft-lb)

– Install and finger tighten 1st - 2nd gear shift fork pinch bolt.

◀ Remove pinch bolts from 3rd - 4th gear shift fork and shift guide. Slide shift fork and shift guide back on selector rod.

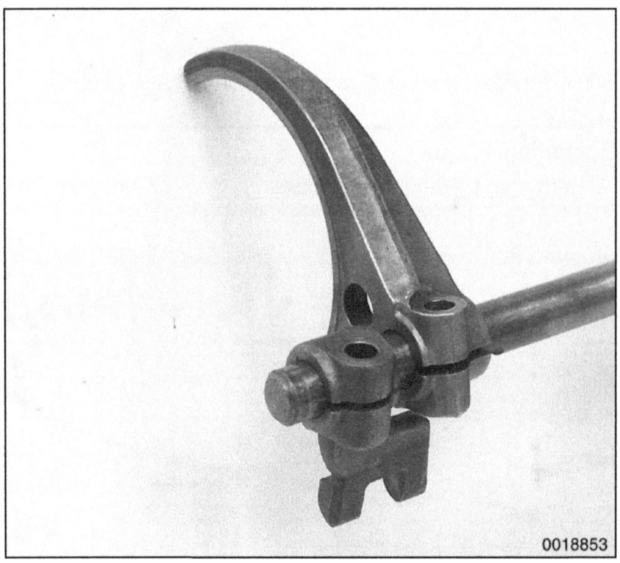

0018854

< Position shift fork onto 3rd - 4th operating sleeve and then slide selector rod into case. Install pinch bolts in forks.

– Insert 3nd - 4th shift rod detent, spring and plug.

Tightening Torque
- Detent plug . . . . 17 Nm (12ft-lb)

0018855

< Install Porsche special tool P 260 a over shift rods and gear shafts.

0018856

< Install 5th gear on input shaft with stepped collar facing gearbox. Install reverse gear on input shaft with marking facing away from gearbox.

◄ Install washer and spacer onto output shaft. Install 5th gear with needle bearing. Install 5th - reverse gear operating sleeve hub and install operating sleeve.

◄ Lock transmission input shaft with Porsche special tool P 37 a or equivalent.

- Engage 5th gear by sliding 5th - reverse gear operating sleeve forward.

NOTE —

With input shaft locking tool (special tool P 37 a) installed and transmission in 5th gear, both shafts should be locked in place.

– Install and torque input and output shaft nuts.

Tightening torqueS

- Input shaft nut . . . . 160 Nm (116 ft-lb)
- Output shaft nut . . . . 250 Nm (181 ft-lb)

– Place 5th gear back in neutral position and remove input shaft lock (special tool P 37 a).

◄ Turn 1st - 2nd gear selector shaft to left until it stops. Turn it back until unmachined flat inner surface of shift guide is almost vertical.

NOTE —

Do not turn shaft back beyond the center point.

- Set 1st - 2nd gear shift fork so that operating sleeve is precisely centered between 1st and 2nd gear synchronizers. Tighten clamping bolt, making sure shift guide is still in vertical position, as described above.

Tightening Torque
- Shift fork pinch bolt . . . . 24 Nm (18 ft-lb)

- ◄ Set 3rd - 4th gear shift fork so that operating sleeve is precisely centered between synchronizers. Install and tighten clamping bolt.

Tightening Torque
- Shift fork pinch bolt . . . . 24 Nm (18 ft-lb)

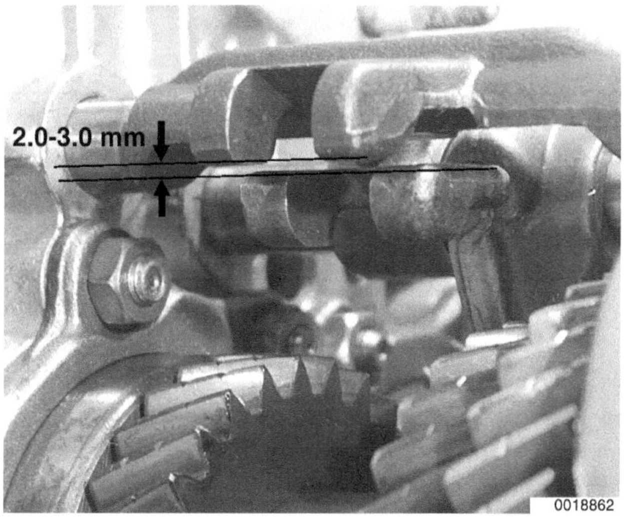

- ◄ Position 3rd - 4th gear shift guide with 2 - 3 mm (0.08 - 0.12 in.) clearance to 1st - 2nd shift guide.
- Tighten 3rd - 4th gear shift guide pinch bolt.

Tightening Torque
- Shift guide pinch bolt . . . . 24 Nm (18 ft-lb)

- Engage 5th gear and lock input shaft using special locking tool P 37 a. Loosen and remove nut from end of output shaft.
- Remove 5th - reverse gear operating sleeve.
- Remove shift fork support plate (Porsche special tool P 260).
- Install new gasket between differential housing and gear housing. Place shifter selector rod in gear housing.

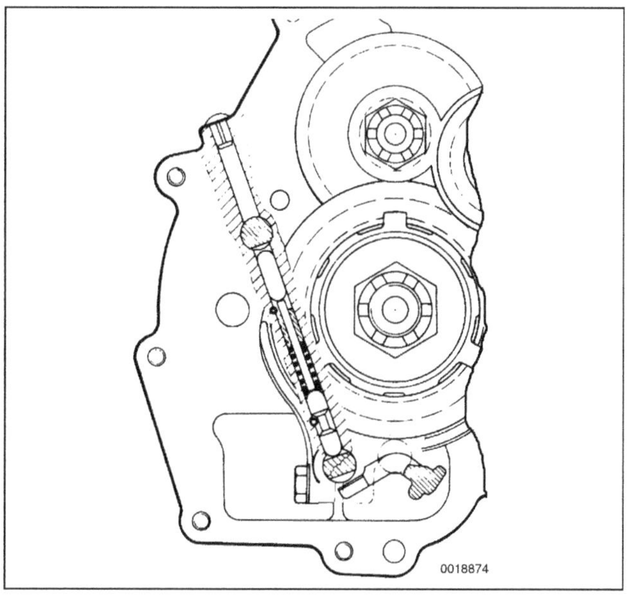

- ◄ Install gear housing, making sure detents in housing are correctly positioned in case. Install and tighten housing nuts with washers.

Tightening Torque
- Gear housing to differential housing . . . . 25 Nm (18 ft-lb)

– Install 5th - reverse gear shift rod into housing with detent cutouts up. Position rod in neutral position.

◄ Install 5th gear with reverse idler shaft:

- Place gear in cutout in idler shaft and place shaft with gear into position. Stepped collar on gear must face housing.
- Notch in shaft end must engage locating pin in housing.

◄ Install reverse gear on input shaft.

- Lock input shaft using Porsche special tool P 37 a.
- Install castle nut and tighten.
- Lock nut in place using roll pin (**arrow**).

Tightening torque
- Input shaft nut. 160 Nm (116 ft-lb)

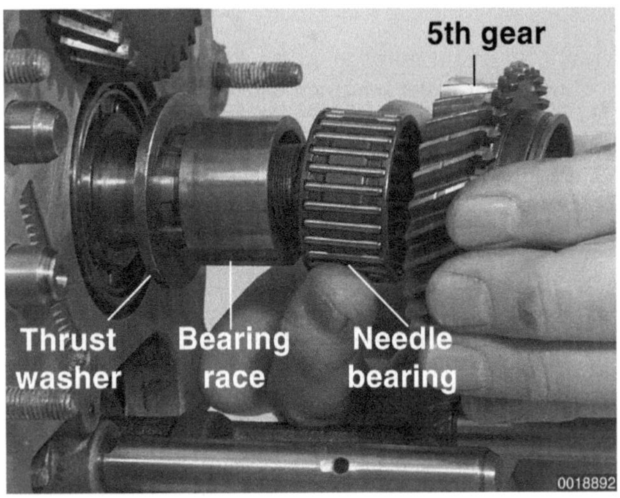

◄ Install thrust washer and 5th gear needle bearing race to output shaft. Install 5th gear with needle bearing.

– Install 5th - reverse gear operating sleeve hub onto output shaft.

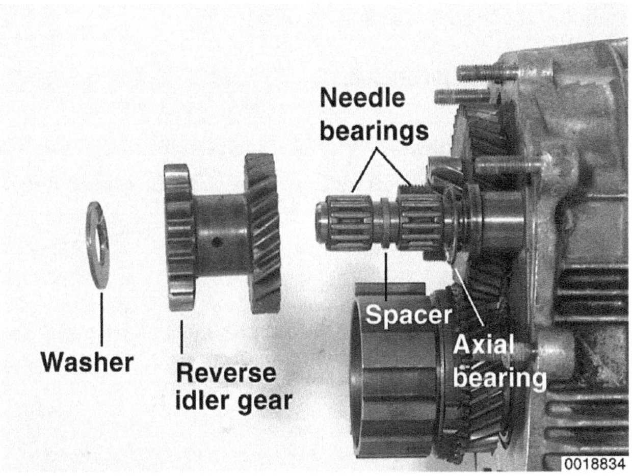

– Unlock input shaft by removing Porsche special tool P 37 a.

◄ Slide reverse idler gear assembly onto shaft.

– Install sealing O-ring to shaft end.

◄ Place 5th-reverse gear shift fork on 5th - reverse gear operating sleeve and gear together. Slide operating sleeve and shift fork on hub and selector shaft.

– Install reverse light roll pin into 5th - reverse gear selector rod with pin flush to back side of shaft.

◄ Lock input shaft with Porsche special tool P 37 a.

- Manually engage 5th gear.
- Install new output shaft nut and tighten.
- Stake nut to shaft (**arrow**),

Tightening Torque

- Output shaft nut . . . . 250 Nm(181 ft-lb)

Push reverse idler against stop and position 5th - reverse shift fork so play between reverse idler gear and reverse operating gear is approximately 1 mm (0.04 in.).

– Tighten shift fork pinch bolt.

Tightening Torques

- 5th gear shift fork pinch bolt . . . . 25 Nm (18 ft-lb)

NOTE —

After tightening shift fork pinch bolt, check the play between reverse operating gear and reverse idler gear by lightly rocking reverse operating gear back and forth. The gears must not contact each other. Readjust shift fork if necessary.

Turn transmission over and install shift cover in place, making sure operating pawl (**arrow**) engages shift rod guides.

Tightening Torques

- Shift cover to transmission housing . . . . 24 Nm (17 ft-lb)

– Turn transmission over and install new O-ring to reverse gear idler shaft.

Install new selector shaft oil seal in end cover. Install reverse light switch operating pin into case hole (**arrow**).

– Using new transmission end cover gasket, slide end cover over selector shaft and gears.

- Install cover nuts and washers and torque.

Tightening Torques

- End cover to gear housing . . . . 25 Nm (18 ft-lb)

– Install new input shaft seal into clutch release guide tube. Install new O-ring to guide tube and reInstall guide tube.

Tightening Torque
- Guide tube to differential housing . . . . 10 Nm (7 ft-lb)

◄ Install reverse light switch with new sealing washer.

Tightening Torque
- Reverse light switch to transmission . . . . 35 Nm (25 ft-lb)

– Reinstall transmission as described in **101 Engine Removal and Installation**. Transmission fluid specification are given below.

Transmission Fluid Specification
- Viscosity . . . . SAE 75W90
- API classification. GL5
- Alternate classification MIL-L2105B

Transmission Fluid Capacity
- Without oil cooler . . . . 3.0 liter (6.34 pt.)
- With oil cooler . . . . 3.1 liter (6.55 pt.)

350 Manual Transmission–Gears and Shafts

Special Tools

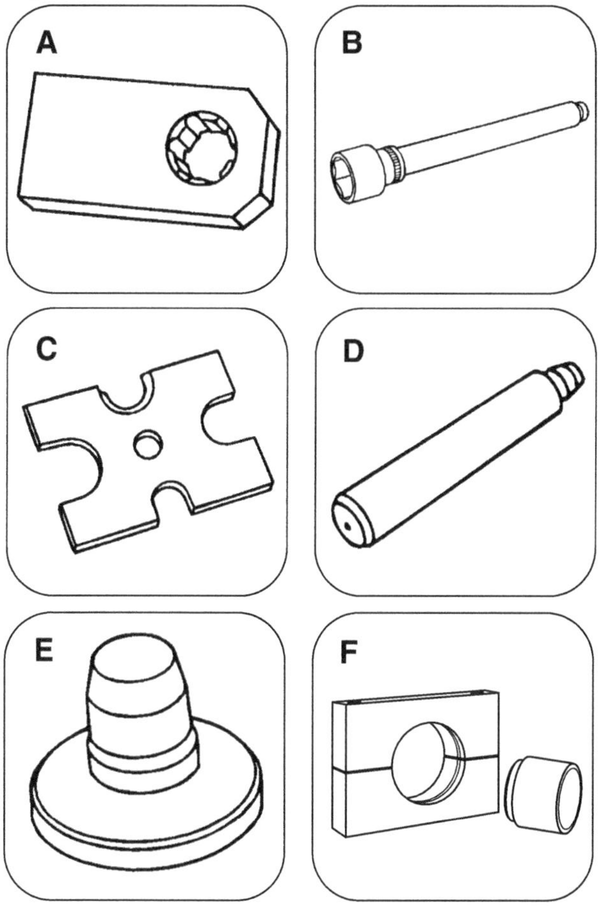

Special tool

- **A** P 355 a . . . . Input shaft mounting tool (source: Baum Tools Unlimited)
- **B** P 252 a . . . . Input shaft collar nut socket (41 mm) (alternate: 9105 Porsche ordering # 000 721 910 50)
- **C** VW 401 or VW 402 . . . . Press support plate
- **D** VW 407 . . . . Input shaft bearing drift
- **E** VW 412 . . . . Input shaft bearing drift
- **F** TL-3100/TL-3150 . . . . Synchronizer gear press tools (source: Automotion)

GENERAL

This section covers the internal repair of the 915 transmission (1984-1986 models). The G50 transmission (1987-1989 models) is not covered here. Other repair groups dealing with transmission removal and repairs are:

- **101 Engine Removal and Installation**
- **301 Clutch**
- **340 Transmission–Controls and Case**
- **390 Differential**

Special tools

Most of the operations and repairs in this section require the use of special tools. If you lack the skills and tools, or a suitable workplace for servicing the transmission, we suggest you leave these repairs to an authorized Porsche dealer or other qualified shop. Completely reconditioned transmissions are available from qualified Porsche repair facilities.

NOTE—
Difficult shifting may be caused by worn out synchronizer teeth, sometimes referred to as "dog" teeth.

INPUT AND OUTPUT SHAFTS

Input shaft, disassembling and assembling

NOTE—
All parts should be installed dry.

– Mount shaft in vise using special tool P 355 a.

– Follow steps in illustration on following page.

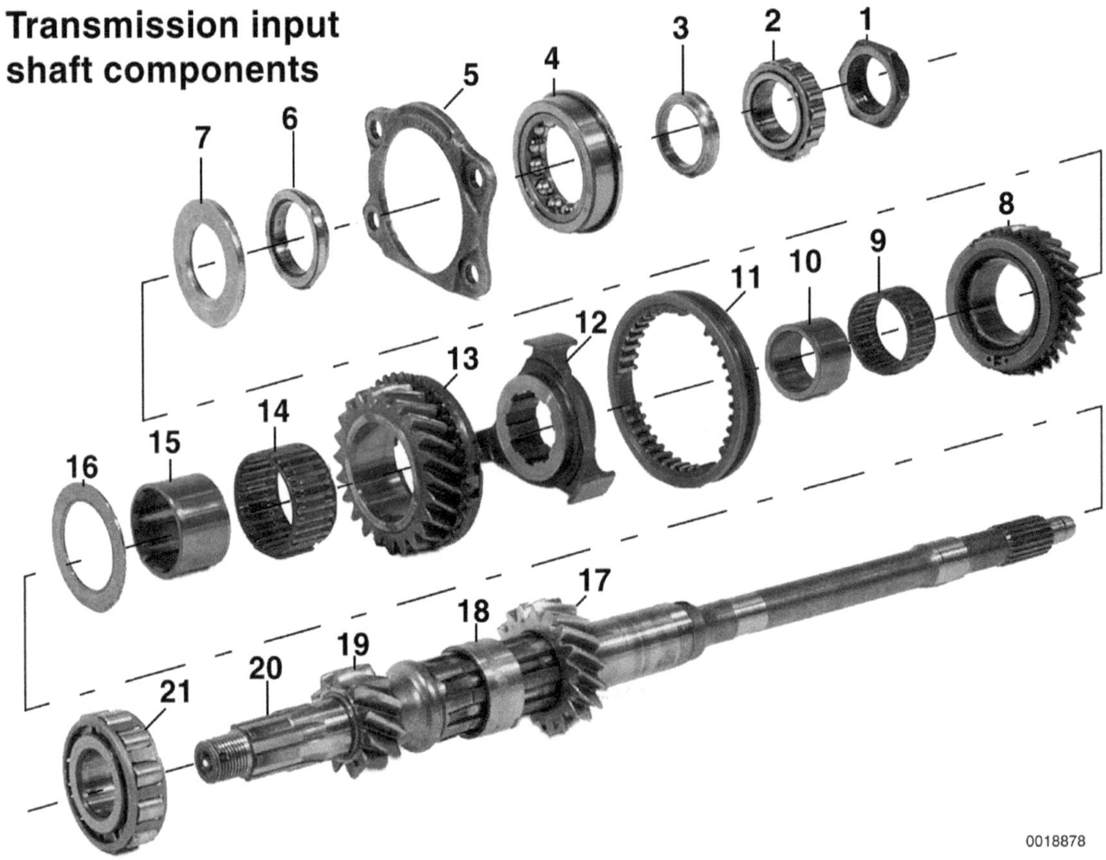

1. Collar nut removal: Use special tool P 252 a (41 mm socket) or equivalent.
   - Installation: Torque to 230 Nm (166 ft-lb).
   - Lock by punching collar. See detail **A**.
2. Roller bearing removal: Press off with special tools VW 401, VW 402, VW 412. When pressing, support under 2nd gear (17).
   - Installation: Heat to approx. 120°C (250°F).
   - Press on with special tools VW 401, VW 454 (press sleeve), VW 407
3. Ball bearing inner race removal: Keep with bearing and other race (**6**) as matched set.
   - Installation: Heat to approx. 120°C (250°F), then press on.
4. Ball bearing
5. Ball bearing flange
6. Ball bearing inner race: same as **3**
7. Thrust washer
8. 4th gear: Check synchronizer as described later.
   - Installation: If necessary, replace matched gears in pairs only.
9. Needle bearing: Keep with gear and race.
10. Needle bearing race: Keep with gear and bearing.
11. 3rd - 4th gear operating sleeve: Check for wear. See detail **B**.
12. Operating sleeve hub: Check for wear.
13. 3rd gear: Check synchronizer as described later.
    - Installation: If necessary, replace matched gears in pairs only.
14. Needle bearing: Keep with gear and race.
15. Needle bearing race: Keep with gear and bearing.
16. Thrust washer
17. 2nd gear: Replace matched gears in pairs only.
    - Installation: Manufacturer's marks face 1st gear. See detail **C**.
18. Spacer
19. 1st gear (integral with shaft)
20. Input shaft
21. Roller bearing removal: Press off with special tool VW 407 and suitable pipe.
    - Installation: Heat to approx. 120°C (250°F).
    - Press on with special tool VW 407 and suitable pipe.

– Always check drive shaft runout after tightening collar nut.

- Remove outer bearing races from transmission case and install them on input shaft.
- Place drive shaft bearings on machinist V-blocks.
- Set dial indicator to measure runout at pilot bearing end of shaft.

Input Shaft Runout

- Maximum runout at pilot bearing end . 0.1 mm (0.004 in.)

NOTE —
A shaft with up to 0.3 mm (0.012 in.) runout can be straightened out cold.

Details

A Stake collar nut to input shaft.

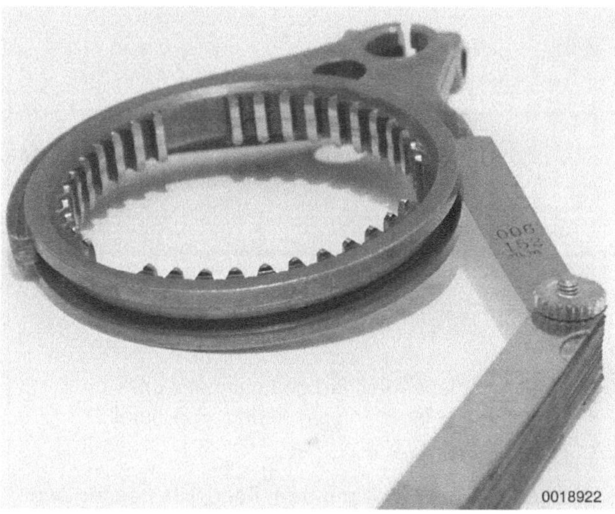

B Check clearance between shift fork and operating sleeve.

Shift Fork Clearance Specification

- Shift fork to operating sleeve. . . . . max 0.5 mm (0.02 in.)

C Install 2nd gear with manufacturer's marks facing 1st gear.

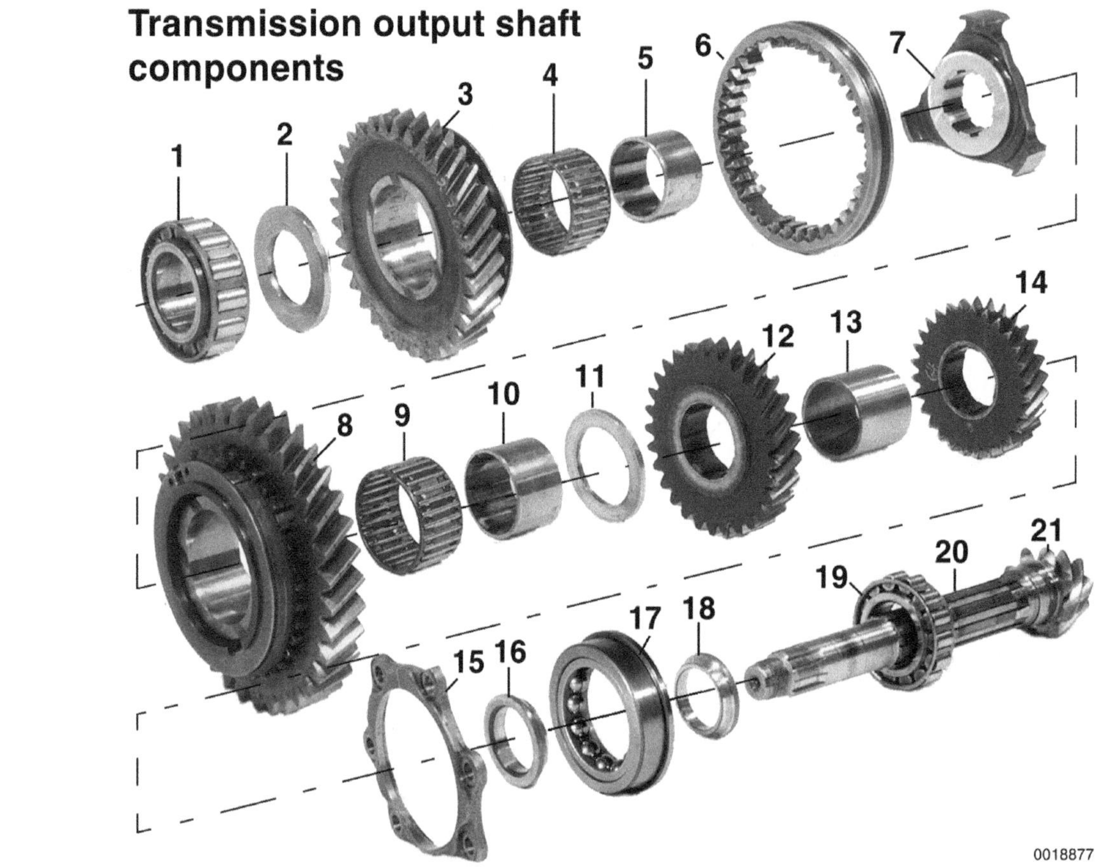

1. Roller bearing removal: Use special tools VW 401, VW 412, P 255 a (press sleeve). Press off by supporting under pinion end roller bearing (19).
   - Installation: Heat to approx. 120°C (250°F).
   - Press on with special tools VW 401, VW 412, VW 244 b (installing drift).
2. Thrust washer
3. 1st gear: Check synchronizer as described later.
   - Installation: First gear is matched to input shaft. If necessary, both must be replaced together.
4. Needle bearing: Keep with gear and race.
5. Needle bearing race: Keep with gear and bearing.
6. 1st - 2nd gear operating sleeve: Check for wear. See detail **A**.
   - Installation: Asymmetric teeth and recess face 1st gear. See detail **B**.
7. Operating sleeve hub: Check for wear.
8. 2nd gear: Check synchronizer as described later.
   - Installation: If necessary, replace matched gears in pairs only.
9. Needle bearing: Keep with gear and race.
10. Needle bearing race: Keep with gear and bearing.
11. Thrust washer
12. 3rd gear installation: Replace matched gears in pairs only.
    - Raised collar faces thrust washer. See detail **C**.
13. Spacer
14. 4th gear: Replace matched gears in pairs only.
    - Raised collar faces thrust bearing. See detail **D**.
15. Ball bearing flange
16. Ball bearing inner race removal: Keep with bearing and other race (**18**) as matched set.
    - Installation: Heat to approx. 120°C (250°F), then press on.
17. Ball bearing
18. Ball bearing inner race: same as 16.
19. Roller bearing removal: Use special tools VW 401, VW 412, P 255 a (press sleeve).
    - Install with collar of needle case facing gear set.
    - Installation: Heat to approx. 120°C (250°F).
    - Press on with special tools VW 407 and VW 415 a (press sleeve).
20. Pinion shaft
21. Drive pinion gear must match ring gear in differential.

Output shaft, disassembling and assembling

NOTE—
- *The output shaft is also referred to as the pinion shaft.*
- *All parts should be installed dry.*

– Follow steps in illustration on facing page. Details are covered below.

–Speedometer gear at forward end of shaft (not illustrated):

- Removal: Use puller
- Installation: Heat to approx. 120°C (250°F) and press on.

Details

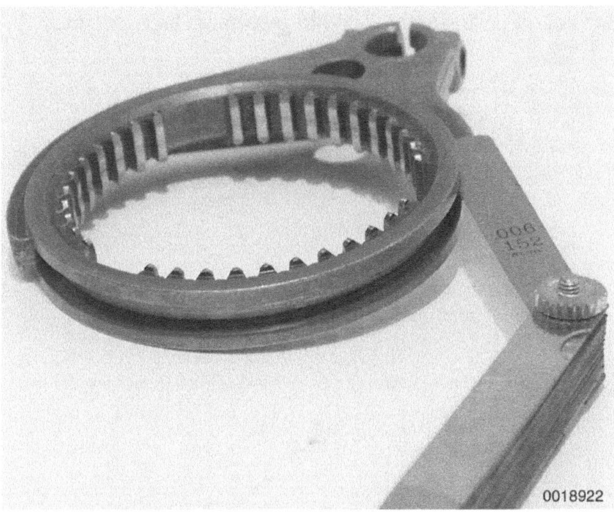

0018922

A Check clearance between shift fork and operating sleeve.

Shift Fork Clearance Specification	
• Shift fork to operating sleeve.	max 0.5 mm (0.02 in.)

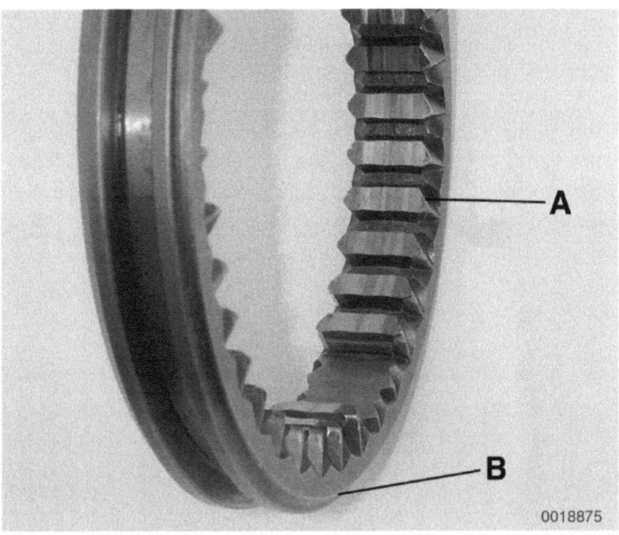

0018875

B Asymmetric teeth (**A**) and recess (**B**) of 1st - 2nd gear operating sleeve must face 1st gear.

0018923

C Raised collar (**arrow**) on 3rd gear must face thrust washer.

0018924

D Raised collar (**arrow**) on 4th gear must face bearing.

SYNCHRONIZERS

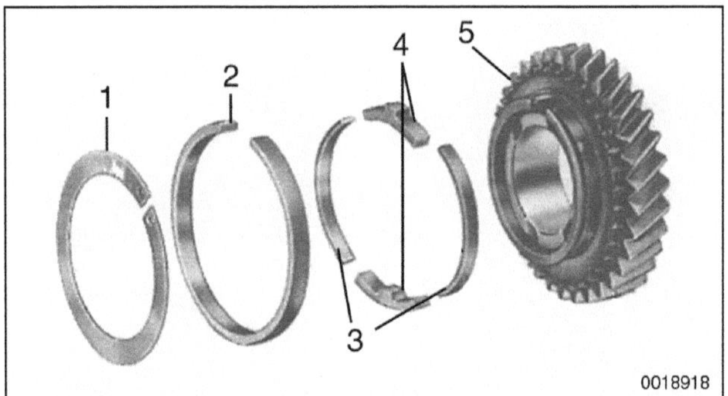

0018918

1. Circlip
2. Synchronizer ring
3. Brake band
4. Thrust block
5. 2nd gear with synchronizer ("dog") teeth

The synchronizer mechanism in the 915 transmission is unlike synchronizers in most other transmissions.

Two of the components, the synchronizer ring and the synchronizer ("dog") teeth, take the brunt of wear and tear. They must be carefully checked during transmission overhaul. Replace as necessary.

Synchronizer, checking

◄ Measure across synchronizer ring with gear and synchronizer fully assembled. If diameter exceeds specifications, replace synchronizer teeth.

Synchronizer Ring Diameter Specification

- 3rd, 4th, 5th gears . . 76.3 ± 0.18 mm (3.004 ± 0.007 in.)
- 1st, 2nd gears. . . . . 86.37 ± 0.17 mm (3.400 ± 0.007 in.)

– Remove circlip. Synchronizer ring should remain coiled inside synchronizer teeth. If not, teeth are worn and should be replaced.

◄ Check synchronizer teeth on gear assembly for damage or wear.

NOTE —
The synchronizer teeth can be worn round or chipped.

Synchronizer ring, removing and installing

◄ Remove circlip holding synchronizer ring in place.

- Remove synchronizer parts and inspect carefully.

CAUTION —
Keep all parts for each synchronizer together.

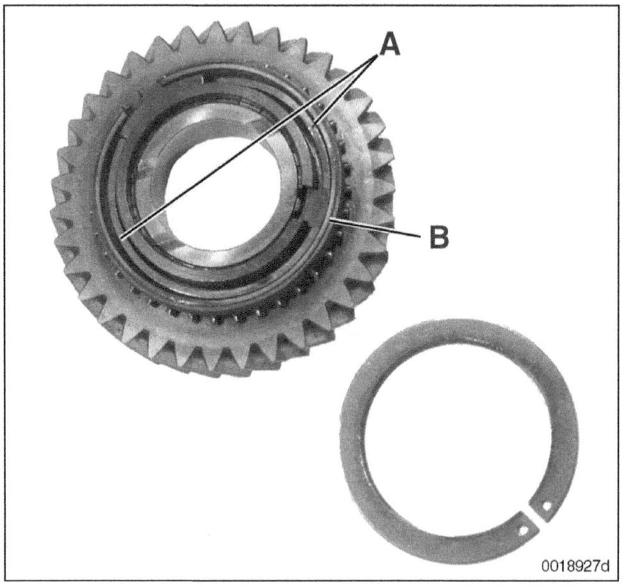

1st gear synchronizer:

- Asymmetric brake bands (**A**).
- Groove in synchronizer ring (**B**).

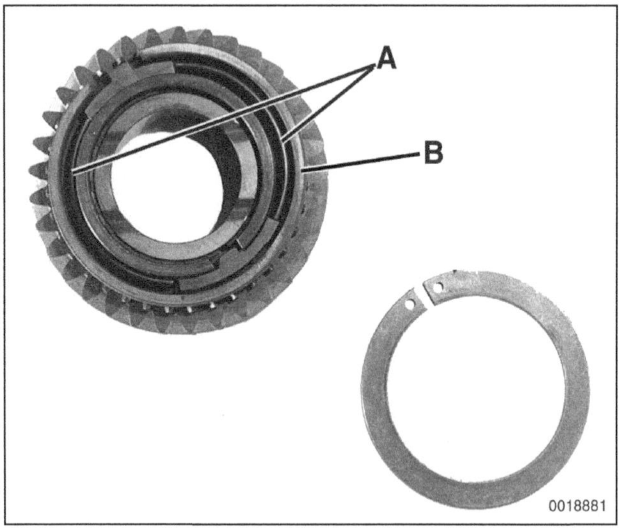

2nd gear synchronizer:

- Symmetrical brake bands (**A**).
- No groove in synchronizer ring (**B**).

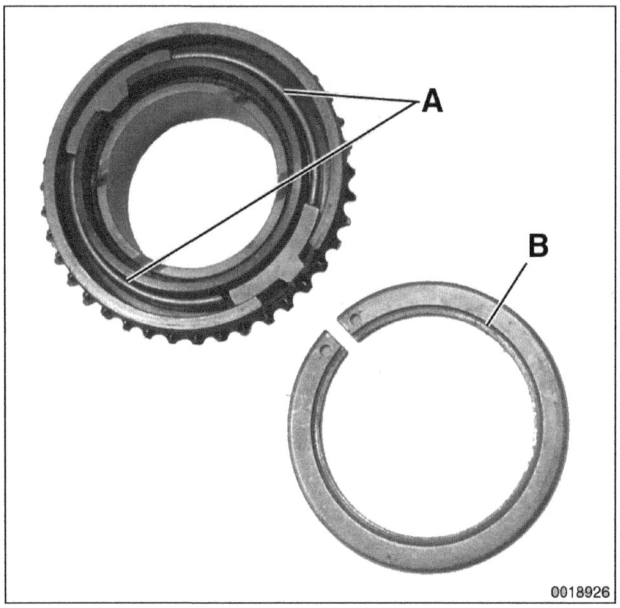

3rd, 4th, 5th gear synchronizers:

- Split brake bands (**A**).

– Install a new synchronizer ring and install circlip.

NOTE —
If the synchronizer ring is reused, install with rough side facing the operating sleeve.

– Install circlip for 3rd, 4th, or 5th gear synchronizer with stepped face (**B**) away from gear.

Synchronizer ("dog") teeth

Synchronizer ("dog") teeth are press fit on each gear. Aftermarket tools and individually available components can be used to replace damaged gear teeth. Consult with the Porsche dealer parts department.

390 Differential

Special Tools

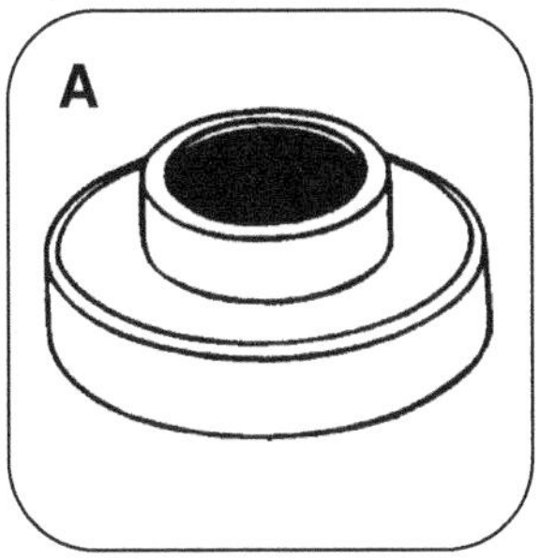

Special tool

A 9252 . . . . Seal installer
(Porsche ordering #000 721 925 00)

GENERAL

The Porsche 911 Carrera transmission assembly houses the forward and reverse gears and the differential, also known as the final drive.

This section covers replacement of the differential flange seals. This does not require disassembly of the differential. Disassembly of the differential requires special tools and knowledge, and is beyond the scope of this manual.

Special tool

The use of a special seal installer is helpful in performing the procedure described below.

Troubleshooting

The source of driveline vibrations and noise can be difficult to pinpoint. Engine, transmission, drive axle or wheel vibrations can be transmitted through the power train mounts to the car body. Noises from the rear of the car may be caused by differential problems or by faulty wheel bearings or constant velocity (CV) joints. Also check that the tires are correctly inflated and are not excessively worn.

To isolate a vibration or noise problem, speed up the engine in the stopped car to the rpm range where the problem occurs. This eliminates the influence of rotating driveshafts and will help indicate if the problem is caused by an engine condition.

Drive the car at the speed where the problem occurs, then shift to different gears to see if the problem changes with road-speed or engine-speed. Road-speed dependent problems usually indicate trouble in the driveline.

Detach the axle shafts at the differential and then operate the engine at the speed and gear position where the problem occurs. If the problem disappears, then it may be in the axle CV joints or wheel bearings.

390

> ***WARNING —***
> *Make sure the car is stable and well supported at all times. Use a professional automotive lift or jack stands designed for the purpose. A floor jack is not adequate support.*

> ***CAUTION —***
> - *Damage to the differential or axle shaft joints can result if the car is run with the wheels off the ground. The wheels hang below their normal height, putting stress on driveline components. Compress the suspension to normal height with jackstands before running the car.*
> - *Never operate the car with one drive wheel turning at a different speed than the other. Damage to the driveline components can result.*

NOTE —
No-load tests with the car stopped or the drive axles disconnected will only give a general idea of the car's performance under the load of normal operation.

For more information on troubleshooting engine or transmission problems, see **100 Engine–General**, **310 Clutch**, **340 Manual Transmission–Controls and Case**, or **350 Manual Transmission–Gears and Shafts**. For information on troubleshooting rear axle, wheel bearing or constant velocity joint problems, see **420 Rear Suspension, Axle Shafts**.

AXLE FLANGE SEAL

Low oil level caused by leakage due to faulty axle flange oil seals may be the cause of noisy transmission operation. The drive flange oil seals can be replaced while the differential is installed in the car.

Do not mistake leaking CV joints for flange seal leaks. It may be helpful to degrease the differential to pinpoint the source of the leak prior to replacing seals.

Axle flange seal, removing and installing

NOTE —
Do not remove both differential drive axle flanges at the same time. If both flanges are removed at the same time you may experience difficulty aligning the drive flange splines.

– Raise car and support in a safe manner.

> ***WARNING —***
> *Make sure the car is stable and well supported at all times. Use a professional automotive lift or jack stands designed for the purpose. A floor jack is not adequate support.*

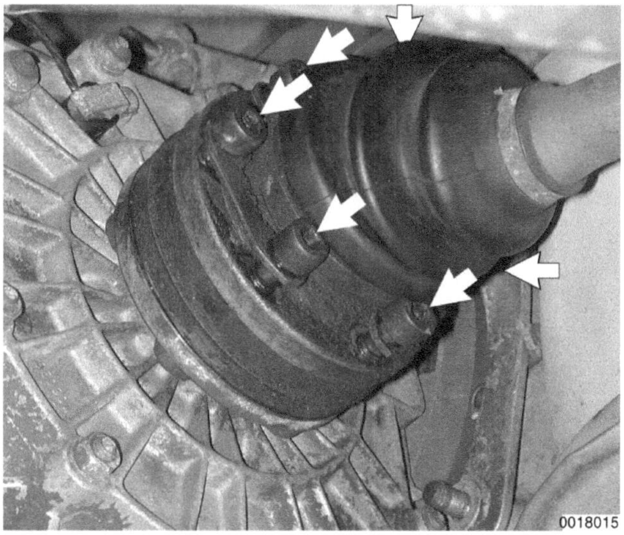

< Remove CV joint bolts (**arrows**) from axle flange.

– Screw 2 bolts into threaded holes in flange.

NOTE —

- *On 915 transmission: Use 2 M8x1.25 bolts.*
- *On G50 transmission: Use 2 M10x1.5 bolts.*

< Counterhold flange using a pry bar between two bolts.

- Remove flange mounting bolt.

– Pull flange out of differential.

– Remove flange seal from differential.

– Install new seal using special tool 9252 or equivalent.

– Reinstall flange and CV joint.

Tightening Torques

• CV joint to flange	
M8	42 Nm (30 ft-lb)
M10	83 Nm (61 ft-lb)
• Flange to differential	44 Nm (32 ft-lb)

400 Suspension, Brakes and Steering–General

TABLES

GENERAL

This section covers general information for front and rear suspension, the braking system, and the steering system. Diagnostic and troubleshooting information on these systems is at the end of the section.

The Porsche 911 Carrera suspension and steering systems are engineered to provide a compromise between taut, responsive handling and ride comfort. Although the front and rear suspension assemblies are independent subsystems, they work together to achieve Porsche's overall combination of handling and comfort.

The suspension has no lubrication points. All suspension joints are permanently lubricated. To maintain good handling and reliability, carry out periodic inspections as listed in **030 Maintenance**. The main items needing attention are:

- Front wheel bearings, checking, adjusting and repacking.
- Wheel nuts, checking tightening torques.
- Constant velocity (CV) joint boots, checking for damage.
- Tire pressures, checking and correcting.

Front suspension and steering

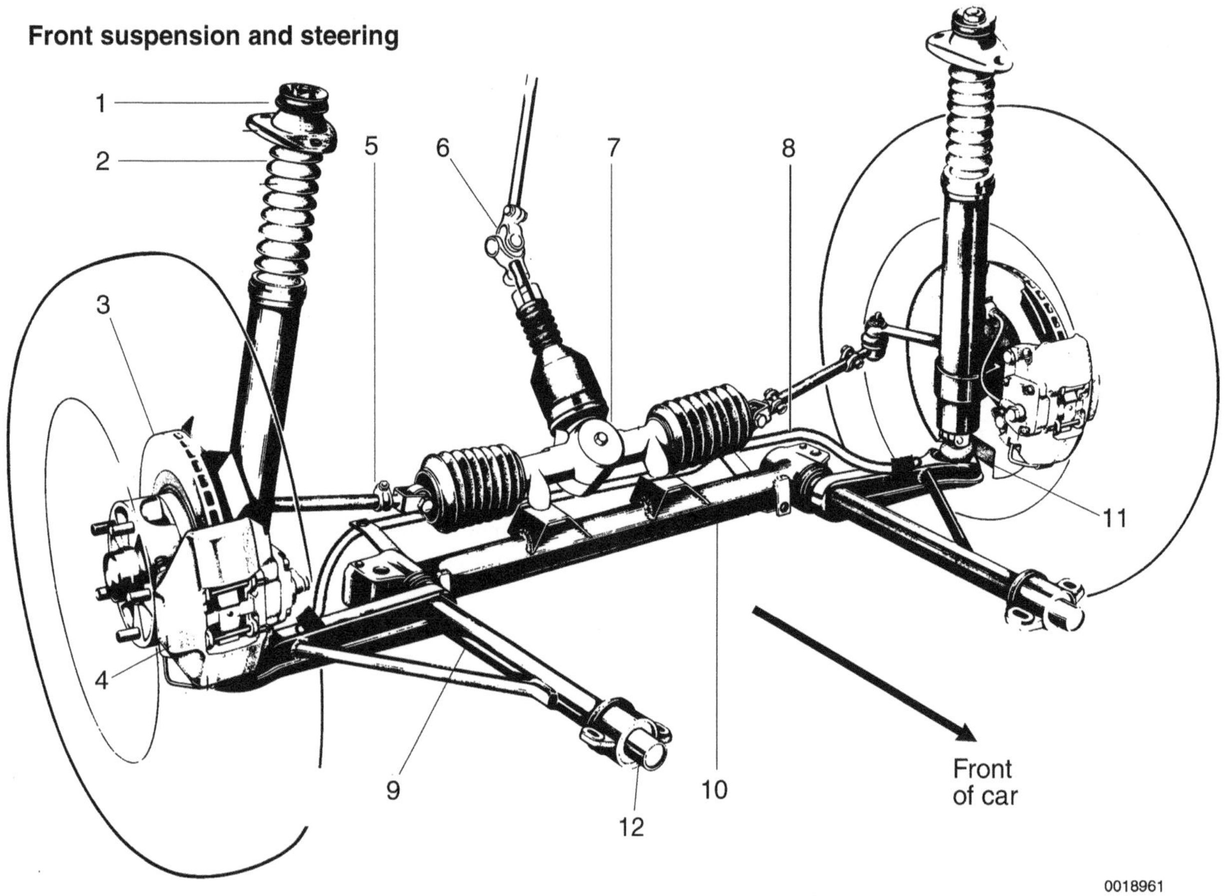

1. Top strut mount
2. Front suspension strut
3. Front brake rotor
4. Front brake caliper
5. Steering tie rod
6. Steering column universal joint
7. Steering rack
8. Stabilizer bar
9. Front control arm
10. Front suspension subframe
11. Ball joint
12. Torsion bar

Front suspension

Each front wheel includes a strut-type shock absorber mounted on a permanently lubricated ball joint. A control arm on each side houses a longitudinal torsion bar and is attached to the suspension subframe.

The torsion bars are adjustable. Suspension travel is limited by progressive rubber buffers (bump stops) installed on each strut piston.

Cornering control is enhanced by a transversely mounted stabilizer bar connected to the control arms and the rear of the suspension subframe with rubber bushings.

For further information or component replacement procedures for the front suspension, see **401 Front Suspension**.

Rear suspension

Rear suspension

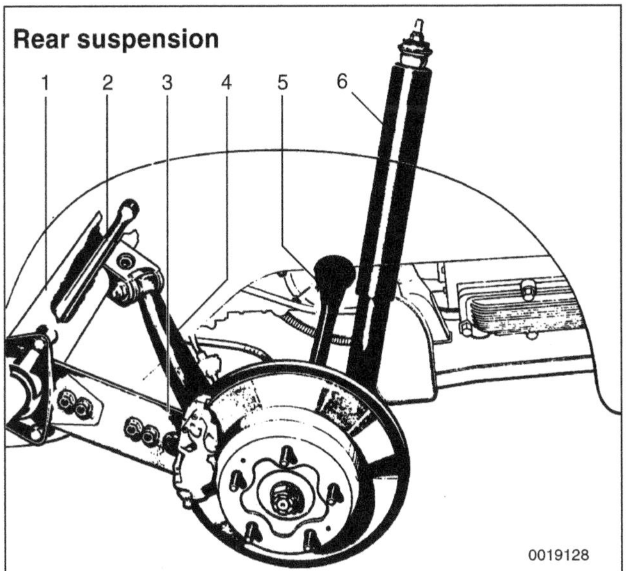

1. Rear suspension subframe (torsion bar carrier)
2. Torsion bar
3. Torsion plate
4. Trailing arm
5. Axle shaft
6. Shock absorber

The main rear suspension components are the trailing arms, torsions plates, torsion bars, and shock absorbers. Torsion plates connect the wheel bearing housings to the transversely mounted torsion bars. The torsion bars, housed within the rear suspension subframe, are adjustable by means of unequal-numbered inner and outer splines.

The rear wheel bearings are housed in the trailing arms. They are sealed and require no lubrication or adjustment.

Half axle shafts with one constant velocity (CV) joint at each end transmit power from the transmission output flanges to the rear wheels. The CV joints are permanently lubricated and need no maintenance, but the covering boots must be inspected and renewed if necessary.

For further information or component replacement procedures for the rear suspension, see **420 Rear Suspension, Axle Shafts**.

Table a. Suspension Specifications

Component	Front suspension	Rear suspension
Suspension	Independent wheel suspension with control arm and ball joint	Independent wheel suspension with alloy trailing arms. Wheel driven by axle shaft
Track: Carrera Turbo-look	 1372 mm (54.016 in) with 6J x 15 rims 1432 mm (56.378 in.) with 7J x 16 rims	 1380 mm (54.331 in.) with 7J x 15 rims 1501 mm (59.095 in.) with 8J x 16 rims
Torsion bars Diameter: 1984 - 1985 1986 - 1989	Longitudinal for each wheel 18.8 mm (0.740 in.) 18.8 mm (0.740 in.)	Transverse for each wheel 24.1 mm (0.949 in.) 25.0 mm (0.984 in.)
Shock absorbers	Strut with replaceable hydraulic or gas-filled cartridge	Double-action hydraulic or gas-filled shock absorber
Stabilizer bar diameter: 1984 - 1985 1986 - 1989	 20 mm (0.787 in.) 22 mm (0.866 in.)	 18.0 mm (0.709 in.) 21.0 mm (0.827 in.)

Steering rack: Side section

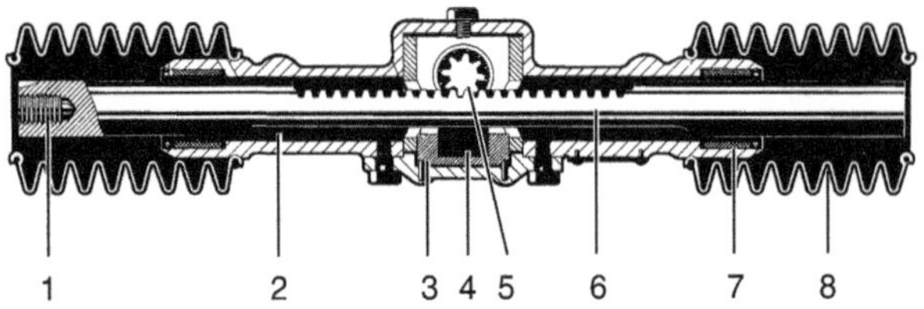

1. Tie rod thread
2. Rack housing
3. Adjusting nut
4. Pressure block
5. Steering pinion
6. Steering rack
7. End bushing
8. Rack boot

Steering Rack: Cross section

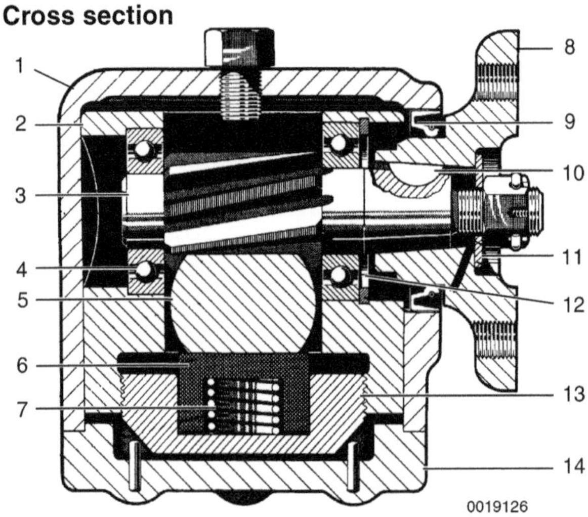

1. Housing
2. Pinion carrier
3. Steering pinion
4. Ball bearing
5. Steering rack
6. Pressure block
7. Spring
8. Steering column flange
9. Oil seal
10. Woodruff key
11. Spacer
12. Lock ring
13. Adjusting nut
14. Base plate

Steering

◄ Rack and pinion steering, used in the Porsche cars, is simple and rugged. There is no power assist to the steering in the cars covered by this manual.

The rack rides in replaceable bushings installed at the ends of the rack housing. The pinion turns in two ball bearings supported in a floating pinion carrier. The rack moves within the confines of the end bushings, pinion gear, a pressure block, spring and adjusting nut. This results in a steering that is very precise with no free play. Torque on the adjusting nut creates a controlled amount of drag in the mechanism.

The steering rack is filled with a special permanent lubricant and needs no further maintenance or lubrication. The protective boots at the tie rod ends of the rack, however, must be inspected periodically and replaced if torn or deteriorated.

For further information or component replacement procedures for the steering system, see **480 Steering**.

Table b. Steering Specifications

Characteristic	Specification
Frictional moment of steering (measured at flange of steering column flange, tie rods detached)	80 - 140 Ncm (113 - 198 in-oz.)
Steering ratio	17.78 : 1
Steering wheel turns lock-to-lock	Approx. 3
Turning circle diameter	10.95 m (35.93 ft.)

Braking system

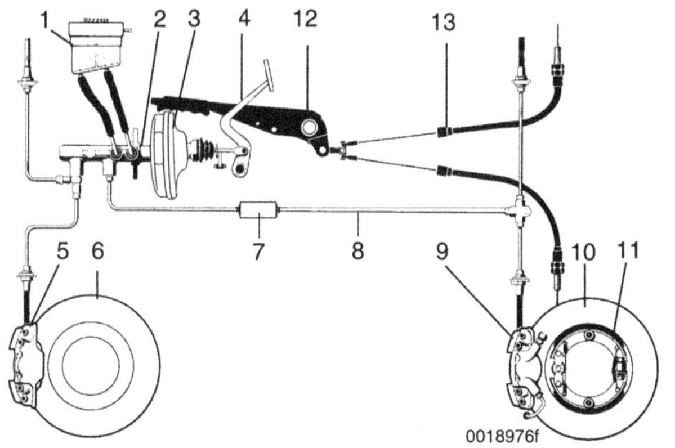

1. Brake fluid reservoir
2. Dual master cylinder
3. Power booster
4. Brake pedal
5. Front brake caliper
6. Front brake rotor
7. Rear brake pressure regulator
8. Brake fluid line
9. Rear brake caliper
10. Rear brake rotor/drum
11. Parking brake shoes
12. Parking brake handle
13. Parking brake cables

Braking system

Cars covered by this manual are equipped with disc brakes on all four wheels. The brake system is a diagonal dual-circuit design, hydraulically actuated by the master cylinder. A brake power booster system is used to reduce pedal effort. All models are also equipped with brake pad wear indicators.

The master cylinder, operated by the brake pedal, creates pressure in the hydraulic system. At the wheels, the hydraulic pressure acts on the calipers, which in turn mechanically apply the brakes. The use of hydraulics makes it possible for the driver to generate high braking forces with a comparatively small amount of effort.

A brake pressure regulator in the brake fluid line leading to the rear brakes prevents rear wheel lock-up during hard braking. The front and rear brake circuits receive equal hydraulic pressure up to approximately 33 bar (479 lb/sq. in.). When brake hydraulic pressure exceeds 33 bar, such as during a panic stop, the pressure regulator biases the brake pressure to the front brakes.

NOTE —

1985 and later Turbo-look cars have different specifications for the brake pressure regulator. See **Table c**.

Large pistons in the rear brake calipers improve braking with low pedal effort.

For further information or component replacement procedures for the braking system, see **460 Brakes–Mechanical** or **470 Brakes–Hydraulic**.

Table c. Braking System Specifications

Description	911 Carrera	911 Carrera Turbo-look (1984)	911 Carrera Turbo-look (1985 and later)
Service brakes	Hydraulic dual circuit, vacuum powered brake booster, vented disc brakes, calipers		
Brake booster: Diameter Boosting factor	 8 in. 2.25		 8 in. 3.0
Master cylinder: Diameter Stroke	 20.64 mm (.813 in.) 20/12 mm (0.787/0.472 in.)	 23.81 mm (0.937 in.) 18/14 mm (0.709/0.551 in.)	
Brake pressure regulator: Switch-over pressure Reducing factor	 33 bar (479 psi) 0.46		 55 bar (798 psi) 0.46
Brake rotor: Diameter: Front Rear Effective disc diameter: Front Rear Thickness (new): Front Rear Thickness (wear limit): Front Rear Min. thickness after machining Front Rear Max. surface roughness after machining Max. thickness variance Max. lateral runout	 282.5 mm (11.122 in.) 290.0 mm (11.417 in.) 228.0 mm (8.976 in.) 244.0 mm (9.606 in.) 24.0 mm (0.945 in.) 24.0 mm (0.945 in.) 22.0 mm (0.866 in.) 22.0 mm (0.866 in.) 22.6 mm (0.890 in.) 22.6 mm (0.890 in.) 0.006 mm (0.0002 in.) 0.02 mm (0.0008 in.) 0.1 mm (0.004 in.)	 304.0 mm (11.969 in.) (perforated) 309.0 mm (12.165 in.) (perforated) 247.0 mm (9.724 in.) 251.0 mm (9.882 in.) 32.0 mm (1.260 in.) 28.0 mm (1.102 in.) 30.0 mm (1.181 in.) 26.0 mm (1.027 in.) 30.6 mm (1.205 in.) 26.6 mm (1.047 in.) 0.006 mm (0.0002 in.) 0.02 mm (0.0008 in.) 0.1 mm (0.004 in.)	
Caliper piston diameter: Front Rear	 48.0 mm (1.890 in.) 42.0 mm (1.654 in.)	 38.0 mm (1.496 in.) (4 pistons/caliper) 30.0 mm (1.181 in.) (4 pistons/caliper)	
Brake pad surface area: Front (each wheel) Rear (each wheel) Total Brake pad thickness: New Wear limit	 76.0 cm^2 (11.78 sq. in.) 52.5 cm^2 (8.14 sq. in.) 257.0 cm^2 (39.84 sq. in.) 10 mm (0.394 in.) 2 mm (0.08 in.)	 94.0 cm^2 (14.57 sq. in.) 94.0 cm^2 (14.57 sq. in.) 376.0 cm^2 (58.28 sq in.) 13 mm (5.118 in.) 2 mm (0.08 in.)	
Brake pedal play (brakes bled, engine stopped)	Approx. 10 mm		
Parking brake	Cable operated shoes, drum brakes on both rear wheels; drums integral with brake rotors		
Brake drum: Diameter (new) Wear limit	 180 mm (7.087 in.) 181 mm (7.126 in.)		
Brake shoes: Width Surface area per wheel Thickness Wear limit	 25 mm (0.984 in.) 85 cm^2 (13.175 sq. in.) 4.5 mm (0.177 in.) 2 mm (0.08 in.)		

Wheels and tires

The Porsche wheels supplied as original equipment are designed for tubeless, radial-ply tires of a specific size. Replacement tires of non-standard size or construction should be installed only if the tire manufacturer specifies them for your model.

Radial-ply tires are installed as standard equipment on all cars covered by this manual. To retain the car's excellent handling characteristics, it is recommended that the tires only be replaced with those of the same (radial-ply) construction, size and speed rating.

Rim and tire sizes recommended by Porsche are listed in **Table d**.

For further information on wheels and tires, see **440 Wheels, Tires, Alignment**.

Table d. Rim and Tire Sizes

Model and year	Standard		Option 1		Option 2	
	Rim size	Tire size	Rim size	Tire size	Rim size	Tire size
Carrera 1984 - 1986						
Front	6 J x 15	185/70 VR 15	7 J x 15	185/70 VR 15	6 J x 16	205/55 VR 16
Rear	7 J x 15	215/60 VR 15	8 J x 15	215/60 VR 15	7 J x 16	225/50 VR 16
Carrera 1987 - 1988						
Front	6 J x 15	195/65 VR 15	7 J x 15	195/65 VR 15	6 J x 16	205/55 VR 16
Rear	7 J x 15	215/60 VR 15	8 J x 15	215/60 VR 15	7 J x 16	225/50 VR 16
Carrera 1989						
Front	6 J x 16	205/55 ZR 16	N-A	N-A	N-A	N-A
Rear	8 J x 16	225/50 ZR 16				
Turbo Look 1984 - 1985						
Front	7 J x 16	205/55 VR 16	N-A	N-A	N-A	N-A
Rear	8 J x 16	225/50 VR 16				
Turbo Look 1986 - 1988						
Front	7 J x 16	205/55 VR 16	N-A	N-A	N-A	N-A
Rear	9 J x 16	245/45 VR 16				
Turbo Look 1989						
Front	7 J x 16	205/55 ZR 16	N-A	N-A	N-A	N-A
Rear	9 J x 16	245/45 ZR 16				

Alignment

Wheel alignment is the precise adjustment of the steering and suspension components to ensure that all wheels are oriented correctly, compared to the other wheels, the chassis, and the direction of travel. Small changes can have a big effect on how the car drives. Proper alignment provides the best compromise between responsiveness, stability, and tire wear.

Alignment specifications differ from model to model. It is important to use the correct alignment specifications for the specific car. Wheel alignment can also be fine-tuned for specific driving conditions. If, for example, the car normally carries only one person or is often heavily loaded, this information may help the alignment technician. Like tire pressure, wheel alignment also has some influence on fuel economy, because of its potential to affect the tires' rolling resistance.

> ***WARNING —***
> *Replacement and/or removal of the trailing arms, torsion bars or torsion plates (rear) disturbs the rear wheel alignment settings. Replacement or adjustment of the front or rear torsion bars or the rear torsion plates also affects vehicle ride height, which directly influences vehicle balance. If ride height is changed, it is strongly recommended that the vehicle balance/weight be checked at all four wheels using suitable wheel-load floor scales. See* **440 Wheels, Tires, Alignment** *for more information.*

TROUBLESHOOTING

This troubleshooting section applies to suspension, steering or braking system problem that affect ride comfort, handling and stability.

Similar problems may be caused by faults in the brakes or steering system. For example, a problem such as consistently pulling to one side may be caused by a bent control arm or by faulty brakes.

Suspension troubleshooting

Stable handling and ride comfort both depend on the integrity of the suspension components. These systems must precisely position the wheels so that the car is stable and controllable, but also allow movement so that the wheels can steer and react to bumps. Any symptom of instability or imprecise road feel may be caused by worn or damaged suspension components.

In addition to inspecting for worn parts, troubleshooting must also consider the condition of tires, wheels, and their alignment. Tire wear and incorrect inflation pressures can dramatically affect handling. Subtle irregularities in wheel alignment angles also affect stability. Mixing different types or sizes of tires, particularly on the same end of the car, can affect alignment and may unbalance a car's handling. See **440 Wheels, Tires, Alignment** for more information.

Table e lists symptoms of suspension problems and their probable causes, and suggests corrective actions. The boldface numbers in the corrective action column refer to headings where the repairs are described.

Diagnostic inspection and testing (suspension)

Suspension problems can usually be isolated and at least partially diagnosed by careful observation of the symptoms and inspection of the components that are the most likely cause.

Torsion bars support the weight of the car and allow the suspension to travel smoothly over bumps and other road irregularities. A torsion bar that is twisted (loaded) by a bump simply unloads in the other direction with nearly the same force that loaded it.

The function of a shock absorber, or damper, is to moderate the torsion bar action. It slows the bounce and helps the torsion bar return to its normal position. Shock absorbers require no routine maintenance and cannot be serviced.

The best evidence of failing shock absorbers is their behavior in normal driving. Worn shock absorbers will allow extra skittishness over bumps, and a less-controlled and wallowing feel after bumps and in corners. When seriously worn, the shock absorbers present little resistance to torsion bar oscillations. Because they so easily reach the limits of their travel, they may knock when going over bumps.

The most common—though not entirely accurate—test of shock absorber function involves vigorously bouncing each end or corner of the car, and then releasing and observing how quickly the bouncing stops. More than one bounce usually suggests that the shock absorbers are not properly damping the torsion bar action and need to be replaced.

NOTE —
Differences from side to side will affect handling and stability. It is strongly recommended that shock absorber cartridges or strut bearings be replaced in pairs.

Tire wear

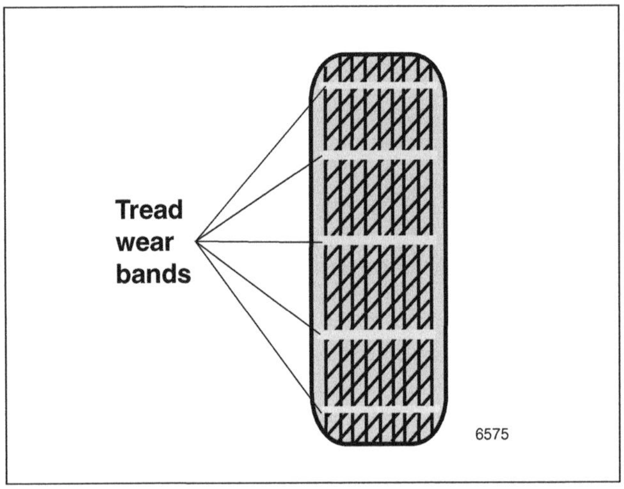

Tire wear is a good indicator of steering and wheel alignment problems. Proper tread wear is difficult to notice, so tires are made with wear-indicator bands that indicate when the tire is nearly worn-out. On an evenly worn tire, these wear-indicator bars will eventually appear as evenly spaced bald "stripes" about ½ in. wide running across the tread surface. The appearance of these tread-wear bands on only one part of the tread indicates uneven wear.

For more information on tire wear, see **440 Wheels, Tires, Alignment**. For information on maintaining proper tire inflation pressures, as well as other general tire maintenance, see **030 Maintenance**.

Isolating pulling symptoms

Consistent pulling to one side in a car driven straight ahead on a level road may be caused either by worn or damaged (bent) suspension components or a faulty tire. In more unusual cases, a brake problem or possibly incorrect torsion bars may be the cause.

To help decide whether the problem is caused by the tires or the suspension, temporarily swap the front wheels (and tires), and then road-test the car. If a tire problem is the cause of the pulling symptom, the problem should switch to the other side of the car when the wheels are switched, and it should now pull to the other side. If the symptom persists, then the problem is probably caused by faulty suspension components.

NOTE —
If no suspension components are found to be faulty, the car may only need a front wheel alignment. See **440 Wheels, Tires, Alignment** *for more information.*

Vibration

Abnormal vibration, if not caused by a mechanical problem, will most often be caused by wheels and tires that are out of balance. Other, less likely, causes of vibration are bent wheels or a tire that has become distorted and out-of-round. These and other tire conditions can be checked by any reputable tire professional.

Before spending the money for balancing, look for an obvious cause of imbalance. Caked-on mud, ice, or snow can dramatically affect wheel balance. Clean the wheels and tires and road-test the car before investigating for more serious causes.

Table e. Suspension, Steering and Alignment Troubleshooting

Symptom	Probable cause	Corrective action
Pull to one side, wandering	a. Incorrect tire pressure b. Defective/unevenly worn tire c. Incorrect wheel alignment d. Faulty brakes (pulls only when braking) e. Faulty tie rod or tie rod end	a. Check and correct tire pressures. See **030 Maintenance**. b. Inspect tires and replace as needed. See **440 Wheels, Tires, Alignment**. c. Check and adjust wheel alignment. See **440 Wheels, Tires, Alignment**. d. See **460 Brakes–Mechanical** or **470 Brakes–Hydraulic**. e. See **480 Steering**.
Steering heavy, poor return to center	a. Worn front upper strut mount(s) b. Incorrect tire pressure	a. Replace front upper strut mounts. See **401 Front Suspension**. b. Check and correct tire pressures. See **030 Maintenance**.
Vibration or shimmy while driving	a. Severely worn front strut cartridges or rear shock absorbers b. Worn ball joints c. Unbalanced or bent wheels/tires d. Loose wheel lug bolts e. Incorrect tire pressure f. Tire(s) out-of-round (radial runout) g. Faulty tie rod(s) or tie rod end(s) h. Faulty steering universal joint	a. Replace strut cartridges or shock absorbers. See **401 Front Suspension** or **420 Rear Suspension, Axle Shafts**. b. Replace ball joints. c. Balance tires. Check tires for uneven wear patterns. Check wheels for damage. See **440 Wheels, Tires, Alignment**. d. Tighten lug bolts to proper torque. See **440 Wheels, Tires, Alignment**. e. Check and correct tire pressure. See **030 Maintenance**. f. Measure tire radial runout and remount or replace as necessary. g. See **480 Steering**. h. See **480 Steering**.
Poor stability, repeated bouncing after bumps, suspension bottoms out easily	a. Worn (weak) strut cartridges in front b. Worn (weak) shock absorbers in rear	a. Replace strut cartridges. See **401 Front Suspension**. b. Replace rear shocks absorbers. See **420 Rear Suspension, Axle Shafts**.
Suspension noise (especially over bumps)	a. Worn front upper strut mounts. b. Loose front strut cartridge c. Faulty rubber shock absorber buffer (bump stop) d. Worn ball joints. e. Loose or worn stabilizer bar rubber mounts f. Loose front suspension subframe	a. Replace front upper strut mounts. See **401 Front Suspension**. b. Disassemble strut, check for damage. See **401 Front Suspension**. c. Replace rubber buffer. d. Replace ball joints. See **401 Front Suspension**. e. Retorque stabilizer bar rubber mounts. Replace rubber mounts as necessary. f. Check for subframe damage. Tighten mounting bolts.
Uneven ride height	a. Incorrectly adjusted or incorrect torsion bars b. Bent or damaged suspension components c. Sagging torsion bars	a. Measure ride height and adjust or replace torsion bars as required. See **401 Front Suspension** or **420 Rear Suspension, Axle Shafts**. b. Inspect and repair/replace as necessary c. Replace torsion bars in pairs. See **401 Front Suspension** or **420 Rear Suspension, Axle Shafts**.
Scraping, growling or rhythmic knocking from wheels while driving, may be more noticeable when turning	a. Worn wheel bearing(s) b. Faulty brakes c. Worn constant velocity (CV) joint	a. Replace wheel bearings. See **401 Front Suspension** or **420 Rear Suspension, Axle Shafts**. b. Repair brakes. See **460 Brakes–Mechanical** c. Inspect CV joint boots for tears. Replace boots or joints as needed. See **420 Rear Suspension, Axle Shafts**
Steering loose, imprecise	a. Incorrect tire pressure b. Loose steering rack mounting bolts c. Worn tie rod(s) or tie rod end(s) d. Faulty front wheel bearing(s) e. Worn or damaged steering rack	a. Check and correct tire pressure. See **030 Maintenance**. b. Inspect and tighten bolts c. See **480 Steering**. d. Replace front wheel bearing(s). See **401 Front Suspension**. e. See **480 Steering**.

Braking system troubleshooting

This heading describes symptoms of trouble with the brakes. The sole function of the brakes is to generate friction to slow or stop the car. Brake problems are usually obvious because they affect the way the car slows and stops. Noise or problems with the car's handling may be caused by the brakes, but may also be caused by faults in the suspension or steering systems. For help in selecting the appropriate repair section, see **Table e** and **Table f**.

Brake diagnostic checks

Component inspection and some general brake system checks can help isolate problems.

Check the brake booster first by pumping the brake pedal approximately 10 times (vacuum-assist systems) or 20 times (hydraulic-assist systems) with the engine off. While holding the pedal down start the engine. The pedal should fall slightly. If not, check for any visible faults before suspecting a faulty brake booster.

Check for leaks in the power booster vacuum line. A faulty check valve or a faulty O-ring between the master cylinder and the vacuum booster will decrease the vacuum assist. Low engine vacuum could also decrease vacuum assist. Test the check valve by removing it from the booster vacuum line and blowing through it. Air should pass through in the direction of the arrow, but not pass through the opposite way.

Brake noise

Occasional groaning or squealing sounds from the disc brakes are usually caused by vibration being transmitted through the brake pads. Brake friction materials that contain little or no asbestos may contribute to brake noise. These noises are normal and rarely indicate a problem.

Although there is no good solution, proper maintenance and repair can help minimize brake noise. Disc brake caliper assemblies include anti-rattle springs to minimize vibration and noise. Brake pads and calipers should be kept clean and free of corrosion. Always resurface or replace brake rotors when changing brake pads.

Table f lists symptoms of problems commonly associated with the brakes, their probable causes, and suggested corrective actions. The boldface numbers in the corrective action column refer to headings where the repairs are described.

Table f. Brake System Troubleshooting

Symptom	Probable cause	Corrective action
Brake squeak or squeal	a. Incorrectly installed brake pads or parking brake shoes b. Brake pad carriers dirty or corroded c. Brake pad anti-rattle springs faulty or missing d. Brake pads heat-glazed or oil-soaked e. Wheel bearings worn (noise most pronounced when turning)	a. Check component installation. See **460 Brakes–Mechanical**. b. Remove brake pads and clean calipers. See **460 Brakes–Mechanical**. c. Install/replace anti-rattle springs. See **460 Brakes–Mechanical**. d. Replace brake pads. Clean rotors. Replace leaking calipers as required. See **460 Brakes–Mechanical** or **470 Brakes–Hydraulic**. e. Replace worn bearings. See **401 Front Suspension** or **420 Rear Suspension, Axle Shafts**.
Pedal goes to floor when braking	a. Brake fluid level low due to system leaks b. Master cylinder faulty	a. Check fluid level and inspect hydraulic system for signs of leakage. Fill and bleed system. See **470 Brakes–Hydraulic**. b. Replace master cylinder. See **470 Brakes–Hydraulic**.
Low pedal even after system bleeding	a. Master cylinder faulty	a. Replace master cylinder. See **470 Brakes–Hydraulic**.
Pedal feels spongy or brakes work only after pedal is pumped	a. Air in brake fluid b. Master cylinder faulty (internal return spring weak) c. Leaking line or hose unions	a. Bleed system. See **470 Brakes–Hydraulic**. b. Replace master cylinder. See **470 Brakes–Hydraulic**. c. Repair or replace lines and hoses. Bleed system. See **470 Brakes–Hydraulic**.
Excessive braking effort	a. Brake pads wet b. Brake pads heat-glazed or oil-soaked c. Vacuum booster or vacuum hose connections to booster faulty	a. Use light pedal pressure to dry pads while driving. b. Replace brake pads. Clean rotors. Replace leaking calipers. See **460 Brakes–Mechanical** or **470 Brakes–Hydraulic**. c. Inspect vacuum lines. Test vacuum booster and replace as required. Test vacuum check-valve for one-way air flow. See **470 Brakes–Hydraulic**.
Brakes pulsate, chatter, or grab	a. Warped brake rotors b. Brake pads worn c. Brake pads heat-glazed or oil-soaked	a. Resurface or replace rotors. See **460 Brakes–Mechanical**. b. Replace brake pads. See **460 Brakes–Mechanical**. c. Replace brake pads. Clean rotors. Replace leaking calipers. See **460 Brakes–Mechanical** or **470 Brakes–Hydraulic**.
Uneven braking, car pulls to one side, rear brakes lock	a. Incorrect tire pressures or worn tires b. Brake pads on one side of car heat-glazed or oil-soaked c. Caliper or brake pads binding d. Worn suspension components	a. Inspect tire condition. Check and correct tire pressures. See **030 Maintenance**. b. Replace brake pads. Clean rotors. Replace leaking calipers. See **460 Brakes–Mechanical** or **470 Brakes–Hydraulic**. c. Clean and recondition brakes. See **460 Brakes–Mechanical** or **470 Brakes–Hydraulic**. d. Inspect for worn or damaged suspension components. See **401 Front Suspension** or **420 Rear Suspension, Axle Shafts**.
Brakes drag, bind, or overheat	a. Brake caliper or brake pads binding b. Master cylinder faulty	a. Clean or replace caliper. See **470 Brakes–Hydraulic**. b. Replace master cylinder. See **470 Brakes–Hydraulic**.

401 Front Suspension

GENERAL

The Porsche 911 Carrera is equipped with independent front suspension. Torsion bars are used at all four wheels. The front end uses strut shock absorbers.

This section covers replacement procedures for front suspension components. For specifications, diagnosis and troubleshooting information see **4 Suspension, Brakes and Steering**. For information on setting ride height and alignment see **440 Wheels, Tires, Alignment**. For information on steering see **480 Steering**.

Special tools

Special Tools

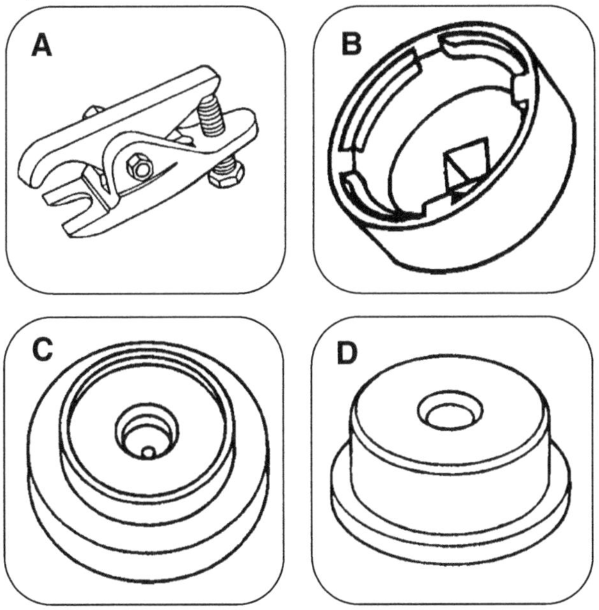

Special tool

A VW 267 a . . . . . Tie-rod end ball joint separator
B P 280 b. . . . . . Ball joint removal spanner (Porsche ordering #000 721 280 20)
C VW 433 . . . . . Front wheel grease seal installer
D VW 447h/447i . . . . . Front wheel outer bearing race installers

Some of the procedures in this group require special tools.

Front suspension

Each front suspension strut assembly includes a replaceable shock absorber cartridge inside a tubular strut housing. The lower end of each strut includes the stub axle for the front wheel assembly.

The front torsion bars are mounted inside the control arms. The spring tension on the front torsion bars is easily adjustable.

The front control arms are mounted to the floor pan. Ball joints are used to connect the control arms to the strut assemblies.

Front suspension assembly

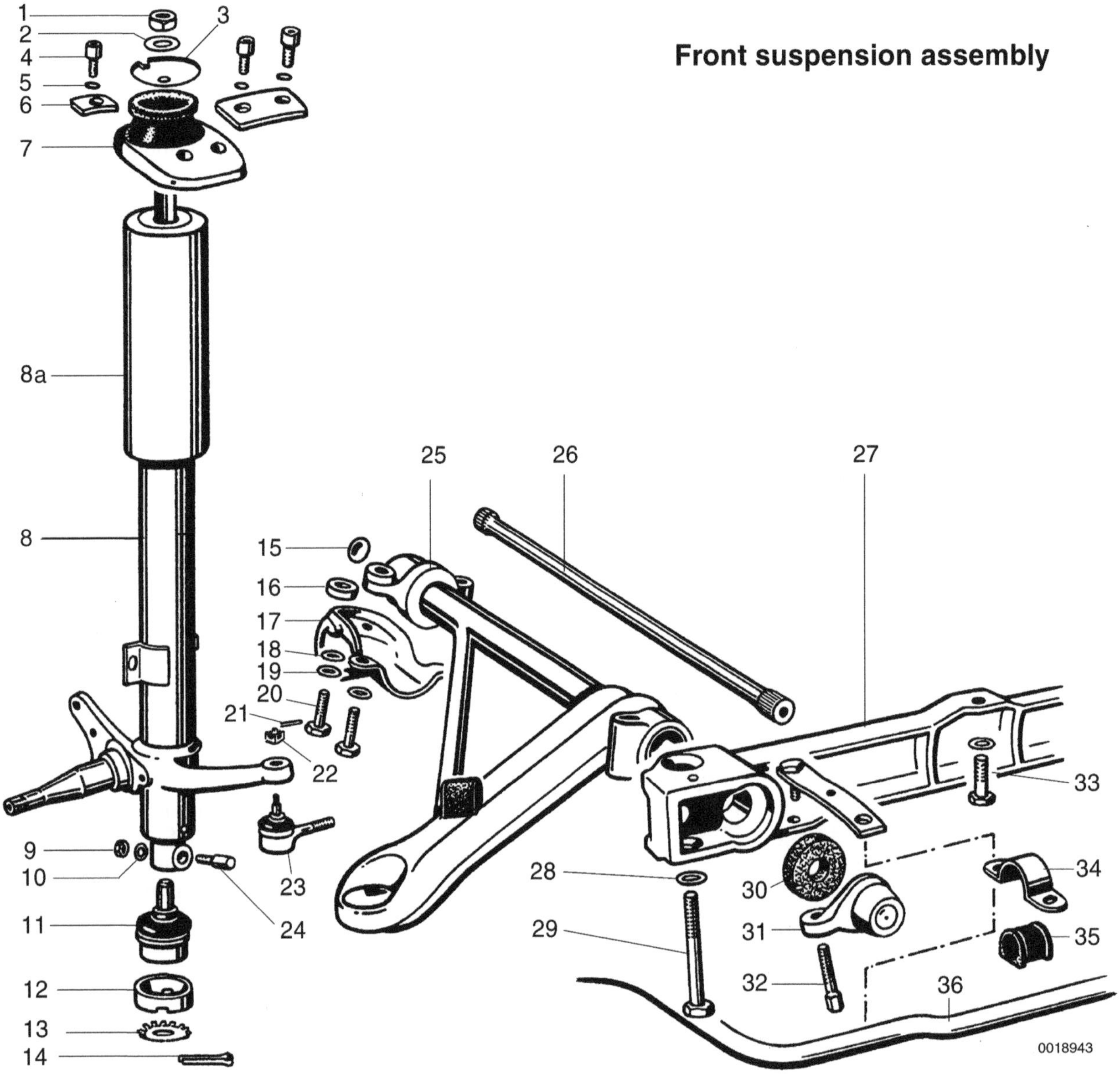

1. Nut (M14 x 1.5)
   tighten to 80 Nm (59 ft-lb)
2. Locking tab: always replace
3. Plate
4. Bolt (M10 x 20 mm)
   tighten to 47 Nm (34 ft-lb)
5. Lock washer (M10)
6. Support plate
7. Top strut mount
8. Strut

8a. Strut dust sleeve

9. Lock nut
   tighten to 22 Nm (16 ft-lb)
10. Washer
11. Ball joint
12. Strut cap
    tighten to 250 Nm (184 ft-lb)
13. Lock plate
14. Cotter pin: always replace
15. Cover
16. Shim
17. Control arm front mounting bracket
18. Washer (M10)
19. Lock washer (M10): always replace

20 Bolt (M10 x 35 mm)
   tighten to 47 Nm (35 ft-lb)

21. Cotter pin (2 x 25): always replace
22. Castle nut (M10 x 1)
    tighten to 45 Nm (33 ft-lb)
23. Tie-rod end
24. Wedge bolt: always replace
25. Control arm
26. Torsion bar
27. Crossmember subframe
28. Lock washer (M12): always replace
29. Bolt (M12 x 1.5 x 85 mm)
    tighten to 90 Nm (66 ft-lb)
30. Foam seal
31. Torsion bar retaining cap
32. Torsion bar adjusting bolt
33. Bolt (M10)
    tighten to 47 Nm (35 ft-lb)
34. Stabilizer bar bracket
35. Stabilizer bar bushing
36. Stabilizer bar

Suspension travel is limited by rubber bump stops on the top of each strut assembly.

A stabilizer bar is mounted to the front control arms and subframe to help reduce body roll when cornering.

When doing any front suspension maintenance or repair, observe the following general cautions.

CAUTION —

- *Do not install bolts and nuts coated with undercoating wax, as correct tightening torque cannot be assured. Always clean the threads with solvent before installation, or install new parts.*
- *Do not attempt to weld or straighten the suspension struts, control arms or subframe. Replace damaged parts.*
- *Do not reuse self-locking nuts or cotter pins. These parts are designed to be used only once.*

FRONT SUSPENSION STRUTS

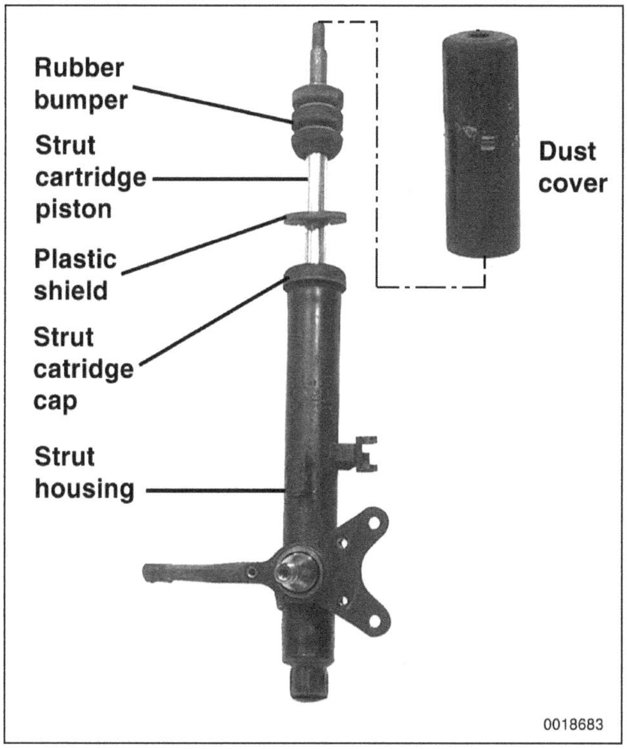

The main components of the front strut assembly are the strut housing, the strut cartridge and the top strut mount. Each of these main components can be replaced separately. The strut housing itself should be replaced if it is damaged.

A variety of struts by different manufacturers, including Boge® and Bilstein®, have been used in Porsche cars. Aftermarket manufacturers, including Koni® and Tokico®, also offer replacement strut cartridges. The strut assembly must be removed from the car and disassembled, as described later, to replace the internal damper cartridge.

Consult with the Porsche dealer parts department or aftermarket supplier when deciding what replacement parts to use.

NOTE —

Any differences in suspension components from side to side will affect handling and stability. It is strongly recommended that struts, strut cartridges or top strut mounts always be replaced in pairs at the same time.

Front suspension strut, removing and installing

NOTE —

- *Shock absorbers should be stored upright. If they are stored horizontally or upside-down for an extended period, they should be placed upright at room temperature with shock absorber rods fully extended for at least 24 hours prior to installation.*
- *The non-gas strut will remain in any desired position when removed. The gas-pressure version will have the damper rod all the way out when removed.*
- *If the upper strut mount is not removed with the strut the wheel alignment will not be disturbed.*

– Loosen front wheel lug nuts. Raise front of car and support it securely.

> ***WARNING —***
> *Make sure the car is stable and well supported at all times. Use a professional automotive lift or jack stands designed for the purpose. A floor jack is not adequate support.*

– Remove front wheel.

– Press brake pedal down about 1 in. and hold in place with a pedal stop.

NOTE —
Holding the brake pedal down will prevent brake fluid from dripping out of open brake lines while the work is being performed.

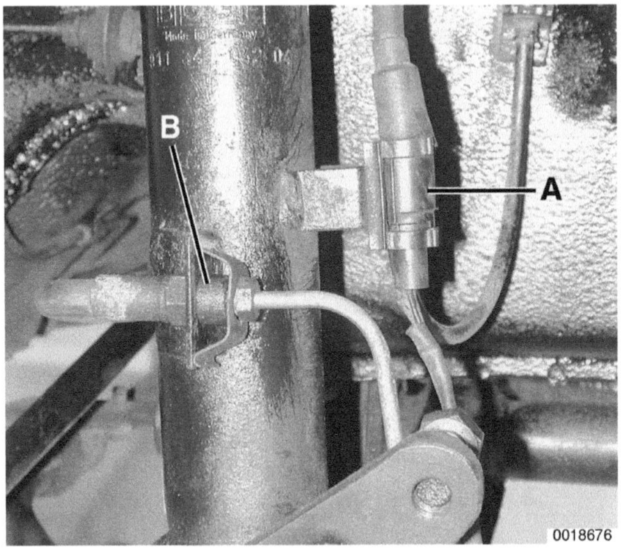

◀ Working in wheel housing, disconnect electrical connector (**A**) for brake pad wear sensor and remove harness from clip on strut body.

– Disconnect rubber brake hose (**B**) from metal brake line near bottom of strut. Plug open lines.

> ***CAUTION —***
> *Always counterhold fitting before loosening or tightening brake lines and hoses.*

– Remove caliper mounting bolts and remove caliper. See **470 Brakes–Hydraulic**.

– Remove wheel bearing dust cover.

0018966

◄ Loosen clamping screw (6 mm allen wrench) in wheel bearing lock nut. Unscrew and remove lock nut from spindle.

– Pull brake rotor with bearing hub and wheel bearings straight off steering spindle.

– Remove cotter pin from tie rod nut and remove castle nut.

– Using special tool (VW 267 a) or equivalent, separate tie rod from strut steering arm.

0018685

◄ Remove ball joint wedge bolt (**arrow**).

– Carefully pry out or tap on control arm to push ball joint down out of strut housing.

0018688

◄ Working in luggage compartment, bend back strut mounting nut locking tab. Remove nut from upper strut mount.

– While supporting strut assembly from below, remove strut assembly. If necessary, use a punch to help push strut piston out of mount.

– Installation is reverse of removal. Always replace ball joint wedge bolt. When installing wedge bolt, line up flat of bolt with cutout in ball joint.

– Repack and adjust wheel bearing as described later.

– Bleed brake hydraulic system. See **470 Brakes–Hydraulic**.

– Check and adjust wheel alignment. See **440 Wheels, Tires, Alignment**.

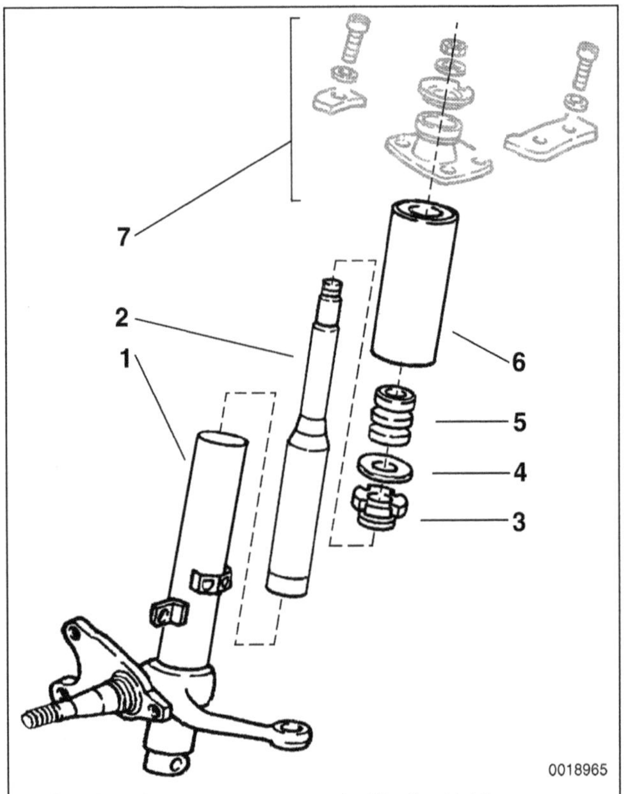

1. Strut housing
2. Strut cartridge (insert)
3. Strut cartridge cap, tighten to 120 + 20 Nm (87 + 14 ft-lb)
4. Plastic shield
5. Rubber bumper
6. Dust cover
7. Upper strut mount assembly

Tightening Torques

- Ball joint wedge bolt (always replace) . . . 22 Nm (16 ft-lb)
- Brake line to brake hose . . . . 12 Nm (9 ft-lb)
- Caliper to strut housing . . . . 70 Nm (51 ft-lb)
- Strut to upper mount (use new locking tab) . . . . 80 Nm (58 ft-lb)
- Tie rod castle nut to tie rod end . . . . 45 Nm (33 ft-lb)
- Wheel bearing lock nut clamp screw (M8) . . . . 15 Nm (11 ft-lb)
- Wheel to wheel hub . . . . 130 Nm (96 ft-lb)

Front strut cartridge, removing and installing

◄ Remove front suspension strut as described earlier. Remove dust cover and rubber bumper from strut piston rod.

– Clamp housing in vise or other strut holding device.

– Using strut nut wrench, remove strut cartridge cap.

- Pull cartridge out of strut housing.
- Empty out oil in strut housing.

– Reassembly is reverse of disassembly.

Tightening Torque

- Strut cartridge cap to strut housing . . . . 120 +20 Nm (87 + 14 ft-lb)

Upper strut mount, replacing

– Loosen front wheel lug nuts. Raise front of car and support it securely.

> ***WARNING —***
> *Make sure the car is stable and well supported at all times. Use a professional automotive lift or jack stands designed for the purpose. A floor jack is not adequate support.*

– Remove front wheel.

◄ Working in luggage compartment, bend back mounting nut locking tab and remove nut from upper strut mount.

- Remove 8 mm allen bolts (**arrows**) holding strut mount to body.

NOTE —

Clean off sealing compound around strut mount plates and mark their locations on the inner fender. This will help in reassembling the front suspension close to its original position.

– Using a long punch, push strut piston rod down out of mount and remove mount from below.

Upper strut mount assembly

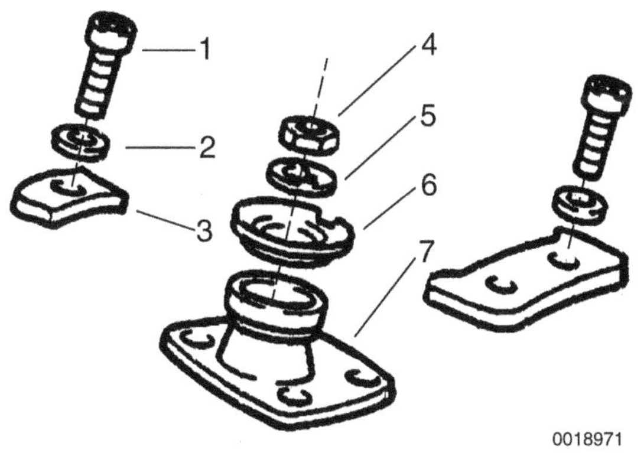

1. Bolt (M10 x 20 mm) tighten to 47 Nm (34 ft-lb)
2. Lock washer (M10)
3. Mounting plate
4. Strut top mounting nut (M10 x 1.5) tighten to 80 Nm (59 ft-lb)
5. Locking tab (always replace)
6. Plate
7. Upper mount

– Installation is reverse of removal.

NOTE —
Use a long M6 stud to reach through the strut mount assembly and into the threads at the top of the strut cartridge piston. Then the piston can be pulled up through the mount.

Tightening Torques

- Brake line to brake hose . . . . 12 Nm (9 ft-lb)
- Strut to upper mount (use new locking tab) . . . . 80 Nm (58 ft-lb)
- Upper strut mount to body . . . . 47 Nm (34 ft-lb)
- Wheel to wheel hub . . . . 130 Nm (96 ft-lb)

– Reset wheel alignment. See **440 Wheels, Tires, Alignment**.

BALL JOINTS, CONTROL ARMS AND TORSION BARS

Ball joint, removing and installing

– Loosen front wheel lug nuts. Raise front of car and support it securely.

WARNING —
Make sure the car is stable and well supported at all times. Use a professional automotive lift or jack stands designed for the purpose. A floor jack is not adequate support.

– Remove front wheel.

< Remove ball joint wedge bolt (**arrow**).

– Carefully pry out or tap on control arm to push ball joint down out of strut housing.

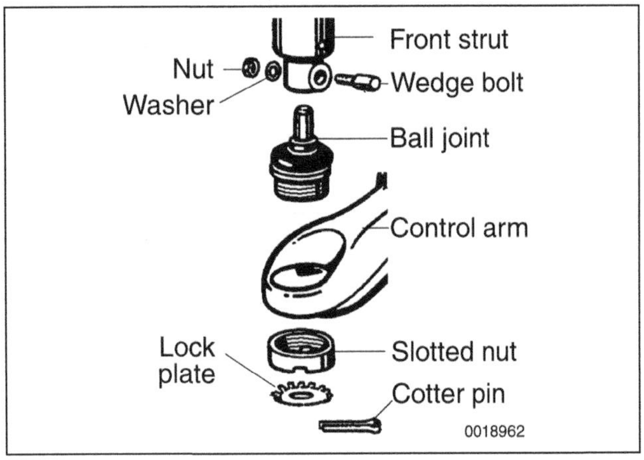

- Remove cotter pin from ball joint lock plate. Remove slotted nut from ball joint using Porsche special tool no. P 280 b.
- Remove ball joint from control arm.
- Installation is reverse of removal.

NOTE —
- *Always replace ball joint wedge bolt.*
- *When installing wedge bolt, line up flat of bolt with cutout in ball joint.*

Tightening torques
- Ball joint slotted nut to ball joint ..... 250 Nm (184 ft-lb)
- Ball joint wedge bolt (always replace)... 22 Nm (16 ft-lb)
- Wheel to wheel hub .... 130 Nm (96 ft-lb)

Front control arm, removing and installing

- Loosen front wheel lug nuts. Raise front of car and support it securely.

WARNING —
Make sure the car is stable and well supported at all times. Use a professional automotive lift or jack stands designed for the purpose. A floor jack is not adequate support.

- Remove front wheels.
- Remove cotter pin from ball joint lock plate. Remove slotted round nut from ball joint using Porsche special tool no. P 280 b.

- Working underneath front of car, remove stone guard mounting bolts (**arrow**). Remove stone guard.

- Unbolt and remove A/C condenser protection bar (**arrow**).
- Remove front control arm mounting bracket bolts (**A**).

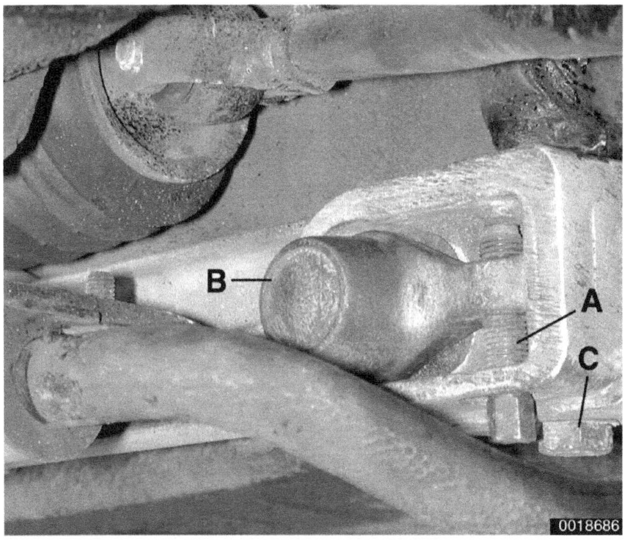

Remove torsion bar adjustment screw (**A**) and retaining cap (**B**).

NOTE —
Count threads on adjustment screw while removing. Record the count. This will help in reassembly.

– Remove single bolt (**C**) at rear of crossmember subframe and pull control arm forward out of subframe.

– With a rubber hammer tap control arm off stabilizer bar.

– Installation is reverse of removal.

NOTE —
- *During installation, use Contifix® or equivalent rubber lubricant when sliding stabilizer bar into control arm stabilizer bushings.*
- *Lubricate torsion bar adjustment screw with multipurpose MoS_2 grease. Install according to number of threads counted during removal.*
- *Be sure to reset front ride height as described in* **440 Tires, Wheels, Alignment**.

Tightening torques
- Ball joint slotted nut to ball joint ..... 250 Nm (184 ft-lb)
- Control arm to body
  Front bracket (M10)................. 47 Nm (35 ft-lb)
  Crossmember subframe (M12)........ 90 Nm (66 ft-lb)
- Wheel to wheel hub .... 130 Nm (96 ft-lb)

Front torsion bar, removing and installing

WARNING —
Replacement and/or removal of the torsion bar affects vehicle ride height, which directly influences vehicle stability. If ride height is changed, it is strongly recommended that the vehicle balance/weight be checked at all four wheels using suitable wheel-load floor scales. See **440 Wheels, Tires, Alignment** *for more information.*

CAUTION —
Torsion bars are pretensioned during production. Right and left bars cannot be exchanged. Bars are marked with a R or L on the ends.

– Raise front of car and support it securely.

WARNING —
Make sure the car is stable and well supported at all times. Use a professional automotive lift or jack stands designed for the purpose. A floor jack is not adequate support.

Working underneath front of car, remove stone guard mounting bolts (**arrow**). Remove stone guard.

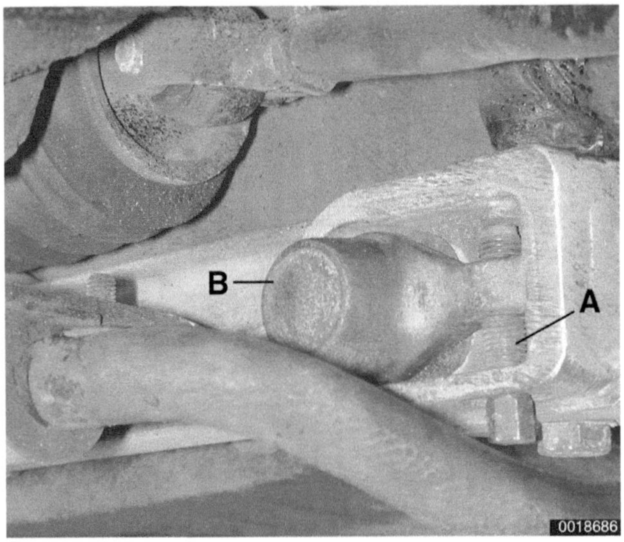

Remove torsion bar adjustment screw (**A**) and retaining cap (**B**).

NOTE —

Count threads on adjustment screw before removing. Record the count. This will help in reassembly.

– Pull torsion bar out toward rear of car.

NOTE —

Pry stabilizer bar down slightly to clear torsion bar as it is being removed.

– Installation is reverse of removal.

NOTE —

- *Lubricate torsion bar adjustment screw with multipurpose MoS_2 grease. Install according to number of threads counted before.*
- *Be sure to check front ride height as described in* **440 Tires, Wheels, Alignment**.

Tightening Torques

- Stone guard/stabilizer rear bracket to body (M10 bolt) . . . . 47 Nm (35 ft-lb)
- Stone guard/stabilizer rear bracket to crossmember subframe (M10 nut) . . . 28 Nm (21 ft-lb)

WHEEL BEARINGS

The front wheel bearings are a tapered-roller design. Periodic repacking and adjustment is part of the normal maintenance routine for the car. See **030 Maintenance**.

When replacing wheel bearings, press equipment should be used to install the bearing races into the bearing hub.

Front wheel bearing assembly

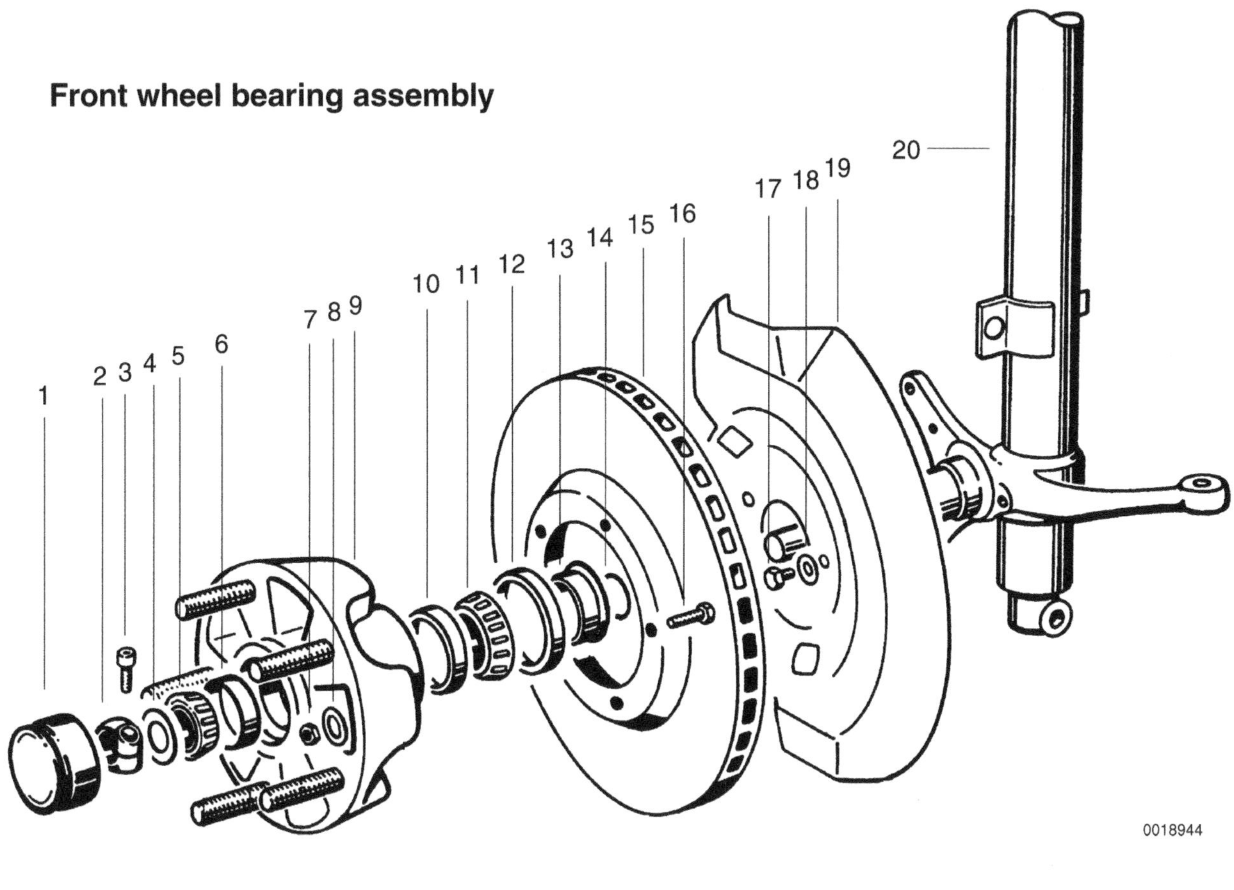

1. Grease cap
2. Wheel bearing lock nut
3. Clamping allen bolt
   tighten to 15 Nm (11 ft-lb)
4. Thrust washer
5. Outer wheel bearing
6. Outer race, outer wheel bearing
7. Nut (M8)
   tighten to 23 Nm (17 Nm)
8. Washer (M8)
9. Bearing hub
10. Outer race, inner wheel bearing
11. Inner wheel bearing
12. Grease seal
13. Spacer
14. O-ring
15. Brake rotor (disc)
16. Bolt (M8)
17. Bolt (M8)
    tighten to 25 Nm (18 ft-lb)
18. Washer (M8)
19. Backing plate
20. Front strut housing

Front wheel bearing play, adjusting

- Loosen front wheel lug nuts. Raise front of car and support it securely.

> ***WARNING —***
> *Make sure the car is stable and well supported at all times. Use a professional automotive lift or jack stands designed for the purpose. A floor jack is not adequate support.*

- Remove front wheel.

- Carefully pry off grease cap in center of bearing hub.

0018966

- ◄ Loosen 6 mm allen clamp screw in wheel bearing lock nut. Loosen bearing lock nut approximately ¼ turn.

0018966a

- ◄ While rotating brake disc and bearing hub, tighten lock nut slowly in small increments until thrust washer can just barely be moved back and forth with a light push on the tip of a screwdriver (do not pry or twist screwdriver).

- Tighten clamping screw to specifications. Recheck tightness of thrust washer.

Tightening Torque
- Wheel bearing lock nut clamp screw (M8) . . . . 15 Nm (11 ft-lb)

- Install grease cap. Reinstall road wheel.

Tightening Torque
- Wheel to wheel hub . . . . 130 Nm (96 ft-lb)

Front wheel bearings, replacing

– Loosen front wheel lug nuts. Raise front of car and support it securely.

> ***WARNING —***
> *Make sure the car is stable and well supported at all times. Use a professional automotive lift or jack stands designed for the purpose. A floor jack is not adequate support.*

– Remove front wheel.

– Press brake pedal down about 1 in. and hold in place with a pedal stop.

NOTE —
Holding the brake pedal down will prevent brake fluid from dripping out of open brake lines while the work is being performed.

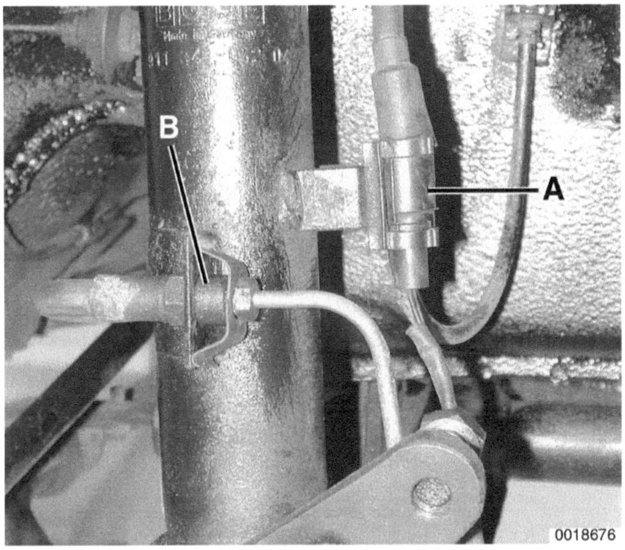

0018676

◄ Working in wheel housing, disconnect electrical connector for brake pad wear sensor (**A**) and remove harness from clip on strut body.

– Disconnect rubber brake hose (**B**) from line near bottom of strut. Plug open lines.

> ***CAUTION —***
> *Always counterhold brake line fittings when loosening or tightening.*

– Remove caliper mounting bolts and remove caliper. See **470 Brakes–Hydraulic**.

– Carefully pry off grease cap in center of bearing hub.

0018966

◄ Loosen 6 mm allen clamp screw in wheel bearing lock nut. Remove lock nut.

– Slide brake rotor and bearing hub straight off axle.

NOTE —
If necessary, rock the brake rotor side-to-side to aid removal of the thrust washer and outer wheel bearing.

- Remove grease seal and inner wheel bearing from bearing hub.

- Using drift, drive wheel bearing outer races from bearing hub.
  - Note cutouts (**arrows**) in bearing hub provided for this purpose.
- Heat bearing hub to 120° - 150°C (250° - 300°F). Press new races into hub.
  - Use special tool VW 447h to press outer bearing race.
  - Use special tool VW 447i to press inner bearing race.
- Pack wheel bearings with lithium-based wheel bearing grease. Lightly coat bearing races and hub with grease.

> ***CAUTION —***
> *Proper packing of wheel bearings is essential for long bearing service life. Use a wheel bearing packing tool if available.*

Front wheel bearings

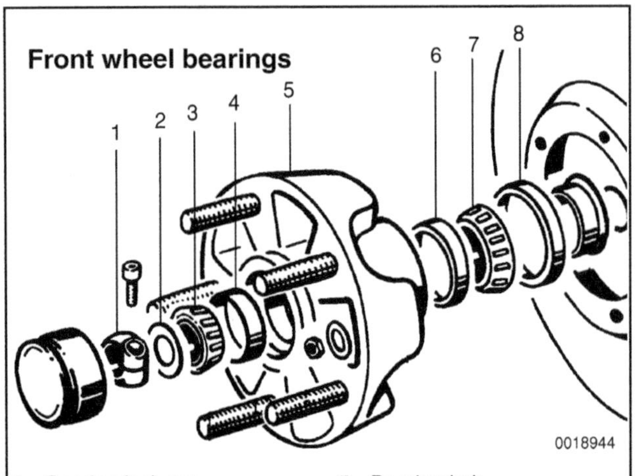

1. Bearing lock nut
2. Thrust washer
3. Outer bearing
4. Inner race, outer bearing
5. Bearing hub
6. Inner race, inner bearing
7. Inner bearing
8. Grease seal

- Place inner wheel bearing in race and press in new inner seal using special tool VW 433.
- Apply grease to gap between sealing lips on inner seal.
- Slide bearing hub with rotor and bearings onto stub axle. Use caution to avoid damaging grease seal.
- Install outer wheel bearing, thrust washer and wheel bearing lock nut. Tighten lock nut while rotating brake rotor.
- Adjust wheel bearing play as described earlier. Tighten wheel bearing lock nut clamp screw to specifications. Recheck tightness of thrust washer.

Tightening Torque

- Wheel bearing lock nut clamp screw (M8) . . . . 15 Nm (11 ft-lb)

- Fill grease cap with 10 grams of grease and install.

– Remainder of installation is reverse of removal.

Tightening Torque
- Brake caliper to strut housing (M12). . . . 70 Nm (52 ft-lb)
- Wheel to wheel hub . . . . 130 Nm (96 ft-lb)

STABILIZER BAR

Front stabilizer bar, removing and installing

– Raise car and support in a safe manner.

CAUTION —
Make sure the car is stable and well supported at all times. Use a professional automotive lift or jack stands designed for the purpose. A floor jack is not adequate support.

◄ Working underneath front of car, remove stone guard mounting fasteners (**arrow**). Remove stone guard.

◄ Loosen stabilizer rear bracket bolt (**A**) on left and right sides.

– Pry stabilizer bracket off crossmember subframe.

– Pull stabilizer backward, first from one control arm, then the other, to remove from car.

– Disassemble stabilizer bracket and inspect rubber bushings for wear. Also inspect bushings in left and right control arm stabilizer brackets. Replace as necessary.

NOTE —
During installation, use Contifix® or equivalent rubber lubricant when sliding stabilizer bar into control arm stabilizer bushings.

Tightening Torques
- Stabilizer bracket to crossmember subframe bracket (M8) . . . . 25 Nm (18 ft-lb)
- Stone guard/ rear stabilizer bracket to body (M10 bolt). . . . . 47 Nm (35 ft-lb)
- Stone guard/rear stabilizer bracket to crossmember subframe (M10 nut) . . . 28 Nm (21 ft-lb)

SUSPENSION SUBFRAME

Front suspension subframe assembly

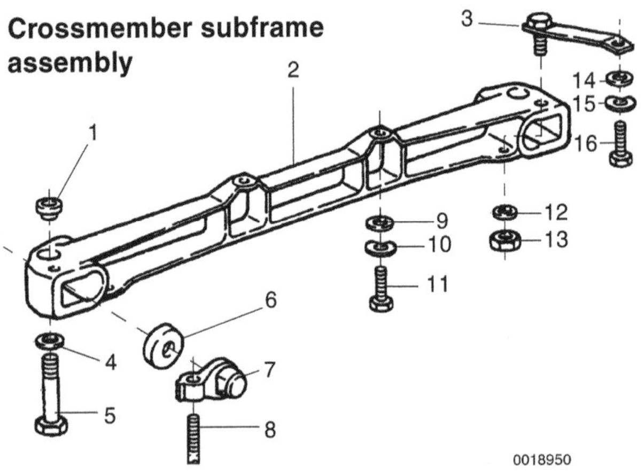

1. Adapter
2. Subframe
3. Stabilizer bar rear mounting bracket
4. Spring washer
5. Bolt (M12 x 1.5) tighten to 90 Nm (66 ft-lb)
6. Foam seal
7. Torsion bar retaining cap
8. Torsion bar adjusting screw (lubricate with MoS_2)
9. Washer
10. Lock washer
11. Steering rack mounting bolt (M10 x 20 mm) tighten to 47 Nm (35 ft-lb)
12. Washer
13. Nut (M10) tighten to 28 Nm (21 ft-lb)
14. Washer
15. Lock washer
16. Bolt (M10) tighten to 47 Nm (35 ft-lb)

The forged light alloy crossmember subframe can be accessed from under the car by removing the stone guard in back of the fuel tank. Details of the subframe assembly are shown in the illustration.

NOTE —

- *Lubricate torsion bar adjustment screw with multipurpose MoS_2 grease.*
- *Be sure to reset front ride height as described in* **440 Tires, Wheels, Alignment**.

Tightening Torques

- Steering rack to subframe (M10 x 20 mm) . . . . 47 Nm (35 ft-lb)
- Stone guard/ stabilizer rear bracket to body (M10 bolt). . . . . 47 Nm (35 ft-lb)
- Stone guard/ stabilizer rear bracket to crossmember subframe (M10 nut) . . . 28 Nm (21 ft-lb)
- Subframe/control arm to body (M12 x 1.5). . . . . 90 Nm (66 ft-lb)

420 Rear Suspension, Axle Shafts

Special Tools

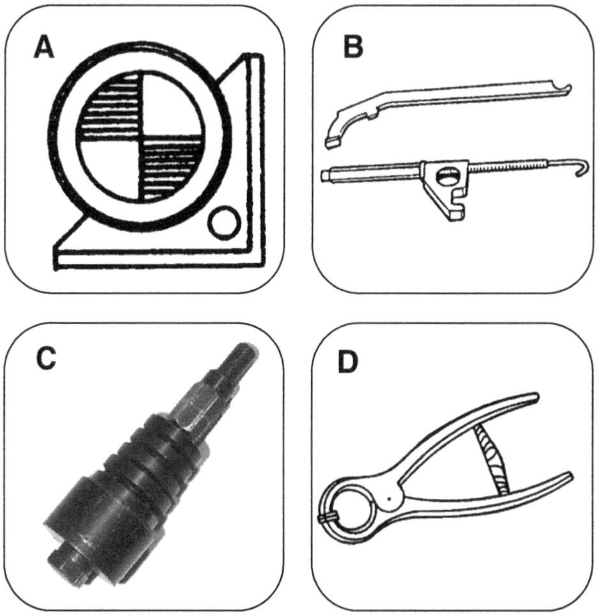

Special tool

A 261/1 (source: Baum Tools Unlimited) . . . . Protractor/level
B P 289 a. . . . . Rear torsion bar preloading tool (Porsche ordering #000721 289 10)
C B 90 . . . . Wheel bearing extractor/installer (source: Baum Tools Unlimited)
D VW 161 a . . . . Circlip pliers

GENERAL

The Porsche 911 Carrera is equipped with independent rear suspension. Torsion bars are used at all four wheels. The rear end uses conventional shock absorbers.

This section covers replacement procedures for rear suspension components. For specifications, diagnosis and troubleshooting information see **4 Suspension, Brakes and Steering**. For information on setting ride height and alignment see **440 Wheels, Tires, Alignment**. For information on transmission output flange seal repairs, see **390 Differential**.

Special tools

< Some procedures found in this repair group require the use of special tools.

Rear suspension

The rear shock absorbers are conventional hydraulic or gas-pressure dampers. Rear torsion bars are mounted transversely inside the rear suspension subframe (torsion bar carrier). Each trailing arm carries a sealed wheel bearing.

A stabilizer bar is mounted to the rear trailing arms to reduce body roll.

NOTE —
The suspension subframe (torsion bar carrier) is welded into the body and is not a replaceable part under normal circumstances.

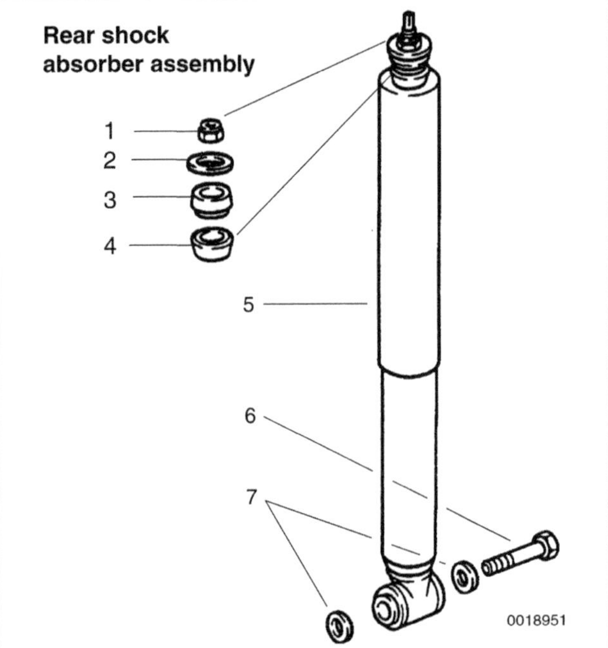

1. Lock nut (M10 x 1), 25 Nm (18 ft-lb)
2. Washer
3. Rubber bumper
4. Rubber bumper
5. Shock absorber
6. Lower mounting bolt (M14 x 1.5) 125 Nm (92 ft-lb)
7. Washer

When doing any rear suspension maintenance or repair, observe the following cautions.

> *CAUTION —*
> - *Do not install bolts and nuts coated with undercoating wax, as correct tightening torque cannot be assured. Always clean the threads with solvent before installation, or install new parts.*
> - *Do not attempt to weld or straighten the torsion plates, trailing arms or subframe. Replace damaged parts.*
> - *Do not reuse self-locking nuts or cotter pins. These parts are designed to be used only once.*

REAR SHOCK ABSORBERS

◄ The rear shock absorbers are conventional dampers.

Rear shock absorber, removing and installing

◄ With weight of vehicle on wheels, remove lower shock bolt.

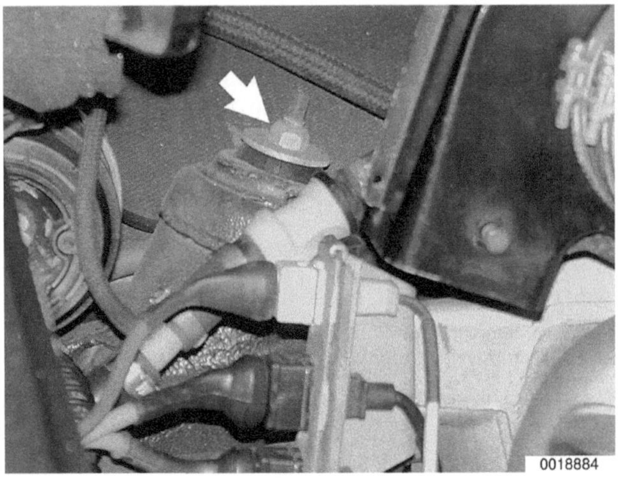

◄ Working in rear of engine compartment, remove shock absorber top mounting nut (**arrow**).

NOTE —
To access the right side top mounting nut, remove air filter cover and, if necessary, housing. See **030 Maintenance**.

- Remove shock absorber.
- Installation is reverse of removal.

Tightening Torques
- Rear shock absorber to trailing arm . . . 125 Nm (92 ft-lb)
- Rear shock absorber to body (M10) . . . . 25 Nm (18 ft-lb)

TRAILING ARMS, TORSION PLATES AND TORSION BARS

> ***WARNING —***
> *Replacement and/or removal of the trailing arm, torsion plate or torsion bar disturbs the rear wheel alignment settings. Replacement or adjustment of the torsion plate or torsion bar also affects vehicle ride height, which directly influences vehicle stability. If ride height is changed, it is strongly recommended that the vehicle balance/weight be checked at all four wheels using suitable wheel-load floor scales. See* **440 Wheels, Tires, Alignment** *for more information.*

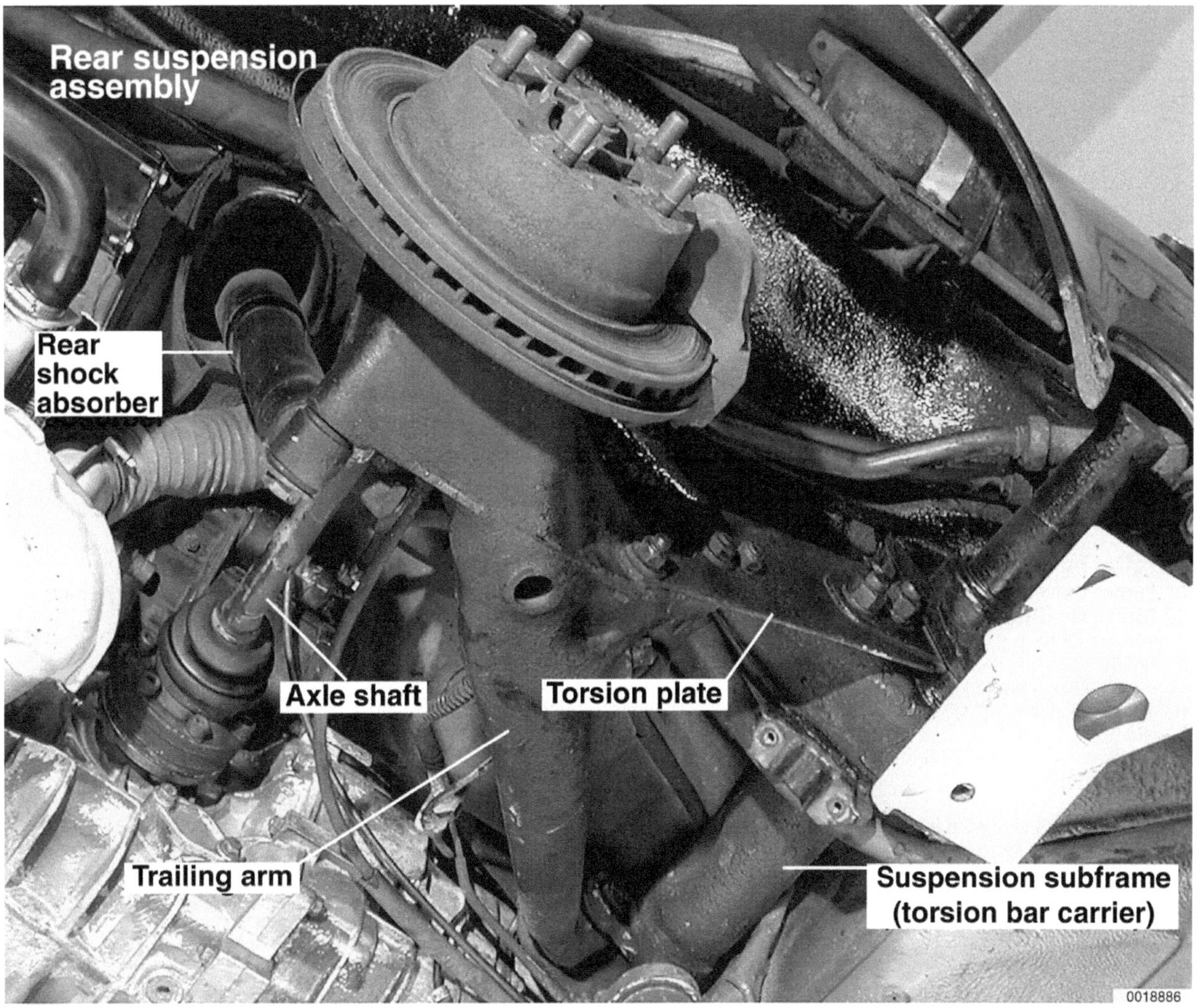

Rear suspension assembly

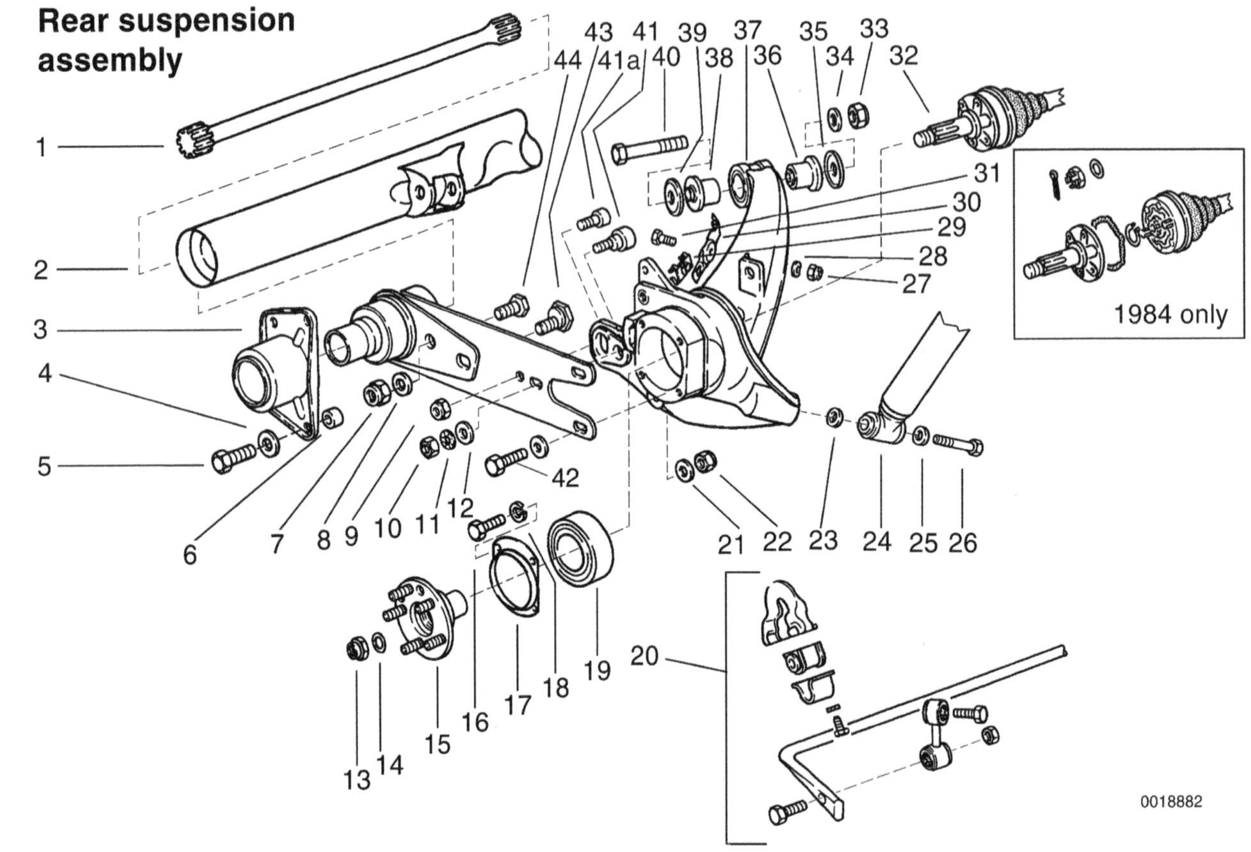

1. Torsion bar
2. Suspension subframe (torsion bar carrier)
3. Torsion bar cover
4. Washer (M10)
5. Bolt (M10)
   - tighten to 47 Nm (34 ft-lb)
6. Spacer
7. Nut (M16)
   - tighten to 245 Nm (177 ft-lb)
8. Washer (M16)
9. Nut (M12)
   - tighten to 85 Nm (61 ft-lb)
10. Nut (M12)
    - tighten to 120 Nm (87 ft-lb)
11. Lock washer (M12)
12. Flat washer (M12)
13. Stub axle nut,
    M20 castellated nut:
    - tighten to 300-320 Nm (217-231 ft-lb)
    M22 self-locking nut:
    - tighten to 460 Nm (339 ft-lb)
14. Washer
15. Wheel hub
16. Bolt (M10)
    - tighten to 47 Nm (34 ft-lb)
17. Wheel bearing cover
18. Washer (M10)
19. Wheel bearing
20. Stabilizer bar assembly
21. Washer (M10)
22. Nut (M12)
    - tighten to 120 Nm (87 ft-lb)
23. Washer (M14)
24. Shock absorber
25. Washer (M14)
26. Bolt (M14)
    - tighten to 125 Nm (92 ft-lb)
27. Nut
28. Washer
29. Clip
30. Clip
31. Bolt
32. Axle shaft
33. Nut (M14)
    - tighten to 100 Nm (72 ft-lb)
34. Washer (M14)
35. Plate
36. Bushing
37. Trailing arm
38. Bushing
39. Plate
40. Bolt (M14)
    - tighten to 100 Nm (72 ft-lb)
41. Eccentric (M12)
    - tighten to 85 Nm (61 ft-lb)

41a. Eccentric (M12)
    - tighten to 85 Nm (61 ft-lb)

42. Bolt (M12)
    - tighten to 125 Nm (90 ft-lb)
43. Eccentric (M16)
    - tighten to 245 Nm (177 ft-lb)
44. Bolt (M16)
    - tighten to 245 Nm (177 ft-lb)

Rear trailing arm, removing and installing

- Apply parking brake and shift transmission into reverse.
- Remove wheel hub center cap. Loosen stub axle nut (32 mm socket wrench) through opening in center of wheel.
- Loosen rear wheel lug nuts.
- Release parking brake and shift car into neutral.
- Press brake pedal down about 1 in. and hold pedal in place with a pedal stop.

NOTE —
This will prevent brake fluid from dripping out of open brake lines while the work is being performed.

- Raise rear of car and support it securely.

WARNING —
- *Make sure the car is stable and well supported at all times. Use a professional automotive lift or jack stands designed for the purpose. A floor jack is not adequate support.*
- *When working on torsion plates and torsion bars, the car should be no higher than 2 feet off the ground.*

0018887

- Remove rear wheel. Remove stub axle nut.
- Working at trailing arm, disconnect electrical connector (**arrow**) for brake pad wear sensor and remove harness from clip on trailing arm.

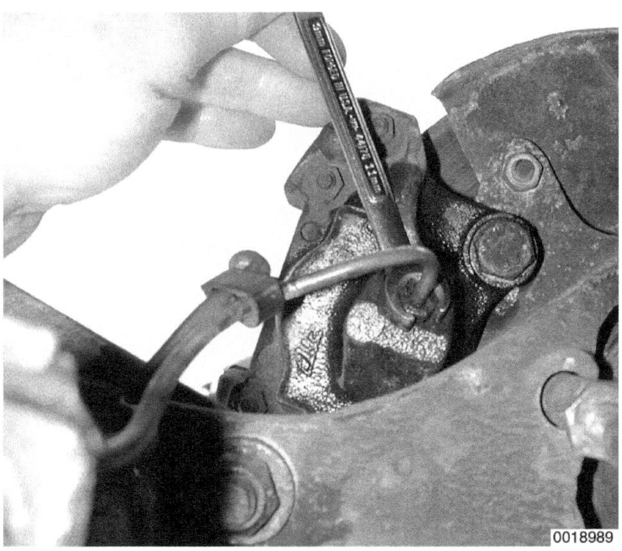

0018989

- Disconnect brake line from brake caliper. Plug hydraulic lines.

CAUTION —
When working with hydraulic brake lines and hoses be sure to counterhold fitting before loosening or tightening.

- Remove brake caliper mounting bolts and remove caliper. See **470 Brakes–Hydraulic**.

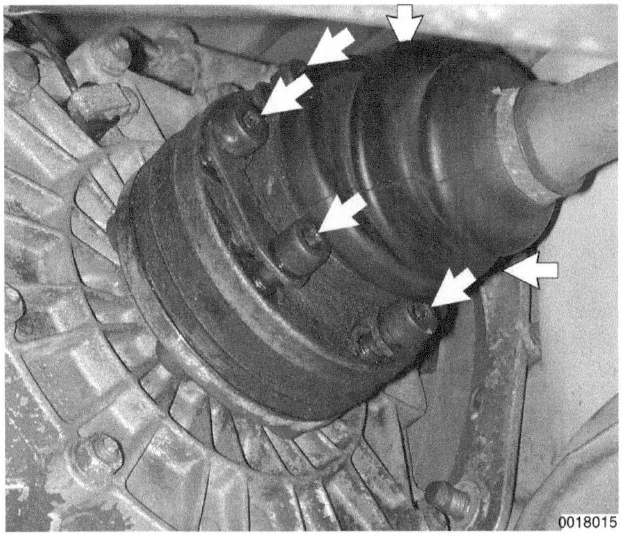

Unbolt CV joint fasteners (**arrows**) from transmission output flange.

- Lower axle shaft. Drive stub axle out of trailing arm.See **Axle shaft, removing and installing**, later in this section.

Remove brake rotor retaining screws (**arrows**) and pull off brake rotor.

Working at parking brake shoes pivot, remove cotter pin and castellated nut (**arrow**).

- Pull parking brake cable out of rear of trailing arm.

◀ Disconnect stabilizer bar link bolt (**arrow**) at trailing arm.

◀ Drive out rear wheel hub using Porsche special tool P 297 a or equivalent drift.

◀ Lower car close to ground level. Use floor jack under trailing arm to relieve suspension load on shock absorber. (Brake backing plate shown removed for clarity.)

- Remove rear shock absorber lower mounting bolt from trailing arm (**arrow**).

NOTE —

Porsche special tool P 289 a may be used to hold rear suspension in a compressed position while working on the rear shock absorber bolt.

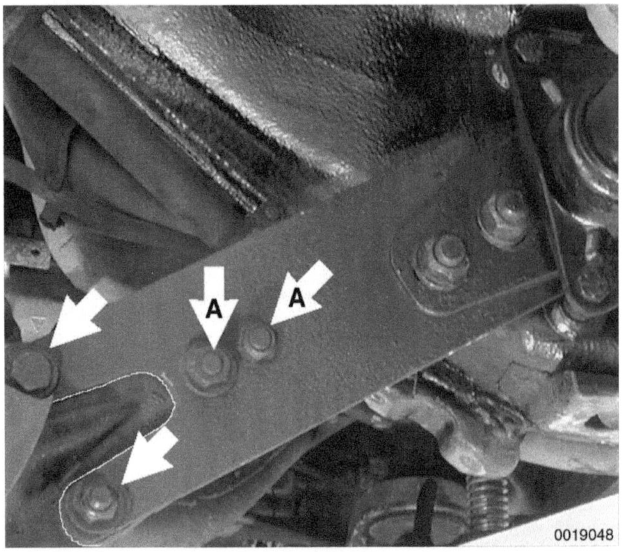

◄ Mark torsion plate position to trailing arm by scribing outline around torsion plate. Remove torsion plate fasteners (**arrows**) and detach trailing arm from torsion plate.

NOTE —

*The torsion plate fastener holes are elongated. Mark relative positions of torsion plate, trailing arm and eccentric bolts (**A**) in order to reassemble suspension close to original alignment setting.*

– Raise car.

◄ Working underneath car, remove trailing arm mounting bolt (**arrow**) from rear suspension subframe (torsion bar carrier). Remove trailing arm.

– Installation is reverse of removal.

- Bleed brakes as described in **470 Brakes–Hydraulic**.
- Reset rear wheel alignment as described in **440 Wheels, Tires, Alignment**.

Tightening torques

- Brake caliper to trailing arm (M12) ..... 60 Nm (43 ft-lb)
- Brake hose to brake line .... 12 Nm (9 ft-lb)
- Brake rotor to wheel hub (M6) .... 5 Nm (44 in-lb)
- CV joint to transmission flange
  M8 bolt .... 42 Nm (30 ft-lb)
  M10 bolt .... 83 Nm (60 ft-lb)
- Rear shock absorber
  to trailing arm (M14) .... 125 Nm (92 ft-lb)
- Stabilizer link to trailing arm (M12) ..... 85 Nm (61 ft-lb)
- Stub axle shaft to hub
  M20 castellated nut...... 300 - 320 Nm (217 - 231 ft-lb)
  M22 self-locking nut .... 460 Nm (339 ft-lb)
- Trailing arm to suspension subframe
  (torsion bar carrier)(M14) .... 100 Nm (72 ft-lb)
- Torsion plate to trailing arm (eccentric) . 85 Nm (61 ft-lb)
- Torsion plate to trailing arm (M12) .... 120 Nm (87 ft-lb)
- Wheel to wheel hub .... 130 Nm (96 ft-lb)

Rear torsion bar, removing and installing

> ***WARNING —***
> *If torsion bars are replaced or reset, it is strongly recommended that the vehicle balance/weight be checked at all four wheels using suitable wheel-load floor scales. See **440 Wheels, Tires, Alignment** for more information.*

– Loosen rear wheel lug nuts. Raise rear of car and support it securely.

> ***WARNING —***
> - *Make sure the car is stable and well supported at all times. Use a professional automotive lift or jack stands designed for the purpose. A floor jack is not adequate support.*
> - *When working on torsion plates and torsion bars, the car should be no higher than 2 feet off the ground.*

– Remove rear wheel.

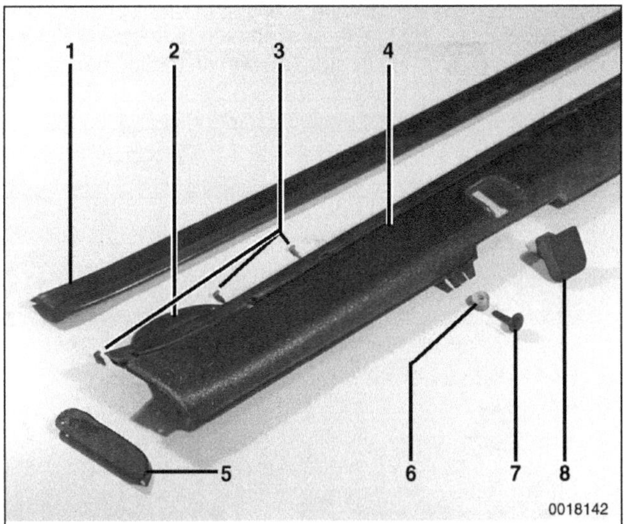

1. Rubber protector
2. Torsion bar access plug
3. Rocker panel mounting screws
4. Rocker panel
5. Rocker panel end cover
6. Plastic spacer
7. Rocker panel bottom mounting screw
8. Jack hole cover

◄ Remove rocker panel.

- Remove rear rocker panel end cap screws and remove cap.
- Working from one end of rocker panel, peel off rubber rocker panel protector.
- Remove top and bottom mounting screws and remove rocker panel.
- Remove rocker panel torsion bar access plug from fender. On 1984-1986 cars, plug is retained to body from behind using nut and bar clamp.

◄ Lower car close to ground level. Use floor jack under trailing arm to relieve suspension load on shock absorber. (Brake backing plate shown removed for clarity.)

- Remove rear shock absorber lower mounting bolt from trailing arm (**arrow**).

NOTE —
Porsche special tool P 289 a may be used to hold rear suspension in a compressed position while working on the rear shock absorber bolt.

– Lower and remove floor jack.

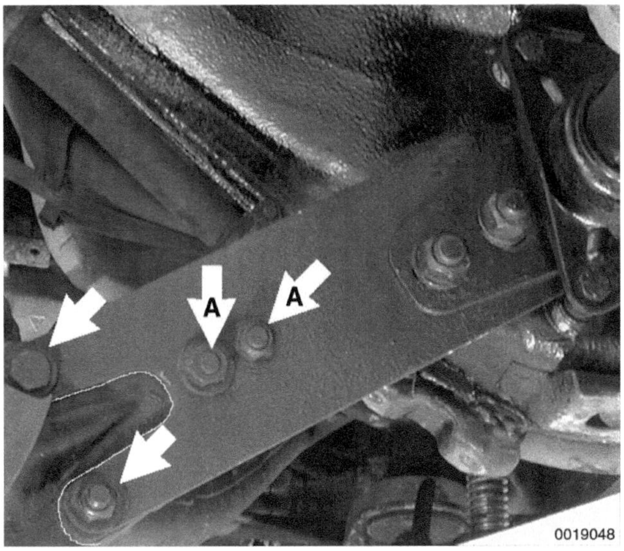

◄ Mark torsion plate position to trailing arm by scribing outline around torsion plate. Remove torsion plate fasteners (**arrows**) and detach trailing arm from torsion plate.

NOTE —
*The torsion plate fastener holes are elongated. Mark relative positions of torsion plate, trailing arm and eccentric bolts (**A**) in order to reassemble suspension close to original alignment setting.*

0019050

With car still close to ground and torsion plate supported from below using floor jack, raise torsion plate just until torsion plate comes off lower stop (**arrow**) on torsion bar cover.

0019052

Loosen 3 torsion bar cover bolts (**A**) and remove stop bolt and sleeve (**B**).

- Slowly lower floor jack to relieve torsion bar tension on torsion plate.

0019053

Raise vehicle on lift. Using spirit level in special protractor tool 261/1 or equivalent, measure inclination of vehicle body at door sill (measurement **A**).

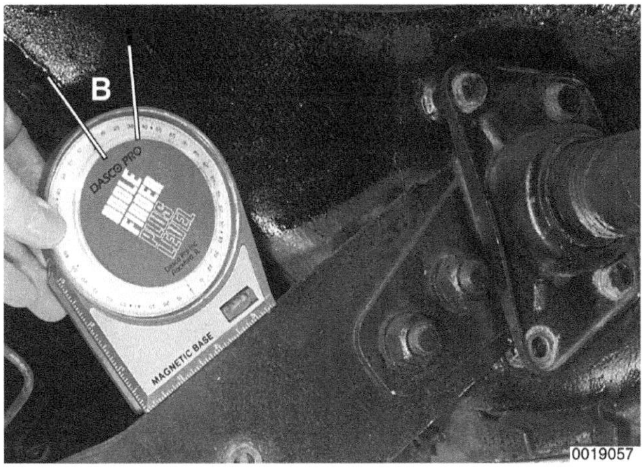

∢ Measure inclination of torsion plate (measurement **B**).

– Combine measurements **A** and **B** (add or subtract body inclination measurement) to find value of torsion plate inclination in relation to vehicle body. Record measurement.

Example

Body inclination (**A**)	–2°
Measured torsion plate inclination (**B**)	34°
Actual torsion plate inclination	**32°**

– Remove torsion bar cover mounting bolts and remove torsion bar cover by prying gently off body.

> ***CAUTION —***
> *When backing off the torsion plate, check that the plate does not hang up on the body mounting boss.*

– Remove torsion bar.

NOTE —
- *Do not interchange rear torsion bars. Bars are marked on the ends "R" and "L" for right and left.*
- *If a torsion bar is broken on a car built before 1987, the broken part that is stuck in the subframe (torsion bar carrier) can be driven out from the opposite side after removing the other torsion bar.*
- *If the car was built in 1987 or later, the transmission must be removed before the broken torsion bar can be driven out of the subframe tube. Replug the hole that has been provided for this purpose after the job is completed.*

– Prior to installation, carefully clean and check all parts for signs of damage. Replace as needed.
- Check torsion plate for corrosion, chipped paint, scrapes.
- Check torsion bar for cracks.

– Coat torsion bar with multipurpose grease and install.

NOTE —
The inner and outer ends of rear torsion bars have a different number of teeth. Be sure to install the bar with the "L" or the "R" on the end facing out.

– Coat rubber seals on torsion bar cover with a rubber lubricant such as Contifix® or Capella®.

∢ Temporarily install 3 bolts (**arrows**) without spacers to hold cover in place. Leave out lower stop bolt.

◄ Using special protractor tool 261/1 or equivalent, measure inclination of vehicle body at door sill (measurement **A**).

◄ Measure inclination of torsion plate (measurement **B**).

– Combine measurements **A** and **B** (add or subtract body inclination measurement) to find value of torsion plate inclination in relation to vehicle body inclination.

Table a. Rear Torsion Bar Specifications

Model	Torsion bar diameter	Torsion bar splines inside/ outside	Torsion plate inclination (see text)
1984 - 1985:			
Carrera with gas struts	24.1 mm	40/44	34°
Carrera w/out gas struts	24.1 mm	40/44	35°
Turbo look	26 mm	40/44	36°
1986:			
Carrera	25 mm	40/44	32°
Turbo look	26 mm	40/44	34°
1987 - 1989:			
Carrera	25 mm	46/47	32°
Turbo look	26 mm	46/47	34°

Example

Body inclination (A)	–2°
Measured torsion plate inclination (B)	34°
Actual torsion plate inclination	**32°**

– If torsion plate inclination is incorrect:

- Remove torsion bar cover fasteners and slide cover with torsion plate off body.
- Reposition torsion bar in torsion plate and reinstall cover using 3 fasteners.
- Recheck inclination and readjust as necessary.

NOTE —

It may be necessary to try several different positions for inner and outer teeth on the torsion bar to achieve the correct setting.

◀ Once torsion bar inclination is correct, use floor jack to raise torsion plate just enough to install lower stop bolt with spacer (**arrow**).

- Tighten all four torsion bar cover mounting bolts.

Tightening Torque
- Torsion bar cover to body (M10) . . . . 47 Nm (34 ft-lb)

– Remainder of installation is reverse of removal.

- Reset rear wheel alignment and check ride height as described in **440 Wheels, Tires, Alignment**.

> ***WARNING —***
> *Replacement and/or removal of the torsion plate or torsion bar disturbs the rear wheel alignment settings and also affects vehicle ride height, which directly influences vehicle stability. If ride height is changed, it is strongly recommended that the vehicle balance/weight be checked at all four wheels using suitable wheel-load floor scales. See* **440 Wheels, Tires, Alignment** *for more information.*

Tightening torques
- Rear shock to trailing arm (M14) . . . . . 125 Nm (92 ft-lb)
- Torsion plate to trailing arm (eccentric) . 85 Nm (61 ft-lb)
- Torsion plate to trailing arm (M12 bolt) . 120 Nm (87 ft-lb)
- Wheel to wheel hub . . . . 130 Nm (96 ft-lb)

Rear torsion bar, setting

> ***WARNING —***
> *If torsion bars are replaced or reset, it is strongly recommended that the vehicle balance/weight be checked at all four wheels using suitable wheel-load floor scales. See* **440 Wheels, Tires, Alignment** *for more information.*

The initial position of each rear torsion bar is set by measuring torsion plate inclination in its unloaded position.

◀ With car close to ground level, use floor jack under trailing arm to relieve suspension load on shock absorber. (Brake backing plate shown removed for clarity.)

- Remove rear shock absorber lower mounting bolt from trailing arm (**arrow**).

> ***NOTE —***
> *Porsche special tool P 289 a may be used to hold rear suspension in a compressed position while working on the rear shock absorber bolt.*

– Lower and remove floor jack.

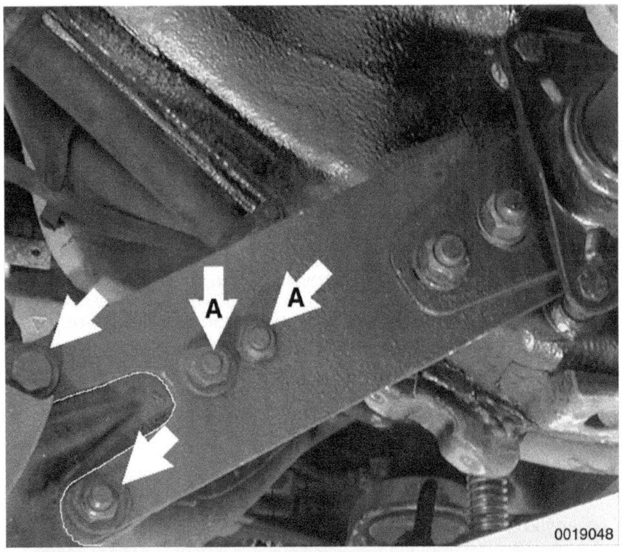

Mark torsion plate position to trailing arm by scribing outline around torsion plate. Remove torsion plate fasteners (**arrows**) and detach trailing arm from torsion plate.

NOTE —
*The torsion plate fastener holes are elongated. Mark relative positions of torsion plate, trailing arm and eccentric bolts (**A**) in order to reassemble suspension close to original alignment setting.*

With car still close to ground and torsion plate supported from below using floor jack, raise torsion plate just until plate comes off lower stop (**arrow**) on torsion bar cover.

Loosen 3 torsion bar cover bolts (**A**) and remove stop bolt and sleeve (**B**).

- Slowly lower and remove floor jack to unload torsion bar tension.

◄ Using spirit level in special protractor tool 261/1 or equivalent, measure inclination of vehicle body at door sill (measurement **A**).

◄ Measure inclination of torsion plate (measurement **B**).

– Combine measurements **A** and **B** (add or subtract body inclination measurement) to find value of torsion plate inclination in relation to vehicle body. Record measurement.

Example	
Body inclination (**A**)	–2°
Measured torsion plate inclination (**B**)	34°
Actual torsion plate inclination	**32°**

NOTE —

There are unequal numbers of splined teeth on the inner and outer ends of each torsion bar. This allows for small incremental changes to be made to the torsion plate inclination.

Rear Torsion Bar Variants

	1984 - 1986	198 7- 1989
Number of splines Inside Outside	 40 44	 46 47
Smallest possible change in torsion plate inclination	0°50′	0°10′
Smallest possible change in vehicle height	Approx. 6.5 mm (0.256 in.)	Approx. 1.4 mm (0.055 in.)

Rear Torsion Bar Specifications

Model	Torsion bar diameter	Torsion bar splines inside/ outside	Torsion plate inclina-tion (see text)
1984-1985:			
Carrera with gas struts	24.1 mm	40/44	34°
Carrera w/out gas struts	24.1 mm	40/44	35°
Turbo look	26 mm	40/44	36°
1986:			
Carrera	25 mm	40/44	32°
Turbo look	26 mm	40/44	34°
1987-1989:			
Carrera	25 mm	46/47	32°
Turbo look	26 mm	46/47	34°

– If torsion plate inclination is incorrect:
  - Remove torsion bar cover fasteners and slide cover with torsion plate off body.
  - Reposition torsion bar in torsion plate and reinstall cover using 3 fasteners.
  - Recheck inclination and readjust as necessary.

NOTE —
It may be necessary to try several different positions for inner and outer teeth on the torsion bar to achieve the correct setting.

– Once torsion bar inclination is correct, use floor jack to raise trailing arm/torsion plate just enough to install lower stop bolt with spacer.

  - Tighten all four torsion bar cover mounting bolts.

Tightening Torque
- Torsion bar cover to body (M10) . . . . 47 Nm (34 ft-lb)

– Remainder of installation is reverse of removal.
  - Reset rear wheel alignment and check ride height as described in **440 Wheels, Tires, Alignment**.

WARNING —
Replacement and/or removal of the torsion plate or torsion bar disturbs the rear wheel alignment settings and also affects vehicle ride height, which directly influences vehicle stability. If ride height is changed, it is strongly recommended that the vehicle balance/weight be checked at all four wheels using suitable wheel-load floor scales. See **440 Wheels, Tires, Alignment** *for more information.*

Tightening torques
- Rear shock to trailing arm (M14) . . . . . 125 Nm (92 ft-lb)
- Torsion plate to trailing arm (eccentric) . 85 Nm (61 ft-lb)
- Torsion plate to trailing arm (M12 bolt) . 120 Nm (87 ft-lb)
- Wheel to wheel hub . . . . 130 Nm (96 ft-lb)

WHEEL BEARINGS

NOTE —
Rear wheel bearings are serviced by replacement only, as they are usually damaged when removed.

The rear wheel bearings are permanently sealed and lubricated and do not require any routine maintenance. Removing and installing the bearing on the car requires a special puller. If the necessary tools are not available, this work is best left to an authorized Porsche dealer or other qualified repair shop.

Check the wheel bearing with the wheel off the ground and the car properly supported. Spin the wheel and listen for any roughness. Alternately, apply pressure to either side of the wheel while checking for play. If any faults are found and no suspension components are found to be at fault, the wheel bearing should be replaced.

Rear wheel bearing, removing and installing

NOTE —
Wheel bearing replacement can be performed by removing the rear trailing arm as described earlier and pressing the bearing out on the bench. The following procedure describes utilizing a special bearing removal and installation tool that eliminates the need to remove the trailing arm from the car.

– Apply parking brake and shift transmission into reverse.

– Remove wheel hub center cap. Loosen stub axle nut (32 mm socket wrench) through opening in center of wheel.

– Loosen rear wheel lug nuts.

– Release parking brake and shift car into neutral. Raise rear of car and support it securely.

> ***WARNING —***
> *Make sure the car is stable and well supported at all times. Use a professional automotive lift or jack stands designed for the purpose. A floor jack is not adequate support.*

– Remove rear wheel. Remove stub axle nut.

– Press brake pedal down about 1 in. and hold in place with a pedal stop.

NOTE —
This will prevent brake fluid from dripping out of open brake lines while the work is being performed.

0018887

◀ Working at trailing arm, disconnect electrical connector (**arrow**) for brake pad wear sensor and remove harness from clip on trailing arm.

Disconnect brake line from brake caliper. Plug hydraulic lines.

> ***CAUTION —***
> *When working with hydraulic brake lines and hoses be sure to counterhold fitting before loosening or tightening.*

– Remove brake caliper mounting bolts and remove caliper. See **470 Brakes–Hydraulic**.

Lower car close to ground level. Use floor jack under trailing arm to relieve suspension load on shock absorber. (Brake backing plate shown removed for clarity).

- Remove rear shock absorber lower mounting bolt from trailing arm (**arrow**).

NOTE —
Porsche special tool P 289 a may be used to hold rear suspension in a compressed position while working on the rear shock absorber bolt.

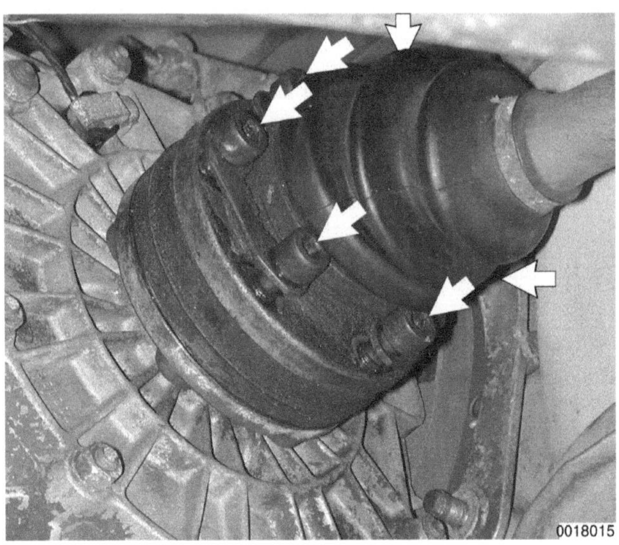

Unbolt CV joint fasteners (**arrows**) from transmission output flange.

- Disconnect axle shaft from drive flange. Drive stub axle out of trailing arm.
- See **Axle shaft, removing and installing**, later in this section.

Remove brake rotor retaining screws and pull off brake rotor.

Working at parking brake shoes pivot, remove cotter pin and castellated nut (**arrow**).

Drive out rear wheel hub using Porsche special tool P 297 a or equivalent.

Remove wheel bearing cover plate bolts (**arrows**). Remove plate with parking brake shoe assembly.

- Remove brake rotor backing plate bolts (**arrows**). Remove backing plate.

Use wheel bearing tool to remove old bearing and install new bearing following manufacturer's instructions.

– Remainder of installation is reverse of removal.

- Bleed brakes as described in **470 Brakes–Hydraulic**.

Tightening torques

- Brake caliper to trailing arm (M12) . . . . . 60 Nm (43 ft-lb)
- Brake hose to brake line . . . . 12 Nm (9 ft-lb)
- Brake rotor to wheel hub (M6) . . . . 5 Nm (44 in-lb)
- CV joint to transmission flange
  M8 bolt . . . . 42 Nm (30 ft-lb)
  M10 bolt . . . . 83 Nm (60 ft-lb)
- Rear shock absorber
  to trailing arm (M14) . . . . 125 Nm (92 ft-lb)
- Stub axle shaft to hub
  M20 castellated nut. . . . . . 300 - 320 Nm (217 - 231 ft-lb)
  M22 self-locking nut . . . . 460 Nm (339 ft-lb)
- Wheel bearing cover plate
  to bearing housing (M8) . . . . 25 Nm (18 ft-lb)
- Wheel to wheel hub . . . . 130 Nm (96 ft-lb)

STABILIZER BAR

The rear stabilizer bar attachment points at the body have been upgraded. The new stronger braces must be welded to the underbody. This work is best left to an authorized Porsche dealer or other qualified repair shop.

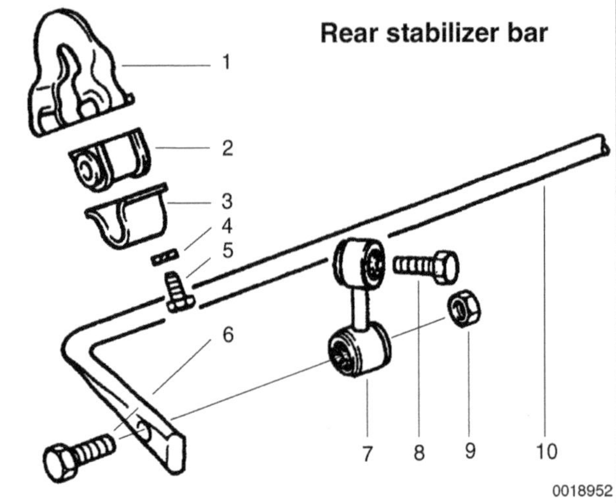

1. Attaching bracket (welded to body)
2. Rubber mount
3. Mounting bracket
4. Washer (M8)
5. Bolt (M8) tighten to 25 Nm (18 ft-lb)
6. Bolt (M12 x 1.5 x 60), tighten to 85 Nm (61 ft-lb)
7. Stabilizer bar link
8. Bolt (M12 x 1.5 x 70), tighten to 85 Nm (61 ft-lb)
9. Nut (M12)
10. Stabilizer bar

Axle shaft assembly

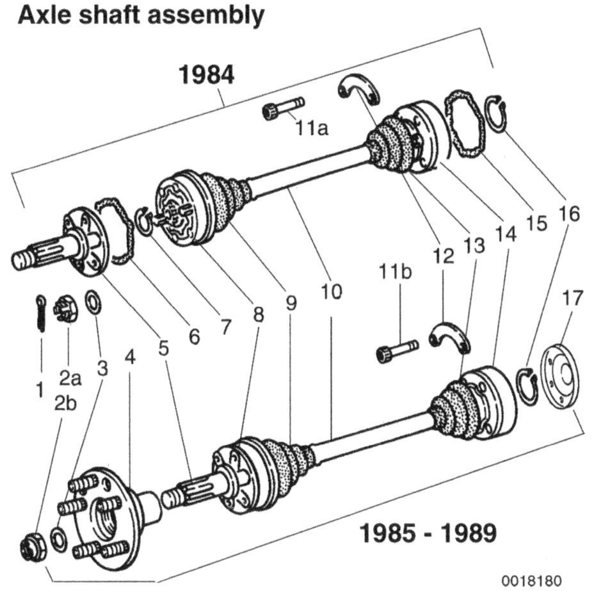

1. Cotter pin
2a. Castellated nut (M20), tighten to 300-320 Nm (217-231 ft-lb)
2b. Self-locking nut (M22), tighten to 460 Nm (339 ft-lb)
3. Washer
4. Wheel hub
5. Stub axle
6. Gasket
7. Circlip
8. Outer CV joint
9. Outer CV boot
10. Axle shaft
11a. Bolt (M8) tighten to 42 Nm (30 ft-lb)
11b. Bolt (M10) tighten to 83 Nm (60 ft-lb)
12. Plate
13. Inner CV boot
14. Inner CV joint
15. Gasket
16. Circlip
17. Inner dust cover

AXLE SHAFTS

In a car with independent rear suspension the drive axles must be able to move with the suspension while delivering power to the wheels. To accomplish this, a constant velocity (CV) joint is located at each end of each axle shaft.

A CV joint is similar in function to the more familiar universal joint. The CV joint allows for the smooth flow of power to the drive wheels even though the final drive, axles and wheels are not lined up with each other. The advantage of the more complicated CV joint is that it operates more smoothly and can accept more radical drive axle angles than a universal joint.

NOTE —
CV joints were increased in size from 100 mm to 108 mm diameter in 1985. The axle shaft assembly design was also changed. Consult with an authorized Porsche dealer parts department or other experienced Porsche repair shop for upgrade information.

CV joint boots should be inspected periodically to see that they are in good condition. A damaged boot will let in dirt, and quickly destroy the joint.

Axle shaft assembly, removing and installing

– Apply parking brake and shift transmission into reverse.

– Remove wheel hub center cap. Loosen stub axle nut (32 mm socket wrench) through opening in center of wheel.

– Loosen rear wheel lug nuts.

– Release parking brake and shift transmission into neutral. Raise rear of car and support it securely.

> ***WARNING —***
> *Make sure the car is stable and well supported at all times. Use a professional automotive lift or jack stands designed for the purpose. A floor jack is not adequate support.*

– Remove rear wheel. Remove stub axle nut.

With car close to ground level use floor jack under trailing arm to relieve suspension load on shock absorber. (Brake backing plate shown removed for clarity.)

- Remove rear shock absorber lower mounting bolt from trailing arm (**arrow**).

NOTE —
Porsche special tool P 289 a may be used to hold rear suspension in a compressed position while working on the rear shock absorber bolt.

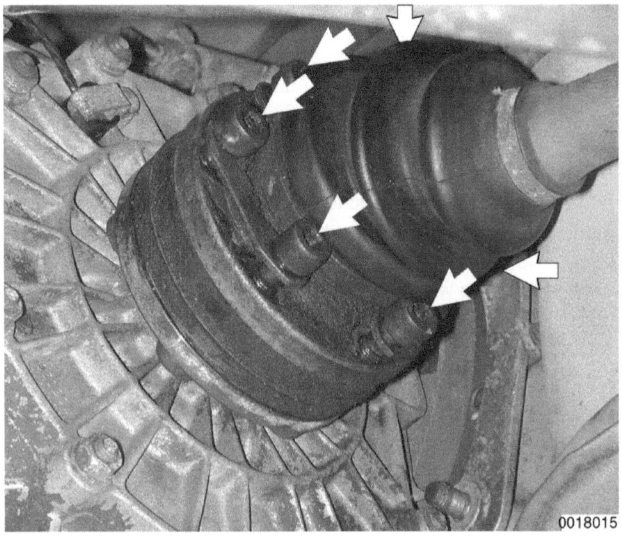

◄ Unbolt CV joint fasteners (**arrows**) from transmission output flange.

– Disconnect axle shaft from drive flange. Drive stub axle out of trailing arm.

– Installation is reverse of removal.

Tightening torques

- CV joint to transmission flange
  - M8 bolt .... 42 Nm (30 ft-lb)
  - M10 bolt .... 83 Nm (60 ft-lb)
- Rear shock absorber to trailing arm (M14) .... 125 Nm (92 ft-lb)
- Stub axle shaft to hub
  - M20 castellated nut...... 300 - 320 Nm (217 - 231 ft-lb)
  - M22 self-locking nut .... 460 Nm (339 ft-lb)
- Wheel to wheel hub .... 130 Nm (96 ft-lb)

CV joint and boot kit

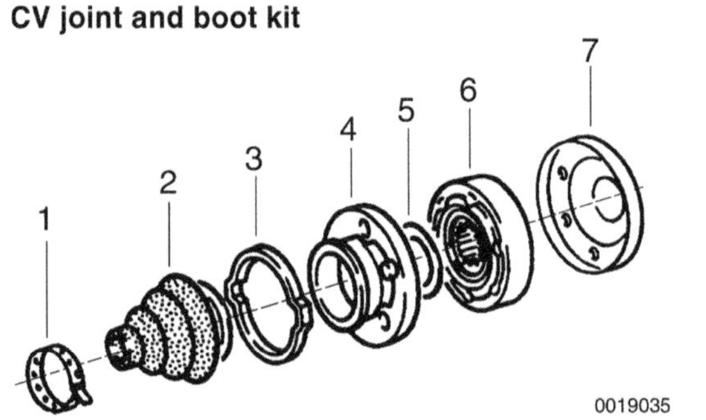

1. Small clamp
2. CV joint boot
3. Large clamp
4. Boot flange
5. Concave washer (install with concave facing CV joint
6. CV joint
7. Inner dust cover

CV joint and boot, removing and installing

NOTE —

On all but 1984 models only the inner CV joint is replaceable separately. The outer joint must be replaced with the axle shaft.

◄ Remove axle shaft assembly as described earlier and place in shop vice with soft jaws.

– Cut off small and large CV boot clamps.

◄ Drive off dust cover and boot flange with punch (**arrows**).

- Slide flange and boot away from inner CV joint.

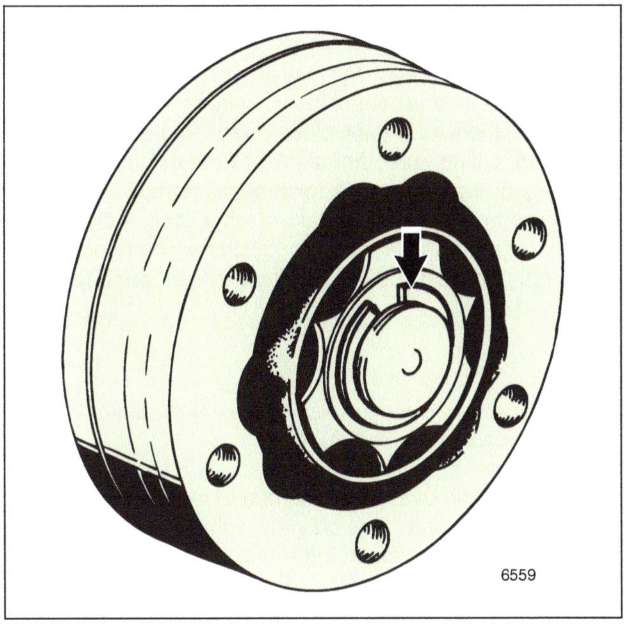

Use special tool VW 161 a or equivalent to spread retaining circlip (**arrow**) and remove from shaft. Discard circlip.

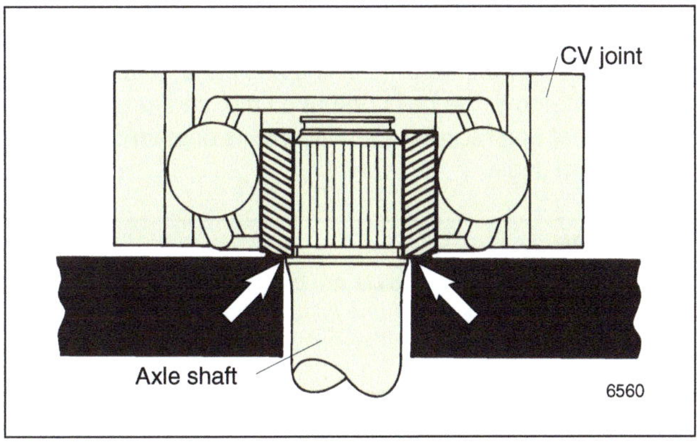

Support CV joint on inner hub (**arrows**) while pressing off axle shaft.

- Slide both boots off shaft.
- Clean all old lubricant off shaft and joint splines.

NOTE —
Inspect the axle shaft and splines for wear or damage. If the shaft is damaged, the shaft and the outer CV joint will have to be replaced as an assembly.

- Inspect CV joints carefully, as described below. If reusing CV joints, clean off old grease and repack with fresh.
- Slide on new boots and secure with new clamps.

NOTE —
A new CV joint boot kit includes new boot, clamps, lubricant and inner CV joint circlip.

- Press inner CV joint on shaft while supporting inner hub of joint. Secure with new circlip.
- Pack CV joint and boot with Optimoly® grease or equivalent.

CV Joint Lubricant Capacity
- Per joint (inner or outer) .... 80 gram (2.8 oz.)

- Fit covers on inner CV joint, taking care to line up bolt holes. Reinstall in car as described earlier.

CV joint, inspecting

The components of each CV joint are precisely matched during manufacture and cannot be serviced or replaced individually. To inspect a CV joint, clean away the grease and look for galling, pitting and other signs of wear or physical damage. Any of these is cause for replacement. Discoloration due to overheating indicates lack of lubrication. A discolored joint that is otherwise in good condition may be reinstalled if thoroughly cleaned and repacked with the proper amount of new grease.

NOTE —

Polished interior surfaces or visible ball tracks alone are not necessarily cause for replacement.

The balls and grooves allow the hub to move, but the parts should fit snugly and move only with some effort. A joint with obvious free play between inner hub, balls and outer cage should be replaced.

Disassembly of the CV joints is not normally recommended. If the joint must be disassembled for cleaning, mark the relative positions of the inner hub, the ball cage, and the outer housing. Remove the balls one at a time, and keep track of them so that each can be reinstalled in its original position in the hub and cage.

CAUTION —

- *The cage, housing, and balls are precisely matched. When disassembling more than one joint, do not intermix components.*
- *The CV joint should go together firmly but smoothly. Heavy force should not be required. If in doubt, start over and recheck the alignment. A joint that is forced together may lock and not come apart again.*

440 Wheels, Tires, Alignment

TABLES

GENERAL

This section covers wheels, tires and alignment information. Troubleshooting information is in **400 Suspension, Brakes and Steering**. For detailed information relating to suspension or steering components, refer to the following sections:

- **401 Front Suspension**
- **420 Rear Suspension, Axle Shafts**
- **480 Steering**

Special tools and equipment

Much of the work described in this section requires special tools and experience. If you lack the skills, the tools, or a suitable workplace for tire or alignment work, we suggest you leave such repairs to an authorized Porsche dealer or other qualified repair shop.

Special Tools

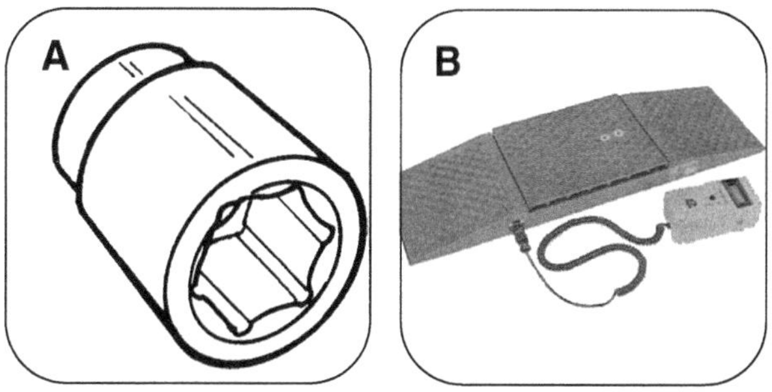

Special tool

A P 300 . . . . 19 mm wheel bolt socket
(Porsche ordering #000 721 300 00)
B Maha RWA 2 U . . . . Wheel load scale

Wheels, tires and alignment

Tires greatly influence the handling, stability and cornering performance of a Porsche. In order to gain the best possible performance from a Porsche all four wheels must be aligned correctly. If the torsion bars or torsion bar components have been replaced, the vehicle's corner weights (weight at each wheel) must be checked before an alignment is performed.

For stability and control, wheels and tires must be in good condition and properly aligned. Tire inflation pressures, tire wear and wheel alignment will all influence how the car feels and responds on the road.

Precise wheel alignment can only be accomplished when the tires, the steering, and the suspension are in good condition. Uneven tire wear, different size tires, or worn suspension and steering parts all affect wheel alignment. Reputable wheel alignment technicians will always inspect the suspension and steering for worn parts before an alignment, and will recommend that any necessary repairs be made first.

Correct tire pressure information can be found on an information label (**arrow**) on the rear engine compartment apron and in the owner's manual.

Tire and wheel specifications are given in **Table a**.

Wheels and tires, general information

- Check for tire directional markings (arrow on sidewall) before mounting a tire.
- Always replace tire valves when mounting new tires. Replace stick-on valve stem supports if missing.
- Use only rubber lubricant on the tire beads when mounting tires. Do not use silicone.
- Clean lubricant from bead area before inflating the tire to reduce the possibility of the tire slipping on the wheel.
- Fast acceleration and hard braking should be avoided for the first 150 miles after mounting a new tire.
- Inflate tire to 4 bar (56 psi) to seat tire bead correctly on the wheel. When inflating, tire bead should seat at no greater pressure than 3.3 bar (48 psi) to avoid damage to the tire bead.
- Wheel locks should be placed on the stud closest to the tire valve.
- Lubricate wheel lugs and shoulders with Optimoly®, Porsche part number 000 043 020 00.
- Tighten wheel lug bolts to 130 Nm (96 ft-lbs).
- Check tire pressures only when tires are cold. The tire pressure increases when hot.

Table a. Wheel and Tire Sizes

Model and year	Standard		Option 1		Option 2	
	Wheel size	Tire size	Wheel size	Tire size	Wheel size	Tire size
Carrera 1984 - 1986:						
Front	6 J x 15	185/70 VR 15	7 J x 15	185/70 VR 15	6 J x 16	205/55 VR 16
Rear	7 J x 15	215/60 VR 15	8 J x 15	215/60 VR 15	7 J x 16	225/50 VR 16
Carrera 1987 - 1988						
Front	6 J x 15	195/65 VR 15	7 J x 15	195/65 VR 15	6 J x 16	205/55 VR 16
Rear	7 J x 15	215/60 VR 15	8 J x 15	215/60 VR 15	7 J x 16	225/50 VR 16
Carrera 1989						
Front	6 J x 16	205/55 ZR 16				
Rear	8 J x 16	225/50 ZR 16				
Turbo Look 1984 - 1985						
Front	7 J x 16	205/55 VR 16				
Rear	8 J x 16	225/50 VR 16				
Turbo Look 1986 - 1988						
Front	7 J x 16	205/55 VR 16				
Rear	9 J x 16	245/45 VR 16				
Turbo Look 1989						
Front	7 J x 16	205/55 ZR 16				
Rear	9 J x 16	245/45 ZR 16				

WHEELS

The Porsche wheels supplied as original equipment are designed for tubeless, radial-ply tires of a specific size. Replacement tires of non-standard size or construction should be installed only if the tire manufacturer specifies them for your specific make and model.

The wheels are sized according to diameter and width. When fitting a non-standard wheel with different dimensions, there is danger of wheels or tires interfering with the body or with suspension parts. Even if there is no interference, incorrectly sized wheels may place additional loading on wheel bearings or adversely affect steering geometry. For best results, rely on the advice of your Porsche dealer or reputable wheel and tire professionals.

Wheel lug bolts should always be tightened using Porsche special tool P 300 (19 mm socket) and a torque wrench. Too little torque is obviously dangerous as the wheel may come loose, but excessive torque is also a problem. Over-tight wheel lug bolts can warp the hub, causing vibration and other problems affecting stability and braking.

Tightening Torque
- Wheel to wheel hub . . . . 130 Nm (96 ft-lb)

0019141

< Damaged wheels may also be a source of vibration problems. Alloy wheels are constructed of a soft aluminum-magnesium alloy and can be easily damaged.

Check for a bent wheel by measuring lateral or radial runout at the part of the wheel where the tire bead meets the wheel.

Alloy Wheel Runout Specification
- Maximum allowable runout . . . . 1.0 mm (0.04 in.)

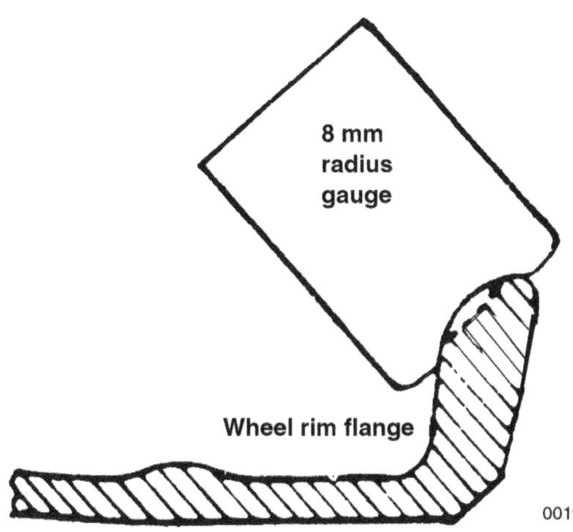

0019142

< Check alloy wheel flanges for wear. The inside flange is more subject to wear. Remove sharp edges and burs, then use an 8 mm radius gauge to check for wear.

TIRES

Radial-ply tires are installed as standard equipment on all cars covered by this manual. To retain the car's excellent handling characteristics, it is recommended that the tires only be replaced with those of the same (radial-ply) construction, size and speed rating.

Tire wear

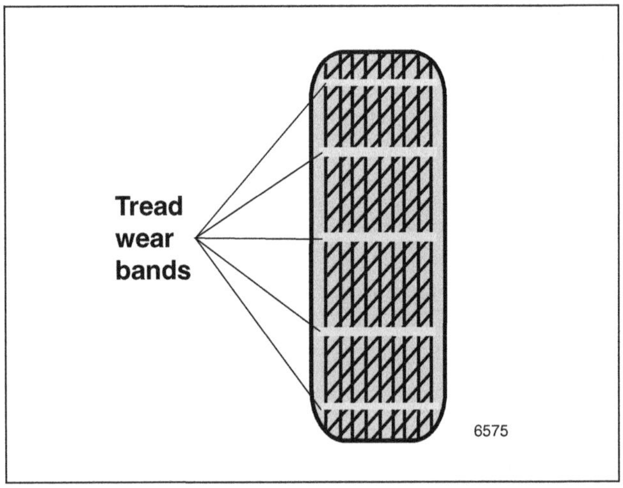

Tire wear is a good indicator of steering and wheel alignment problems. Proper tread wear is difficult to notice, so tires are made with wear-indicator bands that indicate when the tire is nearly worn-out. On an evenly worn tire, these wear-indicator bars will eventually appear as evenly spaced bald "stripes" about ½ in. wide running across the tread surface. The appearance of these tread-wear bands on only one part of the tread indicates uneven wear.

Uneven tire wear is usually caused by improper tire inflation pressures or misalignment. Other kinds of uneven tread wear may be the result of faulty components or improper alignment. The following figures illustrate the influence of these factors on tire wear.

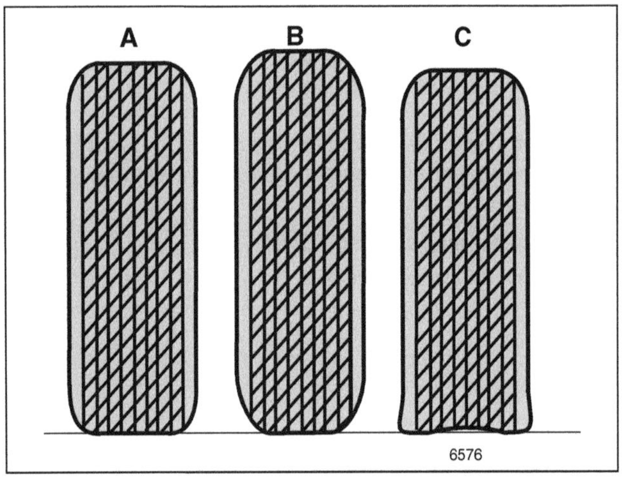

Effect of tire inflation pressure on wear pattern. Condition (**A**) is normal. Overinflation (**B**) causes increased wear in center of tire tread. Underinflation (**C**) causes increased wear on outer edges of tire tread.

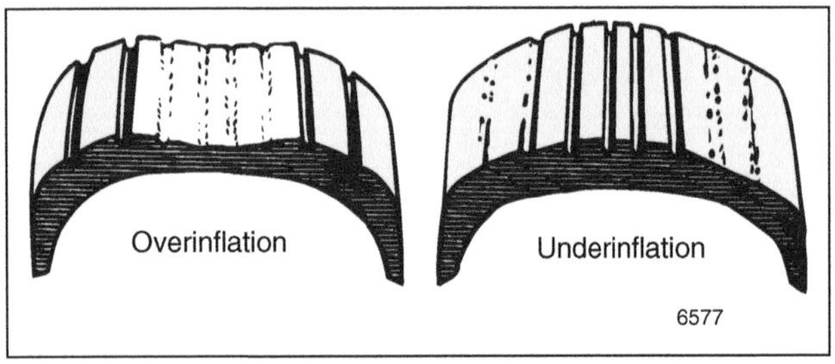

Examples of uneven tread wear caused by overinflated and underinflated tires. (Drawings courtesy of Goodyear Tire and Rubber Co.)

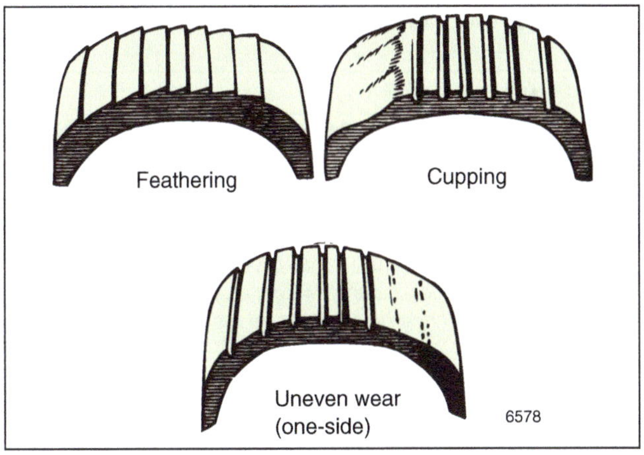

Examples of uneven tread wear caused by misalignment and/or worn or damaged components:

- Feathering suggests incorrect alignment.
- Uneven wear on one side suggests incorrect alignment or repeated hard cornering.
- Cupping results from vibration, often caused by unbalanced tire or faulty shock absorber. (Drawings courtesy of Goodyear Tire and Rubber Co.)

Tire wear that is uneven across the tread—one side worn more than the other, or unusual wear of individual tread ribs—indicates an alignment problem, and perhaps worn suspension or steering components. Cupping or scalloping—wear that is uneven around the circumference of the tire—is a telltale sign of an unbalanced tire or a worn-out shock absorber.

Winter Tires

Though inferior to regular tires for dry-road wear and handling, winter (mud and snow) tires offer a big improvement on snow-covered winter roads. Studded winter tires improve traction on icy surfaces, but may be more dangerous than non-studded tires on dry roads. They should be used only if icy conditions predominate. Check local laws. Studded tires are restricted or completely prohibited in many states.

CAUTION —

If winter tires are to be installed on only two of the four wheels, they should be installed on the rear driving wheels. To avoid creating dangerous handling characteristics, they must be of the same type (radial-ply construction) as those on the front of the car.

Tire storage

- The rubber in tires is sensitive to environmental conditions and foreign substances. Ozone, fuels, greases and oil all can have a destructive effect on tires.
- Store tires indoors where the temperature is between 14° F and 68° F and the air is dry.
- Wash tires before storing them.
- Keep tires out of sunlight or fluorescent lighting.
- Store tires on wheels with full tire pressure.
- Do not hang tires on hooks or stack them on top of each other.

WHEEL ALIGNMENT

> ***WARNING —***
> - *Replacement and/or removal of front or rear suspension components, such as front or rear torsion bars or rear torsion plates, disturbs the wheel alignment settings. Replacement or adjustment of the torsion bars, rear torsion plates or rear trailing arms can affect vehicle ride height, which directly influences vehicle stability, chassis balance and even braking at all wheels. Vehicle balance/weight must be checked at all four wheels, and ride height must be reset. See* **Extended alignment** *later in this group.*
> - *While performing alignment procedures, make sure the car is stable and well supported at all times. Use a professional automotive lift or jack stands designed for the purpose. A floor jack is not adequate support.*

Depending on the circumstances, there are two procedures for wheel alignment:

- **Basic alignment** is required:
  –when replacing suspension components which cause no alteration in ride height
  –In case of excessive or uneven tire wear
  –After a new vehicle has "settled"
  –In case of handling complaints

- **Extended alignment** is required:
  –If suspension replacement components alter vehicle height
  –In case of incorrect vehicle height

Basic alignment requires the use of alignment analyzer equipment. Minor resetting of alignment specifications is permitted.

Extended alignment requires checking and resetting vehicle ride height and corner weights with the use of wheel load scales and then performing the basic alignment.

Basic alignment specification for the front and rear wheels are in **Table b** and **Table c**.

Pre-alignment Checklist (repair groups in bold)
- Fuel tank full
- Tool kit and spare tire in luggage compartment
- Ball joint and suspension parts checked. See **401**.
- Wheel bearing play set and OK. See **401**.
- Tire pressures set to specifications.
- Steering wheel and steering rack in center position. See **480**.
- Check and, if necessary, reset front and rear ride height. See **Extended alignment** in this group.

Basic alignment

– Always start a four wheel alignment procedure by beginning at the rear wheels. Make sure the car is loaded with a normal load. See below.

Normal Load

- Each front seat . . . . .68 kg (150 lb.)
- Center of rear seat .68 kg (150 lb.)
- Luggage compartment . . . . .21 kg (46 lb.)
- Fuel tank. . . . . .full

Table b. Front Wheel Alignment

	Specification	Max. deviation
Toe angle (total)	+15' ± 5'	
Toe difference angle (inside wheel@ 20° lock)	0° to +30"	Can be changed only by replacing steering arm
Camber	0° ± 10'	10'
Caster	6°5' ± 15'	30'
Ride height	108 mm (4.25 in.)	± 5 mm (±0.02 in.)
Steering wheel lock to lock	Approx. 3 turns	
Turning circle diameter	10.35 m (34 feet)	
Track width (7 x 15 wheel)	1374 mm (54.1 in.)	

Table c. Rear Wheel Alignment

	Specification	Max. deviation
Toe per wheel	+10' ± 10'	20'
Camber	- 1° ±10'	20'
Ride height	16 mm ± 5 mm	8 mm
Torsion plate setting		
1984-85 models	35°	0.5°
1986-89 models	32°	0.5°
Track width (7 x15 wheel)	1380 mm	

Extended alignment

After a basic alignment if it is determined that, despite good tires and correct inflation, the handling characteristics of the vehicle are not correct, or if ride height is incorrect, or if the braking is uneven, wheel load scales must be used to set correct wheel load and ride height.

Follow the equipment manufacturer's instructions for measuring and setting wheel load

NOTE —

- *Before setting wheel load and vehicle height, bounce vehicle a few times to let the suspension settle into the normal ride height position.*
- *On cars fitted with Bilstein shocks, the "normal ride height position" can vary with the car in the rest position. Pulling the suspension up or pushing it down at the bumper can result in different ride height positions. When checking or adjusting the ride height, manually position the car in the 'mean' or middle position.*

Wheel load, checking

Changing ride height at any wheel changes wheel loading at all wheels. Increasing the initial tension of a torsion bar raises the vehicle at the corresponding wheel and increases wheel load at that wheel. Decreasing the initial tension of a torsion bar lowers the vehicle at the corresponding wheel and decreases wheel load at that wheel.

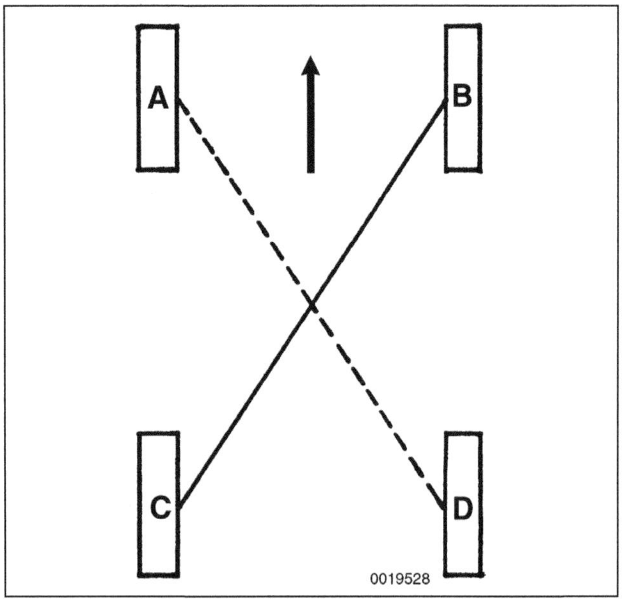

Changing load on one wheel directly affects the wheel at a diagonal from it, while it has the opposite effect on the wheels on the other diagonal.

- For example, increasing load at wheel **A** increases load at wheel **D**.
- Increasing load at wheel **A** decreases load at wheels **B** and **C**.

Wheel Load Specification (front or rear axle)
- Maximum left-right load difference ...... 20 Kg (44 lbs.)

– Check wheel load using Maha wheel load scales RWA 2 U or equivalent equipment.

– Adjust wheel load by setting front and rear ride height, as described later.

NOTE —
- *Any adjustment of a torsion bar will have a wheel load effect on the other three torsion bars.*
- *Increasing the initial torsion bar tension on one side (raising vehicle) will increase wheel load on that wheel.*
- *Decreasing the initial torsion bar tension on one side (lowering vehicle) will decrease wheel load on that wheel.*

Front ride height, setting

Set front ride height after replacing front suspension components such as control arms or torsion bars. The car must have a full fuel tank and the spare tire must be in luggage compartment for correct adjustment.

– Place car on a level floor.

NOTE —
- *Before setting wheel load and vehicle height, bounce vehicle a few times to let the suspension settle into the normal ride height position.*
- *On cars fitted with Bilstein shocks, the "normal ride height position" can vary with the car in the rest position. Pulling the suspension up or pushing it down at the bumper can result in different ride height positions. When checking or adjusting the ride height, manually position the car in the 'mean' or middle position.*

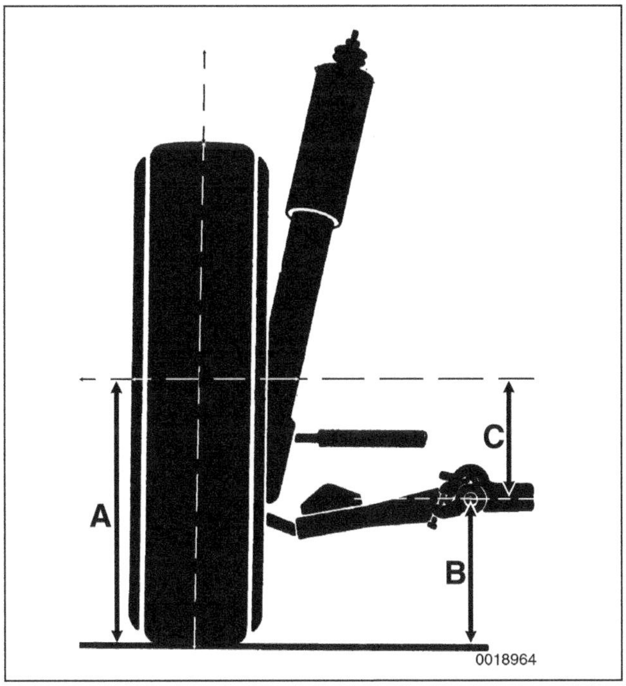

Measure from floor to center of front wheel dust cap to obtain dimension **A**.

- Measure from floor to center of torsion bar adjusting cap to obtain dimension **B**.
- Difference (**A** – **B** = **C**) is ride height.

Front Ride Height

- Dimension **C** . . . . 108 ± 5 mm (4.25 ± 0.20 in.)
- Max. difference between left and right . . . 5 mm (0.20 in.)

Adjust to specifications using front torsion bar adjusting screw (**arrow**).

NOTE —

- *Height-adjustment specifications apply to new vehicles. Older vehicles can be up to 10 ± 5 mm (0.4 ± 0.2 in.) lower.*
- *If a vehicle is slightly low in front it can be raised without causing an excessive difference in wheel load. To do this, evenly turn the front torsion bar adjustment screws clockwise. (This will have an effect on the vehicles toe setting.)*

Rear ride height, setting

Set rear ride height after replacing rear suspension components such as trailing arms, torsion plates or torsion bars. Initial setting of rear torsion bars is covered in **420 Rear Suspension, Axle Shafts**.

– Place car on a level floor.

NOTE —
- *Before setting wheel load and vehicle height, bounce vehicle a few times to let the suspension settle into the normal ride height position.*
- *On cars fitted with Bilstein shocks, the "normal ride height position" can vary with the car in the rest position. Pulling the suspension up or pushing it down at the bumper can result in different ride height positions. When checking or adjusting the ride height, manually position the car in the 'mean' or middle position.*

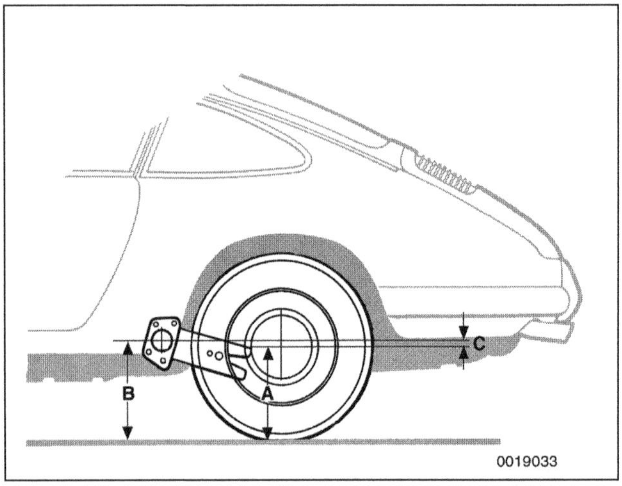

Measure from floor to center of rear wheel dust cap to obtain dimension **A**.

- Measure from floor to center of torsion bar cover to obtain dimension **B**
- Difference (**B - A = C**) is ride height.

Adjust to specifications using height-adjusting eccentric.

Rear Ride Height
- Dimension **C** . . . . 16 mm ± 5 mm (0.6 ± 0.2 in.)
- Max. difference between left and right . . . . 8 mm (0.3 in.)

NOTE —
- *Height-adjustment specifications apply to new vehicles. Older vehicles can be up to 10 ± 5 mm (0.4 ± 0.2 in.) lower.*
- *The two-piece torsion plate is set at the factory so that the vehicle can be raised with the height adjustment eccentric, but not lowered.*

Tightening Torque
- Clamping or eccentric bolt . . . . 245 Nm (181 ft-lb)

Rear wheel camber and toe, adjusting

WARNING —
While performing alignment procedures, make sure the car is stable and well supported at all times. Use a professional automotive lift or jack stands designed for the purpose. A floor jack is not adequate support.

CAUTION —
Changing camber at one rear wheel will cause a change in the load at that wheel and consequently at all other wheels. Be sure to check vehicle corner loads as a final check. Changing camber at one rear wheel will also affect the toe setting at that wheel.

NOTE —
A change in the rear toe by adjusting the toe eccentric will have very little effect on camber, wheel load or ride height.

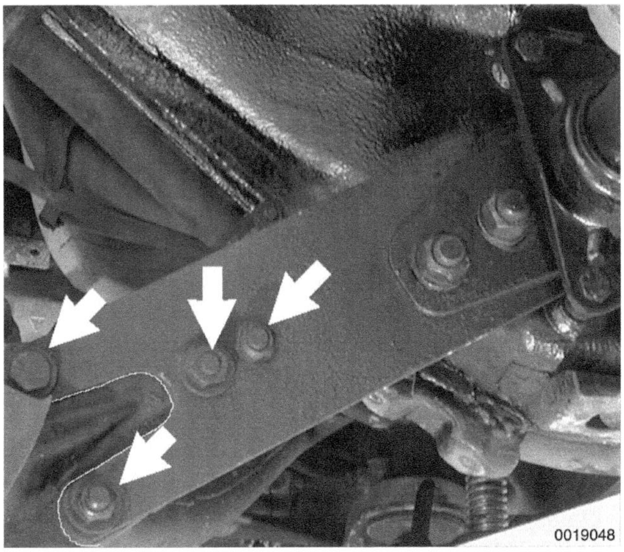

◄ Mark torsion plate position to trailing arm by scribing outline around torsion plate.

- Loosen torsion plate fasteners (**arrows**).

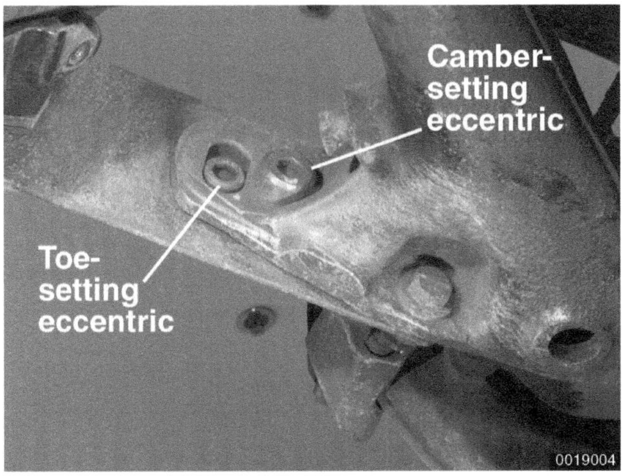

◄ Set rear wheel camber and toe using an alignment analyzer.

Tightening torque
- Torsion plate to trailing arm (eccentric). . 85 Nm (63 ft-lb)
- Torsion plate to trailing arm (M12) . . . . 120 Nm (89 ft-lb)

0019529a

Front wheel camber and caster, adjusting

◀ Working in luggage compartment, clean sealing compound away from front upper strut mounts and bolts (**arrows**).

– Loosen strut mount bolts slightly.

0019529b

◀ Set front wheel camber and caster using an alignment analyzer.

Tightening torque

- Upper strut mount to body . . . . 47 Nm (35 ft-lb)

NOTE —
Reseal upper strut mounts with sealing compound.

Front wheel toe, adjusting

– Check to see if steering rack is in its center position when steering wheel is centered:

- Starting from center position, turn steering wheel in one direction to stop, counting turns.
- Turn to stop in opposite direction while counting turns.
- If necessary, remove and center steering wheel at half the number of turns from stop to stop.

0019136

◀ Loosen 22 mm inner and outer tie rod locking nuts (**arrows**).

- Adjust tie rod by turning center tube to change length.
- To keep steering wheel centered, adjust both tie rods equal amounts.

– When final toe adjustments have been made tighten tie rod locking nuts.

NOTE —
The toe difference angle cannot be adjusted. If the difference is excessive, steering or suspension parts should be inspected for damage or deformation.

460 Brakes–Mechanical

TABLES

GENERAL

This repair group covers replacement procedures for mechanical brake components including brake pads, brake rotors and parking brake components.

For general specifications, diagnosis and troubleshooting information see **400 Suspension, Brakes and Steering–General**. Repairs to the brake master cylinder, power booster, calipers and associated components are covered in **470 Brakes–Hydraulic**. Pedal cluster repairs are covered in **301 Clutch**.

Special tools

Some procedures require special tools.

Special Tools

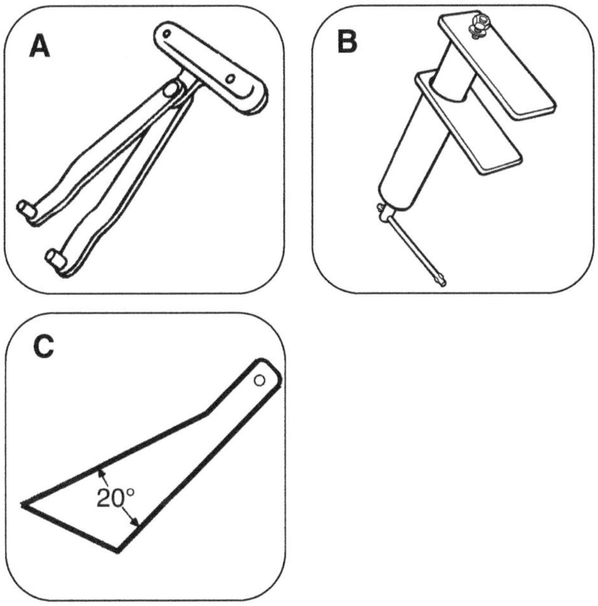

Special tool

A P 86 . . . . Brake pad removal tool
(Porsche ordering #000 721 086 00)
B 1023/4 . . . . Caliper piston retractor
(source: Baum Tools Unlimited)
C P 84 . . . . Caliper piston 20° gauge
(source: Baum Tools Unlimited)

> ***WARNING —***
>
> - *Brake friction materials such as brake linings or pads may contain asbestos fibers which can lead to illnesses. Do not create dust by grinding, sanding, or cleaning the pads with compressed air. Avoid breathing any asbestos fibers or dust.*
> - *Brake fluid is poisonous, highly corrosive and dangerous to the environment. Wear safety glasses and rubber gloves when working with brake fluid. Do not siphon brake fluid with your mouth. Immediately clean away any fluid spilled on painted surfaces and wash with water. Dispose of brake fluid properly.*
> - *Do not reuse self-locking nuts, or fasteners. They are designed to be used only once and may fail if reused. Always replace them with new self-locking nuts.*

Braking system

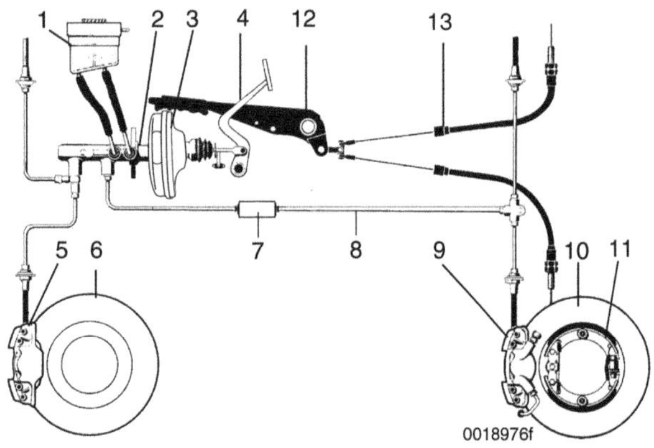

1. Brake fluid reservoir
2. Dual master cylinder
3. Power booster
4. Brake pedal
5. Front brake caliper
6. Front brake rotor
7. Rear brake pressure reducer
8. Brake fluid line
9. Rear brake caliper
10. Rear brake rotor/drum
11. Parking brake shoes
12. Parking brake handle
13. Parking brake cables

Braking system

Porsche 911 Carrera models are equipped with power-assisted four-wheel disc brakes.

A brake booster power-assists a dual-circuit master cylinder. A vacuum boost system is used in all models.

Dual-piston calipers act on vented front and rear rotors. A brake pad wear sensor for each wheel indicates when pads need replacement.

The parking brake is mechanically operated via dual cables. The parking brake system is independent of the rear disc brake system. The drum-style parking brake shoes act on a small drum, which is integral with the rear brake rotor.

Brake noise

Occasional groaning or squealing sounds from the disc brakes are usually caused by vibration being transmitted through the brake pads. Brake friction materials that contain little or no asbestos may contribute to brake noise. These noises are normal and rarely indicate a problem.

Although there is no single solution to brake noise problems, proper maintenance and repair can help minimize noise. Disc brake caliper assemblies include anti-rattle springs to minimize vibration and noise. Brake pads and calipers should be kept clean and free of corrosion. Always resurface or replace brake rotors when changing brake pads.

Brake pad wear warning system

The brake pad wear warning system warns the driver with a light on the instrument panel when the pads are worn. This warning system should not be ignored. Failure to replace the pads soon after the light comes on may cause caliper or rotor damage that will add to the cost of repairs.

A brake pad warning sensor is used on the inner pad of each brake caliper. The pad wear sensors are plugged into the pad wear warning system. The plug for each front sensor is clipped to the side of the front strut. The plug for each rear sensor is located in a clip attached to the trailing arm.

With the ignition on and the engine stopped, the pad warning indicator light is activated via the oil pressure light circuit. The light comes on if:

- There is a short to ground anywhere in the warning circuit.
- A brake pad is worn, causing the sensor to make contact with the brake rotor while braking.
- The wire is damaged and touches ground.

The indicator light then stays on until the ignition is turned off, even if there is no longer a ground contact. If the ignition is turned off and back on, the light stays off until a ground short turns it back on again.

The plastic body of the pad wear sensor will be ground away when the brake pad wear below 2 mm (0.08 in.). Replacement of the warning sensors can be avoided by replacing the brake pads when pad thickness is 2.5 mm (0.1 in.).

NOTE —
Due to the extra space required by the pad wear sensor harness at each caliper, all 14 inch and most 15 in wheels cannot be fitted to these cars.

Troubleshooting

Brake performance is mainly affected by three things: the level and condition of the brake fluid, the system's ability to create and maintain hydraulic pressure, and the condition of the friction components.

Air in the brake fluid will make the brake pedal feel spongy during braking or will increase the brake pedal force required to stop. Fluid contaminated by moisture or dirt can corrode the system. Inspect the brake fluid inside the reservoir. If it is dirty or murky, or is over a year old, the fluid should be replaced.

Visually check the hydraulic system starting at the master cylinder. To check the function of the master cylinder hold the brake pedal down hard with the car stopped and the engine running. The pedal should feel solid and stay solid. If the pedal slowly falls to the floor, either the master cylinder is leaking internally, or fluid is escaping from other points in the system. If no leaks can be found, the master cylinder is faulty and should be replaced. Check all brake fluid lines and couplings for leaks, kinks, chafing or corrosion.

Check the brake booster by pumping the brake pedal approximately 10 times with the engine off. Then hold the pedal down and start the engine. The pedal should fall slightly. If not, check for any visible faults before suspecting a faulty brake booster, and check the non-return valve for one-way flow.

NOTE —
- *The brake booster non-return valve is in the vacuum hose at the left side manifold, in the engine compartment.*
- *See* **400 Suspension, Brakes and Steering–General** *for a brake troubleshooting table.*
- *See* **470 Brakes–Hydraulic** *for repair information for hydraulic brake components or the brake power-booster.*

Worn or contaminated brake pads will cause poor braking performance. Oil-contaminated or glazed pads cannot produce as much friction, and stopping distances will increase. Inspect the rotors for glazing, discoloration, scoring or oiliness.

Steering wheel vibration while braking at speed is often caused by warped brake rotors, but can also be caused by worn suspension parts.

Maintenance

– Porsche specifies the maintenance steps below to be carried out at particular time or mileage intervals for proper maintenance of the brakes.

- Check brake fluid. See **030 Maintenance**.
- Inspect brake hoses, lines and wheel calipers for leaks.
- Check brake pads and shoes for wear.
- Replace brake fluid. See **470 Brakes–Hydraulic**.

BRAKE PADS AND ROTORS

Brake pads can be replaced without disconnecting the brake fluid line from the caliper or having to bleed the brakes.

Brake rotor removal and replacement requires that the caliper be disconnected from the brake line in order to avoid bending the metal line. The rear rotors can be replaced without dissembling the wheel hub and bearing, but front rotor replacement requires that the wheel bearing hub be removed with the rotor and disassembled on a work bench.

> ***CAUTION —***
> - *Always machine or replace brake rotors in pairs.*
> - *Always replace brake pads in sets (one set = one axle).*

Brake pads, inspecting

– Raise car and remove wheels.

> ***WARNING —***
> *Make sure the car is firmly supported on jack stands designed for the purpose. Place the jack stands beneath a structural chassis point. Do not place jack stands under suspension parts.*

◄ Visually inspect brake pads for thickness and compare to specifications. Replace as necessary.

Brake Pad Thickness Specification
- Wear limit **A** (front or back) . . . . 2 mm (0.08 in.)

Brake pads, replacing

> ***CAUTION—***
> *To prevent brake fluid overflow when caliper pistons are pushed back, siphon some fluid out of brake master cylinder reservoir.*

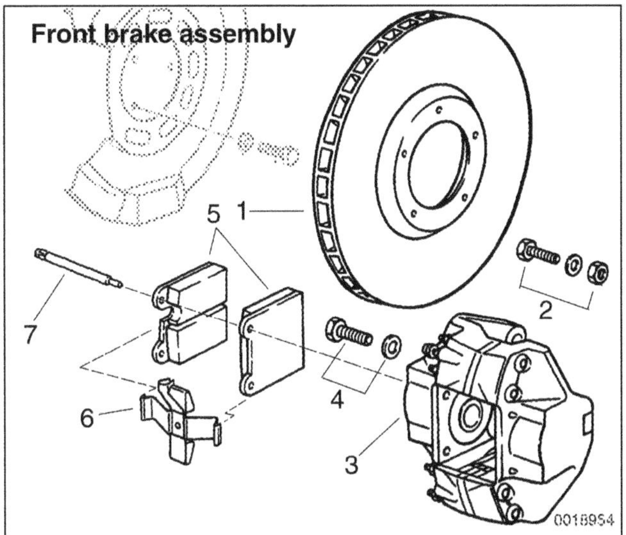

1. Brake rotor
2. Brake rotor to bearing hub fasteners (M8)
   Tighten to 23 Nm (17 ft-lb)
3. Brake caliper
4. Brake caliper to strut housing fastener (M12)
   Tighten to 70 Nm (52 ft-lb)
5. Brake pads
6. Anti-rattle cross spring
7. Retaining pin

◀ This procedure is applicable to both front and rear brake pads.

– Raise car and remove wheels.

> ***WARNING—***
> *Make sure the car is firmly supported on jack stands designed for the purpose. Place the jack stands beneath a structural chassis point. Do not place jack stands under suspension parts.*

◀ Front brakes: Working in wheel housing, disconnect electrical connector (**A**) for brake pad wear sensor and remove harness from clip on strut body.

Rear brakes: Working at trailing arm, disconnect electrical connector (**A**) from brake pad wear sensor and remove harness from clip on trailing arm.

Use drift to drive brake pad retaining pins out while depressing anti-rattle cross-spring.

- Remove anti-rattle cross spring. Detach brake pad warning sensor wire from spring.

Remove brake pads from caliper with Porsche special tool P 86 or equivalent.

NOTE —

If brake pads are to be reused, mark original position.

◀ Pry out brake pad warning contact from inner pad.

- Clean around contact and pry out carefully to avoid damage to sensor contact.

◀ Push pistons back into caliper body with Porsche special tool P 83 or equivalent (Baum tool 1023/4 shown).

NOTE—
If the special tool in not available, a hard wood block can be used to push the piston back.

– Use coarse file or wire brush to clean brake pad contact surfaces in calipers.

◀ Inspect caliper piston dust boots (**arrows**). Replace caliper with damaged boots.

NOTE—
Brake caliper rebuilding procedures are beyond the scope of this manual. Remanufactured calipers are available from the Porsche dealer parts department, as well as a variety of aftermarket sources.

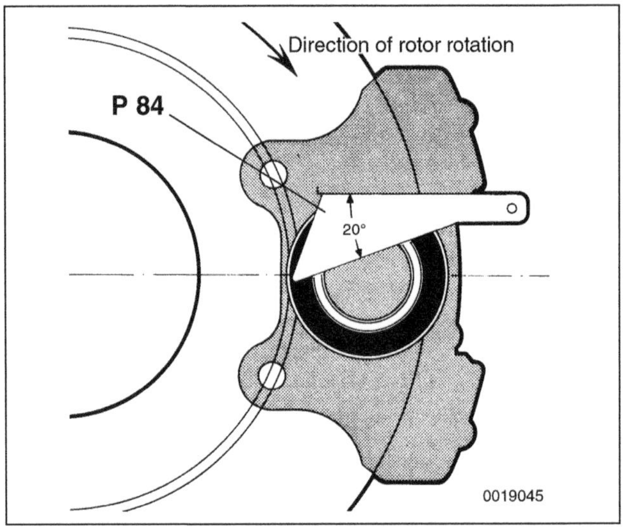

◀ Set caliper piston cutout alignment using Porsche special tool P 84.

– Check brake rotor for wear as described later. Replace or reface if necessary.

– Inspect brake pad sensor for wear or damage. Replace if necessary.

– Install new brake pads in caliper.

NOTE —
- *Apply thin coat of Optimoly® HT or equivalent anti-seize paste to sliding edges of pads.*
- *Install pad with sensor cutout in inside position of caliper.*

– Install brake pad sensor into inside pad.

– Install anti-rattle cross spring and retaining pins.

◀ Route pad sensor harness under anti-rattle cross spring (**arrow**).

– Plug sensor harness into connector on strut body (front) or trailing arm (rear) and reinstall harness retaining clips.

– Remainder of installation is reverse of removal.
- Before driving car after brake work has been performed, depress brake pedal several times to seat pads against brake disc.
- Recheck brake fluid level.

Tightening Torque
- Wheel to wheel hub . . . . 130 Nm (96 ft-lb)

Brake rotor, checking

Check the brake rotors for wear any time the brakes are serviced. Rotors that are scored with sharp ridges, warped, worn irregularly or cracked should be replaced.

Brake rotors should always be resurfaced in pairs, with an equal amount of material removed from both sides of each rotor. **Table a** lists the dimensions for resurfacing rotors. Rotors which fail to meet these requirements should be replaced.

NOTE —
On original equipment rotors, the minimum thickness dimension can be found stamped into the rotor's hub.

Table a. Brake Rotor Reconditioning

	Front and rear
Thickness when new	24.0 mm (0.945 in.)
Minimum thickness after machining	22.6 mm (0.890 in.)
Wear limit (minimum permissible thickness)	22.0 mm (0.866 in.)
Thickness tolerance (difference between any two measurements)	0.02 mm (0.0008 in.)
Maximum lateral runout	0.1 mm (0.004 in.)

Use a micrometer to measure rotor thickness at eight to ten positions on rotor braking surface. Rotor runout should be measured using a dial indicator setup.

If a low speed shimmy goes away when the brakes are released, excessive rotor runout is probably the cause of the shimmy. If the rotor's runout is within limits, and the shimmy still persists, check the front wheel bearings as described in **401 Front Suspension**.

Brake rotors can be resurfaced by most local automotive machine shops.

Front brake rotor, removing and installing

CAUTION—
Always replace or machine brake rotors in pairs.

- Press brake pedal down about 1 in. and hold in place with a pedal stop.

NOTE—
This will prevent brake fluid from dripping out of open brake lines while the work is being performed.

- Raise car and remove front wheels.

WARNING—
Make sure the car is firmly supported on jack stands designed for the purpose. Place the jack stands beneath a structural chassis point. Do not place jack stands under suspension parts.

Working in wheel housing, disconnect electrical connector (**A**) for brake pad wear sensor and remove harness from clip on strut body.

- Disconnect rubber brake hose (**B**) from metal brake line near bottom of strut. Plug hydraulic lines.

CAUTION—
When working with hydraulic brake lines and hoses be sure to counterhold fitting before loosening or tightening.

– Remove caliper mounting bolts and remove caliper. See **470 Brakes–Hydraulic**.

– Remove wheel bearing dust cover.

< Loosen 6 mm allen clamping screw in wheel bearing lock nut and unscrew and remove lock nut from spindle.

– Pull brake rotor with bearing hub and wheel bearings straight off steering spindle.

< Remove 8 mm bolts (**arrows**) mounting brake rotor to hub.

- Tap hub out of rotor with a soft hammer.

< Clean rotor mounting surface on hub with a wire brush.

– Install new rotor on hub and torque bolts.

Tightening Torque

- Brake rotor to bearing hub (M8) . . . . 23 Nm (17 ft-lb)

– Reinstall hub and rotor on spindle and install outer wheel bearing.

- Adjust wheel bearing as described in **401 Front Suspension**.
- Install bearing dust cap.

Tightening Torque
- Wheel bearing lock nut clamp screw (M8) . . . . 15 Nm (11 ft-lb))

– Remainder of assembly is reverse of disassembly.
- Bleed brakes as described in **470 Brakes–Hydraulic**.

Tightening Torques
- Brake caliper bleeder screw
  7 mm. . . . . 3 Nm (27 in-lb)
  Turbo-look. 8 - 11 Nm (6 - 8 ft-lb)
- Brake line to brake hose . . . . 12 Nm (9 ft-lb)
- Caliper to strut housing. . . . . 70 Nm (52 ft-lb)
- Wheel to wheel hub . . . . 130 Nm (96 ft-lb)

Rear brake rotor, removing and installing

WARNING —
Always replace or machine brake rotors in pairs.

– Press brake pedal down about 1 in. and hold in place with a pedal stop.

NOTE —
This will prevent brake fluid from dripping out of open brake lines while the work is being performed.

– Raise car and remove rear wheels.

WARNING —
Make sure the car is firmly supported on jack stands designed for the purpose. Place the jack stands beneath a structural chassis point. Do not place jack stands under suspension parts.

0018887

Working at trailing arm, disconnect electrical connector (**arrow**) for brake pad wear sensor and remove harness from clip on trailing arm.

0018989

◄ Disconnect metal brake line from brake caliper. Plug hydraulic fittings.

– Remove brake caliper mounting bolts and remove caliper. See **470 Brakes–Hydraulic**.

0019047

◄ Remove two screws (**arrows**) mounting brake rotor to hub and remove rotor.

NOTE —

If rotor hangs up and will not slide off, back off parking brake adjuster. See **Parking brake shoes, adjusting**, *later in this section.*

– Clean hub face and install new rotor.

– Remainder of assembly is reverse of disassembly.

- Adjust Parking brake shoes as necessary.
- Bleed brakes as described in **470 Brakes–Hydraulic**.

Tightening torques

- Brake caliper bleeder screw
  - 7 mm . . . . . 3 Nm (27 in-lb)
  - Turbo-look . . . . . 8 - 11 Nm (6 - 8 ft-lb)
- Brake rotor to hub screws . . . . . 5 Nm (3.5 ft-lb))
- Caliper to trailing arm . . . . . 60 Nm (44 ft-lb))
- Hydraulic brake line to caliper . . . . . 14 Nm (10 ft-lb)
- Wheel to wheel hub . . . . . 130 Nm (96 ft-lb)

Parking brake assembly

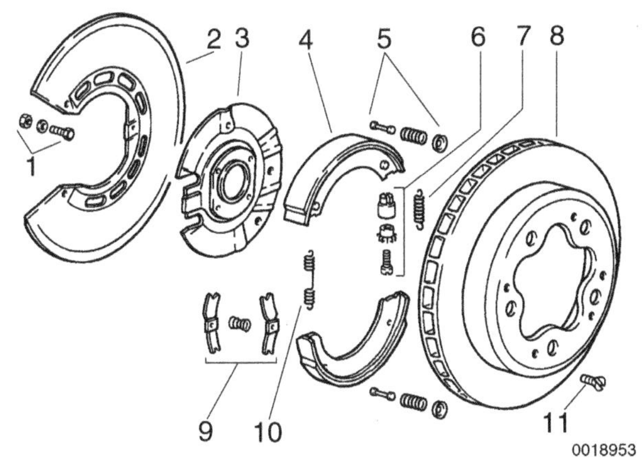

1. Bolt, washer, nut (M8) tighten to 25 Nm (18 ft-lb)
2. Backing plate
3. Brake shoes carrier
4. Brake shoe
5. Shoe hold-down assembly
6. Star-wheel adjuster assembly (Right side shown. Left side is installed upside down from illustration.)
7. Return spring
8. Brake rotor/drum
9. Shoe expander assembly
10. Return spring
11. Brake rotor retaining screw tighten to 5 Nm (3.6 ft-lb)

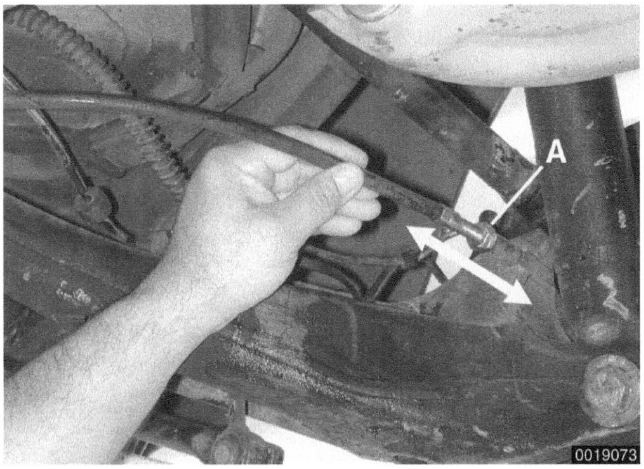

PARKING BRAKE

The parking brake uses a small brake drum inside the rear rotor. The parking brake shoe assembly is fitted in the drum. The mechanical parking brake operates only on the rear wheels and is independent of the main hydraulic brake system.

Each of the two parking brake cables is connected to an expander assembly on each of the brake shoe assemblies. When the parking brake is applied, the brake shoes are forced out against the brake drum.

Parking brake shoes, adjusting

There is no automatic adjusting mechanism for the parking brake. The parking brake shoes need periodic adjustment to compensate for wear. The parking brake should also be adjusted if a cable, brake rotor, or parking brake shoes are replaced. Check to see that the parking brake functions properly whenever the cables have been detached for rear brake or rear suspension service.

– Raise car and remove rear wheels.

WARNING —
Make sure the car is firmly supported on jack stands designed for the purpose. Place the jack stands beneath a structural chassis point. Do not place jack stands under suspension parts.

NOTE —
- *Parking brake should be fully released for this procedure.*
- *Push brake pads away from brake rotor so that rear rotor turns freely.*

Working underneath car, check for axial (in and out) play in brake cable at trailing arm.

- There should be slight movement in brake cable.
- Adjust locking nuts (**A**) as necessary.
- Repeat on other side.

– Turn brake rotor until brake shoe adjuster (star wheel) is visible through adjustment/inspection hole.

0019074

< Use a screwdriver to turn adjuster until brake rotor can no longer be turned by hand.

- Loosen adjuster by 4 clicks and check rotor for free movement.
- Loosen more if necessary.
- Adjust other side.

NOTE —

On both wheels, the screwdriver handle must be pulled backward (toward back of car) to tighten and forward to loosen adjuster. The two star wheel assemblies, though identical, are installed upside down from each other on the two sides of the car.

– With parking brake lever pulled up two teeth, both rear wheels should be difficult to turn.

– Release parking brake lever and check wheels. Both should turn freely.

Parking brake cable, adjusting

Parking brake cable adjustment is necessary if:

- Parking brake lever needs to be pulled up more than 4 clicks (max. 5 clicks) to lock rear wheels.
- One wheel stops before the other when parking brake is applied.

NOTE —

If parking cable was replaced, pull up and release the parking brake handle several times to stretch and set the cable.

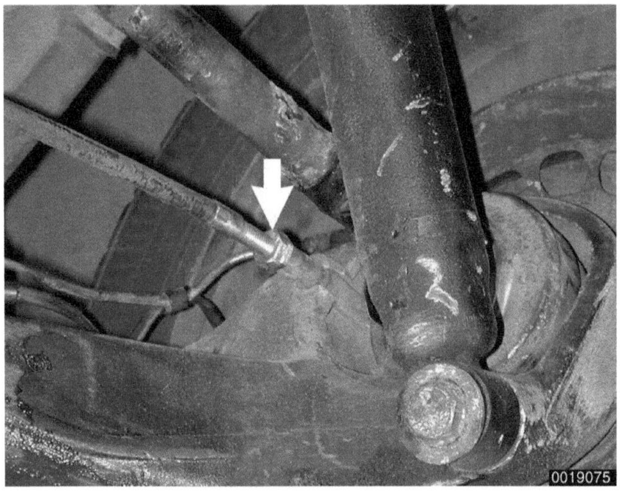
0019075

< Working underneath rear of car and with parking brake fully released, adjust cables at locking nuts (**arrow**) so that equal amounts of thread show on each cable.

- Check to be sure there is slight axial (in and out) play in both cables at the trailing arms.
- Check to be sure both wheels turn freely with parking brake released.

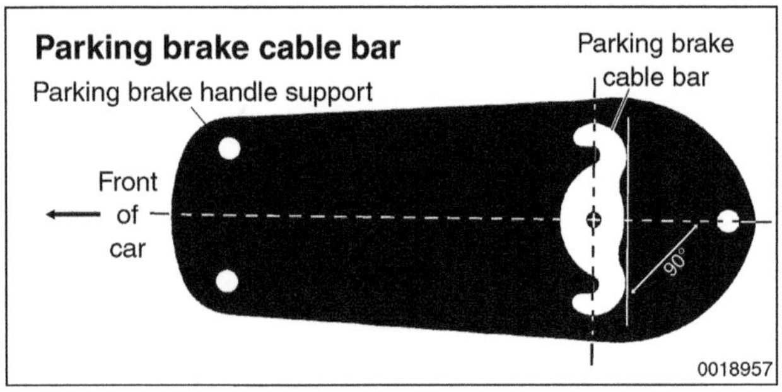

0018957

< Access parking brake handle assembly as described later under **Parking brake cable, replacing**.

- Check to make sure cable bar is exactly transverse (90° angle) to car.
- Adjust cables as necessary.

Parking brake cable, replacing

– Press brake pedal down about 1 in. and hold in place with a pedal stop.

NOTE —
This will prevent brake fluid from dripping out of open brake lines while the work is being performed.

– Raise car and remove rear wheel.

> ***WARNING —***
> *Make sure the car is firmly supported on jack stands designed for the purpose. Place the jack stands beneath a structural chassis point. Do not place jack stands under suspension parts.*

0018887

◄ Working at trailing arm, disconnect electrical connector (**arrow**) for brake pad wear sensor and remove harness from clip on trailing arm.

0018989

◄ Disconnect metal brake line from brake caliper. Plug hydraulic fittings.

– Remove brake caliper mounting bolts and remove caliper. See **470 Brakes–Hydraulic**.

Remove two screws (**arrows**) mounting brake rotor to hub and remove rotor.

NOTE —

If rotor hangs up and will not slide off, back off parking brake adjuster as described earlier.

Remove cotter pin (**arrow**) from castellated nut on end of brake cable and remove nut.

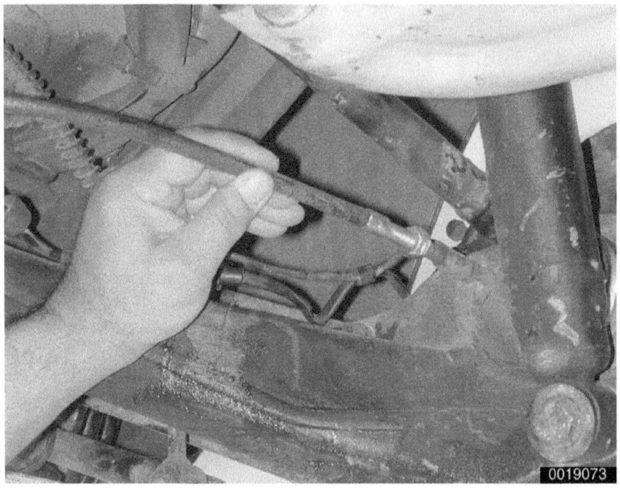

Working underneath car, pull parking brake cable out of brake backing plate and trailing arm.

– Remove passenger seat to gain access to heater controls and parking brake handle assembly.

– Remove access panel from right side of heater control.

◀ On cars with automatic heating, disconnect heater control linkage rod (**arrow**).

– Remove heater control trim screws from floor. Lift heater control and lay aside.

– Pull carpet up around parking brake handle.

◀ Remove bolts (**arrows**) holding parking brake handle support to floor.

Parking brake handle components

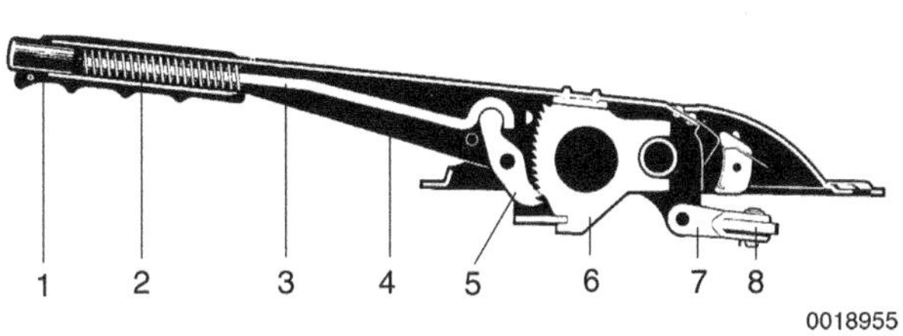

1. Release button
2. Spring
3. Push rod
4. Parking brake handle
5. Pawl
6. Toothed sector
7. Link
8. Cable bar

◀ Lift up parking brake handle mechanism and disconnect link at base of handle by removing clevis pin from cable bar.

– Detach cables from cable bar.

– Pull out old cable and replace with new.

NOTE—
With the old cable removed, blow out cable conduit in body with compressed air.

– Attach new cable eye to cable bar at handle mechanism.
- Reinstall handle mechanism.

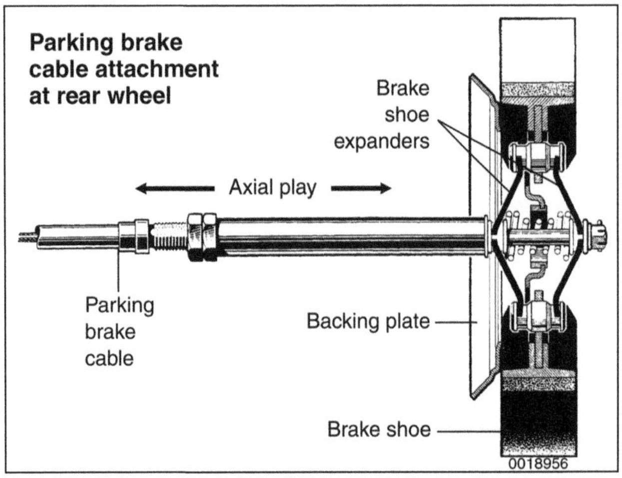

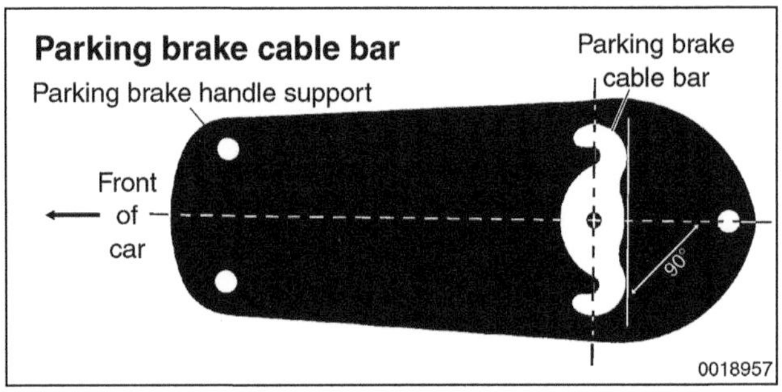

– Push cable through trailing arm and backing plate.
  • Install castellated nut.
  • Use new cotter pin.

◄ Check brake cable housing axial play. If play is found, adjust nuts on cable until slight axial play is felt. Lock adjusting nuts together.

– Reinstall brake rotor and caliper.

Tightening Torques
- Brake caliper to trailing arm (M12) ..... 60 Nm (43 ft-lb)
- Brake rotor to hub (M6)............... 5 Nm (3.5 ft-lb)
- Hydraulic brake line to caliper......... 14 Nm (10 ft-lb)

– Adjust parking brake shoes adjuster as described earlier.

– Pull parking brake handle up and release 4 or 5 times to stretch and set new cable.

◄ Working inside car, check to make sure cable bar below parking brake handle is exactly transverse (90° angle) to car.
  • Adjust cables as necessary.

– Raise parking brake handle 2 clicks. With lever in this position both rear wheels should be difficult to turn.

– Bleed brakes as described in **470 Brakes–Hydraulic**.

– Install rear wheel.

Tightening Torque
- Wheel to wheel hub .... 130 Nm (96 ft-lb)

– Replace rug over parking handle housing. Reinstall heater controls and passenger seat.

Parking brake shoes, replacing

– Press brake pedal down about 1 in. and hold in place with a pedal stop.

NOTE —
This will prevent brake fluid from dripping out of open brake lines while the work is being performed.

– Raise car and remove rear wheel.

WARNING —
Make sure the car is firmly supported on jack stands designed for the purpose. Place the jack stands beneath a structural chassis point. Do not place jack stands under suspension parts.

Working at trailing arm, disconnect electrical connector (**arrow**) for brake pad wear sensor and remove harness from clip on trailing arm.

Disconnect metal brake line from brake caliper. Plug hydraulic fittings.

- Remove brake caliper mounting bolts and remove caliper. See **470 Brakes–Hydraulic**.

Remove two screws (**arrows**) mounting brake rotor to hub and remove rotor.

NOTE —

If rotor hangs up and will not slide off, back off parking brake adjuster as described earlier.

- Remove cotter pin (**arrow**) from castellated nut on end of brake cable. Remove nut.

– Remove shoe expander and spring.

Parking brake assembly

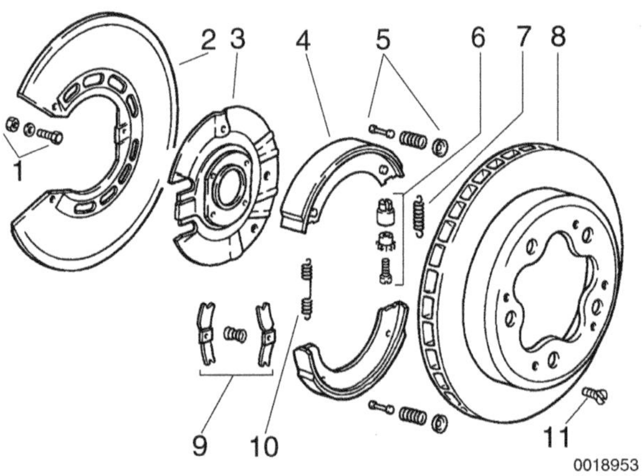

1. Bolt, washer, nut (M8) tighten to 25 Nm (18 ft-lb)
2. Backing plate
3. Brake shoes carrier
4. Brake shoe
5. Shoe hold-down assembly
6. Star-wheel adjuster assembly (Right side shown. Left side is installed upside down from illustration.)
7. Return spring
8. Brake rotor/drum
9. Shoe expander assembly
10. Return spring
11. Brake rotor retaining screw tighten to 5 Nm (3.6 ft-lb)

- Remove brake shoe hold-down springs and cups.

– Remove brake shoe adjuster (star wheel) and rear return spring.

NOTE —

Take note of brake shoe adjuster orientation:
*–The left side adjuster slotted head points **up**.*
*–The right side adjuster slotted head points **down**.*

– Pull shoes forward and remove with front return spring.

– Clean and inspect all parts. Replace as necessary.
  - Replace oily, worn or cracked shoes.
  - Lubricate adjuster to make sure it is functioning properly.

– Installation is reverse of removal.
  - Adjust parking brake as described earlier.
  - Bleed brakes as described in **470 Brakes–Hydraulic**.

Tightening Torques

- Brake caliper to trailing arm (M12) . . . . . 60 Nm (43 ft-lb)
- Brake rotor to hub (M6). . . . . 5 Nm (3.5 ft-lb)
- Hydraulic brake line to caliper. . . . . 14 Nm (10 ft-lb)
- Wheel to wheel hub . . . . 130 Nm (96 ft-lb)

470 Brakes–Hydraulic

GENERAL

This repair group covers brake bleeding as well as replacement procedures for hydraulic brake components, including brake master cylinder, brake power booster and brake calipers.

For general specifications, diagnosis and troubleshooting information see **400 Suspension, Brakes and Steering–General**. Repairs to the brake pads and rotors or to the parking brake system are covered in **460 Brakes–Mechanical**. Pedal cluster repairs are covered in **301 Clutch**.

Special tools

Some procedures require special tools.

Special Tools

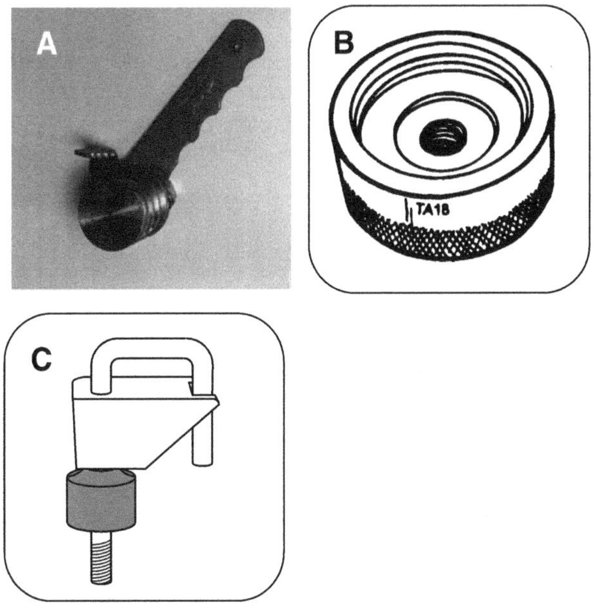

Special tool no.

A . . . . Tubing bender
B TA18. . . . ATE brake master cylinder pressure bleeder adapter
C 647 b . . . . Flexible line clamp tool (source: Baum Tools Unlimited)

> ***WARNING —***
> - *Brake friction materials such as brake linings or pads contain abrasive fibers which can lead to illnesses. Do not create dust by grinding, sanding, or cleaning the pads with compressed air. Avoid breathing any brake friction fibers or dust.*
> - *Brake fluid is poisonous, highly corrosive and dangerous to the environment. Wear safety glasses and rubber gloves when working with brake fluid. Do not siphon brake fluid with your mouth. Immediately clean away any fluid spilled on painted surfaces and wash with water. Dispose of brake fluid properly.*
> - *Do not reuse self-locking nuts, bolts or fasteners. They are designed to be used only once and may fail if reused. Always replace them with new self-locking nuts.*

Braking system

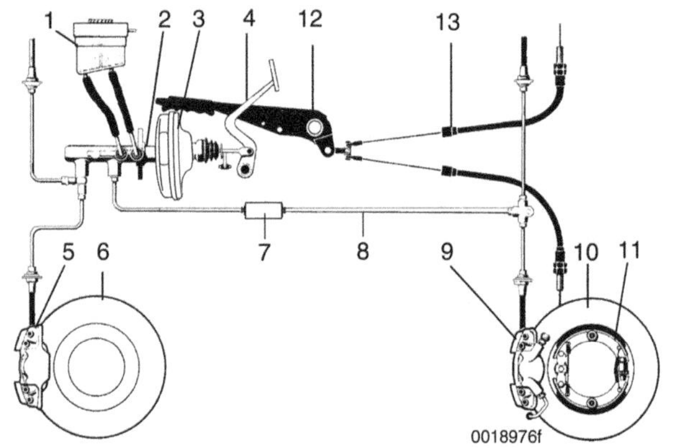

1. Brake fluid reservoir
2. Dual master cylinder
3. Power booster
4. Brake pedal
5. Front brake caliper
6. Front brake rotor
7. Rear brake pressure regulator
8. Brake fluid line
9. Rear brake caliper
10. Rear brake rotor/drum
11. Parking brake shoes
12. Parking brake handle
13. Parking brake cables

Braking system

All Porsche 911 Carrera models are equipped with power-assisted four-wheel disc brakes. A brake booster power-assists a dual-circuit master cylinder. A vacuum boost system is used in all models.

Dual-piston calipers act on vented front rotors and rear rotors. A brake pad wear sensor for each wheel indicates when pads need replacement. The dual-drum parking brake is integrated with the rear brake rotors.

In order to prevent rear wheel lock-up and loss of steering control during hard braking, hydraulic pressure to the rear calipers is reduced by a rear brake pressure regulator (**arrow**) installed in the front-to-rear brake line in the luggage compartment, under a small cover.

Replacing brake fluid

Porsche recommends that the brake fluid be replaced at least once every two years. This is due to the fact that brake fluid readily absorbs moisture. Moisture in the brake fluid can adversely affect braking performance and may also damage the system, leading to costly repairs.

Replace the brake fluid using one of the procedures described later to expel the old fluid. Remove the filter/strainer from the brake fluid reservoir and clean it in new, unused brake fluid. Using new, unused brake fluid, pump at least 1 pint (500 cc) of brake fluid through each caliper to completely flush the system and expel the old fluid. Then refill the reservoir and bleed the brakes as described above. See **030 Maintenance** for brake fluid specifications. Use only Porsche-recommended brake fluid.

NOTE —

The brake fluid reservoir also supplies the hydraulic clutch master cylinder. It is a good idea to also flush the clutch slave cylinder when replacing the brake fluid. See **301 Clutch.**

Flushing brake system

Do not rely on flushing alone to clean a system contaminated with dirt or corrosion. The flushing procedure may actually force dirt in the lines into the calipers. To do the job thoroughly, the system must be disassembled and the parts individually cleaned. Use only brake fluid to flush the lines. Alcohol must not be used since it will encourage the accumulation of water in the system.

When preparing to bleed the brakes, be sure to read the following warnings and cautions.

> ***WARNING —***
> *When working underneath the car, make sure the car is firmly supported on jack stands designed for the purpose. Place the jack stands beneath a structural chassis point. Do not place jack stands under suspension parts.*

> ***CAUTION —***
> - *Brake fluid is highly corrosive and dangerous to the environment. Dispose of it properly.*
> - *When adding or replacing brake fluid always add new brake fluid from an unopened container.*
> - *It is important to bleed the entire brake system whenever any part of the brake hydraulic system has been opened.*

Brake lines

Only straight metal brakes lines are available as replacements. To replace a brake line, first remove the old line from the car and measure its length. Using a new straight line of the same length, bend it to match the pattern of the old line. Each bend radius should be no smaller than about 15 mm (9/16 in.). Check that the protective coating on the brake line is not damaged after bending.

Make sure the new brake line has sufficient clearance from the body and suspension when installed, to allow for suspension bottoming and steering travel.

Brake hoses must be checked for rubber deterioration, cracking or other damage and replaced as necessary.

Tightening Torques
- Brake line to brake component . . 10 - 16 Nm (7 - 11 ft-lb)
- Brake line to brake hose . . . . . . 13 - 16 Nm (10 - 12 ft-lb)

> ***CAUTION —***
> *The use of a tubing bender is highly recommended for shaping replacement brake lines. Bending brake lines by hand can cause kinks that weaken the line and restrict fluid flow. Inexpensive tubing benders are available at most auto parts stores.*

Troubleshooting

Brake performance is mainly affected by three things: the level and condition of the brake fluid, the system's ability to create and maintain hydraulic pressure, and the condition of the friction components.

Air in the brake fluid will make the brake pedal feel spongy during braking or will increase the brake pedal force required to stop. Fluid contaminated by moisture or dirt can corrode the system. Inspect the brake fluid inside the reservoir. If it is dirty or murky, or is over a year old, the fluid should be replaced.

Visually check the hydraulic system starting at the master cylinder. To check the function of the master cylinder hold the brake pedal down hard with the car stopped and the engine running. The pedal should feel solid and stay solid. If the pedal slowly falls to the floor, either the master cylinder is leaking internally, or fluid is escaping from other points in the system. If no leaks can be found, the master cylinder is faulty and should be replaced. Check all brake fluid lines and couplings for leaks, kinks, chafing or corrosion.

Check the brake booster by pumping the brake pedal approximately 10 times with the engine off. Then hold the pedal down and start the engine. The pedal should fall slightly. If not, check for any visible faults before suspecting a faulty brake booster, and check the one-way check-valve for air flow.

NOTE —
- *See* **400 Suspension, Brakes and Steering–General** *for a brake troubleshooting table.*
- *See* **470 Brakes–Hydraulic** *for hydraulic brake components or brake power-booster repair information.*

Maintenance

– Porsche specifies the maintenance steps below to be carried out at particular time or mileage intervals for proper maintenance of the brakes.
  - Check brake fluid. See **030 Maintenance**.
  - Inspect brake hoses, lines and wheel calipers for leaks.
  - Check brake pads and shoes for wear.
  - Replace brake fluid (described in this section).

BLEEDING BRAKES

Brake bleeding is usually done for one of two reasons; either to replace old brake fluid as part of routine maintenance or to expel trapped air in the system that resulted from opening the brake hydraulic system during repairs.

Brake system bleeding can be done with a pressure bleeder, a vacuum bleeder or manually. Pressure bleeding is recommended as the most efficient method.

When bleeding the brakes, start at the wheel farthest from the master cylinder and progress in the following order:

- Right rear brake
- Left rear brake
- Right front brake
- Left front brake.

Pressure bleeding brakes

– Using clean syringe, remove old brake fluid from reservoir. Clean out reservoir with a lint-free cloth.

< Using special tool 647 b or equivalent, clamp off vent line from master cylinder reservoir.

< Top off brake fluid and connect pressure bleeder (**arrow**) to reservoir.

– Pressurize system to approximately 1 bar (14.5 psi).

> ***CAUTION —***
> *Do not exceed a pressure of 2 bar (29 psi) when pressure bleeding the brake system. Excessive pressure will damage the brake fluid reservoir.*

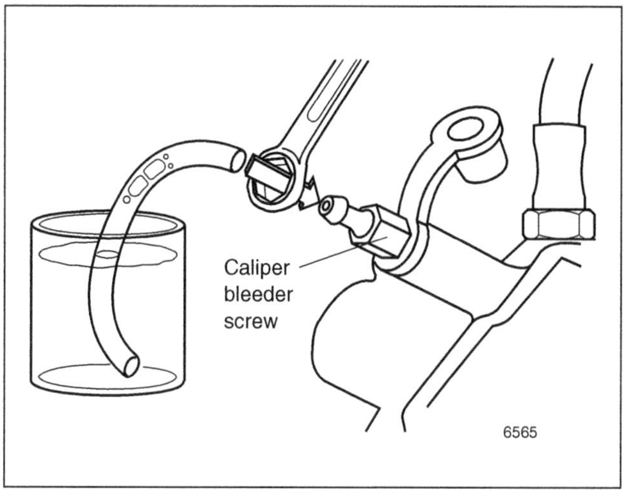

< Connect bleeder hose and waste bottle to right rear caliper bleeder screw.

– Have helper hold brake pedal down.

◀ Open bleeder screw (**arrow**). Have helper slowly pump brakes *at least 12 times* with bleeder screw open, holding pedal down on last pump.

- Close bleeder screw and release brake pedal.
- When finished, replace rubber bleeder cap.

Tightening Torques

- Bleeder screw (7 mm). . . . . 3 Nm (27 in-lb)
- Bleeder screw (Turbo-look). . . . . 8 -11 Nm (6 - 8 ft-lb)

– Refill brake fluid reservoir and proceed to next wheel in order given above.

Vacuum bleeding brakes

– Using clean syringe, remove old brake fluid from reservoir. Clean out reservoir with a lint-free cloth.

– Top off brake fluid and replace reservoir cap.

– Connect vacuum pump to right rear brake bleeder screw. Route discharge hose into waste bottle.

> ***CAUTION —***
> *Vacuum bleeding requires an air-tight seal between the pump and the bleeder screw. Any leakage will result in faulty bleeding.*

– Open bleeder screw and withdraw brake fluid until it is clean and without air bubbles.

NOTE —
Fill the brake fluid reservoir level frequently to prevent it from emptying during brake bleeding.

– Close bleeder screw.

- Replace rubber bleeder cap.

Tightening Torques

- Bleeder screw (7 mm). . . . . 3 Nm (27 in-lb)
- Bleeder screw (Turbo-look). . . . . 8 -11 Nm (6 - 8 ft-lb)

– Refill brake fluid reservoir and proceed to next wheel in order given above.

Manually bleeding brakes

– Using clean syringe, remove old brake fluid from reservoir. Clean out reservoir with a lint-free cloth. Top off brake fluid and replace reservoir cap.

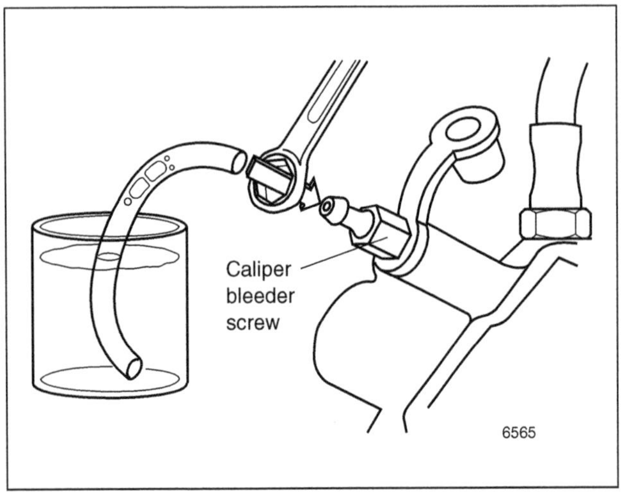

◀ Connect bleeder hose and waste bottle to right rear bleeder screw.

– Have helper pump brakes several times. Then hold pedal down firmly on last pump.

– Open right rear caliper bleeder screw and allow fluid to expel. Close bleeder screw.

– Repeat previous two steps at least 12 times, or until discharged brake fluid is clean and without air bubbles.

NOTE —
Fill the brake fluid reservoir level frequently to prevent it from emptying during brake bleeding.

– Close bleeder screw.

- Replace rubber bleeder cap.

Tightening Torques
- Bleeder screw (7 mm). . . . . 3 Nm (27 in-lb)
- Bleeder screw (Turbo-look). . . . . 8 -11 Nm (6 - 8 ft-lb)

– Refill brake fluid reservoir and proceed to next wheel in order given above.

BRAKE MASTER CYLINDER AND POWER BOOSTER

The master cylinder is mounted to the rear of the brake booster. The master cylinder, power booster and brake fluid reservoir are mounted in the luggage compartment.

The brake pedal operates the master cylinder via a pedal pushrod that passes through the firewall and operates a reversing lever. The reversing lever in turn operates the master cylinder/brake booster push rod. The master cylinder piston acts on the brake fluid to create pressure in the system.

The master cylinder and brake booster can be removed and installed with ordinary hand tools. Be sure to have an adequate supply of new, unopened brake fluid on hand, as bleeding the brakes, described earlier, will be necessary after installation.

Although master cylinder parts and rebuilding kits are available for most models, replacing the master cylinder as a complete unit is preferable. Whether or not it can be successfully rebuilt depends on its internal condition, which can only be determined after it is disassembled.

The brake master cylinder and power booster are removed and installed as one unit. Once removed from the car and placed on a work bench, the two components can be separated and replaced as necessary.

Brake master cylinder/power booster, removing and installing

– Remove driver's side floor mats and pedal cluster cover.

Working above and forward of pedal cluster, remove locking clip (**arrow**) from brake pedal push rod and pull out pin.

Working above pedal cluster, remove master cylinder mounting bolt (**arrow**) above driver's knees.

Working in luggage compartment, disconnect power booster vacuum hose at connection in luggage compartment floor (**arrow**).

Detach and remove air duct hoses (**arrows**) behind master cylinder/power booster assembly.

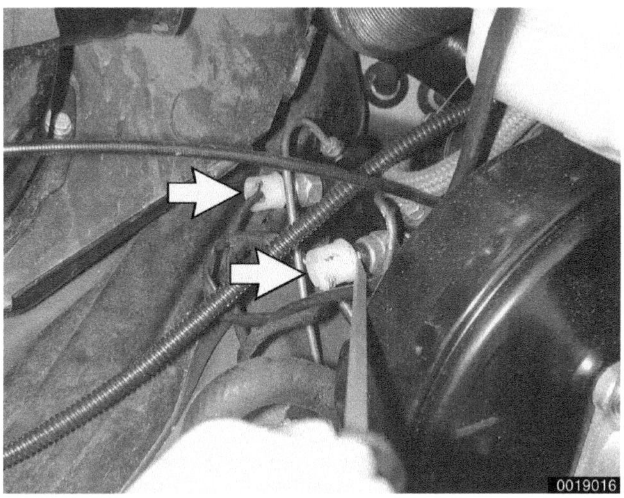

Disconnect both stop light switch plugs (**arrows**) at master cylinder.

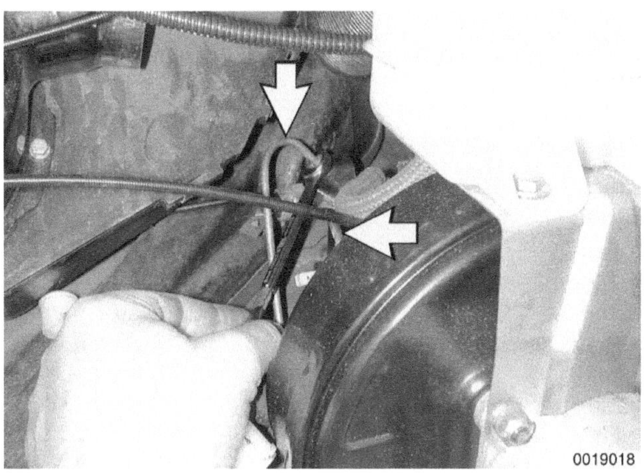

Loosen and remove brake lines (**arrows**) from master cylinder.

– Drain master cylinder reservoir using a syringe.
  - Detach master cylinder feed lines.
  - Remove fluid reservoir and set aside.

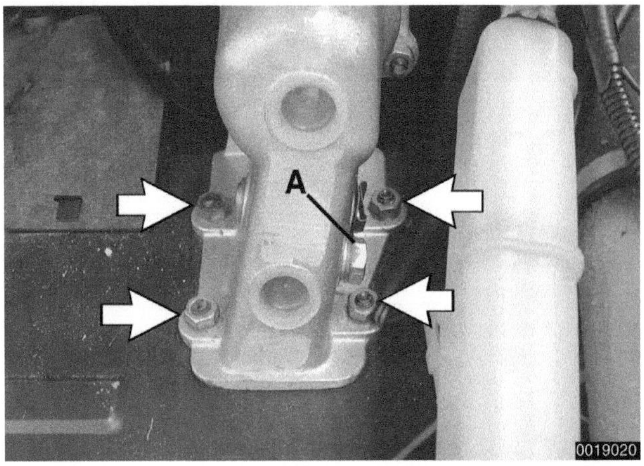

◀ Remove pedal cluster brace bolt (**A**) from side of power booster mounting bracket.

– Remove power booster bracket mounting nuts (**arrows**). Remove master cylinder and power booster as a unit.

◀ With booster and master cylinder on work bench, remove master cylinder mounting nuts (**arrows**) and pull cylinder out of power booster.

– Installation is reverse of removal.

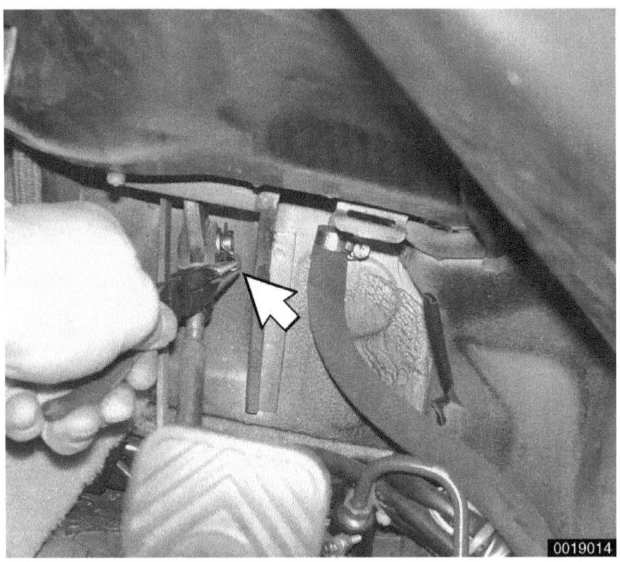

◀ When attaching brake pedal push rod to reversing lever, be sure to insert pivot pin so that locking clip (**arrow**) can be installed from the right side.

– Be sure to bleed complete brake system as described earlier.

– On 1987 or later models: Bleed clutch slave cylinder.

Tightening torques

- Bleeder screw
  7 mm . . . . 3 Nm (27 in-lb)
  Turbo-look . . . . 8 -11 Nm (6 - 8 ft-lb)
- Brake line to master cylinder . . . . 14 Nm (10 ft-lb)
- Brake light switch to master cylinder . . . . 15 Nm (11 ft-lb)
- Master cylinder to booster . . . . 25 Nm (18 ft-lb)
- Power booster to
  luggage compartment floor . . . . 25 Nm (18 ft-lb)
- Pedal cluster brace to power booster . . . 46 Nm (33 ft-lb)
- Wheel to wheel hub . . . . 130 Nm (94 ft-lb)

BRAKE CALIPERS

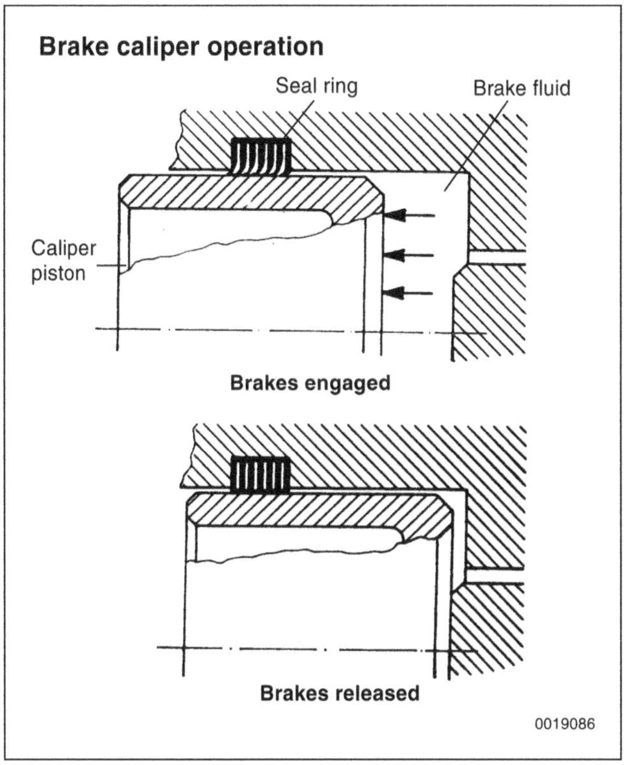

Brake calipers are self-adjusting units. When brake fluid pressure is applied to the caliper piston, the seal ring in the caliper wall deforms slightly, allowing the piston to move out and engage the brakes. When brake fluid pressure is reduced, the seal ring returns to its original shape, pulling the piston back and disengaging the brakes.

As brake pads or rotors wear, the deformation of the seal ring reaches its limit and the piston actually slides out further before regaining a stable position again.

Brake fluid seeping or leaking from around the caliper piston is the result of a failed or damaged piston seal ring. A damaged seal ring is usually caused by corrosion, scoring, or pitting of the piston or caliper bore. The seal can be replaced separately, but a damaged piston will quickly damage the new seal. To remedy a leaking caliper piston seal and avoid future problems, complete replacement of the caliper is recommended.

Brake caliper, removing and installing

This procedure is applicable to both front and rear brake calipers.

NOTE—
It is recommended brake calipers be replaced in pairs.

- Raise car and remove wheels.

WARNING—
Make sure the car is firmly supported on jack stands designed for the purpose. Place the jack stands beneath a structural chassis point. Do not place jack stands under suspension parts.

- Depress brake pedal approximately 1 inch and hold in place with a pedal depressor.

NOTE—
This will prevent brake fluid from dripping out of open brake lines while the work is being performed.

Front brakes: Working in wheel housing, disconnect electrical connector (**A**) for brake pad wear sensor and remove harness from clip on strut body.

- Disconnect rubber brake hose (**B**) from metal brake line near bottom of strut. Plug hydraulic lines.

> ***CAUTION —***
> *When working with hydraulic brake lines and hoses, be sure to counterhold fitting before loosening or tightening.*

◄ Rear brakes: Working at trailing arm, disconnect electrical connector (**arrow**) for brake pad wear sensor and remove harness from clip on trailing arm.

◄ Disconnect metal brake line from brake caliper. Plug hydraulic fittings.

◄ Remove caliper mounting bolts (**arrows**) and remove caliper. (Front caliper shown. Rear is similar).

– Inspect brake rotor for heat damage due to caliper pistons hanging up.

– Installation is reverse of removal.

– Bleed brakes as described earlier.

Tightening Torques

- Bleeder screw
  - 7 mm . . . . 3 Nm (27 in-lb)
  - Turbo-look . . . . 8 -11 Nm (6 - 8 ft-lb)
- Brake line to caliper . . . . 14 Nm (10ft-lb)
- Front caliper to strut . . . . 70 Nm (52 ft-lb
- Rear caliper to arm . . . . 60 Nm (43 ft-lb)
- Wheel to wheel hub . . . . 130 Nm (94 ft-lb)

480 Steering

TABLE

GENERAL

The Porsche 911 Carrera is equipped with manual rack and pinion steering.

This repair group covers servicing of the steering system, including steering rack replacement, rack boot and tie rod replacement and repairs to the steering column universal joints.

Troubleshooting information is in **400 Suspension, Brakes and Steering**. For information on the suspension or alignment, see the following repair groups:

- **401 Front Suspension**
- **420 Rear Suspension**
- **440 Wheels, Tires, Alignment**

Special tools

Some procedures require special tools.

Special Tool

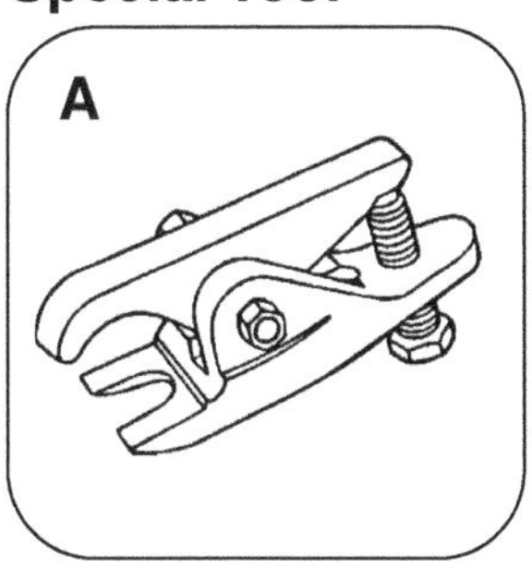

Special tool

A VW 267 a . . . . . Tie rod removal tool

> ***WARNING —***
> - *Do not reuse self locking nuts. They are designed to be used only once and may fail if reused. Always replace them with new locking nuts.*
> - *Do not install bolts and nuts coated with undercoating wax, as correct tightening torque cannot be assured. Always clean the threads with solvent before installation, or install new parts.*
> - *Do not attempt to weld or straighten any steering components. Always replace damaged parts.*

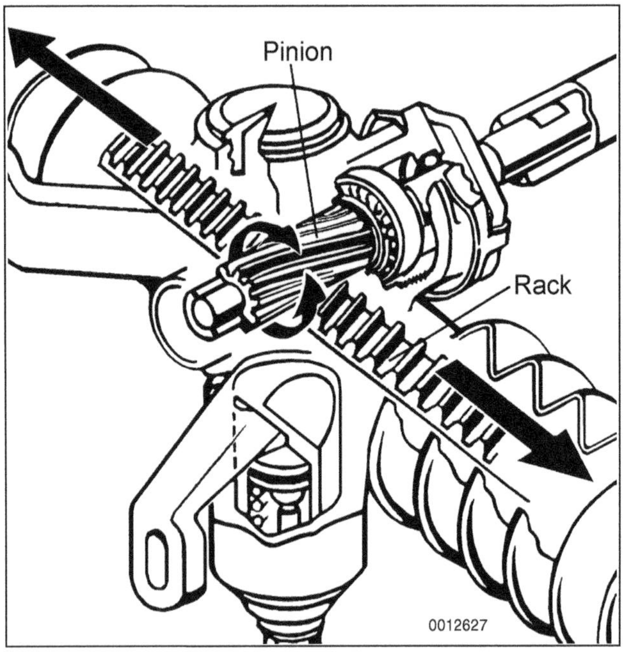

Steering system

The steering wheel and column are connected to the steering rack by a short shaft and two universal joints. Turning the steering wheel turns the rack pinion gear which in turn moves the rack and steering linkage.

Wear and excessive play or clearance anywhere in the steering system will cause sloppy, loose steering. On cars with high mileage, tie rod ends are prone to wear.

If the steering requires too much effort or lacks precise feel, the steering mechanism may not be to blame. Worn suspension parts may also cause or contribute to the symptoms. Before checking for any steering problems, check tires, tire inflation pressures front suspension components.

There are no provisions for lubrication of the steering rack or linkage. In general, the steering system is serviced mainly by the replacement of worn parts.

Table a. Steering Specifications

Characteristic	Specification
Frictional moment of steering (measured at flange of steering column flange, tie rods detached)	80 - 140 Ncm (113 - 198 in.-oz)
Steering ratio	17.78 : 1
Steering wheel turns lock-to-lock	Approx. 3
Turning circle diameter	10.95 m (35.93 ft.)

STEERING WHEEL AND COLUMN

NOTE —
Removal and installation of the steering column switches and the ignition switch is covered in **960 Electrical Switches, Interior Lighting**.

Steering wheel, removing and installing

The steering wheel must be removed for access to the steering column switches. The steering wheel is held in place by a 27 mm nut.

– Disconnect negative (–) battery cable.

> ***CAUTION —***
> *Prior to disconnecting the battery, read the battery disconnection cautions given at the front of this manual on page vii.*

Pull horn pad straight back to release locking clips.

NOTE —
There are three locking clips, one at top center of pad and one at each lower corner.

◄ Disconnect horn wire from horn pad.

– Make sure front wheels are pointed straight ahead. Center steering wheel.

– Remove steering wheel center nut and washer.

– Lightly punch mark or scribe position of steering wheel to steering column shaft.

– Pull steering wheel straight off steering shaft.

– Installation is reverse of removal.

- Align horn pad with steering wheel locking clips. Then press horn pad down on steering wheel until clips snap into place.
- Check operation of horn and steering wheel.

Tightening torques

- Steering wheel to column . . . . 50 Nm (36 ft-lb)

Steering wheel, centering

If the steering rack was disturbed while the steering wheel was removed, or if a new unmarked steering wheel is being installed, the steering rack will have to be centered prior to installing the wheel in a straight ahead position.

To center the steering rack, temporarily position the steering wheel on the column and turn until the road wheels point straight ahead. Reposition steering wheel and install lock washer and nut. Back car up a short distance and drive forward a short distance to see if the car is moving in a straight line with the steering wheel centered. Torque steering wheel nut, install horn pad. Road test to check steering wheel position.

Steering column universal joints, removing and installing

Access to the steering column universal joints is through the floor of the luggage compartment. The fresh air plenum and blower must first be removed.

– Disconnect negative (–) battery cable.

> ***CAUTION —***
> *Prior to disconnecting the battery, read the battery disconnection cautions given at the front of this manual on page vii.*

– Unsnap luggage compartment floor carpet and lift off.

Remove fresh air blower cover retaining fasteners (**arrows**). Remove cover.

Remove screws (**arrows**) holding fresh air vent trim in cowl.

– Remove vent trim and screen.

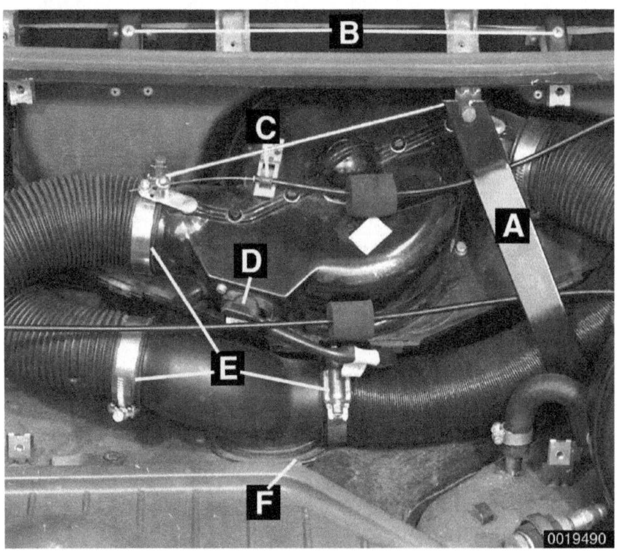

Working inside luggage compartment, detach fresh air plenum and blower connections:

- Remove diagonal brace (**A**).
- Remove screws (**B**) holding fresh air blower housing in place.
- Disconnect heater/defroster cable (**C**) from housing.
- Disconnect fresh air hoses (**D**) from housing.
- Disconnect electrical harness connector (**E**).
- Remove retaining screw (**F**) and pull off large plastic air hose T-fitting.
- Remove blower.

< Remove drain hose (**arrow**) at bottom of fresh air plenum.

– Remove fresh air plenum.

NOTE —

If plenum sticks to cowl, pry carefully to avoid damage to plenum housing.

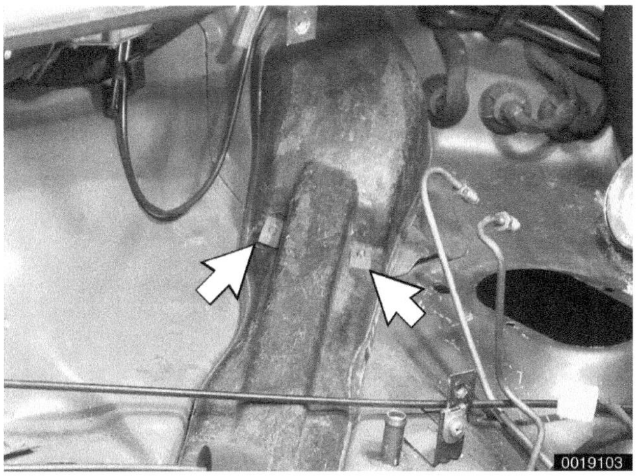

< Pry off speed nuts (**arrows**) used to hold cover in place over steering shaft. (Brake master cylinder/power booster unit shown removed for clarity.)

- Remove cover.

< Open access door in floor behind fuel tank.

Remove locking nut and bolt (**arrow**) from upper universal joint.

Remove locking nut and bolt (**arrow**) from lower universal joint.

– Remove steering shaft with both universal joints.

NOTE —

*It may be necessary to remove allen bolts (**A**) from steering rack shaft support and to remove bracket.*

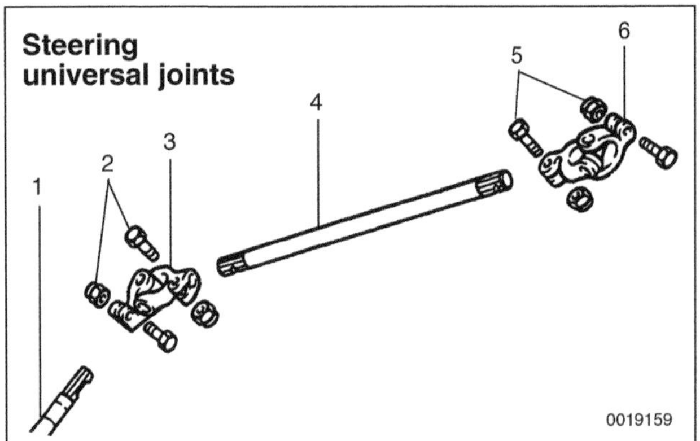

1. Steering rack shaft
2. Universal joint pinch bolt/nut tighten to 20 Nm (15 ft-lb)
3. Lower universal joint
4. Steering shaft
5. Universal joint pinch bolt/nut tighten to 20 Nm (15 ft-lb)
6. Upper universal joint

Inspect steering universal joints. Disassemble universal from shaft and replace as necessary.

NOTE —

Always replace universal joint lock nuts.

– Installation is reverse of removal.

Tightening Torque

- Steering rack shaft support to body (6 mm allen) . . . . 25 Nm (18 ft-lb)
- Steering universal pinch bolt (M8) . . . . . 20 Nm (15 ft-lb)

Steering column, removing and installing

– Disconnect negative (–) battery cable.

> ***CAUTION —***
> *Prior to disconnecting the battery, read the battery disconnection cautions given at the front of this manual on page vii.*

– Working inside luggage compartment, detach upper steering column universal as described earlier.

– Working in passenger compartment, remove steering wheel and steering column switches as described earlier.

0019134

Working under dashboard, remove knee trim panel screw (**arrow**) located above luggage compartment lid release handle.

0019133

Remove left outer A/C vent from dash.

- Remove screw (**A**) from trim panel on left side of dash.
- While pulling on A/C vent, release locking tabs (**arrows**) at top and bottom of vent with a small screwdriver.

0019108

Remove knee trim:

- Working through A/C vent opening, remove trim mounting nut and washer from left stud (**A**).
- Working just below heater controls in center of dashboard, remove trim mounting nut and washer from right stud (**B**).

– Use a small screwdriver to carefully pry tachometer out of dashboard.

  • Disconnect wiring from back of tachometer while removing.

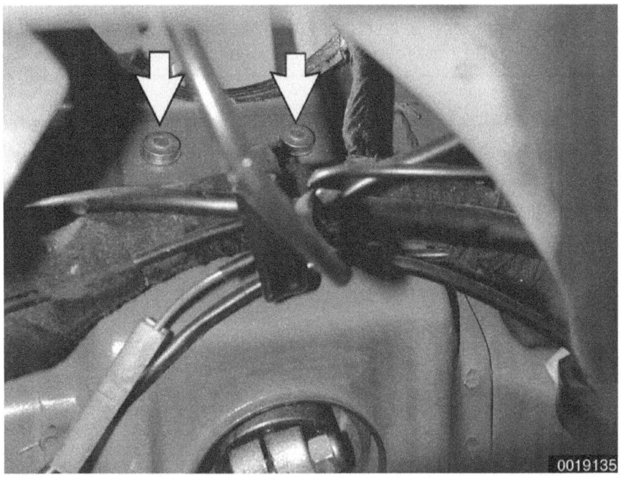

< Use a 1/8 in. bit in an angle drill to drill holes in each steering column shearbolt located behind tachometer.

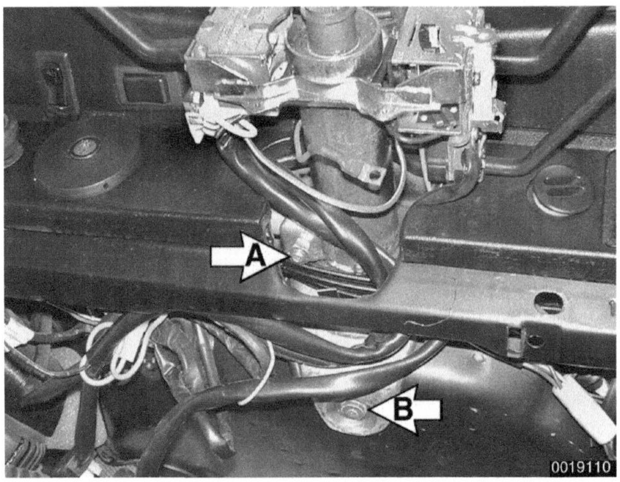

< Loosen lock nut and allen bolt (**A**) holding ignition lock arm in column support.

– Drill 1/8 in. hole in shearbolt (**B**) located above driver's knees.

– Use an easy-out to remove all 3 shearbolts.

– Remove column by sliding it to right to release steering lock.

– Installation is reverse of removal.

NOTE —
Always use shearbolts to attach steering lock to steering column.

Tightening torques

• Steering shaft universal joint pinch bolt	20 Nm (15 ft-lb)
• Steering wheel to column	50 Nm (36 ft-lb)

Ignition lock, removing and installing

– Disconnect negative (–) battery cable.

> ***CAUTION —***
> *Prior to disconnecting the battery, read the battery disconnection cautions given at the front of this manual on page vii.*

◄ Working under dashboard, remove knee trim panel screw (**arrow**) located above luggage compartment lid release handle.

◄ Remove knee trim:

- Working from below, remove trim mounting nut and washer from left stud (**A**).
- Working just below heater controls in center of dashboard, remove trim mounting nut and washer from right stud (**B**).

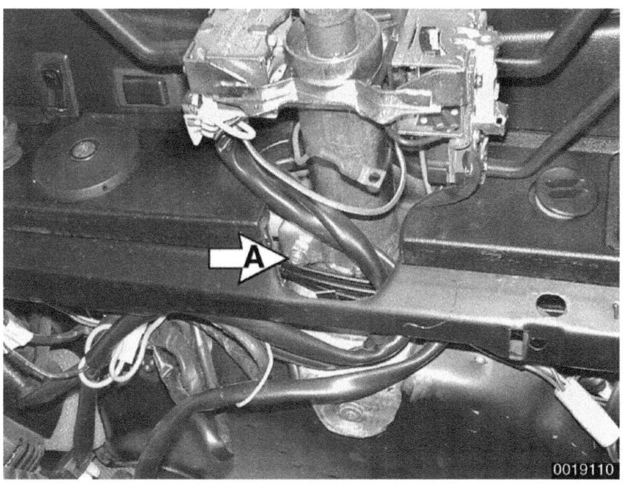

◄ Loosen lock nut and allen bolt (**A**) holding ignition lock arm in column support.

- Working from rear of ignition lock, disconnect harness connector from ignition switch.
- Remove ignition switch rubber trim.
- ◄ Drill out shear bolts (**arrows**) mounting lock to dashboard. Use a screw extractor to remove shear bolts from ignition lock.
- Push ignition lock through dash and remove from behind.
- Installation is reverse of removal. Use new shear bolts and tighten until bolt heads shear off.

STEERING RACK

◄ The illustration shows the basic components of a rack-and-pinion steering gear.

The steering gear and linkage require no maintenance other than alignment and periodic inspection for worn components.

Steering rack assembly

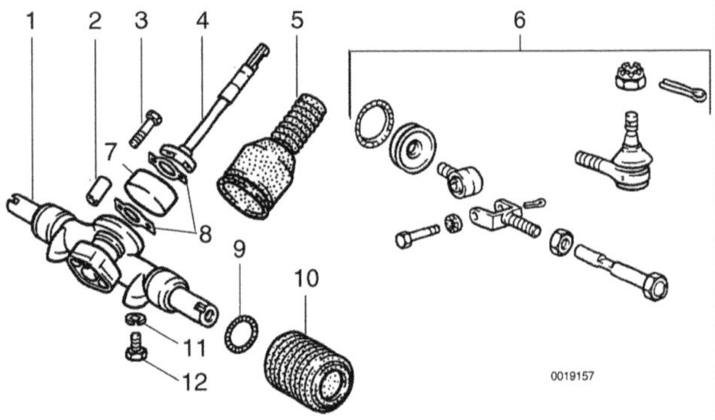

1. Steering rack
2. Spacer bushing
3. Bolt (M8)
   tighten to 25 Nm (18 ft-lb)
4. Steering shaft
5. Bellows
6. Tie rod assembly
7. Steering shaft coupling
8. Lock plate
9. Spring clamp
10. Rack boot
11. Washer (M10)
12. Bolt (M10)
    tighten to 47 Nm (35 ft-lb)

Steering rack, removing and installing

- Raise car and remove front wheels.

> ***WARNING —***
> *Make sure the car is firmly supported on jack stands designed for the purpose. Place the jack stands beneath a structural chassis point. Do not place jack stands under suspension parts.*

- Disconnect negative (–) battery cable.

> ***CAUTION —***
> *Prior to disconnecting the battery, read the battery disconnection cautions given at the front of this manual on page vii.*

◄ Working underneath front of car, remove stone guard mounting fasteners (**arrow**). Remove stone guard.

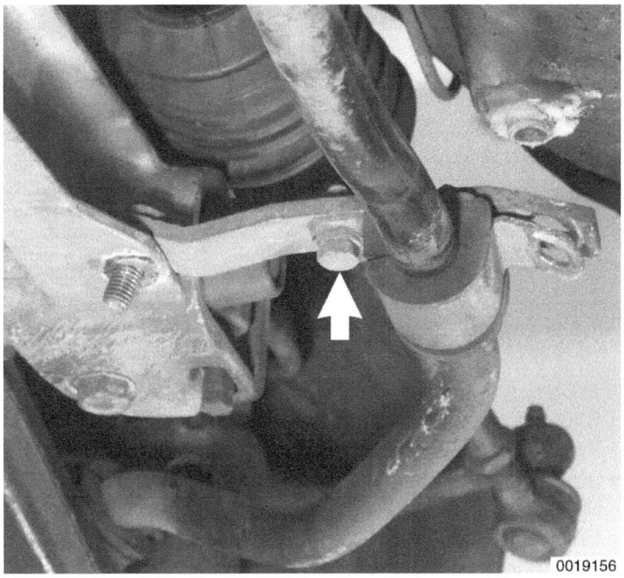

Remove front stabilizer bar mounting clamp bolts (**arrow**). (Right side shown.)

- Remove stabilizer bar.

NOTE —
Remove one end of the stabilizer bar at a time by tapping with a rubber hammer.

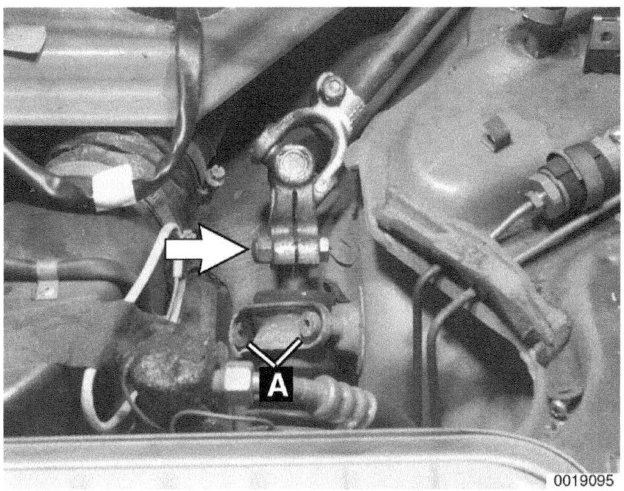

Access steering shaft universal joints as described earlier.

- Remove locking nut and bolt (**arrow**) connecting lower universal joint to steering rack.
- Pull universal off steering rack.
- Remove allen bolts (**A**) from steering rack shaft support. Remove bracket.

– Remove cotter pins and castle nuts from both outer tie rod ends.

Use special tool VW 267 a or equivalent to release tie rod from steering arm. Repeat for other side.

– Remove horn and mounting bracket from front suspension subframe.

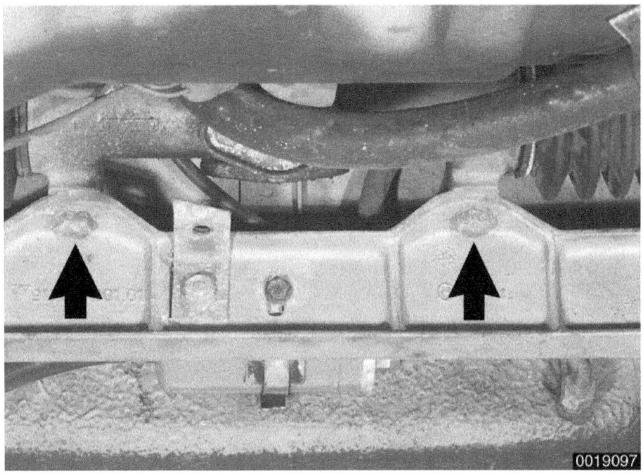

Remove steering rack mounting bolts (**arrows**).

- Remove steering rack, being careful not to damage brake lines and A/C lines.
- Installation is reverse of removal.

NOTE —
Always replace universal joint lock nuts.

Tightening Torques

- Steering rack shaft support to body (6 mm allen) . . . . 25 Nm (18 ft-lb)
- Steering rack to suspension subframe . . 47 Nm (35 ft-lb)
- Steering universal pinch bolt. . . . . 20 Nm (15 ft-lb)
- Stone guard to body
  - M10. . . . . 45 Nm (33 ft-lb)
  - M8. . . . . 25 Nm (18 ft-lb)
- Tie rod to steering arm . . . . 45 Nm (33 ft-lb)
- Wheel to wheel hub . . . . 130 Nm (96 ft-lb)

Steering rack boot, removing and installing

- Raise car and remove front wheel on side to be repaired.

WARNING —
Make sure the car is firmly supported on jack stands designed for the purpose. Place the jack stands beneath a structural chassis point. Do not place jack stands under suspension parts.

- Remove cotter pin and castle nut from outer tie rod end.

Use special tool VW 267 a or equivalent to release tie rod from steering arm.

Remove spring clamps by sliding over end of boot.

◀ Slide boot off over tie rod.

– Installation is reverse of removal.

Tightening torques
- Tie rod to steering arm . . . . 45 Nm (33 ft-lb)
- Wheel to wheel hub . . . . 130 Nm (94 ft-lb)

TIE RODS

◀ The Porsche 911 Carrera tie rods have two replaceable ends. The outer ball joint end connects to the steering arm with a castle nut. The inner end is a metal yoke that attaches to a rubber bushing on the rack end.

It is possible to replace the outer tie rod end, but not the inner, without removing the tie rod from the car

Tie rod assembly

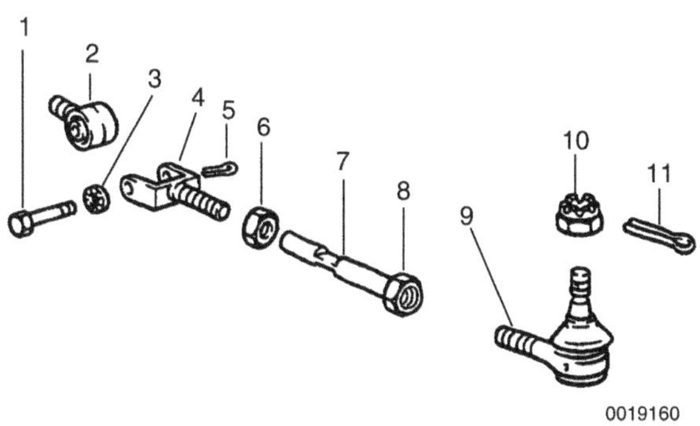

1. Bolt (M10) tighten to 47 Nm (34 ft-lb)
2. Rack end bushing
3. Lock washer (M10)
4. Inner tie rod end yoke
5. Cotter pin
6. Inner tie rod adjuster lock nut (left-hand thread) tighten to 45 Nm (33 ft-lb)
7. Tie rod shaft
8. Outer tie rod adjuster lock nut (right-hand thread) tighten to 45 Nm (33 ft-lb)
9. Outer tie rod
10. Castle nut tighten to 45 Nm (33 ft-lb)
11. Cotter pin

Tie rod, removing and installing

NOTE —

To replace a worn tie rod end, first measure and record the overall length of the tie rod. This will ensure that the alignment will be disturbed as little as possible because of removal and installation.

– Raise car and remove front wheel on side to be repaired.

> ***WARNING —***
>
> *Make sure the car is firmly supported on jack stands designed for the purpose. Place the jack stands beneath a structural chassis point. Do not place jack stands under suspension parts.*

◀ Loosen inner and outer tie rod adjuster lock nuts.

NOTE —

The inner tie rod adjuster threads are left-handed.

– Remove cotter pin and castle nut from outer tie rod end.

◀ Use special tool VW 267 a or equivalent to release tie rod from steering arm.

◀ Remove cotter pin from inner tie rod end bolt. Remove bolt.

– Either tie rod end can now be replaced by clamping tie rod shaft in shop vise.

NOTE —

Clean all threads and coat with anti-seize paste.

– Installation is reverse of removal.

- Reset tie rod total length to measurement obtained earlier.
- Have car aligned professionally.

Tightening Torques

- Inner tie rod to rack bushing . . . . 47 Nm (34 ft-lb)
- Tie rod adjuster nut to tie rod shaft (M14). . . . . 45 Nm (33 ft-lb)
- Tie rod to steering arm . . . . 45 Nm (33 ft-lb)
- Wheel to wheel hub . . . . 130 Nm (94 ft-lb)

Outer tie rod end, removing and installing

NOTE —
- *Measure and record the overall length of the tie rod. This will ensure that the alignment will be disturbed as little as possible because of removal and installation.*
- *Another technique for maintaining the basic alignment setting is to count the number of turns as the tie rod end is unscrewed and reinstall the new end with exactly the same number of turns.*

– Raise car and remove front wheel on side to be repaired.

> ***WARNING —***
> *Make sure the car is firmly supported on jack stands designed for the purpose. Place the jack stands beneath a structural chassis point. Do not place jack stands under suspension parts.*

◄ Loosen outer tie rod lock nut (B).

– Remove cotter pin and castle nut from outer tie rod end.

◄ Use special tool VW 267 a or equivalent to release tie rod from steering arm.

– Unscrew old tie rod end and replace with new.

NOTE —
Clean all threads and coat with anti-seize paste.

– Reassembly is reverse of removal.
- If necessary, loosen inner tie rod adjuster nut and reset tie rod length to measurement obtained earlier.
- Have car aligned professionally.

Tightening Torques
- Tie rod adjuster nut to tie rod shaft (M14) . . . . 45 Nm (33 ft-lb)
- Tie rod to steering arm . . . . 45 Nm (33 ft-lb)
- Wheel to wheel hub . . . . 130 Nm (94 ft-lb)

500 Body Assembly–General

GENERAL

The 1984 - 1989 911 Carrera's all-steel and galvanized bodyshell was available in three body styles; Coupe, Targa and Cabriolet.

Targa and Cabriolet models

Both the Targa and Cabriolet models are all-season vehicles with a high level of functionality, excellent interior comfort, and acoustic refinement.

Early versions of the Cabriolet model used a manually operated top, with the fully automatic convertible top being introduced in the beginning of the 1986 model year.

All pillars and frame members are of thin wall sheet metal and have been deep-drawn for maximum strength. This allows for control cables, linkages and heater lines to be accommodated. The outer skin extending from the windshield pillar to the tail end is welded to the floor assembly, the wheel arches, and pillars to form a rigid unit.

Front fenders

The front fenders are attached with bolts. The fuel filler flap is located in the left front fender.

Body dimensions

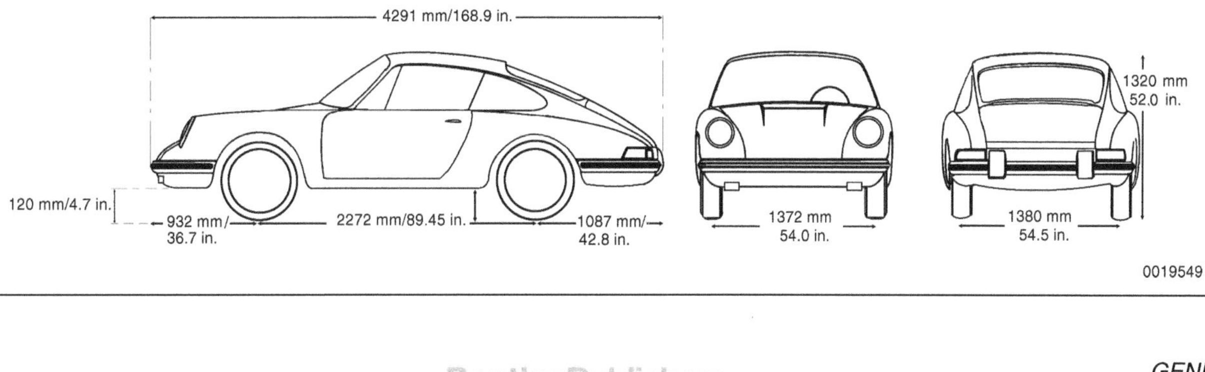

Bumpers

The bumpers are detachable. They are painted in the same color as the car and have rubber trim strips.

Doors

The doors are fixed to the front pillars of the body with internally fitted hinges. A unique feature of the door is the built-in alarm lock cylinder (**arrow**).

Lids

The rear lid release handle is located on the drivers door post (B-pillar) and for safety reasons can only be accessed when the door is open. The luggage compartment lid is double-latched. Gas support struts enable the lids to be open in any position. The luggage compartment is under the front lid and contains the tool kit, the spare wheel, the battery and the fusebox/relay panel. The rear lid covers the engine compartment and carries the air intake grille for the engine.

Seats

Front seats are adjustable backward and forward. The backrest angle can be adjusted to position down to horizontal. The front seat backrest has an additional catch which prevents the seat from tilting forward.

Luggage compartment

Luggage space is provided in the front compartment under the front lid. The spare wheel, tool kit, battery and fuse/relay panel are also located here. Additional luggage space can be made in the back of the driving compartment if the seat backs of the rear seats are folded forward.

Interior fittings

All interior surfaces are covered with padded leather in many colors and patterns. Perforated leather is used for the headliner. Carpeting covers the interior floor and console as well as the luggage compartment.

For sound proofing and insulation all hollow components are sprayed or painted with rust inhibitor before welding. All surfaces which are not painted in the body color, in particular the floor group with the wheel arch panels and the interior compartment are sprayed with an efficient sound deadening compound which at the same time provides rust protection. For further sound proofing the interior compartment together with the front and rear compartments are covered on special instructions with bitumen board, bitumen felt, jute felt and cork felt mats.

Heated air from the engine is blown by the engine compartment blower through heater ducts into the car. The air volume is controlled by an automatic heating control unit mounted in the center console between the seats.

Instruments and controls

Everything in the passenger compartment essential to the driver is logically grouped and easy to reach. All instruments and controls have been arranged ergonomically and are fully integrated into the overall design of the vehicle.

510 Fenders

GENERAL

This repair group covers replacement of the front fenders.

FRONT FENDERS

The front fenders are among the few body panels that are bolted on and can be replaced separately.

For corrosion protection, the galvanized fenders are assembled with a sealing strip between the top mating surfaces to prevent moisture from becoming trapped. This strip should be replaced any time the fenders are removed.

Front fender assembly

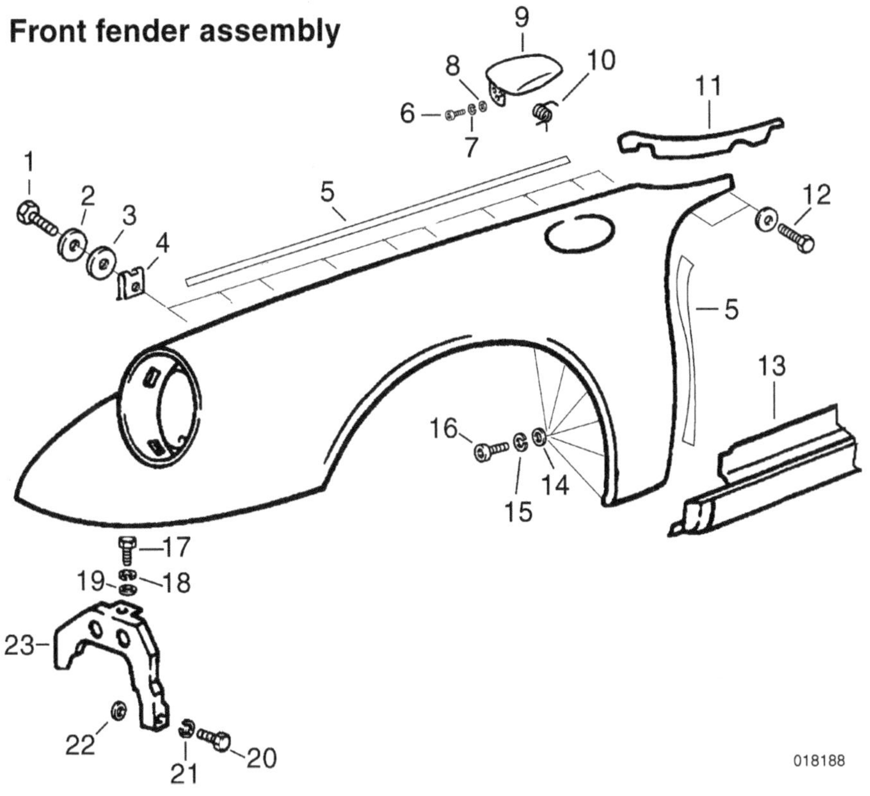

1. Bolt
2. Washer
3. Sealing washer
4. Speed nut
5. Seal
6. Bolt
7. Lock washer
8. Sealing washer
9. Fuel filler door
10. Spring
11. Gasket
12. Bolt
13. Rocker panel
14. Sealing washer
15. Lock washer
16. Bolt
17. Bolt
18. Lock washer
19. Sealing washer
20. Bolt
21. Lock washer
22. Nut
23. Bracket

Fender, removing and installing

- Raise front hood.
- Remove front bumper. See **630 Bumpers.**
- Peel up rubber hood seal on side to be repaired.

< Remove screw securing headlight ring (**arrow**) and remove ring.

< Remove headlight assembly screws (**arrows**). Disconnect electrical plugs and remove headlight assembly.

- Remove wiring connectors from headlight harness plug.
- Withdraw wiring harness from headlight bucket.
- Working inside headlight bucket, remove fender mounting bolt (M8).

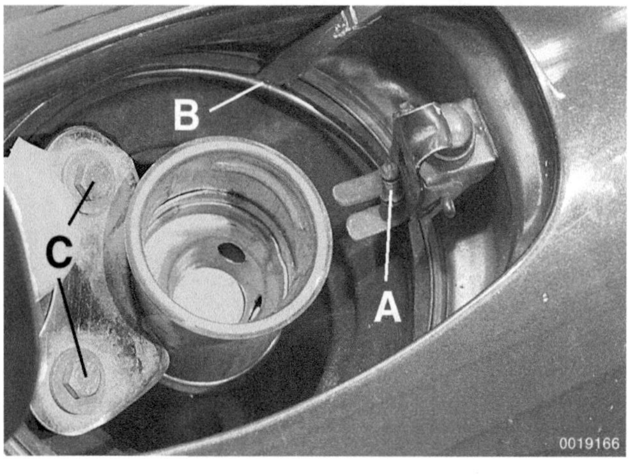

< Left fender only:

- Disconnect cable from lid release (**A**).
- Pry out retaining ring from fuel filler neck sealing bellows (**B**).
- Remove filler neck bracket bolts (**C**).
- Working at underside of fender, disconnect washer fluid hose from hose tube.

– Working inside wheel housing, remove mounting bolts (qty. 6) along fender vertical.

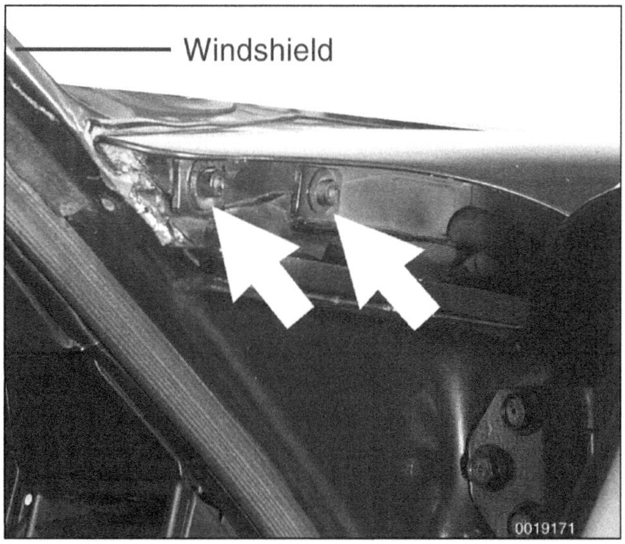

◀ Open door fully. Remove two screws below windshield (**arrows**).

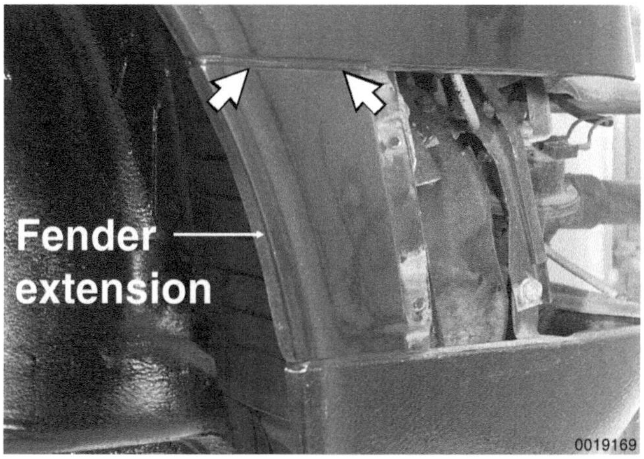

◀ Working from inside wheel housing, remove upper screws securing fender extension to fender (**arrows**).

– Remove screws securing rubber bumper cover to fender.

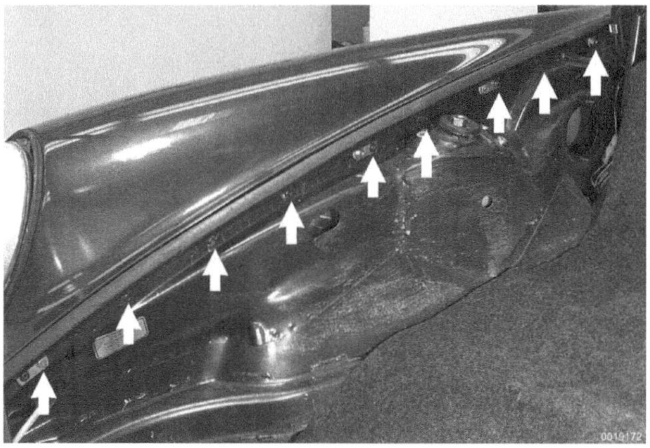

◀ Working inside luggage compartment, remove mounting bolts along fender edge (**arrows**).

– When removing right side fender, remove upper bolt from the oil cooler bracket.

– Carefully separate fender from body and remove.

– When installing new fender, clean off old sealant and protective coating from body mounting surface.

– Replace all gaskets and seals.

– Position fender in place and loosely install all mounting bolts. Align fender correctly with door and windshield pillar (A-pillar), then tighten bolts.

– Remainder of installation is the reverse of removal. Align panel so that even gaps exists at all joints. Repair any paint damage and paint any exposed metal.

Fender gaps

- Fender to door . . . . .4 mm (0.16")
- Fender to hood . . . . .3 mm (0.12")
- Wheel housing to top edge of fender . . . . .17 mm (0.67")

550 Engine Lid, Hood

GENERAL

This repair group covers removal and installation of the hood and engine lid.

FRONT HOOD ASSEMBLY

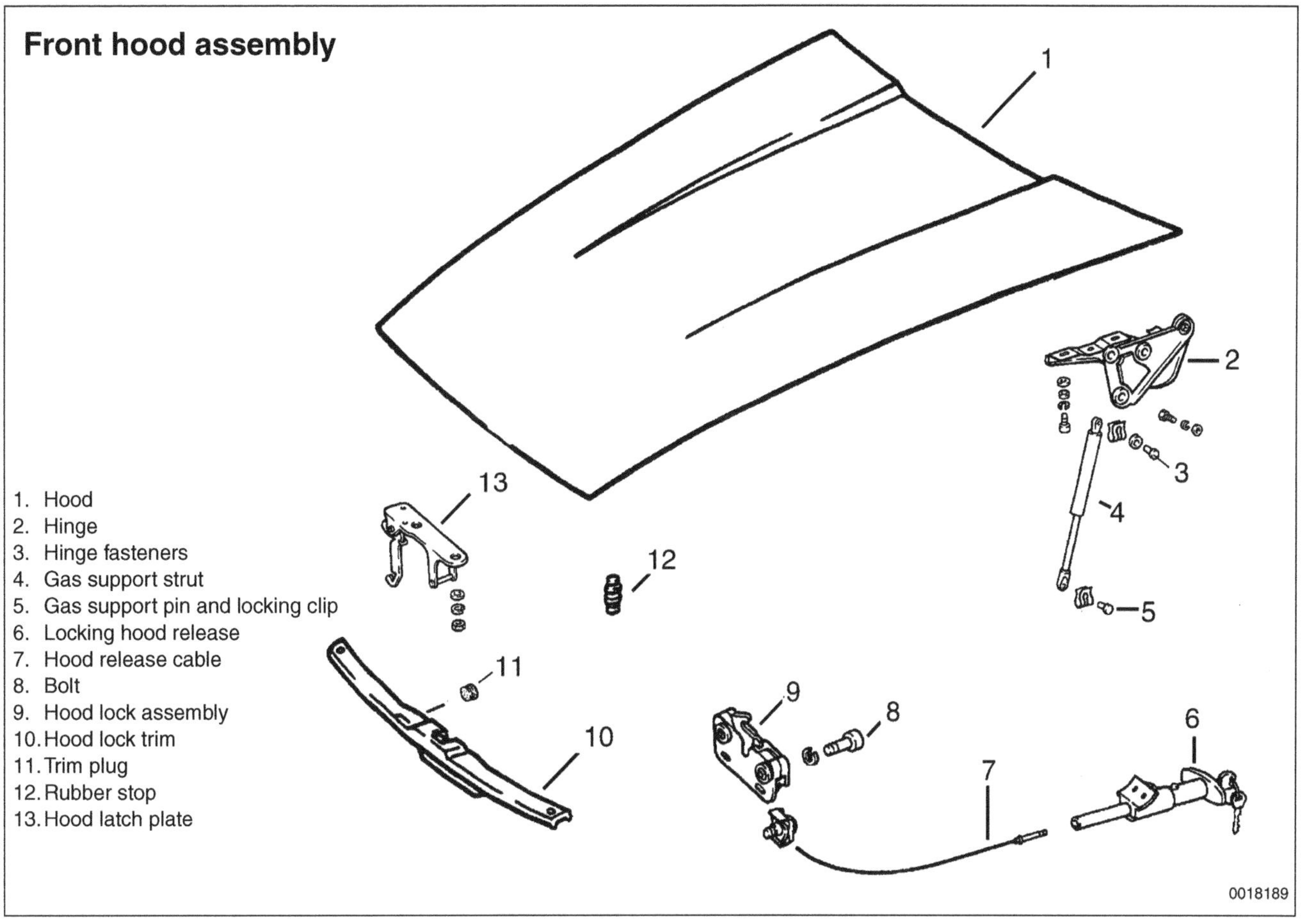

Front hood, removing and installing

- Raise hood.
- ◄ Disconnect harness connector for luggage compartment light. Unclip harness from hood hinge. Remove light switch retaining nut from switch.
- Precisely mark hinge location on hood panel.

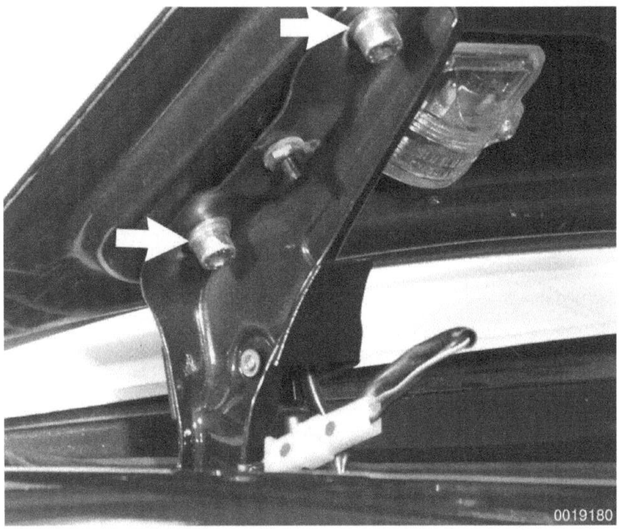

- ◄ Loosen hood mounting bolts on left and right sides (**arrows**).
- With aid of a helper, support hood while removing mounting bolts. Remove hood, guiding harness through hinge.
- Installation is reverse of removal.
  - Repair any paint damage and paint any exposed metal.
  - Check hood alignment as described next.

Hood, aligning

When installing the hood, align hinges as close to the original position as possible.

- Align hood so that gaps are even on sides and rear.

Hood gaps
- Hood to cowl panel . . . . 4 - 5 mm (0.16 - 0.20")
- Side gaps . . . . 4 mm (0.16")

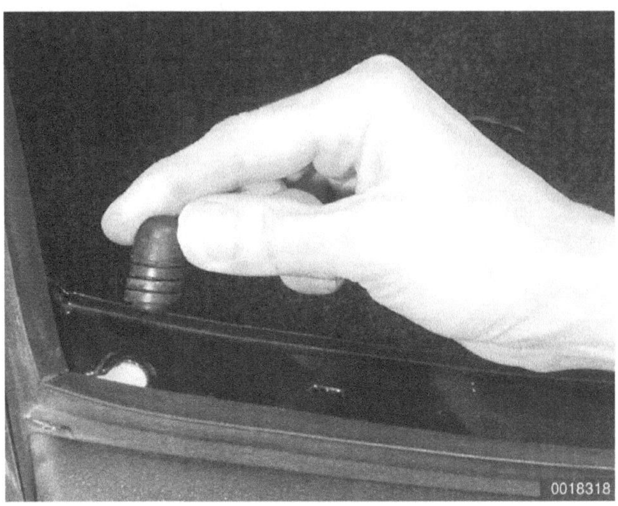

- ◄ Set hood to same height as fender joint edge by adjusting threaded rubber stops on either side of hood latch support.

Hood latch, adjusting

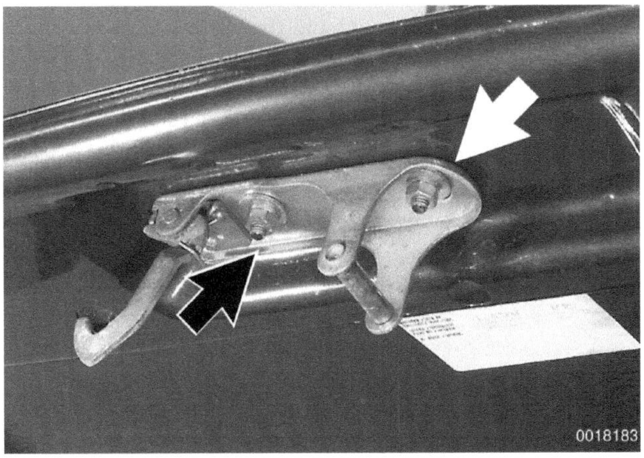

Before adjusting hood latch, be sure the hood is aligned evenly to fenders and cowl panel.

The latch must locate in the center of the lock to close properly. The height at the front of the hood can be altered by sliding the lock assembly up or down on its mounting bolts.

< Loosen latch plate bolts (**arrows**). Tighten bolts finger tight.

– Lower, **but do not latch**, hood so that latch plate pin centers itself in latch.

> ***CAUTION —***
> *Do not let hood lock with latch plate bolts loose.*

– Open hood without altering position of latch plate. Tighten latch plate mounting bolts.

Hood cable, replacing and adjusting

– Raise hood.

– Remove broken cable.

< Install new cable and thread end through latch release arm and into cable clamp (**arrow**).

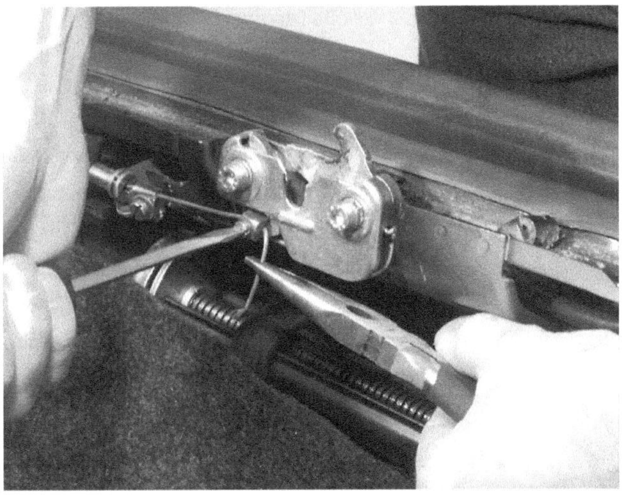

< While holding cable tight using pliers, tighten cable clamping screw.

– Check for proper hood latch operation before closing hood. While a helper pulls hood release, check that latch releases smoothly and returns to the locked position. If necessary readjust cable.

Hood, emergency opening with broken cable

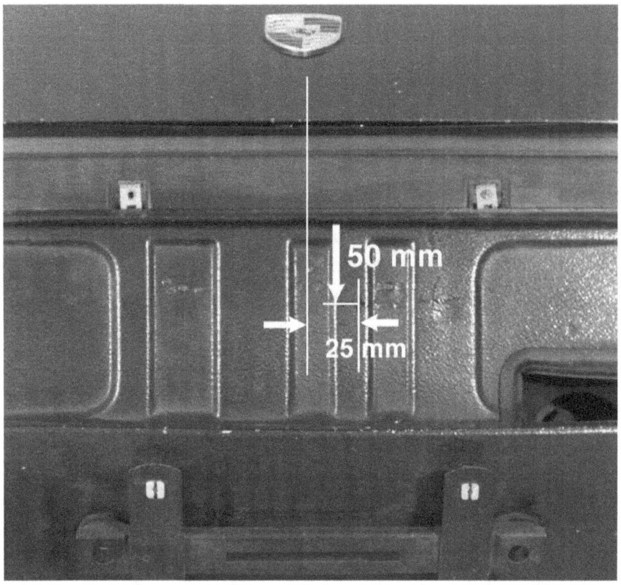

– Remove front bumper assembly. See **630 Bumpers**.

< Locate and punch mark position 50 mm (2 in.) down from bottom of bumper trim panel and 25 mm (1 in.) to the left of body center line. Drill an 8 mm (5/16 in.) hole through front body panel.

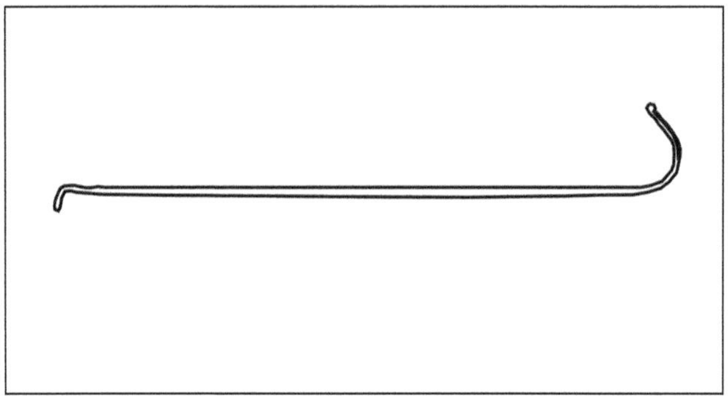

< Using a piece of stiff steel wire (6 mm or 1/4 in.) approximately 40 cm (16 in.) long, make a 90° bend 15 mm (0.6 in.) from wire end.

– Insert tool upward through hole in panel to catch hood release. Turn tool to right to release lock.

– Plug drilled hole (use Porsche part no. 999 703 044 50).

– Reinstall front bumper.

Hood hinge, replacing

– Raise hood and support in up position.

– Mark hinge location on hood.

NOTE —
When replacing both hood hinges, do one hinge at a time.

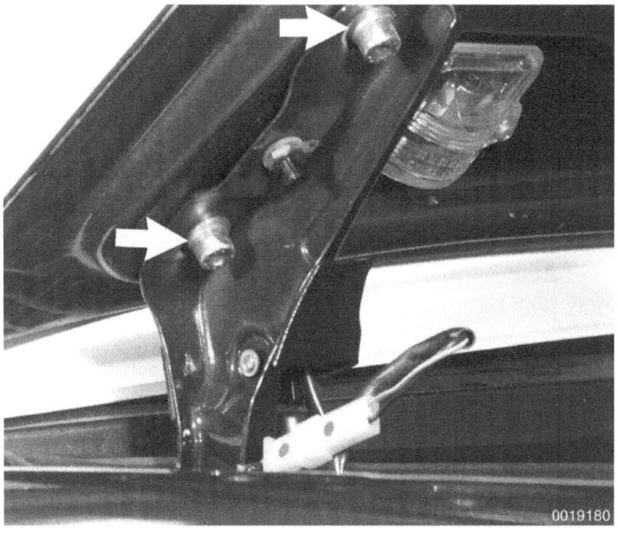

< Remove hood mounting bolts (**arrows**).

– Disconnect hood support strut from hinge.

• Remove upper hood strut pin locking clip and remove pin.

◀ Remove hinge mounting bolts (**arrows**) and remove hinge.

– Installation is reverse of removal.

- Repair any paint damage and paint any exposed metal.
- Check hood alignment as described earlier.

Hood support struts, replacing

When replacing both hood support struts, do only one strut at a time.

– Raise hood and support in up position.

– Remove locking clip and clevis pin from support strut upper mounting (**upper arrow**).

– Remove locking clip and clevis pin from support strut lower mounting (**lower arrow**).

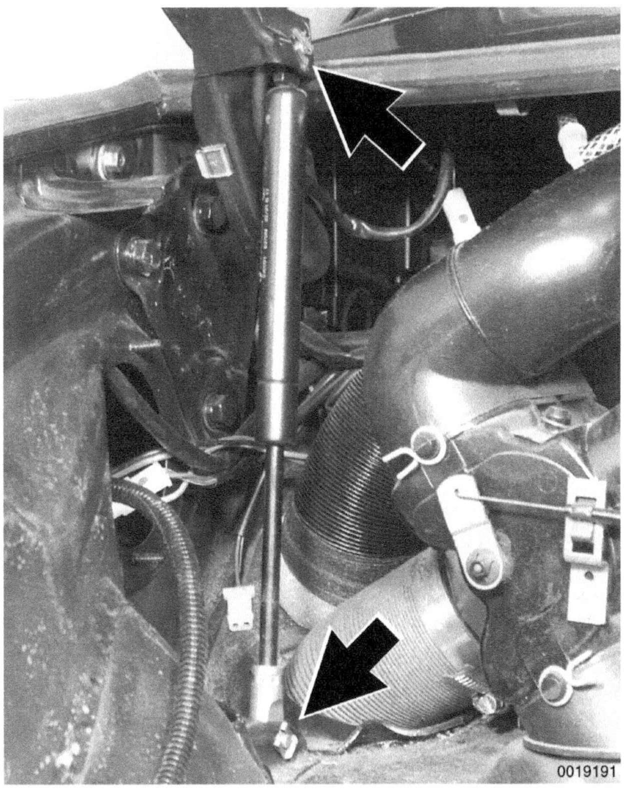

0019191

◀ Place new strut in position and install lower pin and locking clip.

– While compressing strut slightly, install upper pin and locking clip.

ENGINE LID

Engine lid assembly

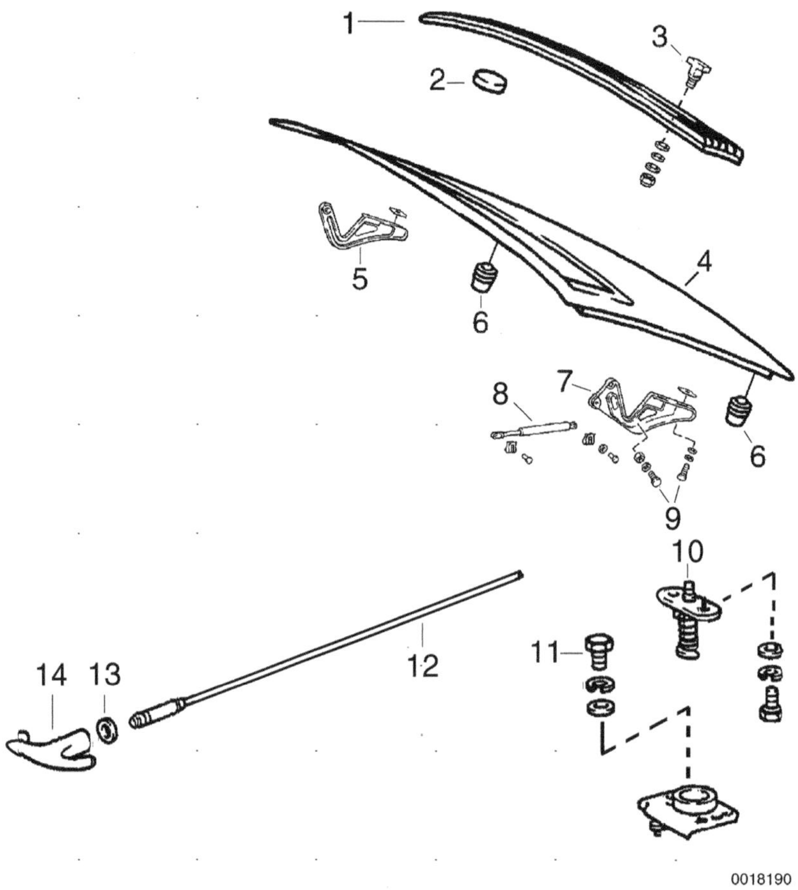

1. Grill
2. Cap
3. Screw
4. Engine lid
5. Right hinge
6. Lid stop
7. Left hinge
8. Gas support pin
9. Hinge bolts
10. Lock
11. Bolt
12. Lid release cable
13. Washer
14. Handle

Engine lid, removing and installing

– Raise engine lid.

◄ Remove 2 screws from A/C condenser center bracket and remove bracket (**arrow**).

> ***CAUTION—***
> *Do not loosen or disconnect A/C refrigerant lines from condenser.*

– Disconnect engine compartment light wiring harness connector and unclip harness from right hinge.

– Mark location of hinges on engine lid.

◄ Loosen engine lid mounting bolts on left (**arrows**) and right sides.

– With aid of a helper, support lid and remove hinge mounting bolts, noting position of any shims. Remove lid.

– Installation is reverse of removal. Install shims in same place as originally installed. Align lid to hinges using marks made during removal. Align engine lid so that gaps are even on sides and rear.

Engine lid, aligning

– Open engine lid.

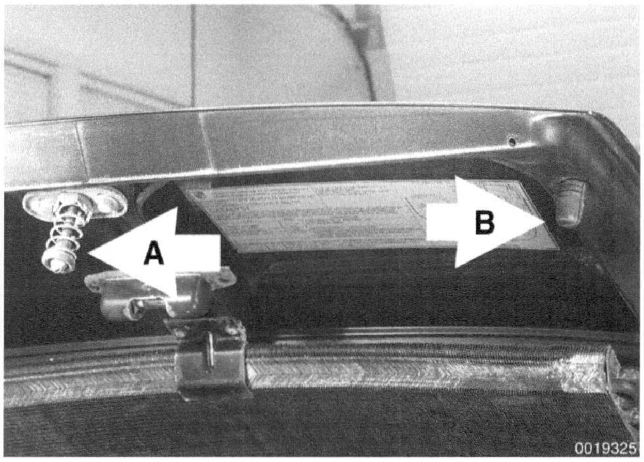

– Loosen engine lid hinge bolts and then tighten finger tight.

– Close lid but do not latch and align lid evenly to opening in body.

Engine lid gaps

- Top . . . . 4 -5 mm (0.16 - 0.20")
- Sides. . . . . .4 mm (0.16")

– Open engine lid and tighten hinge mounting bolts.

◄ Set engine lid to the same height as fender joint edge by adjusting threaded latch pin (**A**) and threaded rubber stops on left and right sides of lid (**B**).

Engine lid grill, removing and installing

– Remove engine lid as described earlier.

NOTE —
Do not disconnect A/C lines from condenser.

◄ Remove grill mounting nuts from rear side of lid and remove grill from front.

– Installation is reverse of removal.

Engine lid strut, removing and installing

- Raise engine lid.
- Support lid with hood in up position.
- Remove clips and pins from support strut.
- Install pins and clips in new strut.

NOTE —
Models with rear spoiler may have two struts.

570 Doors

GENERAL

This repair group covers door repair information, including removal and installation of the door and the interior door panel.

NOTE —
- *For information on the power windows and door glass replacement, see* **640 Windows**.
- *For information on the power door locking system, see* **950 Central Locking**.

Door assembly

1. Door handle
2. Door lock cylinder
3. Base bracket
4. Washer
5. Nut
6. Door
7. Plug
8. Vapor barrier
9. Door check
10. Nut
11. Cover
12. Bushing
13. Shim
14. Hinge
15. Bolt
16. Hinge pin
17. Gasket

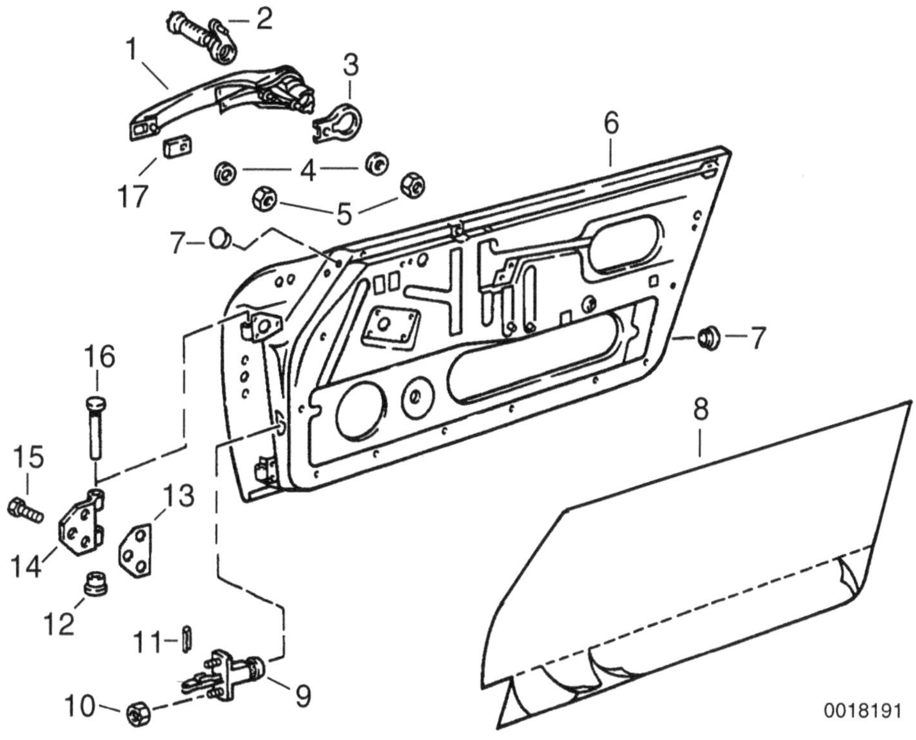

DOORS

Door, removing and installing

– Disconnect battery negative (–) cable.

> ***CAUTION —***
> *Prior to disconnecting the battery, read the battery disconnection cautions given at the front of this manual on page viii.*

0019320a

◄ Drive out door check pivot pin (**arrow**).

– Remove door panel. See **Door panel, removing and installing**.

– Label and disconnect wiring from all switches. Arrange wiring so that it can be easily withdrawn through door once door is unbolted.

Table a. Window switch wiring connections

Switch Terminal	Left window driver's door switch	Right window driver's door switch	Right window passenger door switch
1	Blue	Red / white	Blue
2	Blue / black	Red / black	Blue / black
3	Single red / blue	Single red / blue	Red / white
4	Brown	Brown	Brown
5	Double red / blue	Double red / blue	Red / black

0019263

◄ While supporting door, remove upper and lower hinge bolts on body. Carefully withdraw wiring harnesses when removing door.

– Installation is reverse of removal, noting the following:

- Align door so that panel gaps are equal on either side, as described later. If necessary adjust hinges as described later.
- Adjust door striker so that trailing edge of door is slightly higher than front edge.
- Repair any paint damage and paint any exposed metal.

Door panel, removing and installing

0019197

◄ Remove trim screw and trim piece (**right arrow**) at front door edge. Remove two screws behind trim piece.

– Pry off concealing plug at rear of trim panel and remove screw behind plug (**left arrow**).

– Remove upper trim panel by lifting up and over door lock pin.

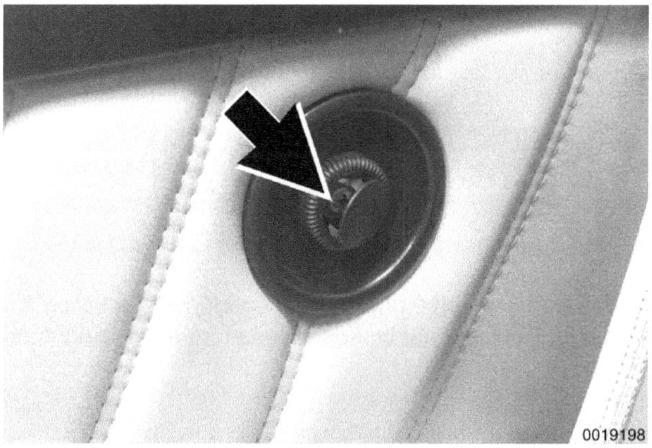

- Pry off cover from door lock knob (**arrow**). Remove knob mounting screw and knob.

- Remove lock knob trim by turning counterclockwise.

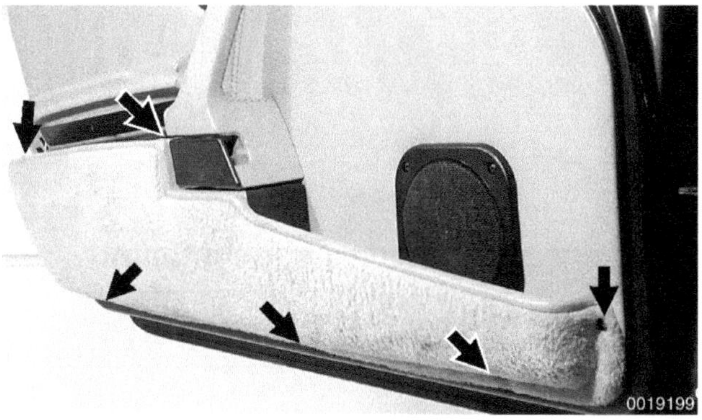

- Remove screws from map pocket perimeter (**arrows**) and remove carpeted map pocket.

- Remove two screws from pocket door hinge and remove map pocket door.

- Remove speaker grill screws and remove grill and speaker.

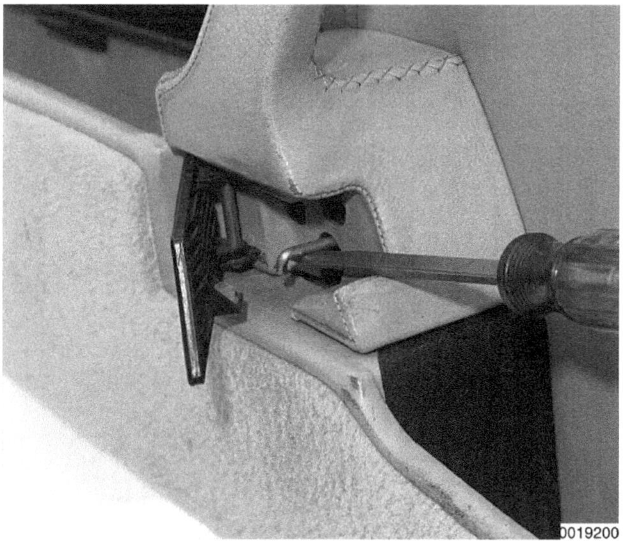

- Pull door handle out and pry out actuating rod from handle.

◀ Remove bolts at bottom and top of door pull (**arrows**).

– Pry out power window switches from door panel.

NOTE —
Label wires on switches before disconnecting.

– Carefully unclip door panel at perimeter and remove.

– Installation is reverse of removal. Check door trim clips for damage. Replace any broken clips before reinstalling door panel.

Door check, removing and installing

– Disconnect battery negative (–) cable.

> ***CAUTION —***
> *Prior to disconnecting the battery, read the battery disconnection caution given at the front of this manual on page viii.*

– Remove interior door panel as described earlier.

◀ Drive out door check pivot pin (**arrow**).

– Remove door check mounting bolts at door edge (**arrows**). Withdraw door check from inside door.

– Installation is reverse of removal.

Door window frame, removing and installing

– Remove door panel. See **Door panel, removing and installing**.

– Remove windows glass. See **640 Windows**.

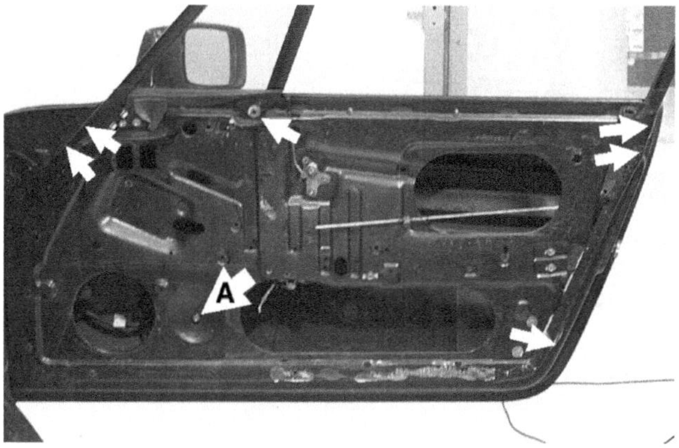

< Remove window frame mounting bolts (**arrows**).

– Working inside door, remove bracket from lower end of window channel (**A**).

– Slide window frame up an out of door.

– Installation is the reverse of removal.

- Align marks on window frame with door.
- Check that window seats squarely in opening and frame closes evenly on the door gasket.
- Adjust window frame if necessary. The frame tilt (in-out) can be adjusted at rear and middle channel bracket. Frame height (up-down) can be adjusted at two upper bolts at front of door.

DOOR LOCK MECHANISM

Door lock assembly

1. Locking knob
2. Locking rod
3. Lock control rod
4. Support
5. Socket
6. Latch assembly
7. Washer
8. Screw
9. Return spring
10. Concealing plug
11. Screw
12. Washer
13. Rotary Knob
14. Trim cover
15. Screw
16. Locking swivel
17. Socket
18. Washer
19. Gasket
20. Latch release swivel
21. Release cable
22. Support
23. Support
24. Release rod

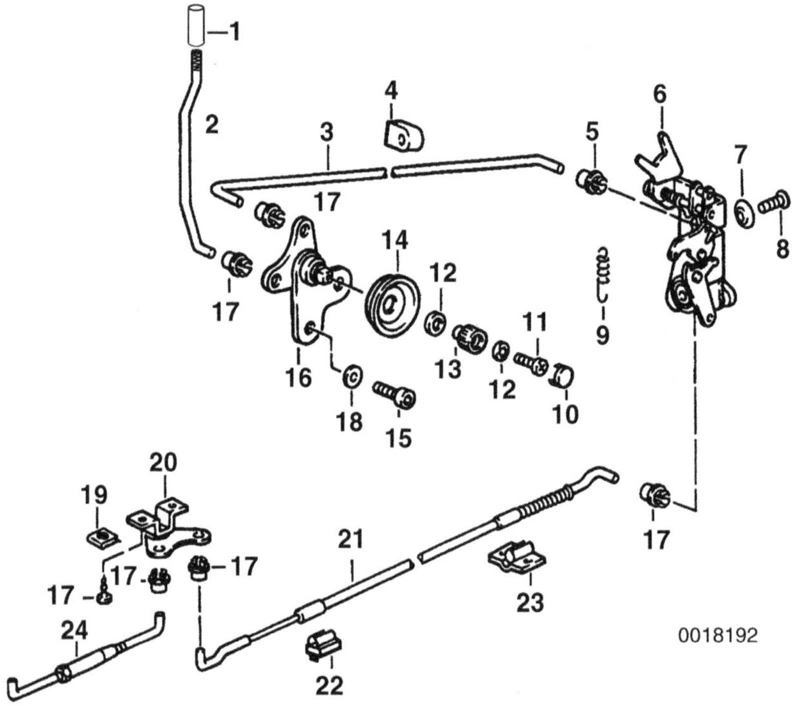

Door latch, removing and installing

- Remove door panel. See **Door panel, removing and installing**.
- Remove window glass. See **640 Windows**.
- Remove window frame. See **Door window frame, removing and installing**.
- Working inside door, disconnect release cable and power door lock control rod from latch assembly.

◄ Disconnect locking rod from locking swivel on door. Remove rod support from door (**arrows**).

Working at door exterior, remove latch bolts (**arrows**) and remove latch from inside door.

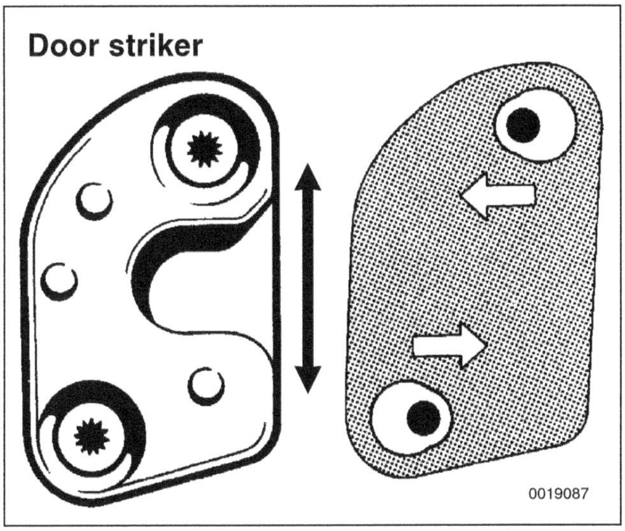

Installation is reverse of removal. If necessary, adjust striker position and tighten bolts.

Door outside handle, removing and installing

- Close window completely.
- Remove door panel. See **Door panel, removing and installing**.
- Disconnect lock wiring harness from connection at bottom of door.

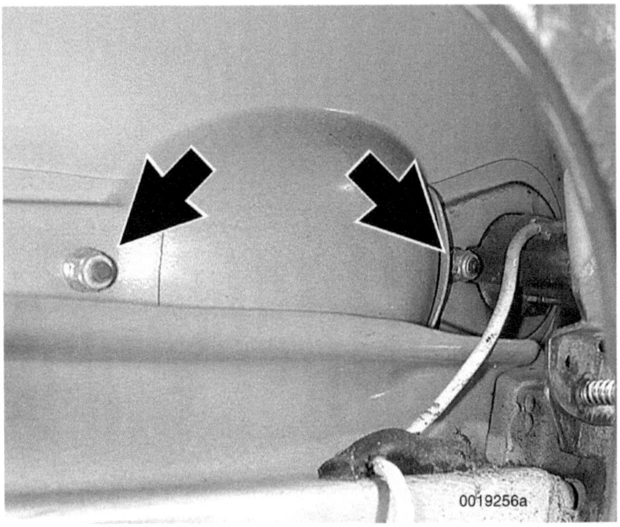

- Working from inside door, remove door handle securing nuts (**arrows**).
- Carefully pull door handle out of door.
- Installation is reverse of removal. Use new gaskets between door and body when installing handle.

Door striker, replacing

- Open door. Mark position of striker on body.

- Remove bolts securing striker to body.
- Remove striker and striker shims (where applicable).
- Installation is reverse of removal.

601 Sunroof

GENERAL

The electric sunroof is operated by a motor drive and drive cables. The sunroof panel can be adjusted with out removing it from the car. Replacement of the sunroof liner or components such as the cable assembly require that the sunroof panel be removed. The sun roof drive motor is protected from damage by a slip coupling fitted between the motor and the drive.

NOTE —
Be sure to check the front and rear drains in the corners of the sunroof for blockage. If water is entering through the sunroof one of more of these drains are usually plugged.

Sunroof assembly

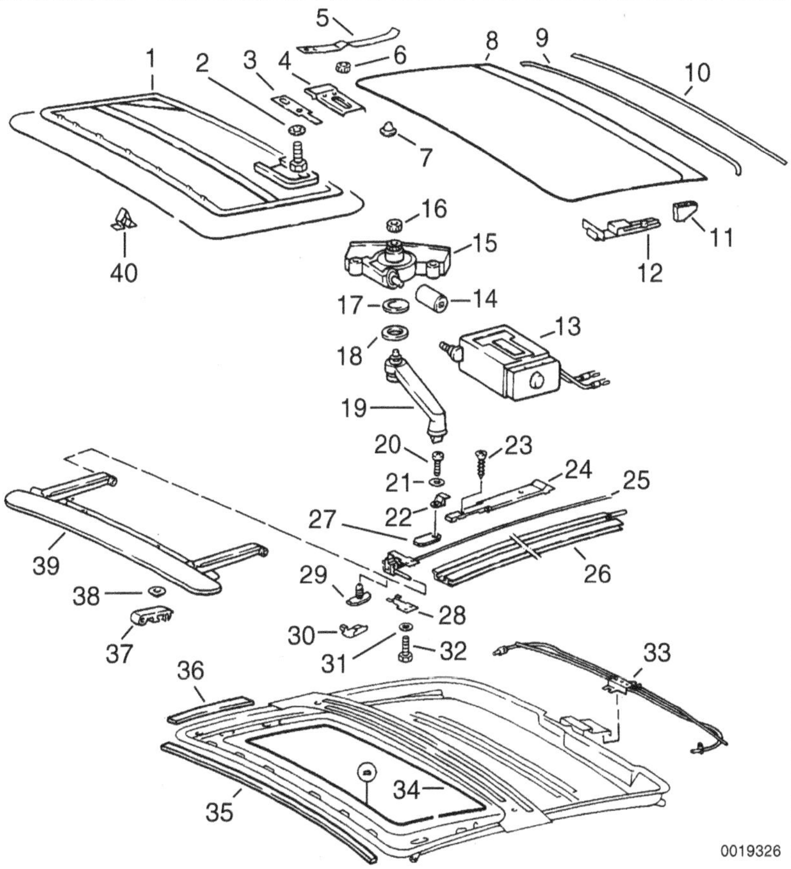

1. Frame
2. Washer
3. Lock plate
4. Guide
5. Leaf spring
6. Knurled nut
7. Plug
8. Cover
9. Velvet strip
10. Sealing rubber
11. Guide shoe
12. Support
13. Motor
14. Clutch
15. Transmission
16. Drive pinion
17. Cap
18. Washer
19. Hand crank
20. Screw
21. Washer
22. Terminal piece
23. Screw
24. Intermediate piece
25. Cable
26. Guide rail
27. Threaded plate
28. Reinforcement
29. Plug
30. Guide piece
31. Lock washer
32. Bolt
33. Cable guide
34. Beading
35. Front seal
36. Side seal
37. Clip
38. Protection cap
39. Wind deflector
40. Clamp

SUNROOF ASSEMBLY

The motor and drive unit are located in the roof just ahead of the rear window. Access to the motor and drive is a zipper opening in the headliner.

Sunroof panel, emergency closing

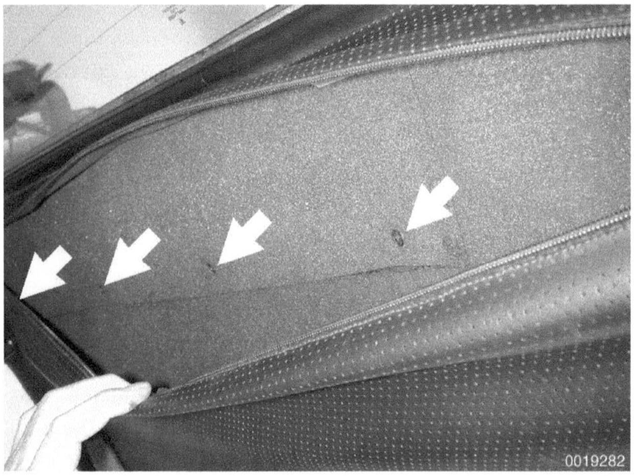

- Working in rear headliner, just forward of top of rear window, open zipper in headliner.

- Remove sunroof drive assembly access panel screws (**arrows**). Remove panel.

- Remove screw from drive unit (**arrow**). Makes sure spacer(s) come off with screw.

NOTE —
The hand crank for closing the sunroof can usually be found in the glove box or in the tool kit.

- Insert handcrank into drive unit while aligning drive keys in handcrank with drive slots in drive unit. Tighten knurled screw to secure crank to drive unit.

- Turn hand crank to close sunroof.

Sunroof panel, adjusting height

The sunroof panel should be adjusted:

- If the top of the closed sunroof becomes misaligned with the roof of the car;
- If it does not close squarely;
- if there are wind noises at speed;
- If there are water leaks;
- If the sunroof has been removed.

< Open sunroof approximately 100 mm (4").

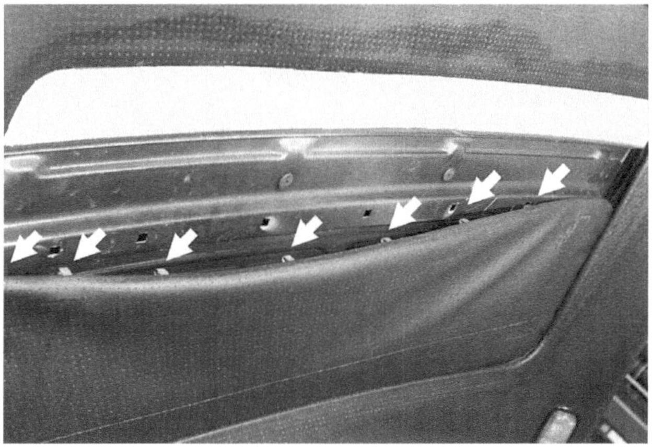

< Carefully pull down on front of trim panel to free securing clips.

– Slide headliner panel back into roof cavity.

– Close sunroof.

< Remove plastic caps covering front guide screws on left and right sides. Left side is shown (**arrow**).

To adjust front height of sunroof panel:

- Remove front guides.
- Turn knurled nuts to adjust height (**arrow**).
- Reinstall guides and check roof adjustment.

NOTE —
Adjust front height so that panel is on same plane as roof or just slightly lower.

To adjust rear height of sunroof panel:

- Loosen slotted height adjustment screw (**arrow**).
- Adjust ratchet (**inset**) so that sunroof panel is on same plane as roof or just slightly higher.
- Tighten screws.

– Open sunroof approximately 100 mm (4").

– Slide headliner trim panel forward and snap clips into place.

– Check operation of sunroof.

Sunroof panel, removing and installing

◀ Open sunroof lid approximately 100mm (4").

- Carefully pull down on front of trim panel to free securing clips.
- Slide sunroof headliner panel back into roof.
- Close sunroof.

◀ Remove guide bracket covers from front guide brackets (left and right sides). Remove front guide bracket mounting bolts (**arrows**).

NOTE —
Place a strip of masking tape cross front of sunroof opening to protect the paint when removing sunroof parts.

◀ Working a rear of panel, lift trim panel leaf spring off pin and turn to side (left and right sides).

- Remove screws from rear guide brackets (left and right sides).
- Carefully push front of sunroof panel up and slide forward and out of body.
- Installation is reverse of removal. Check sunroof adjustment as described earlier.

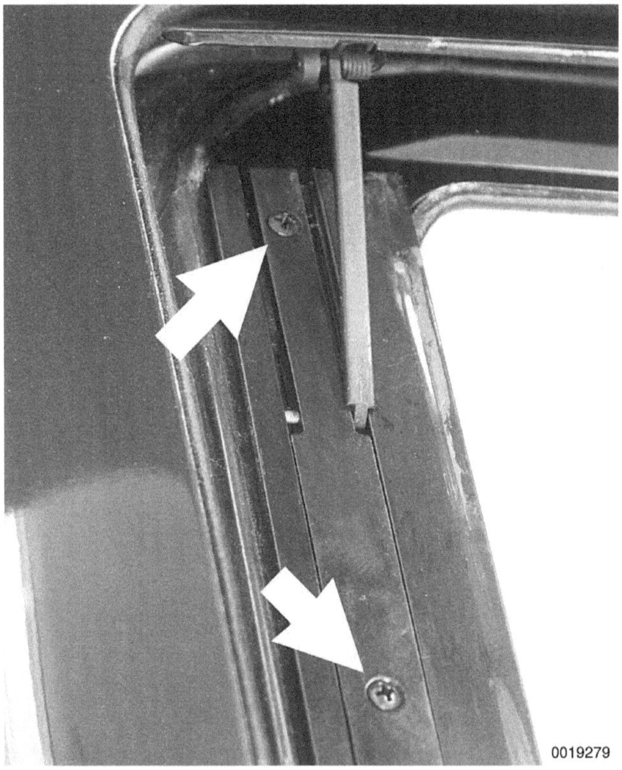

0019279

Wind deflector, removing and installing

- Open sunroof.

< Remove intermediate plate screws (**arrows**) and remove plate (left and right sides).

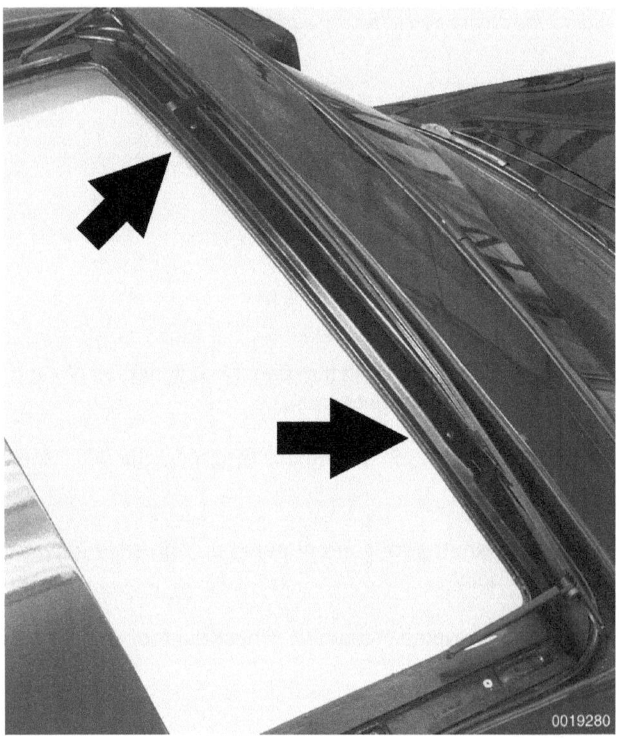

0019280

< Remove mounting screws from trim cover (**arrows**).

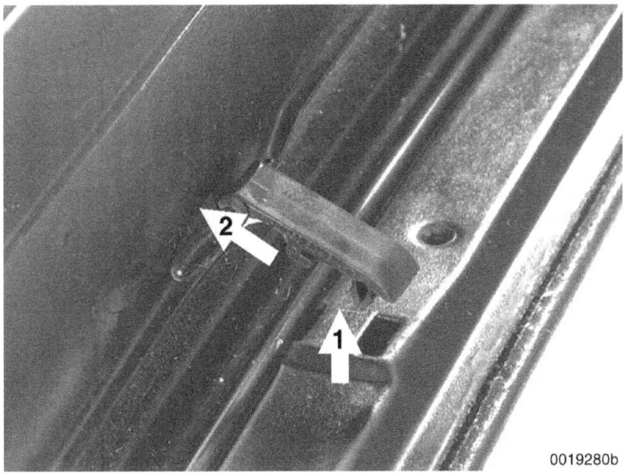

0019280b

⮜ While lifting up cover, carefully pry up on rear of deflector clip (**1**) and slide forward clip off cover (**2**). Remove wind deflector with clips.

– Installation is reverse of removal.

Sunroof cables, replacing

– Remove sunroof panel, described later.

– Remove sunroof drive motor, described earlier.

– Remove wind deflector, described earlier.

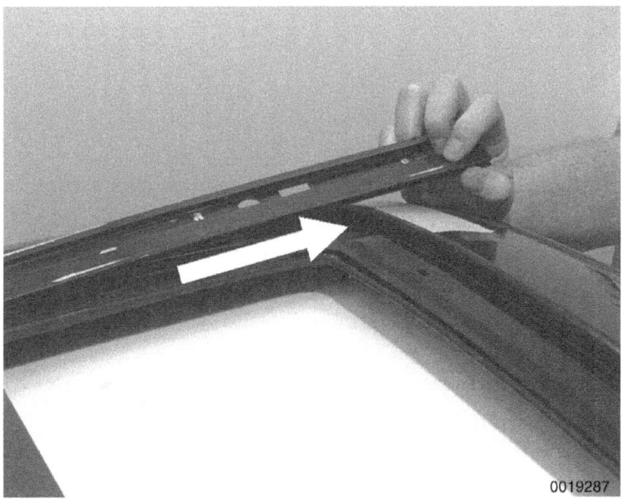

0019287

⮜ Pull sunroof guide rails straight forward and out of sunroof opening to remove.

– Pull sunroof cables from guides.

– Clean guide rails with hot soapy water and wipe dry.

NOTE —
Do not use a lubricant on sunroof cables.

– Installation is reverse of removal

- Inspect all parts for wear and replace as necessary.
- Adjust sunroof height as described earlier.

Sunroof motor and drive unit, removing and installing

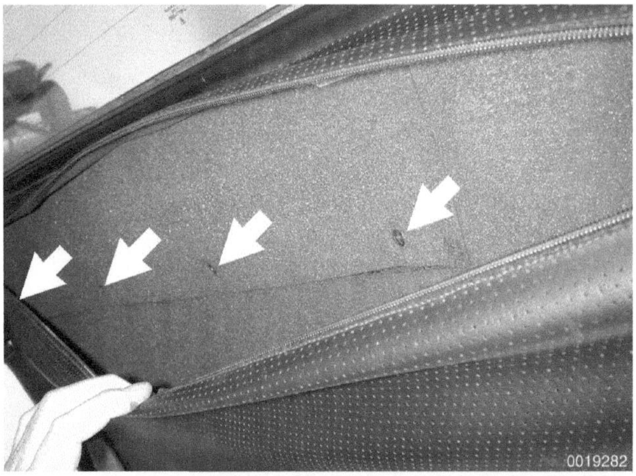

- Working in rear headliner, just forward of top of rear window, open zipper in headliner.
- Remove sunroof drive assembly access panel screws (**arrows**). Remove panel.

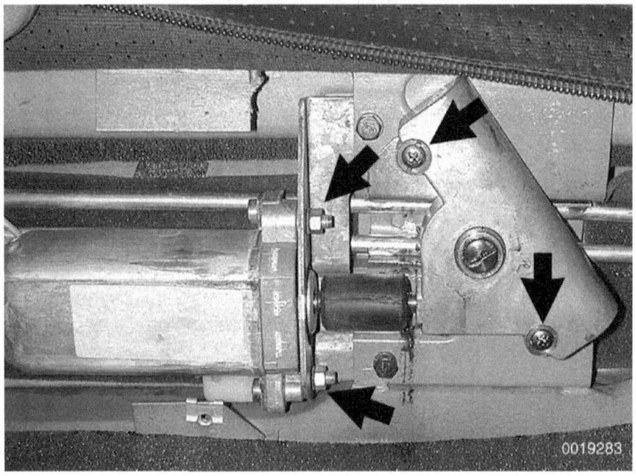

- Disconnect electrical harness connector. Remove sunroof drive assembly mounting screws (**arrows**). Lower drive assembly with motor.
- Installation is reverse of removal.

Sunroof guide rails, removing and installing

- Remove sunroof panel, sunroof drive unit and wind deflector as described earlier.

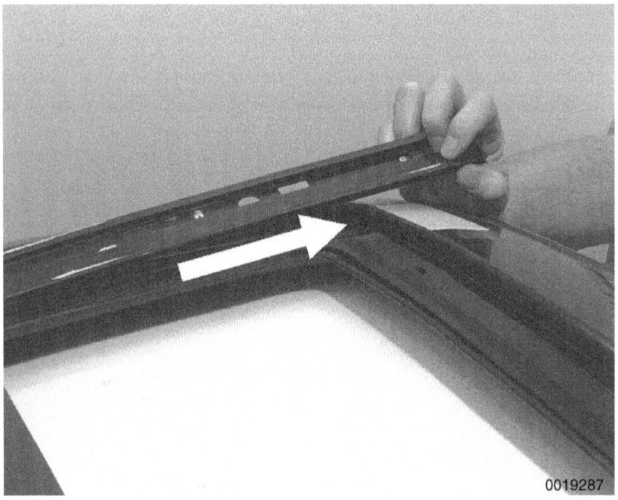

- Pull sunroof guide rails straight forward out of sunroof opening to remove.
- Installation is reverse of removal.

612 Targa Top

GENERAL

The targa top is secured to the roof bar by means of three pins, and to the windshield frame by two locks, facilitating easy removal and installation.

Targa top assembly

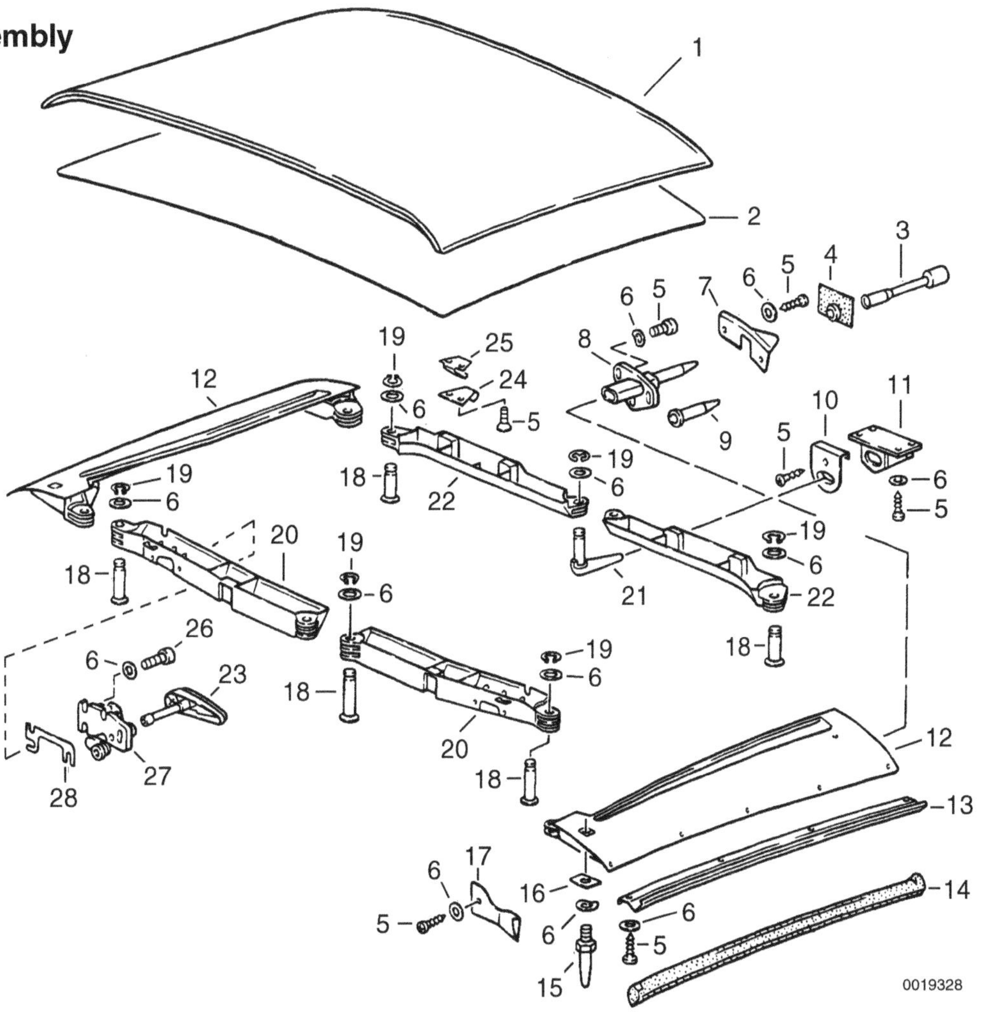

1. Outer lining
2. Inner lining
3. Silent block
4. Sealing
5. Screw
6. Washer
7. Trim
8. Silent block
9. Covering
10. Cover plate
11. Mounting
12. Threaded plate
13. Water strip
14. Sealing
15. Fixing bolt
16. Spring washer
17. Trim
18. Bolt
19. Lock ring
20. Roof frame, front
21. Locking bolt
22. Roof frame, rear
23. Handle
24. Latch
25. Clamp
26. Screw
27. Catch
28. Shim

TARGA TOP

Prior to removal, clean the targa roof panel to prevent road dust or dirt from soiling clothes or the luggage compartment carpet. When storing the top in the luggage compartment, protect it from objects that could damage the panel surfaces.

NOTE —
If the top is to be left removed from the vehicle for an extended period, store it away from the vehicle in an open, not folded, position.

Targa top, removal and installation

Special targa top removal handles are provided in the tool kit.

- Working inside car in front seat, insert both release handles in lock receptacles in front roof support bow of targa top (above sun visors). Start with handles pointed left.
- Turn handles counterclockwise until they are at the 4 o'clock position.
- Continue turning left handle while gently lifting left side of top until it snaps out of lock.
- Perform same operation on opposite side of top.
- Till front of top up and slide forward to remove from car.
- Installation is reverse of removal.

Targa top, folding and unfolding

Once it is removed from the car, the targa top must be folded in order to be stored in the luggage compartment. To fold or unfold, support the rear edge of the roof on a soft surface.

NOTE —
The toes of your shoes work well in this operation.

- **Folding:** Apply pressure simultaneously to middle of both hinged bows. Finish folding by applying light pressure to sides.
- **Unfolding:** Simultaneously push both hinged bows outward to locked position.

Targa top latches, adjusting

The front latch assemblies can be adjusted in or out by means of shims. Remove targa top and plastic latch covers. Loosen latch mounting screws and add or remove shims as necessary.

630 Bumpers

GENERAL

This repair group covers removal and installation of the front and rear bumper and trim.

FRONT BUMPER

Front bumper assembly

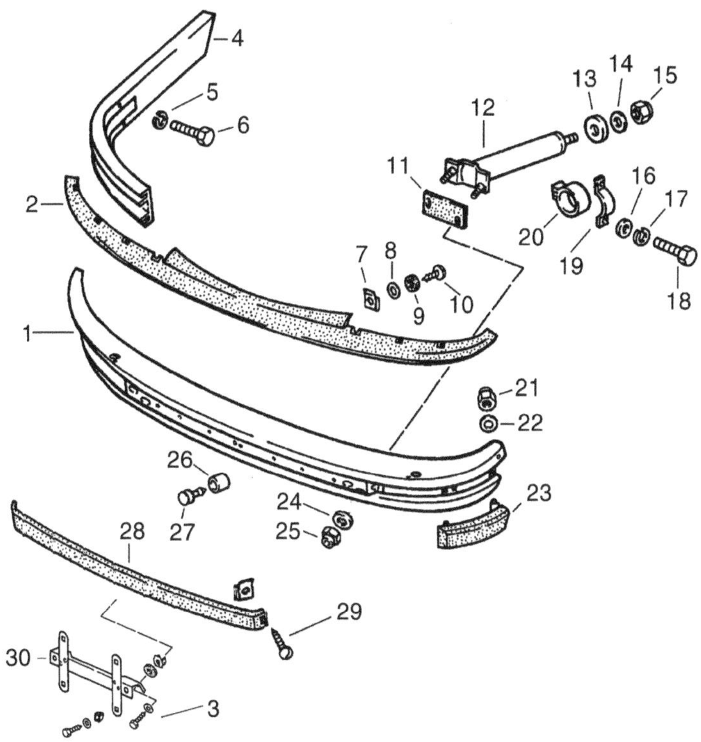

1.	Bumper frame	16.	Washer
2.	Lip seal	17.	Lock washer
3.	License plate mounting hardware (kit)	18.	Bolt
		19.	Clamping bracket
4.	Reinforcing bracket	20.	Insulator
5.	Washer	21.	Nut
6.	Bolt	22.	Washer
7.	Clip	23.	Protective strip
8.	Washer	24.	Washer
9.	Washer	25.	Nut
10.	Bolt	26.	Insert
11.	Gasket	27.	Clip
12.	Impact absorber	28.	Protective strip
13.	Fender washer	29.	Screw
14.	Washer	30.	License plate frame
15.	Nut		

Front bumper, removing and installing

- Remove parking light lens.

< Remove screw securing reflectors to parking light lens assembly. Disconnect wiring harness from reflectors and remove reflectors from left and right sides.

- Remove fog lights. See **940 Exterior Lighting**.
- Working through foglight openings, withdraw parking light wiring harnesses from rear of parking light housings.

< Remove screw from end of protective bumper strip on left (**arrow**) and right sides.

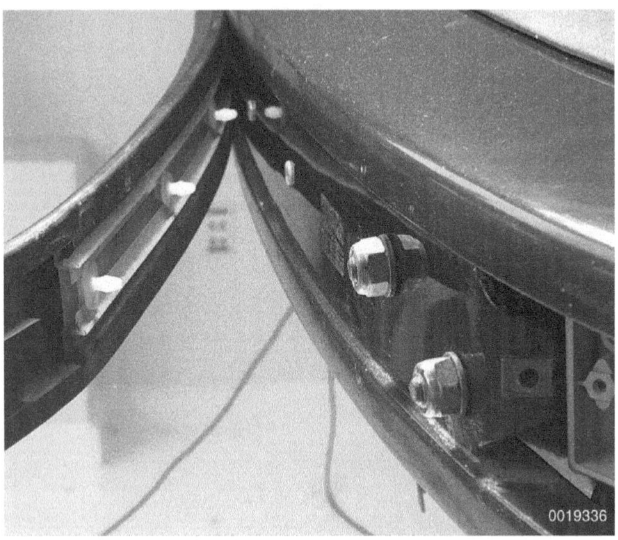

< Remove protective strip by carefully prying out clips on rear of strip.

NOTE —

Have 7 trim snaps and plugs on hand in case they break when being removed.

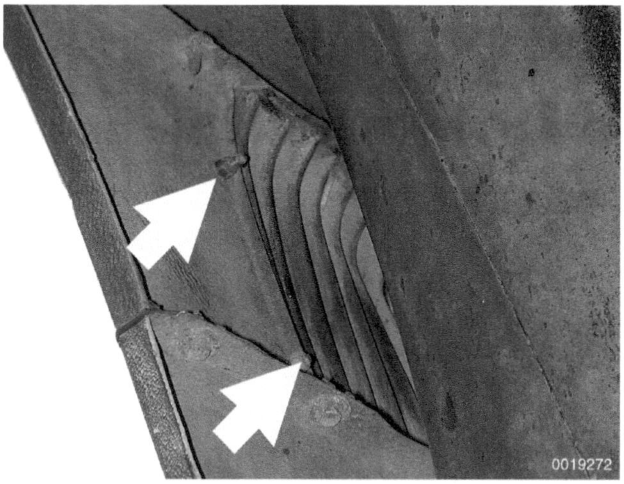

◄ Working behind bumper, remove nuts (**arrows**) securing bumper bellows to fenders on left and right sides.

- Pull bumper bellows out so that studs clear fender. Wrap studs with tape to prevent damaging painted surfaces.

◄ Remove bumper securing nuts and washers from left and right sides. Left side shown (**arrows**).

- Remove bumper by sliding straight off.
- Installation is reverse of removal.

Front bumper impact absorber, replacing

- Remove front bumper as described earlier.
- Left side: remove washer fluid reservoir tank.
  Right side: remove horn mounting bracket with horns.

◄ Remove mounting nut (**1**) from rear of impact absorber.

- Remove bolts (**2**) holding side clamping bracket in place.
- Remove impact absorber unit.
- Installation is reverse of removal.

REAR BUMPER

Rear bumper assembly

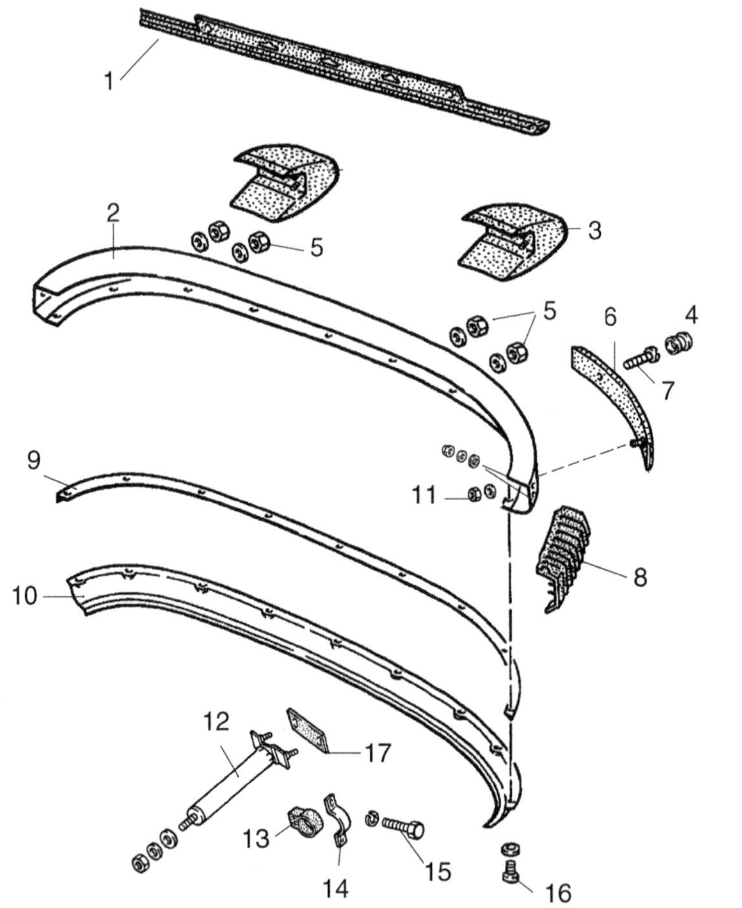

1. Trim seal
2. Bumper frame
3. Buffer
4. Concealing cap
5. Nut and washer
6. Protective strip
7. Screw
8. Bellows
9. Valence sealing strip
10. Valence
11. Nut and washer
12. Impact absorber
13. Insulator
14. Clamping bracket
15. Bolt
16. Screw
17. Gasket

Rear bumper, removal and installation

- Working from behind, remove nuts and washers securing bumper bellows to body (left and right sides of bumper).
- Carefully pull bellows away from bumper and body.
- Remove license plate light assemblies from bumper buffers and disconnect wiring harnesses.
- Working from behind, remove 3 nuts securing bumper buffers to bumpers. Pull buffers off of bumper on left and right sides.

◄ Remove concealing cap from bumper protective strip and remove screw behind cap.

◀ Remove bumper mounting nuts (**arrows**) at left and right sides. (Left side shown).

– Remove bumper by sliding it straight off of body.

– Installation is reverse of removal.

Rear bumper impact absorber, replacing

– Remove front bumper as described earlier.

◀ Remove mounting nut from rear of impact absorber (**1**).

– Remove bolts holding side clamping bracket in place (**2**).

– Remove impact absorber unit.

– Installation is reverse of removal.

640 Windows

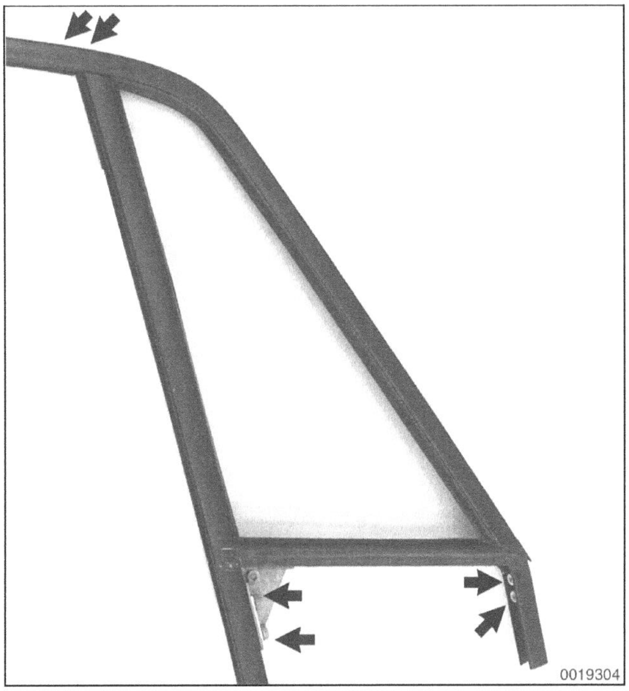
0019304

GENERAL

This repair group covers replacement of the vent, door and quarter panel glass. Also covered here is the replacement of the power window regulator and power window motor. For removal and installation of the doors, door panels, and window frame, see **570 Doors**.

WINDOW SERVICE

Vent window, removing and installing

- Remove door trim panel. See **570 Doors**.
- Remove window frame from door. See **570 Doors.**

◄ Remove screws (**arrows**) from frame assembly.

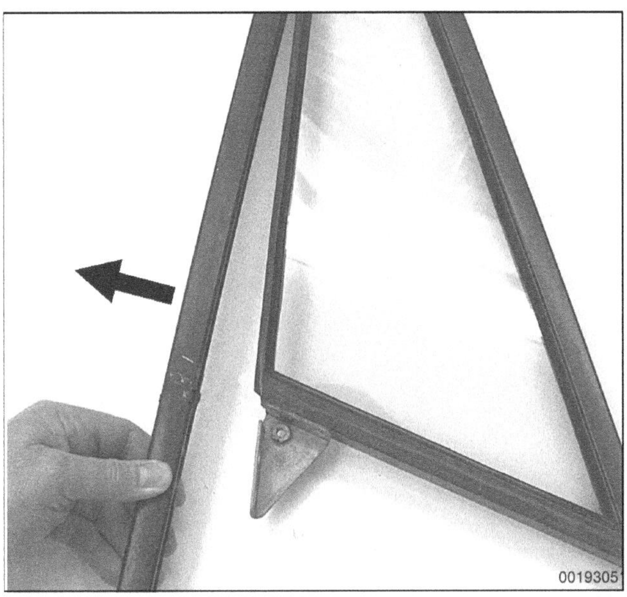
0019305

◄ Carefully separate center window guide rail from window frame.

- Remove vent window with rubber seal from window frame.
- Remove rubber seal from vent window.
- Installation is reverse of removal.

640

Door window, removing and installing

– Position window about halfway down in door.

– Remove door trim panel. See **570 Doors**.

– Remove window frame from door. See **570 Doors**.

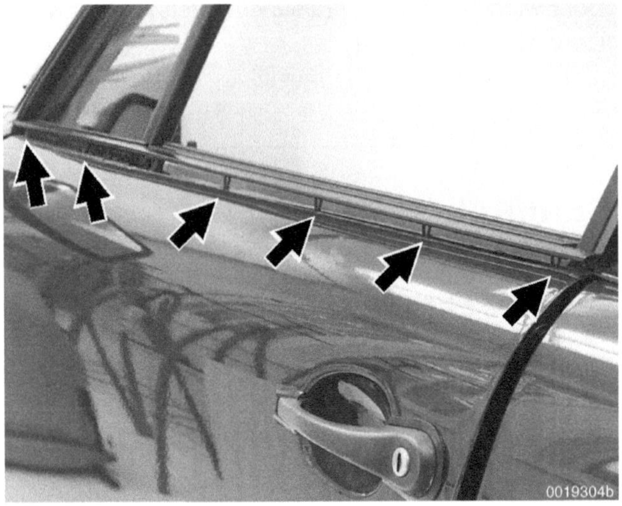

NOTE—

- *Be careful not to drop two bolts located at the top front of the door. If bolts do drop into the channel, use a magnet to fish them out through the drain hole in the bottom of the door.*
- *When removing the door frame, the window glass stays in the door.*

◀ Carefully pry outer window weather-strip from top of door at retaining clips (**arrows**).

– Carefully peel off inner weather strip.

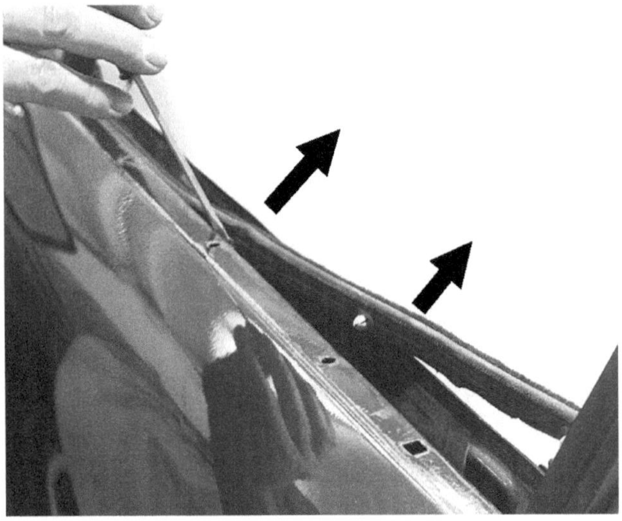

◀ Carefully pry out felt weather-strip from inside of door.

0019308

Remove bolts (**arrows**) from rear regulator arm support.

0019309

Tilt window up at rear and pull out of door while sliding both guide wheels out of front of window guide channel.

NOTE —

The guide wheels will not come out through the rear of the window guide rail.

– Installation is reverse of removal.

CAUTION —

If replacing a broken glass be very sure to clean all the glass from window channels.

Door window regulator, removing and installing

– Position window about halfway down.

– Remove door trim panel. See **570 Doors**.

< Remove regulator arm support bolts (**arrows**).

– Disconnect electrical plug from window motor.

< Remove bolt securing regulator to door (**arrow**).

– Slide both regulator guide wheels out of front window guide rail.

NOTE —

The guide wheels will not come out through the rear of the window guide rail.

Push window into closed position and tape in place.

> ***CAUTION —***
> *Be sure tape holds well. If the window drops into the door unexpectedly, it will most likely shatter.*

0019348

Lower regulator into door to access window motor. Unbolt window motor (**arrows**) from regulator and set inside door.

0019350

Remove regulator through hole in door.

Door window, adjusting (Coupe models)

Whenever the window regulator is removed, window adjustment should be checked. The glass should contact the top of the window frame squarely and should seat against the window seal uniformly. If the window stop is not adjusted correctly the door may become hard to close or wind noise may be a problem.

The window regulator has two adjustments, one is for in-out tilt and the other is for front/rear height parallelism.

– Remove door trim panel. See **570 Doors**.

◄ Loosen bolts securing rear regulator guide rail to door panel.

– Move rear of arm up or down in elongated slot (small arrow) to change window parallelism.

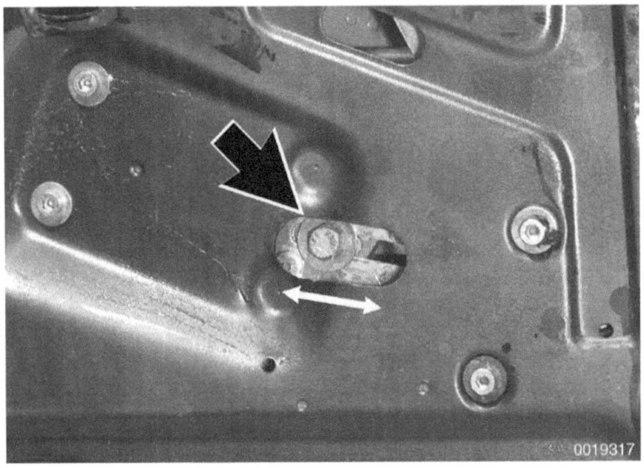

◄ Loosen height adjusting bolt (**arrow**).

– Reposition window stop left or right in elongated slot (small arrow) and check window height adjustment.

- Move stop to rear to decrease amount of window travel.
- Move stop to front to increase the amount of window travel.
- Tighten bolt when adjustment is correct.

– Install door trim panel. See **570 Doors**.

Door window, adjusting (Targa and Cabriolet models)

Whenever the window regulator is removed, the window adjustment should be checked. The glass should contact the top of the window frame squarely and should seat against the window seal uniformly. If the window stop is not adjusted correctly the door may become hard to close or wind noise may be a problem.

The window regulator used in the targa and convertible uses to stops for setting window travel.

– Remove door trim panel. See **570 Doors**.

◄ Lower window until stop bracket is aligned with access hole in door (**arrow**).

– Loosen bolt securing stop to regulator.

- Move stop forward to raise window height.
- Move bolt backward to lower window height.
- Tighten bolt when adjustment is correct.

◄ Loosen bolts (**arrows**) securing rear regulator guide rail to door panel.

- Move rear of arm up or down in elongated slot (**small arrow**) to change window parallelism. Tighten bolts when adjustment is correct.

– Install door trim panel.

660 Exterior Trim

TABLE

Special Tools

Special tools

A 9700/9701/9702 . . . Wiring connector removal tool (source: Baum Tools Unlimited)

GENERAL

This repair group covers removal and installation of the outside rear view mirrors and the exterior trim.

For front and rear bumper and bumper trim removal, see **630 Bumpers**.

Special tools

Some of the procedures require special tools.

Outside rear view mirror assembly (1987-1989)

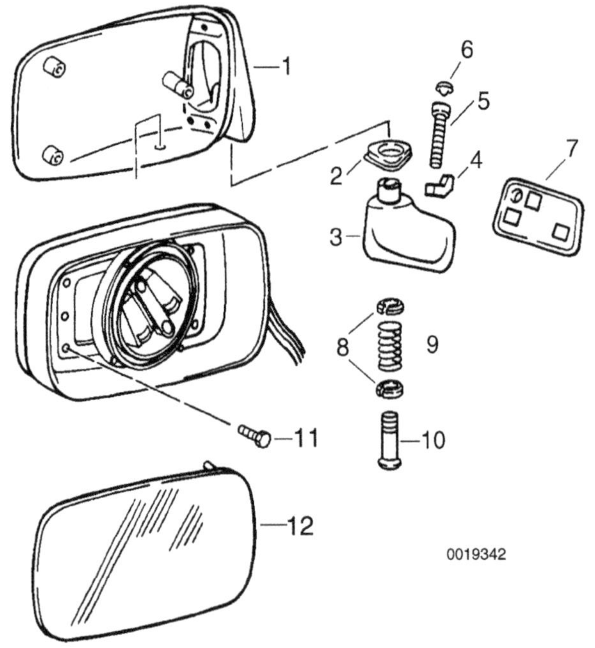

1. Mirror housing
2. Seal
3. Mount
4. Brace
5. Bolt M5 x 35 mm
6. Cap
7. Base gasket
8. Adjusting washers
9. Adjusting spring
10. Guide tube
11. Bolt M5 x 10 mm
12. Mirror glass

OUTSIDE REAR VIEW MIRRORS

Starting with 1987 models, the remote controlled outside mirrors are electrically heated. The mirror glass, mirror drive motor and mirror housing are available separately from an authorized Porsche dealer. Note that two variants of the outside power mirror was used in the Carrera series: The early variant was used on 1984 - 1986 cars and the later style (as shown) was used on 1987 - 1989 cars.

NOTE —

- *The main difference between the early style mirror and the later style mirror is the way the glass is mounted to the motor assembly. On 1983-1986 cars, the glass is retained by pins at all four corners. On 1987 and later cars, the glass is retained by a circular locking ring. The procedures given are for the 1987 - 1989 mirror assembly, the early assembly is similar.*
- *Mirror heating is controlled by the rear window heater circuit. See* **970 Electrical Wiring Diagrams**.

Outside mirror glass, replacing (1987-1989 cars)

On 1987 - 1989 cars, the mirror glass is held in place by a locking ring. Turning the locking ring left or right locks and unlocks the glass.

CAUTION —
The mirror glass should be at or above room temperature before removal. Otherwise, small plastic parts or glass will break.

– Press mirror glass in at bottom.

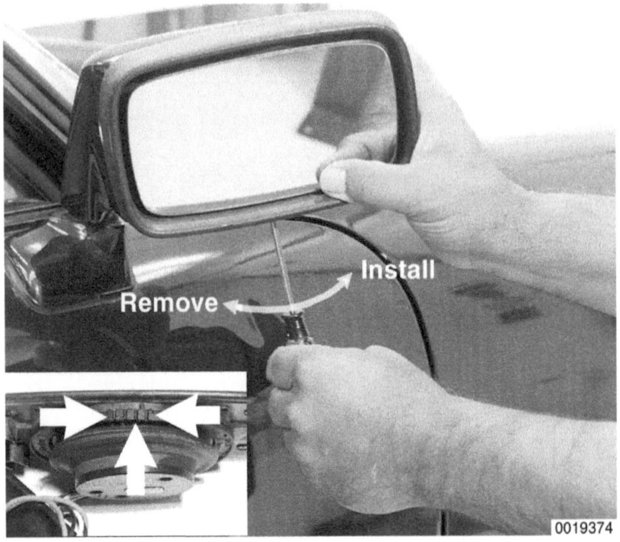

– Insert small screwdriver through hole in bottom of mirror housing and locate locking ring slots.

- Turn locking ring counterclockwise to release mirror glass.

NOTE —
Inset shows slots in bottom of locking ring.

– Disconnect heating element electrical connectors from back of mirror glass.

– Install new glass into position by pressing firmly into housing and turning locking ring clockwise. Check that glass is locked in place.

Outside mirror drive motor, removing and installing

– Remove mirror glass as described earlier.

◀ Remove three bolts (**arrows**) securing motor to housing.

– Pull motor out of housing and disconnect wiring harness plug.

– Installation is reverse of removal.

Outside mirror housing, removing and installing

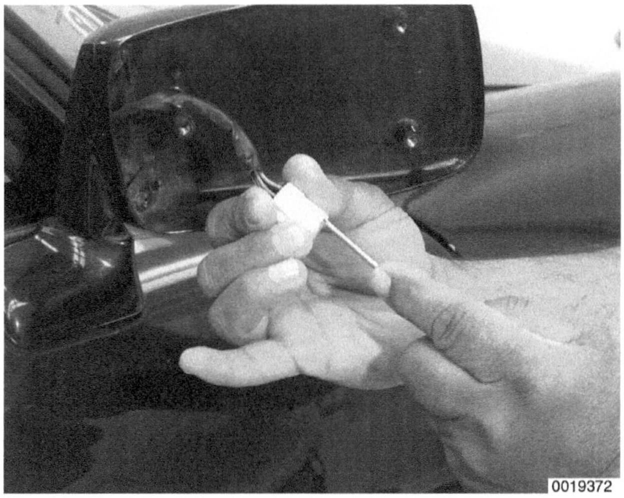

– Remove mirror glass and motor as described earlier.

◀ Use wiring connector removal tool to push wire terminals out of mirror motor harness connector.

◀ Remove allen mounting bolt (**arrow**) at base of mirror housing.

- Remove mirror while feeding wiring harness through mirror base.

– Installation is reverse of removal. Wiring colors are shown in **Table a**.

Table a. Outside Mirror Drive Motor Wire Colors

Terminal	1984-1985		1986-1989	
	Left	Right	Left	Right
1	white/green	white/green	blue	black
2	white/brown	white/blue	black	blue
3	black	black/blue	grey/green	grey/green
4	green/white/green	green/white/green	brown	brown
5	grey/green	grey/green	brown	brown
6	brown	brown	white	white

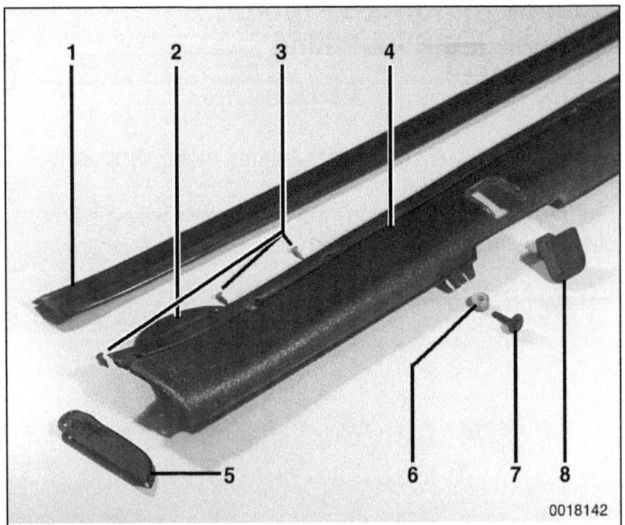

1. Rubber protector
2. Rear torsion bar end cover
3. Rocker panel mounting screws
4. Rocker panel
5. Rocker panel end cover
6. Plastic spacer
7. Rocker panel bottom mounting screw
8. Jack hole cover

EXTERIOR TRIM

Rocker panel, removing and installing

- Pull off rubber rocker panel protector.
  - Remove top and bottom mounting screws and remove rocker panel.

NOTE —
*In early production cars the rear torsion bar end cover (**2**) is one piece with the rocker cover. In later years it comes as a separate piece.*

- Remove screws securing end cover and remove covers.
- Installation is reverse of removal.

NOTE —
To reinstall the rubber protective strip use a rubber lubricant and slide the strip in from one end.

Rear reflector panel, removing and installing

- Open engine compartment lid.
- Lift seal in back of engine compartment up to expose reflector panel plastic fasteners.
  - Release fasteners by turning each a half turn.

NOTE —
When the fastener is released it will pop-up.

- Tilt reflector panel out at top and lift out of retaining clips at bottom.
- Installation is reverse of removal.

Porsche emblem

- Two speed nuts (**arrows**) are used to mount the Porsche emblem to the luggage compartment hood.
- Use a deep 8 mm socket to remove nuts. Use new nuts when installing emblem.

680 Interior Trim

GENERAL

This repair group covers interior trim removal and installation procedures. For removal and installation of interior switches, including interior lighting equipment, see **960 Electrical Switches, Interior Lighting**.

CENTER CONSOLE

The hazard warning switch, the rear defogger switch, the central locking main switch, as well as some of the heating/A/C controls are installed in the center console. Access to these, as well as removal of the shifter boot, makes it necessary to partially or completely disassemble the center console.

Center console, removing and installing

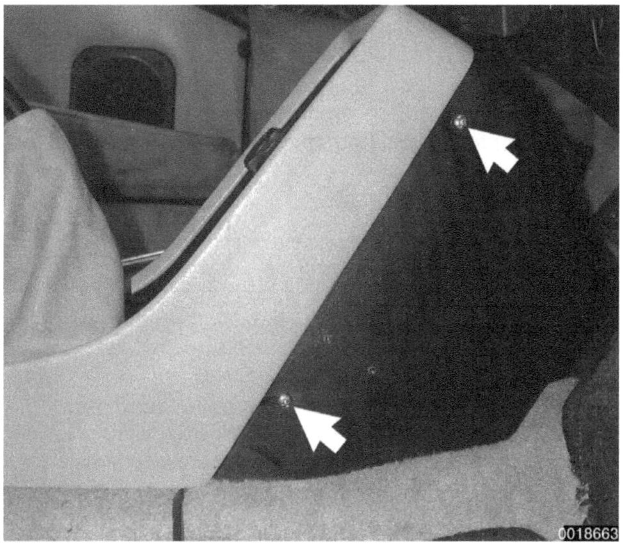

Remove left and right side screws (**arrows**) securing front cover to center console. (Right side shown; left is similar.) Slide cover forward (toward front of car) off center console.

680

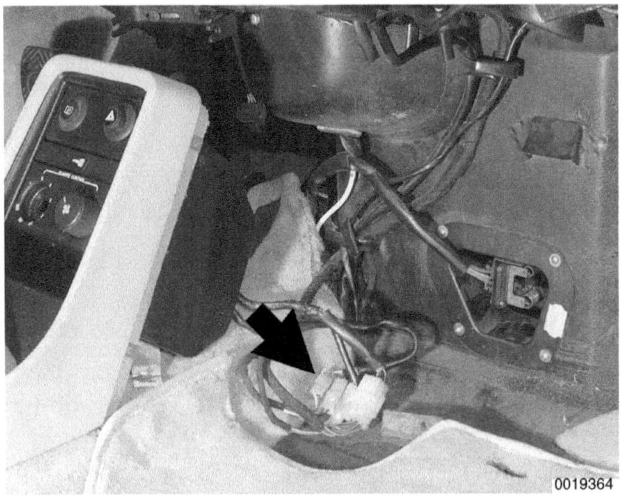

◄ Working in front of center console, detach emergency flasher, rear window defogger and power door lock harness connectors (**arrow**).

NOTE —
Harness connectors are at the front bulkhead to the right of the center tunnel.

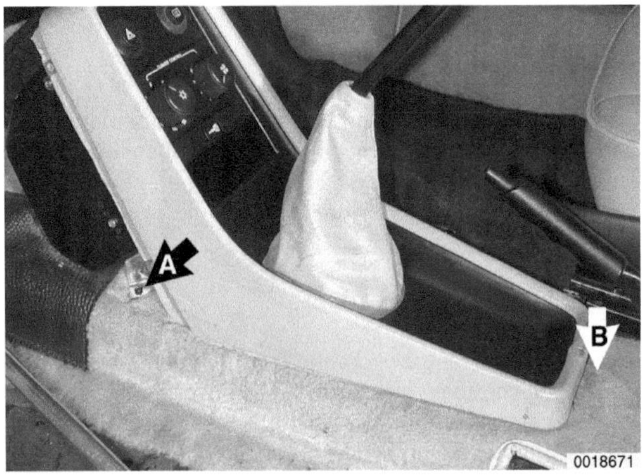

◄ Remove center console mounting screws.

- Screw on left shown (**A**). Right side is similar.
- Pull back rug slightly to reveal screw in middle of rear (**B**).

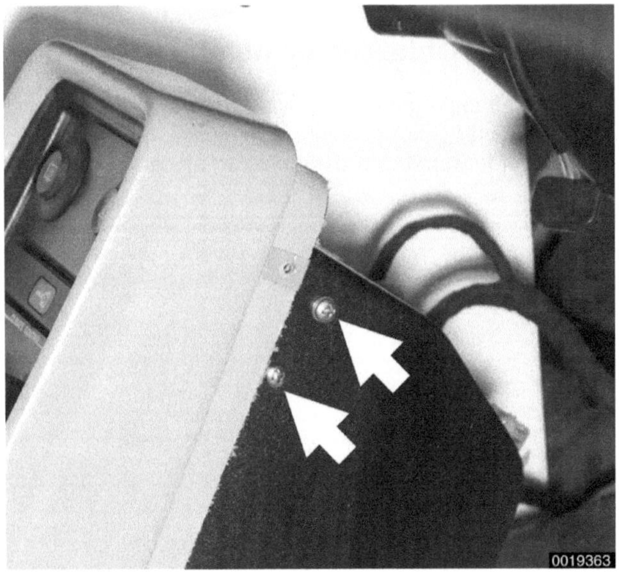

◄ Remove screws (**arrows**) securing switch supports to center console. (Right side shown; left is similar.)

– Pull off A/C temperature control knob, then remove slotted nut in back and push A/C switch forward out of support.

– Pull switch supports and switches out of console, carefully feeding harnesses into front of console housing.

> ***CAUTION —***
> *Do not pull on the A/C temperature control switch. This can damage the temperature sensor conduit.*

NOTE —
4 out of 5 switches can be removed together with the switch support assembly. The A/C temperature control switch is attached to the temperature sensor conduit which is inserted into the evaporator in the luggage compartment. The A/C switch must be removed through the front of the console housing.

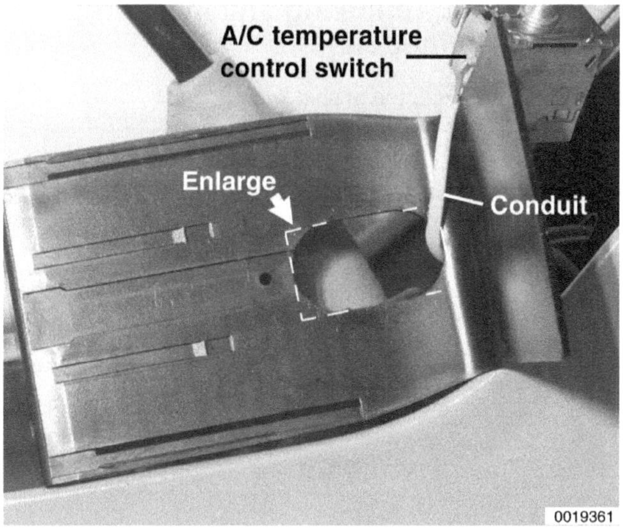

⮜ To get temperature control switch out of switch support, opening in support may have to be enlarged (**dotted lines**).

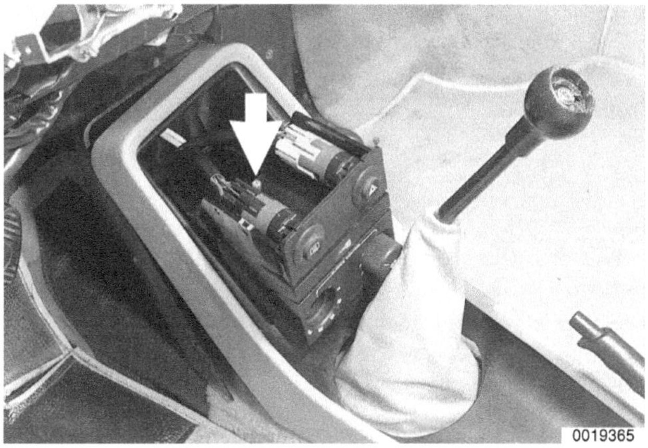

⮜ Remove switch support retaining screw (**arrow**).

- Detach wiring harness connectors from switches. Remove switch support assembly from console

⮜ Switch supports can be slid apart from each other (**arrows**) to access and remove different switches.

- Lift console over shifter and remove.
- Installation is reverse of removal.

Shifter boot, removing and installing (1987-1989 models)

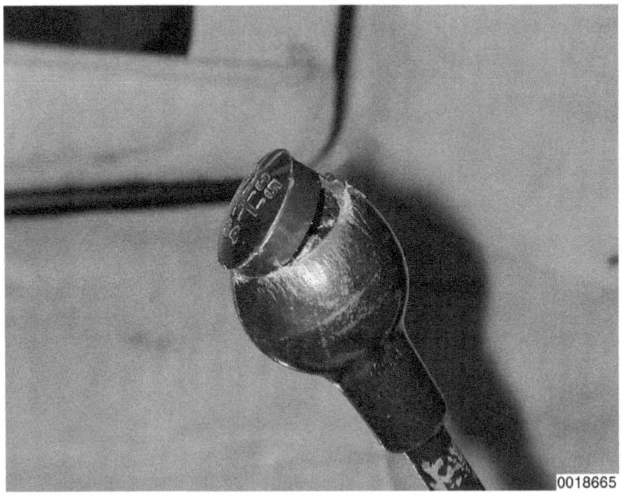

◀ Gently pry cap from top of shift knob.

NOTE —
Cap locking tabs are at 4, 8 and 12 o'clock.

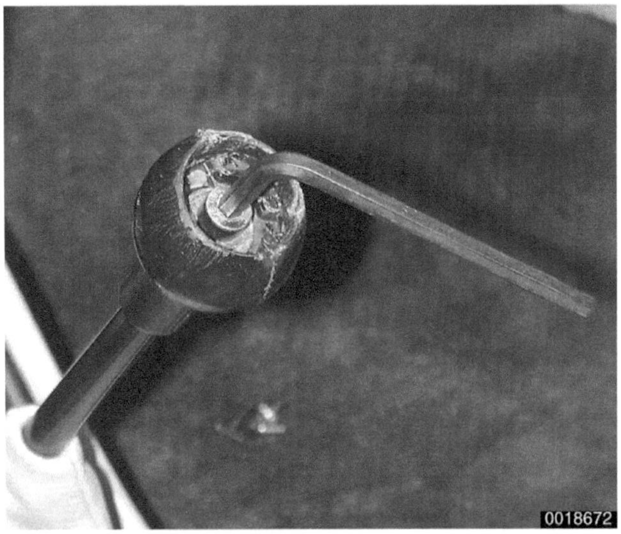

◀ Remove allen bolt holding knob in place. Remove knob.

– Remove center console as described earlier.

◀ Detach leather shifter boot from foam rubber cover (**arrow**).

0018664

Pull shifter boot up to expose rubber o-ring (**arrow**) securing boot to shift lever. Slide O-ring down off boot and remove boot.

DASHBOARD

Most gauges and instruments are removed from the dashboard cover by gentle prying. See **901 Instruments**.

A cracked or damaged dashboard cover can be replaced using the procedure given below.

Dashboard cover, removing and installing

- Remove windshield.
- Use a sharp blade to slice off the lip of the windshield gasket inside the car. Then gently tip windshield forward and out.

0019417

Remove five screws (**arrows**) securing defroster outlet duct to dashboard cover. Remove duct.

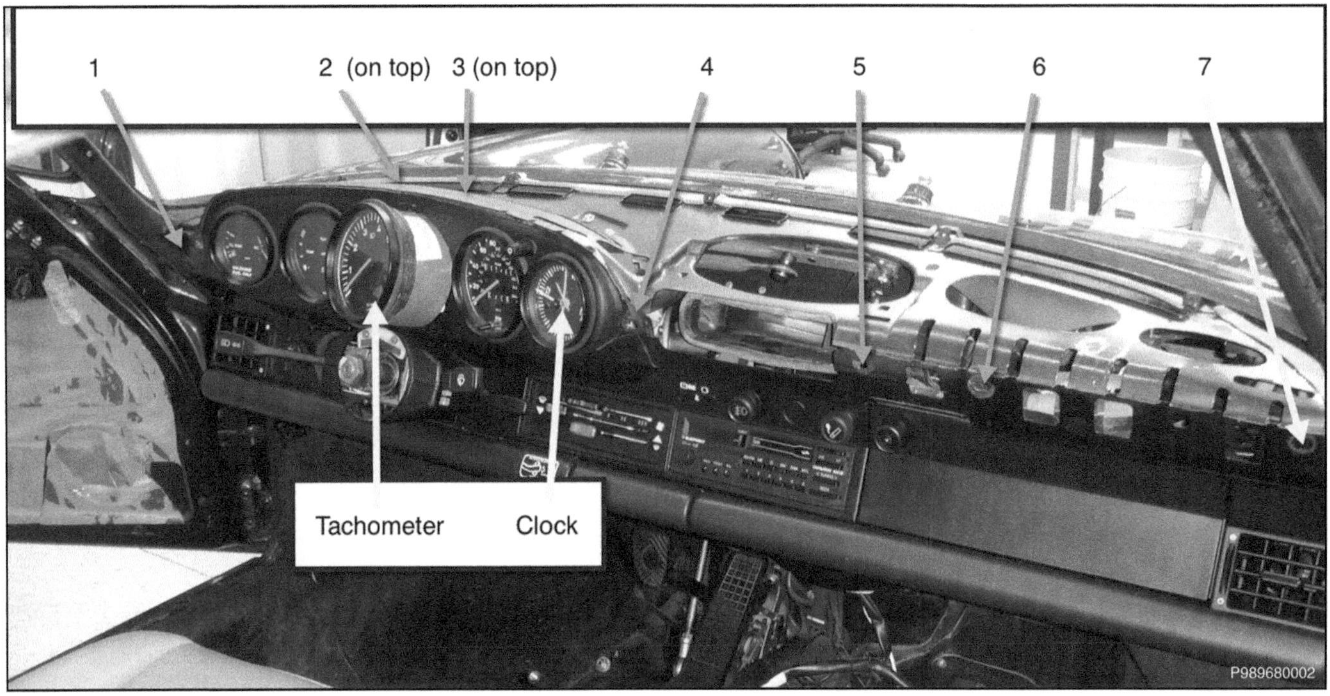

NOTE —
Above photo shows dashboard pad removed.

- Pry out and remove tachometer and clock from dashboard.
- Working through instrument openings, remove three nuts from dashboard mounting studs.
  - Tachometer opening: remove two nuts at top dashboard cover (**2**, **3**).
  - Clock opening: remove nut (**4**) on right side.
- Working from inside luggage compartment, locate and remove 4 lock nuts (**1**, **5**, **6**, and **7**) securing dashboard cover to dashboard.
- If applicable, lift dashboard cover to release snap clips located above gauges and work dashboard cover away from dashboard.

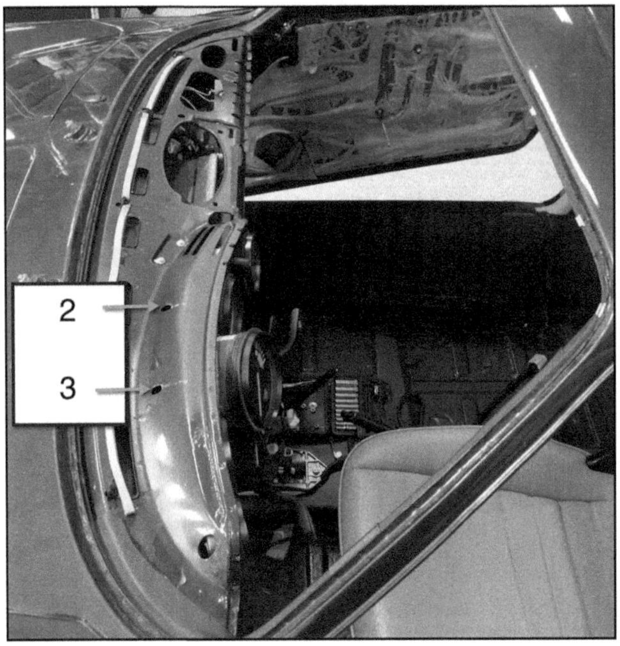

Installation is reverse of removal. When reinstalling cover, locate upper studs in openings **2** and **3** when fitting cover to dashboard.

Knee bar, removing and installing

◄ Working under dashboard, remove knee bar screw (**arrow**). (Left side shown. Right is similar.)

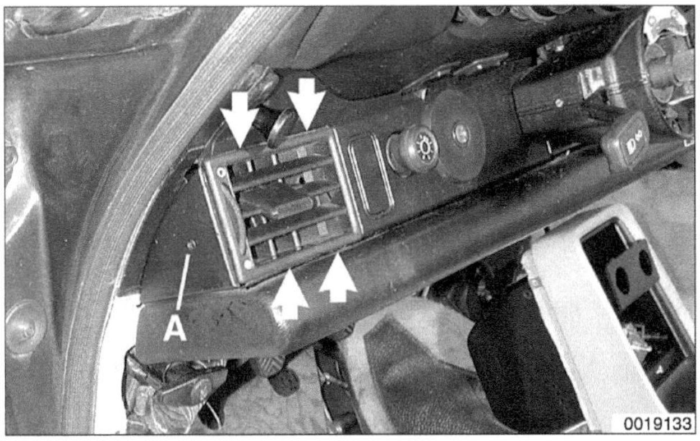

◄ Remove outer A/C vent from dash.

- Remove screw (**A**) from trim panel on side of dash.
- While pulling on A/C vent, release locking tabs (**arrows**) at top and bottom of vent with a small screwdriver.

◄ Remove knee bar:

- Working through A/C vent opening, remove knee bar mounting nut and washer from stud (**A**).
- Working just below heater controls or radio in center of dashboard, remove knee bar mounting nut and washer from stud (**B**).
- Right side: Remove radio as described in **910 Radio**. Remove knee bar mounting nut and washer through radio opening.

– Installation is reverse of removal.

GLOVE COMPARTMENT

Glove compartment, removing and installing

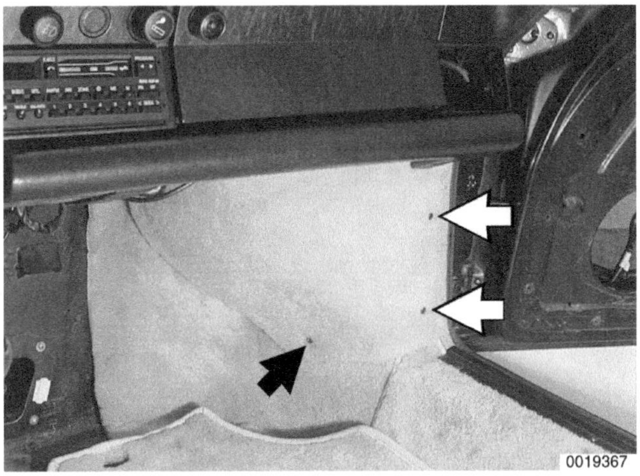

◄ Remove right side kickpanel mounting screws (**arrows**) and remove panel.

– Loosen clamps and remove blower hose.

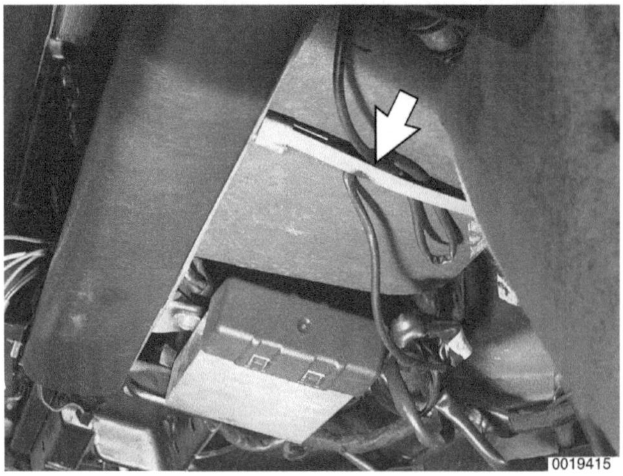

Remove screw from glove compartment mounting strap (**arrow**).

- Remove glove compartment.
- Installation is reverse of removal.

Glove compartment door, removing and installing

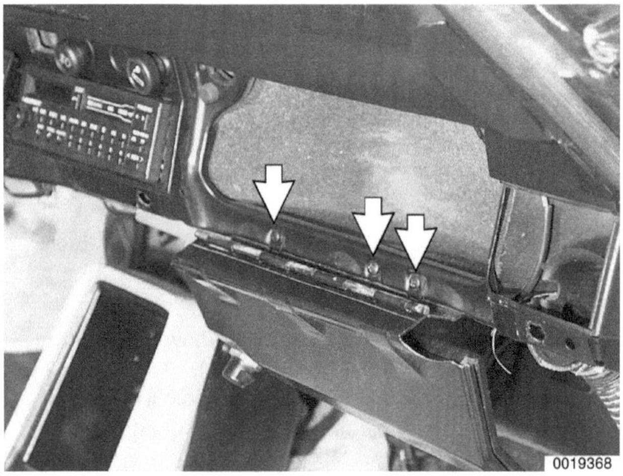

- Remove right knee bar as described earlier.
- Open glove compartment door.

Remove hinge screws (**arrows**) and remove door.

- Installation is reverse of removal.

SUN VISORS

Visor tension, adjusting

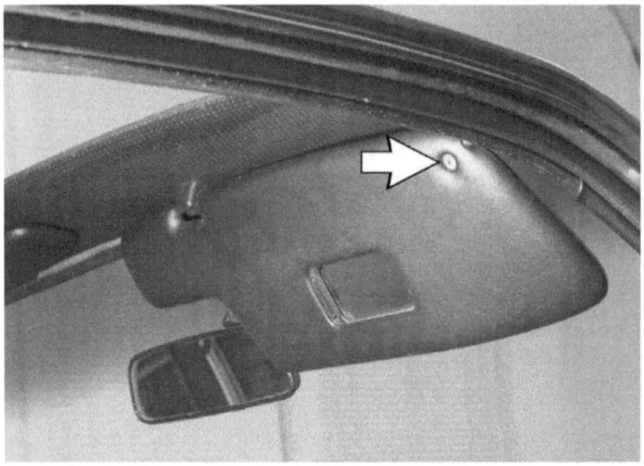

Visor tension can be changed by loosening or tightening the adjusting screw (**arrow**).

Visor, removing and installing

- Snap visor out of center support and turn 90° to side.
- Remove 2 screws securing visor bracket to headliner and remove visor.
- Installation is reverse of removal.

720 Seats

GENERAL

A variety of seat styles including leatherette, leatherette/leather and all-leather versions are installed in the models covered by this manual. There are also manual, semi-electric and fully electrically powered versions. Some versions are equipped with heated seat and backrest.

This repair group covers removal and installation of the front seats. For electrically powered seats, replacement of the seat height adjustment servo motor and cable is also covered.

Upholstery repair and repairs to seat heating components are beyond the scope of this manual.

SEAT REMOVAL

Manual and electrically powered seat removal procedures are similar.

Front seat, removing and installing

– On model with power or heated seat: Disconnect negative (–) cable from battery.

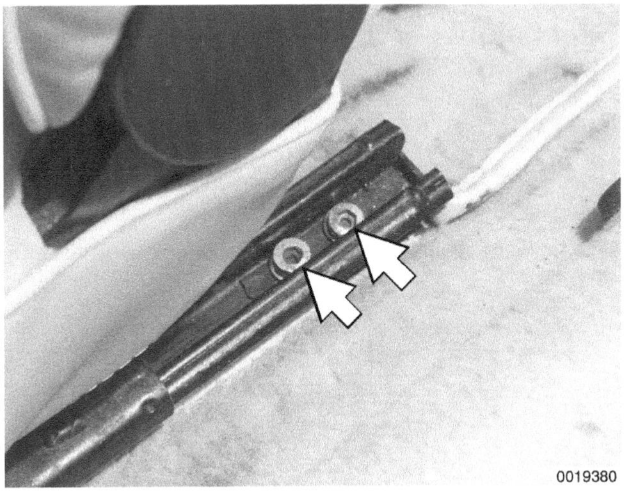

0019380

◄ Move seat to full forward position. Remove mounting bolts (**arrows**) at rear of each seat rail. (Left rail shown. Right is similar.)

Slide seat back to rear position. Remove mounting bolts (**arrow**) at front of seat rails.

NOTE —
- *Use a blanket to protect door sill from scuffing during seat removal.*
- *If car is equipped with a power driver seat and a manual passenger seat, the passenger seat will have spacers between seat rail and seat.*

– On model with power or heated seat: Tilt seat back to access and unplug electrical harness connectors under seat cushion.

– Cut wire ties as necessary.

– Remove seat from car.

– Installation is reverse of removal.
- Use wire ties or equivalent means to keep seat harness wiring from damage.
- Install all seat mounting bolts finger-tight before torquing.

Tightening Torque
- Seat to seat rail.................... 23 Nm (17 ft-lb)

SEAT REPAIR

There are two versions of power seats used in the 911 Carrera: Partially power adjustable or fully power adjustable. The partially adjustable seat has servo motors to adjust front and rear seat height only. The fully adjustable seat is equipped with servo motors for horizontal adjustment, height adjustment, backrest adjustment and lumbar adjustment.

Repairs to the servo motors can be carried out once the seat has been removed as described earlier.

Seat height servo motor and cable, removing

– Remove seat as described earlier and turn upside down.

Detach seat height control cable from servo motor by prying spring clip from end of cable.
- Remove mounting screws (**A**) and detach motor from seat frame.

< Angle motor out of seat frame and detach electrical harness connectors (**arrows**).

- Installation is reverse of removal.

800 Heating, Ventilation, Air Conditioning–General

TABLES

GENERAL

This section covers general information for the heating, ventilation and air conditioning (A/C) systems. Diagnostic and troubleshooting information for these systems is at the end of the section.

The Porsche flow-through heating and ventilation system provides draft-free air-flow while driving. Outside air flows through the grille in front of the windshield and into the passenger compartment via the windshield defroster, dashboard and footwell vents. The flow-through air is heated or cooled according to settings on the heating and A/C controls.

Heating,ventilation, A/C system components

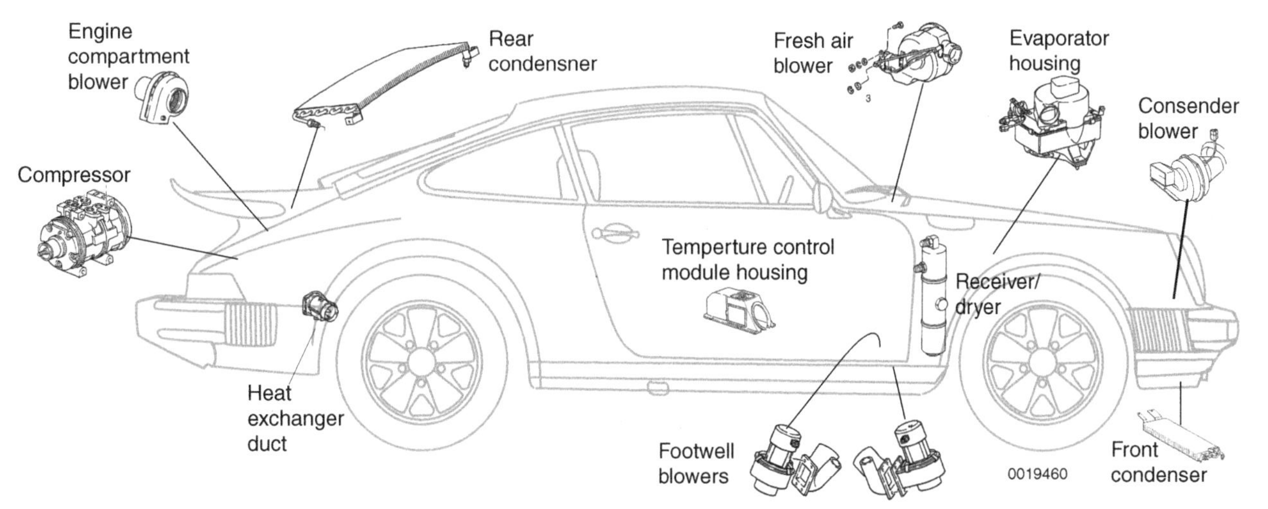

Fresh air and heating footwell/defrost flap controls

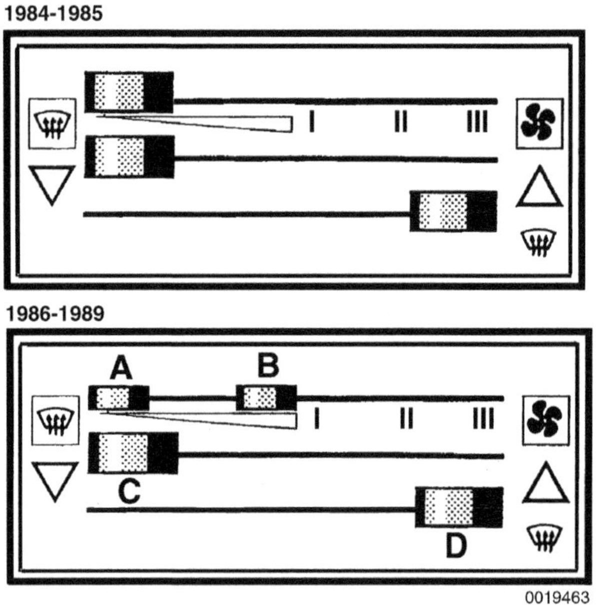

Fresh and heated air distribution

The air distribution levers control fresh air and heated air to the footwells and the defroster outlets.

NOTE —
The slider controls have no control over air distribution to the center dash vent or the side dash vents. This center vent is used for A/C only. The left and right side vents are for A/C or heated air and are turned on and off by the vertical knurled control knob.

- **Top lever(s)**: Controls fresh air intake from outside the vehicle. When the lever(s) is in the extreme left position, air intake from the outside is closed off. When the lever(s) is in the extreme right position, maximum ventilation of the car interior is obtained.
- **Middle lever**: Fresh air footwell/defrost blend door. This lever directs fresh air to the footwells (extreme left position) or the defroster (extreme right position) or a blend of the two.
- **Bottom lever**: Heated air footwell/defrost blend door. This lever directs heated air to the footwells (extreme left position) or the defroster (extreme right position) or a blend of the two. On cars with automatic heating, heated air temperature is controlled by the temperature control knob between the seats.

NOTE —
- *For information on heating and defrosting component replacement, see* **801 Heating**.
- *For information on ventilation system component replacement, see* **850 Ventilation**.

Heater control module assembly

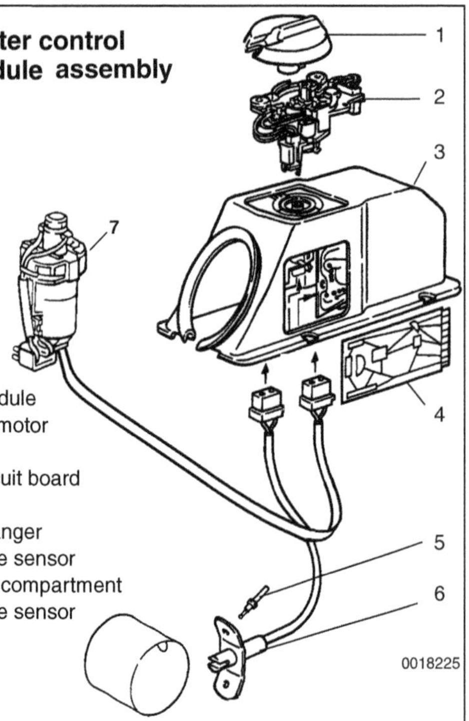

1. Knob
2. Control module with servo motor
3. Housing
4. Printed circuit board
5. Rivet
6. Heat exchanger temperature sensor
7. Passenger compartment temperature sensor

Automatic heating system

The automatic heating system controls the passenger compartment temperature based on the setting of the temperature control knob.

By comparing the passenger compartment temperature to the setting at the temperature control knob, the control module servo motor automatically opens or closes the heater control box valves to maintain the desired temperature.

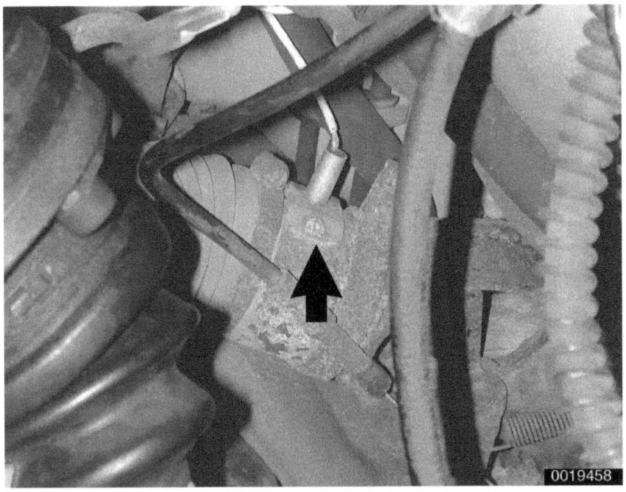

0019458

The **heat exchanger temperature sensor** (**arrow**) is used as part of the automatic heater system to automatically control heat distribution into the passenger compartment. The sensor is installed inside the left heater duct next to the transmission and prevents operation of the footwell blower motors when the engine is cold.

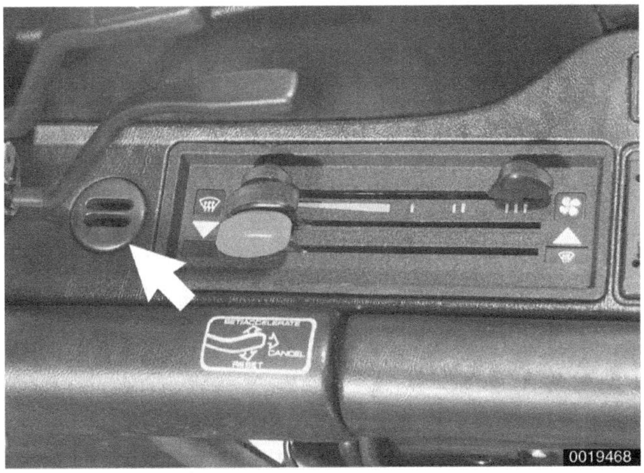

0019468

The **passenger compartment temperature sensor** monitors the passenger compartment temperature. On 1984 and 1985 cars, the sensor is located behind a trim cover in the headliner between the sun visors. On 1986 and later cars, the sensor is fitted in the dash, behind a small grille (**arrow**).

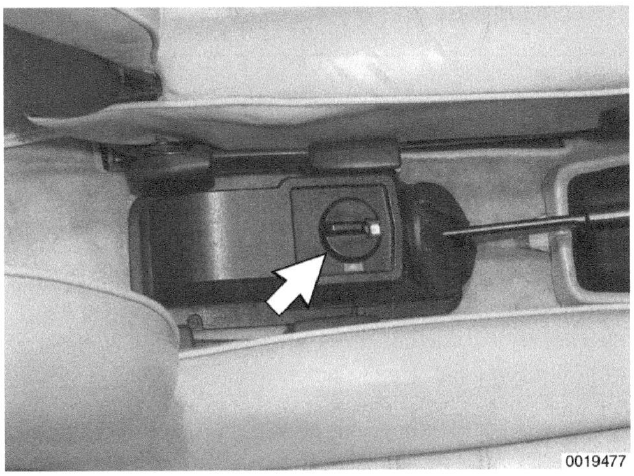

0019477

The **temperature regulating control module** is operated through the temperature control knob (**arrow**) and regulates the opening and closing of the heater control box valves via an integrated servo motor. The control module is located below the handbrake console.

The **engine compartment blower** forces air through the heat exchangers and into the passenger compartment when heat is called for at the temperature control knob. The blower is mounted on top of the engine.

Footwell blower motors provide supplemental heated air flow to the air distribution ducting. There are two footwell blower motors, located ahead of the left and right A-pillars.

Air Conditioning

With the A/C on, the air is cooled and dehumidified by the evaporator to maintain the desired temperature.

Cooling efficiency of the air conditioner system is dependent on the engine rpm. The greater the rpm the higher the output from the air conditioner. The air conditioner works only with the engine running. The rear lid must be closed for the A/C system to operate properly.

The schematic view shown below is a typical A/C system. In the Porsche system there are two condensers in the vehicle, one in front, cooled by a blower, and another in the rear.

A/C components

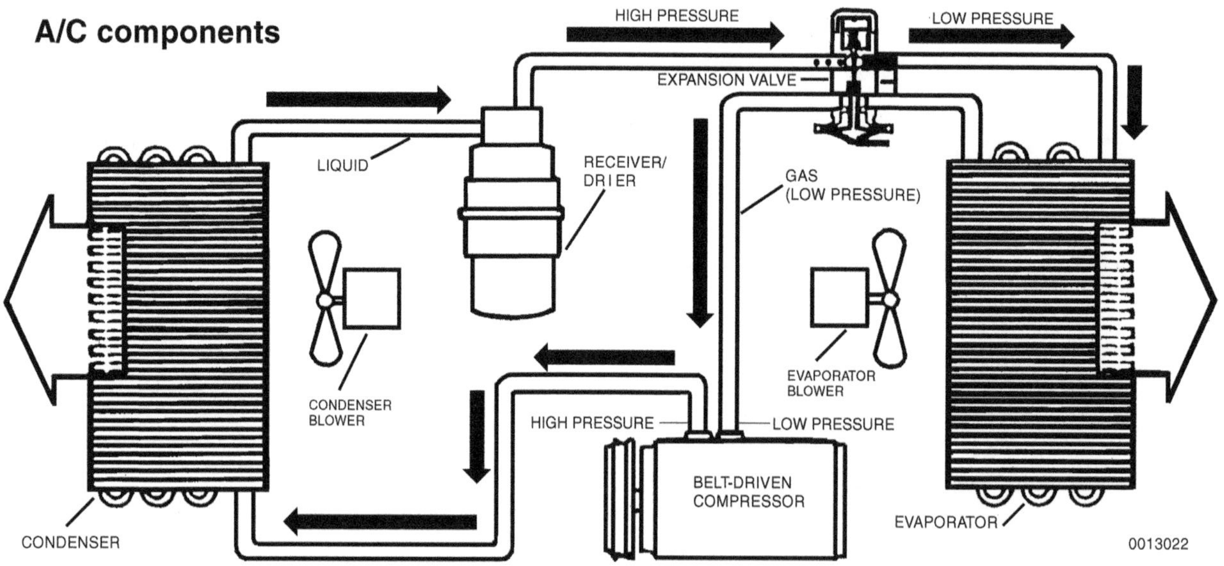

The A/C system removes heat and moisture from the passenger compartment. It accomplishes this through the application of four physical principles: materials absorb heat as they change from a liquid to a gas (evaporate); materials give off heat as they change from a gas to a liquid (condense); the boiling point of a liquid varies with its pressure; and, heat always flows from hot to cold. For example, the first principle is demonstrated by wetting your hand and then blowing on it. As the water evaporates, it takes some heat with it and your hand feels cooler.

In the A/C system, the heat from the passenger compartment boils (evaporates) the refrigerant (R-12) in the **evaporator**, causing heat to be absorbed by the R-12. This heat is then released into the atmosphere when the R-12 is cooled and condensed into a liquid in the condensers.

Moisture is removed at the evaporator in the same way that water drops form on a cold glass. The moisture drips beneath the evaporator and onto the ground. This is the reason a water puddle may often be seen under the car when the A/C is operating.

The **A/C compressor** forces refrigerant through the system and at the same time pressurizes it, raising the refrigerant boiling point to force it to condense more easily. The compressor is engaged by an electro-magnetic clutch that is actuated when the A/C is turned on. The compressor clutch disengages when the temperature in the passenger compartment reaches the level set on the operating controls.

The **condensers**, which look like small radiators, are located in front and rear of the car. The front condenser is located below the front of the car, protected by a perforated pan. It has a separate blower, accessible through the front of the luggage compartment. The rear condenser is integral with the engine cover.

The **receiver/drier**, located in the left front wheel housing, removes small amounts of moisture and dirt from the system.

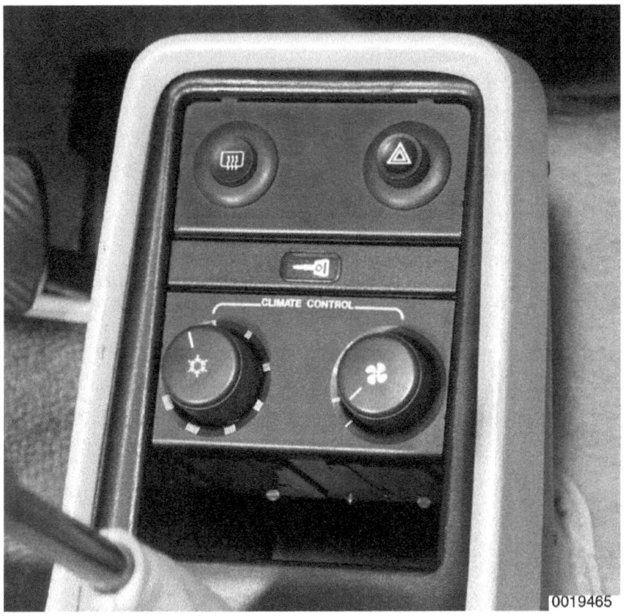

◀ Two switches in the center console operate the A/C system.

The **temperature control knob** (left hand switch) regulates the temperature of cooling air coming from the A/C vents in the dash.

- Fully left. . . . . low cooling
- Fully right . . . . full cooling

The **blower switch** (right hand switch) also turns on the compressor clutch.

- Position I . . . . no cooling, blower off
- Position II to IIII . . . .A/C on, increasing blower speed

Vents. Air is drawn in from the passenger footwell and blown through the air conditioner evaporator where it is cooled. Air conditioned air is then emitted from the center and side dash vents.

NOTE —
For maximum cooling, the console A/C control switches should be turned to the extreme right.

For information on A/C system component replacement, see **870 Air Conditioner**.

AUTOMATIC HEATING TROUBLESHOOTING

For all practical purposes, the heating system is separate from the A/C system. Therefore, problems should fall within one of two categories: heating faults or air conditioning faults. Each system has its own causes and corrective actions.

Troubleshooting of the heater electronic controls begins by removing the heater control module housing as described in **801 Heater**. Heater control module troubleshooting is summarized in **Table a**. Use the photos that follow for connector and test terminals referred to in **Table a**.

Table a. Temperature Control Troubleshooting

Symptom	Probable cause	Corrective action (refer to photos)
Temperature control knob in position 1, control servo moves to maximum heating	Open circuit in passenger compartment temperature sensor	Test sensor resistance at connector **B**. Passenger compartment temperature sensor resistance specification: 1.9 kΩ at 20°C (68°F).
	Open circuit in temperature control potentiometer	Test resistance at temperature control potentiometer terminals with wires connected and switch in position 5. Potentiometer resistance specification: 950 ± 20 Ω.
Control module regulates to maximum heating in defrost switch position only	Short circuit in passenger compartment temperature sensor	Test sensor resistance at connector **B**. Passenger compartment temperature sensor resistance specification: 1.9 kΩ at 20°C (68°F).
	Short circuit in temperature control potentiometer	Test resistance at temperature control potentiometer terminals with wires connected and switch in position 5. Potentiometer resistance specification: 950 ± 20 Ω.
Too much heat, does not regulate correctly.	Open circuit in heat exchanger temperature sensor	Test sensor resistance at connector **C**. Heat exchanger temperature sensor resistance specification: 1.7 kΩ at 20°C (68°F).
	Temperature control potentiometer set incorrectly	Set potentiometer switch to position 5 and remove knob. Connect ohmmeter to potentiometer. Hold shaft of switch with pliers and adjust potentiometer with a screwdriver to specification: 950 ± 20 Ω.
Too little heat, does not regulate correctly.	Short circuit in heat exchanger temperature sensor	Test sensor resistance at connector **C**. Heat exchanger temperature sensor resistance specification: 1.7 kΩ at 20°C (68°F).
	Temperature control potentiometer set incorrectly	Set potentiometer switch to position 5 and remove knob. Connect ohmmeter to potentiometer. Hold shaft of switch with pliers and adjust potentiometer with a screwdriver to specification: 950 ± 20 Ω.
Control module not regulating	Open circuit in power supply	Check for battery voltage at connector **A** terminal 15. Check for continuity to ground at connector **A** terminal 31.
	Printed circuit board defective	Replace circuit board.
	Control module defective	Replace control module.

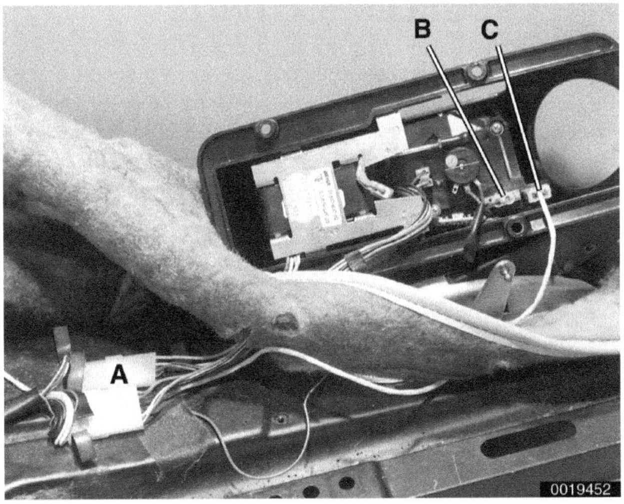

◄ Test connections at temperature control module:

- Harness connector **A** (power supply).
- Harness connector **B** (passenger compartment temperature sensor).
- Harness connector **C** (heat exchanger temperature sensor).

◄ Test connections (**arrows**) at temperature control potentiometer.

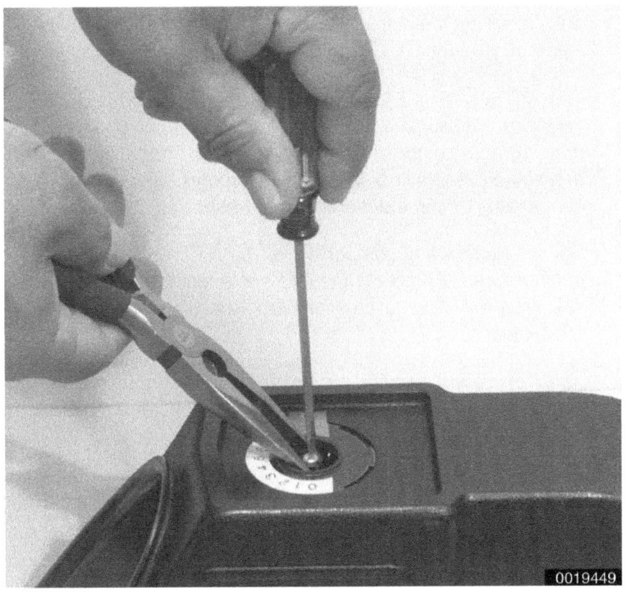

◄ Adjust temperature sensor potentiometer:

– Set potentiometer switch to position 5 and remove knob. Connect ohmmeter to potentiometer (shown above). Hold shaft of switch with pliers and adjust potentiometer with a screwdriver to specification: 950 ± 20 Ω

A/C TROUBLESHOOTING

Although detailed A/C diagnosis is beyond the scope of this manual, this troubleshooting section may help to pinpoint common air conditioning problems.

Before starting to troubleshoot the A/C system, check the following:

- Evaporator blower motor: do all three speeds work?
- Is the A/C compressor drive belt tight?
- Is the A/C compressor clutch engaging when switched on?

A/C warnings and cautions

Heeding the warnings and cautions will help prevent personal injury or damage to the system.

WARNING —

- *Work in a well ventilated area. Avoid breathing refrigerant vapors. Exposure may irritate eyes, nose, and throat. Switch on exhaust/ventilation systems when working on the refrigerant system.*

- *Always be careful that refrigerant does not come in contact with your skin. Always wear hand and eye protection (gloves and goggles) when working around the A/C system. If refrigerant does come in contact with your skin or eyes:*
  -Do not rub skin or eyes
  -Immediately flush skin or eyes with cool water for 15 minutes
  -Rush to a doctor or hospital.
  -Do not attempt to treat yourself.

- *Refrigerant gases are heavier than air, displace oxygen and may cause suffocation in areas of poor circulation, for example under a car.*

- *Do not discharge/charge the A/C system without proper equipment and training. Damage to the vehicle and personal injury may result.*

- *The A/C system is filled with refrigerant gas which is under pressure. Pressurized refrigerant in the presence of oxygen may form a combustible mixture. Never introduce compressed air into any container (full or empty), capped off A/C component, or piece of service equipment.*

- *Do not expose any component of the A/C system to high temperatures (Above 80°C/176°F) or open flames. Excessive heat will cause a pressure increase which could burst the system.*

- *Keep refrigerant away from open flames. Poisonous gas will be produced if it burns. Do not smoke near refrigerant gases for the same reason.*

> ***CAUTION —***
> * *Immediately plug open connections on A/C components to prevent dirt and moisture contamination.*
> * *Do not steam clean condensers or evaporators. Use only cold water or compressed air.*
> * *It is recommended that all service to the A/C system refrigerant components be left to an authorized Porsche dealer or other qualified A/C service facility. If any of the refrigerant components are disconnected or removed, special equipment, supplies, and new parts will be needed to remove moisture from the system and recharge the refrigerant before it is closed back up.*
> * *As of January 1, 1992 any person who services a motor vehicle air conditioner must, by law, be properly trained and certified, and use approved refrigerant recycling equipment. Technicians must complete an EPA-approved recycling course to be certified*
> * *Do not top-off a partially charged refrigerant system. Discharge system, evacuate and then recharge system.*
> * *State and local governments may have additional requirements regarding air conditioning servicing. Always comply with state and local laws.*

A/C inspection and tests

Periodic inspections will help keep the A/C operating at its peak. There are almost always small leaks in the system that will require that it eventually be recharged with R-12. The V-belt that drives the compressor is subject to wear. The condenser fins can become bent or covered with debris, reducing air flow and raising system pressure to damaging levels.

NOTE —
Run the air conditioning system for a few minutes every few weeks to keep the seals lubricated.

Clean any debris or bugs from the front of the condenser. Straighten any bent fins using a fin comb. Check the fresh air intake for obstructions. Inspect the compressor, the hoses, and all visible components for any oil leaks. These leaks are often seen at the bottom side of the fittings and components. Inspect the wiring to the compressor clutch. Check the compressor mountings for tightness, and check the condition and tension of the V-belt as described later.

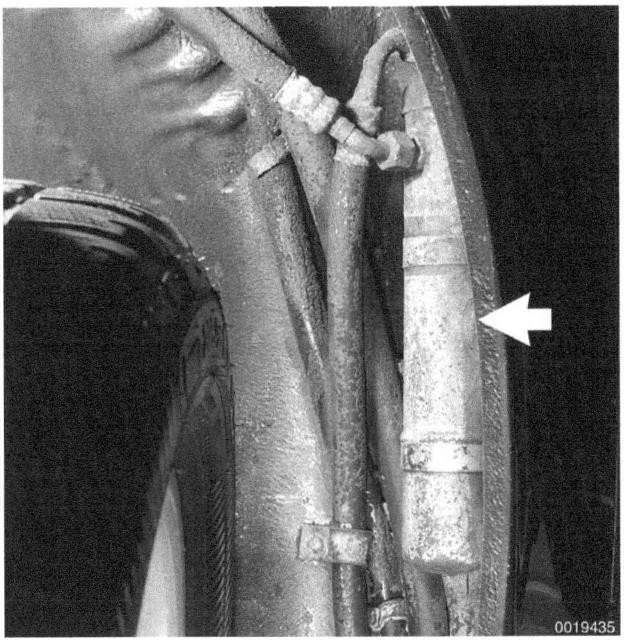

Refrigerant charge, checking

The refrigerant (R-12) charge is quickly checked by looking at the sight glass on the receiver/dryer (**arrow**), located in the left front wheel housing.

NOTE —
It may be necessary to loosen the receiver/dryer mounting clamps in order to rotate the sight glass into view.

– Start engine and turn A/C on to its maximum setting. With compressor running (clutch cycled on) view sight glass. There should be few or no bubbles visible in the glass. A constant foaming indicates that system charge is low. Streaks on interior of glass indicate that system is discharged.

If using an aftermarket refrigerant product to recharge the system, follow the manufacturer's directions closely. To prevent component damage, a totally discharged system must be evacuated (sometimes called pulling a vacuum) using special equipment before recharging. This removes any harmful moisture from lines and components.

A/C system pressure, testing

Test system pressure at an engine speed of 2,000 rpm and an outside temperature of approximately 77°F (25° C). Keep engine lid closed.

A/C System Pressure

- Low side . . . . 0.81 - 2.0 bar (12 - 30 psi)
- High side. . . . . 7.93 - 14.13 bar (115 - 205 psi)

WARNING —
Wear eye protection when testing the system. A/C refrigerant at normal atmospheric pressures can evaporate and freeze anything it contacts.

Table b lists A/C refrigerant system capacities.

Table b. A/C System Capacities

Compontent	Oil capacity
Refrigerant (R12): Total system capacity	 1350 gr. (47 oz.)
A/C lubricating oil: Compressor Evaporator Condensers Receiver/dryer	 50 gr. (2.0 oz) 40 gr. (1.6 oz.) 20 gr. (0.8 oz.) 10 gr. (0.4 oz.)

A/C evaporator blower motor, testing

– To test evaporator blower, first remove luggage compartment floor mat.

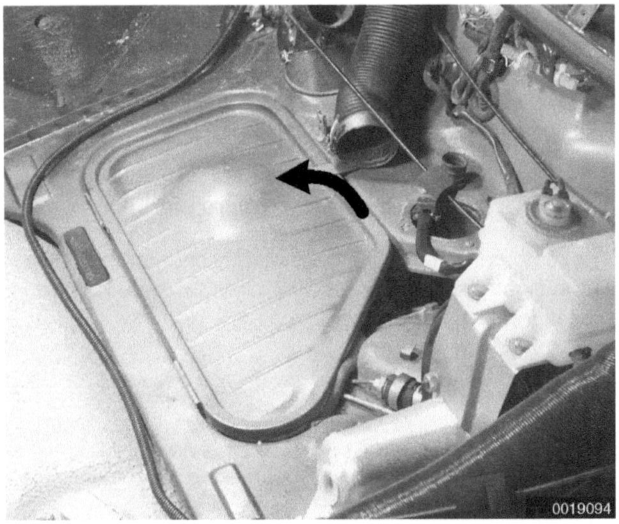

◄ Open access door in floor behind fuel tank.

Detach connector (**arrow**) for A/C evaporator blower motor and turn ignition and A/C blower switch on (position IIII, maximum setting).

- Check for battery voltage at connector red/white wire.
- If voltage is present and blower does not operate, blower motor is faulty and should be replaced.
- If voltage is not present, check for a faulty fuse (luggage compartment fuse box 3, fuse 2), damaged wires or connectors, or a faulty A/C blower switch.

A/C blower resistor pack, testing

The A/C blower resistor pack, located in a compartment behind the passenger side footrest, is also referred to as the bimetallic switch.

– If the evaporator blower works at the maximum setting, but not in A/C blower switch positions II and III, test resistor pack.

- Detach harness connector at resistor pack.
- With ignition switch on and blower switch in position II, there should be battery voltage at harness connector red/green wire.
- With ignition on and blower switch in position III there should be battery voltage at harness connector red/black wire.
- If voltage is present and blower does not operate in the lower speeds, blower resistor pack is faulty and should be replaced.
- If voltage is not present in either switch setting, check for damaged wires or connectors or a faulty A/C blower switch.

A/C compressor clutch, testing

To quickly check the operation of the compressor clutch, turn the ignition on and turn the temperature regulating knob to its full cold position. Turn the A/C switch on and listen for the compressor clutch to click on.

If no click is heard, the compressor clutch may not be receiving voltage or the clutch itself may be faulty.

- Check for battery voltage at single-wire connector (**arrow**) leading out of compressor.
- If voltage is present, compressor clutch is faulty and should be replaced.
- If no voltage is present, check for a faulty fuse (luggage compartment fuse box 3, fuse 2), or damaged wiring.

Table c suggests some corrective actions for additional A/C fault symptoms. Most other testing of the A/C system requires specialized knowledge and equipment and is beyond the scope of this manual. Servicing by an authorized Porsche dealer or other qualified air conditioning shop is highly recommended.

Table c. A/C Troubleshooting

Symptom	Probable cause	Corrective action
Cooling has failed completely	Burst seal on receiver/drier due to A/C system overheating	Check front condenser blower fuse, relay and motor. See **971 Electrical Component Locations**. Replace receiver drier. See **870 Air Conditioner**.
Low pressure in both high and low sides	Leak in system	Check all hoses, lines and the compressor seal for leaks. Replace any seal found leaking.
Low side pressure to high High side pressure to low	Faulty compressor	Replace A/C compressor. See **870 Air Conditioner**.
Low side pressure normal High side pressure too high	Refrigerant overfilled	Discharge system and recharge.
Low side pressure too low High side pressure too low	Refrigerant low	Charge system. Check for leaks.
Low side pressure too high High side pressure normal	Expansion valve faulty	Replace expansion valve. See **870 Air Conditioner**.

801 Heater

GENERAL

This group covers automatic heating system component replacement. Refer to the following groups for further information:

- **800 Heating, Ventilation, Air Conditioning–General**
- **850 Ventilation**
- **870 Air Conditioner**

NOTE —

- *Some of the cars covered by this manual may be fitted with a manual heating control system and not the optional automatic system (M424 option code). For information on the ducting and ventilation controls for these cars, see* **850 Ventilation**.
- *Automatic heating system troubleshooting is covered in* **800 Heating, Ventilation, Air Conditioning–General**.

Special tools

Some of the procedures require special tools.

Special Tools

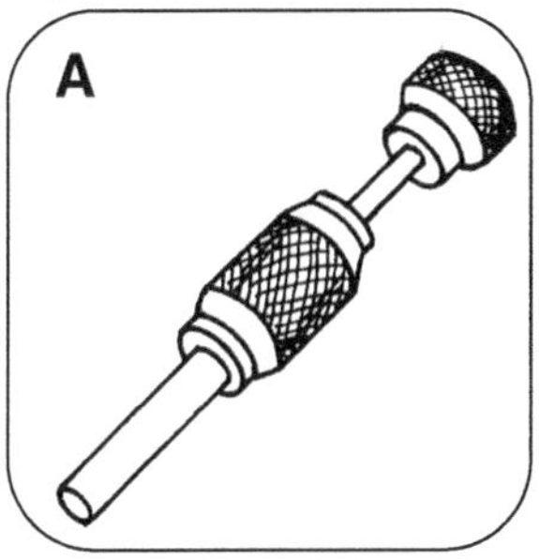

Special tools

A Baum 9700/9701/9702. . . . . Wiring connector removal tool (source: Baum Tools Unlimited)

Automatic heating system

The automatic heating system controls the passenger compartment temperature based on the setting of the temperature control knob (**arrow**). By comparing the passenger compartment temperature to the setting at the temperature control knob, the control module servo motor automatically opens or closes the heater control box valves to maintain the desired temperature. For a description of the automatic heating system, see **800 Heating, Ventilation, Air Conditioning–General**.

Heated air is blown through the exhaust system heat exchangers by an engine-mounted electric fan and supplemented by two blower motors in the left and right footwells in the passenger compartment.

Temperature is controlled by automatic regulation of the flaps on the heat exchanger valve housings. Control of the heat exchanger valves is via the servo motor in the control module housing.

NOTE —
Heat exchanger replacement is covered in **260 Exhaust System**.

Heating system components

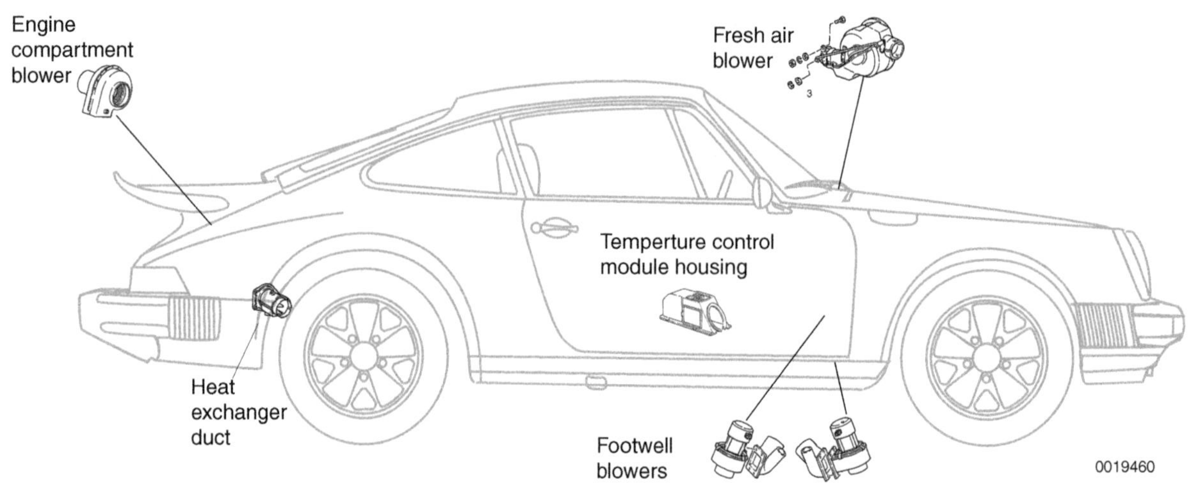

Automatic Heating System Component Replacement

Control module housing, removing and installing

- Remove passenger seat to gain access to heater controls and parking brake handle assembly.
- Remove access panel from right side of heater control housing.
- ◄ Disconnect heater valve servo control rod (**arrow**) from lever.
- Remove heater control housing trim screws from floor.
- Remove rubber sleeve covering parking brake lever.

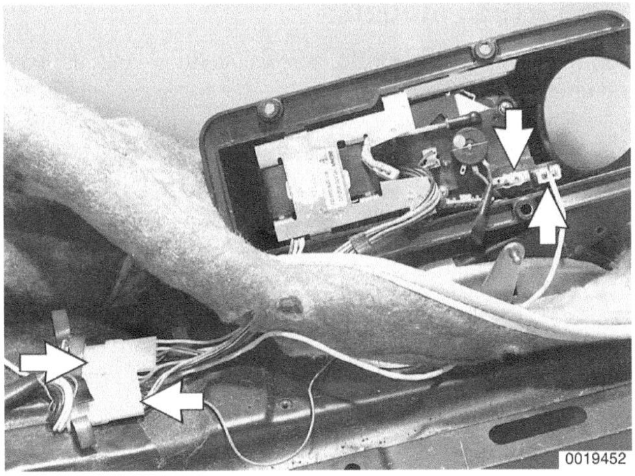

- Lift control module housing up and turn over.
  - Lift rug behind and under housing.
  - Disconnect electrical harness connectors (**arrows**).
- With all wiring harnesses disconnected remove control module housing.

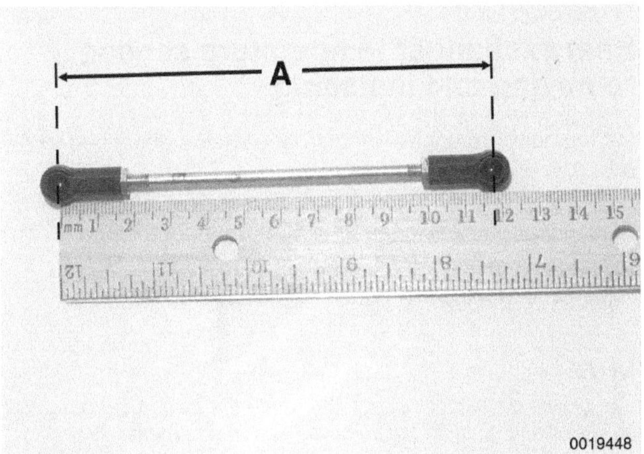

- If replacing servo control rod, set length (**A**) to specification below.

Servo Control Rod

- Length **A** . . . . 118 ± 2 mm (4.646 ± 0.079 in.)

- Installation is reverse of removal.

Passenger compartment temperature sensor, removing and installing

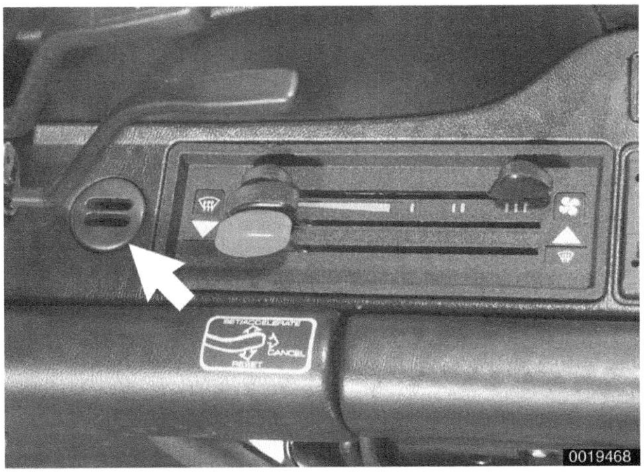

The passenger compartment temperature sensor monitors the passenger compartment temperature. On 1984 and 1985 cars, the sensor is located behind a trim cover in the headliner between the sun visors. On 1986 and later cars, the sensor is fitted in the dash, behind a small grille (**arrow**).

1984 - 1985 models:

- Remove screws holding sensor trim cover to headliner and remove cover.
- Pull sensor out of trim cover and unplug.
- Installation is reverse of removal.

1986 - 1989 models:

- Remove ashtray mounting screws, ashtray and ashtray mount.

- ∢ Remove mounting screws (**arrows**) from sensor housing.
- Disconnect hose from blower.
- Disconnect electrical harness connector and remove sensor.
- Installation is reverse of removal.

Heat exchanger temperature sensor, removing and installing

- Lift temperature control module housing as described earlier.

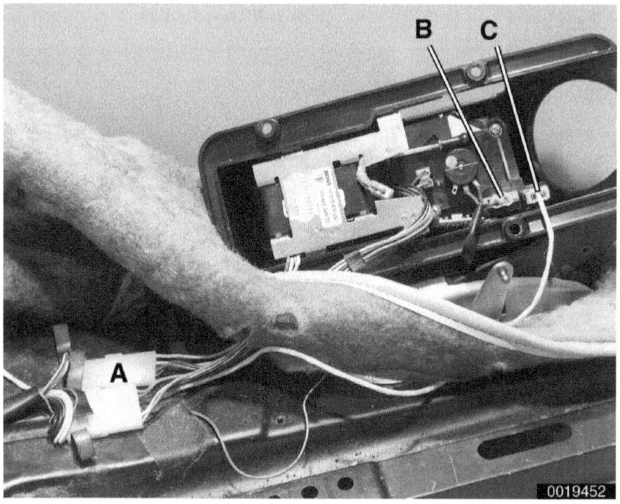

- ∢ Disconnect harness connector **C** at front of module.
    - Use special wire connector tool to push wire ends out of plastic connector.

NOTE —
The plastic connector will not fit through the hole in the rear bulkhead and must first be removed from the harness.

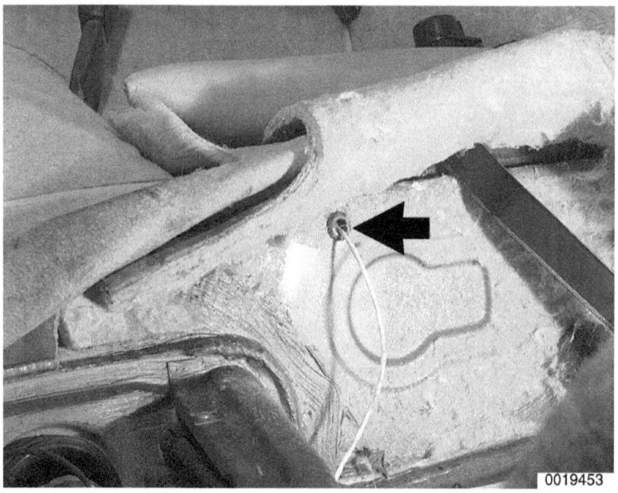

- ∢ Follow white wiring harness back to bulkhead (behind driver's seat) and push rubber grommet and harness through bulkhead.

– Remove heater hose from left heat exchanger valve housing.

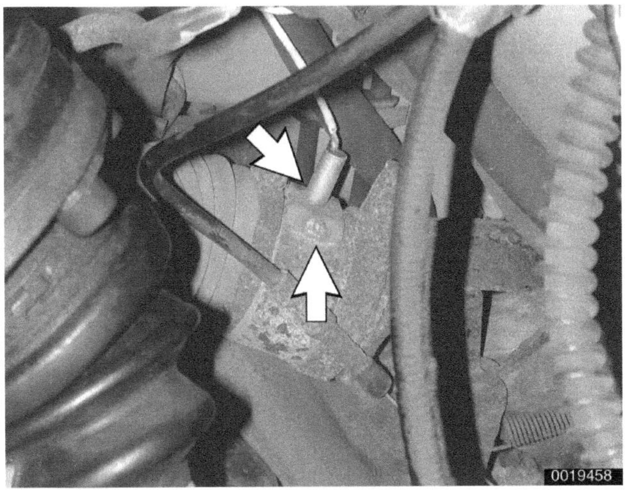

⮜ Loosen clamp bolt and remove heat sensor sleeve.

- Drill out rivets (**arrows**) and remove sensor from sleeve.

– Rivet new sensor to sensor sleeve. The remainder of installation is the reverse of removal.

Footwell blower, removing and installing

The 2-speed footwell blowers are located behind the left and right side kickpanels.

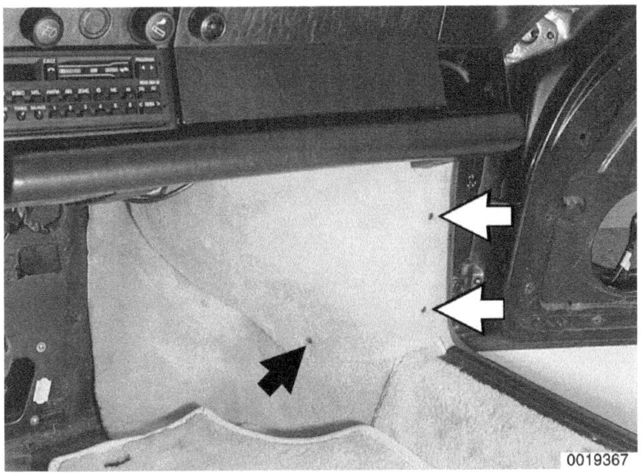

⮜ Remove kickpanel mounting screws (**arrows**) and remove panel.

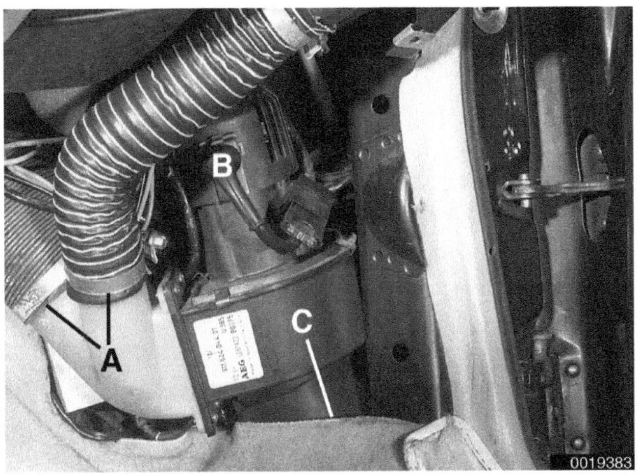

⮜ Loosen hose clamps and remove air hoses (**A**).

– Disconnect electrical harness connector (**B**).

– Remove blower from air intake tube (**C**).

– Installation is reverse of removal.

Footwell blowers series resistor, removing and installing

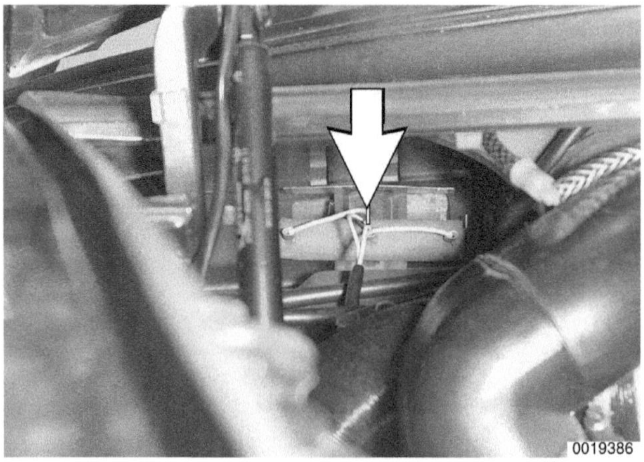

◄ The series resistor for the footwell blowers is located in the luggage compartment at the base of the windshield on the right side (**arrow**).

– Working inside luggage compartment, remove fresh air plenum cover.

– Disconnect resistor harness connector.

– Bend end tabs and remove resistor.

– Installation is reverse of removal.

Heater blower motor, removing and installing

The heater blower motor is located on top of the engine on the left side.

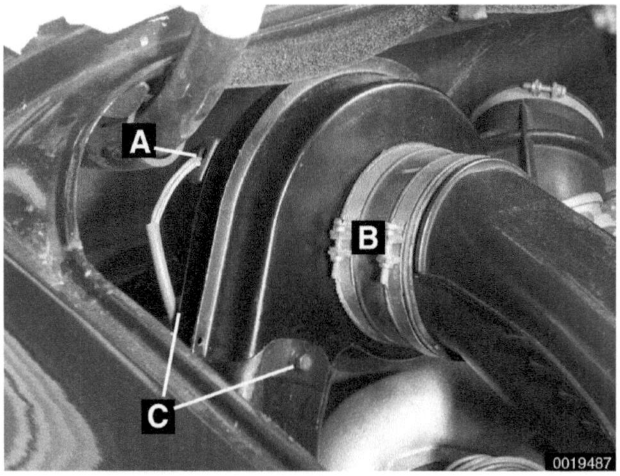

◄ Disconnect electrical harness connector (**A**) from blower.

- Detach blower air intake duct (**B**).
- Remove two bolts (at front and rear) (**C**) from output elbow and detach elbow.

◄ Remove blower motor mounting fasteners (**arrows**). Remove blower motor.

– Installation is reverse of removal.

Heat exchanger control cable, replacing and adjusting

The heat exchanger valve on each side is held open by a spring. Closing of the valves is controlled by a cable attached to the temperature control module or control lever.

– Raise rear of car and support it securely..

> ***WARNING —***
> *Make sure the car is stable and well supported at all times. Use a professional automotive lift or jack stands designed for the purpose. A floor jack is not adequate support.*

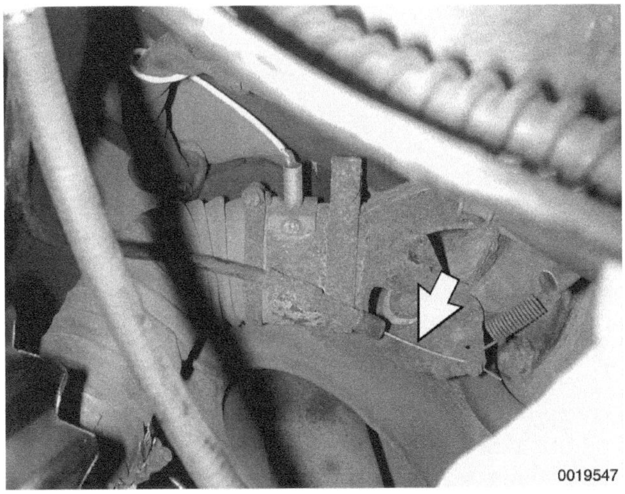

0019547

◄ Working underneath car, disconnect cable ends (**arrow**) from heat exchanger valves. (Left side shown. Right is similar.)

NOTE —
One cable is looped and used to connect the control module to both heat exchanger valves.

– Working inside passenger compartment, remove heater control module housing as described earlier.

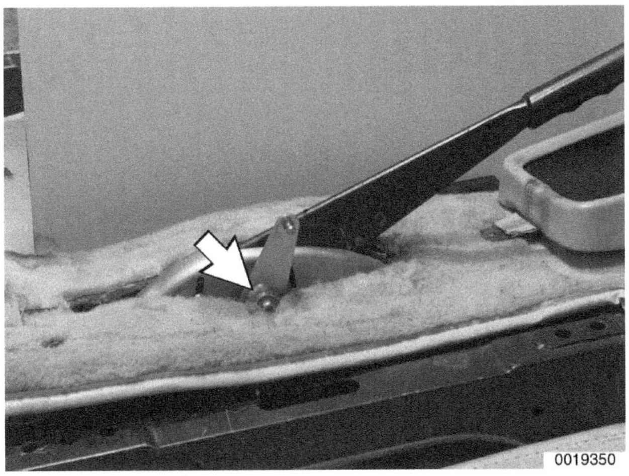

0019350

◄ Remove lock nut (**arrow**) from heater lever mounting nut at base of parking brake lever.

– Remove heater lever and pull heater cable out of guide tubes.

– Install new cable on heater control lever.

– Install cable into right and left guide tubes.

– Reinstall heater control lever and nut.

– Push heater control lever forward to stop.

0019547

- Install cable ends into heat exchanger valve attachment pins (**arrow**) and tighten set screws on left and right sides.
- Reinstall heater control module housing and attach servo control rod.
- Check heater operation. Reinstall seat.

850 Ventilation

GENERAL

This group covers component replacement and adjustment procedures for the ventilation system. Refer to the following groups for further information:

- **800 Heating, Ventilation, Air Conditioning–General**
- **801 Heater**
- **870 Air Conditioner**

Fresh and heated air distribution

Fresh air and heating footwell/defrost flap controls

1984-1985

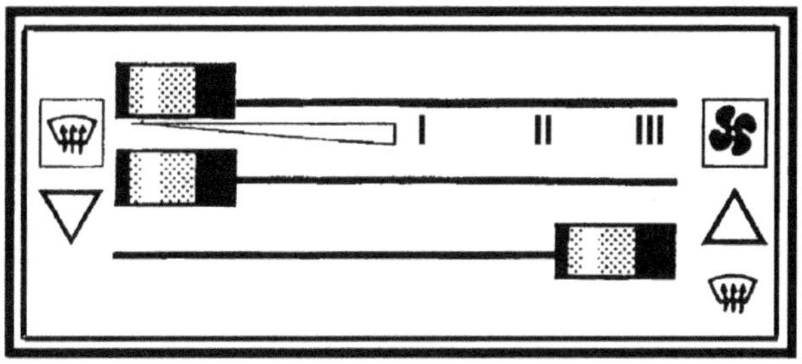

1986-1989

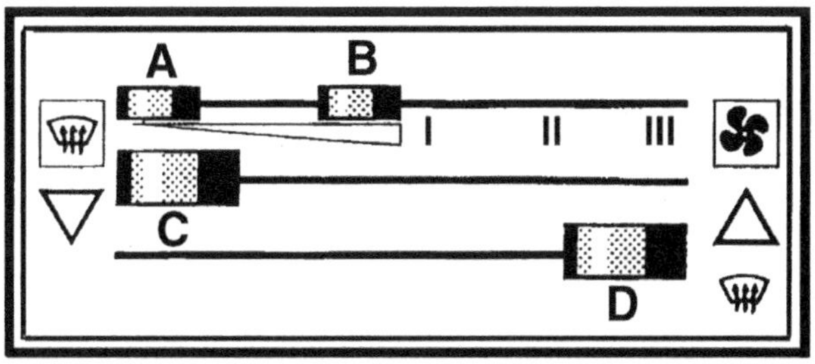

0019463

The air distribution levers use cables to control fresh air and heated air to the footwells and the defroster outlets. A fresh air blower is located in the luggage compartment below the windshield.

NOTE—
The ventilation controls have no control over air distribution to the center dash vent or the side dash vents. This center vent is used for A/C only. The left and right side vents are for A/C or heated air and can are turned on and off by the vertical knurled control knob.

- **Top lever(s)**: Controls fresh air intake from outside the vehicle. When the lever(s) is in the extreme left position, air intake from the outside is closed off. When the lever(s) is in the extreme right position, maximum ventilation of the car interior is obtained.
- **Middle lever**: Fresh air footwell/defrost blend door. This lever directs fresh air to the footwells (extreme left position) or the defroster (extreme right position) or a blend of the two.
- **Bottom lever**: Heated air footwell/defrost blend door. This lever directs heated air to the footwells (extreme left position) or the defroster (extreme right position) or a blend of the two. On cars with automatic heating, heated air temperature is controlled by the temperature control knob between the seats.

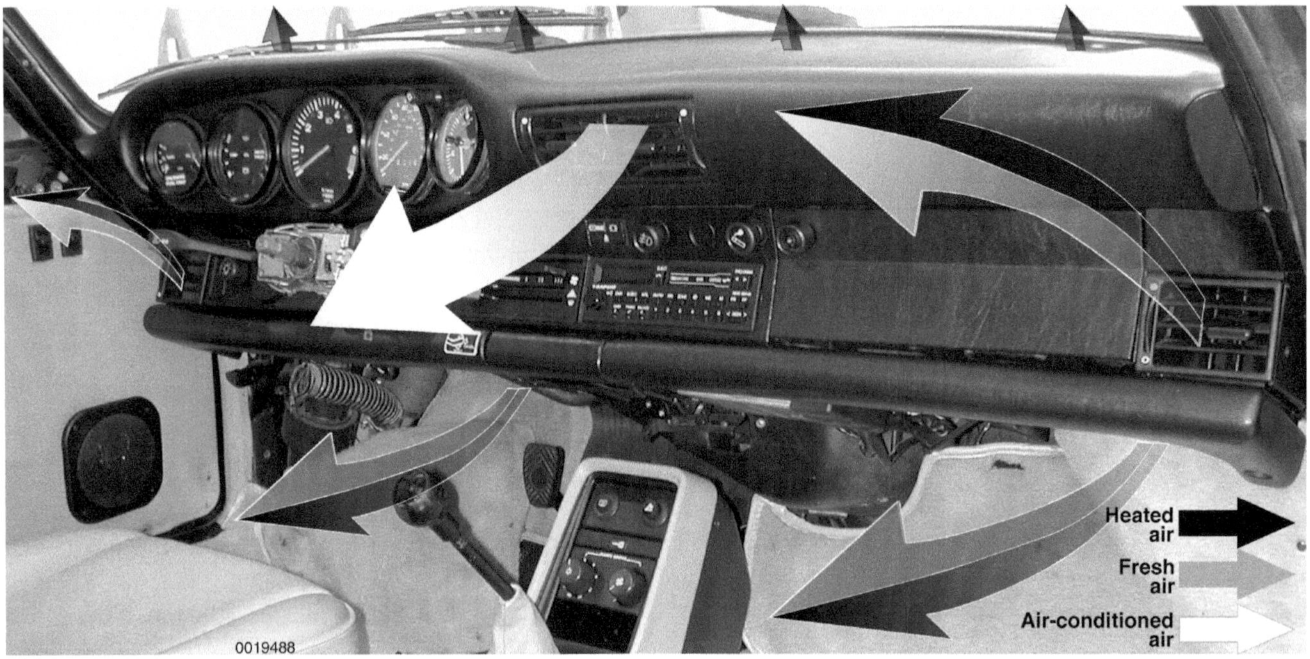

NOTE —
The HVAC system uses several other blowers for heating and A/C air distribution:

- *Heater blower motor, in engine compartment. See* **801 Heater**.
- *Footwell blowers. See* **801 Heater**.
- *A/C evaporator blower, in luggage compartment. See* **870 Air Conditioner**.

FRESH AIR BLOWER

The fresh air blower is located in the luggage compartment below the windshield. It circulates fresh air to the footwell vents and the windshield defroster outlets.

Fresh air blower, removing and installing

– Disconnect negative (–) battery cable.

– Unsnap luggage compartment floor carpet and lift off.

< Remove fresh air plenum cover retaining fasteners (**arrows**). Remove cover.

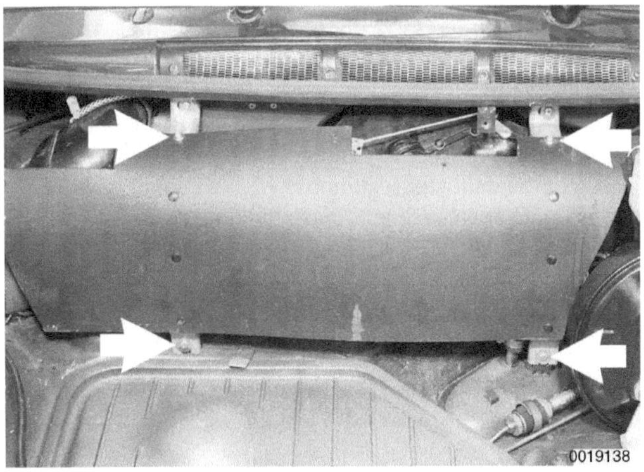

Remove fresh air vent trim screws (**arrows**) in cowl.

- Remove vent trim and screen.

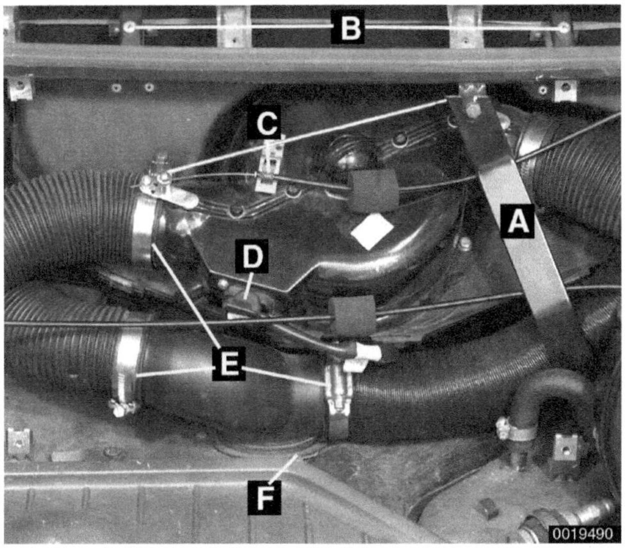

Working inside luggage compartment, detach fresh air plenum and blower connections:

- Remove diagonal brace (**A**).
- Remove screws (**B**) holding fresh air blower housing in place.
- Disconnect heater/defroster cable (**C**) from housing.
- Disconnect electrical harness connector (**D**).
- Disconnect fresh air hoses (**E**) from housing.
- Remove retaining screw (**F**) and pull off large plastic air hose T-fitting.
- Remove blower.

- Installation is reverse of removal.

VENTILATION CONTROLS

In order to access the ventilation controls and operating cables, first remove fresh air plenum cover retaining fasteners (**arrows**) inside luggage compartment. Remove cover.

Ventilation controls, removing and installing

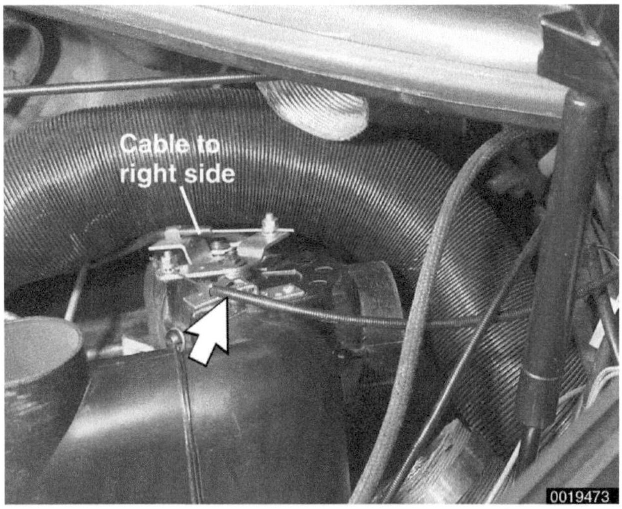

- Remove fresh air blower assembly as described earlier.
- Working behind brake master cylinder inside luggage compartment, detach control cables (**arrow**) from left fresh air distribution valve housing. (Heated air cable shown. Fresh air cable is similar.)

NOTE—

- *If necessary, lift housing off lower duct to access heated air cable.*
- *On each side of left distribution housing, one cable is attached to the controls while the other is attached to the right side air distribution housing.*

- Working under dashboard, just ahead of ashtray:
  - Remove bolt securing ventilation control assembly to firewall.
  - Disconnect ground wire to left of ashtray.
- Remove ventilation controls (with control cables attached) through dashboard opening, detaching harness connector.
- Installation is reverse of removal.

Fresh air valve cable, adjusting

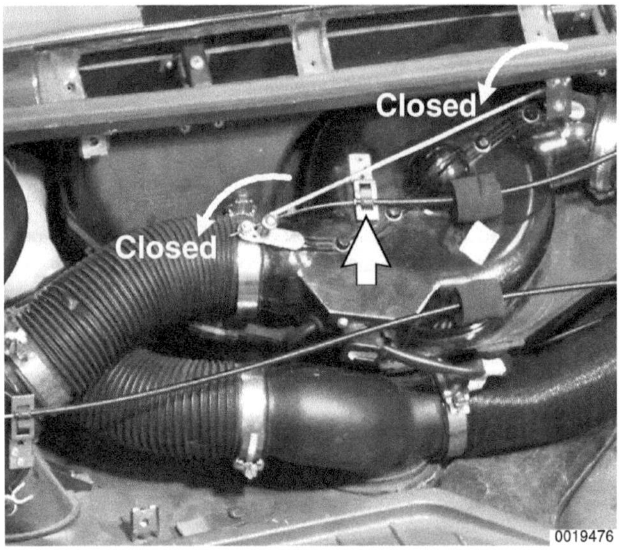

- Access fresh air blower as described earlier.
- Remove bowden cable retaining spring clip (**arrow**).
- Close fresh air flaps.
- Working at dashboard ventilation controls, slide fresh air lever to extreme left.
- Reinstall retaining spring clip. Check that flap operates smoothly and correctly.

Fresh air distribution cable, adjusting

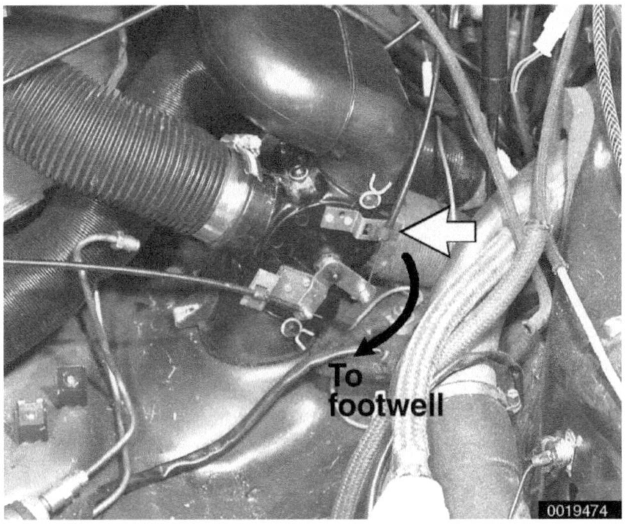

◄ Working in luggage compartment behind brake master cylinder, remove bowden cable retaining spring clip (**arrow**) at left fresh air distribution valve housing.

- Set valve on distribution box so air flows to footwell.

NOTE —
Brake master cylinder and booster removed for clarity.

– Working at dashboard ventilation controls, slide fresh air lever to left.

– Reinstall retaining spring clip.

Heated air distribution cable

The procedure for adjusting the heated air distribution cable is similar to the fresh air cable adjustment. However, the linkage for the cable is hidden between the valve housing and the dashboard bulkhead. Lift off the valve housing to access and adjust the cable.

870 Air Conditioner

GENERAL

This section covers removal and installation of air conditioning components. A/C refrigerant discharge, evacuation and recharge methods, techniques and equipment are not covered here.

> ***WARNING —***
> *When discharging and charging R-12 refrigerant, refer to* **A/C warnings and cautions** *later in this group.*

A/C system components

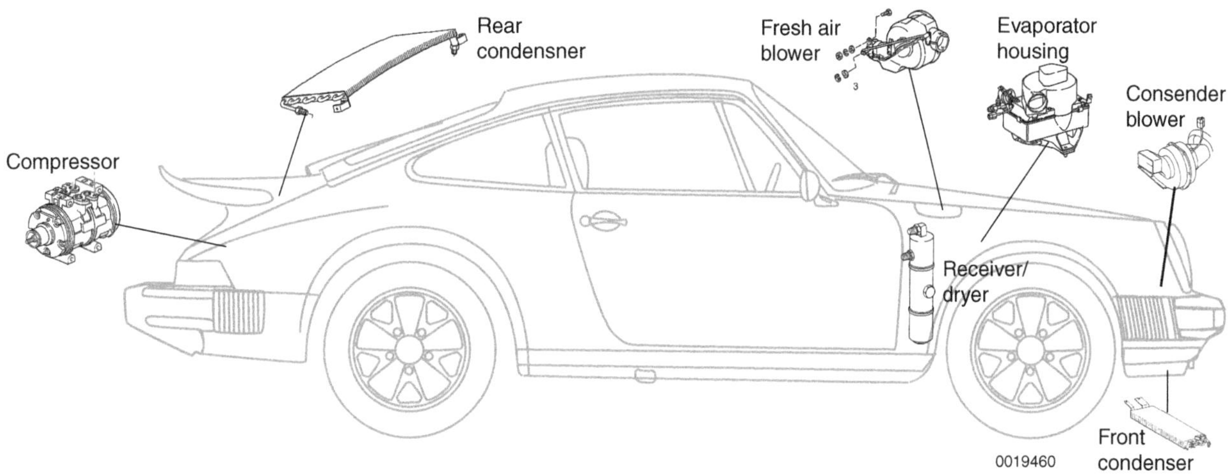

A/C service

NOTE —

- *Removal and installation of A/C console mounted switches is covered in* **680 Interior Trim**.
- *For further information on the ventilation system, refer to the following groups:* **801 Heater** *and* **850 Ventilation**.
- *The A/C system is separate from the heating and ventilation system and uses its own blower motor.*

A/C service and repair requires special equipment and knowledge. Incorrect procedures may not only damage the system, but also may be hazardous. Pressures in excess of 300 psi are created in the system when it is operating. The refrigerant used (R-12) is not poisonous, but in vapor form it can accumulate in areas with poor ventilation and cause suffocation. Also, in vapor or liquid form, R-12 can immediately freeze anything it contacts, including eyes and skin.

Although tests can determine A/C efficiency, it is recommended that all service to the system be left to an authorized Porsche dealer or other qualified repair shop. Some diagnostic and troubleshooting information for A/C components is provided in **800 Heating, Ventilation, Air Conditioning–General**.

Dismounting of the A/C compressor and the condenser, without disconnecting the hoses, is covered as part of the engine removal procedure in **101 Engine Removal and Installation**. If any of the hoses or components are disconnected and the system is opened, special equipment will be needed to remove moisture from the system before it can be closed up and refilled with refrigerant and refrigerant oil.

A/C warnings and cautions

Heeding the warning and cautions on the following pages will help prevent personal injury or damage to the system.

WARNING —

- *Work in a well ventilated area. Avoid breathing refrigerant vapors. Exposure may irritate eyes, nose, and throat. Switch on exhaust/ventilation systems when working on the refrigerant system.*
- *Always be careful that refrigerant does not come in contact with your skin. Always wear hand and eye protection (gloves and goggles) when working around the A/C system. If refrigerant does come in contact with your skin or eyes:*
  -Do not rub skin or eyes;
  -Immediately flush skin or eyes with cool water for 15 minutes;
  -Immediately go to a doctor or hospital;
  -Do not attempt to treat yourself.
- *Refrigerant gases are heavier than air, displace oxygen and may cause suffocation in areas of poor circulation, for example under a car.*
- *Do not discharge/charge the A/C system without proper equipment and training. Damage to the vehicle and personal injury may result.*
- *The A/C system is filled with refrigerant gas which is under pressure. Pressurized refrigerant in the presence of oxygen may form a combustible mixture. Never introduce compressed air into any container (full or empty), capped off A/C component, or piece of service equipment.*
- *Do not expose any component of the A/C system to high temperatures (above 80°C/176°F) or open flames. Excessive heat will cause a pressure increase which could burst the system.*
- *Keep refrigerant away from open flames. Poisonous gas will be produced if it burns. Do not smoke near refrigerant gases for the same reason.*

CAUTION —

- *Immediately plug open connections on A/C components to prevent dirt and moisture contamination.*
- *Do not steam clean condensers or evaporators. Use only cold water or compressed air.*
- *It is recommended that all service to the A/C system refrigerant components be left to an authorized Porsche dealer or other qualified A/C service facility. If any of the refrigerant components are disconnected or removed, special equipment, supplies, and new parts will be needed to remove moisture from the system and recharge the refrigerant before it is closed back up.*
- *As of January 1, 1992, any person who services a motor vehicle air conditioner must, by law, be properly trained and certified, and use approved refrigerant recycling equipment. Technicians must complete an EPA-approved recycling course to be certified.*
- *State and local governments may have additional requirements regarding air conditioning servicing. Always comply with state and local laws.*
- *Do not top off a partially charged refrigerant system. Remove refrigerant, evacuate and then recharge system.*

A/C COMPRESSOR AND RECEIVER/DRYER

A/C compressor, replacing

- Following manufacturer's instructions, connect an approved refrigerant recovery/recycling/recharging unit to A/C system and remove refrigerant.
- Disconnect negative (–) battery cable.
- ≺ Disconnect compressor clutch electrical connection (**A**).
- Disconnect A/C refrigerant lines (**B**) and plug openings.

- ≺ Loosen compressor mounting plate bolts and nut (**arrows**); do not remove fasteners. Loosen V-belt adjuster lock nut and screw (**A**). Slide compressor to left and remove V-belt.
- Remove compressor mounting bolts (**B**) and remove compressor from adjusting plate.
- Installation is reverse of removal.
  - Be sure there is sufficient oil in compressor. See **Table a**.
  - Before recharging, evacuate system for at least 30 minutes.
  - Recharge system following equipment manufacturer's instructions.

> ***WARNING —***
> *A new A/C compressor from an authorized Porsche parts department is shipped pressurized with refrigerant. Use care when removing sealing caps and follow all manufacturer's instructions.*

> ***CAUTION —***
> *Always replace O-rings when reconnecting refrigerant lines.*

Table a. A/C System Capacities

Component	Oil capacity
Refrigerant (R-12):	
Total system capacity	1350 gr. (47 oz.)
A/C lubricating oil:	
Compressor	50 gr. (2.0 oz)
Evaporator	40 gr. (1.6 oz.)
Condensers	20 gr. (0.8 oz.)
Receiver/dryer	10 gr. (0.4 oz.)

Tightening Torques

- A/C compressor adjusting plate to mount . . . . 22 Nm (17 ft-lb)
- A/C compressor to adjusting plate . . . . . 22 Nm (17 ft-lb)
- Refrigerant lines to A/C compressor
  3/4 in. fitting . . . . 20 - 28 Nm (14 - 20 ft-lb)
  7/8 in. fitting . . . . 29 - 37 Nm (21 - 27 ft-lb)

A/C compressor clutch, replacing

The compressor clutch can be replaced without disturbing the sealed refrigerant system.

A/C compressor assembly

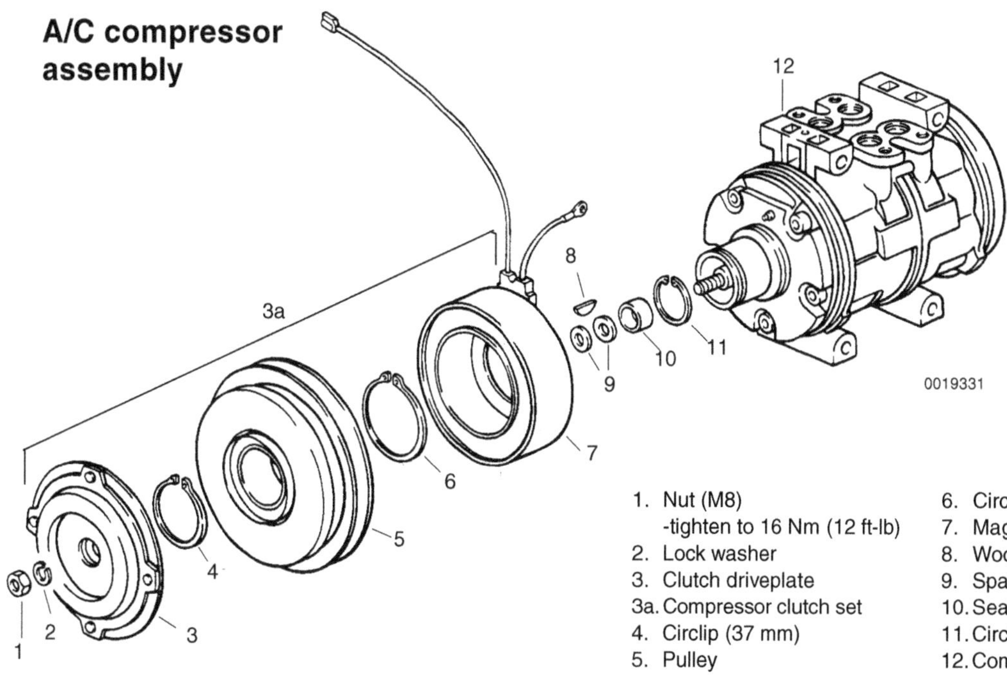

1. Nut (M8)
   -tighten to 16 Nm (12 ft-lb)
2. Lock washer
3. Clutch driveplate
3a. Compressor clutch set
4. Circlip (37 mm)
5. Pulley
6. Circlip (45.8 mm)
7. Magnetic coil
8. Woodruff key
9. Spacers
10. Seal
11. Circlip
12. Compressor

- Disconnect compressor clutch electrical connection.
- Loosen V-belt adjuster, slide A/C compressor to left, and remove V-belt.
- Hold A/C compressor clutch driveplate stationary with strap tool or equivalent to allow removal of retaining nut.
- Remove drive plate and pulley to access clutch magnetic winding.
- Remove spacers.
- Remove circlip and slide magnetic coil off support.
- If reassembling with old parts, test magnetic coil resistance.

NOTE—
Magnetic coil resistance is 3.8 ± 0.2 ohms.

– During installation be sure to reinstall all spacers. The correct spacing between clutch drive plate and pulley is 0.4 - 0.7 mm (0.016 - 0.028 in).

Tightening Torques

- A/C compressor adjusting plate to mount . . . . 22 Nm (17 ft-lb)
- A/C compressor clutch to compressor shaft . . . . 16 Nm (12 ft-lb)

A/C receiver/drier, replacing

The A/C receiver/drier is mounted in the rear of the left front wheel housing.

> ***CAUTION —***
> *The receiver/drier must be replaced when;*
> *–There is dirt in the A/C system;*
> *–The compressor has seized or has been replaced for any reason;*
> *–The condenser or evaporator are replaced for any reason;*
> *–The A/C system is leaking and there is no more refrigerant;*
> *–The A/C system was open for 24 hours or more.*

– Following manufacturer's instructions, connect an approved refrigerant recovery/recycling/recharging unit to A/C system and remove refrigerant.

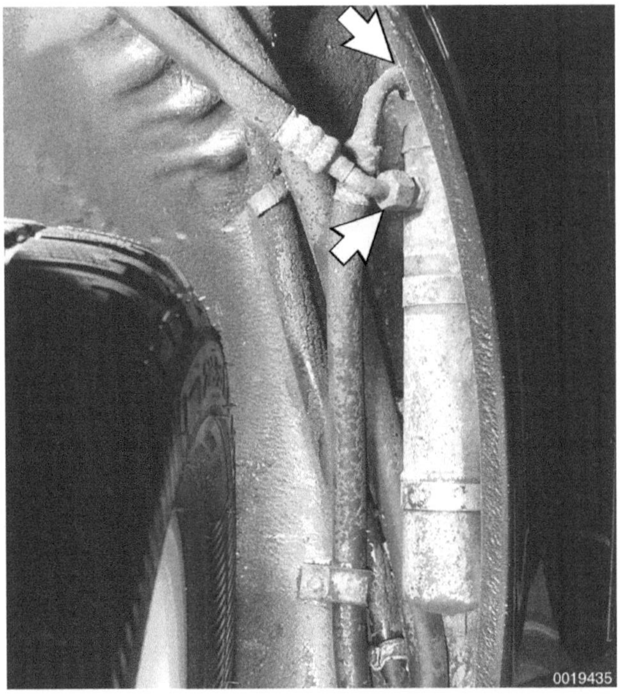

◀ Disconnect and plug refrigerant lines (**arrows**).

– Remove hose clamps used to mount receiver/drier to inner fender panel.

– Remove receiver/drier.

– Installation is reverse of removal.

- Be sure there is sufficient oil in receiver/dryer. See table below.
- Before recharging, evacuate system for at least 30 minutes.
- Recharge system following equipment manufacturer's instructions.

> ***CAUTION —***
> *Always replace O-rings when reconnecting refrigerant lines.*

Tightening Torque

- Refrigerant line to receiver/dryer:
  5/8 in. fitting . . . . 14 - 20 Nm (10 - 14 ft-lb)

A/C System Capacities

Component	Oil capacity
Refrigerant (R-12):	
Total system capacity	1350 gr. (47 oz.)
A/C lubricating oil:	
Compressor	50 gr. (2.0 oz)
Evaporator	40 gr. (1.6 oz.)
Condensers	20 gr. (0.8 oz.)
Receiver/dryer	10 gr. (0.4 oz.)

A/C CONDENSERS

A/C condenser (front), removing and installing

– Following manufacturer's instructions, connect an approved refrigerant recovery/recycling/recharging unit to A/C system and remove refrigerant.

> ***WARNING —***
> *Do not discharge/charge the A/C system without proper equipment and training. Damage to the vehicle and personal injury may result. Read warnings and cautions given earlier.*

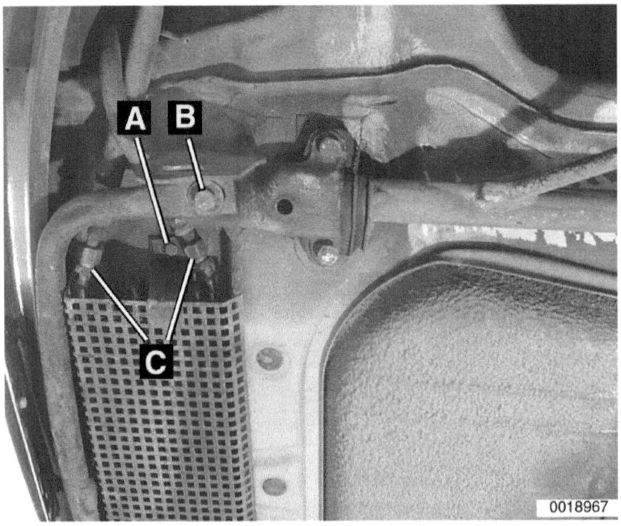

◄ Remove front condenser protective bar and pan mounting bolts (**A** and **B**). (Left side shown. Right is similar.)

– Disconnect refrigerant lines (**C**) from left side of condenser. Plug lines and fittings immediately.

– Remove A/C condenser and protective pan.

– Installation is reverse of removal.

- Be sure there is sufficient oil in condenser. See table below.
- Before recharging, evacuate system for at least 30 minutes.
- Recharge system following equipment manufacturer's instructions.

> ***CAUTION —***
> - *Always replace O-rings when reconnecting refrigerant lines.*
> - *When installing a new condenser always replace the receiver/drier unit.*

Tightening Torque

- Refrigerant lines to front condenser
  5/8 in. fitting . . . . 14 - 20 Nm (10 - 14 ft-lb)

A/C System Capacities

Component	Oil capacity
Refrigerant (R-12):	
Total system capacity	1350 gr. (47 oz.)
A/C lubricating oil:	
Compressor	50 gr. (2.0 oz)
Evaporator	40 gr. (1.6 oz.)
Condensers	20 gr. (0.8 oz.)
Receiver/dryer	10 gr. (0.4 oz.)

A/C condenser (rear), removing and installing

The rear A/C condenser is mounted to the underside of the engine lid.

– Following manufacturer's instructions, connect an approved refrigerant recovery/recycling/recharging unit to A/C system and remove refrigerant.

> ***WARNING —***
> *Do not discharge/charge the A/C system without proper equipment and training. Damage to the vehicle and personal injury may result. Read warnings and cautions given earlier.*

< Working at underside of engine lid, disconnect refrigerant lines (**arrows**) from sides of condenser. Plug lines and fittings immediately.

– Remove condenser mount (**A**) from center of engine lid.

– Remove refrigerant line support clamps on hinges.

– Installation is reverse of removal.

- Be sure there is sufficient oil in condenser. See table below
- Before recharging, evacuate system for at least 30 minutes.
- Recharge system following equipment manufacturer's instructions.

> ***CAUTION —***
> *Always replace O-rings when reconnecting refrigerant lines.*

Tightening Torque

- Refrigerant lines to rear condenser
  3/4 in. fittings. . . . . 20 - 28 Nm (14 - 20 ft-lb)

A/C System Capacities

Component	Oil capacity
Refrigerant (R-12):	
Total system capacity	1350 gr. (47 oz.)
A/C lubricating oil:	
Compressor	50 gr. (2.0 oz)
Evaporator	40 gr. (1.6 oz.)
Condensers	20 gr. (0.8 oz.)
Receiver/dryer	10 gr. (0.4 oz.)

A/C condenser blower (front), removing and installing

– Disconnect negative (–) battery cable. See **030 Maintenance**

– Remove luggage compartment floor mat.

– Remove battery.

– Remove spare tire.

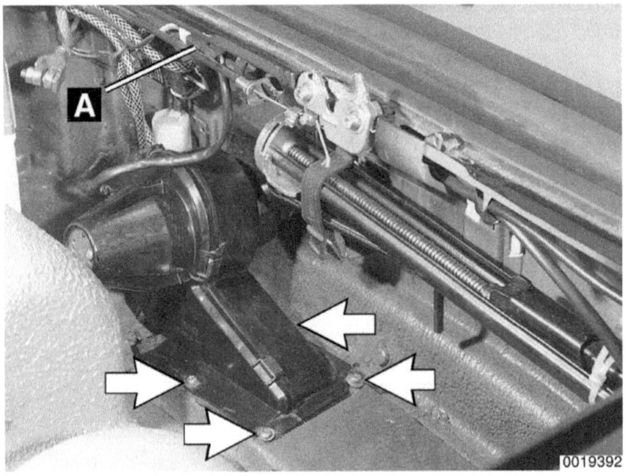

◄ Disconnect blower electrical harness connector (**A**).

– Remove screws (**arrows**) securing blower to floor.

– Slide blower back and lift out.

– Installation is reverse of removal.

A/C EVAPORATOR ASSEMBLY

The A/C evaporator assembly includes the following:

- **Evaporator** and **evaporator blower**
- **Expansion valve** attached to the outside of the evaporator housing
- **A/C temperature control switch** in the center console, connected to the **evaporator temperature probe** in the evaporator housing.
- **Evaporator blower resistor** in the evaporator blower air intake port in the passenger compartment

NOTE —
The blower resistor is also referred to as the bimetallic switch.

Expansion valve, replacing

– Following manufacturer's instructions, connect an approved refrigerant recovery/recycling/recharging unit to A/C system and remove refrigerant.

> ***WARNING —***
> *Do not discharge/charge the A/C system without proper equipment and training. Damage to the vehicle and personal injury may result. Read warnings and cautions given earlier.*

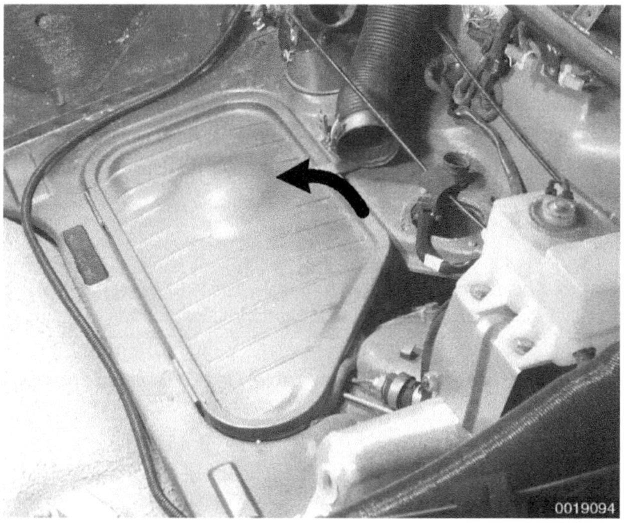

◄ Remove luggage compartment floor mat and open access panel in compartment floor.

◄ Remove evaporator temperature sensor duct support from expansion valve and gently push duct away from valve.

– Remove insulating tape (**A**) from expansion valve.

– Disconnect refrigerant connections (**B**) from expansion valve and remove valve.

– Installation is reverse of removal.

  - Be sure there is sufficient oil in evaporator. See table below.
  - Before recharging, evacuate system for at least 30 minutes
  - Recharge system following equipment manufacturer's instructions.

> ***CAUTION—***
> *Always replace O-rings when reconnecting refrigerant lines.*

Tightening Torques

- Refrigerant lines to expansion valve or evaporator
  - 5/8 in. fitting . . . . 14 - 20 Nm (10 - 14 ft-lb)
  - 3/4 in. fitting . . . . 20 - 28 Nm (14 - 20 ft-lb)
  - 7/8 in. fitting . . . . 29 - 37 Nm (21 - 27 ft-lb)

A/C System Capacities

Component	Oil capacity
Refrigerant (R-12):	
Total system capacity	1350 gr. (47 oz.)
A/C lubricating oil:	
Compressor	50 gr. (2.0 oz)
Evaporator	40 gr. (1.6 oz.)
Condensers	20 gr. (0.8 oz.)
Receiver/dryer	10 gr. (0.4 oz.)

A/C evaporator housing, removing and disassembling

– Following manufacturer's instructions, connect an approved refrigerant recovery/recycling/recharging unit to A/C system and remove refrigerant.

> ***WARNING —***
> *Do not discharge/charge the A/C system without proper equipment and training. Damage to the vehicle and personal injury may result. Read warnings and cautions given earlier.*

– Disconnect negative (–) battery cable. See **030 Maintenance**.

– Remove expansion valve as described earlier.

– Loosen hose clamp and detach large air duct at top of evaporator housing.

< Disconnect evaporator blower harness connector (**1**).

– Remove both support brackets (**2**).

– Gently pull evaporator temperature probe (**3**) out of base of evaporator and lay aside without damaging duct.

– Remove evaporator assembly from well.

– Remove spring clips securing lower evaporator housing to upper housing. Remove upper housing.

– Remove, service and replace blower motor or evaporator, as needed.

– Installation is reverse of removal.

- Be sure there is sufficient oil in evaporator. See table below.
- Before recharging, evacuate system for at least 30 minutes.
- Recharge system following equipment manufacturer's instructions.

> ***CAUTION —***
> *Always replace O-rings when reconnecting refrigerant lines.*

Tightening Torques

- Refrigerant lines to expansion valve or evaporator
  - 5/8 in. fitting .... 14 - 20 Nm (10 - 14 ft-lb)
  - 3/4 in. fitting .... 20 - 28 Nm (14 - 20 ft-lb)
  - 7/8 in. fitting .... 29 - 37 Nm (21 - 27 ft-lb)

A/C System Capacities

Component	Oil capacity
Refrigerant (R-12):	
Total system capacity	1350 gr. (47 oz.)
A/C lubricating oil:	
Compressor	50 gr. (2.0 oz)
Evaporator	40 gr. (1.6 oz.)
Condensers	20 gr. (0.8 oz.)
Receiver/dryer	10 gr. (0.4 oz.)

A/C temperature control switch, replacing

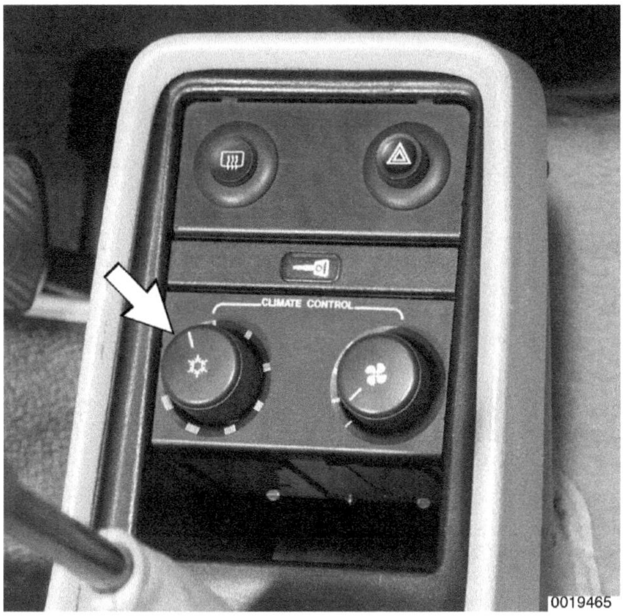

0019465

The A/C temperature switch (**arrow**) in the center console is connected via a capillary tube to the temperature probe in the evaporator housing.

- Access evaporator in compartment below luggage compartment.

0019436

Remove temperature probe from hole in evaporator housing.

- Remove temperature control switch from center console. See **680 Interior Trim**.
- Carefully feed temperature sensor tube and probe through opening in bulkhead into passenger compartment.
- Installation is reverse of removal.

Evaporator blower resistor, replacing

The evaporator blower resistor, also referred to as the bimetallic switch, is located in the intake port of the evaporator, in front of the passenger foot-rest.

- Detach resistor harness connector (**arrow**).
- Remove intake port mounting screws (**A**) and remove resistor and port.

– Installation is reverse of removal.

900 Electrical System–General

TABLES

Special Tools

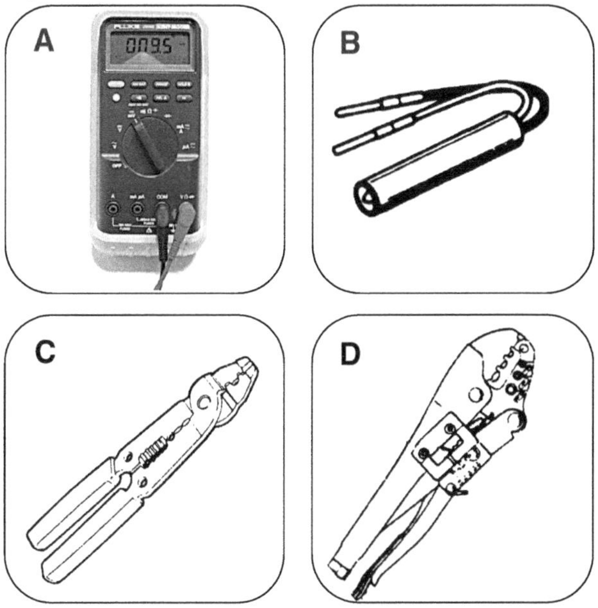

Special tool

A Fluke 87 . . . . Automotive digital multimeter
B 1115 . . . . LED tester with thin spade probes
(source: Baum Tools Unlimited)
C T746 . . . . Wire stripper/crimper
(source: Baum Tools Unlimited)
D T748 . . . . One-step crimping pliers
(source: Baum Tools Unlimited)

GENERAL

A brief description of the principal parts of the electrical system is presented here. Also covered here are basic electrical system troubleshooting tips.

Special tools

Many procedures and tests in this partition of the manual require the use of some electronic test equipment. Read the procedure through before starting work.

Electrical system safety precautions

Please read the following warnings and cautions before doing any work on your electrical system.

WARNING —

- *The ignition system of the car operates at lethal voltages. People with pacemakers or weak hearts should not expose themselves to the ignition system. Extra caution must be taken when working on the ignition system or when servicing the engine while it is running or the key is on. See* **280 Ignition System** *for additional ignition system warnings and cautions.*
- *Before operating the starter without starting the engine (as when making a compression test), disable the ignition system by removing harness connector from the DME main/fuel pump relay. See* **240 Fuel Injection**.

CAUTION —

- *Always turn off the engine and disconnect the negative (–) cable from the battery before removing any electrical components.*
- *Connect and disconnect ignition system wires, multiple connectors, and ignition test equipment leads only while the ignition is off.*
- *Disconnecting the battery cables erases engine control module (ECM) adaptive memory. It may be necessary to drive the car for approximately 10 minutes after reconnectiung the battery to reset ECM adaptive memory and restore normal engine performance.*
- *Do not disconnect the battery with engine running.*
- *Do not quick-charge the battery (for boost starting) for longer than one minute, and do not exceed 16.5 volts at the battery with the boosting cables attached. Wait at least one minute before boosting the battery a second time.*
- *Do not use a test lamp that has a normal incandescent bulb to test circuits containing electronic components. The high electrical consumption of these test lamps may damage the components.*
- *Do not use an analog meter. Use only a digital multimeter.*
- *Solid-state modules are static sensitive. Static discharge will permanently damage them. Always handle the modules using proper static prevention equipment and techniques.*
- *To avoid damaging harness connectors or relay panel sockets, use jumper wires with flat-blade connectors that are the same size as the connector or relay terminals.*
- *Always switch a digital multimeter to the appropriate function and range before making test connections.*
- *Do not try to start the engine of a car which has been heated above 176°F (80°C), (for example, in a paint drying booth). Allow it to cool to normal temperature.*
- *Disconnect the battery before doing any electric welding on the car.*
- *Do not wash the engine while it is running or any time the ignition is switched on.*

Voltage and polarity

The vehicle electrical system is a 12-volt direct current (DC) negative-ground system. A voltage regulator controls system voltage at 12.6 - 14.5 volts. All circuits are grounded by direct or indirect connection to the negative (–) terminal of the battery. A number of ground connections throughout the car connect the wiring harness to chassis ground. These circuits are completed by the battery cable or ground strap between the body and the battery negative (–) terminal.

Wiring, fuses and relays

Nearly all parts of the wiring harness connect to components of the electrical system with keyed, push-on connectors that lock into place. Notable exceptions are the heavy battery cables and the starter wiring. The wiring is color-coded for circuit identification.

With the exception of the battery charging system, most electrical power is routed from the ignition switch or the battery through the main fuse/relay panel, located in the left rear corner of the luggage compartment. Fuses are color coded to indicate current capacities.

The relays and control units/modules are mounted in various places throughout the vehicle. See **971 Electrical Component Locations**.

Electrical test equipment

Many of the electrical tests described in this manual call for measuring voltage, current or resistance using a digital multimeter (DMM). Digital meters are preferred for precise measurements and for electronics work because they are generally more accurate than analog meters. The numerical display is also less likely to be misread, since there is no needle position to be misinterpreted by reading at an angle.

An LED test light is a safe, inexpensive tool that can be used to perform many simple electrical tests that would otherwise require a digital multimeter. The LED indicates when voltage is present between any two test-points in a circuit.

CAUTION —

- *Use a digital multimeter with at least 10 megaohm input impedance, or an LED test light. An analog meter (swing-needle) or a test light with a normal incandescent bulb may draw enough current to damage sensitive electronic components.*
- *An ohmmeter must not be used to measure resistance on solid state components such as control units or time delay relays.*
- *Disconnect battery before making resistance (ohm) measurements on the circuit.*

WIRING DIAGRAMS

The wiring diagrams shown in **970 Electrical Wiring Diagrams** have been specially designed to enable quick and efficient diagnosis and troubleshooting of electrical malfunctions.

Wiring codes and abbreviations

A lot of information is included in each wiring diagram if you know how to read them. Wire colors in the diagrams are abbreviated. Combined color codes indicate a multi-colored wire. For example the code BLU/RED indicates a blue wire with a red stripe. See **Table a**.

Many electrical components, connectors, fuses, and ground locations are identified using a unique number. Each of these numbers correspond to a particular part in the circuit commonly found in **970 Electrical Wiring Diagrams**.

NOTE —
Sometimes the color of an installed wire may be different than the one on the wiring diagram. Do not be concerned. Just be sure to confirm that the wire connects to the proper terminals.

Table a. Wire Color Codes

BLU	Blue
BRN	Brown
YEL	Yellow
GRN	Green
GRY	Gray
ORG	Orange
RED	Red
BLK	Black
VIO	Violet
WHT	White

Most terminals are identified by numbers on the components and harness connectors. The terminal numbers for major electrical connections are shown in the diagrams. Though many terminal numbers appear only once, several other numbers appear in numerous places throughout the electrical system and identify certain types of circuits. Some of the most common circuit numbers are listed below in **Table b**.

Table b. Terminal and Circuit Numbers

Number	Circuit description
1	Low voltage switched terminal of coil
4	High voltage center terminal of coil
X or +X	Originates at ignition switch. Supplies power when the ignition switch is in the PARK, RUN or START position
15	Originates at ignition switch. Supplies power when ignition switch is in RUN or START position
30	Battery positive (+) voltage. Supplies power whenever battery is connected. (Not dependent on ignition switch position, unfused)
31	Ground, battery negative (–) terminal
50	Supplies power from battery to starter solenoid when ignition switch is in START position only
+54	Originates at ignition switch. Supplies power when ignition switch is in the RUN position only
85	Ground side (–) of relay coil
86	Power-in side (+) of relay coil
87	Relay actuated contact
D	Alternator warning light and field energizing circuit

Additional abbreviations shown in the wiring diagrams are given below.

Abbreviations

- A/C .... air conditioning
- DME .... digital motor electronics
- ECM .... electronic control module

ELECTRICAL TROUBLESHOOTING

Four things are required for current to flow in any electrical circuit: a voltage source, wires or connections to transport the voltage, a load or device that uses the electricity and a connection to ground. Most problems can be found using a digital multimeter (volt/ohm/amp meter) to check for voltage supply, for breaks in the wiring (infinite resistance/no continuity), or for a path to ground that completes the circuit.

Electric current is logical in its flow, always moving from the voltage source toward ground. Electrical faults can usually be located through a process of elimination. When troubleshooting a complex circuit, separate the circuit into smaller parts. The general tests outlined below may be helpful in finding electrical problems. The information is most helpful when used with the wiring diagrams.

Be sure to analyze the problem. Use the wiring diagrams to determine the most likely cause. Get an understanding of how the circuit works by following the circuit from ground back to the power source.

When making test connections at connectors and components, use care to avoid spreading or damaging the connectors or terminals. Some tests may require jumper wires to bypass components or connections in the wiring harness. When connecting jumper wires, use blade connectors at the wire ends that match the size of the terminal being tested. The small internal contacts are easily spread apart, and this can cause intermittent or faulty connections that can lead to more problems.

Voltage and voltage drop

The wires, connectors, and switches that carry current are designed with very low resistance so that current flows with a minimum loss of voltage. A voltage drop is caused by higher than normal resistance in a circuit. This additional resistance actually decreases or stops the flow of current. A voltage drop can be noticed by problems ranging from dim headlights to sluggish wipers. Some common sources of voltage drops are corroded or dirty switches, dirty or corroded connections or contacts and loose or corroded ground wires and ground connections.

A voltage drop test is a good test to make if current is flowing through the circuit, but the circuit is not operating correctly. A voltage drop test will help to pinpoint a corroded ground strap or a faulty switch. Normally, there should be less than 1 volt drop across most wires or closed switches. A voltage drop across a connector or short cable should not exceed 0.5 volts.

NOTE —

- *A voltage drop test is generally more accurate than a simple resistance check because the resistances involved are often too small to measure with most ohmmeters. For example, a resistance as small as 0.02 ohms would results in a 3 volt drop in a typical 150 amp starter circuit. (150 amps x 0.02 ohms =3 volts).*
- *Keep in mind that voltage with the key on and voltage with the engine running are not the same. With the ignition on and the engine off (battery voltage), voltage should be approximately 12.6 volts. With the engine running (charging voltage), voltage should be approximately 14.0 volts. Measure voltage at the battery with the ignition on and then with the engine running to get exact measurements.*

Voltage, measuring

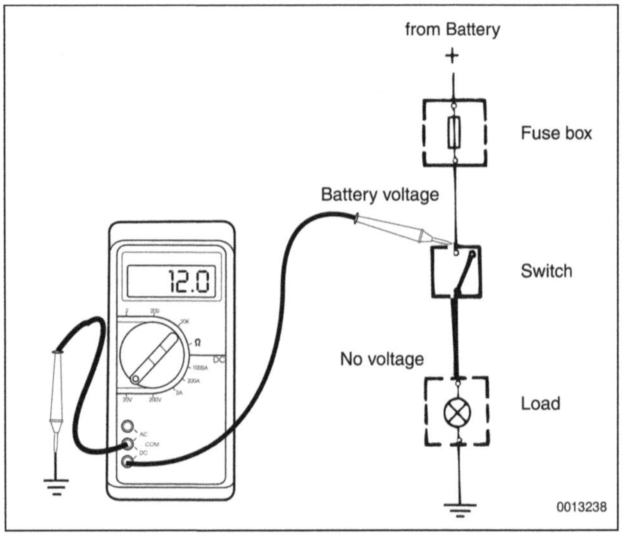

– Connect digital multimeter negative lead to a reliable ground point on car.

NOTE —
The negative (–) battery terminal is always a good ground point.

◄ Connect digital multimeter positive lead to point in circuit you wish to measure. If a reading is obtained, current is flowing through circuit.

NOTE —
The voltage reading should not deviate more than 1 volt from the voltage at the battery. If the voltage drop is more than this, check for a corroded connector or loose ground wire.

Voltage drop, testing

Voltage drop can only be checked when current is running through the circuit, such as by operating the starter motor or turning on the headlights. A digital multimeter should be used to ensure accurate readings.

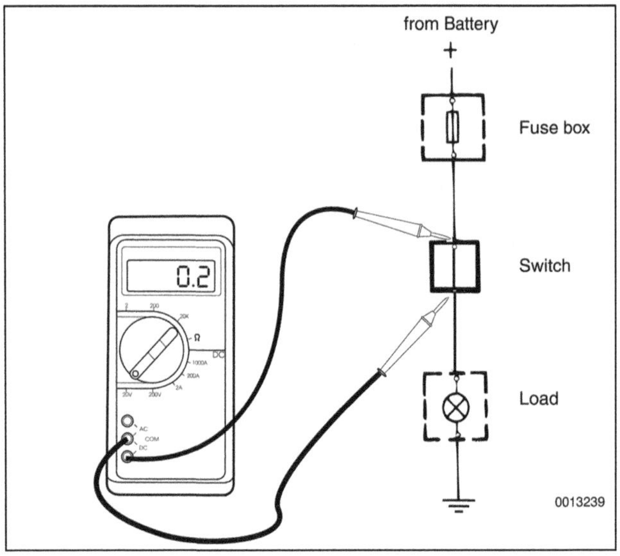

– Connect digital multimeter positive lead to positive end of cable or switch being tested.

◄ Connect digital multimeter negative lead to other end of cable or switch being tested.

– With power on and circuit working, meter shows voltage drop (difference between two points). This value should not exceed 1 volt.

NOTE —
The maximum voltage drop in an automotive circuit, as recommended by the Society of Automotive Engineers (SAE), is as follows: 0 volts for small wire connections; 0.1 Volts for high current connections; 0.2 volts for high current cables; and 0.3 volts for switch or solenoid contacts. On longer wires or cables, the drop may be slightly higher. In any case, a voltage drop of more than 1.0 volt usually indicates a problem.

Continuity, checking

The continuity test can be used to check a circuit or switch. Because most automotive circuits are designed to have little or no resistance, a circuit or part of a circuit can be easily checked for faults using an ohmmeter. An open circuit or a circuit with high resistance will not allow current to flow. A circuit with little or no resistance allows current to flow easily.

> **CAUTION —**
> *Use only a high quality digital ohmmeter having high input impedance when checking electronic components. The internal power source used in most analog (swing-needle) meters can damage solid state components.*

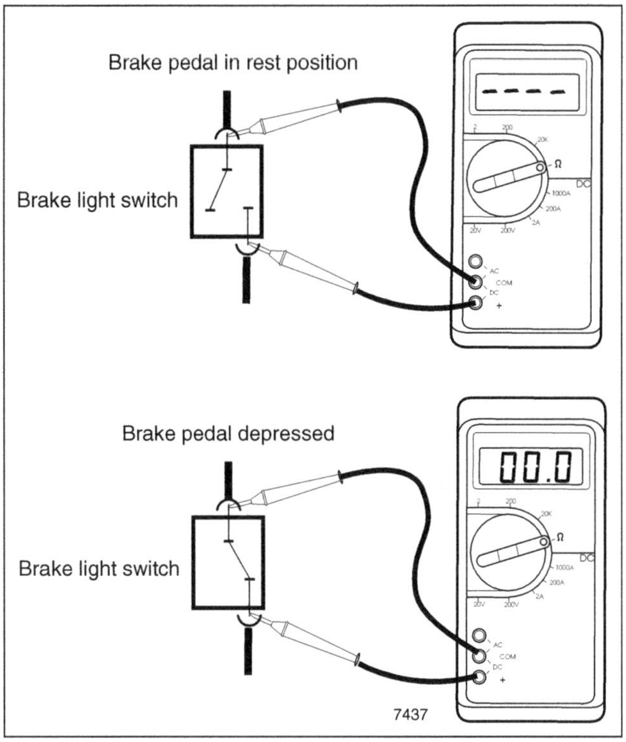

Brake light switch being tested for continuity.

- With brake pedal in rest position (switch open) there is no continuity (infinite ohms). With pedal depressed (switch closed) there is continuity (zero ohms).
- When checking continuity, the ignition should be off. On circuits that are powered at all times, the battery should be disconnected. Using the appropriate wiring diagram, a circuit can be easily tested for faulty connections, wires, switches, relays, and engine sensors by checking for continuity.

Short circuits

A short circuit is exactly what the name implies. The circuit takes a shorter path than it was designed to take. The most common short that causes problems is a short to ground where the insulation on a positive (+) wire wears away and the metal wire is exposed. When the wire rubs against a metal part of the car or other ground source, the circuit is shorted to ground. If the exposed wire is live (positive battery voltage), a fuse will blow and the circuit may be damaged.

Shorts to ground can be located with a digital multimeter. Short circuits are often difficult to locate and may vary in nature. Short circuits can be found using a logical approach based on the current path.

> **CAUTION —**
> - *On circuits protected with high rating fuses (25 amp and greater), the wires or circuit components may be damaged before the fuse blows. Always check for damage before replacing fuses of this rating.*
> - *When replacing blown fuses, use only fuses having the correct rating. Always confirm the correct fuse rating printed on the fuse/relay panel cover.*

Short circuit, testing with ohmmeter

– Remove blown fuse from circuit and disconnect cables from battery.

– Disconnect harness connector from circuit's load or component.

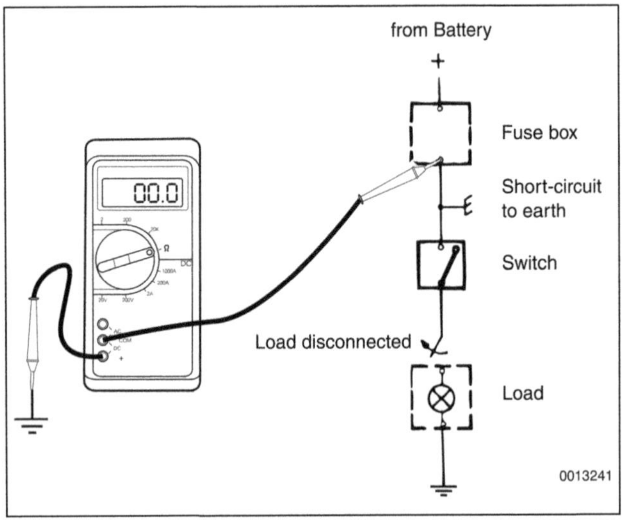

◄ Using an ohmmeter, connect one test lead to load side of fuse terminal (terminal leading to circuit) and the other test lead to ground.

– If there is continuity to ground, there is a short to ground.

– If there is no continuity, work from wire harness nearest to fuse/relay panel and move or wiggle wires while observing meter. Continue to move down harness until meter displays a reading. This is the location of short to ground.

Visually inspect the wire harness at this point for any faults. If no faults are visible, carefully slice open the harness cover or the wire insulation for further inspection. Repair any faults found.

Short circuit, testing with voltmeter

– Remove blown fuse from circuit.

– Disconnect harness connector from circuit's load or component.

NOTE —

Most fuses power more than one component. Be sure all components are disconnected when checking for a short circuit.

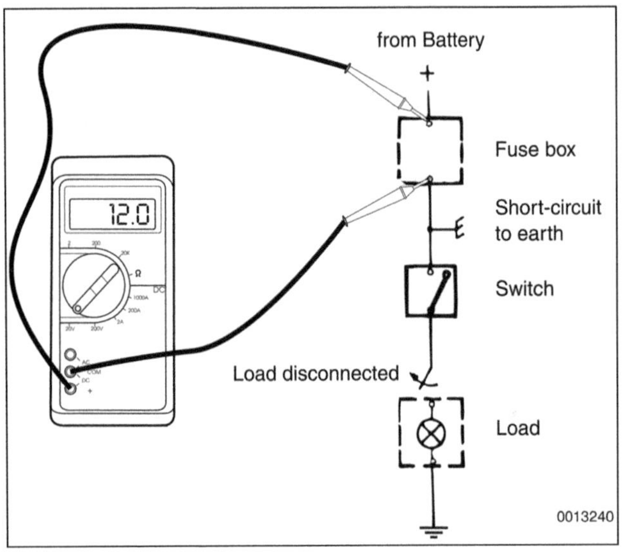

◄ Using a digital multimeter, connect test leads across fuse terminals. Make sure power is present in circuit. If necessary turn key on.

– If voltage is present at voltmeter, there is a short to ground.

– If voltage is not present, work from wire harness nearest to fuse/relay panel and move or wiggle wires while observing meter. Continue to move down harness until meter displays a reading. This is the location of short to ground.

– Visually inspect wire harness at this point for any faults. If no faults are visible, carefully slice open harness cover or wire insulation for further inspection. Repair any faults found.

901 Instruments

Special Tools

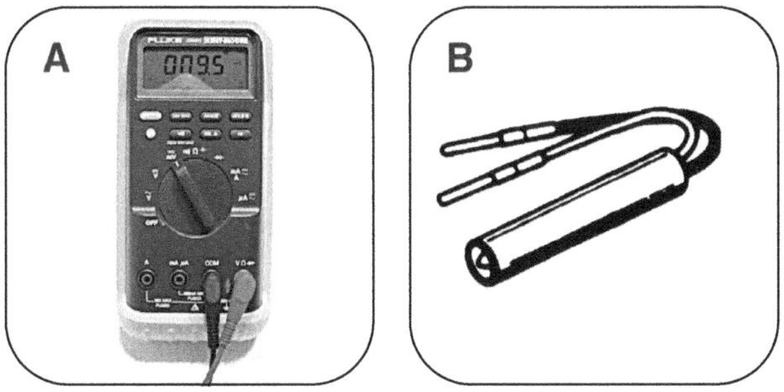

Special tools

A Fluke 87 .... Automotive digital multimeter
B 1115 .... LED tester with thin spade probes (source: Baum Tools Unlimited)

GENERAL

This repair group covers the removal and installation of the gauges and warning lights. Bulb replacement information for the gauges is also included.

Special tools

Many procedures and tests in this partition of the manual require the use of some electronic test equipment. Read the procedure through before starting work.

INSTRUMENTS

The instrument gauges are mounted in rubber retaining rings in the dashboard on the drivers side. Five gauges are used, warning lights are also used. All gauges are electronic.

Disconnect the negative (–) cable and cover battery terminal cable from accidentally contacting battery terminal.

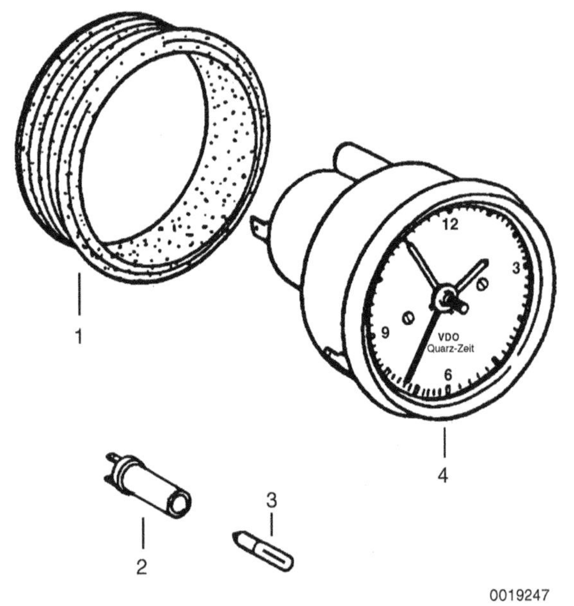

1. Rubber ring
2. Bulb holder
3. Bulb
4. Clock

Instruments, removing and installing

- Working inside passenger compartment, use a plastic pry tool or other blunt tipped object to pry the gauge out of its rubber retaining ring.

> **CAUTION —**
> *Use care not to damage the dashboard or gauge.*

- Label and disconnect wires.
- Installation is reverse of removal.

Instrument bulbs, replacing

- Remove instrument as described earlier.
- Pull bulb holder from back of instrument.
- Replace bulb.
- Installation is reverse of removal.

Seatbelt/brake warning lamp, removing and installing

The seatbelt/brake warning lamp is contained in a plastic housing with a control unit and circuit board. The bulbs are soldered directly to the circuit board and are not replaceable without special tools.

- Remove clock. See **Instruments, removing and installing**.

- Disconnect harness connector from back of seatbelt/brake warning lamp (**A**).
- Press warning lamp out from behind dashboard (**B**).

> **CAUTION —**
> *Do not pry on front of seatbelt/brake warning lamp housing. The lens is brittle and may break.*

- Installation is the reverse of removal.

Heating and ventilation control lamps, replacing

- Remove heating and ventilation control mounting bolt as described in **801 Heater**.
- Slide heating and ventilation control out of dash to access bulb sockets.
- Remove socket from control and replace bulb.

> ***CAUTION —***
> *Replace bulbs with proper bulbs only. See Porsche parts for the latest information.*

- Installation is reverse of removal.

HORNS

The components of the horn system are the dual horns, horn relay, and the horn buttons on the steering wheel. A brush and slip ring assembly makes the electrical contact between the rotating steering wheel and the steering column. The horns are located below the right headlight and are accessible by removing the front bumper.

The horns are connected to ground. Depressing the horn button activates the horn relay to supply the horns with positive voltage.

Horns, removing and installing

- Remove front bumper. See **630 Bumpers**.

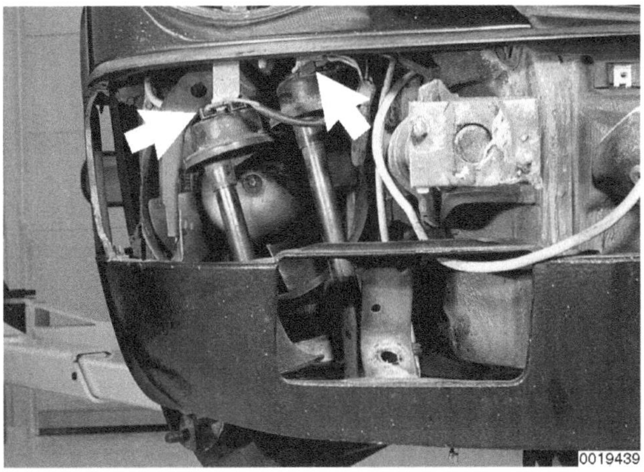
0019439

- ◄ Unplug wiring harnesses from horns (**arrows**).

NOTE —
Fog light is shown removed to illustrate horn removal.

- Remove mounting bracket nut and both horns with mounting bracket.
- Installation is reverse of removal.

SPEEDOMETER

The electronic speedometer uses the signal from the speedometer sensor to display the road speed. The speedometer sensor is mounted in the final drive above the drive axle.

NOTE —
On 1984 - 1986 cars the speedometer sensor is on the right side of the final drive unit. On 1987 - 1989 cars the sensor is on the left side of the final drive unit. A two wire sensor is used on all cars.

Speedometer sensor, removing and installing

- Working inside passenger compartment, remove floor mats at rear of center tunnel.

NOTE —
Slide driver's seat forward to gain access to rear center tunnel.

- Remove four securing screws and tunnel access cover.

0018009

- Disconnect sensor wires (**arrows**) in tunnel.
- Raise vehicle to allow access to transmission.

> ***WARNING —***
> *Be sure vehicle is stable and fully supported on the manufacturer's lift points as shown in* **010 General**. *Use proper jack stands to support the vehicle at all four corners.*

- Remove retaining bolt (**arrow**) and sensor holder from transmission.

NOTE —
Illustration at left shows sensor on 1987 - 1989 cars located on left side of transmission. Sensor on 1984 - 1986 cars is located on right side of transmission.

- Remove wires from retaining clips on transmission and pull wires from tunnel.
- Remove sensor from transmission.
- Installation is reverse of removal.

909 Alarm System

GENERAL

This repair group covers factory installed alarm. Aftermarket anti-theft systems are not covered.

Special tools

Special Tools

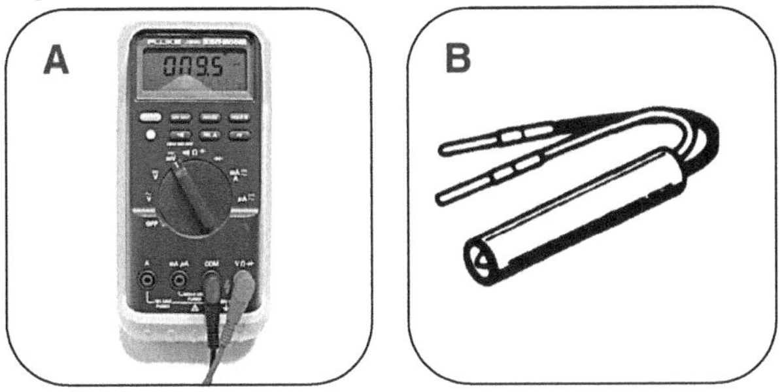

Special tools

A Fluke 87 . . . . Automotive digital multimeter
B 1115. . . . . LED tester with thin spade probes (source: Baum Tools Unlimited)

Many procedures and tests in this partition of the manual require the use of some electronic test equipment. Read the procedure through before starting work.

Alarm System

The alarm system uses its own key for activation. Keys are not available separately. If the alarm key is lost the alarm switch must be replaced.

When activated, the alarm will sound a horn and will interrupt power to the fuel pump. Even with the alarm off, a faulty alarm control unit may interrupt power to fuel pump.

The switch for the alarm (**arrow**) is located in the driver's door behind the door lock. The door must be partially opened to access the switch. To activate the alarm insert key and turn 90° clockwise. To deactivate the alarm insert key and turn 90° counter-clockwise.

> ***CAUTION—***
> *Always remove alarm key from door before closing.*

When activated, the alarm will be set off by opening the drivers door, passenger door or front hood, or by turning on the ignition switch.

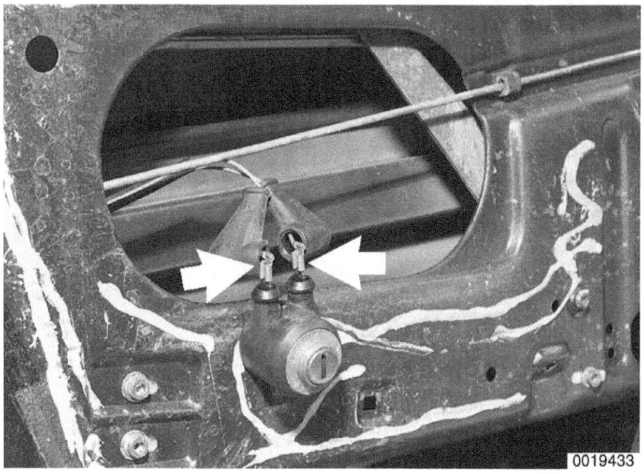

TROUBLESHOOTING

Alarm system troubleshooting requires a digital multimeter, 12V (max. 3W) test lamp and the appropriate wiring diagrams.

Alarm system switch, testing

– Alarm switch in off position. Ignition switch in off position.

– Remove alarm switch as described later.

Measure resistance between terminals (**arrows**).

- Alarm switch on, (key horizontal): 2-3 KΩ
- Alarm switch off, (key vertical): 4-5.5 KΩ

– Replace switch if resistance not as specified.

Alarm circuit, testing

> ***CAUTION—***
> *Alarm circuit is tested with battery connected. Use care when connecting and disconnecting plugs. DO NOT allow any wires to touch vehicle body.*

– Remove fresh air plenum in luggage compartment to access alarm system control module.

Detach harness connectors from alarm control module (**1**).

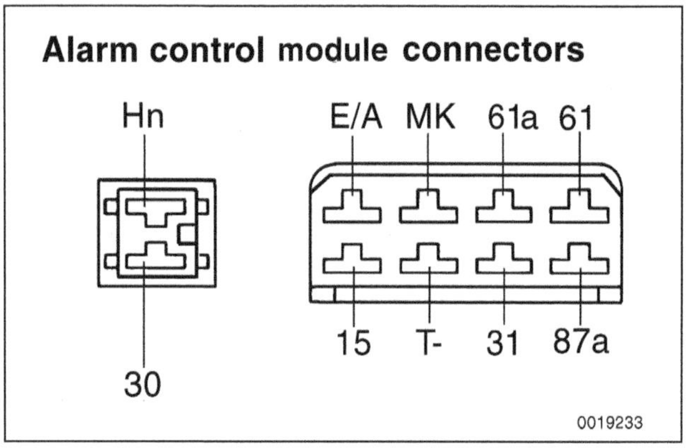

Test harness connectors using **Table a** with a test lamp and digital multimeter.

Table a. Alarm control unit connector tests

Connect Terminals	Result
30 and 87a	Test lamp will light
Hn and 30	Test lamp will light
E/A and 31	Alarm switch on (key horizontal): 2 - 3 kΩ Alarm switch off (key vertical): 4 - 5.5 kΩ
MK and 30	Test lamp on, press trunk light contact switch, lamp goes out
61 and 30	Test lamp on, turn ignition switch on, lamp goes out
15 and 31	Test lamp will go out with ignition switch turned on
T- and 30	Open door (driver or passenger), test lamp comes on close door, lamp goes out
31 and 30	Test lamp will light

Alarm control unit, testing

Testing the alarm control unit requires a 12V test light (max. 3w).

- Unplug 2-pin connector from alarm control unit, 8-pin connector remains connected.
- Depress contact switch for trunk and secure with tape or clamp.

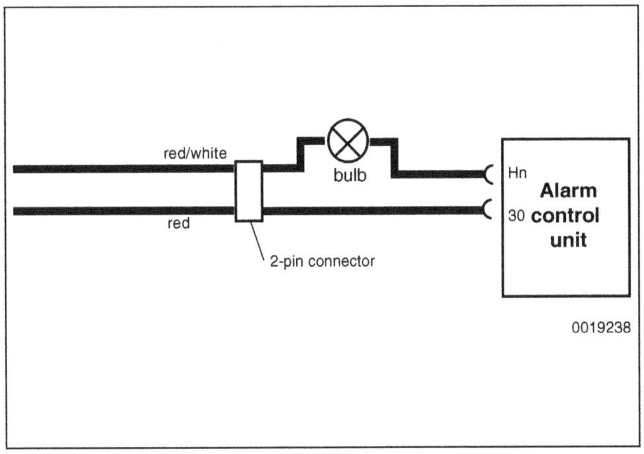

< Connect jumper wire between terminal 30 in control unit and terminal 30 in connector. Connect test light between terminal Hn in control unit and terminal Hn in connector, as shown.

- Turn on alarm. Open driver side door fully and close door. Light should flash for approx. 30 seconds.
- After light stops flashing turn on ignition switch. Light should flash. Attempt to start engine, engine should not start.
  - Open passenger door fully. Light should flash.
  - Remove tape or clamp from trunk contact switch. Light should flash.
- Disconnect jumper harness and reconnect 2-pin plug.
- Installation is reverse of removal.

COMPONENT REPLACEMENT

This section covers removal and installation of alarm system components. Always disconnect battery before connecting or disconnecting electrical components.

Alarm switch, removing and installing

- Raise driver's side window.
- Disconnect negative (–) battery cable and cover battery terminal to keep cable from accidentally contacting battery terminal.
- Remove inner door panel from driver's door. See **570 Doors**.

< Remove plastic nut cover and alarm switch mounting nut (**arrow**).

- Remove switch from inside door and unplug wires.
- Installation is reverse of removal.

Alarm control unit, removing and installing

- Disconnect negative (–) battery cable and cover battery terminal to keep cable from accidentally contacting battery.
- Remove fresh air blower. See **850 Ventilation**.
- < Push control unit (**1**) back towards dashboard until clip is free of bracket.
- Disconnect multi pin connectors and remove control unit.
- Installation is reverse of removal.

Alarm system horn, removing and installing

- The alarm horn is bolted to front suspension cross-member, behind fuel pump.
- Remove stone guard from front suspension.
- < Remove mounting bolt (**arrow**) and disconnect wires.
- Installation is reverse of removal.

910 Radio

GENERAL

This section covers the Porsche factory installed sound system. Radio and speaker removal are covered. Servicing of components is beyond the scope of this manual and should be referred to an authorized agent of the equipment in question.

The basic sound system uses a radio and four speakers. The audio option pack uses a radio, a separate power amplifier and may have additional speakers. The antenna is incorporated into the windshield and utilizes its own amplifier to boost the signal.

Aftermarket radios, amplifiers, speakers, etc. are beyond the scope of this manual. Servicing of aftermarket sound equipment should be referred to an authorized agent of the equipment in question.

Special tools

Special Tools

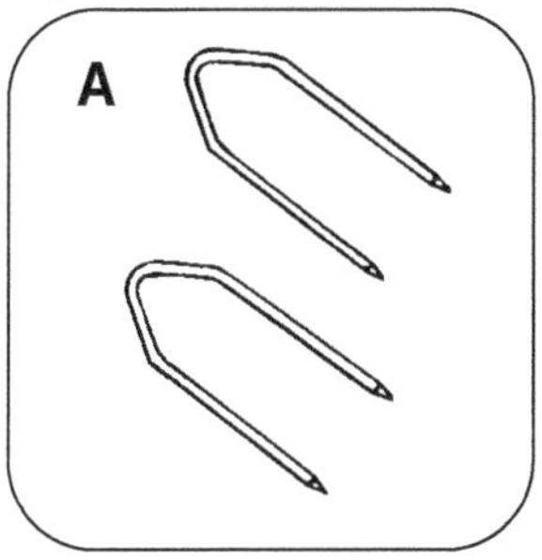

Special tool

A PNA 721 021 . . . Radio removal tool

Removal of the radio requires the use of a special radio removal tool to release the retaining clips of the pull-out radio unit.

Radio, removing and installing

> ***CAUTION—***
> - *Before beginning work on the sound system verify that the sound system is original equipment and the wiring harness has not been modified.*
> - *Do not remove the radio or disconnect the battery without obtaining the anti-theft code.*
> - *The radio is held in with tamper resistant clips. Special tools are necessary to remove radio without damage.*

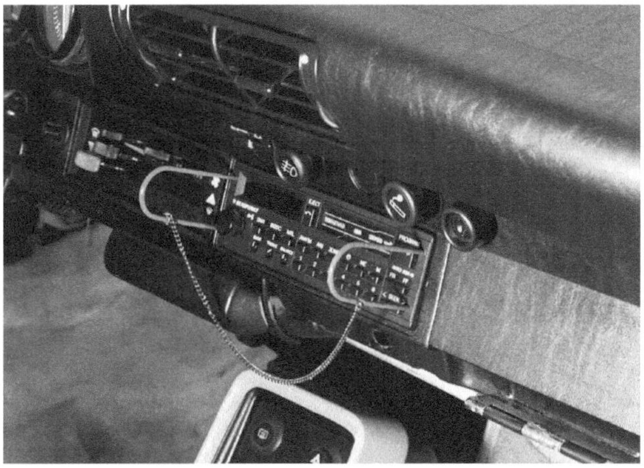

Insert radio removal tools into holes in faceplate and gently pull radio out.

– Label and disconnect wires and antenna. Remove radio.

– Install radio by sliding into dash until retaining clips engage.

ANTENNA AND AMPLIFIER

The radio antenna is integrated into the windshield. In the event of antenna failure the windshield must be replaced.

The antenna amplifier is located below the right lower corner of the windshield and is accessible through the luggage compartment.

Antenna amplifier, removing and installing

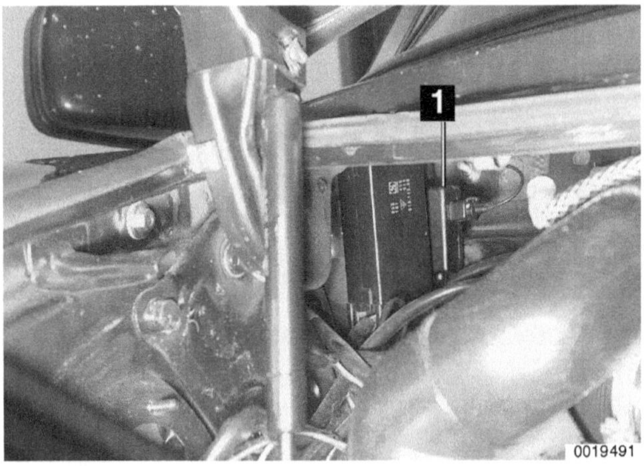

- Remove antenna amplifier (**1**) mounting nut and washer from top of amplifier.
- Disconnect wire from antenna to amplifier.
- Remove radio as described above.
- Push antenna harness and rubber grommet into luggage compartment.
- Remove antenna amplifier with wire harness.
- Installation is the reverse of removal.

FRONT SPEAKERS

The front speakers are mounted in the door panels. Sound systems without audio option pack use one full range speaker per side. Sound systems with audio option package may use separate high and low frequency speakers on each side.

Front speaker, removing and installing

Front speaker assembly

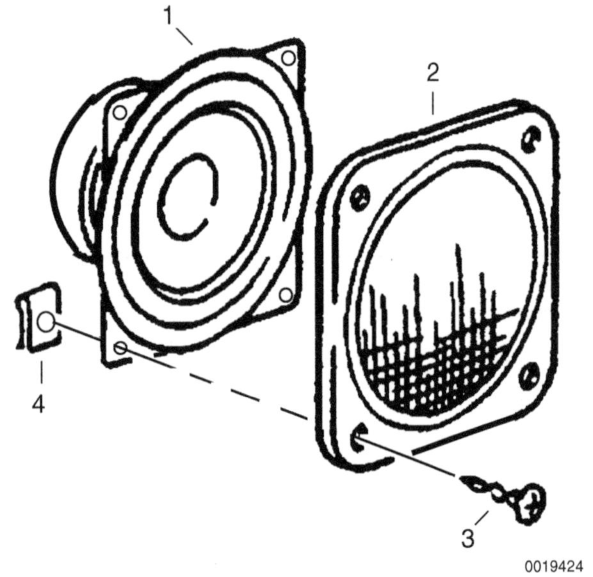

1. Front speaker
2. Grille
3. Mounting screw
4. Speed nut

- Remove speaker mounting screws.

NOTE —

If speaker mounting screws are not visible, carefully remove speaker grill.

- Remove speaker from door. Label and disconnect wires.
- Installation is reverse of removal.

Rear speaker assembly

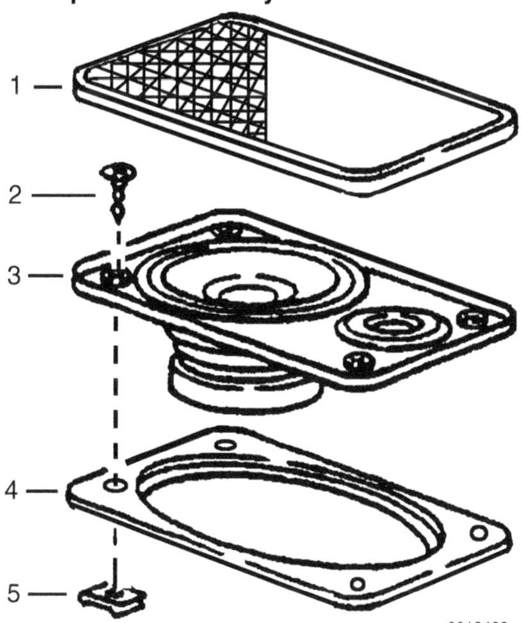

1. Grill
2. Mounting screw
3. Speaker
4. Frame
5. Speed nut

REAR SPEAKERS

Rear speakers are mounted in the shelf below the rear window on the left and right. Sound systems without audio option pack use one full range speaker per side. Sound systems with audio option package may use separate high and low frequency speakers on each side.

Rear speaker, removing and installing

◄ Remove speaker mounting screws.

NOTE —
If speaker mounting screws are not visible, carefully remove speaker grill.

- Remove speaker from shelf. Label and disconnect wires.
- Installation is reverse of removal.

920 Windshield Wipers and Washers

GENERAL

This repair group covers windshield wiper and washer system repair information. Removal and replacement of the windshield wiper/washer stalk switch assembly is covered in **960 Electrical Switches, Interior Lighting**.

Electrical wiring diagrams and relays for the wiper/washer are covered in **970 Electrical Wiring Diagrams**.

WIPER BLADES

Wiper blade symptoms and fixes

Common problems with the windshield wipers include streaking or sheeting, water drops after wiping, and blade chatter.

Streaking is usually caused when wiper blades are coated with road film or car wash wax. Clean the blades using soapy water. If cleaning does not cure the problem, the blades or inserts should be replaced. Insert replacement is the most economical method, although over time the wiper blade carrier itself will become worn.

Drops that remain behind after wiping are usually caused by oil, road film or diesel exhaust residue on the glass. Use a luke warm soap/water solution or a commercial window cleaning product. Sekuriflex windshields may be cleaned with normal drugstore alcohol if very dirty.

> ***CAUTION—***
> *Use only non-abrasive cleaners on windshield and windows.*

Wiper blade chatter may be caused by dirty or worn blades or by wiper arms that are out of alignment. Clean the blades and windshield as described above. Adjust the wiper arm so that there is even pressure along the blade, and so that the blade is perpendicular to the windshield at rest. If the problem persists, the blade carrier and wiper arms should be replaced.

Wiper blades, replacing

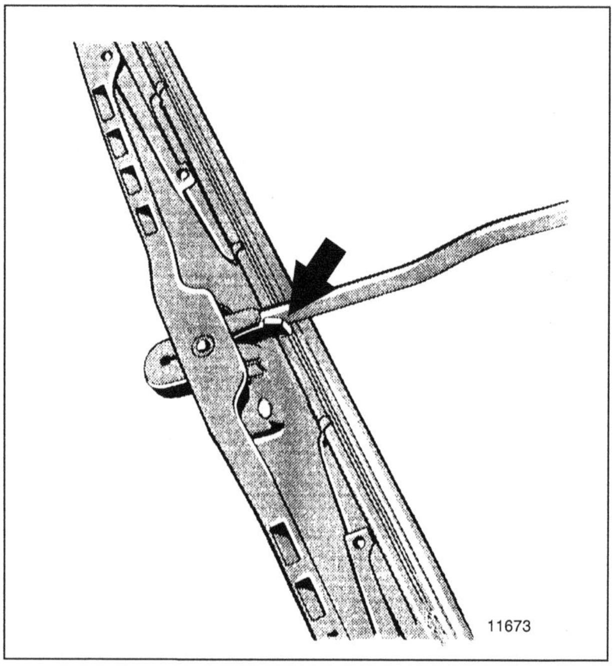

11673

- Pivot wiper arm away from glass.
- Position wiper blade carrier approximately perpendicular to wiper arm.

◀ Remove wiper blade carrier from wiper arm by depressing retaining tab (**arrow**) and sliding blade out of arm.

NOTE—
Some wiper blade versions may have two retaining tabs.

- Installation is reverse of removal. Install wiper blade carrier to wiper arm until it clicks into position.

Wiper blade inserts, replacing

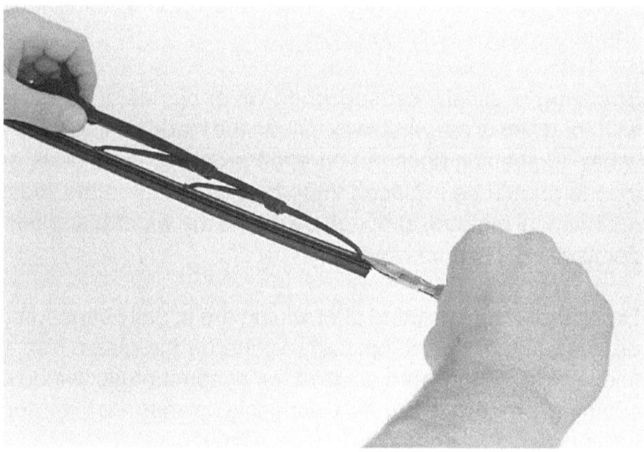

0011784

- Remove wiper blade as described earlier.

◀ Unhook end of wiper blade insert from wiper arm guides.

NOTE—
If necessary, spread guides slightly using needle nose pliers.

- Pull old insert from wiper arm guides.
- Remove metal support strips from old insert and install into slots in new insert, noting installation direction of cutouts in support strips.

NOTE—
The notched cutouts in the retaining strips should engage the molded tangs in the inserts.

- Slide new insert through wiper blade guides. Lock insert in place at end guides.

Wiper arms, removing and installing

- Swing hinged plastic cover upwards from base of wiper arm.
- ◄ Loosen, but do not remove wiper arm retaining nut (**arrow**).

NOTE—
Marking the position of the wiper arm on the shaft will aid installation.

- Pivot wiper arm up slightly to release from tapered mounting shaft.
- Remove mounting nut and take off wiper arm.
- Install driver's side wiper arm first. Wiper arm should not touch bottom part of windshield gasket.

NOTE—
When installing wiper arm, be sure to carefully align end of arm onto tapered shaft before tightening mounting nut.

- Install passenger wiper arm.

NOTE—
Passenger side wiper blade should be positioned approximately 25 mm (1 inch) above driver's side wiper blade when both arms are in park position.

WINDSHIELD WIPER ASSEMBLY

The windshield wiper assembly (linkage and motor) is removed as a single unit. Once the assembly is removed, the wiper motor and other linkage parts can be repaired or replaced.

Windshield wiper assembly, removing and installing

NOTE—
The wiper arms should be in the parked position before removing the wiper assembly. To avoid damaging the wiper arms and pivots, do not manually slide or force the wiper arm across the windshield.

- Disconnect negative (–) battery cable. See **030 Maintenance**.
- Unsnap luggage compartment carpet and lift out.
- ◄ Remove fresh air plenum cover retaining fasteners (**arrows**). Remove cover.
- Remove fresh air blower assembly. See **850 Ventilation**.
- Remove wiper arms as shown previously.

Windshield wiper assembly

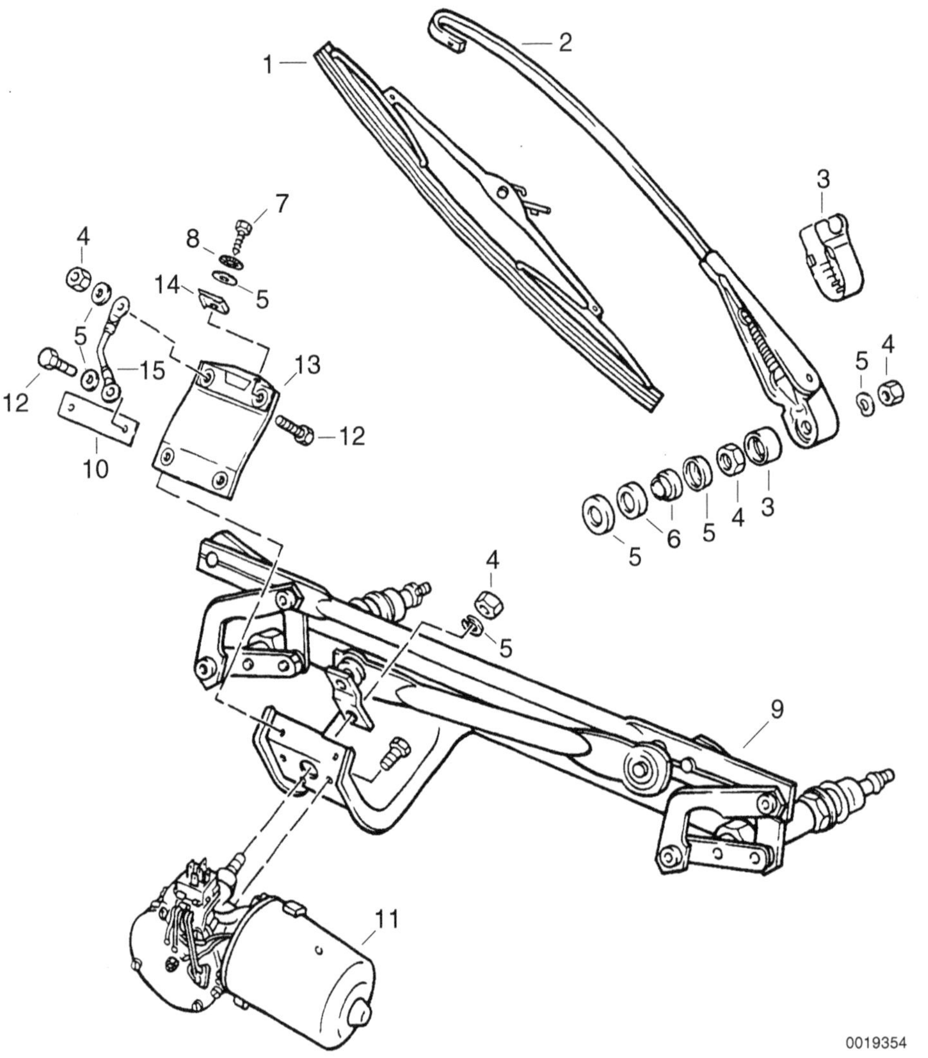

1. Wiper blade
2. Wiper arm
3. Cap
4. Nut
5. Washer
6. Rubber bushing
7. Tapping screw
8. Shake proof washer
9. Crank mechanism
10. Lock plate
11. Motor
12. Bolt
13. Fabric mounting
14. Speed nut
15. Ground cable

– Remove clock and speedometer from dashboard. See **901 Instruments**.

– Working though speedometer opening, unplug wiper motor and remove M6 wiper motor mounting bolts.

– Support windshield wiper assembly and remove plastic caps beneath wiper arms. Remove retaining nuts.

– Lower windshield wiper assembly and remove.

– Installation is reverse of removal.

– To remove wiper motor from crank mechanism, use windshield wiper assembly component illustration above.

> ***CAUTION—***
> - *Be sure ground wire at fabric mount is securely fastened.*
> - *Run wipers and allow to return to parked position before installing wiper arms.*

WASHER SYSTEM

The windshield washer system includes the windshield spray nozzles, the washer fluid pump and the washer fluid reservoir. A separate washer fluid tank, pump and nozzles are used for the intensive washer system.

Cars with the headlight washer system are equipped with an additional fluid pump, hoses and nozzles on top of the front bumper.

Washer spray nozzles can be aimed by using a sewing needle or a similar diameter stiff piece of wire.

Windshield spray nozzle, removing and installing

◀ Working inside luggage compartment, press washer nozzle from the rear (**arrow**) and pivot nozzle up out of the cowl.

NOTE—
It may be necessary to slide washer nozzle slightly toward front of car while pushing from below (arrow).

- Remove clamp from nozzle and remove washer hose.
- Unplug 2-pin connector for heated nozzle (as applicable) and remove nozzle.

NOTE—
2-pin connector for heater washer nozzle is not located at the nozzle itself but in the plenum area.

- Installation is reverse of removal.

NOTE—
Be sure to replace washer nozzle gasket as necessary.

Windshield washer pump, removing and installing

◄ Unplug wires from windshield washer pump (**1**).

– Remove clamps and fluid hoses.

– Remove pump bracket from rubber mounts and remove pump.

– Connectors on pump are labeled plus (+) and minus (-). Brown wire connects to minus (-) connector.

– Installation is reverse of removal.

Washer fluid reservoir, removing and installing

– Raise and safely support vehicle.

> ***WARNING —***
> *Make sure car is stable and well supported at all times. Use a professional automotive lift or jack stands designed for the purpose. A floor jack is not adequate support.*

– Remove left front wheel.

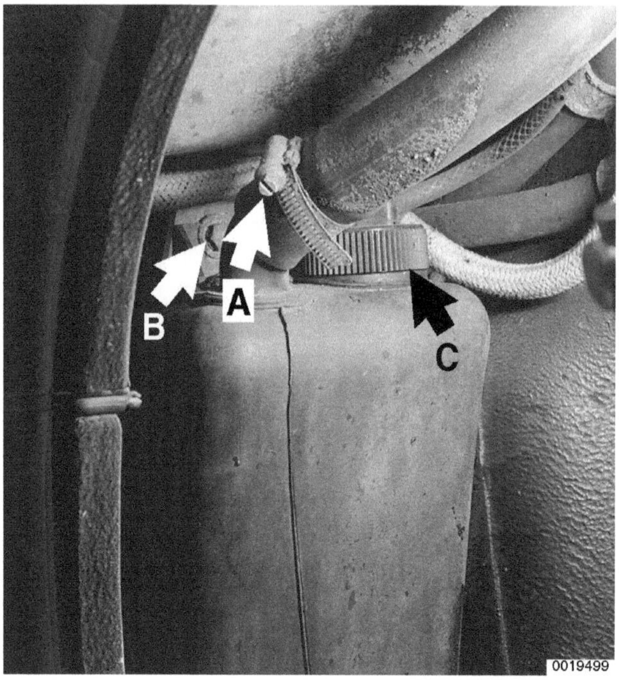

◄ Unscrew filler tube clamp (**arrow A**) and remove tube from reservoir.

– Unscrew collar (**arrow C**) for washer pump feed and return lines and set to side.

– Remove lower mounting bolt and retaining strap from underneath reservoir.

– Remove upper reservoir mounting bolt (**arrow B**) and lower reservoir down.

– Remove washer pump feed and return lines from reservoir. Remove reservoir.

> ***NOTE —***
> *Feed line should be equipped with a small filter at the end.*

– Installation is the reverse of removal.

Windshield wiper/washer switches

The windshield wiper/washer switch is mounted on the right side of the steering column. The wipers have three speeds and an off position. Pushing up on the wiper lever increases the speed. Removal of the wiper lever/switch is covered in **960 Electrical Switches, Interior Lighting**.

The windshield washer system is operated by pulling the lever towards the steering wheel.

Intermittent wiper action switch, removing and installing

The intermittent wiper rheostat is located on the dashboard between the speedometer and the clock. The wiper switch on the steering column must be in the off position (all the way down) to actuate intermittent operation.

– Disconnect negative (–) battery cable. See **030 Maintenance**.

> ***WARNING —***
> *Prior to disconnecting the battery, read the battery disconnection cautions given at the front of this manual on page vii.*

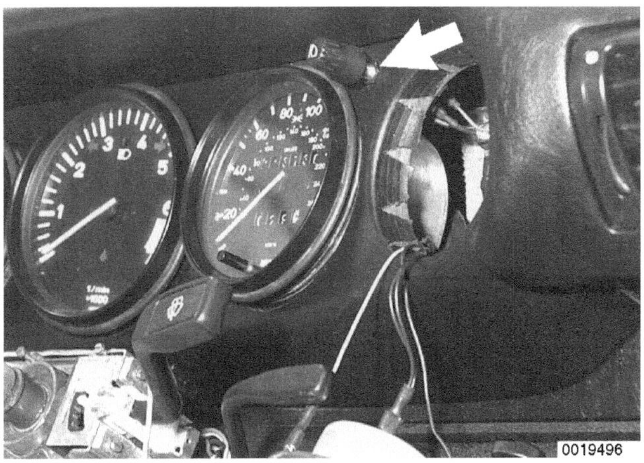

0019496

– Remove clock from the dashboard. See **901 instruments**.

< Pull off knob and remove mounting nut (**arrow**).

– Slide switch back into dash and remove through clock opening.

– Label and disconnect wires.

– Installation is reverse of removal.

940 Exterior Lighting

GENERAL

This section covers exterior light troubleshooting and component replacement.

See **960 Electrical Switches, Interior Lighting** for exterior light switch replacement procedures.

See **970 Electrical Wiring Diagrams** for information on individual circuits.

Test equipment

Electrical and electronic automotive components require the use of low impedance test equipment.

Special Tools

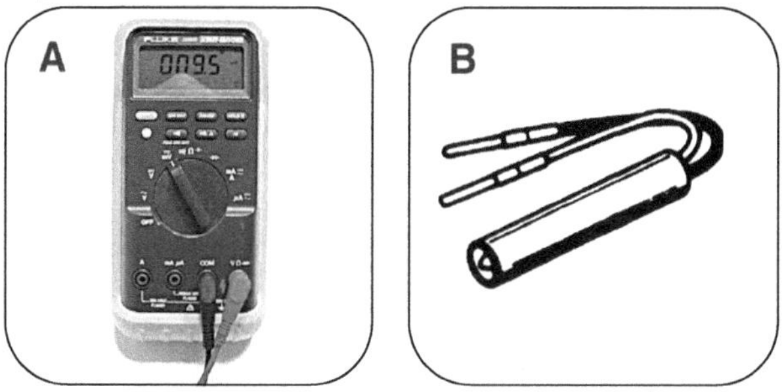

Special tool

A Fluke 87 . . . . Automotive digital multimeter
A 1115 . . . . LED tester with thin spade probes (source: Baum Tools Unlimited)

Troubleshooting

Many lighting problems are due to dirty or corroded sockets, or poorly fitting bulb contacts or connectors. Check that voltage is reaching the bulb and that the socket has a good connection to ground.

Most exterior lighting involves pairs of lights, and this fact is an aid to troubleshooting. If only one of a pair of lights is out then the problem is most likely due to a failed bulb or some other problem with that particular light socket or its wiring. A simple test is to exchange the failed bulb with one that is known to be good.

If a pair of bulbs are both out then the problem is most likely in some part of the system common to both lights. Begin by checking for a failed fuse. Test switches using simple continuity checks made with a digital ohmmeter. Check the switch connectors for voltage and continuity to ground. Always check the bulbs as they are the most common cause of lighting problems.

For a general description of electrical test procedures, see **900 Electrical System–General**.

NOTE —
Replacement exterior bulbs are listed in **Table a**.

> ***CAUTION —***
> - *When working on electrical systems, detach negative (–) battery cable. See* **030 Maintenance**.
> - *Prior to disconnecting the battery, read the battery disconnection cautions given at the front of this manual on page viii.*

Table a. Bulb Specifications

Location	Rating	DIN (SAE) number
Back-up light	32 cp	(1073)
Brake/taillight	32/3 cp	(1034)
Front fog light H3	55 W	PK 22 S
Headlight		
Sealed beam	60/50 W	(H6014)
H4	60/55 W	P43t-38
License plate	2 cp	(1895)
Side marker	2 cp	(1895)
Sign light	2 cp	2.9 x 4.6
Turn signal/side marker	32/3 cp	(1034)

NOTE —
In the table above, "cp" stands for candle power.

FRONT LIGHTING

> ***CAUTION —***
> *Do not touch halogen bulbs with your bare hands. Dirt and skin oils may cause premature bulb failure. If necessary clean bulb using a clean cloth dampened with rubbing alcohol.*

Sealed beam headlights are standard equipment in 1984 - 1986 cars. Halogen bulbs are used for headlights and fog lights for 1987 and later cars.

Headlights, aiming

> ***CAUTION—***
> *Headlights must be aimed using a headlight aiming tool. DO NOT aim headlights by eye.*

NOTE—
Prior to adjusting headlights, make sure the car has a full fuel tank, correct tire pressures and 75 kg (165 lbs.) in driver's seat.

- Remove rubber plugs from headlight trim ring.

NOTE—
If necessary, remove trim ring to expose headlight adjusting screws.

- Mount headlight aiming tool.
- Working through holes in headlight trim ring (**arrows**), adjust aiming screws.
- Reinstall rubber plugs.

Headlight sealed beam, replacing

- Remove retaining screw and headlight trim.
- Remove three screws and sealed beam retaining ring.
- Remove sealed beam and unplug electrical connector.
- Installation is reverse of removal.

Sealed beam headlight assembly

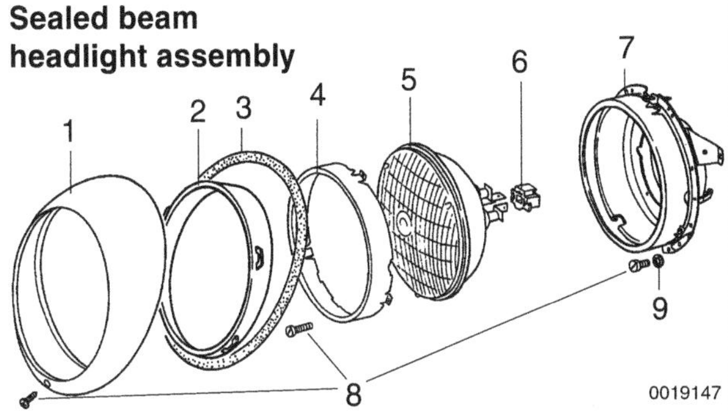

1. Trim ring
2. Intermediate ring
3. Sealing ring
4. Sealed beam retaining ring
5. Sealed beam
6. Electrical connector
7. Sealed beam housing
8. retaining crews
9. Washer

Halogen headlight assembly

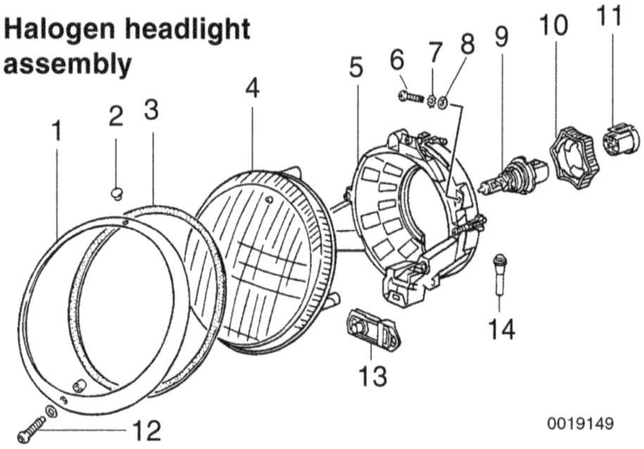

1. Trim ring
2. Rubber plug
3. Sealing ring
4. Reflector housing
5. Headlight housing
6. Aligning screw
7. Lock washer
8. Washer
9. Halogen bulb
10. Plastic retaining nut
11. Electrical connector
12. Retaining screw
13. Humidity absorber
14. Drain plug

Headlight halogen H-4 bulb, replacing

< Remove retaining screw and headlight trim.

- Remove four screws and pull headlight assembly out of fender.

- Unplug harness connector.

- Remove large plastic retaining nut and remove bulb.

> ***CAUTION—***
> - *Bulb has O-ring seal and may be difficult to remove. DO NOT apply excessive force as bulb may shatter.*
> - *Do not touch halogen bulbs with your bare hands. Dirt and skin oils may cause premature bulb failure. If necessary clean bulb using a clean cloth dampened with rubbing alcohol.*

- Installation is reverse of removal.
  - Align grooves in bulb with tabs in lens.

Front fog light assembly, removing and installing

The following procedure includes fog light lens replacement and fog light bulb replacement.

NOTE—
There are several different fog lights. Procedures may vary slightly.

< Remove fog light mounting screws.

- Remove fog light lens assembly.

- Disconnect electrical harness connector from fog light bulb. Remove bulb from rear of fog light assembly.

- Installation is reverse of removal.
  - Check electrical connector for corrosion.
  - Spray with contact cleaner, if necessary.

Front fog light assembly

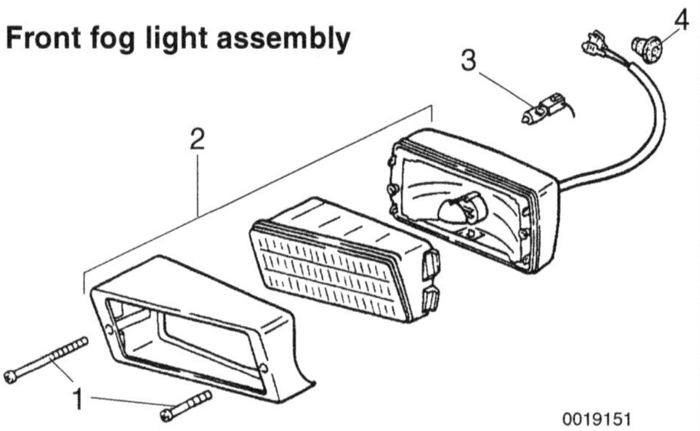

1. Retaining screw
2. Fog light assembly housing
3. Halogen bulb H3
4. Grommet

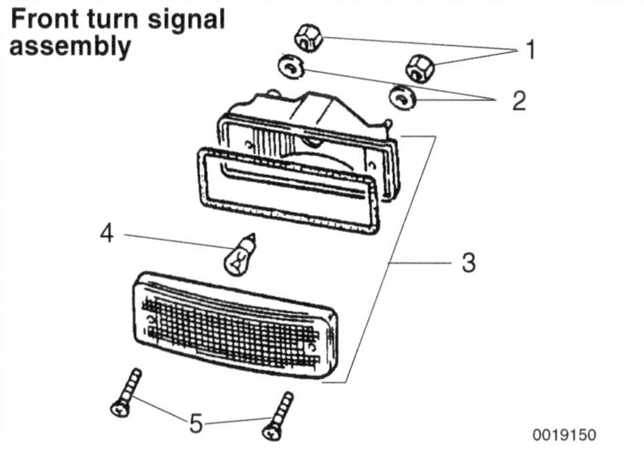

1. Housing retaining nut
2. Washer
3. Turn signal assembly housing
4. Bulb
5. Lens screw

Front turn signal bulb, replacing

◄ Remove front turn signal lens screws. Remove lens.

- Remove bulb.
- Inspect socket for corrosion.
- Installation is reverse of removal.

Front turn signal assembly, removing and installing

- Remove front bumper. See **630 Bumpers**.
- Remove turn signal mounting nuts.
- Unplug wires and remove turn signal assembly.
- Installation is reverse of removal.

REAR LIGHTING

Taillight bulb, replacing

- Remove taillight lens mounting screws and remove lens.
- To remove bulb from bulb socket, push bulb in gently and rotate counterclockwise. Pull bulb from socket.
- Inspect socket for corrosion. Clean or replace as required.
- Installation is reverse of removal.

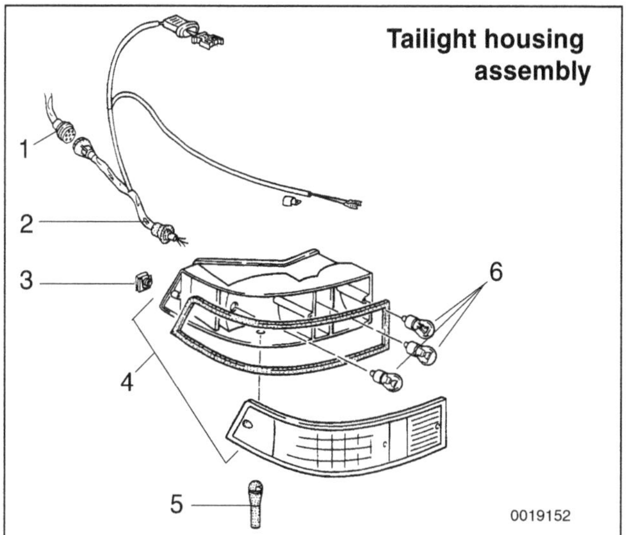

1. Harness connector (7-pin)
2. Harness
3. Nut
4. Tailight housing
5. Drain plug
6. Bulb

Taillight assembly, removing and installing

◄ Remove taillight lens mounting screws and remove lens.

- Remove taillight drain plug.
- Remove taillight housing mounting screws.
- Pull taillight housing from car.
- Disconnect multipin connector, in engine compartment, and remove taillight housing.
- Installation is reverse of removal.

Center brake light bulb, replacing

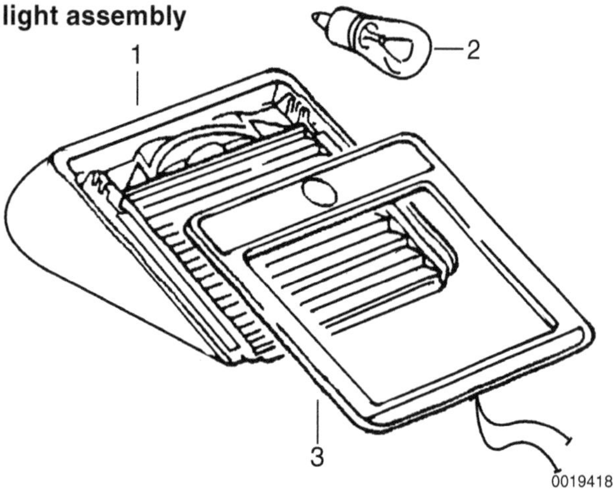

1. Housing
2. Bulb
3. Frame

◄ Working at utility shelf below rear window, depress locking tab in center of center brake light plastic housing and lower housing.

- Pull housing straight out from rear window.
- Press two locking tabs at top of red lens and lift lens off.
- To remove bulb from bulb socket, push bulb in gently and rotate counterclockwise. Pull bulb from socket.
- Installation is reverse of removal.

License plate light, removing and installing

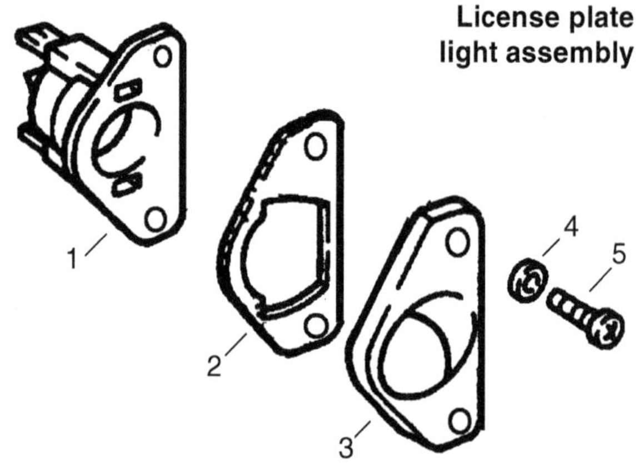

1. Light socket
2. Gasket
3. Lens
4. Washer
5. Screw

◄ Remove screws retaining license plate light assembly to rear bumper pad.

- Pull license plate light assembly from bumper pad and remove lens.
- To remove bulb from bulb socket, push bulb in gently and rotate counterclockwise. Pull bulb from socket.
- Installation is reverse of removal.

950 Central Locking

Special Tools

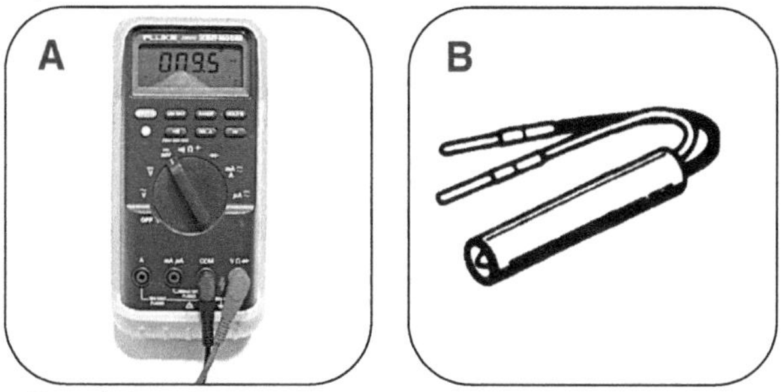

Special tool

A Fluke 87 . . . . Automotive digital multimeter
A 1115 . . . . LED tester with thin spade probes (source: Baum Tools Unlimited)

GENERAL

This repair group covers central locking system troubleshooting and component replacement.

Door handle and interior door panel removal is covered in **570 Doors**.

Removal of the central locking switch, located in the center console, is covered in **680 Interior Trim**.

Special tools and equipment

Electrical and electronic automotive components require the use of low impedance test equipment.

Central locking system

The central locking system uses electrical locking actuators at each door. The locking system can be actuated by either front door lock. When the key is in the lock position, a microswitch engages the central locking system via the central locking control module. Microswitches are operated via the door lock cylinder at both front doors.

Central locking function, checking

Both doors can be locked or unlocked electrically by pressing the central locking switch on the center console with ignition switch in position one or two. The central locking switch will illuminate to indicate that the doors are locked.

Troubleshooting

Check for 12 volts at terminal 6 of central locking module (**1**), found in luggage compartment behind right hinge.

Check condition of central locking system main fuse (luggage compartment fuse/relay panel, fusebox #1, fuse 2.)

950

CENTRAL LOCKING SERVICE

Central locking components

1. Center console
2. Central locking switch
3. Screw
4. Door lock
5. Microswitch
6. Washer
7. Operating rod
8. Pin
9. Locking clip
10. Lock actuator
11. Support
12. Speed nut
13. Locking device
14. Central locking control module
15. Control module bracket

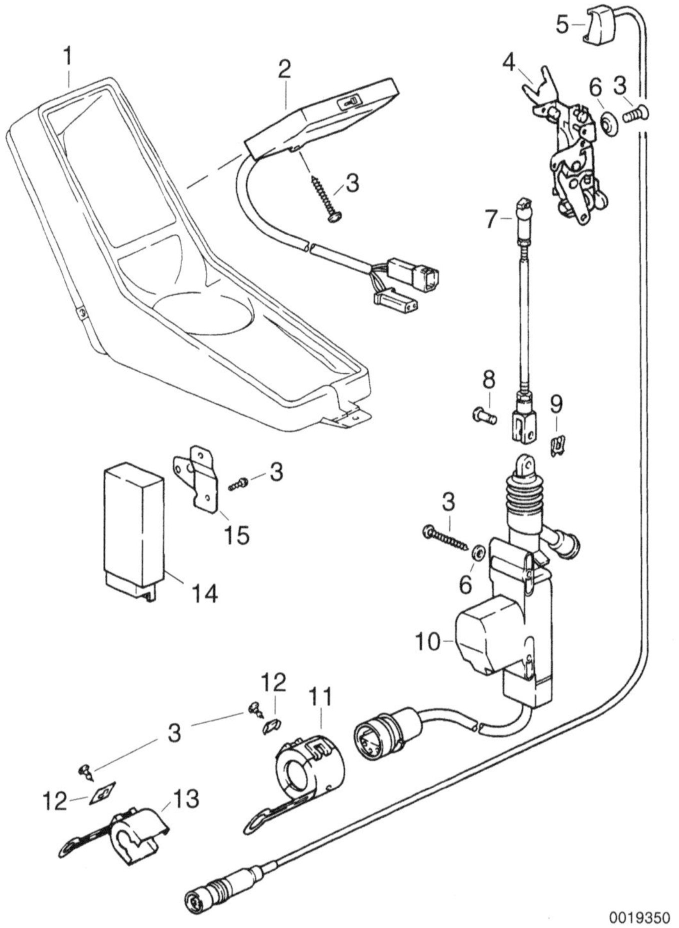

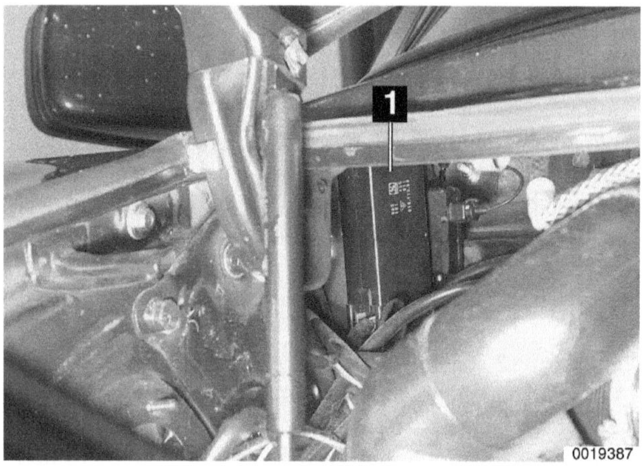

Central locking control module, replacing

The central locking control module (**1**) is mounted in the right rear of the luggage compartment, directly below the windshield post.

- Disconnect negative (–) battery cable.
- Remove nut securing module to inner fender panel.
- Remove module from its mounting stud and disconnect multiple pin connector.
- Installation is reverse of removal.

Door lock cylinder microswitch, removing and installing

The door lock microswitch is attached to the door handle/door lock assembly. It is available as a separate replacement part.

- Close window completely.
- Disconnect negative (–) battery cable.
- Remove door panel as described in **570 Doors**.

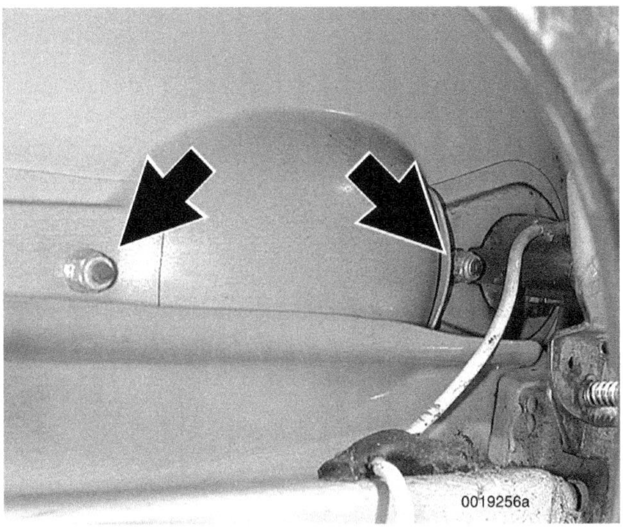

◀ Working from inside door, remove door handle securing nuts (**arrows**).

- Disconnect microswitch wiring harness from connection at bottom of door.

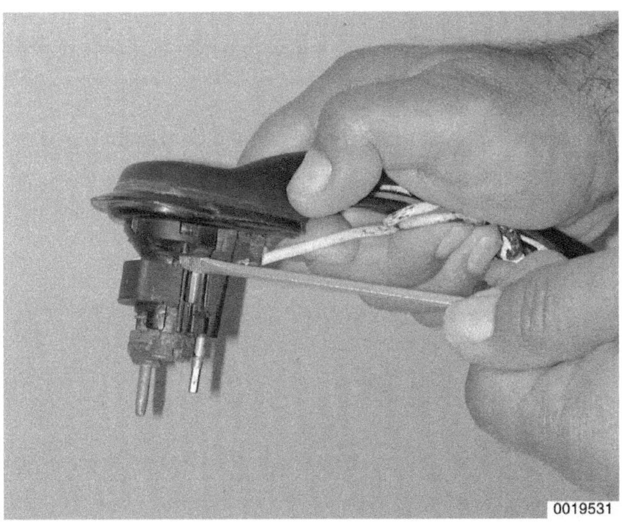

◀ Pull door handle out of door. Use a small screwdriver to carefully pry microswitch off lock mechanism.

- Detach wire tires and remove harness and switch from door.
- Installation is reverse of removal. Use new gaskets between handle and door.

Front door locking actuator, replacing

NOTE —

If just one of the central locking actuators does not operate, the locking linkage could be frozen or stiff. Remove the interior door panel and lubricate the locking linkage.

– Close window completely.

– Disconnect negative (–) battery cable.

– Remove door trim panel. See **570 doors**.

0019425

◀ Disconnect actuator connecting rod from door latch ball.

0019445

◀ Remove actuator mounting screws (**arrow**) in lower part of door.

– Disconnect wiring harness from connector at bottom of door and remove actuator.

– Installation is reverse of removal.

Front door locking actuator, adjusting

- Close window completely.
- Disconnect negative (–) battery cable.
- Remove locking actuator from door as described earlier.

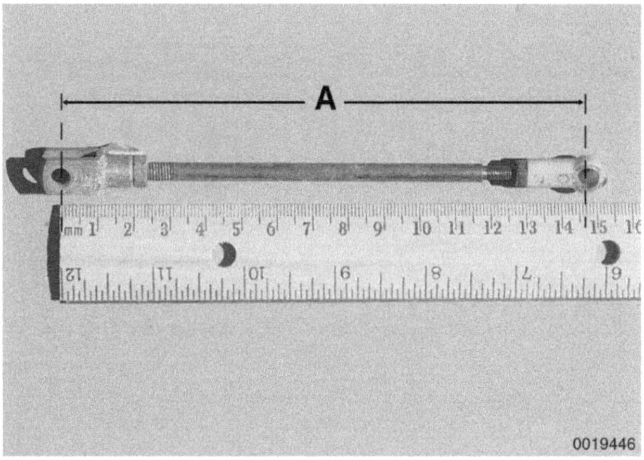

- ◀ Check connecting rod length (**A**) and adjust if necessary.

NOTE—
Measure rod length from pin center to ball center.

Central Locking Actuator Rod	
• Actuator rod length (**A**)	146.5 mm (5.77 in.)

- Mount actuator on door, but leave mounting screws loose.

NOTE—
The bulge of the actuator motor must point toward the front of the car.

- Set door locks to unlocked position.
- Pull actuator all the way down, then slide back up between 2 and 3 mm (0.08 - 0.12 in.). Tighten mounting screws.
- Connect actuator harness connector to locking actuator.
- Check function of door locks.

960 Electrical Switches, Interior Lighting, Cruise Control

TABLES

GENERAL

This section covers replacement of electrical switches and interior lights. The cruise control system is also covered here.

See **901 Instruments** for dash gauge and instrument repair information.

See **940 Exterior Lighting** for exterior light repair information.

See **970 Electrical Wiring Diagrams** for information on individual circuits.

Test equipment

Electrical and electronic automotive components require the use of low impedance test equipment.

Special Tools

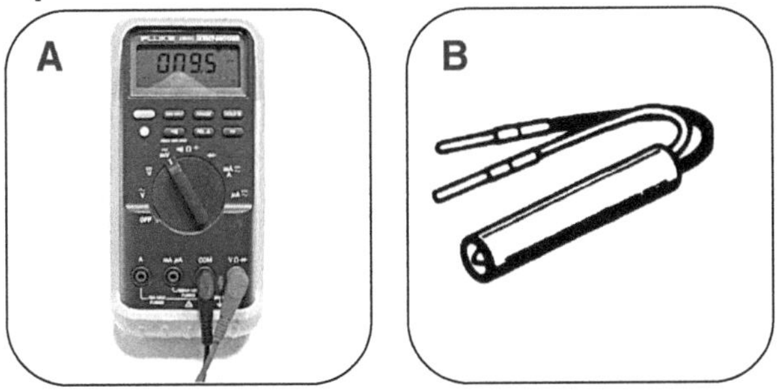

Special tool

A Fluke 87 . . . . Automotive digital multimeter
B 1115 . . . . LED tester with thin spade probes (source: Baum Tools Unlimited)

> ***CAUTION —***
> - *When working on electrical systems, detach negative (–) battery cable. See* **030 Maintenance**.
> - *Prior to disconnecting the battery, read the battery disconnection cautions given at the front of this manual on page viii.*

960

ELECTRICAL SWITCHES

Console mounted switches

The switches mounted in the center console are covered in **680 Interior Trim.** This includes the following switches:

- A/C blower switch
- A/C temperature switch
- Hazard warning switch
- Rear window defroster switch
- Central locking switch

Door contact switch, removing and installing

The door contact switches are located in each front door post.

Door contact switch assembly

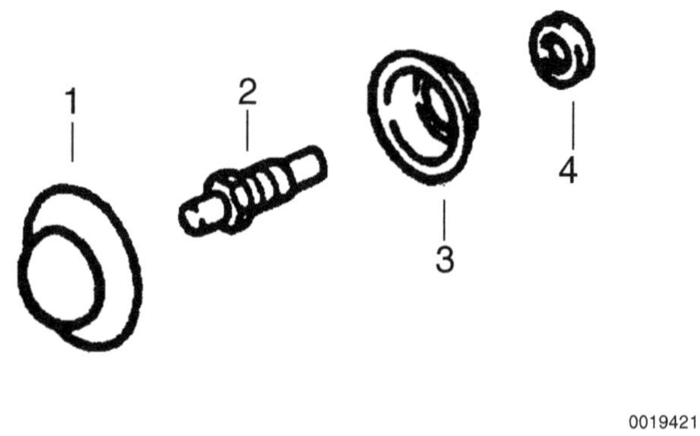

1. Rubber cap
2. Door contact switch
3. Concave washer
4. Washer

– Disconnect negative (–) battery cable.

– Remove rubber boot from switch.

– Using an 11 mm socket, unscrew switch and disconnect wire.

– Installation is reverse of removal.

> *CAUTION —*
> *DO NOT allow wire to fall into door pillar.*

Fog light switch, removing and installing

– Disconnect negative (–) battery cable.

– Remove fresh air blower and plenum, see **850 Ventilation**.

– Unscrew fog light switch knob in counter-clockwise direction.

– Use two-pin spanner wrench to remove switch retaining collar from dashboard.

– Remove switch from behind.

– Label and disconnect wires.

– Installation is the reverse of removal.

Luggage compartment light/alarm switch, removing and installing

The luggage compartment light switch is located on the right side hinge, between the hinge bolts. A copper connector is crimped to the wire and to the switch.

– Disconnect negative (–) battery cable.

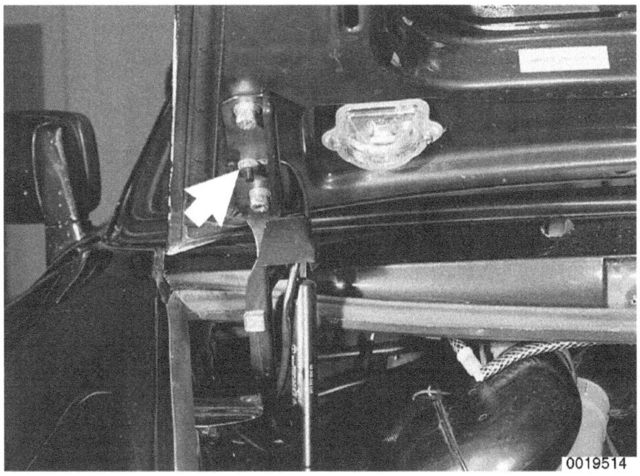
0019514

◄ Unscrew switch retaining nut (**arrow**) from hinge.

– Remove copper connector wire from switch and remove switch.

> ***CAUTION —***
> *DO NOT allow wire to fall into hood.*

– Installation is reverse of removal.

Headlight switch, removing and installing

– Disconnect negative (–) battery cable.

– Pry out plastic center from rubber headlight switch knob.

– Unscrew nut and remove headlight switch knob.

0019430

◄ Using a small spanner in retaining ring holes (**arrows**) remove the retaining ring and push switch out.

– Remove switch from below dashboard.

– Label and disconnect wires.

– Installation is reverse of removal.

Horn contact ring, removing and installing

- Disconnect negative (–) battery cable.
- Remove steering wheel, see **480 Steering**.

- Remove horn contact ring mounting screws (**arrows**) and contact ring.
- Installation is the reverse of removal.

Ignition switch, removing and installing

- Remove left side knee protection trim from dashboard.
- Working under dashboard, remove electrical wiring harness from ignition switch.
- Remove ignition switch mounting screws and remove from ignition/steering lock mechanism.

See **480 Steering** for additional information on the steering lock.

Power mirror selector switch, removing and installing

The power mirror selector switch is located above the ignition switch to the right of the sunroof switch (if applicable). This switch selects which mirror (driver or passenger side) is selected for adjustment with the mirror adjustment switch.

- Disconnect negative (–) battery cable.
- Use a plastic pry tool or other blunt tipped object to pry the temperature/oil pressure gauge out of the dashboard.

0019516

◄ Working in luggage compartment, pry out switch retaining tabs (**arrow**) while pushing switch down out of dashboard.

– Disconnect harness connector.

– Install switch by pushing back up into dashboard.

Power mirror adjustment switch, removing and installing

– Disconnect negative (–) battery cable.

– Remove mirror adjustment switch knob by pulling upward.

0019518

◄ Unscrew switch retaining ring (**arrow**).

– Remove upper door panel trim.

– Disconnect harness connector and remove mirror adjustment switch.

– Installation is reverse of removal.

Power window switch, removing and installing

– Disconnect negative (–) battery cable.

> ***CAUTION—***
> *Prior to disconnecting the battery, read the battery disconnection cautions given at the front of this manual on page vii.*

– Loosen and/or remove enough door panel mounting hardware to pull front part of panel away from door. See **570 Doors** for more information.

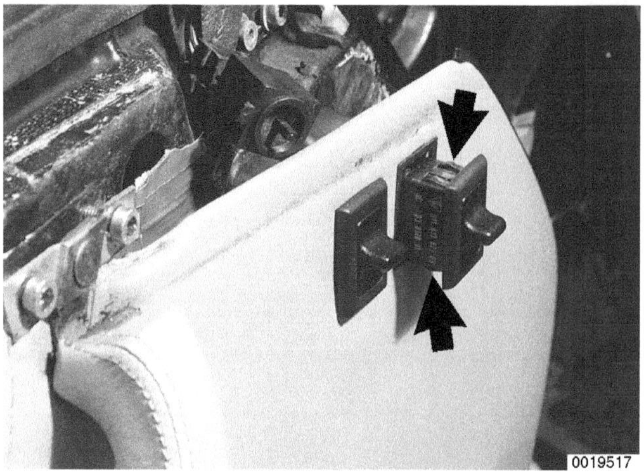

◀ Push out power window switch from behind door panel while depressing spring tabs on top and bottom of switch (**arrows**).

– Pull rubber boot back and disconnect wires from switch.

– Installation is the reverse of removal.

Table a. Window switch wiring connections

Switch Termi-nal	Left window driver's door switch	Right window driver's door switch	Right window passenger door switch
1	Blue	Red / white	Blue
2	Blue / black	Red / black	Blue / black
3	Single red / blue	Single red / blue	Red / white
4	Brown	Brown	Brown
5	Double red / blue	Double red / blue	Red / black

Steering column switches, removing and installing

The turn signal/headlight dimmer switch, windshield wiper/washer switch and cruise control switch are all mounted on the steering column.

– Disconnect negative (–) battery cable.

– Working inside passenger compartment, remove steering wheel as described in **480 Steering**.

◀ Remove steering column cover outer mounting screws (**arrows**) located on sides of cover.

NOTE —
Screws are located behind switch arms.

◀ Remove horn contact plate mounting screws (**arrows**).

- Disconnect ground wire from back of plate.
- Remove plate.

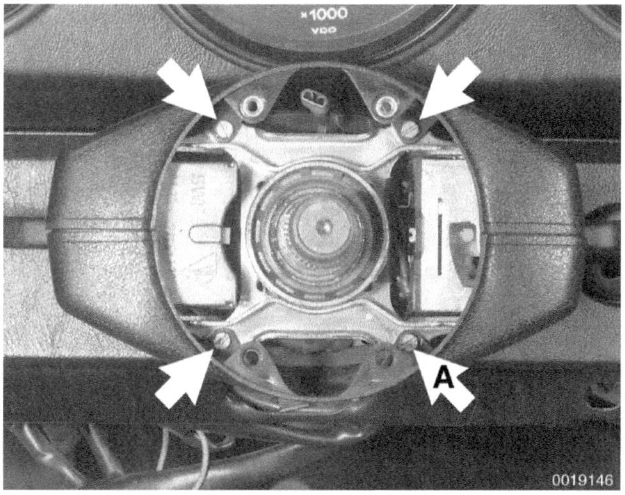

◄ Remove steering column cover inner mounting screws (**arrows**). Remove covers.

NOTE —
*Ground wire connection is on lower right screw (**A**).*

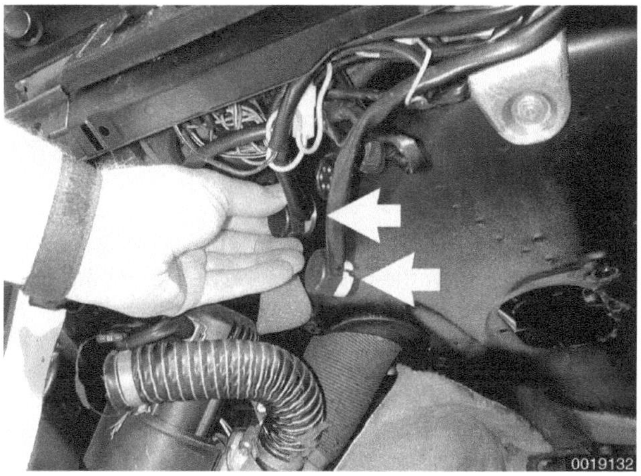

◄ Unplug switch wiring harness (**arrows**) for appropriate switch from underneath dashboard.

NOTE —
Harness connector for cruise control switch is not shown in illustration. Follow wires down to right of steering column to locate harness connector.

◄ To remove turn signal switch from column:

- Cut wire ties.
- Remove small mounting screws (**arrows on left side**).
- Remove switch.

◄ To remove windshield wiper/washer switch or cruise control switch from column:

- Cut wire ties.
- Remove small mounting screws (**arrows on right side**).
- Remove switch.

– Installation is reverse of removal.

- Tighten mounting screws, then lock in place with lacquer or paint.
- Reattach harnesses with new wire tires.
- When installing steering column covers, be sure to anchor ground wire under lower right cover screw.

Tightening Torque
- Steering wheel to steering column . . . . . 50 Nm (36 ft-lb)

Sunroof switch, removing and installing

– Disconnect negative (–) battery cable.

– Use a blunt tipped screwdriver to pry the temperature/oil pressure gauge out of the dashboard.

0019515

< Reach through gauge opening and push sunroof switch (**arrow**) downward out of dashboard.

– Disconnect harness connector.

– Install switch by pushing back up into dashboard.

INTERIOR LIGHTS

General

The dome lights for the coupe are located over each door and can be operated by simply tilting the lens face. The glove compartment light is located in the dash above the glove compartment door. A switch in the dash is released when the glove compartment door is opened that turns on the light. Both the dome and glove compartment light can be set to operate by the door contact switches.

The targa and cabriolet use one interior light located above the windshield between the sun visors. The light has a built-in three position rocker switch which is operated by pressing on the lens.

The various interior light bulb specifications are provided in **Table a**.

Table b. Bulb Specifications

Location	Rating	DIN (SAE) number
Passenger compartment and trunk	10 W	BA 15s
Instruments, ashtray	2 W	BA 7
Glass base indicator lamp	1.2 W	W2 x 4.6d
Sign light (USA only)	1.2 W	2.9 x 4.6

NOTE —
Use only original equipment replacement bulbs.

Engine compartment bulb, replacing

The engine compartment light is located in the center of the engine compartment lid. The sliding bulb cover is also an on/off switch. The engine compartment light operates only when the headlights are on.

– Disconnect negative (–) battery cable.

– Pull housing cover off light assembly at arrows.

NOTE —
It may be necessary to pry off the housing cover with a small screwdriver.

– Rotate bulb cover/switch to expose bulb.

– Remove bulb.

– Installation is reverse of removal.

Glove compartment bulb, replacing

The glove compartment light is located in the dashboard, directly above the glove compartment. The light has a built-in three position rocker switch which is operated by pressing on the lens.

– Disconnect negative (–) battery cable.

– Insert screwdriver in cut-out on light housing and carefully pry out light assembly.

– Remove bulb from between contacts.

– Install new bulb and press light into place.

Interior light bulb, replacing

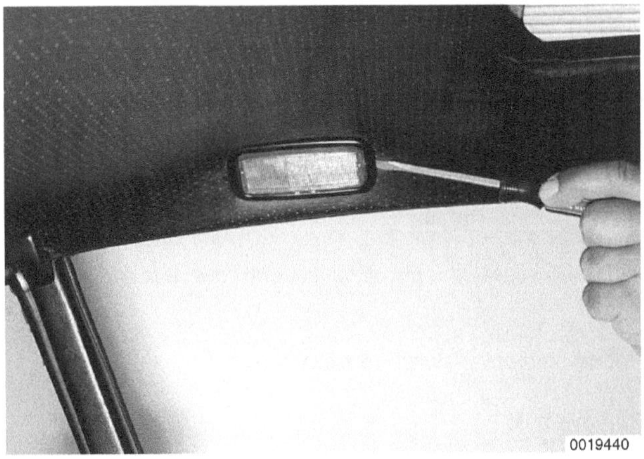

- Disconnect negative (–) battery cable.
- ◄ Insert screwdriver into front of light bulb housing and carefully pry out interior light assembly.
- Remove bulb from between contacts.
- Install new bulb and press interior light into place.

Luggage compartment bulb, replacing

The luggage compartment light is installed in the luggage compartment lid near the windshield, on the right.

- ◄ Remove two mounting screws (**arrows**) and remove light from lid.
- Slide bulb holder out of the lens.
- Remove bulb from socket.
- Installation is reverse of removal.

> ***CAUTION —***
> *Align plastic gasket before installing light.*

CRUISE CONTROL

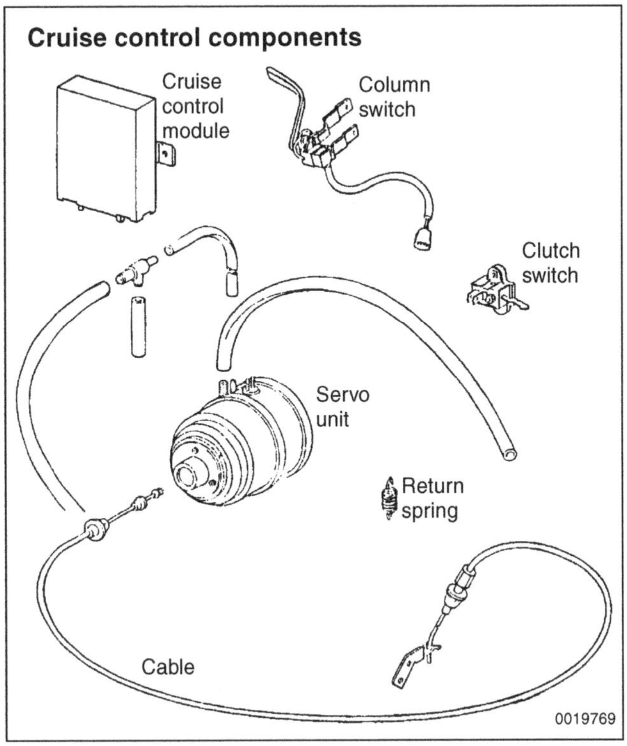

The cruise control system consists of an electronic control unit, a cruise control servo, a control switch on the steering column, and switches at the brake and clutch pedals.

The control module uses a signal from the speedometer to compare the car's true speed to the speed selected by the driver via the column switch. The control unit then signals the servo to open or close the throttle via a cable.

The system will maintain the set speed while going up and down hills within the engine's power range. The set speed can be increased at any time by depressing the accelerator pedal. When the accelerator pedal is released the vehicle speed will return to the previously set speed. If either the clutch or brake pedal are depressed the system will be disengaged. If the vehicle speed drops below 15 mph at any time the system will automatically disengage.

To disengage cruise control when the clutch pedal is depressed, a clutch pedal switch is used. To disengage cruise control when the brakes are used an extra connection at one of the brake light switches is used.

NOTE —
Cruise control switch removal and installation is covered earlier in this group under **Steering column switches, removing and installing**.

Troubleshooting

– When troubleshooting cruise control system, first check for a faulty fuse 6 in the fuse/relay panel in luggage compartment.

– To begin troubleshooting cruise control components and electrical circuit, access cruise control module as described later.

- Remove fresh air plenum in luggage compartment.
- Detach electrical harness connector at cruise control module.
- Proceed to tests in Table c.

– If no faults are found with cruise control components or electrical circuits, replace cruise control module.

Table c. Cruise control troubleshooting

Test (terminals at cruise control module harness connector)	Results	Probable cause
Connect ohmmeter between terminal 3 and ground	Meter reads infinite Ω	None
Depress clutch pedal	Meter reads approx. 10 Ω	None
	Meter reads infinite Ω	Clutch pedal switch or servo unit faulty
Connect ohmmeter between terminals 3 and 7	Meter reads 11 -17 Ω	None
	Meter reads <11 Ω or >17 Ω	Servo unit faulty
Connect ohmmeter between terminal 12 and ground	Meter reads 0 Ω	See wiring diagram in **970 Electrical Wiring Diagrams**
Connect ohmmeter between terminal 11 and ground. Turn rear wheels	Meter deflects	None
	Meter does not deflect; speedometer functions	Wiring between cruise control module and speedometer faulty
	Meter does not deflect; speedometer does not function	Speedometer sensor at transmission faulty or wiring from speedometer sensor faulty
Connect ohmmeter between terminal 6 and ground	Meter reads infinite Ω	None
	Meter reads 0 Ω	Brake light switch faulty
Connect voltmeter between terminal 6 and ground, turn ignition on Depress brake pedal	Meter reads battery voltage	None
	Battery voltage not present	Brake light switch faulty
Connect voltmeter between terminal 9 and ground Push up on column switch	Meter reads battery voltage	None
	Battery voltage not present	Fuse 6 faulty
Connect voltmeter between terminal 8 and ground Pull column switch toward steering wheel	Battery voltage not present	None
	Meter reads battery voltage	Cruise control column switch faulty
Connect voltmeter between terminal 10 and ground Push column switch down	Meter reads battery voltage	None
	Battery voltage not present	Cruise control column switch faulty

Cruise control module, removing and installing

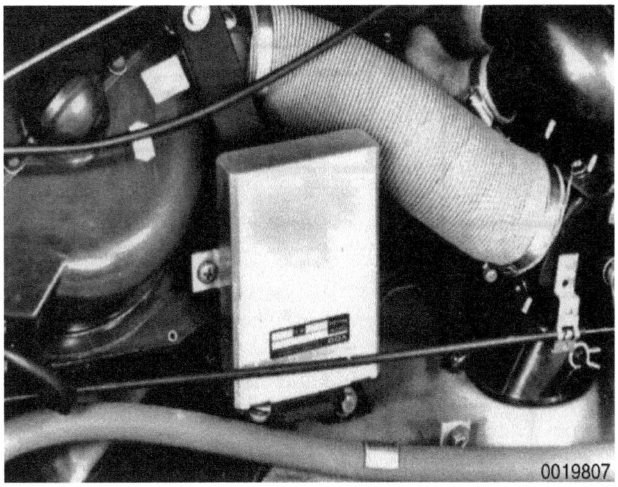

◀ Working in luggage compartment, remove fresh air plenum cover in luggage compartment.

NOTE —

See also page 971-2 for alternate location of cruise control module on late models.

– Pull electrical harness connectors off module.

– Remove phillips head screw (**arrow**) and remove module.

– Installation is reverse of removal.

Clutch pedal switch, removing and installing

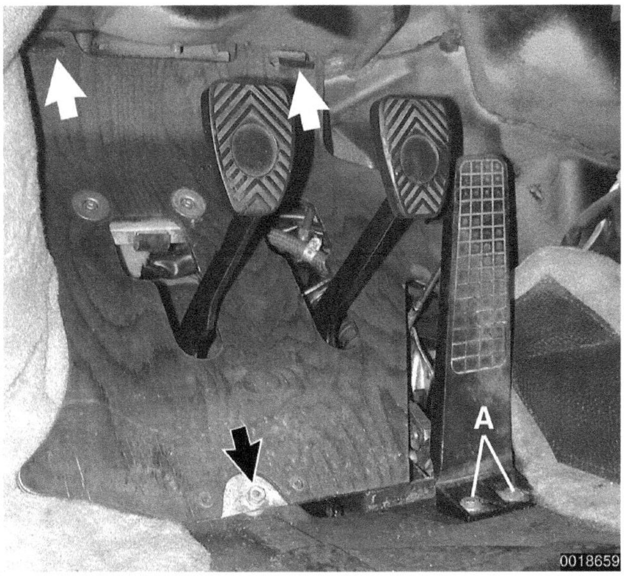

◀ Working inside passenger compartment at the pedals, remove accelerator pedal mounting screws (**A**) and remove pedal.

– Remove pedal cluster cover after removing carpet securing straps at top and fastener at bottom (**arrows**).

◀ Remove clutch pedal switch bracket fasteners, disconnect wires and remove bracket with switch.

– Installation is reverse of removal.

- Make sure switch is adjusted correctly for clutch pedal. Use adjusting slots so that switch is not damaged when pulling back clutch pedal to check clutch play.

0019808

Cruise control servo unit, removing and installing

- Working inside engine compartment on left side of engine, remove hose between heater blower motor and heater distribution pipe.

< Disconnect cruise control servo electrical harness connector (**A**).

- Detach vacuum line (**B**) on servo.

- Working at throttle assembly:
  - Detach cruise control cable attachment bracket from throttle lever.
  - Slide plastic cable housing end out of holder bracket.

- Remove servo mounting screws. Remove servo with cable.

- Installation is reverse of removal.

Cruise control cable, removing and installing

- Working at throttle assembly:
  - Detach cruise control cable attachment bracket from throttle lever.
  - Slide plastic cable housing end out of holder bracket.

- Working inside engine compartment on left side of engine, remove hose between heater blower motor and heater distribution pipe.

- Working at cruise control servo, remove screws holding cable end flange to vacuum diaphragm at cruise control servo.

0018738

< To disconnect cable from servo, squeeze plastic retainer clip and push forward to release cable.

- Remove plastic clip from cable. Remove flange from cable.

- Installation is reverse of removal.
  - Be sure there are no kinks in cable.
  - Make sure there is as little slack as possible in cable at throttle end.
  - With engine off, there should be no tension on cable.
  - Use plastic hexagon nut at throttle housing end of cable to adjust cable tension.

970 Electrical Wiring Diagrams

ELECTRICAL COMPONENT LOCATIONS

Relay, fuse and control module positions, ground locations . . . . see **Repair Group 971**

GENERAL

This section contains wiring diagrams for 1984 to 1989 Porsche 911 Carrera vehicles.

NOTE —
- *The alarm system is also referred to as the anti-theft system.*
- *The warning chimes relay is also referred to as the warning relay or the clock relay.*

CAUTION —
Disconnecting the battery cables erases engine control module (ECM) adaptive memory. It may be necessary to drive the car for approximately 10 minutes after reconnectiung the battery to reset ECM adaptive memory and restore normal performance.

Power Distribution

1984 (1 of 3)

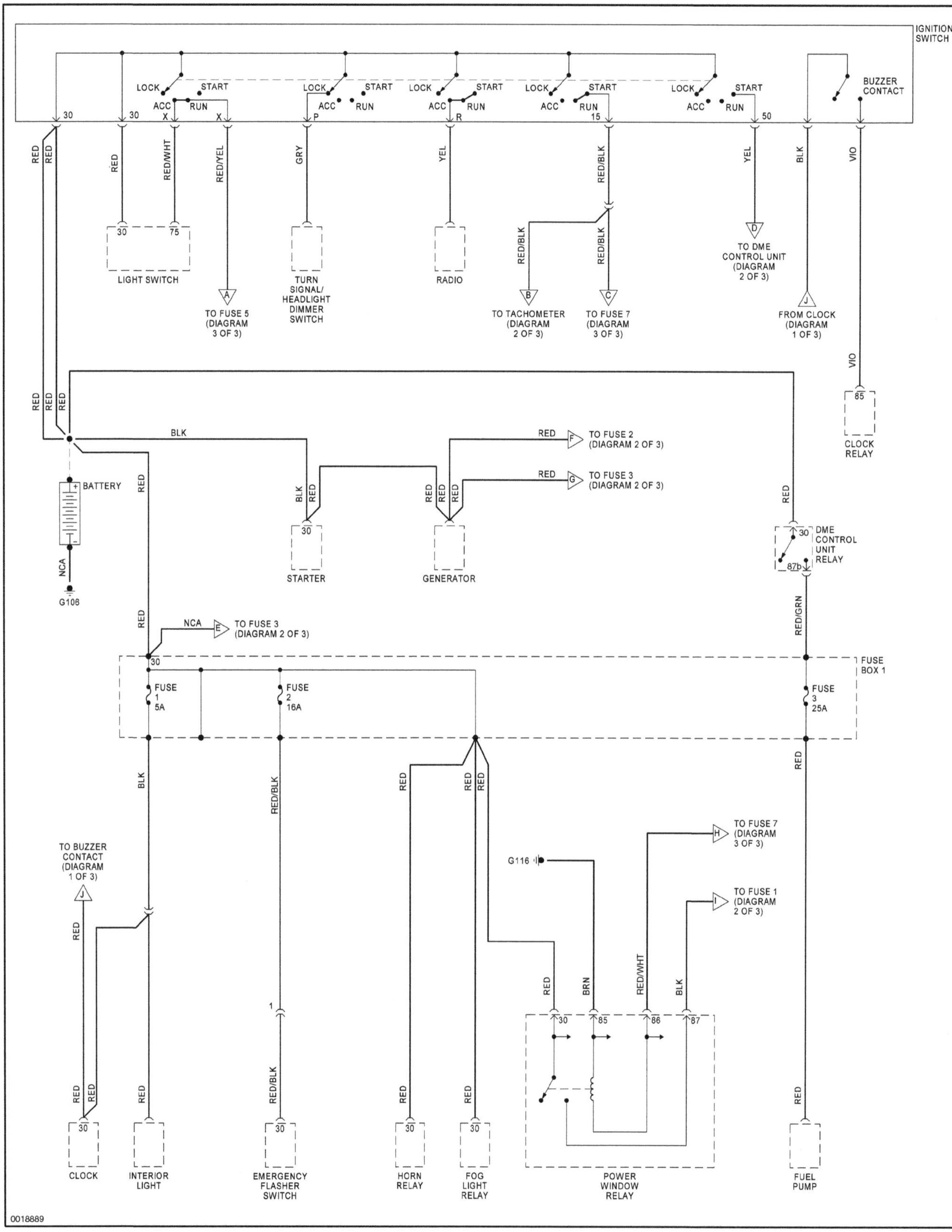

Power Distribution

1984 (2 of 3)

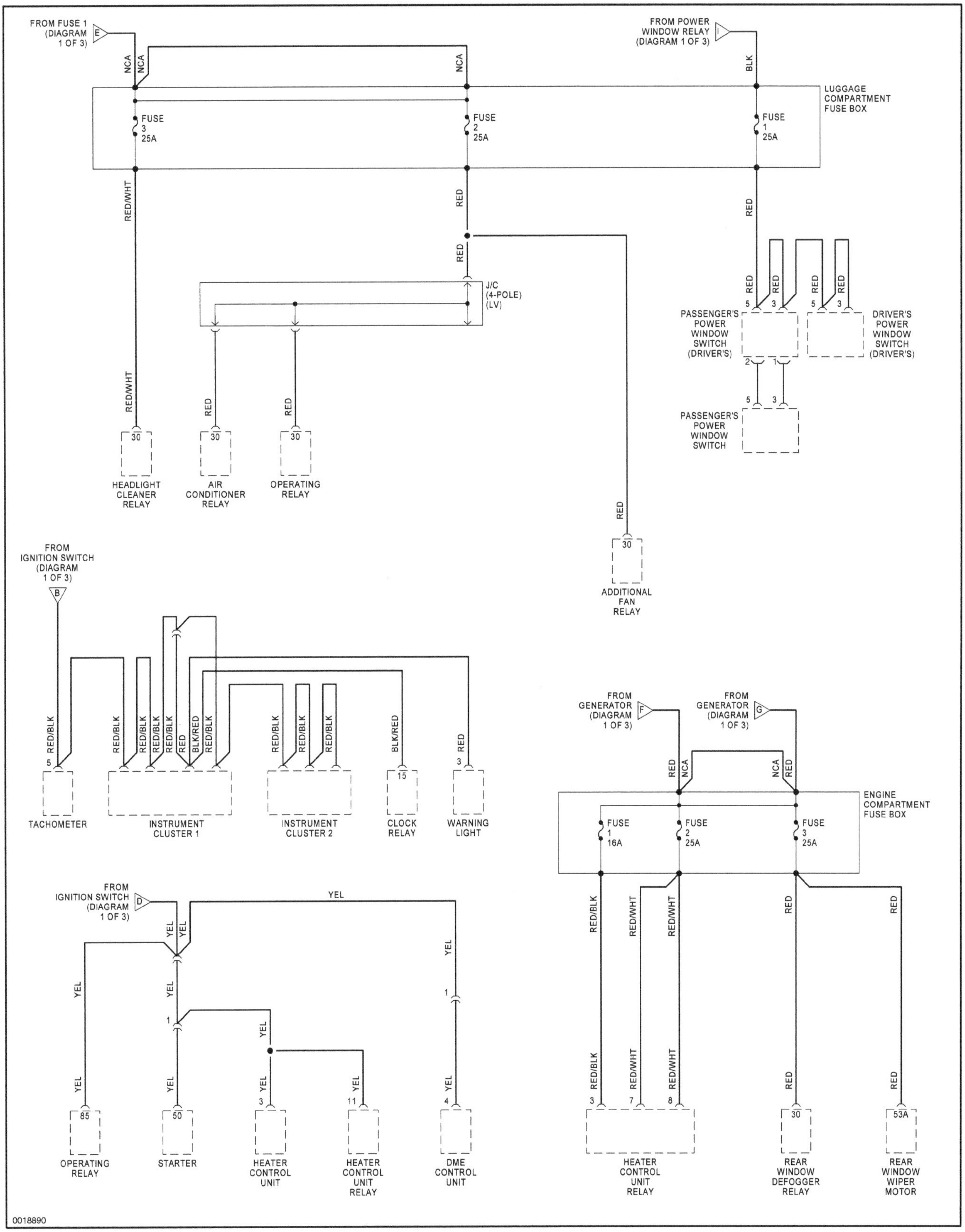

Power Distribution

1984 (3 of 3)

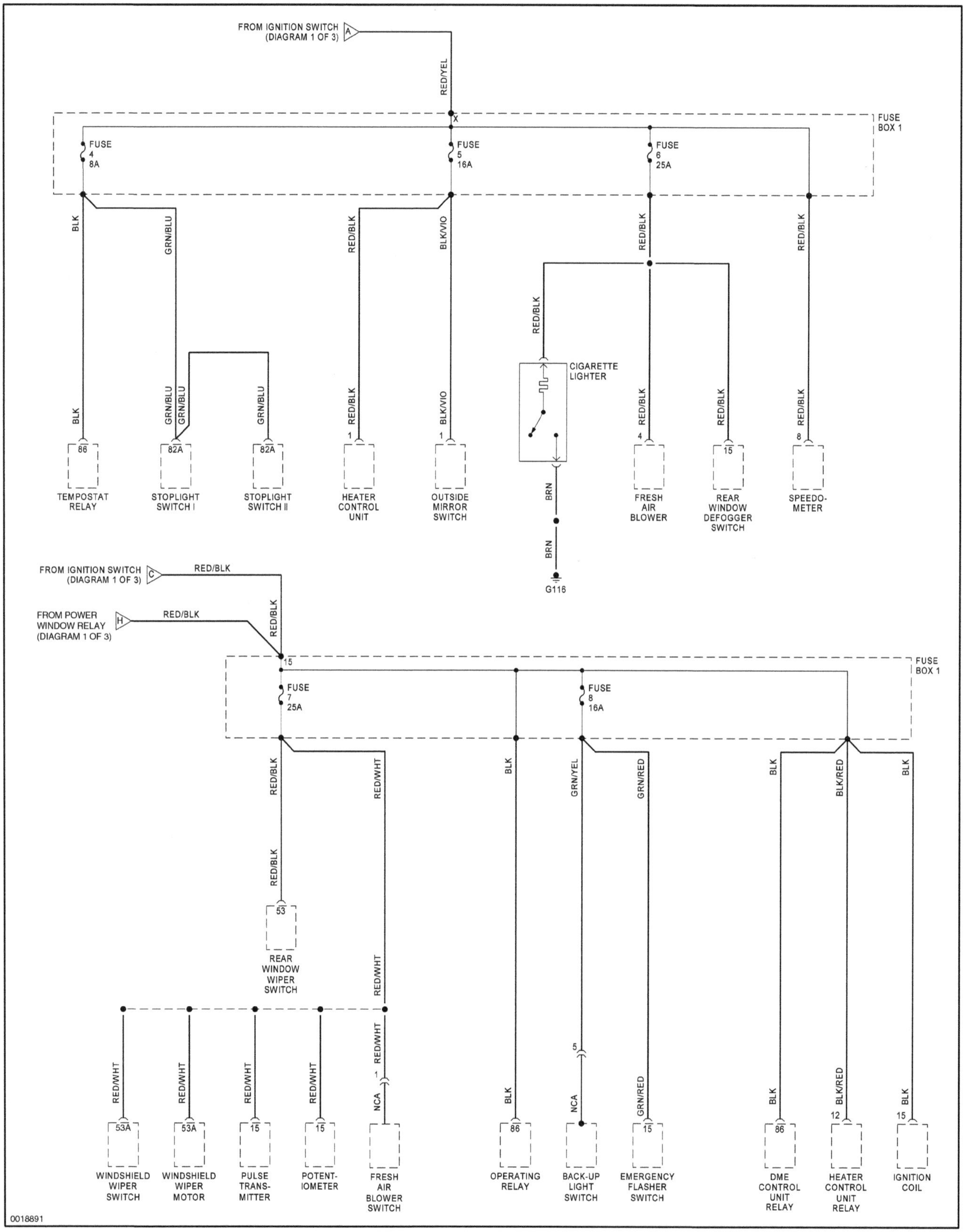

Power Distribution

1985 (1 of 3)

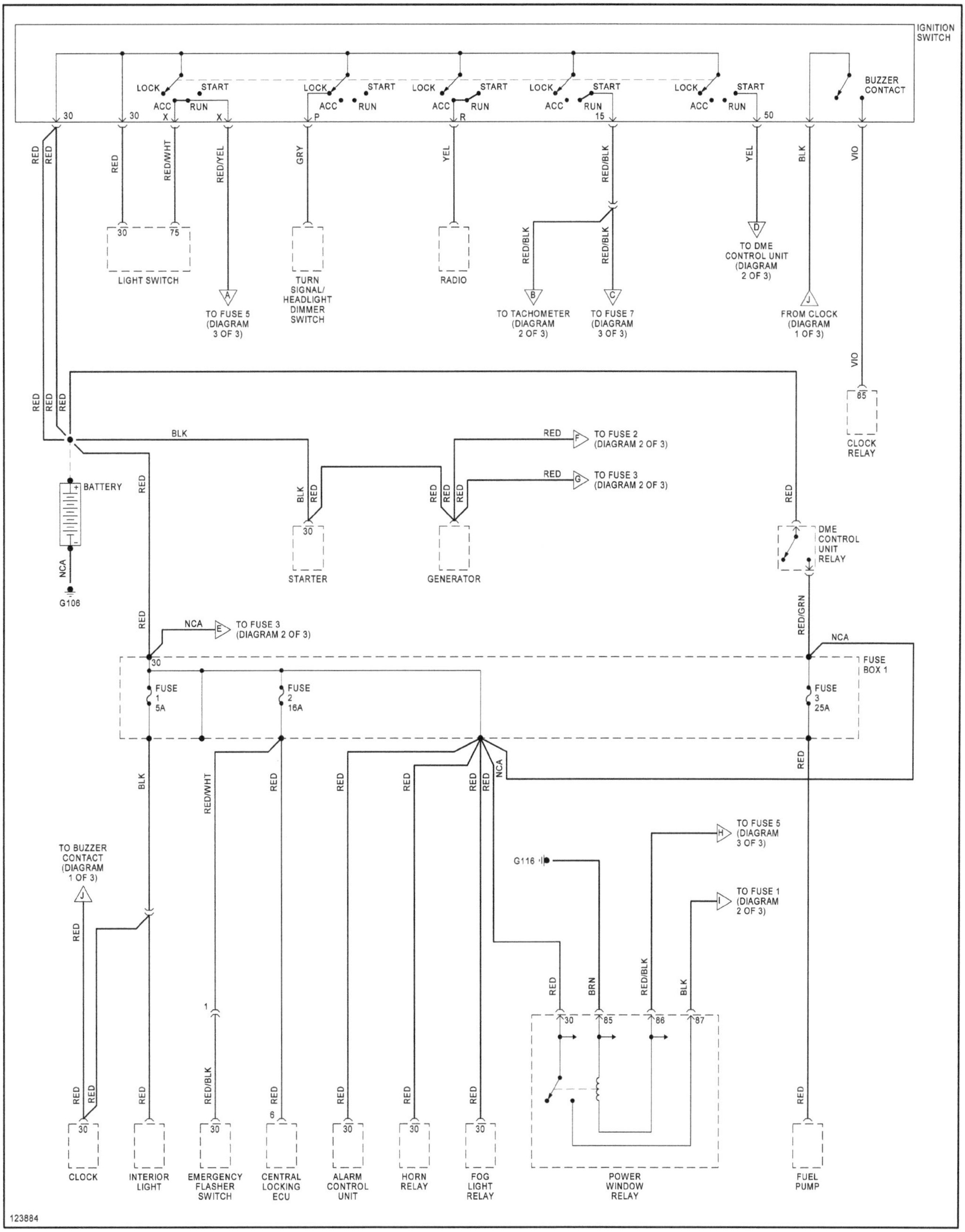

Power Distribution

1985 (2 of 3)

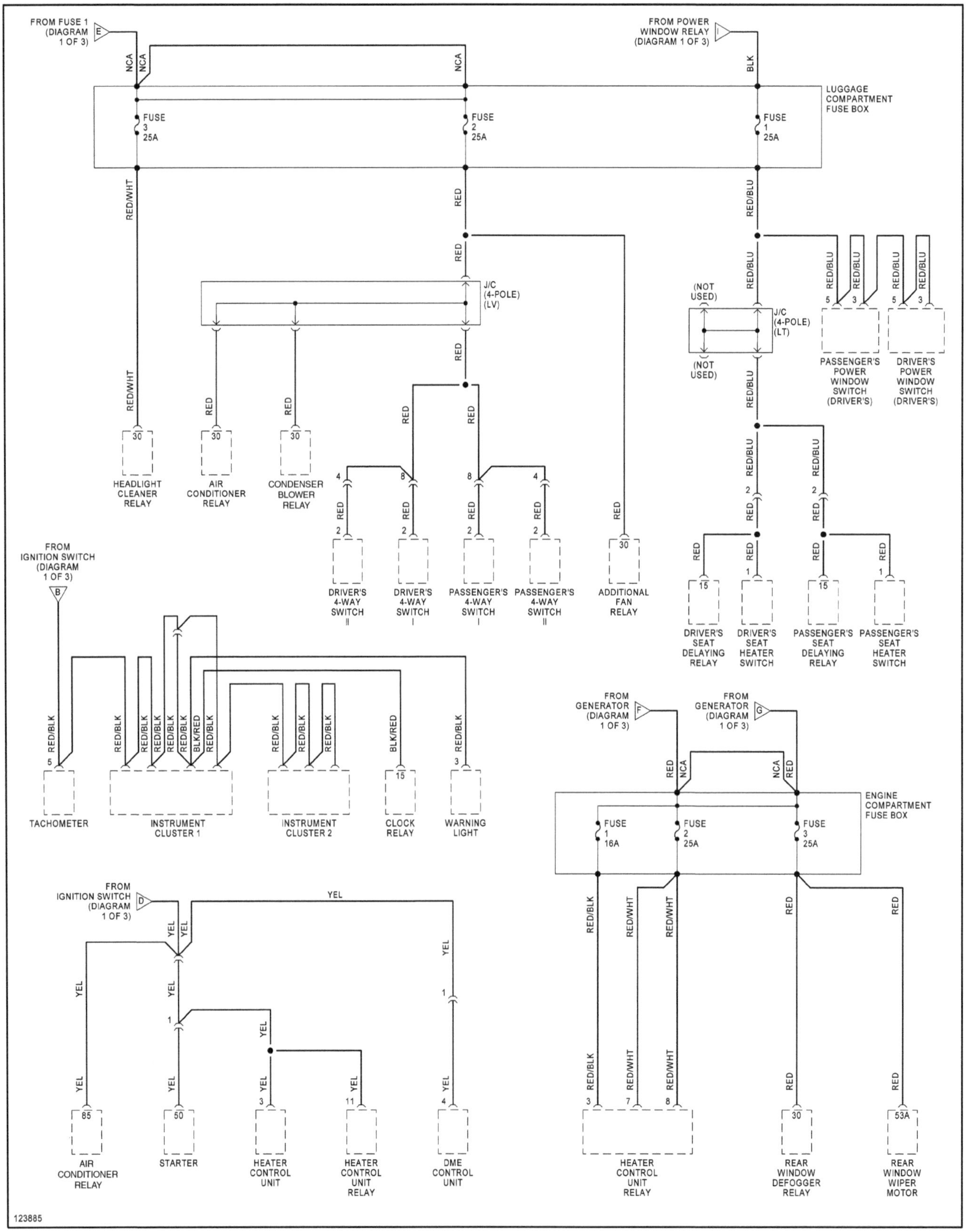

Power Distribution

1985 (3 of 3)

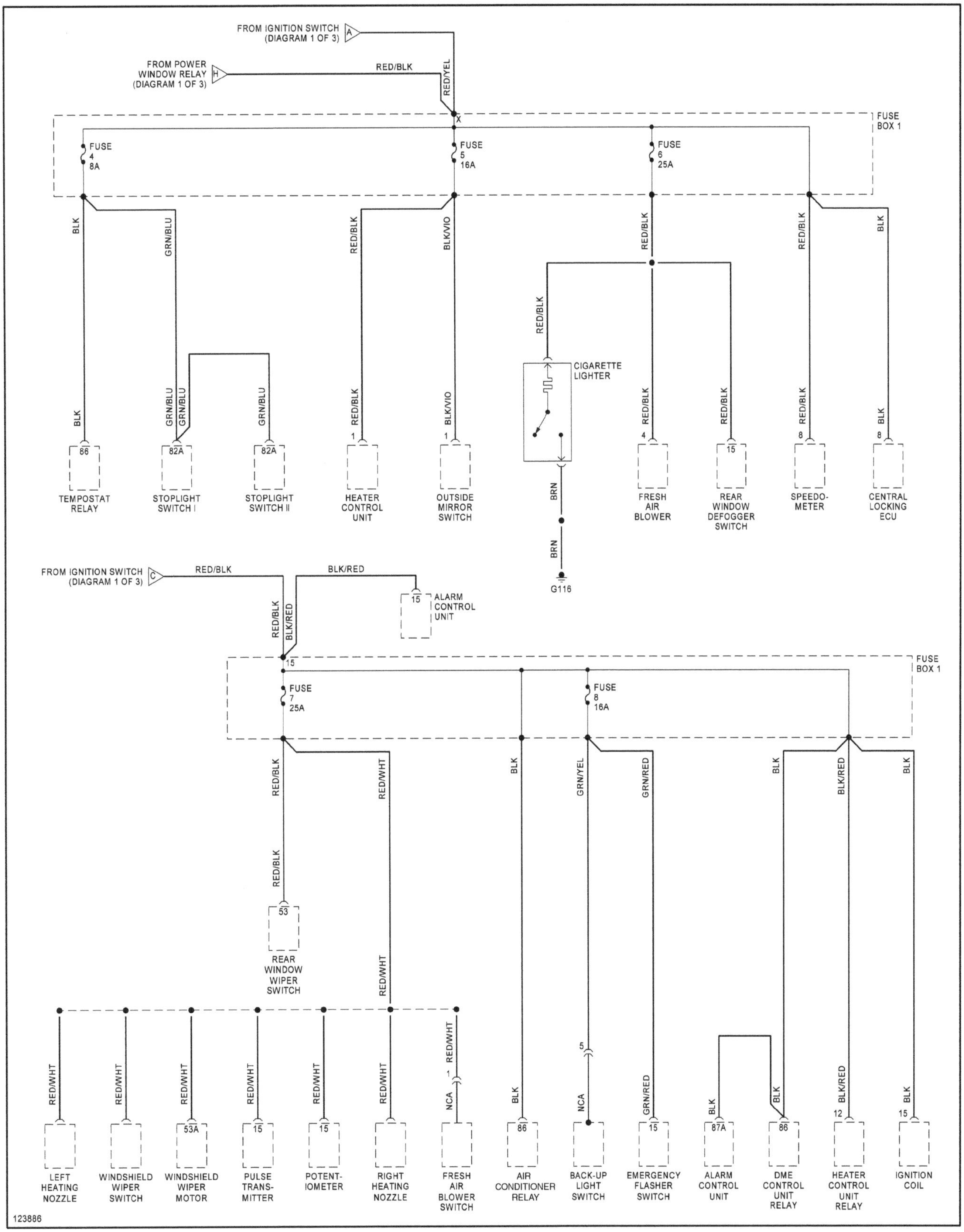

Power Distribution

1986-1987 (1 of 3)

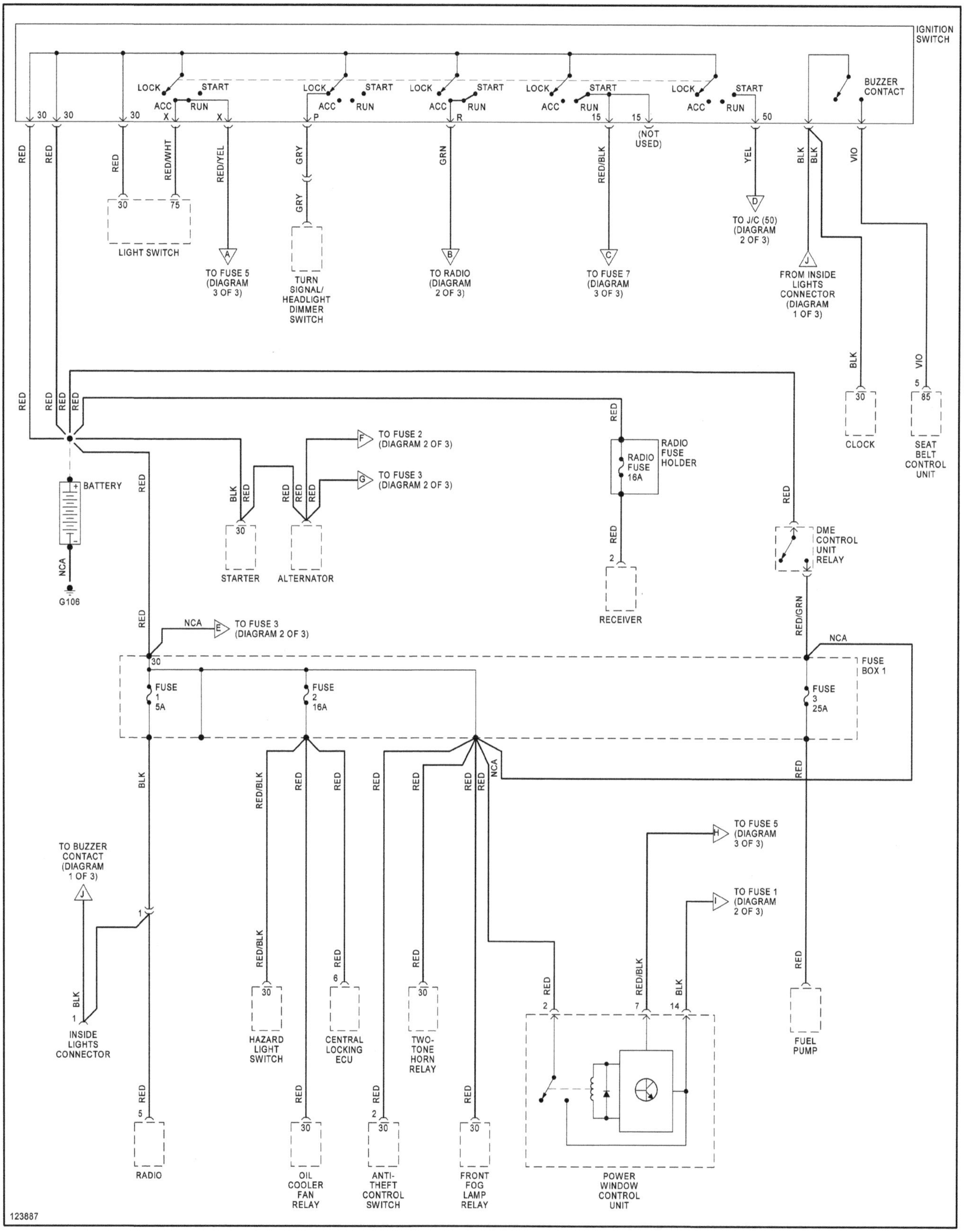

Power Distribution

1986-1987 (2 of 3)

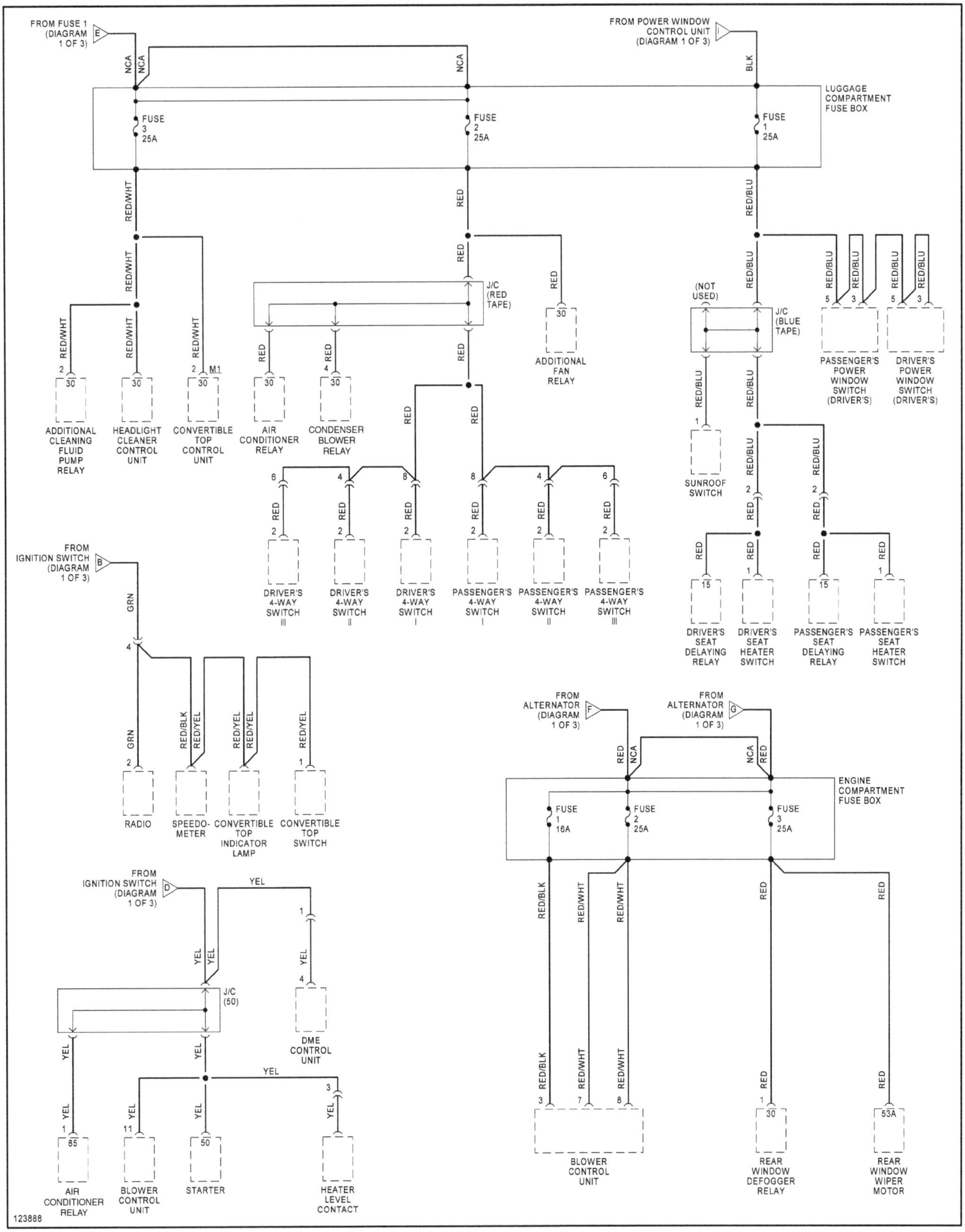

Power Distribution
1986-1987 (3 of 3)

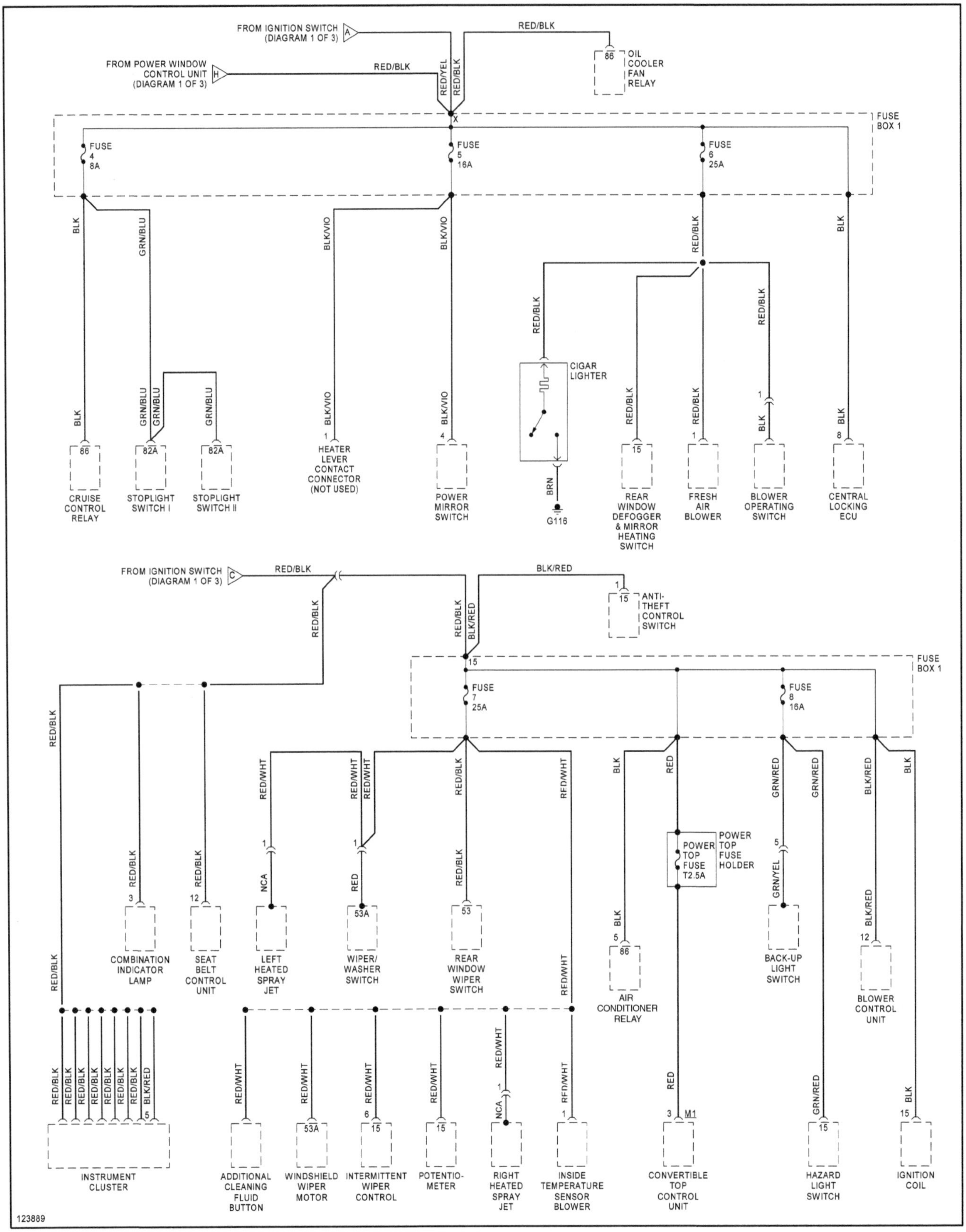

Power Distribution

1988-1989 (1 of 3)

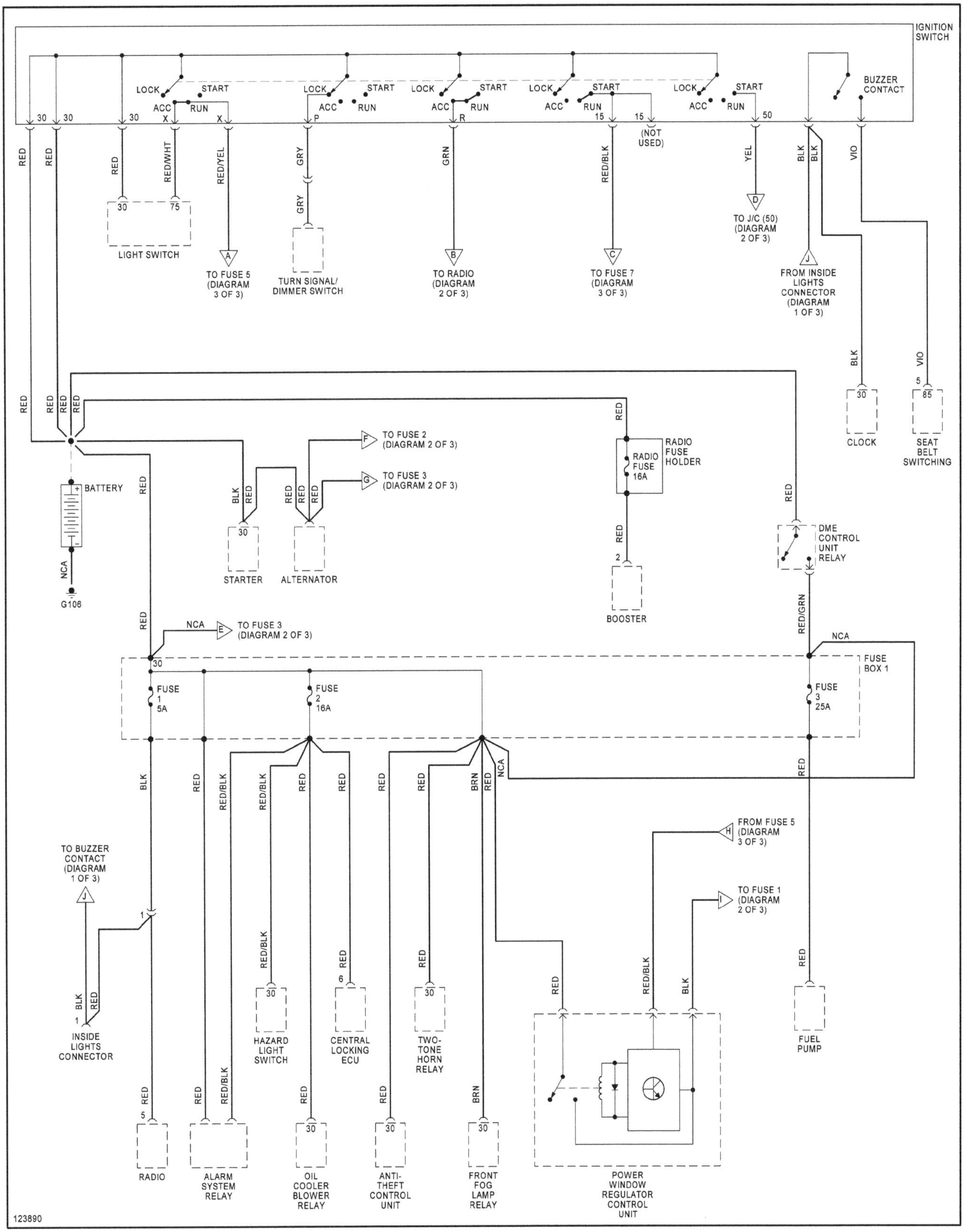

Power Distribution

1988-1989 (2 of 3)

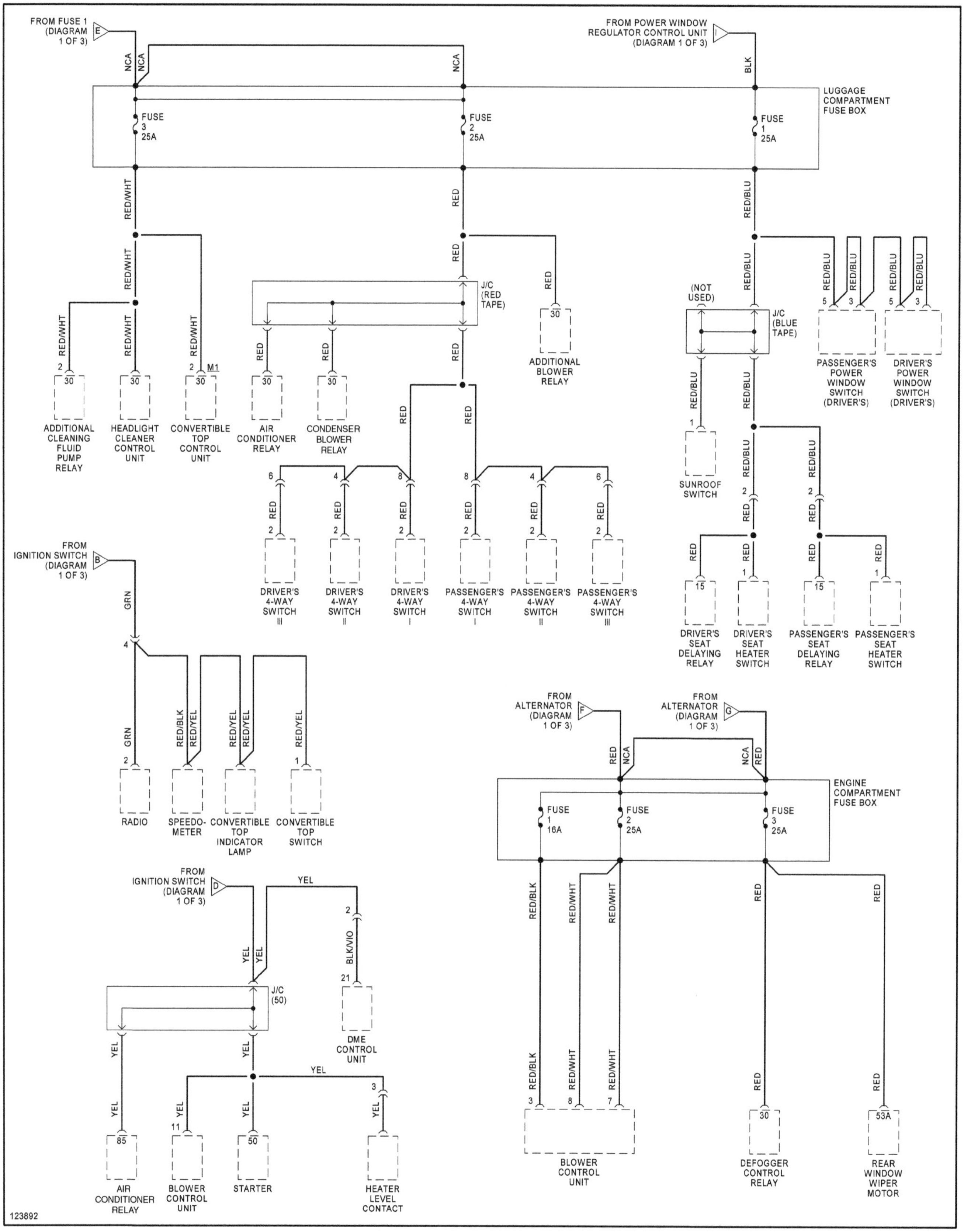

Power Distribution

1988-1989 (3 of 3)

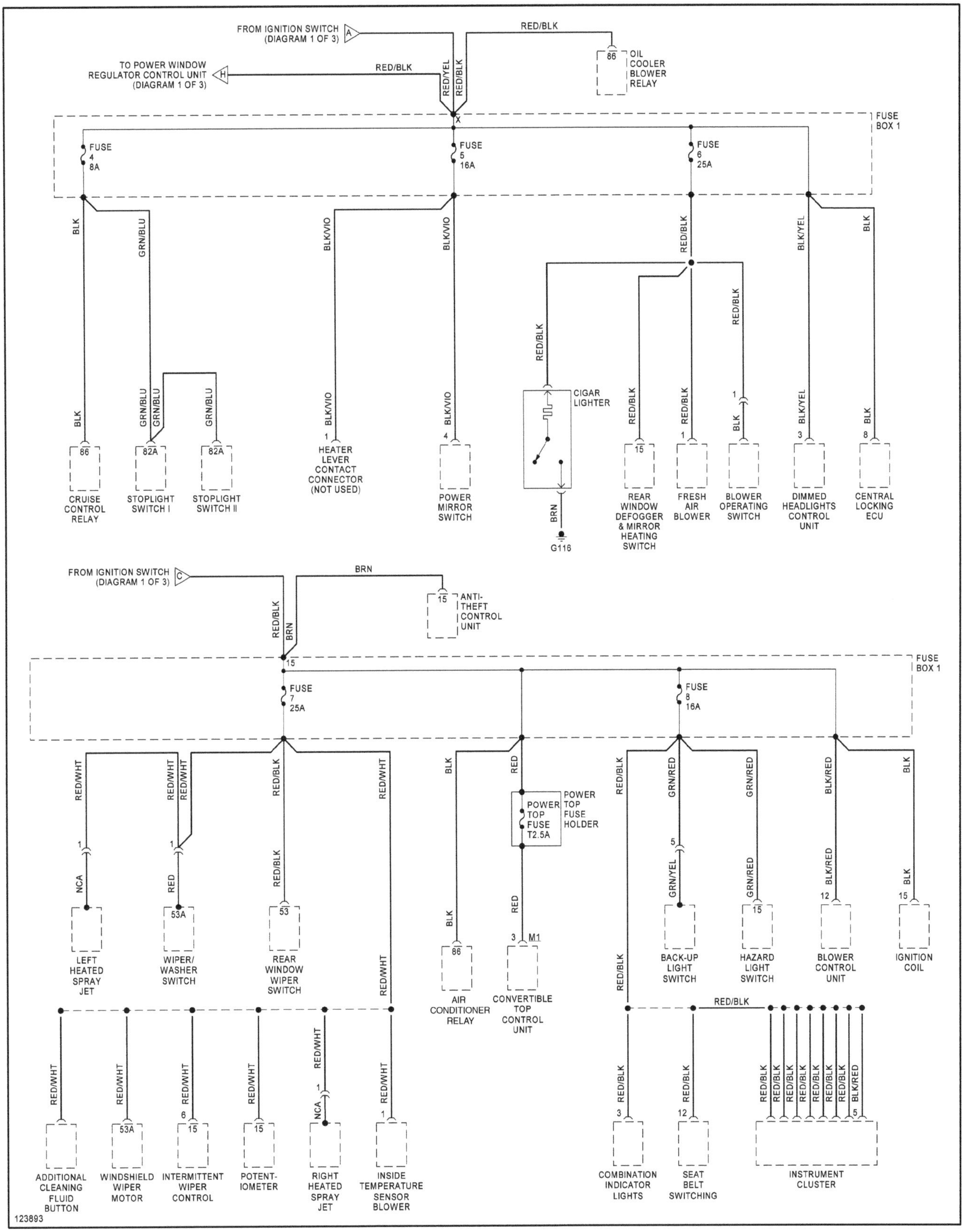

Ground Distribution

1984 (1 of 3)

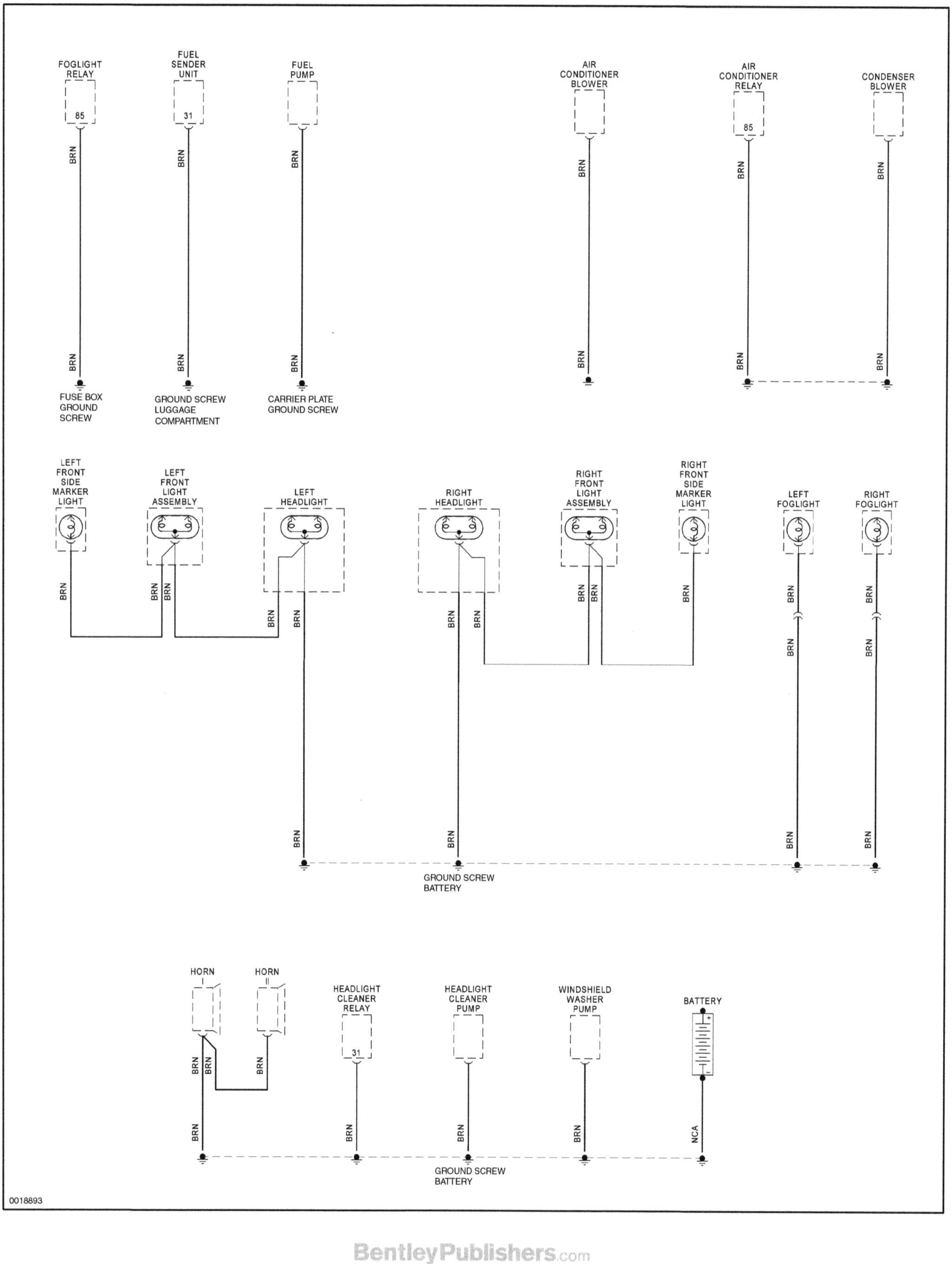

Ground Distribution

1984 (2 of 3)

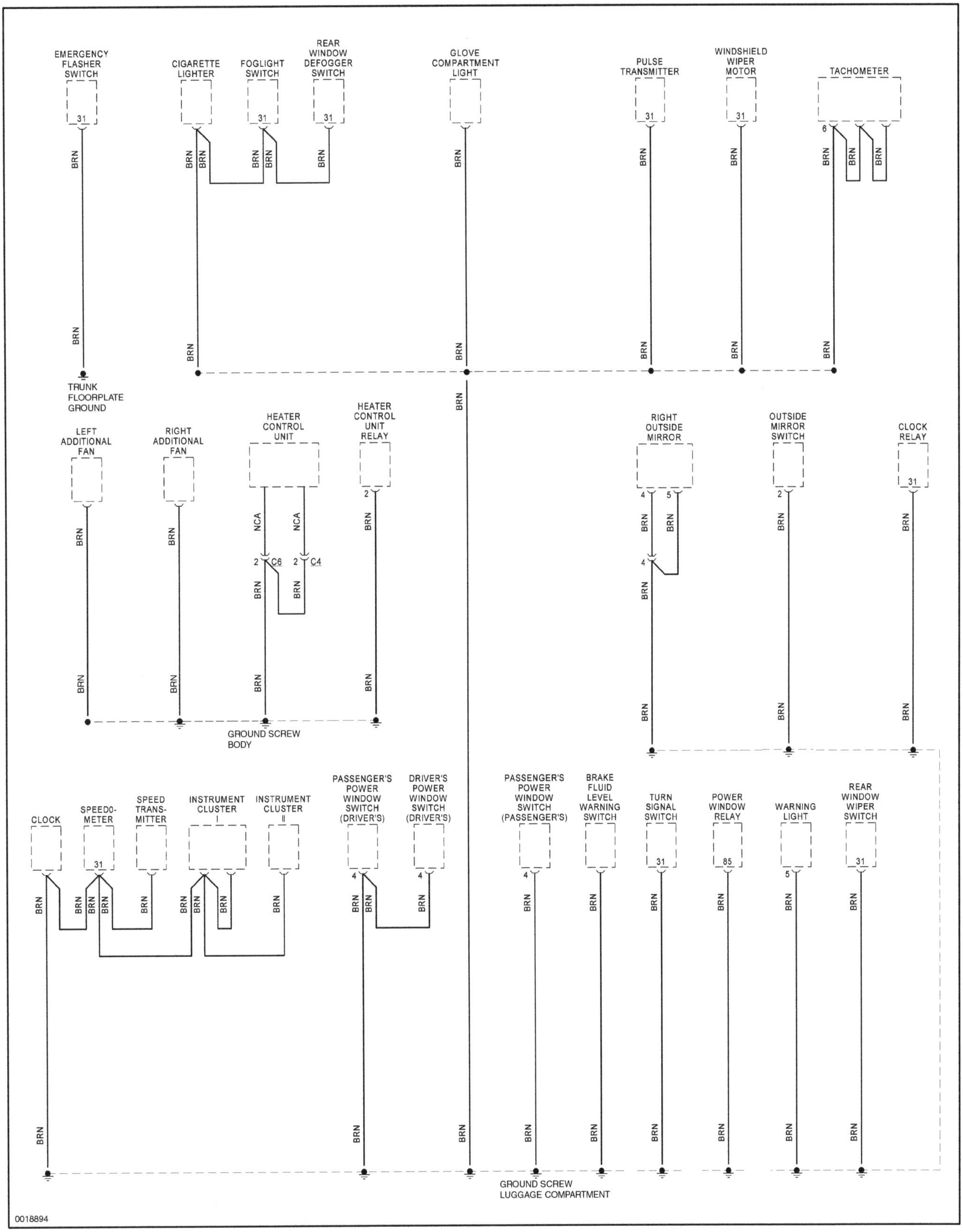

Ground Distribution

1984 (3 of 3)

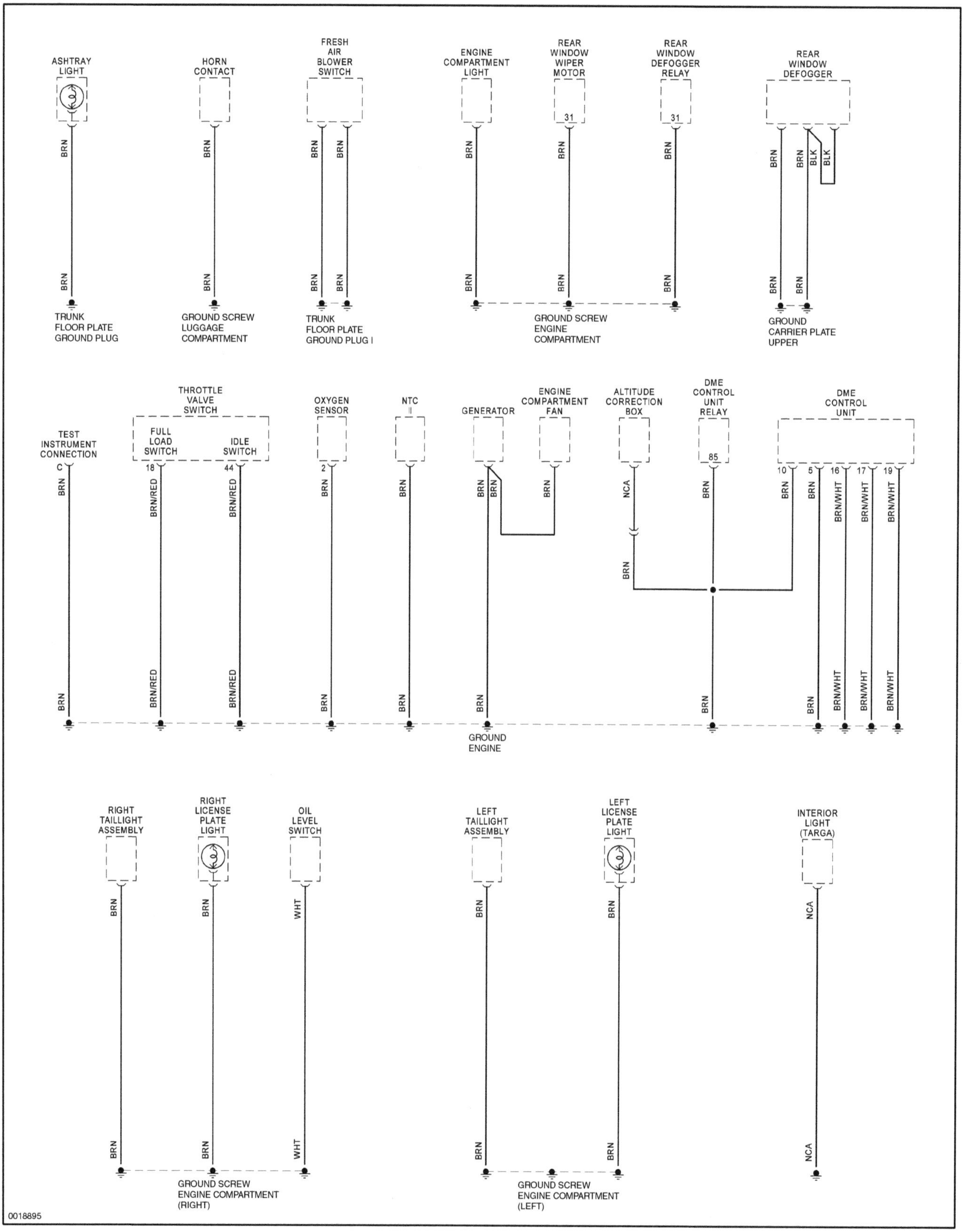

Ground Distribution

1985 (1 of 3)

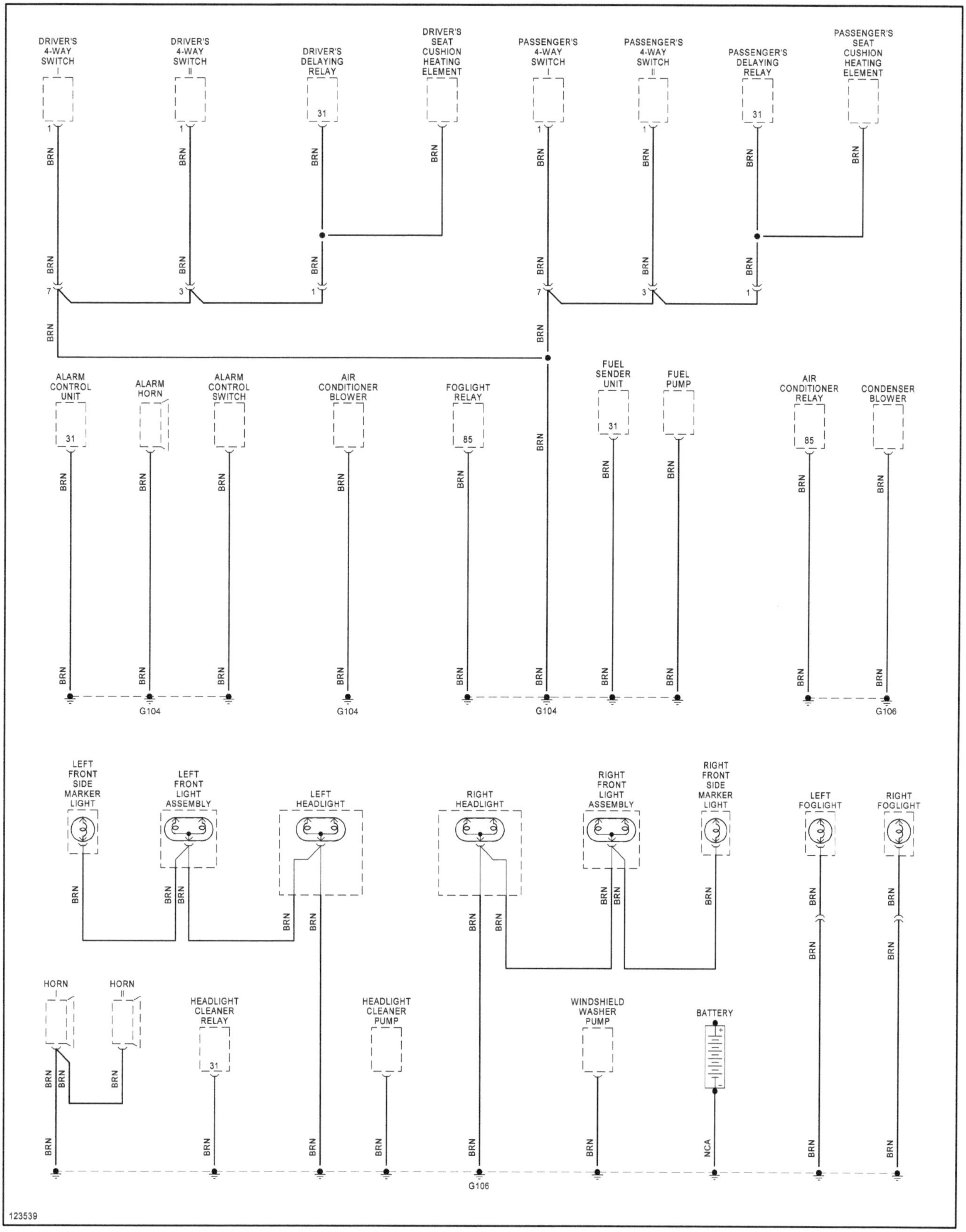

Ground Distribution

1985 (2 of 3)

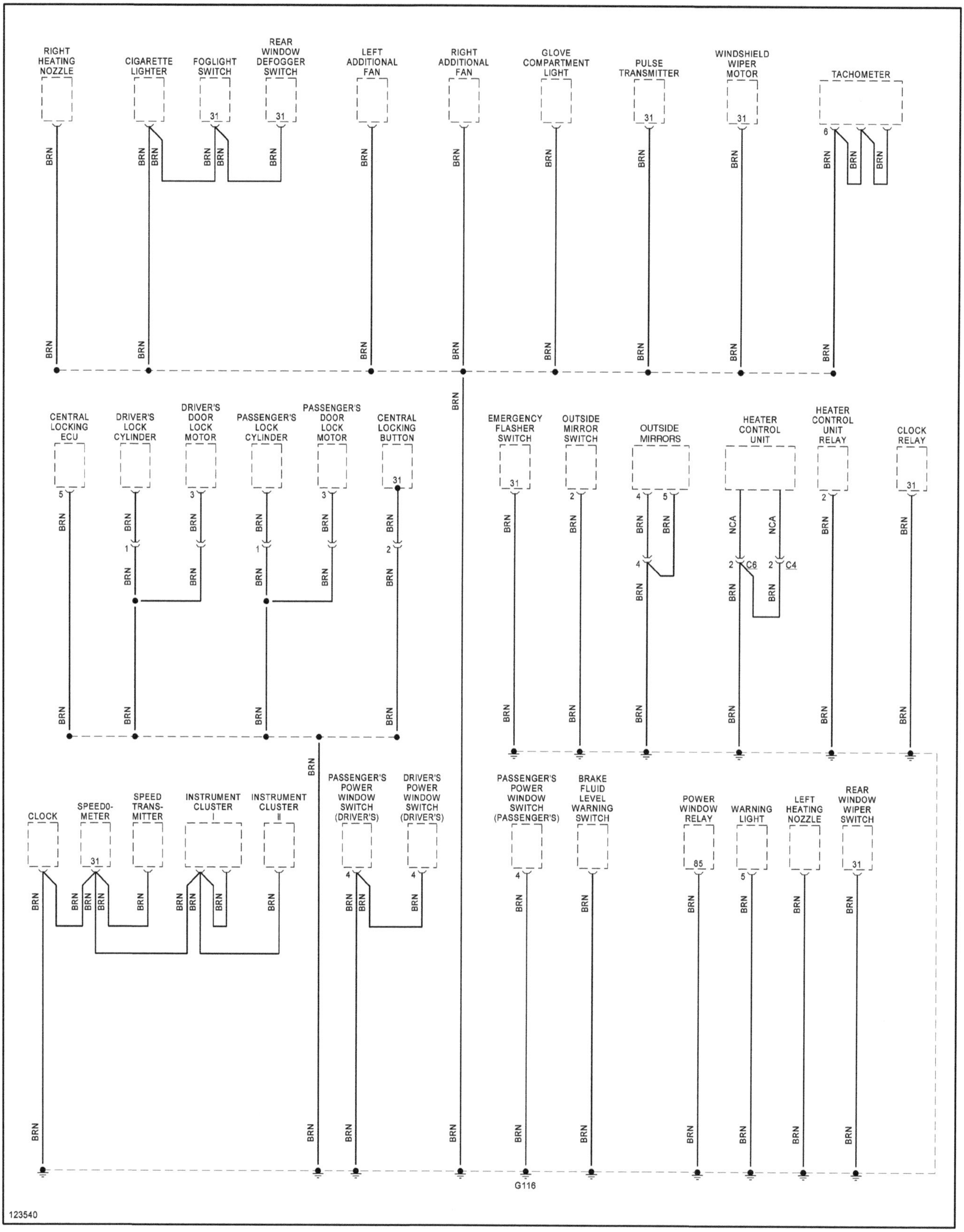

Ground Distribution

1985 (3 of 3)

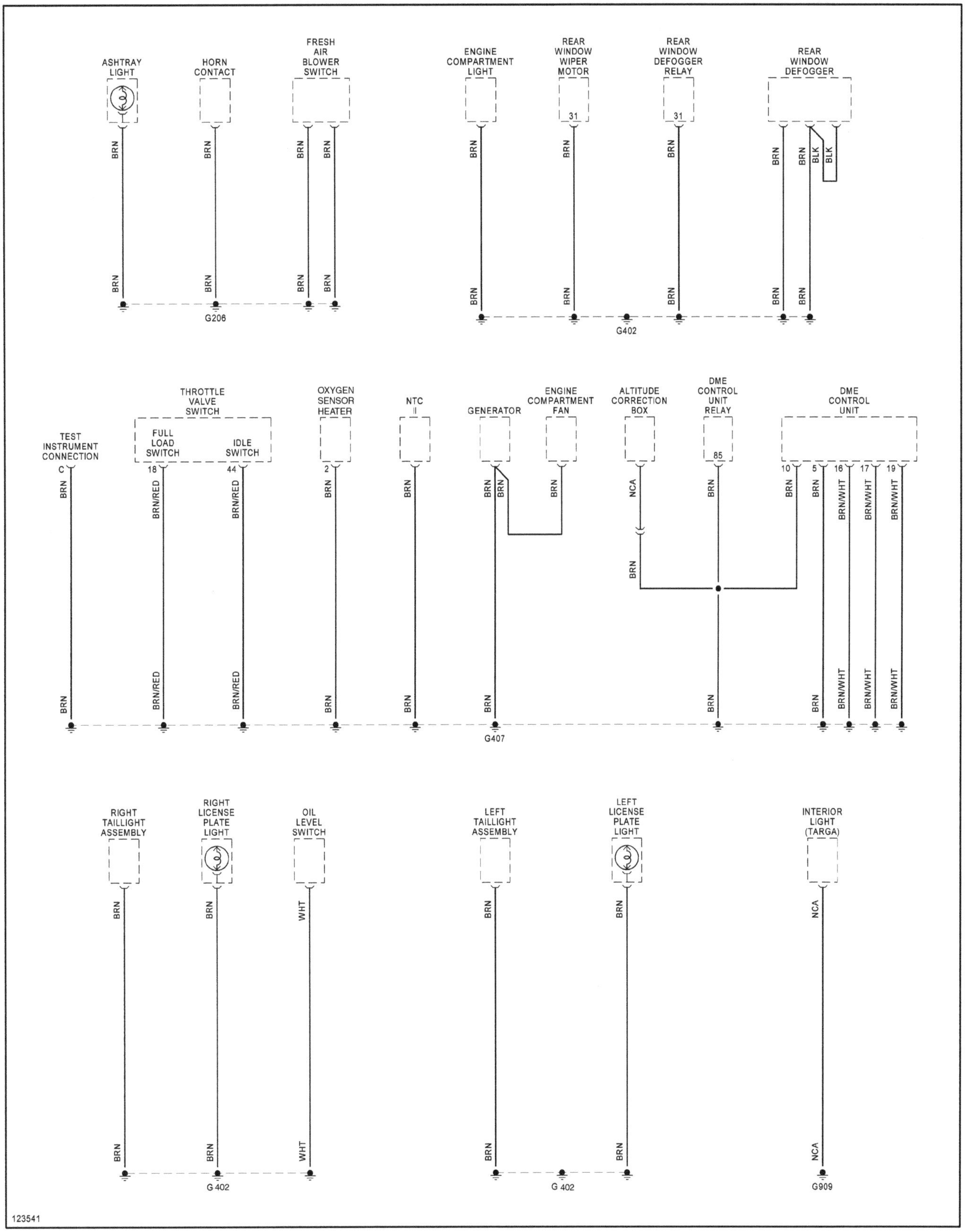

Ground Distribution

1986-1987 (1 of 3)

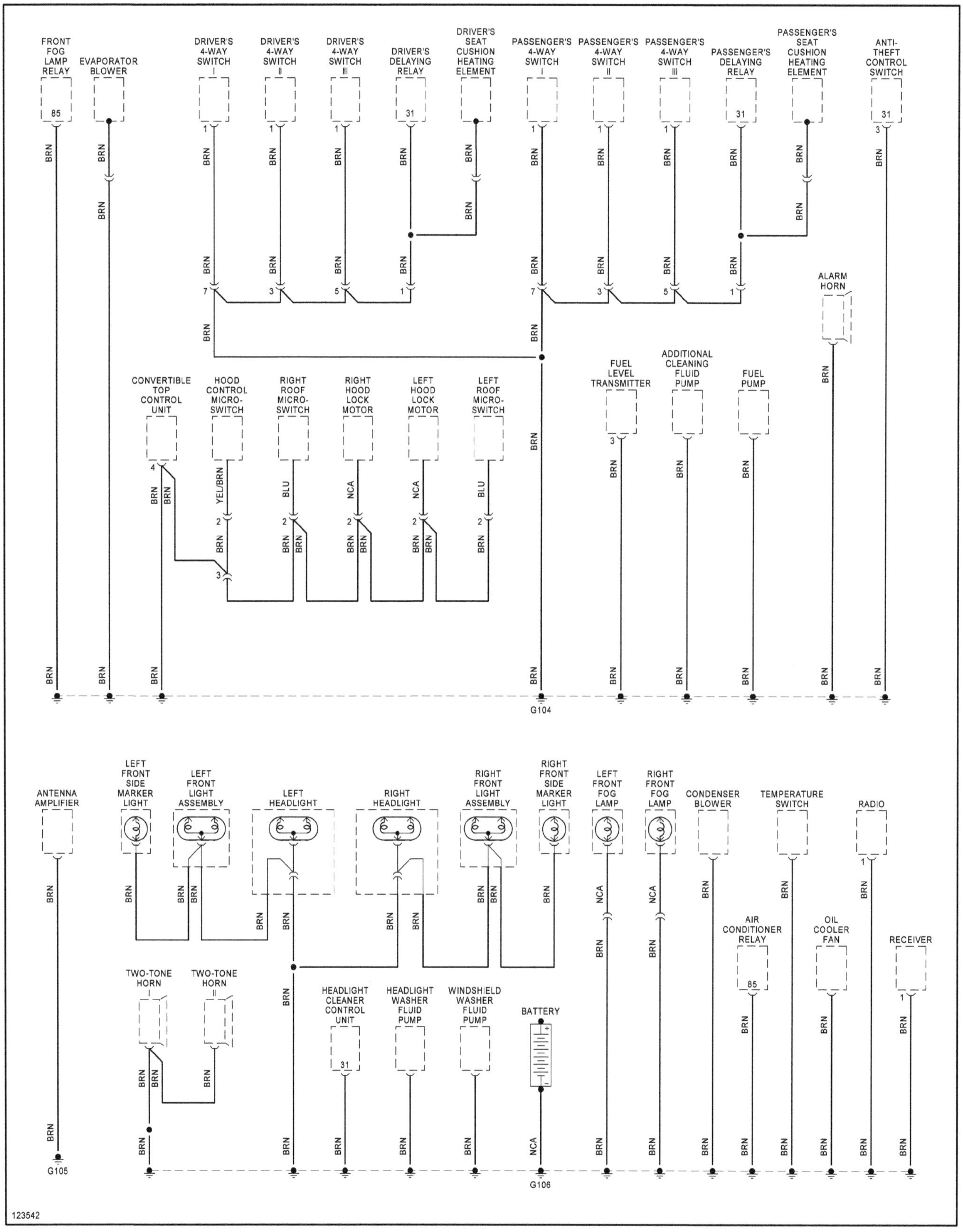

Ground Distribution

1986-1987 (2 of 3)

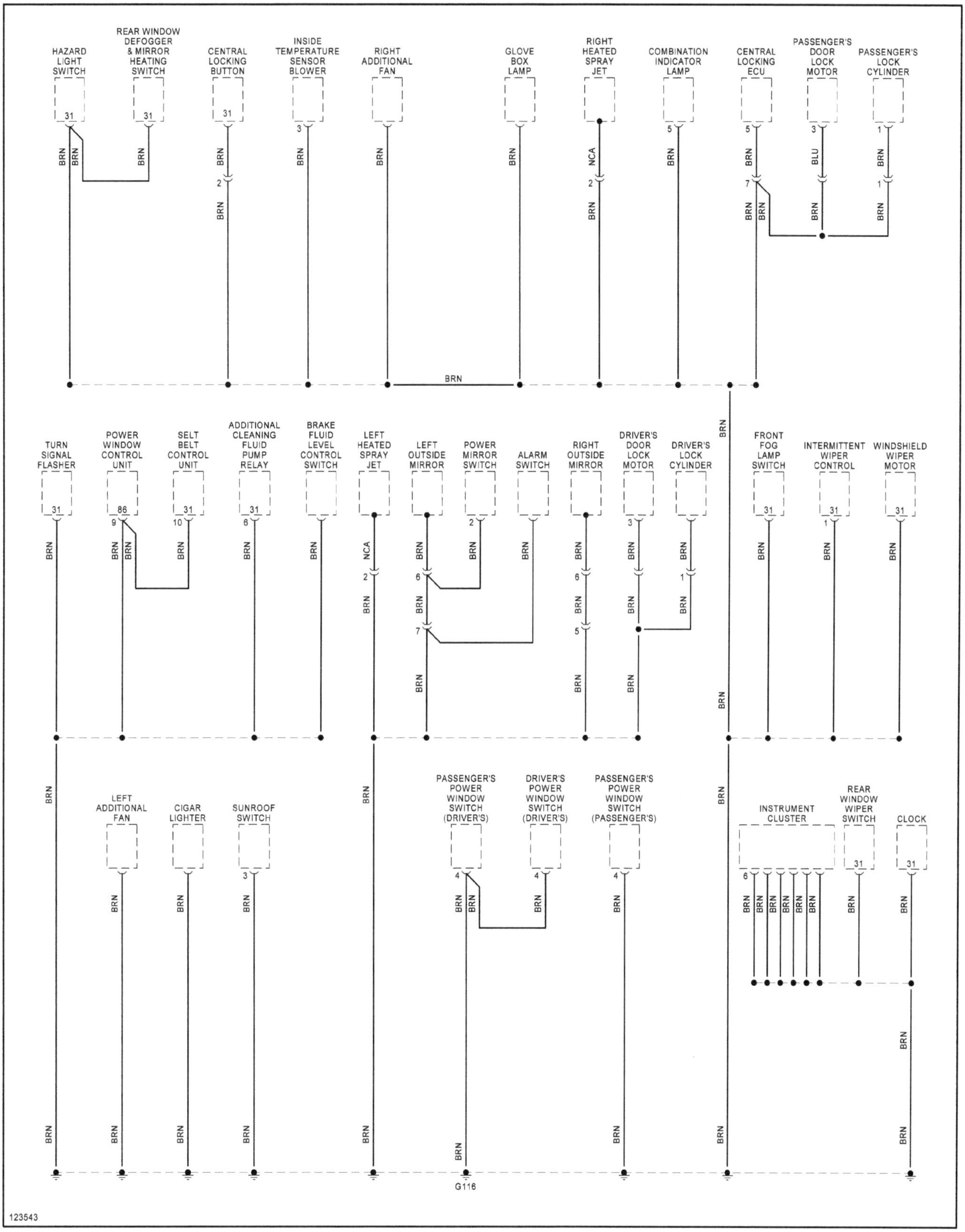

Ground Distribution

1986-1987 (3 of 3)

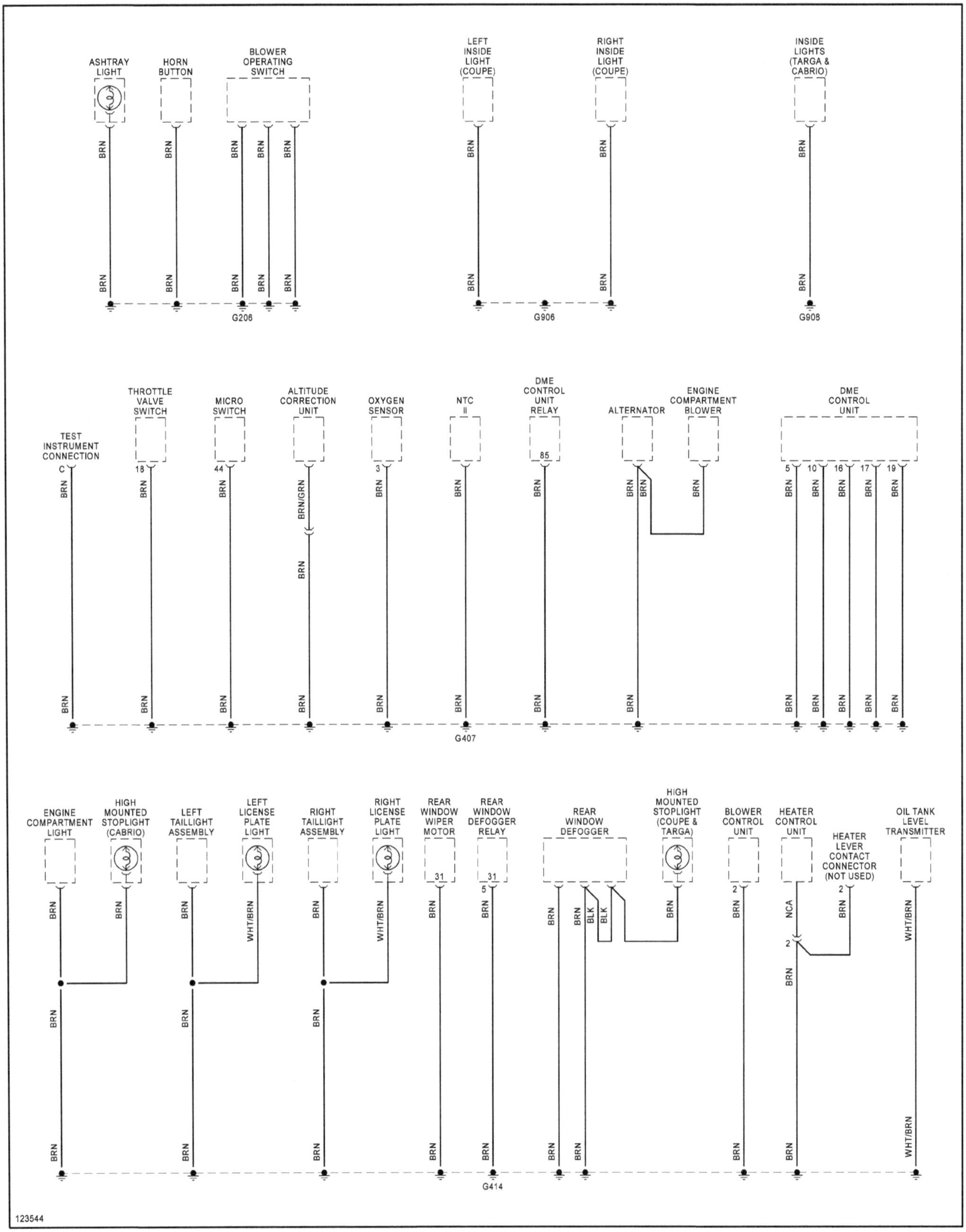

Ground Distribution

1988-1989 (1 of 3)

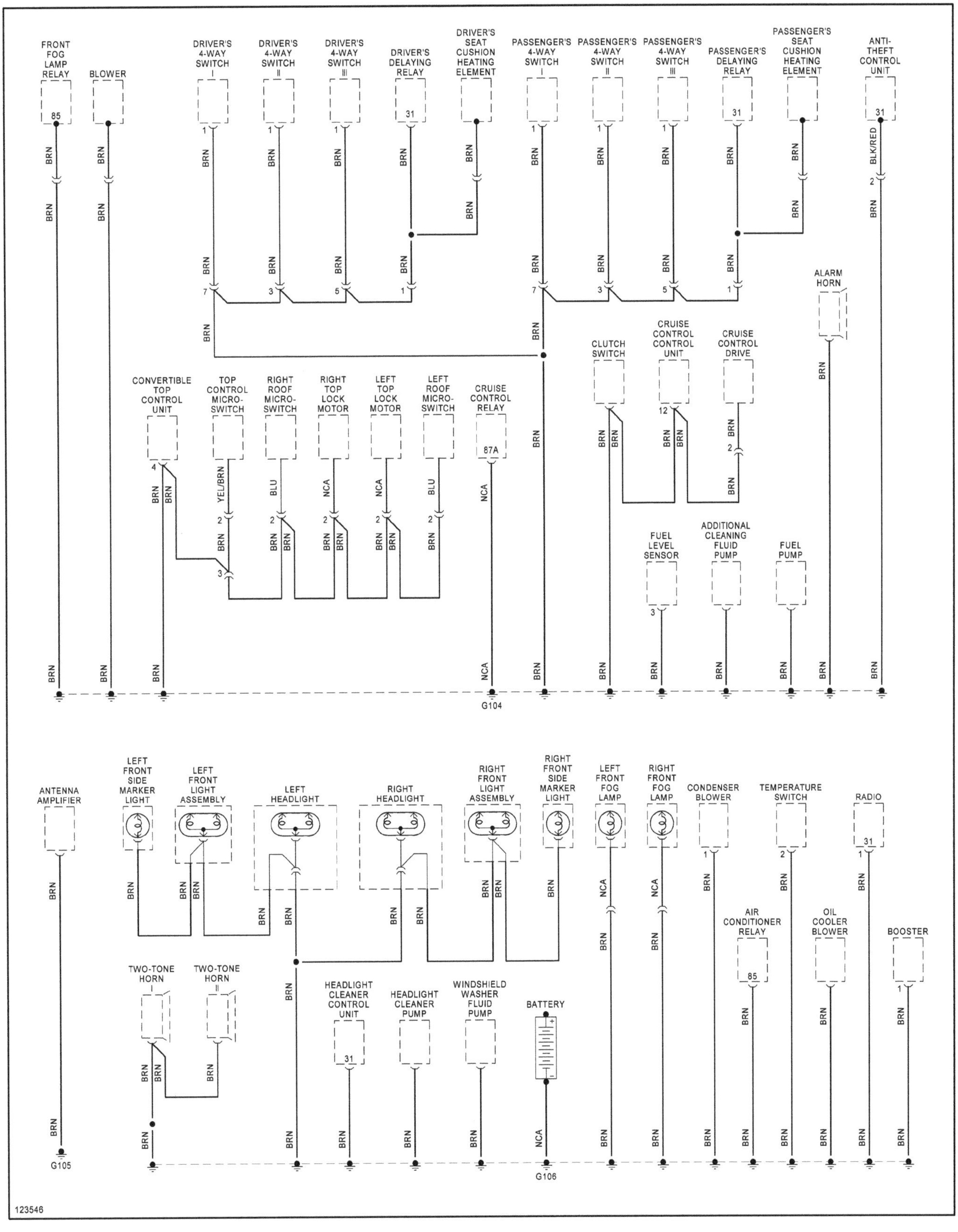

Ground Distribution

1988-1989 (2 of 3)

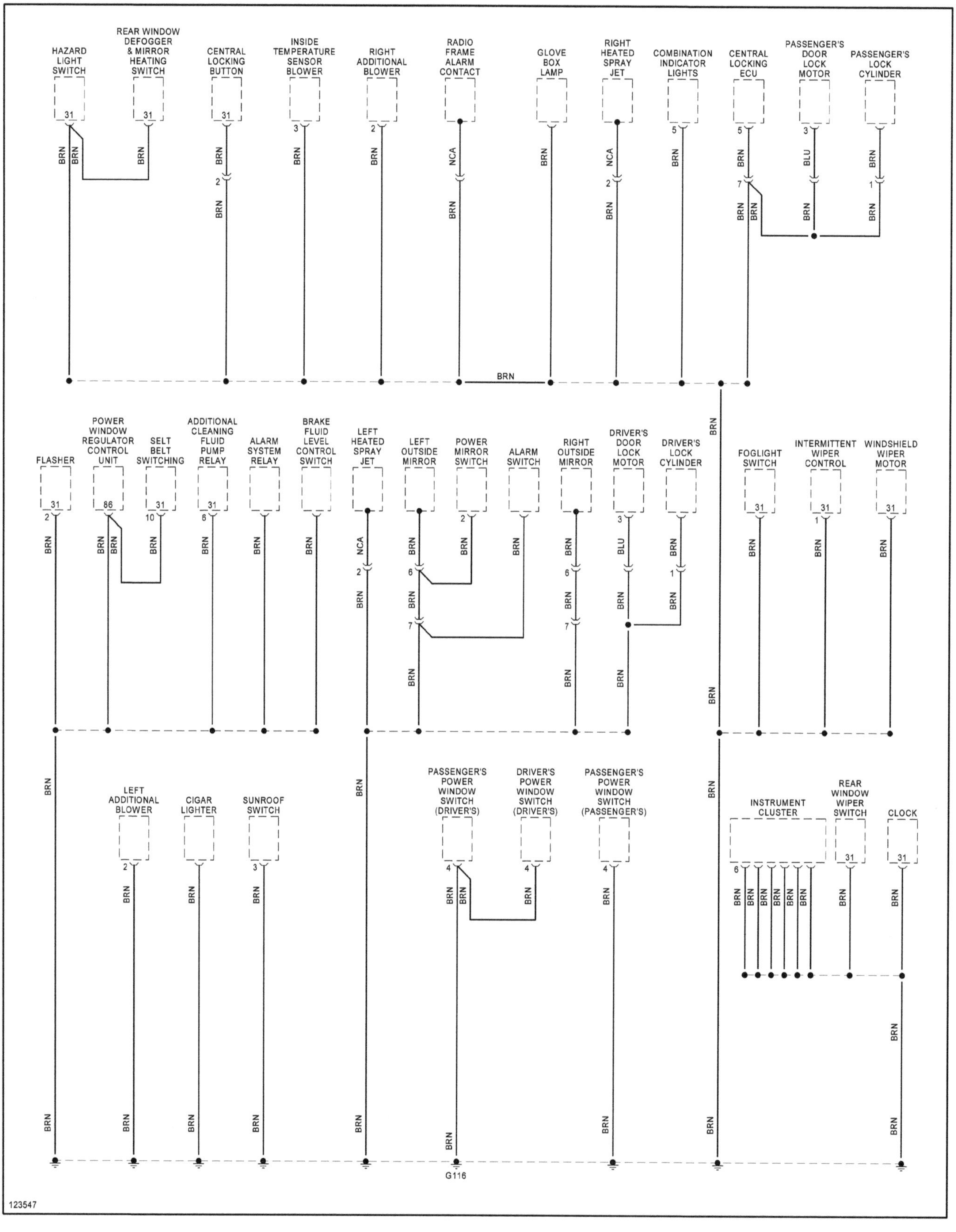

Ground Distribution

1988-1989 (3 of 3)

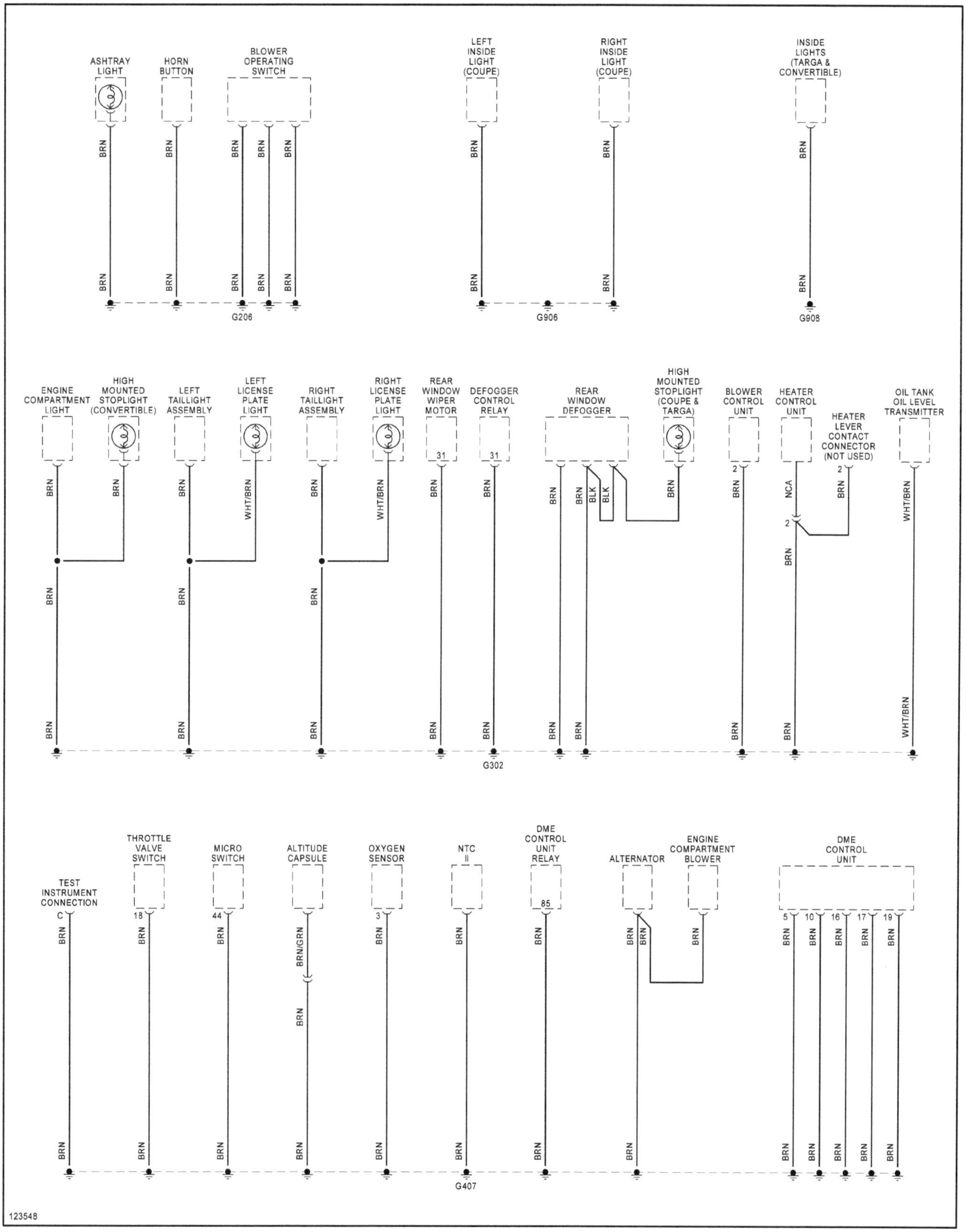

Engine Management
1984-1985 (1 of 2)

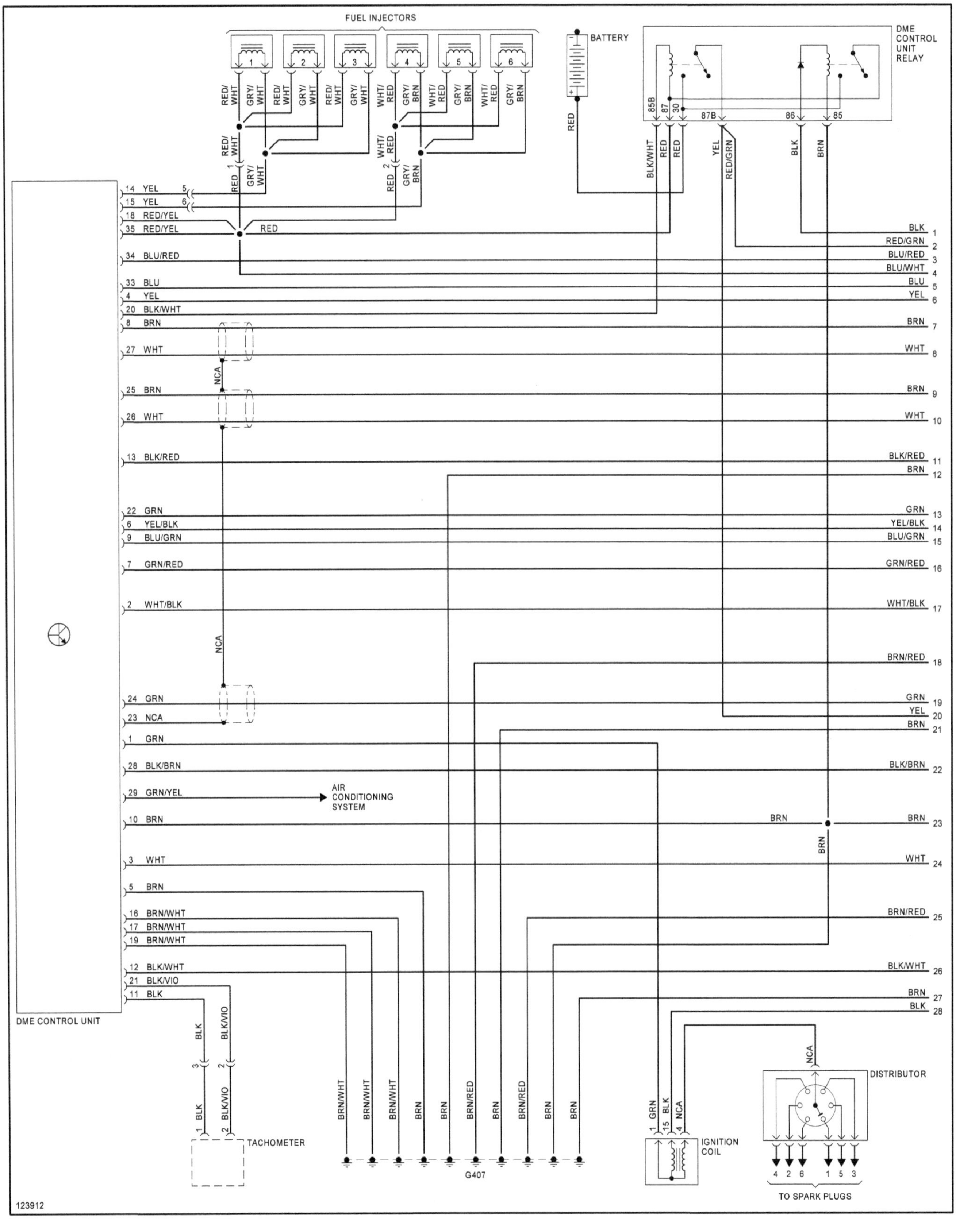

Engine Management

1984-1985 (2 of 2)

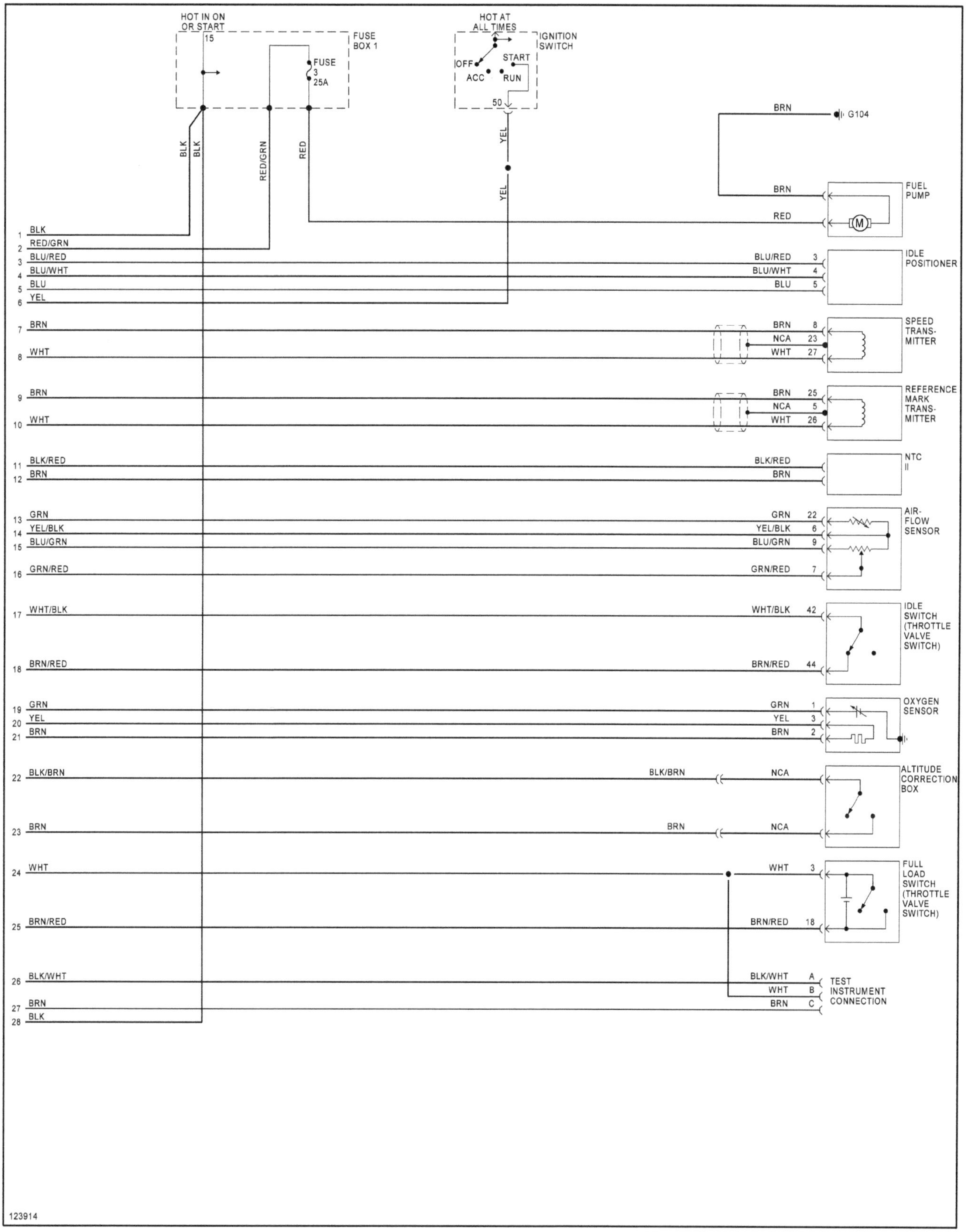

Engine Management

1986-1987 (1 of 2)

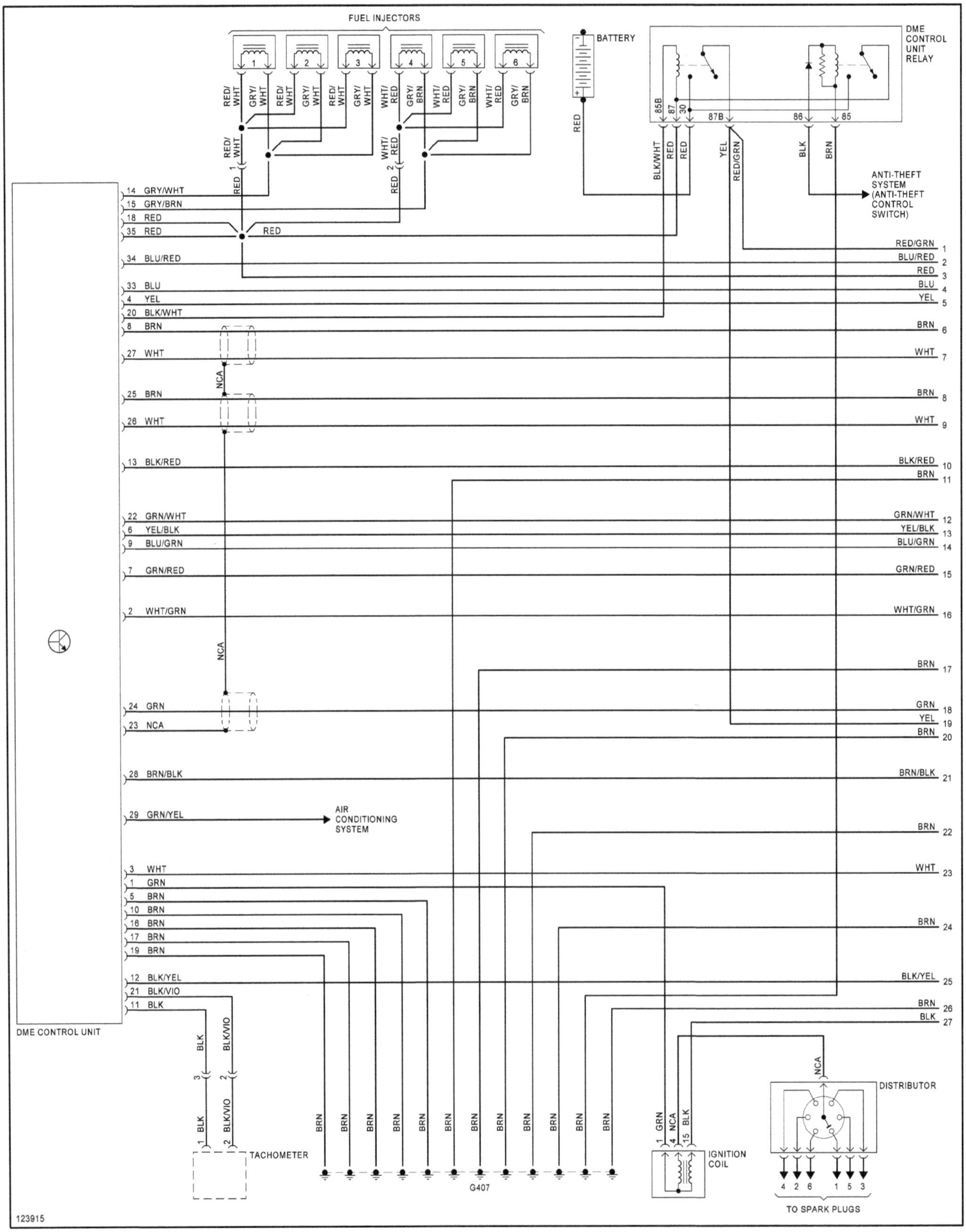

Engine Management

1986-1987 (2 of 2)

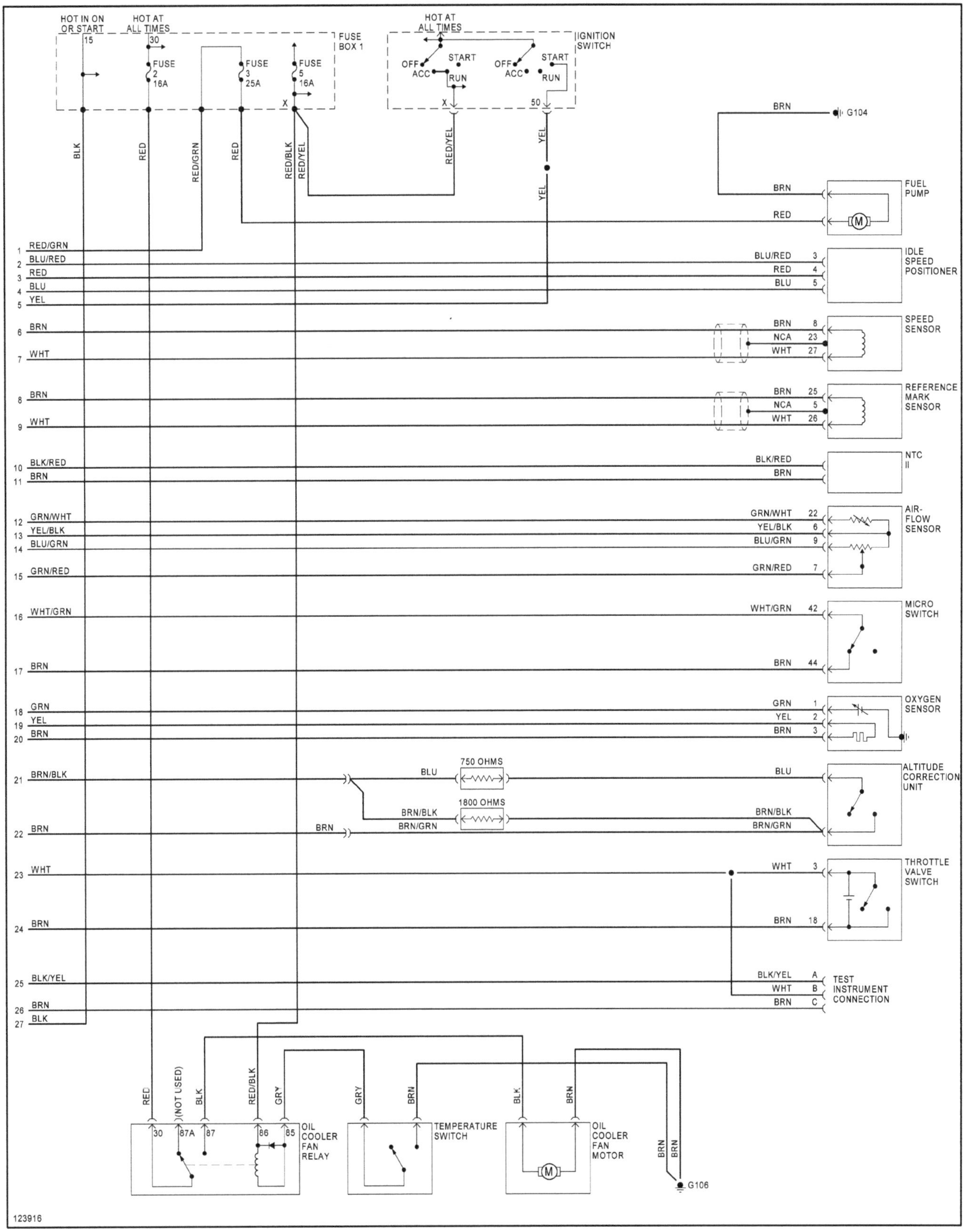

Engine Management
1988-1989 (1 of 2)

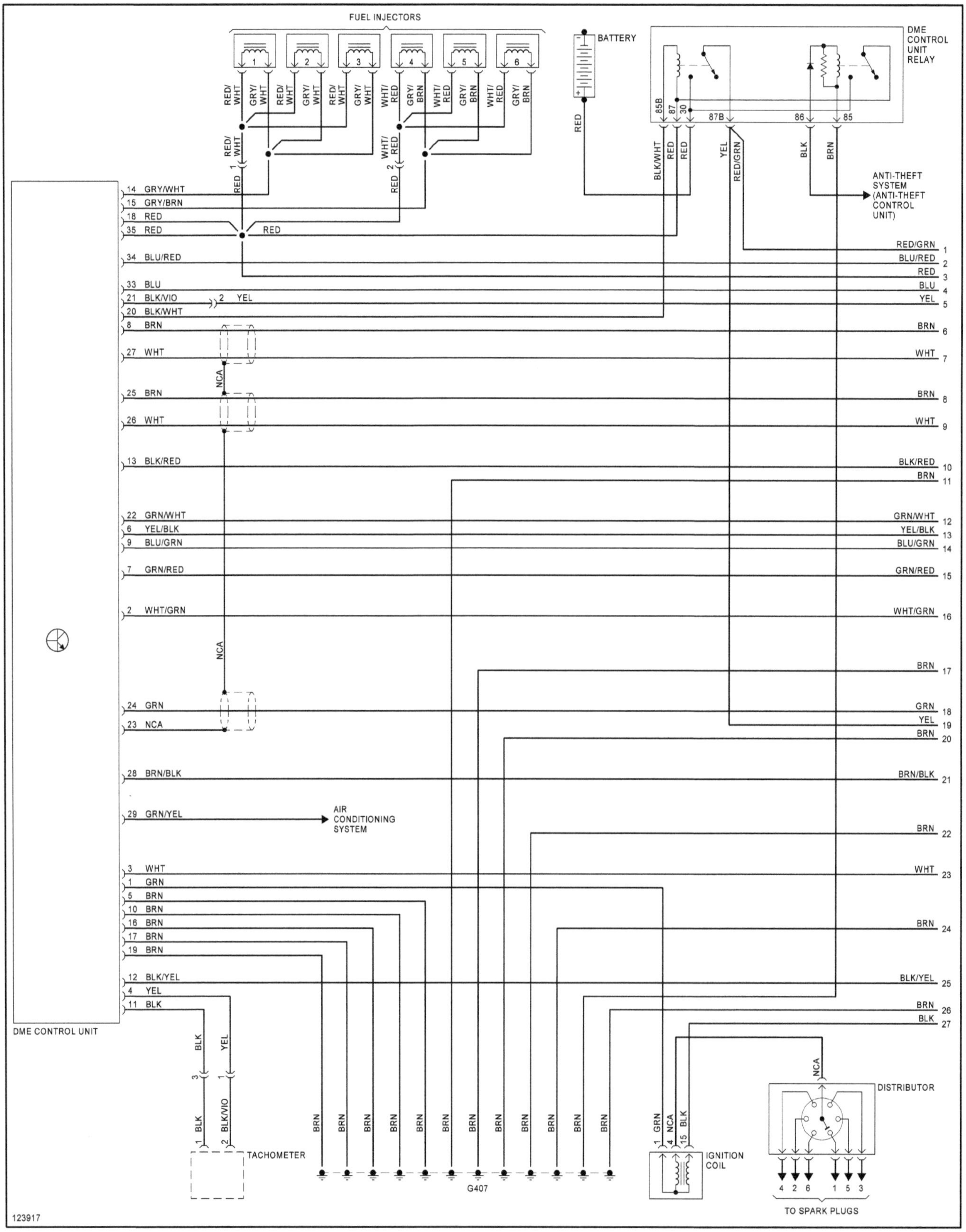

Engine Management

1988-1989 (2 of 2)

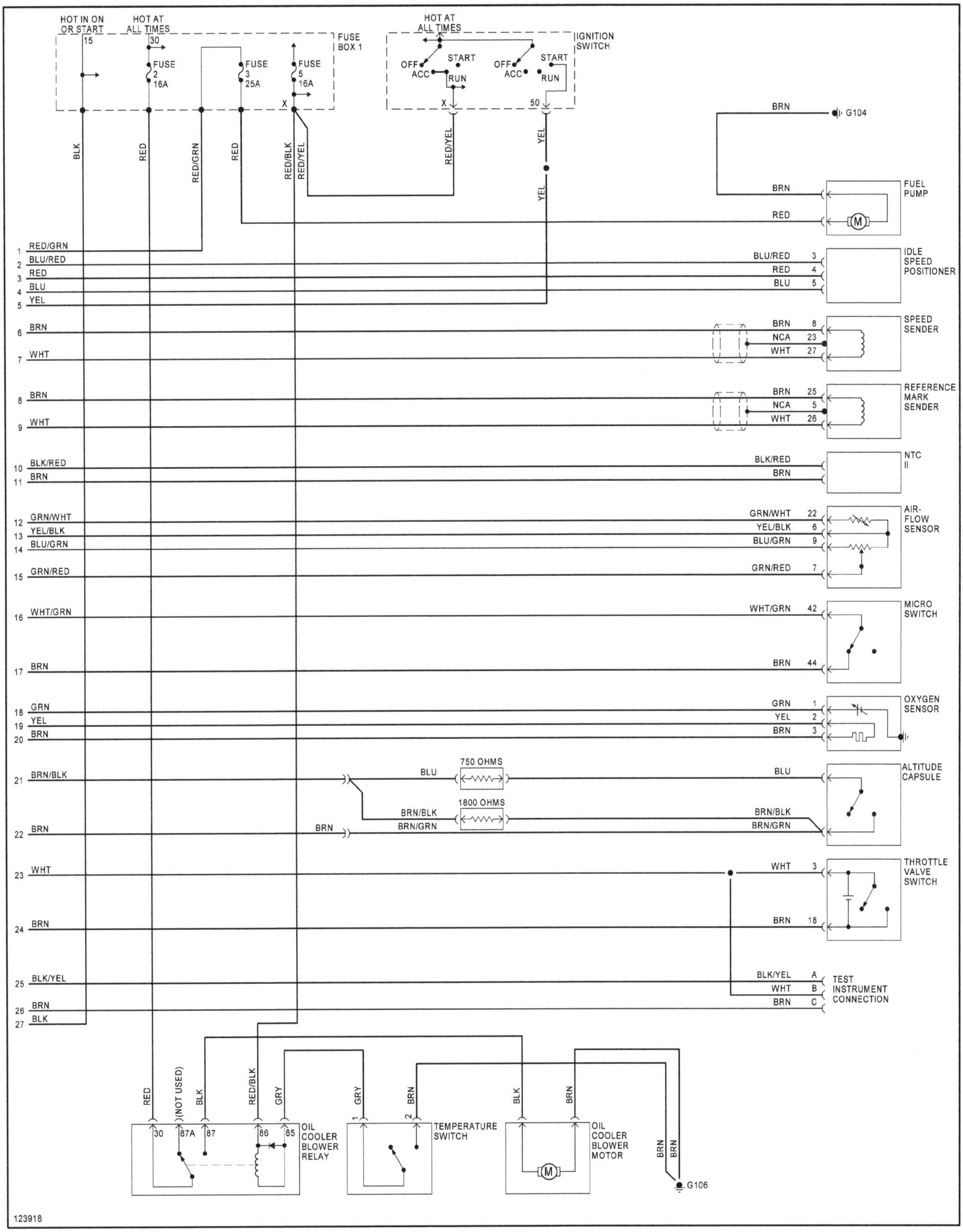

Engine Oil Cooling Fan

1986-1987

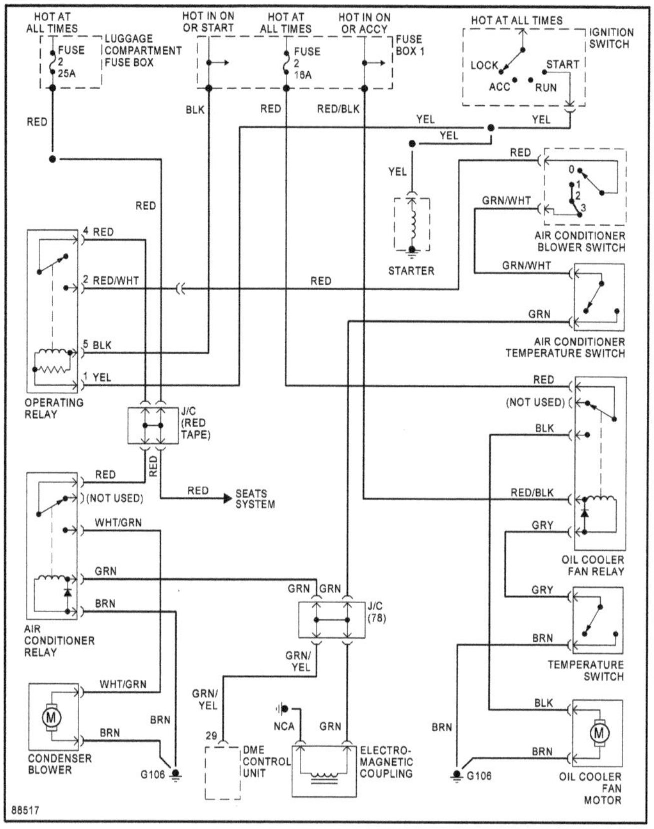

Engine Oil Cooling Fan

1988-1989

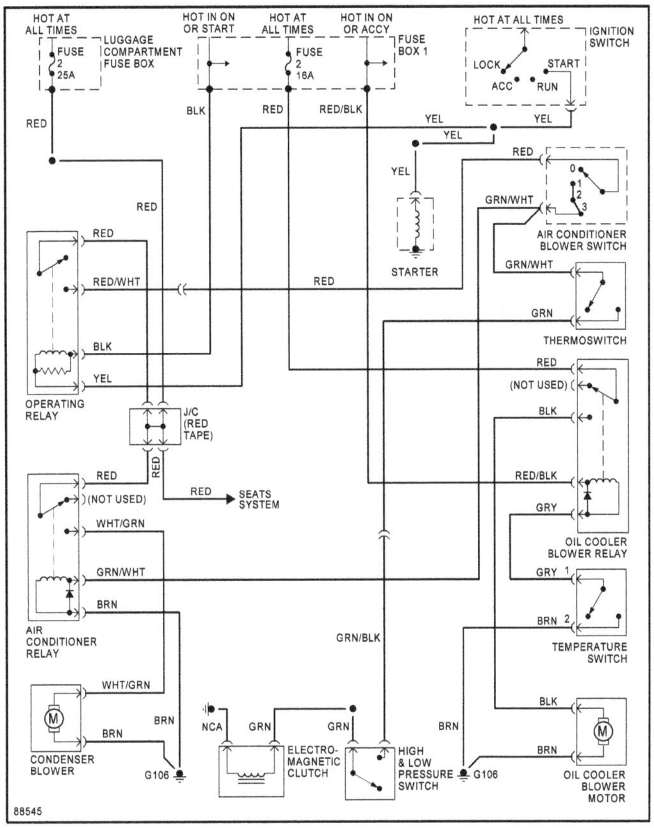

Charging

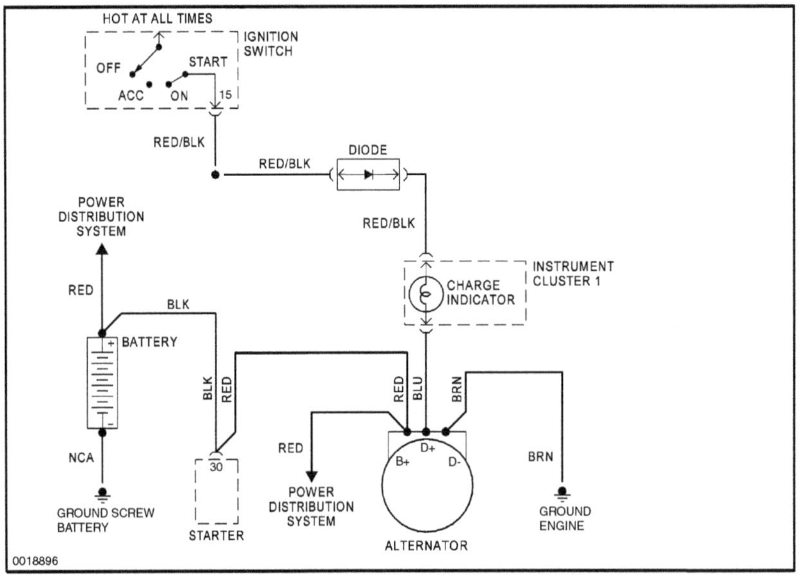

1984 Charging

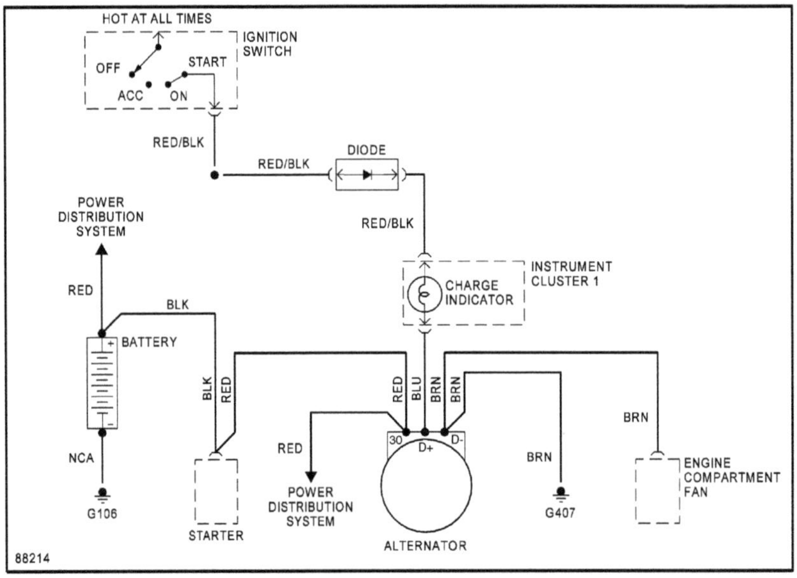

1985 Charging

Charging
1986-1989

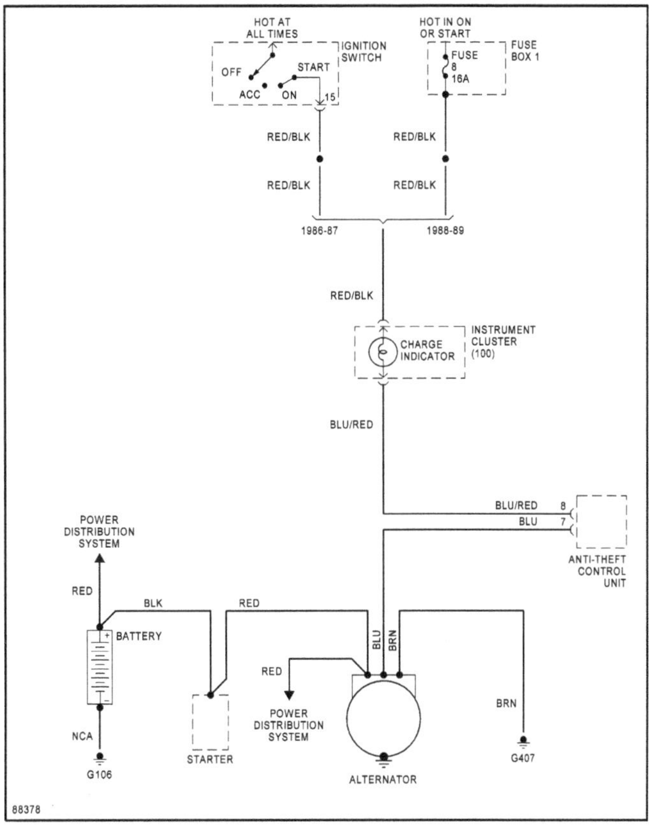

Starting

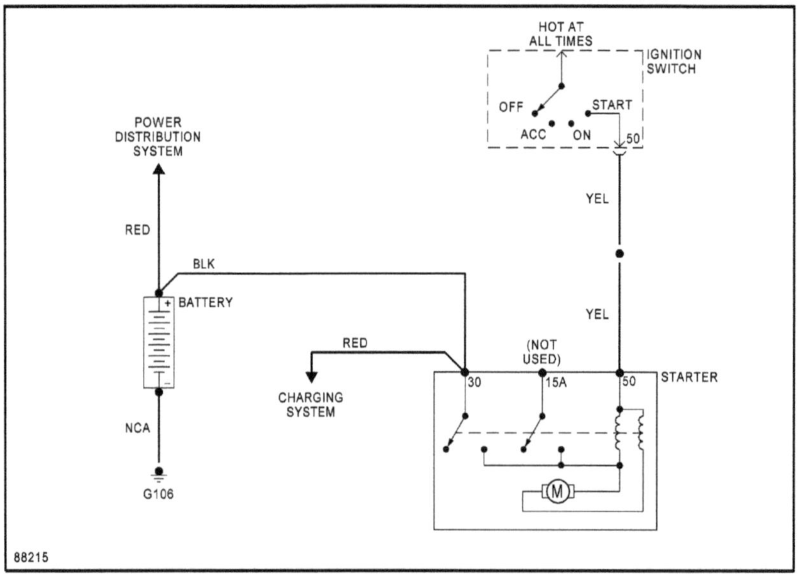

1984-1985 Starting

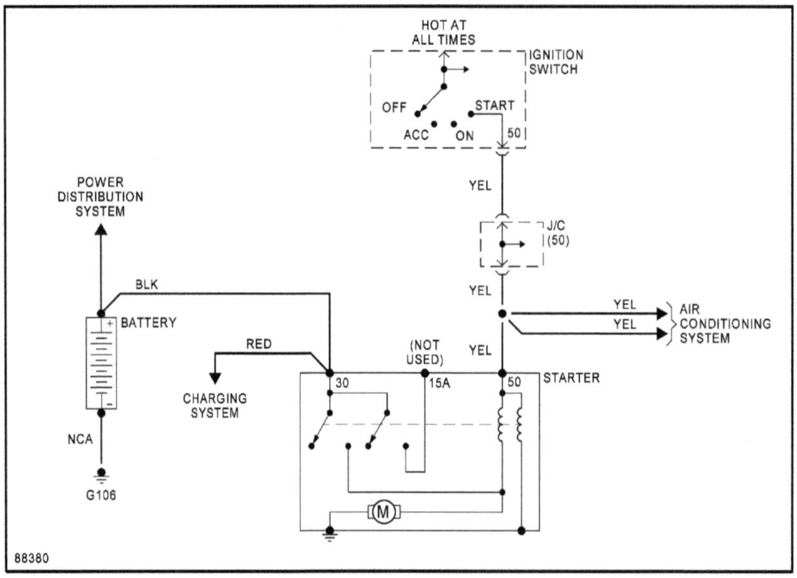

1986-1989 Starting

Central locking
1985-1989

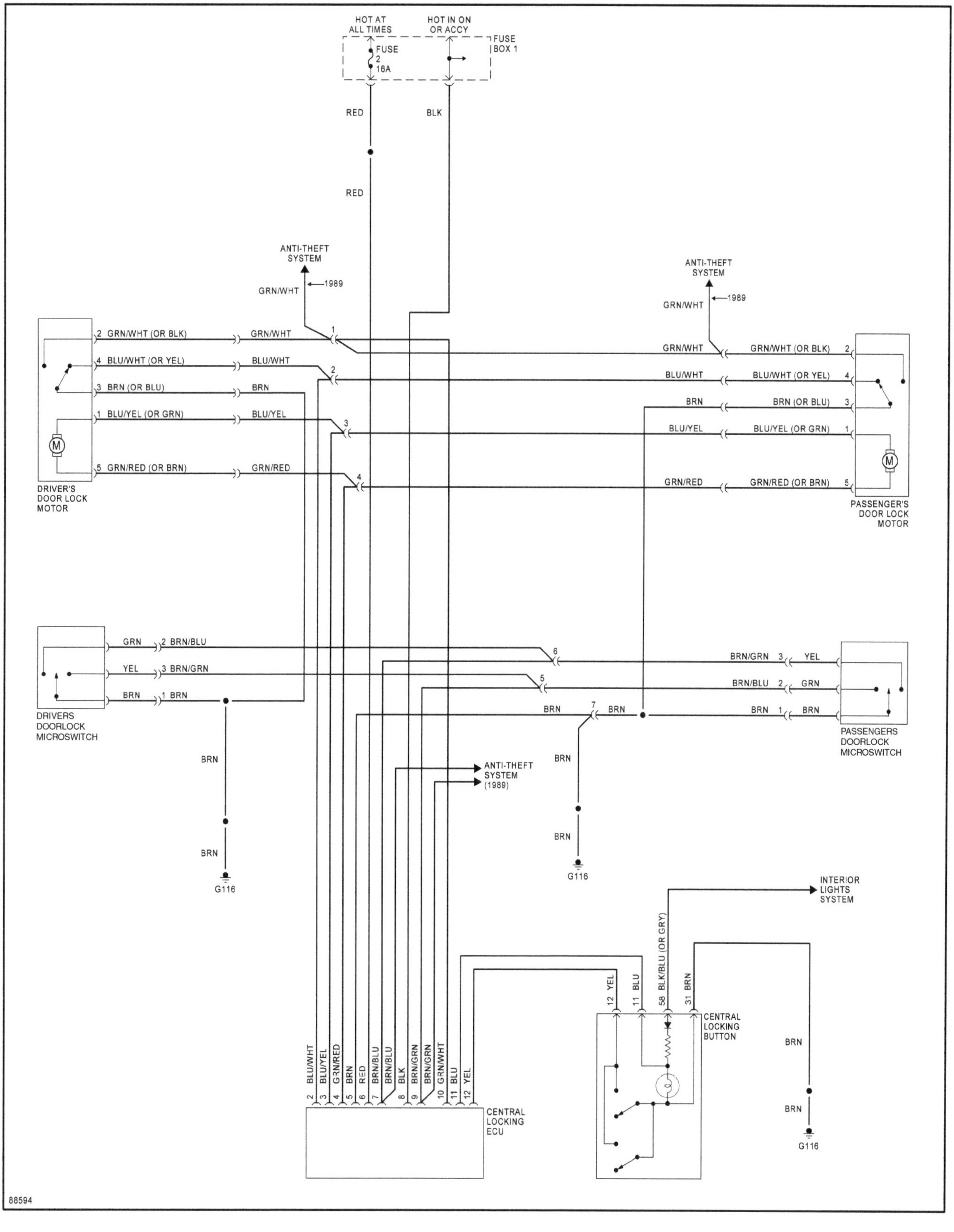

Convertible Top

1986-1989

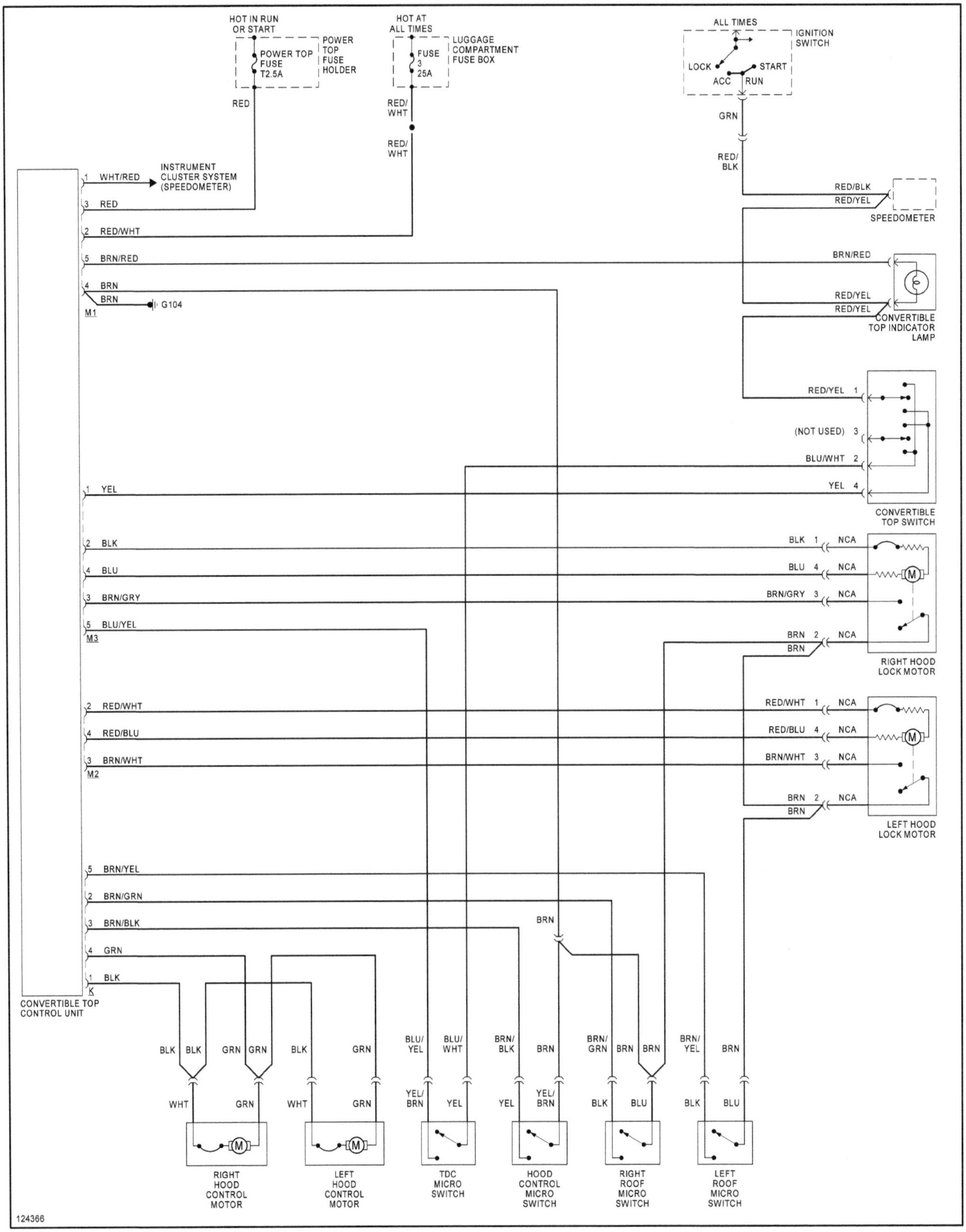

Sunroof
1986-1989

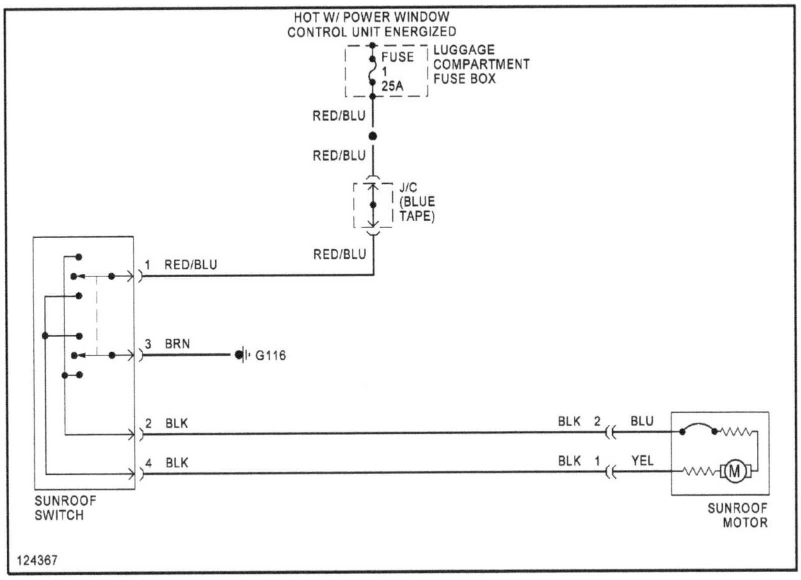

Power Windows

1984

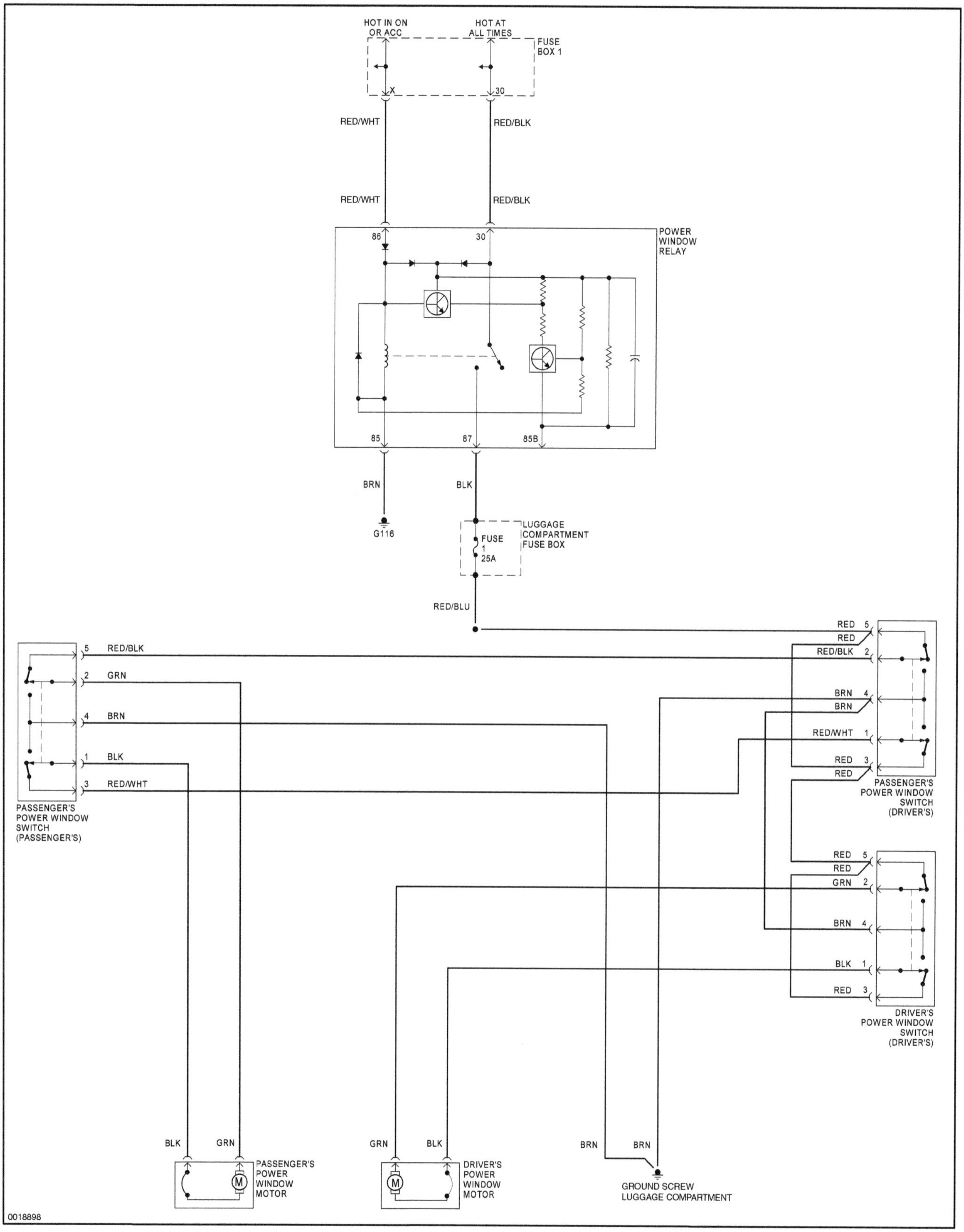

Power Windows
1985

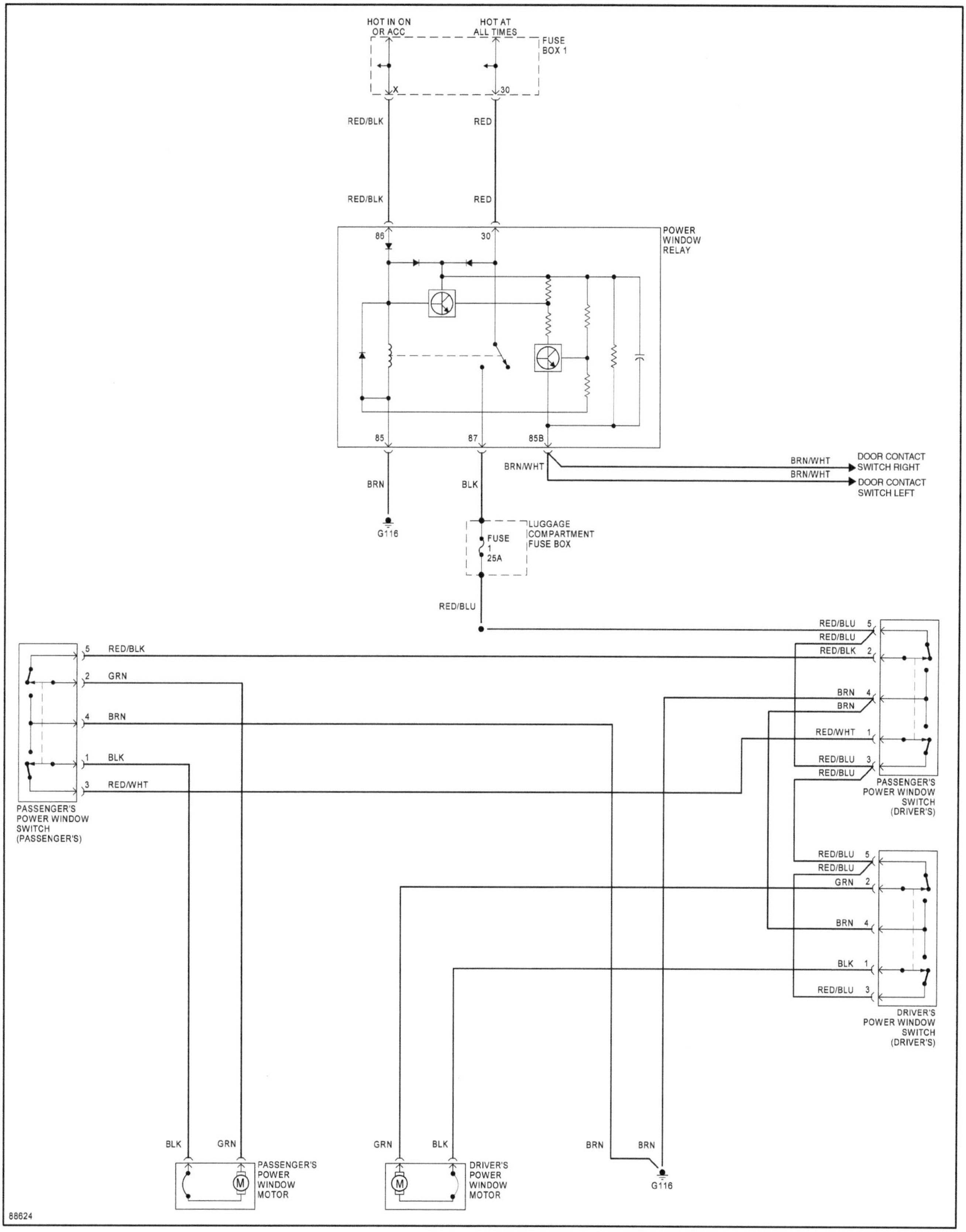

Power Windows

1986-1989

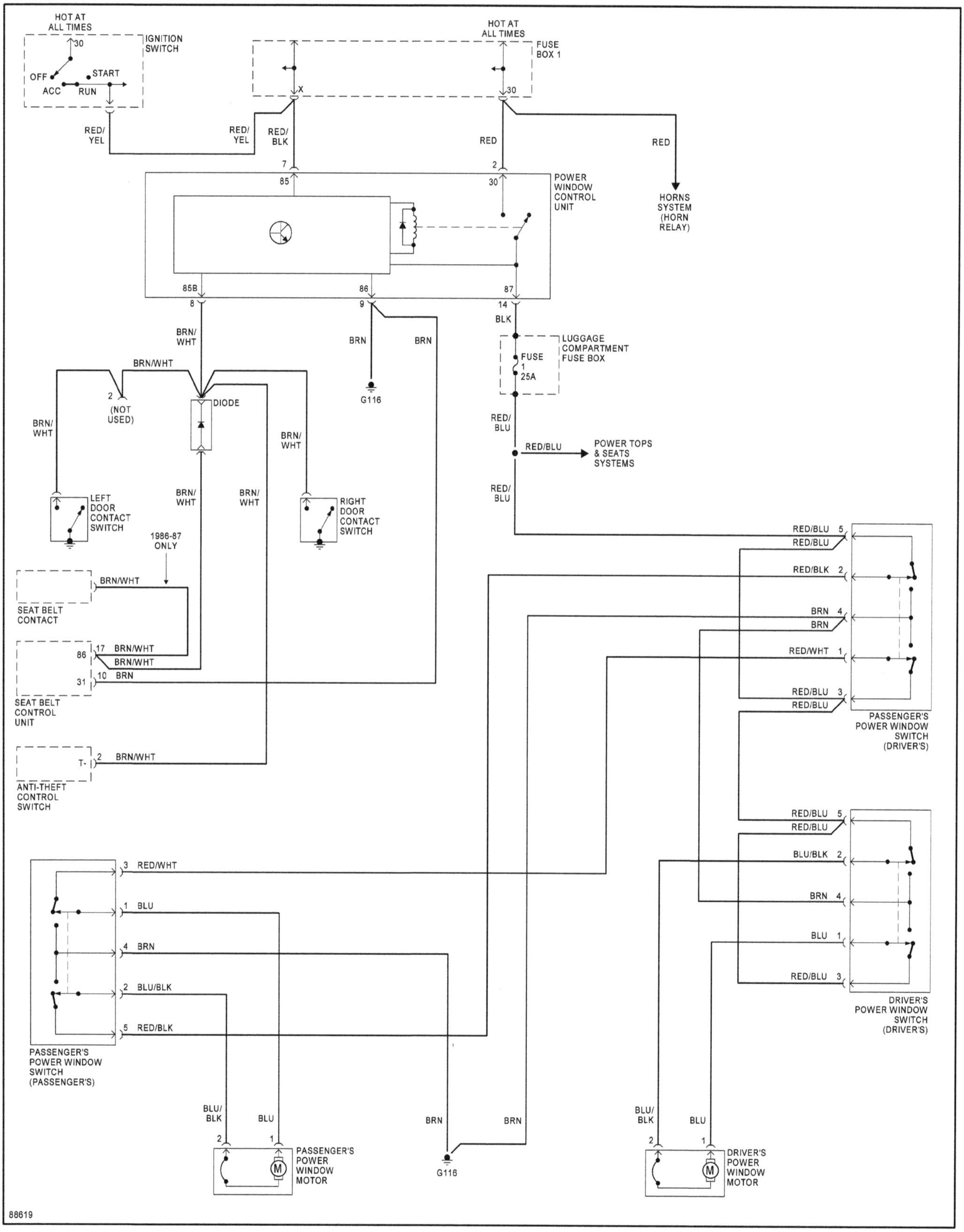

Seats

1985

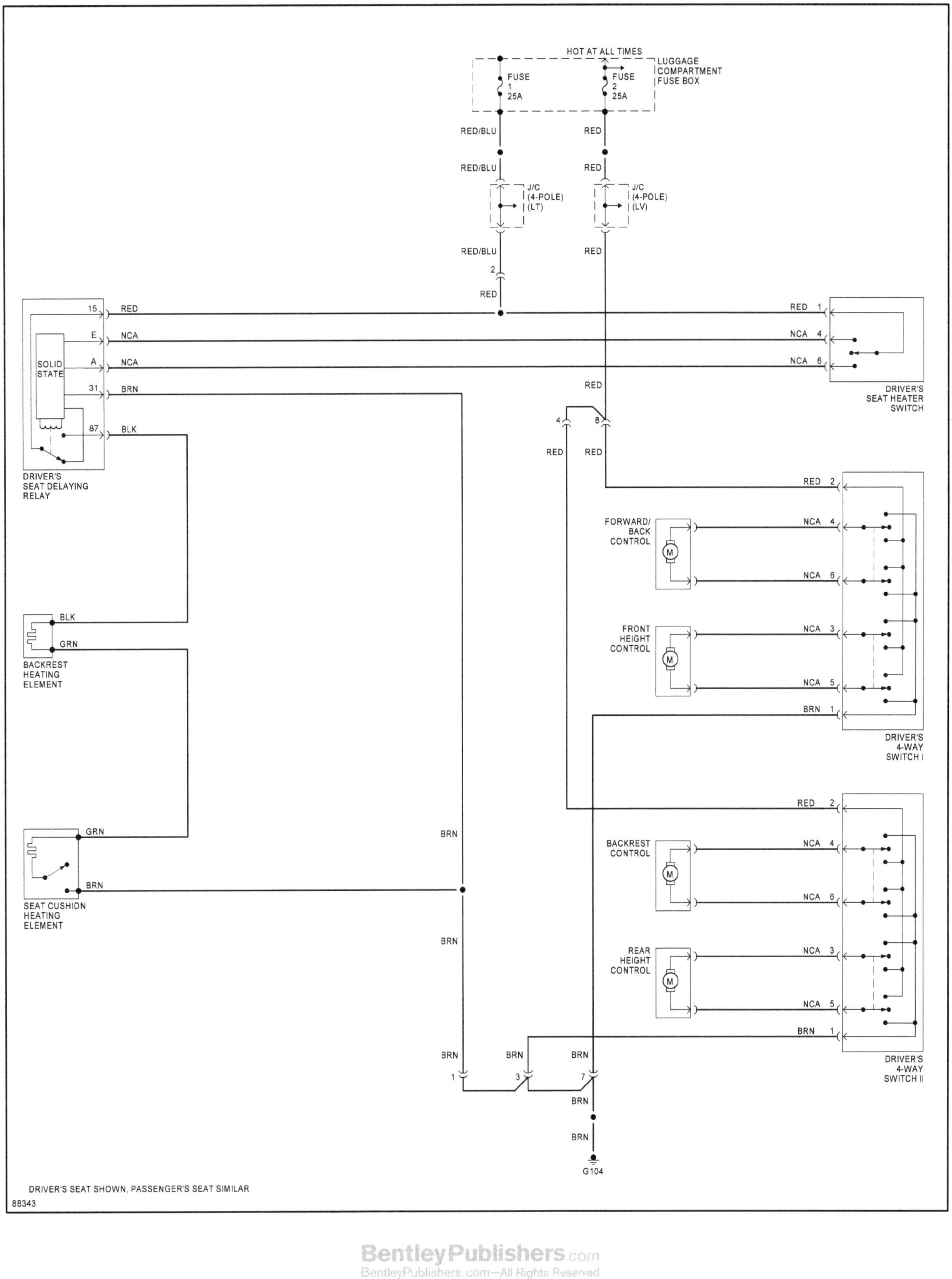
HOT AT ALL TIMES
LUGGAGE COMPARTMENT FUSE BOX
FUSE 1 25A
FUSE 2 25A
RED/BLU
RED
J/C (4-POLE) (LT)
J/C (4-POLE) (LV)
DRIVER'S SEAT DELAYING RELAY
SOLID STATE
15
E
A
31
87
NCA
BRN
BLK
GRN
DRIVER'S SEAT HEATER SWITCH
FORWARD/ BACK CONTROL
FRONT HEIGHT CONTROL
DRIVER'S 4-WAY SWITCH I
BACKREST CONTROL
REAR HEIGHT CONTROL
DRIVER'S 4-WAY SWITCH II
BACKREST HEATING ELEMENT
SEAT CUSHION HEATING ELEMENT
G104
DRIVER'S SEAT SHOWN, PASSENGER'S SEAT SIMILAR
88343

Seats

1986-1989

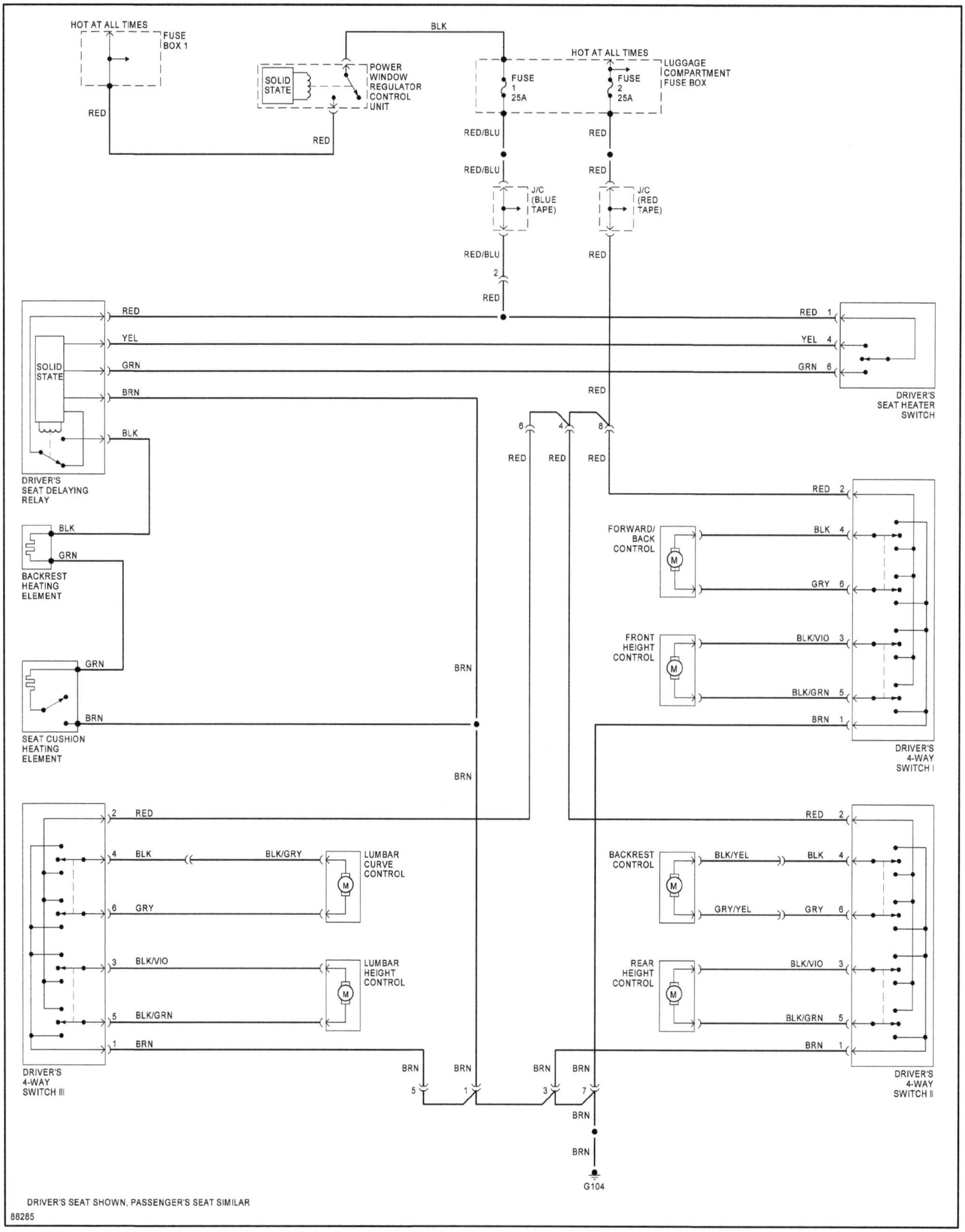

Air Conditioning and Heating

1984 (1 of 2)

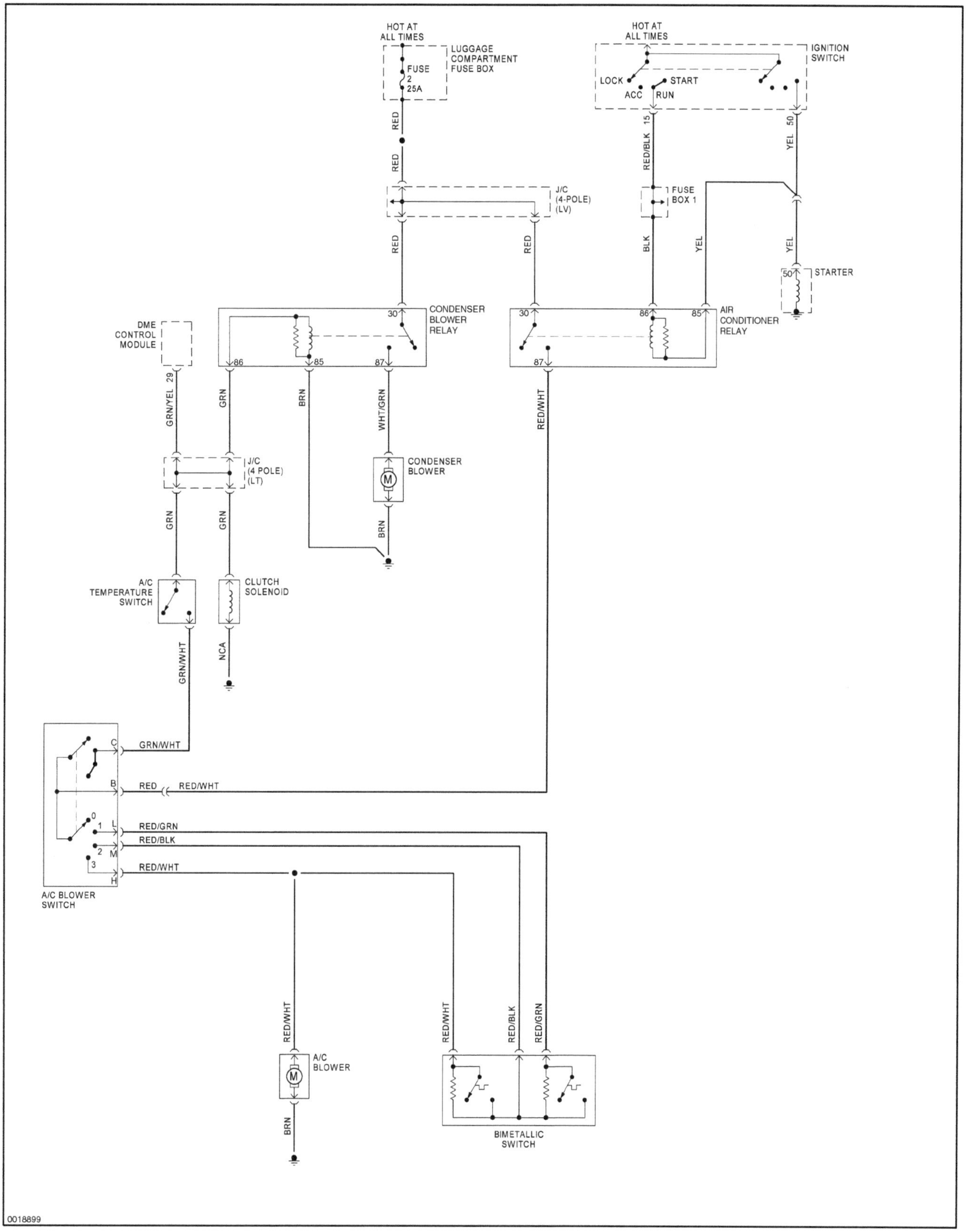

Air Conditioning and Heating

1984 (2 of 2)

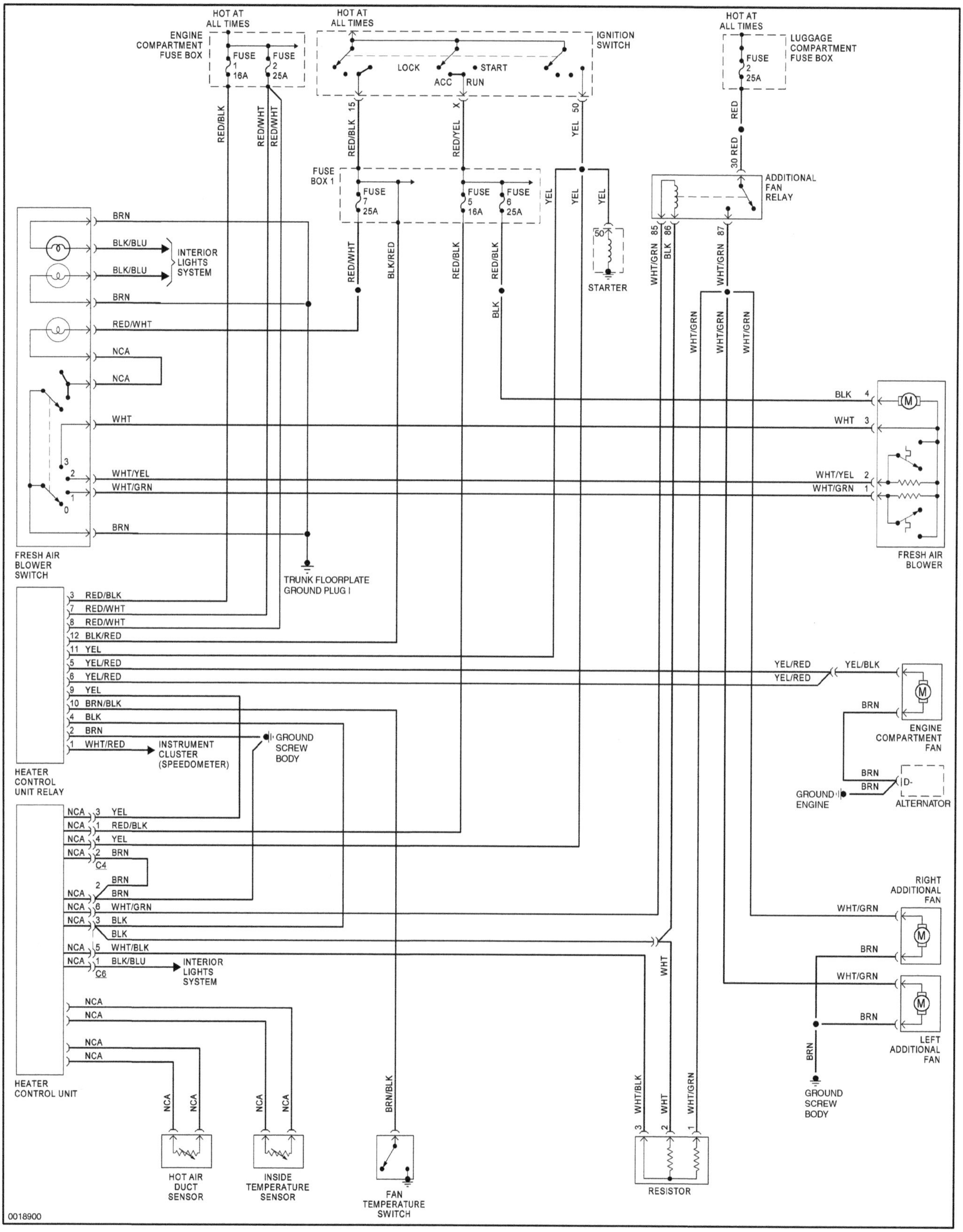

Air Conditioning and Heating

1985 (1 of 2)

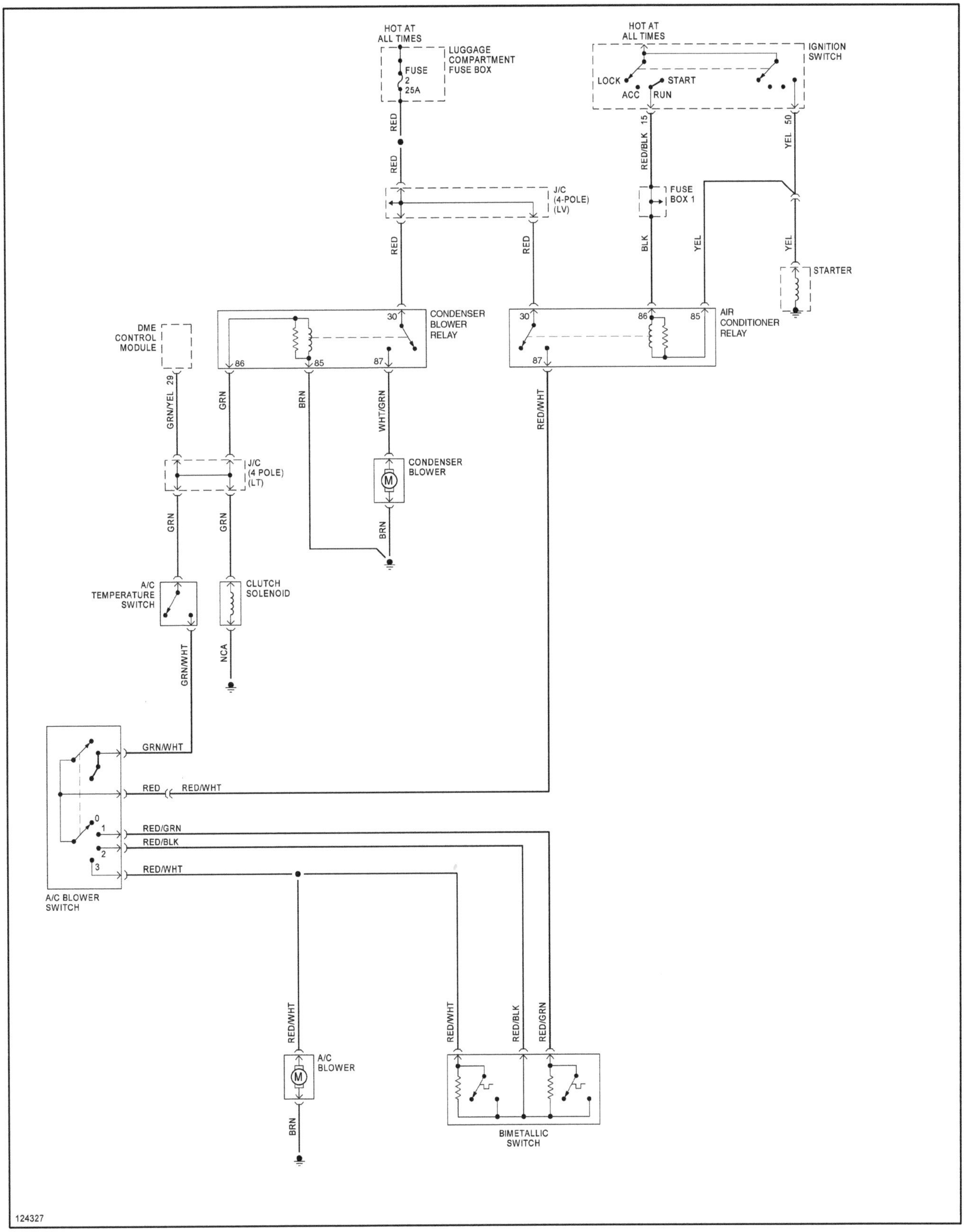

Air Conditioning and Heating

1985 (2 of 2)

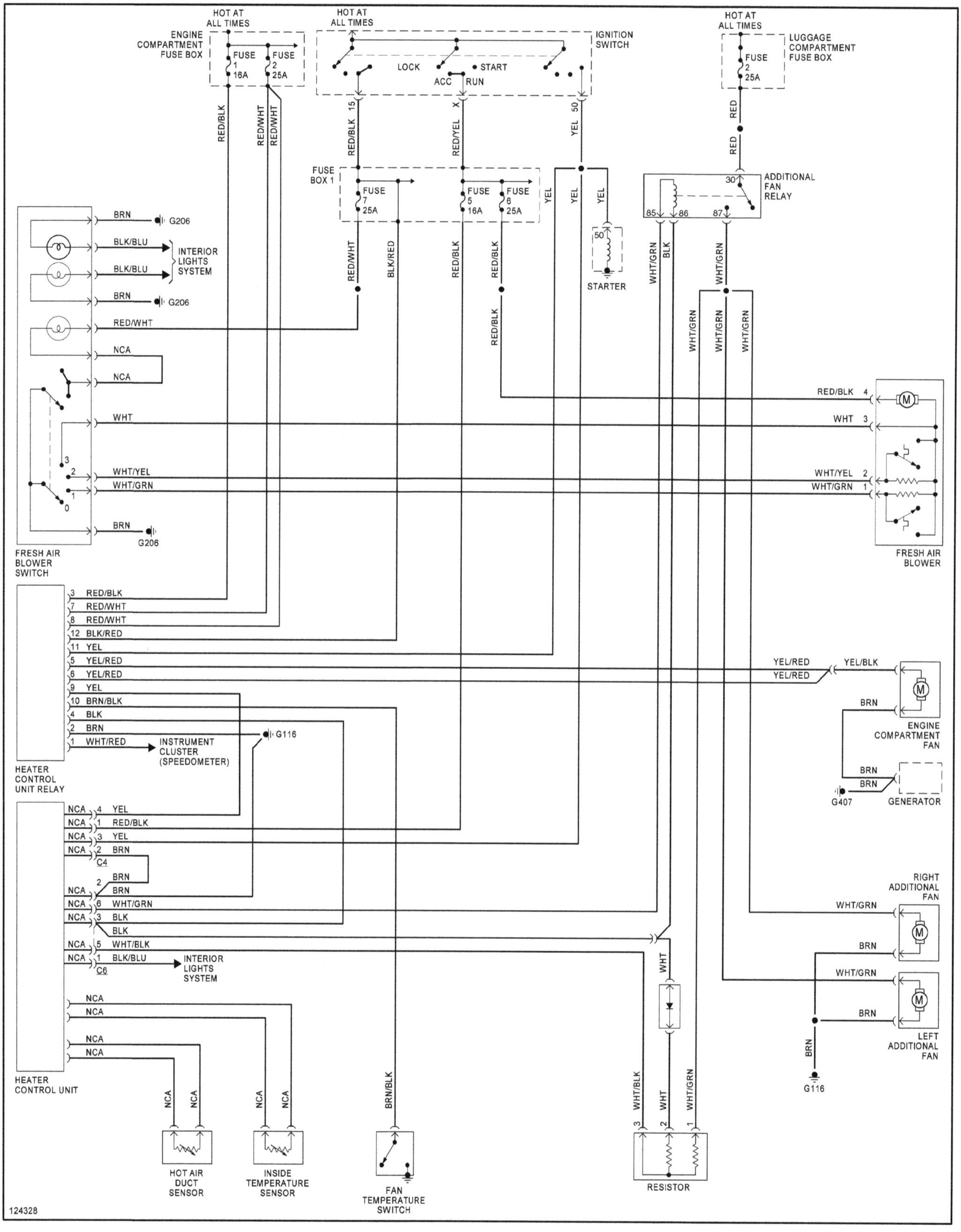

Air Conditioning and Heating

1986-1987 (1 of 2)

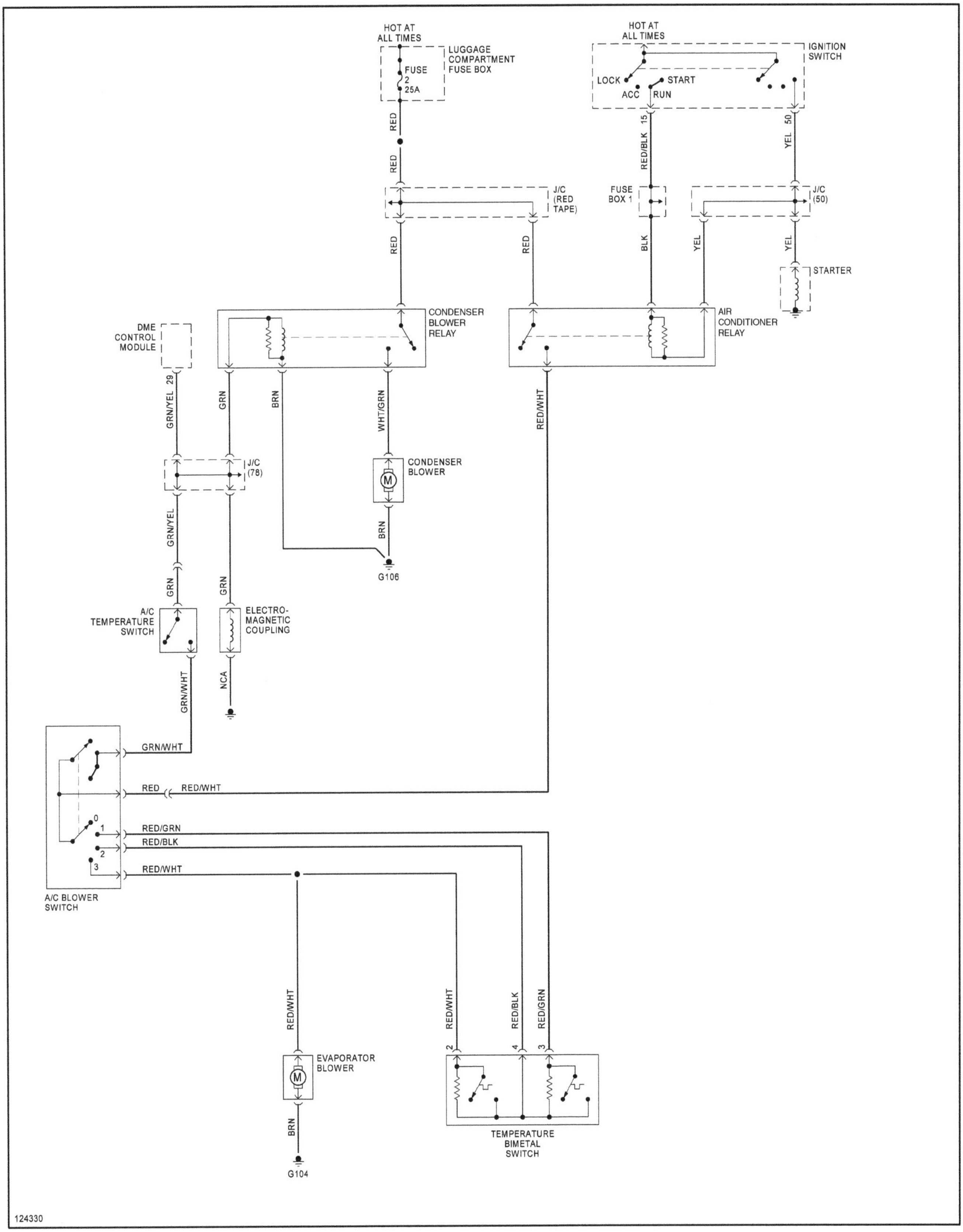

Air Conditioning and Heating

1986-1987 (2 of 2)

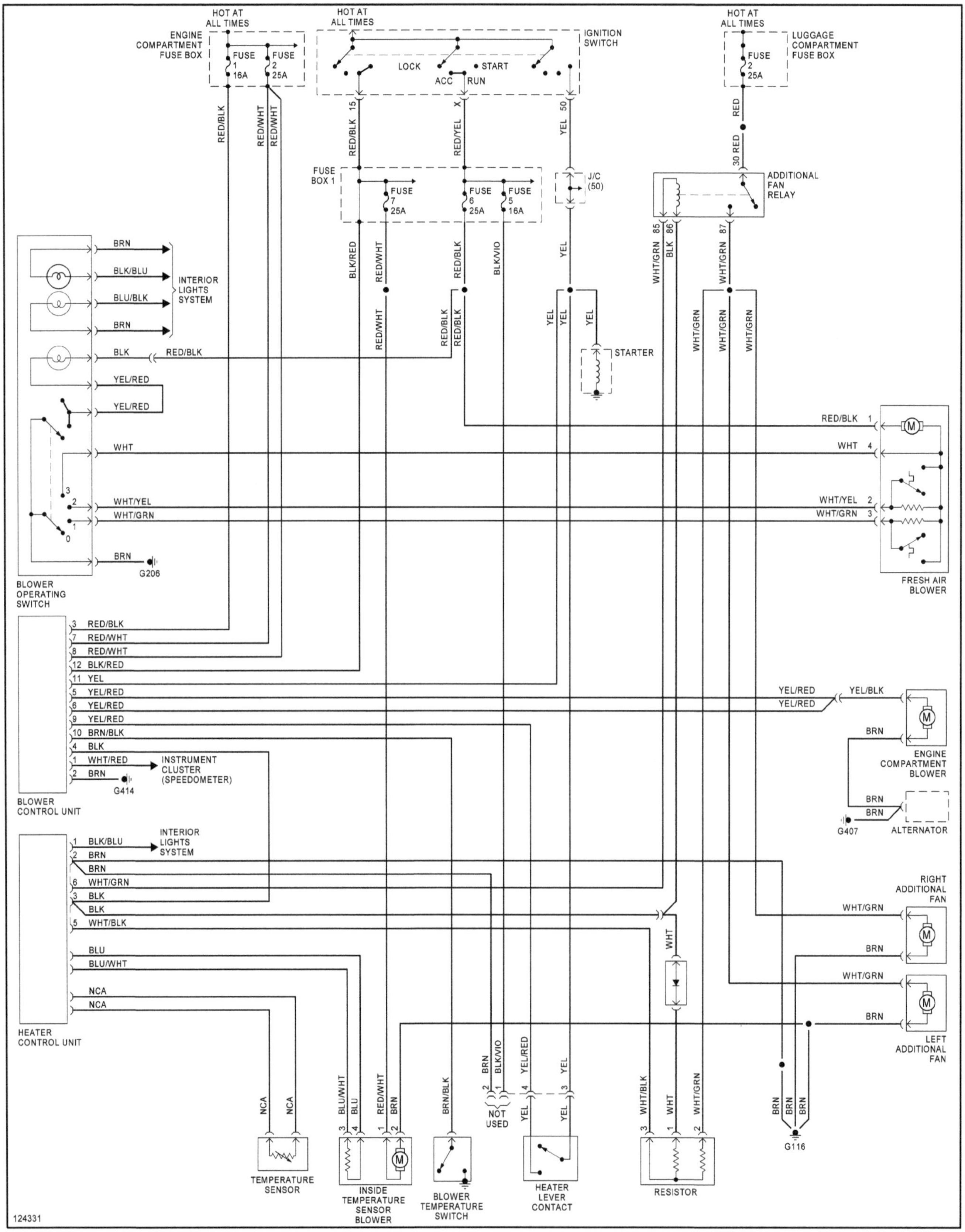

Air Conditioning and Heating

1988-1989 (1 of 2)

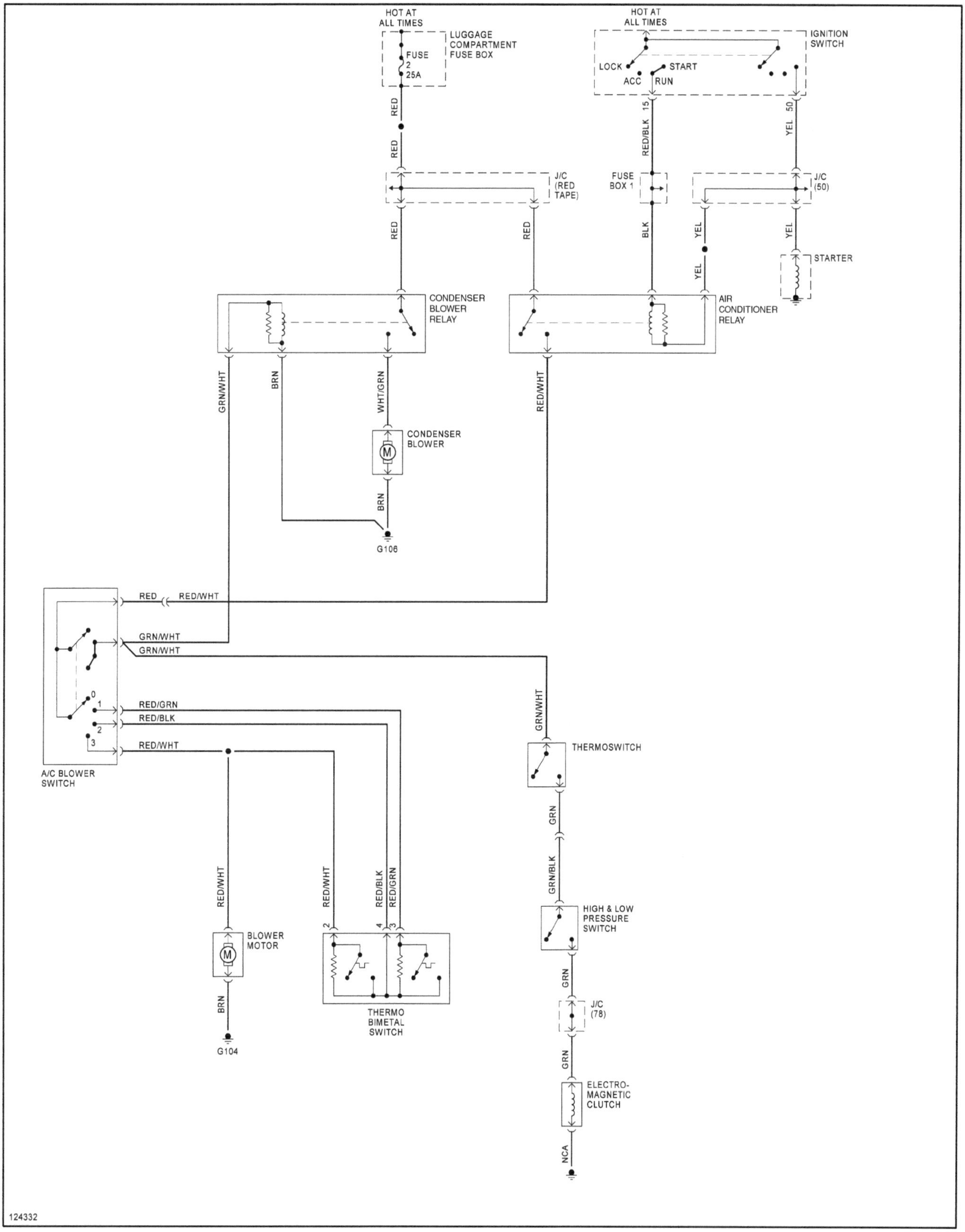

Air Conditioning and Heating

1988-1989 (2 of 2)

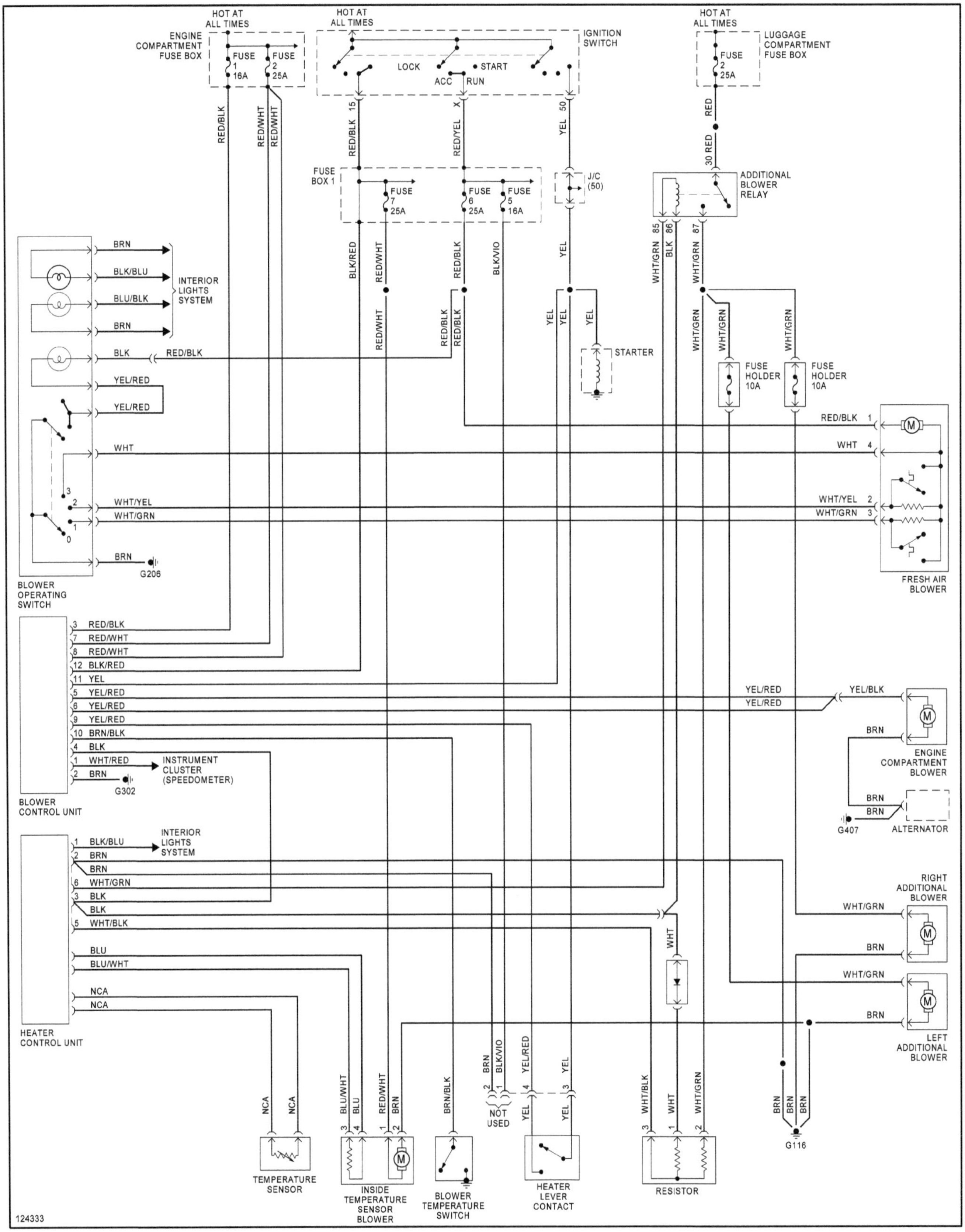

Defogger
1984

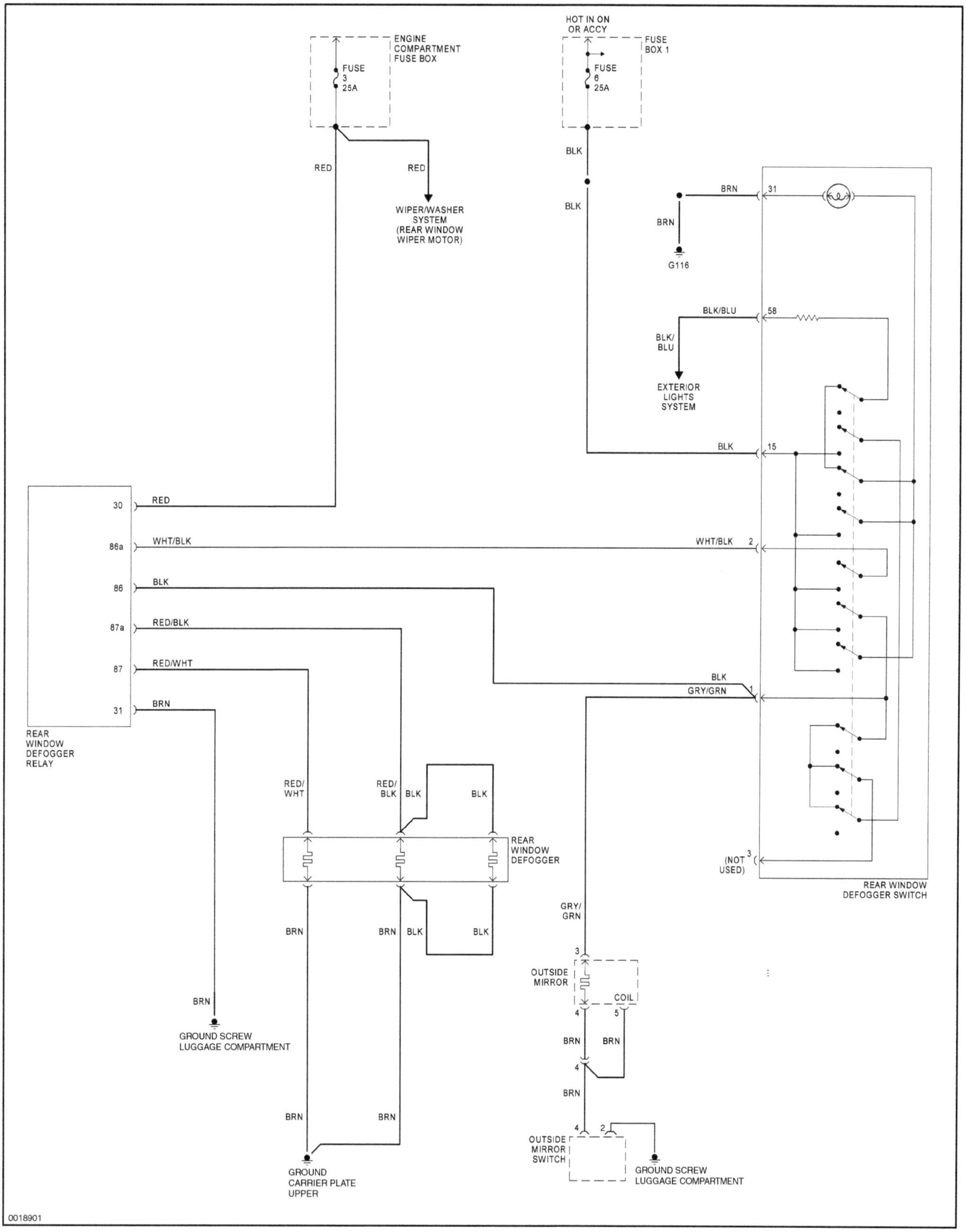

Defogger

1985

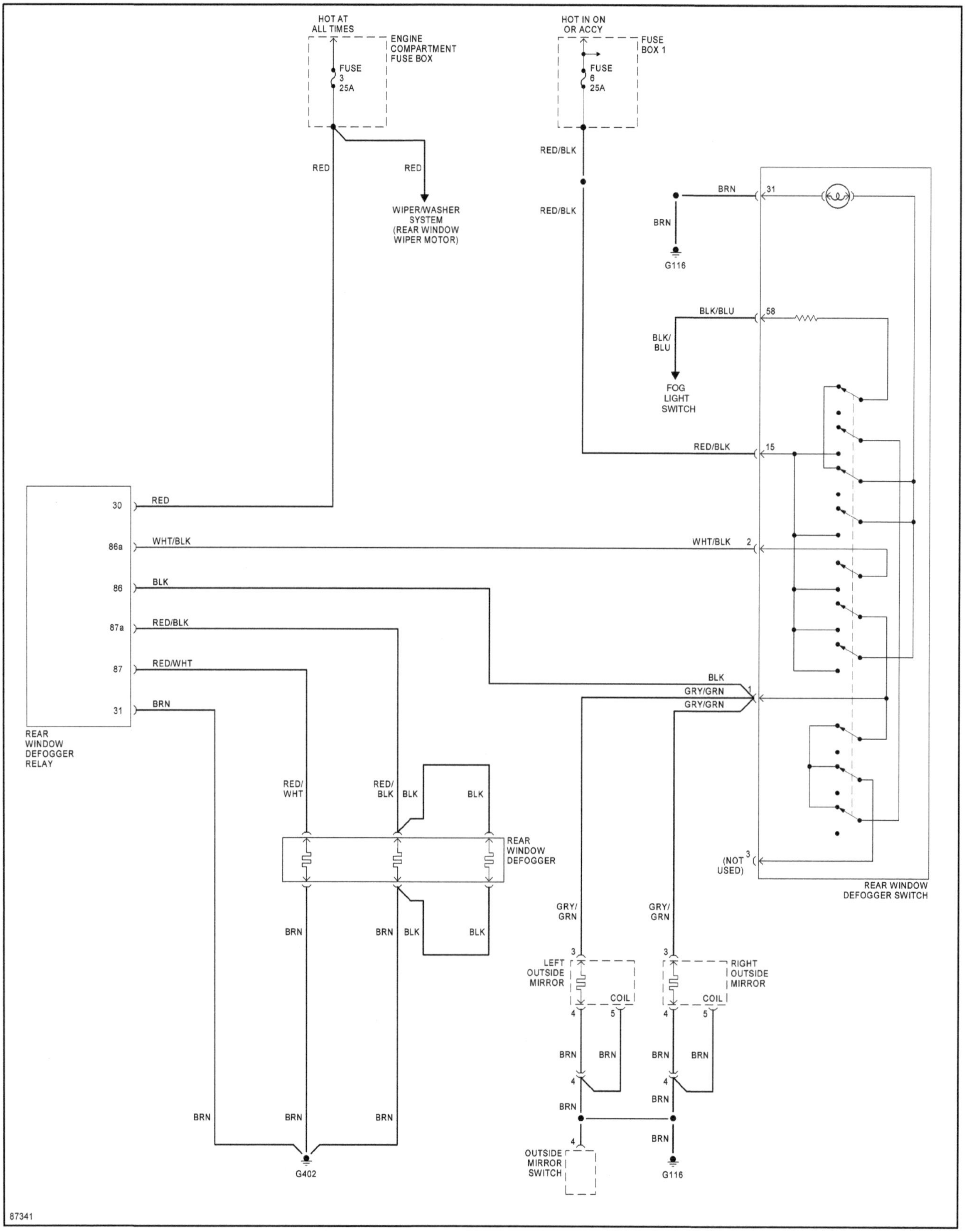

Defogger

1986-1989

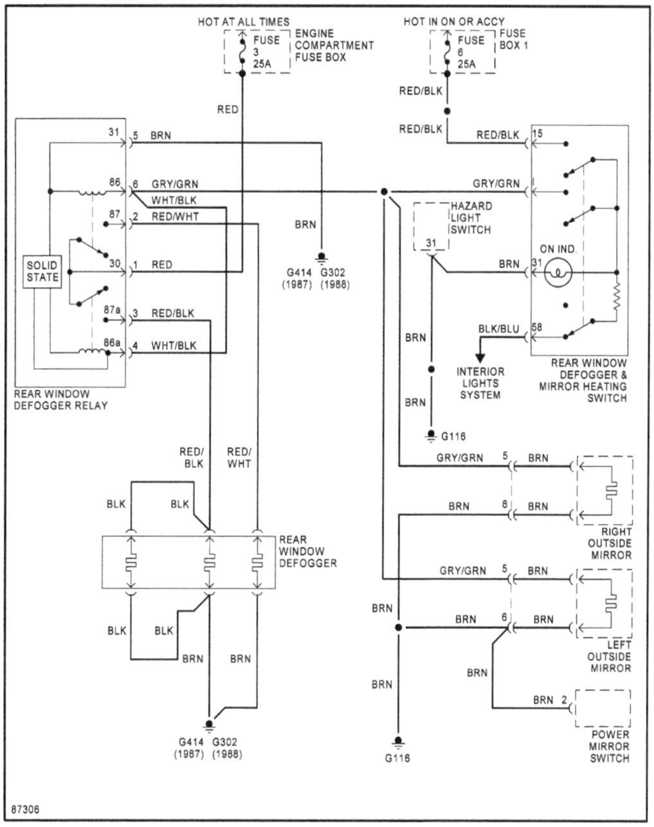

Anti-theft
1985

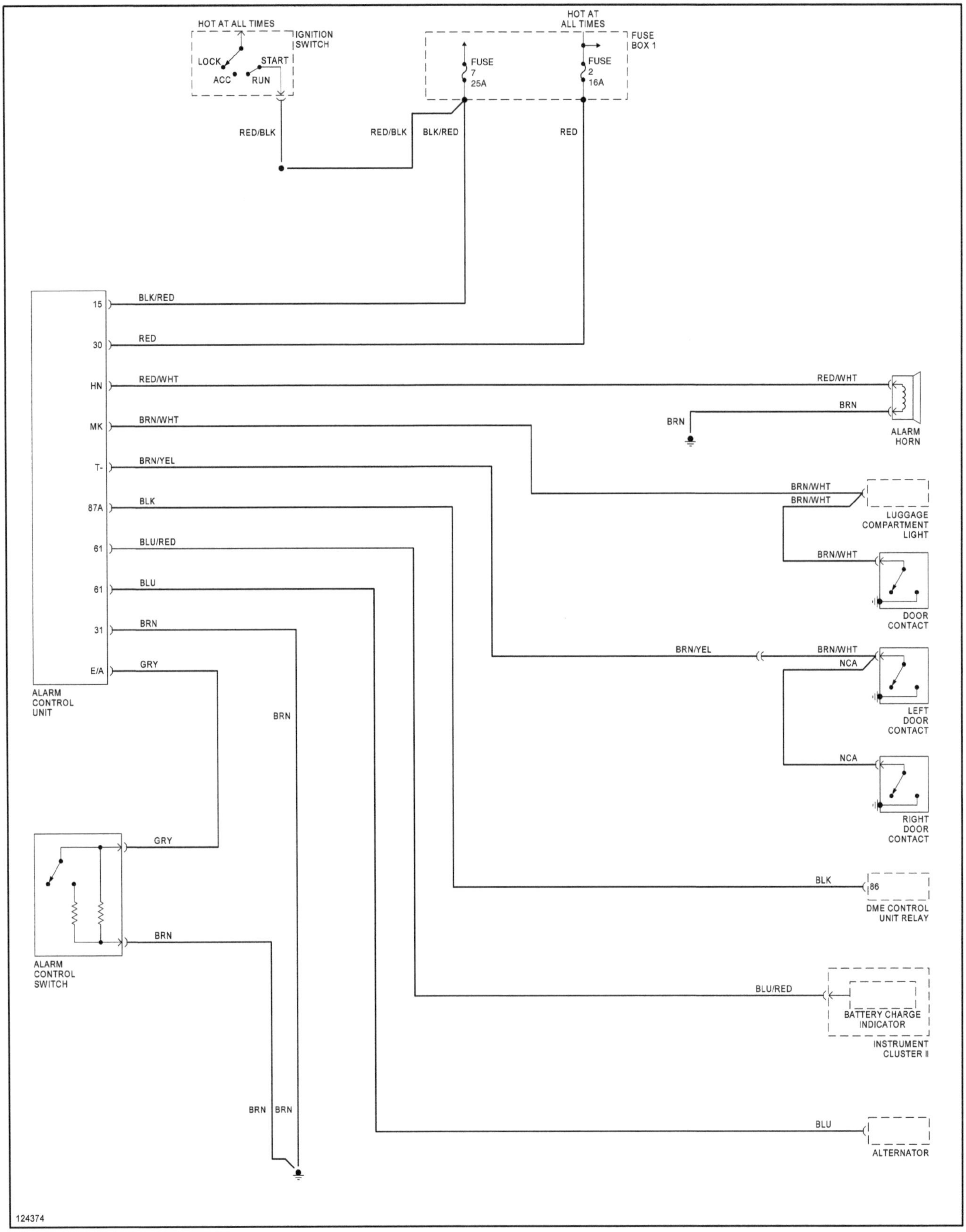

Anti-theft
1986-1987

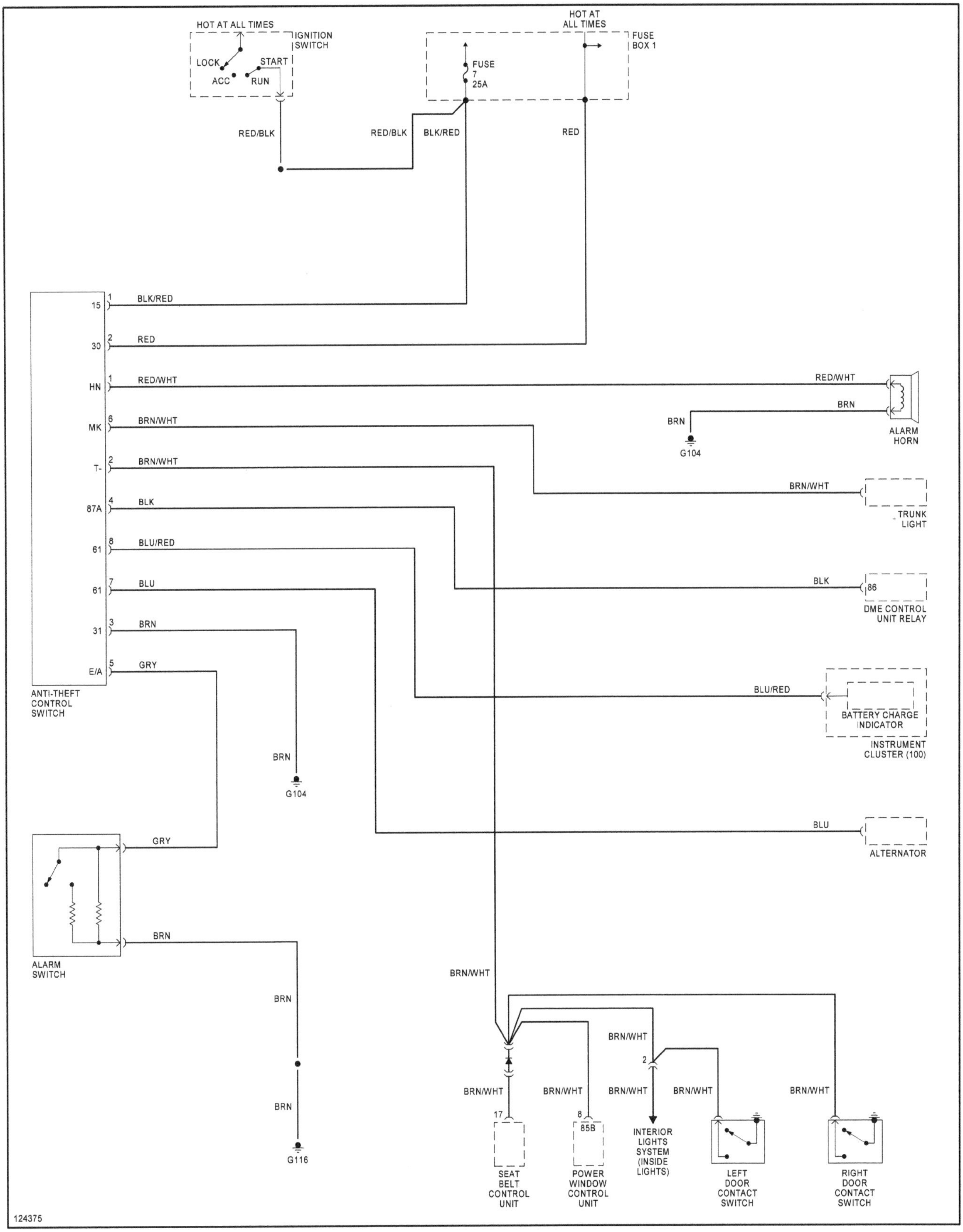

Anti-theft

1988-1989

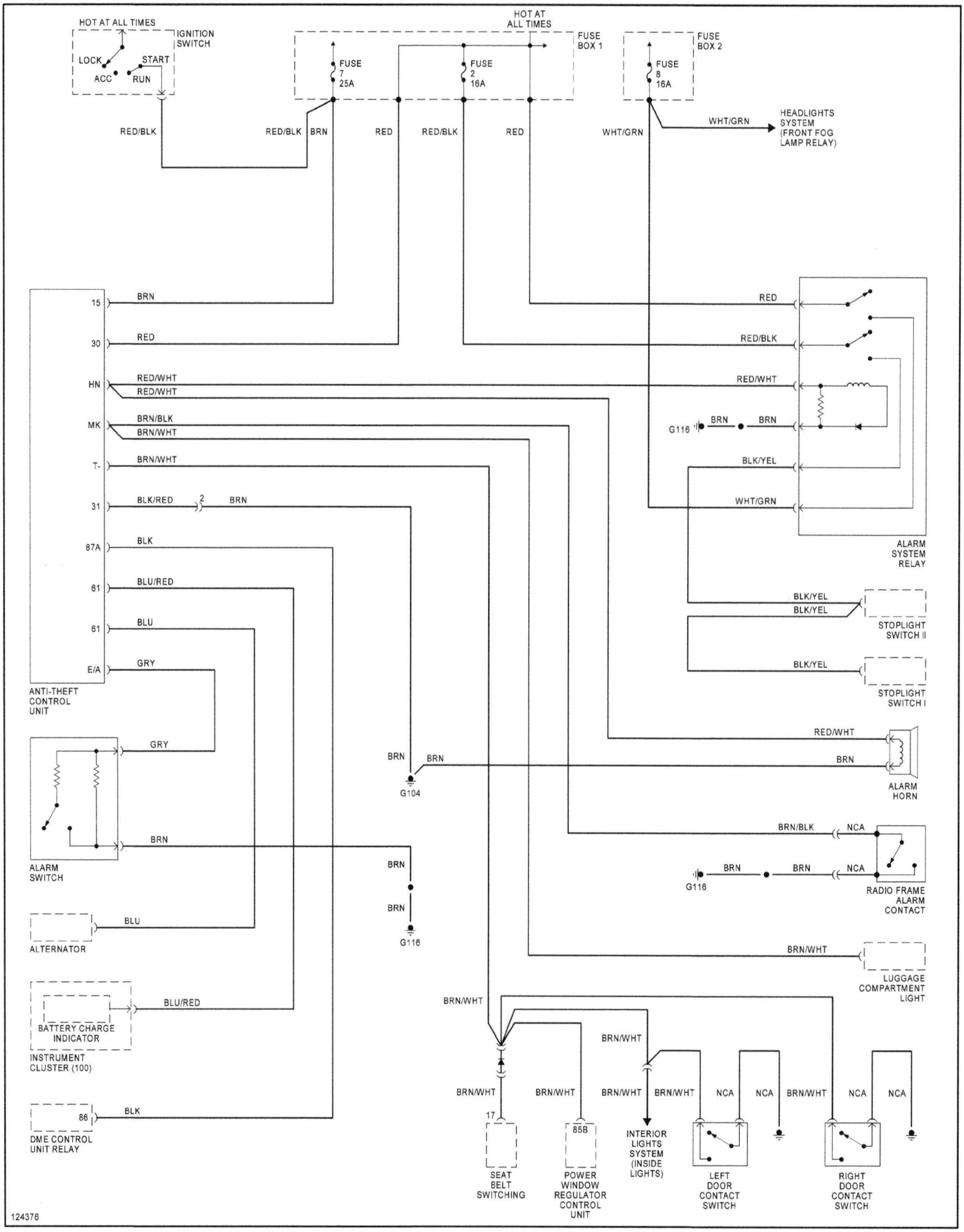

Cruise Control

1984

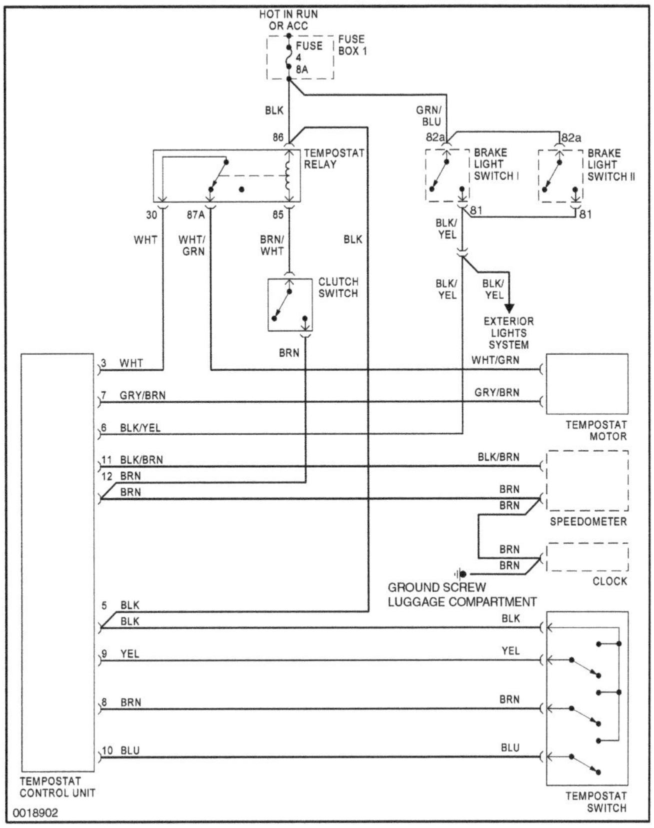

Cruise Control

1985

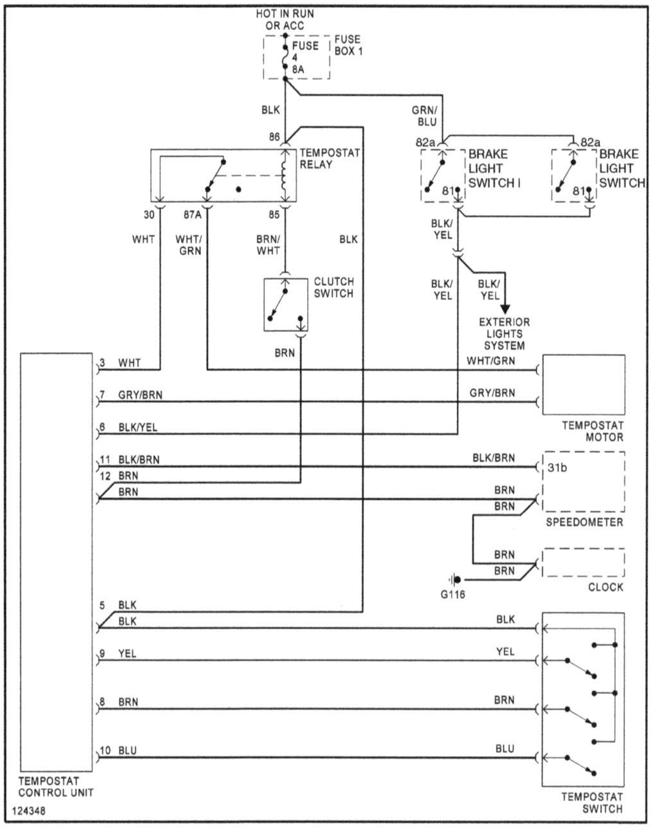

Cruise Control

1986-1987

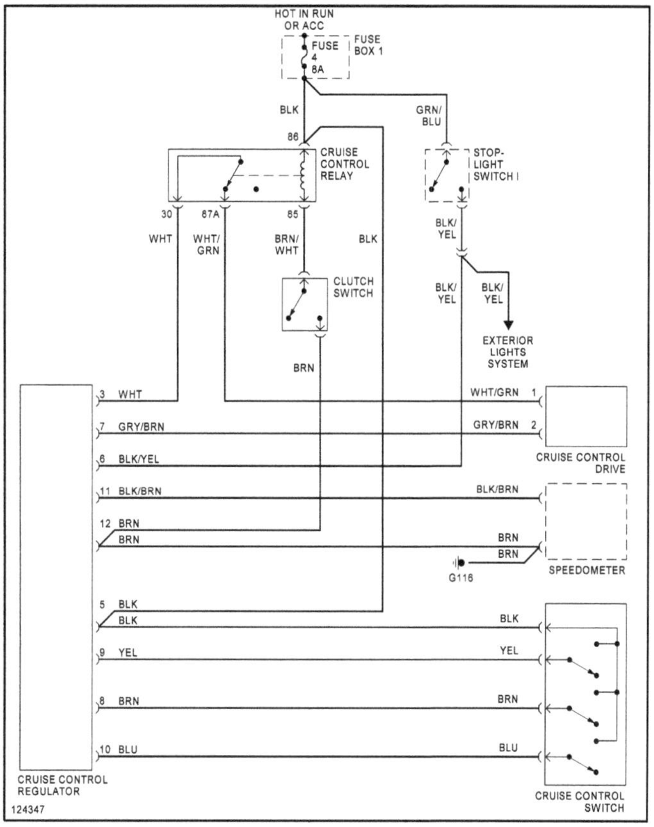

Cruise Control

1988-1989

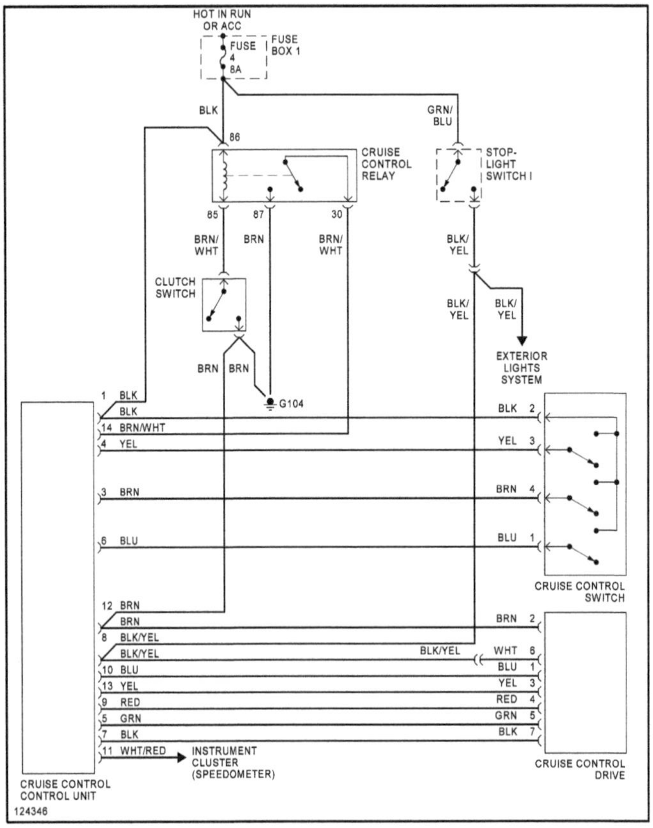

Exterior Lights

1984

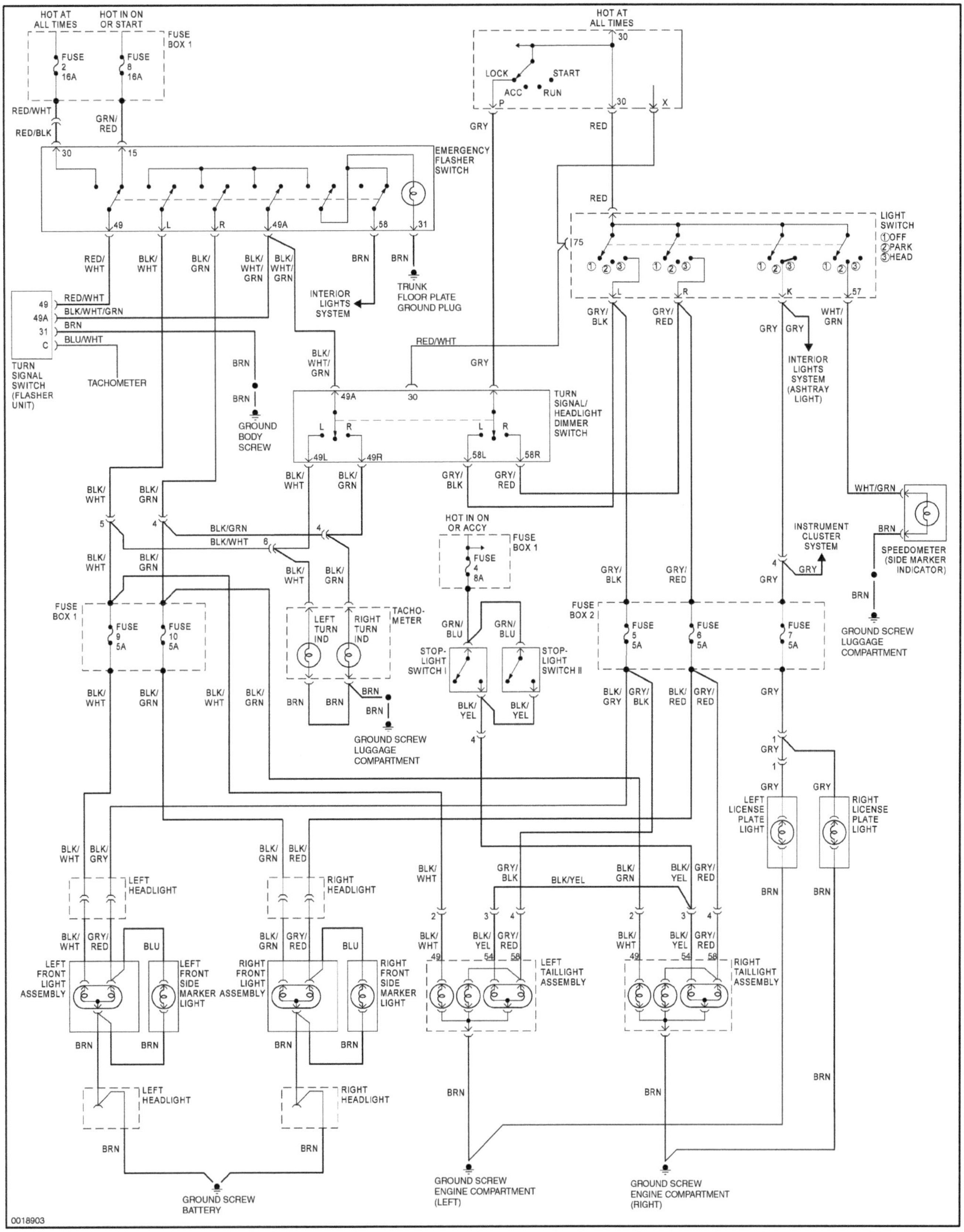

Exterior Lights
1985

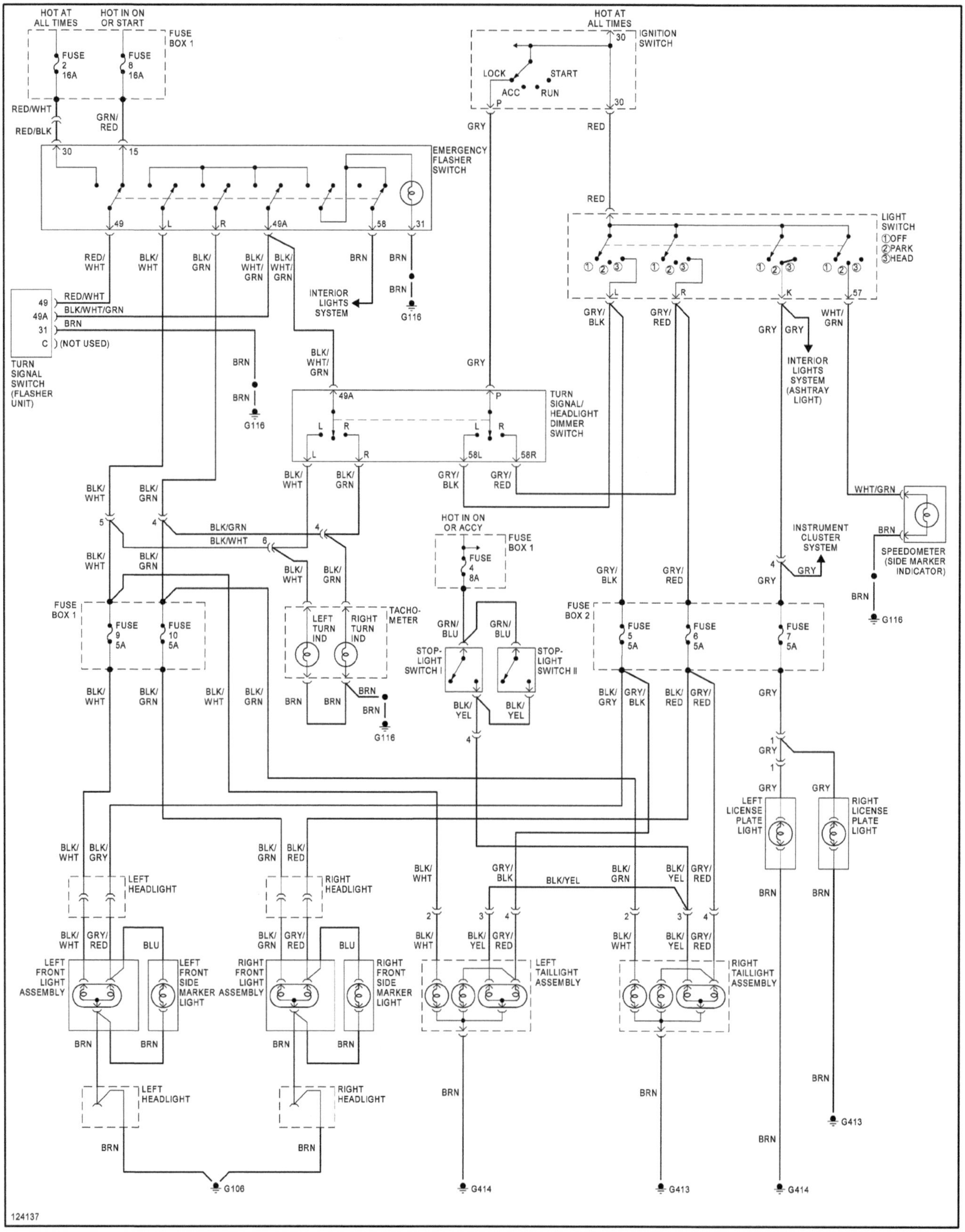

Exterior Lights

1986-1987

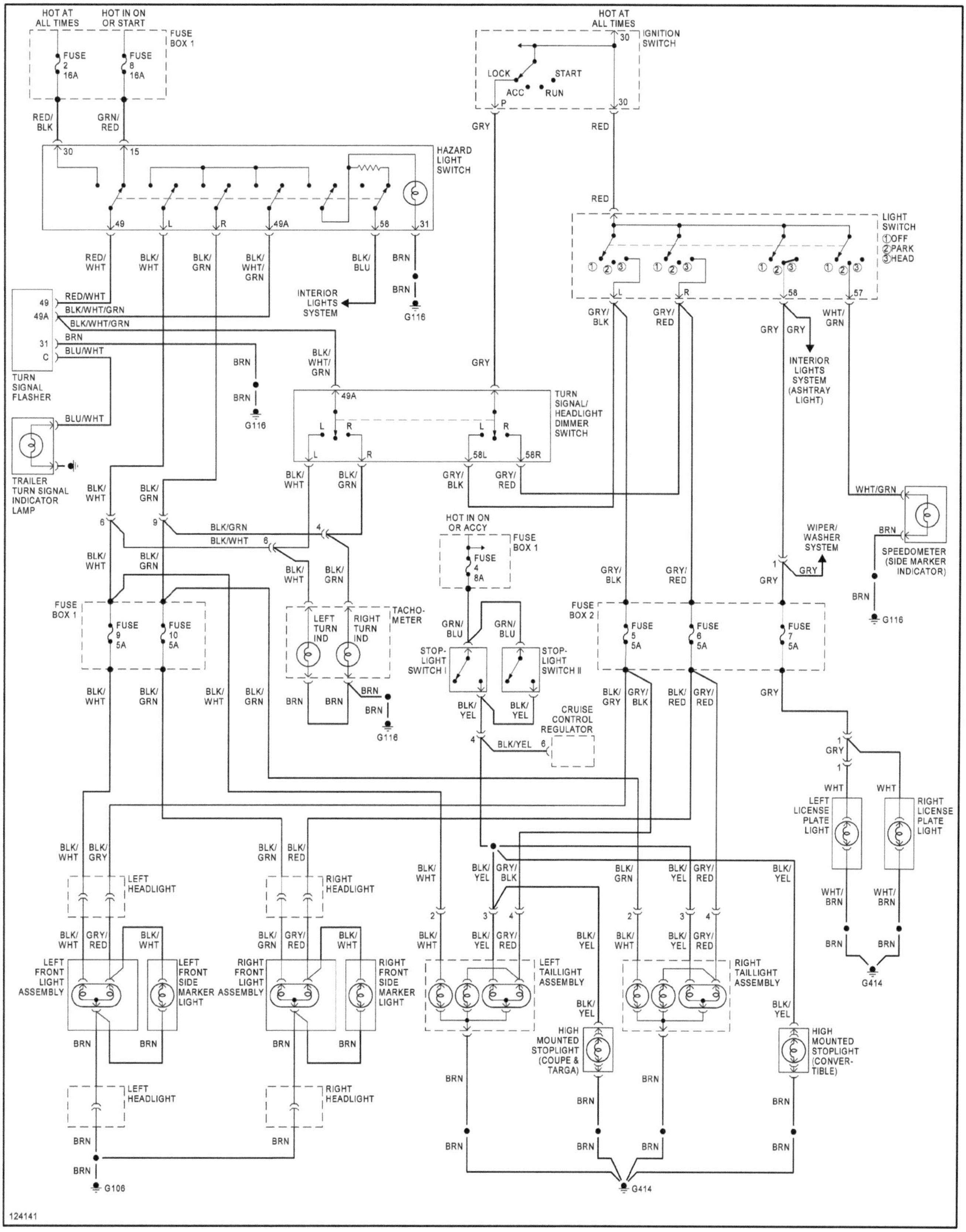

Exterior Lights
1988-1989

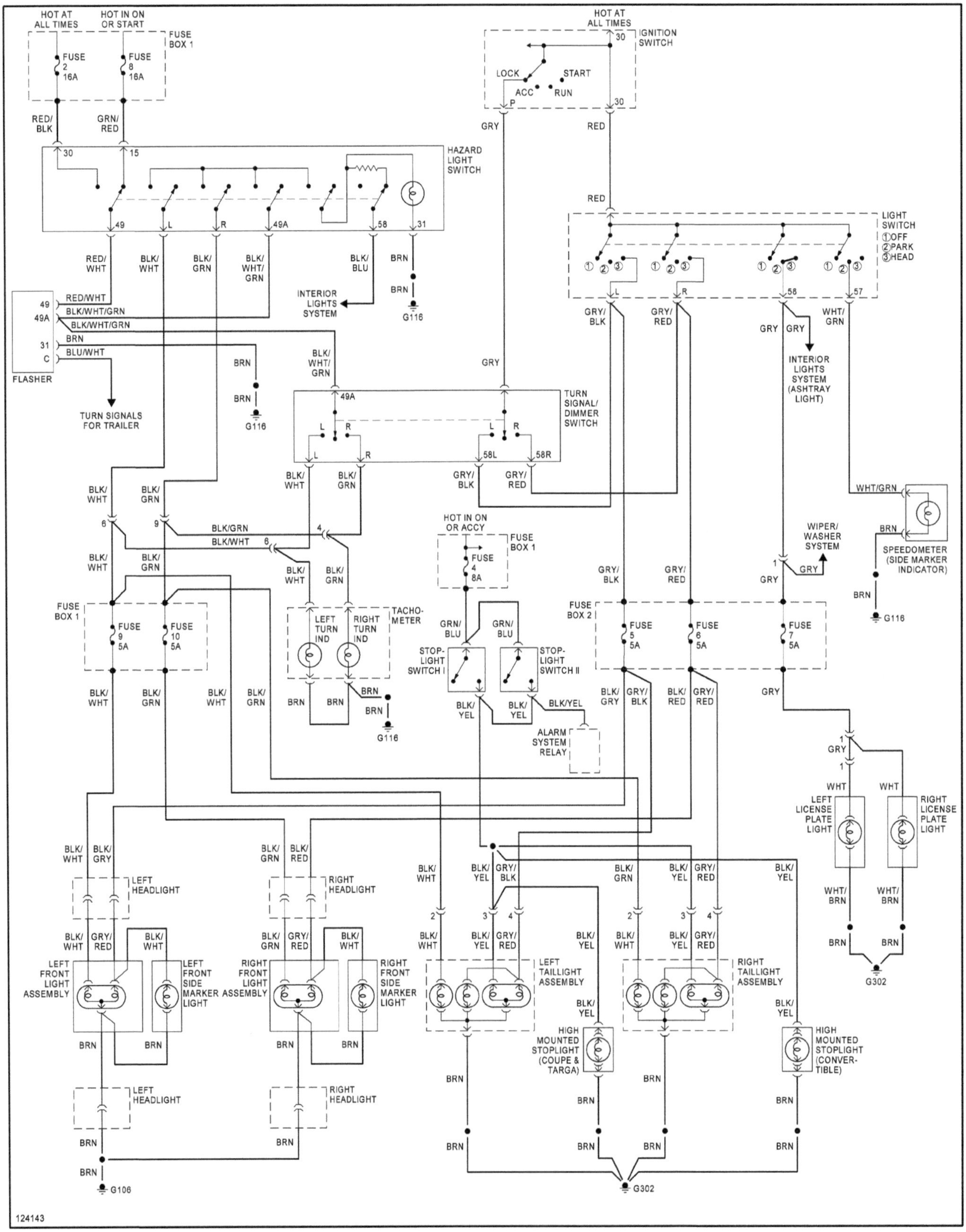

Back-up Lights

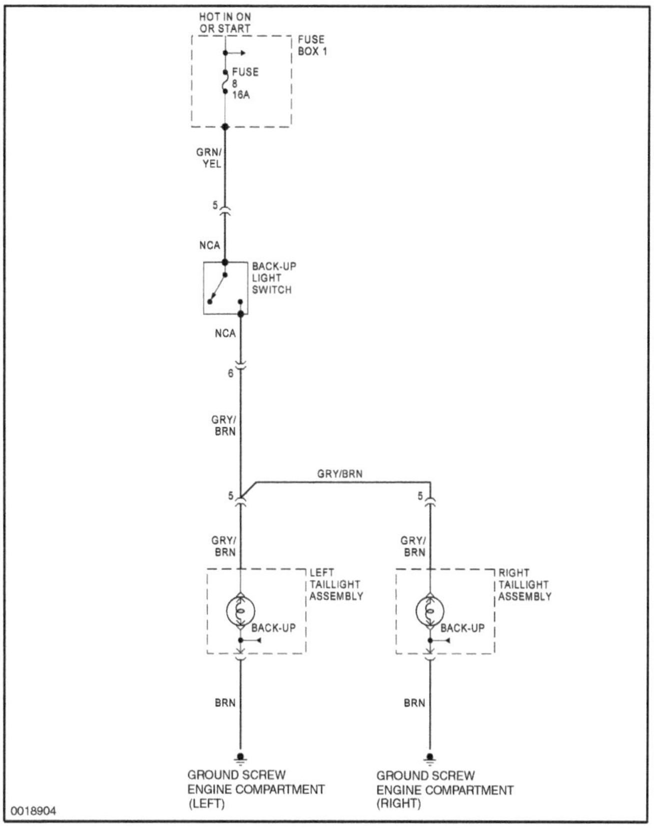

1984 Back-up Lights

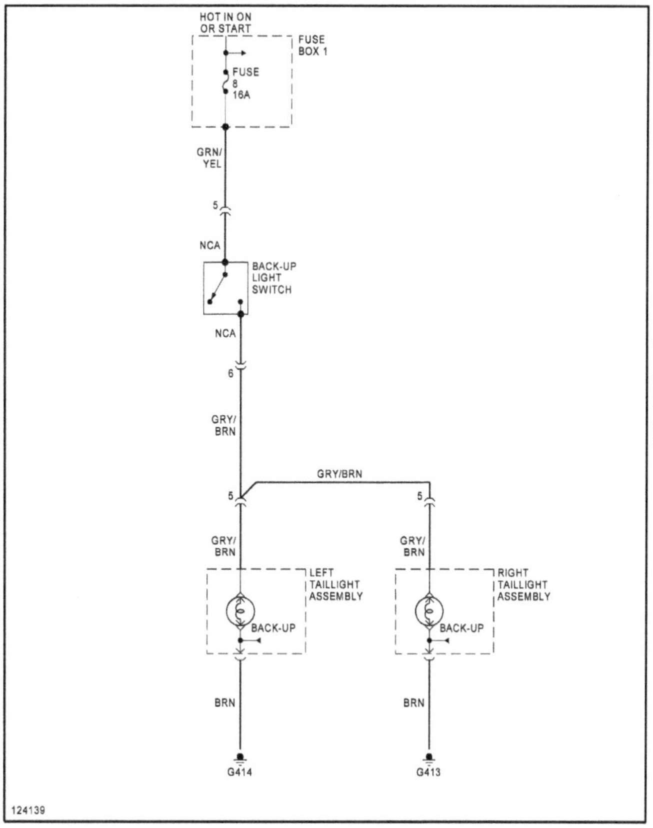

1985 Back-up Lights

Back-up Lights

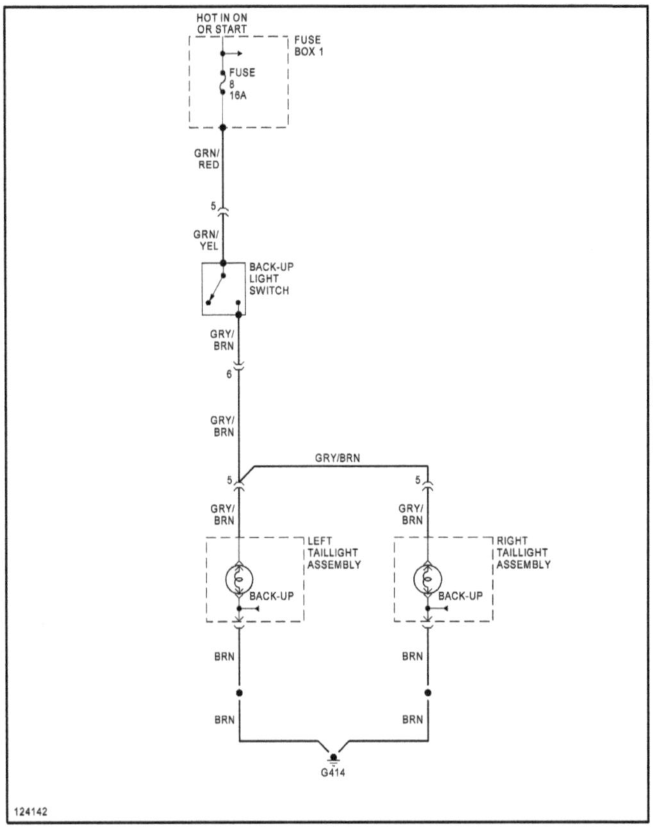

1986-1987 Back-up Lights

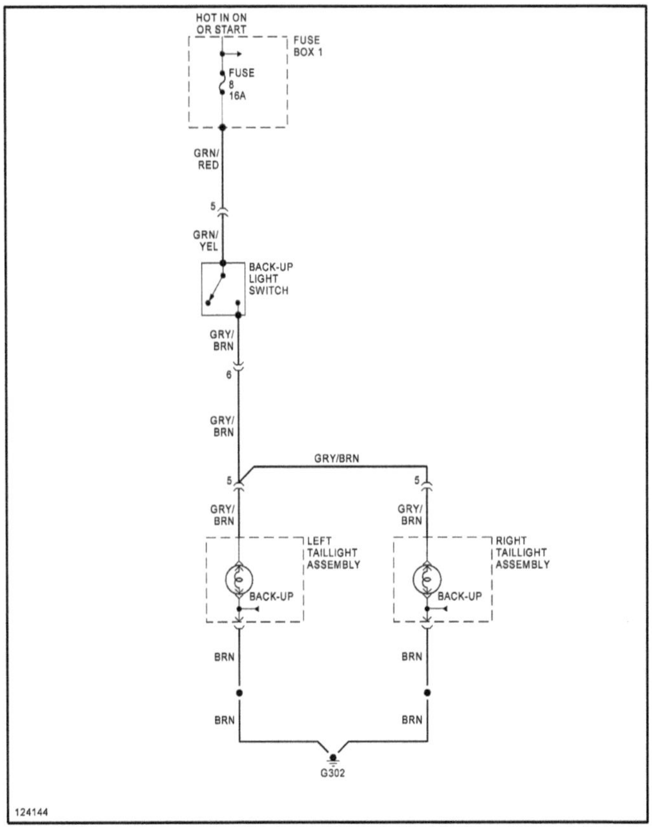

1988-1989 Back-up Lights

Fog Lights

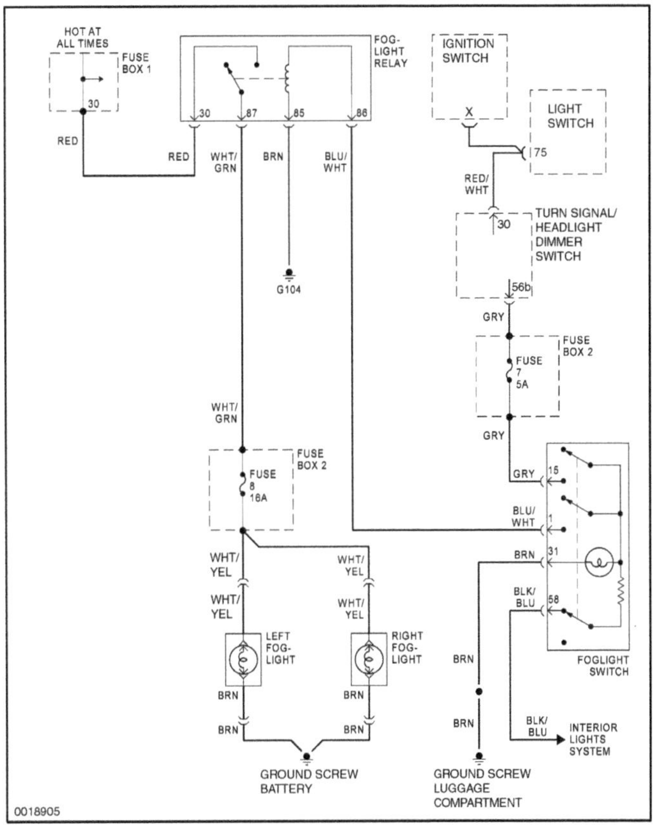

1984 Fog Lights

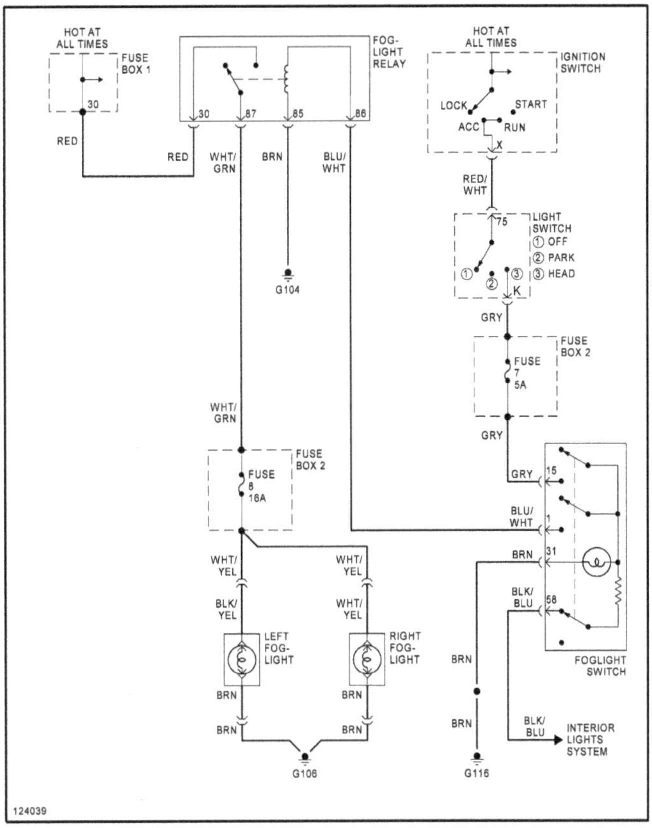

1985 Fog Lights

Fog Lights
1986-1989

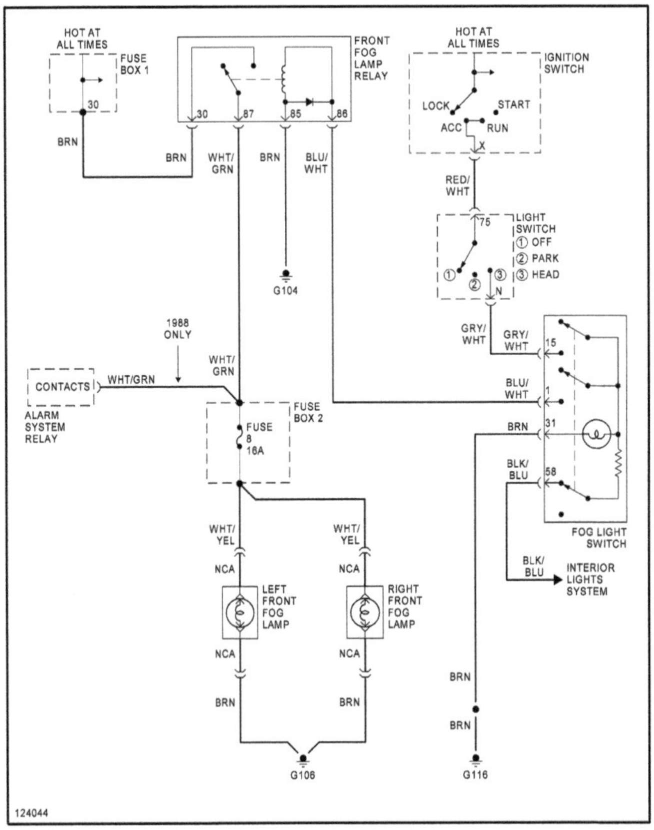

Headlights

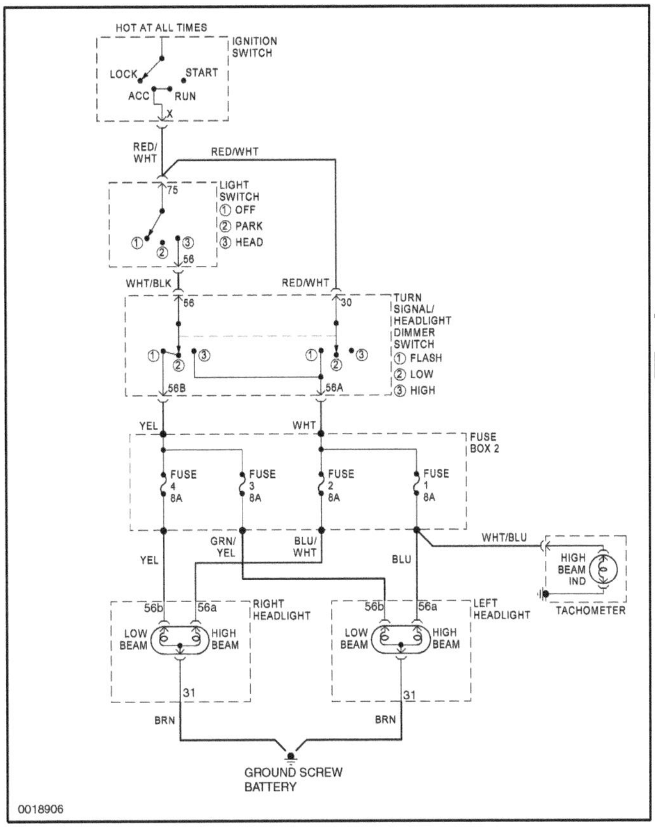

1984 Headlights

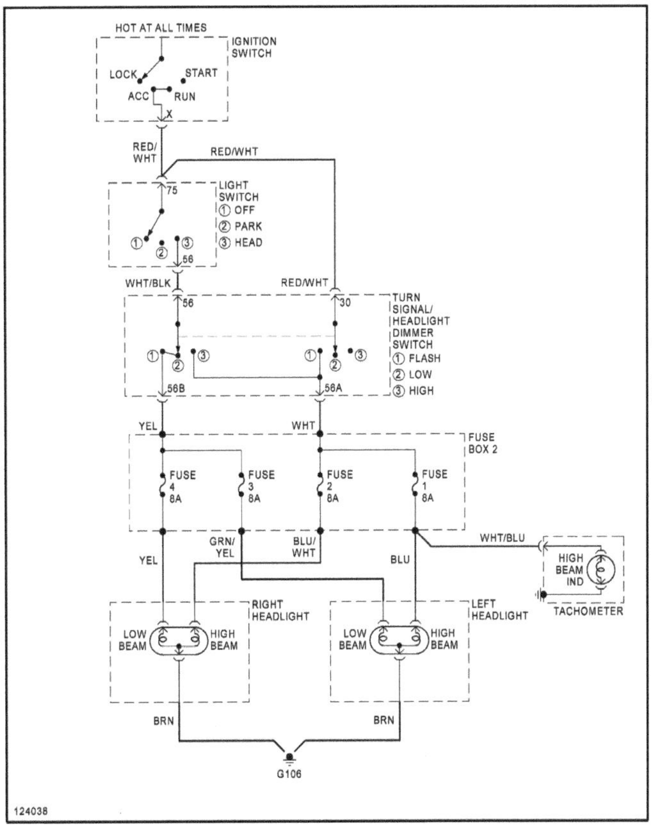

1985 Headlights

Headlights

1986-1989

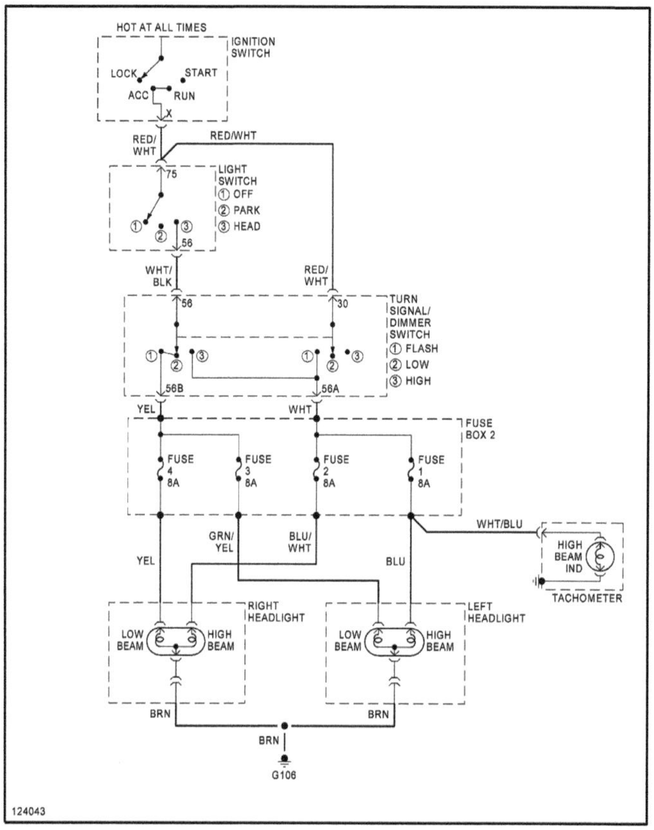

Horns

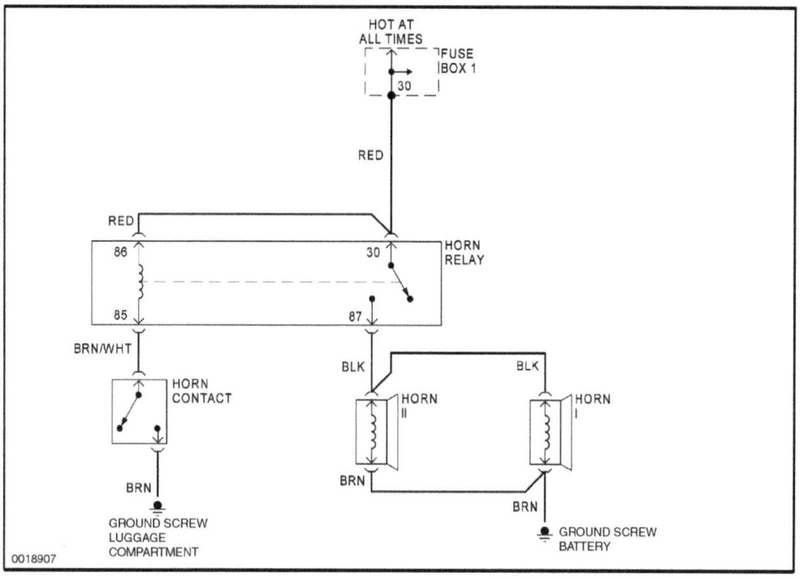

1984 Horns

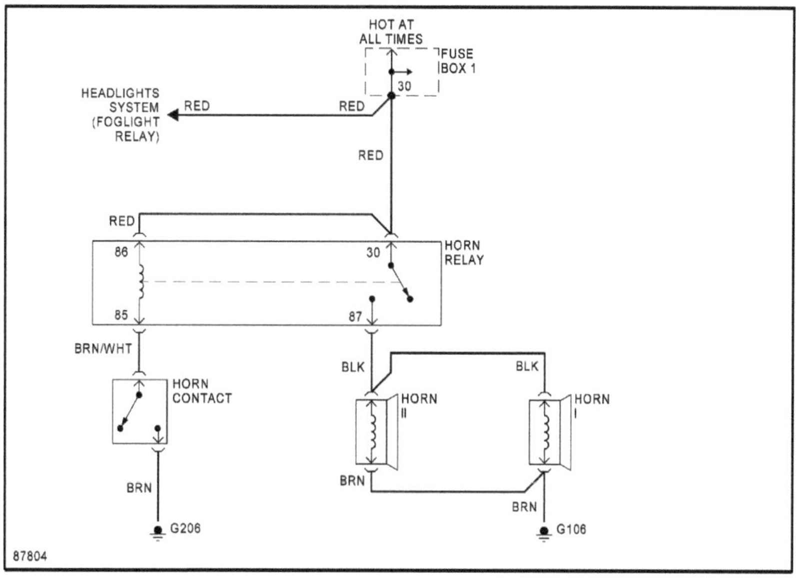

1985 Horns

Horns
1986-1989

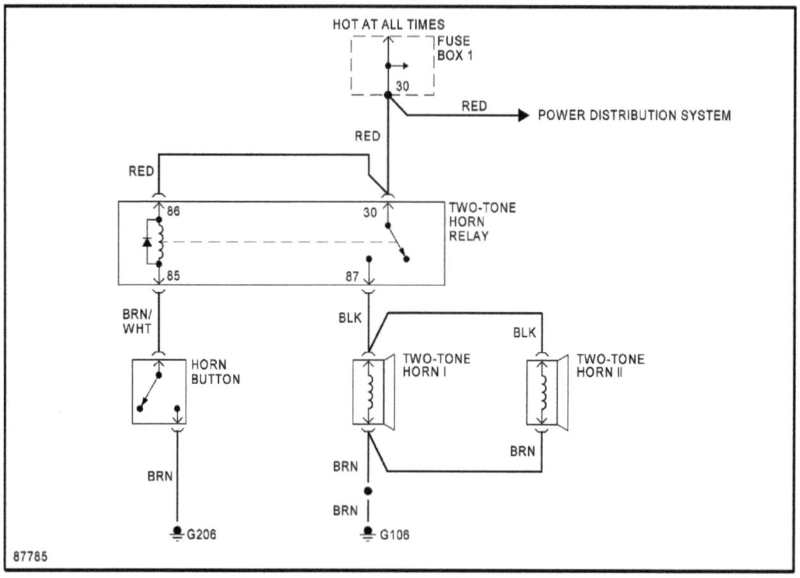

Instrument Panel

1984 (1 of 2)

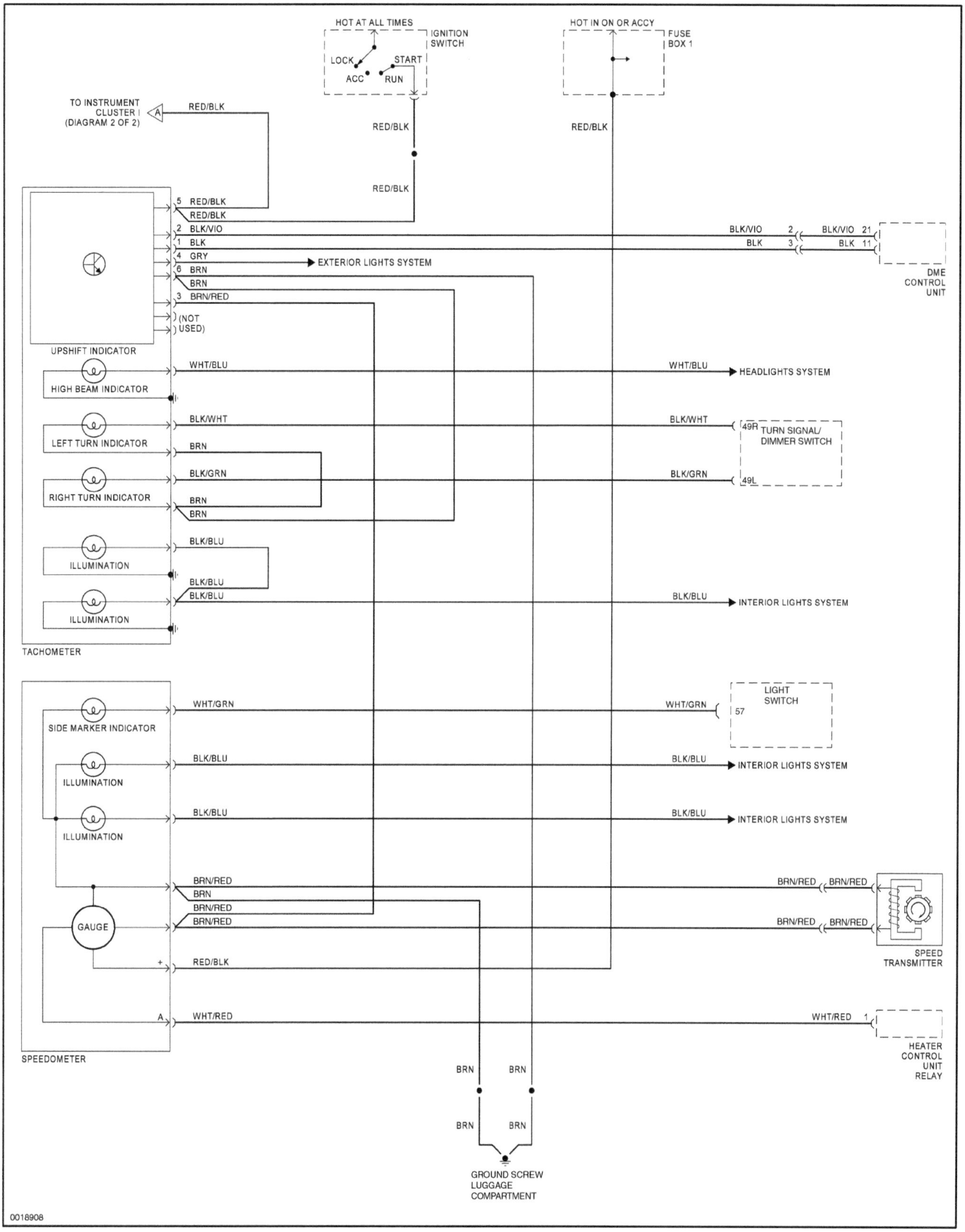

Instrument Panel

1984 (2 of 2)

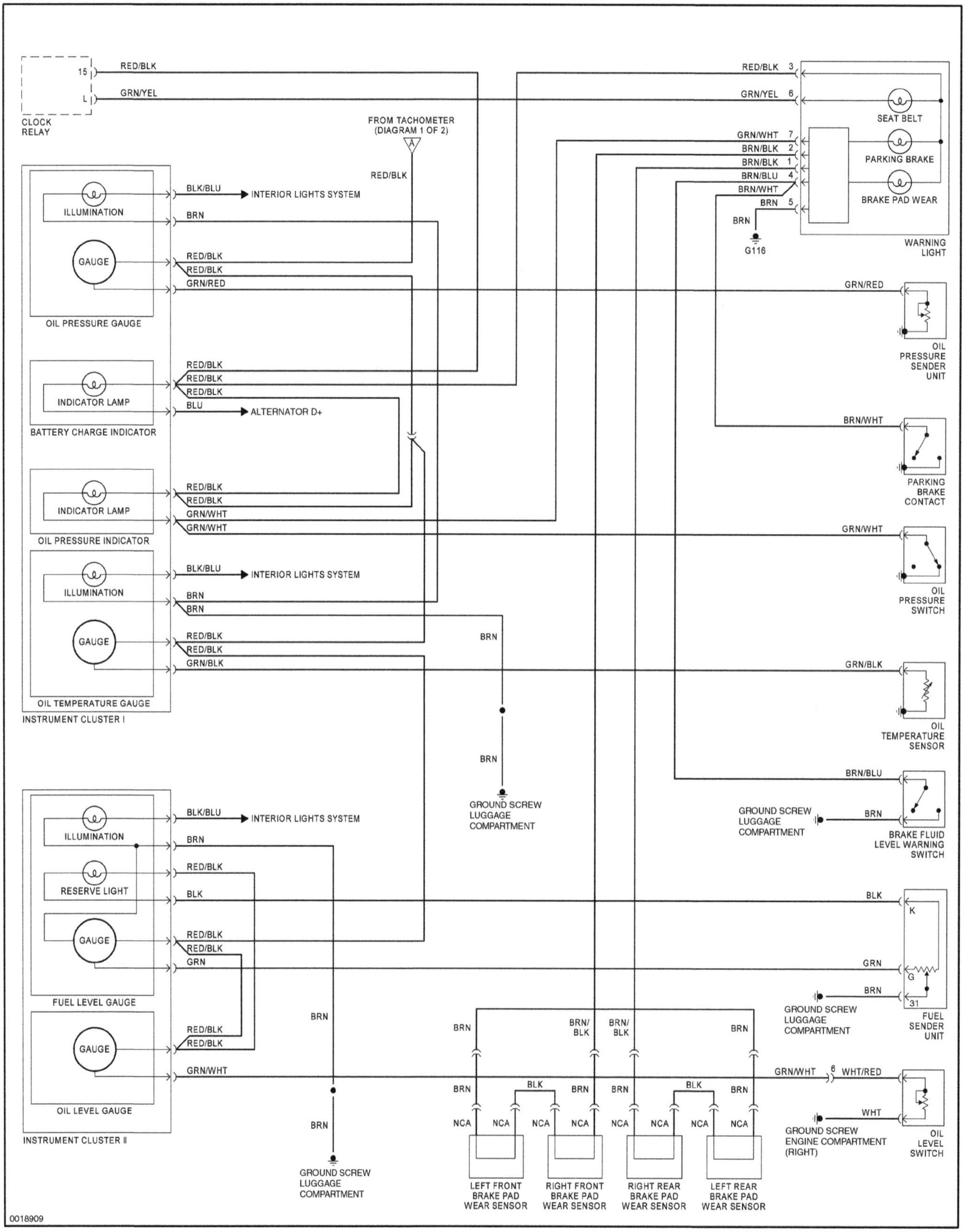

Instrument Panel

1985 (1 of 2)

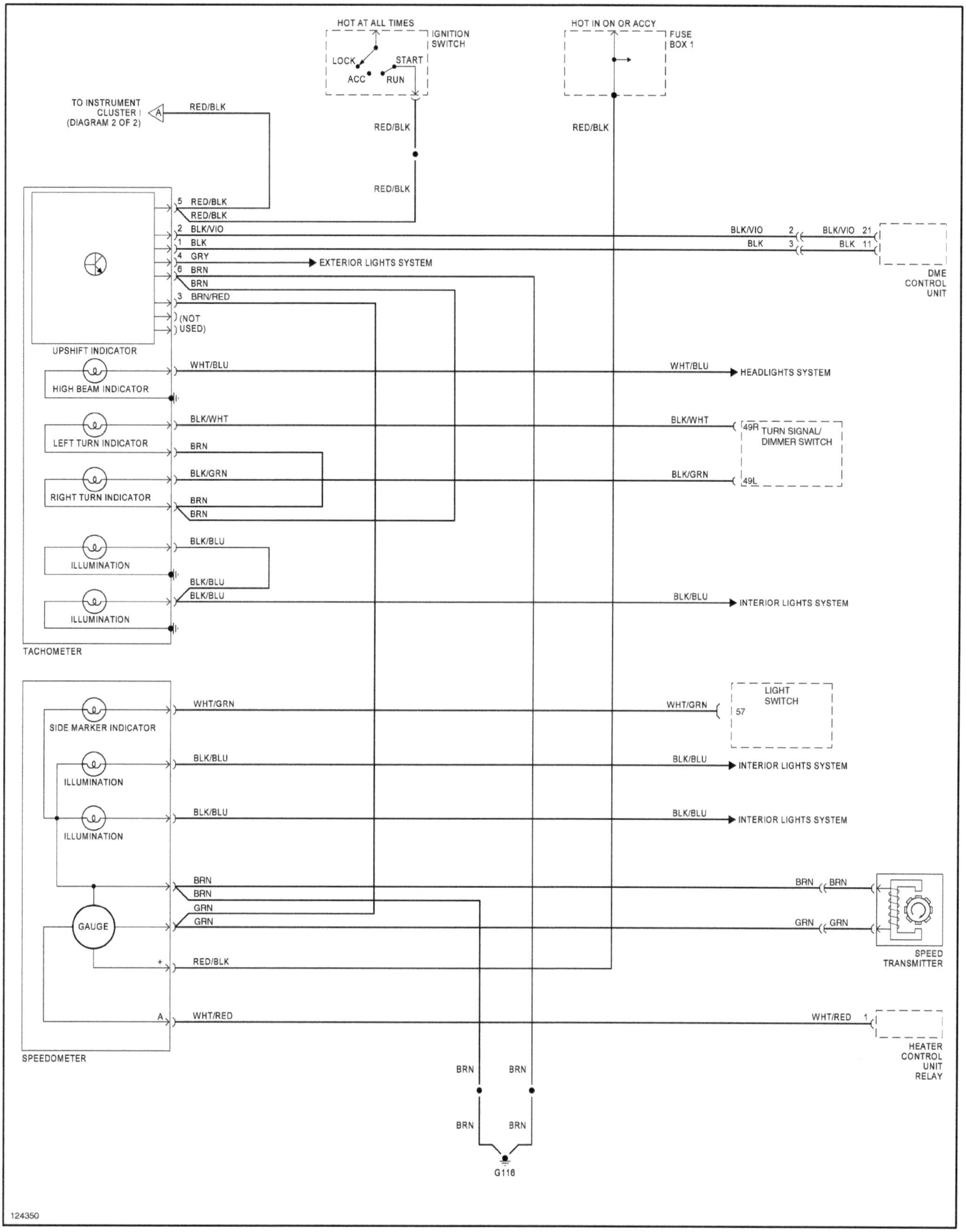

Instrument Panel

1985 (2 of 2)

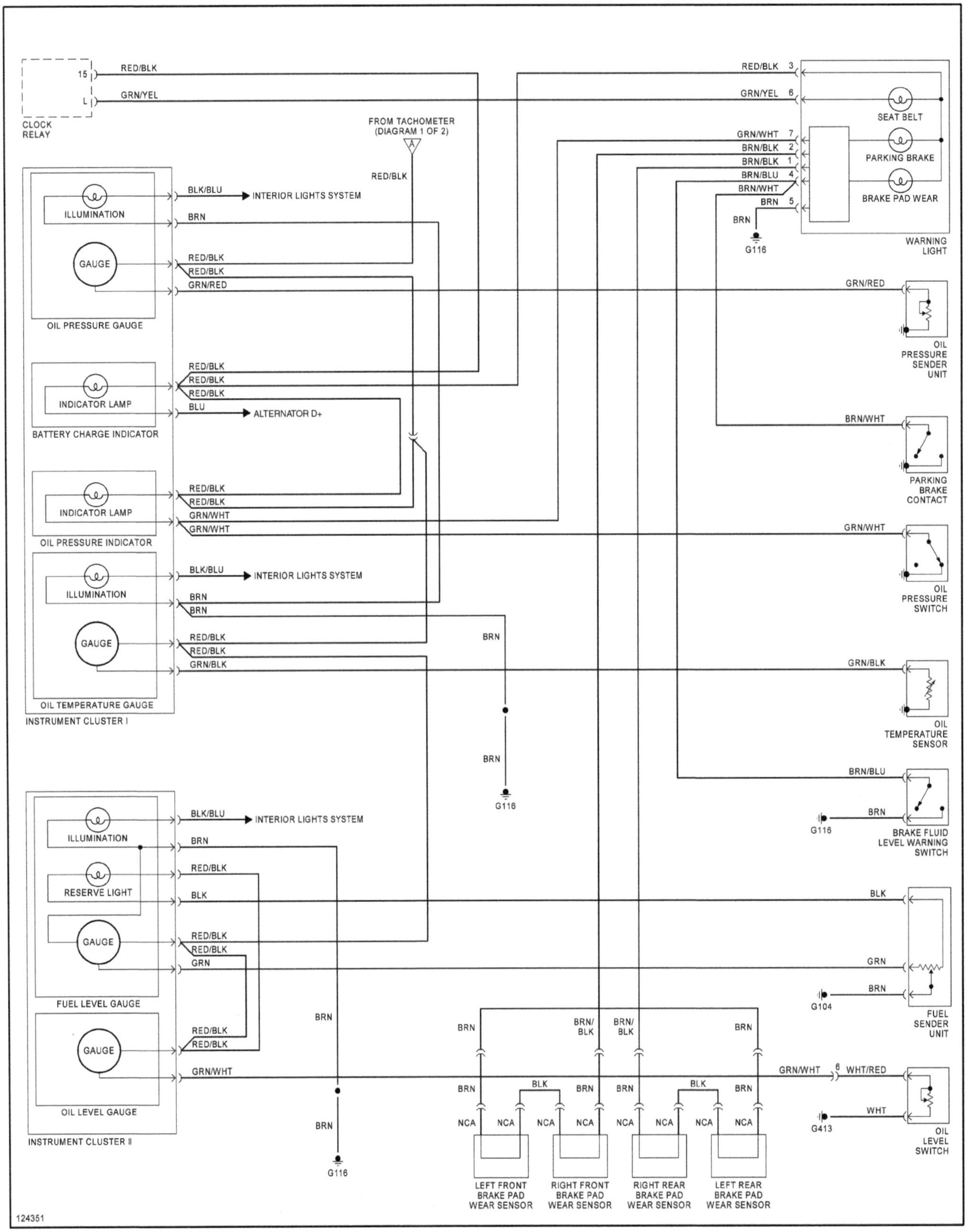

Instrument Panel
1986-1987 (1 of 2)

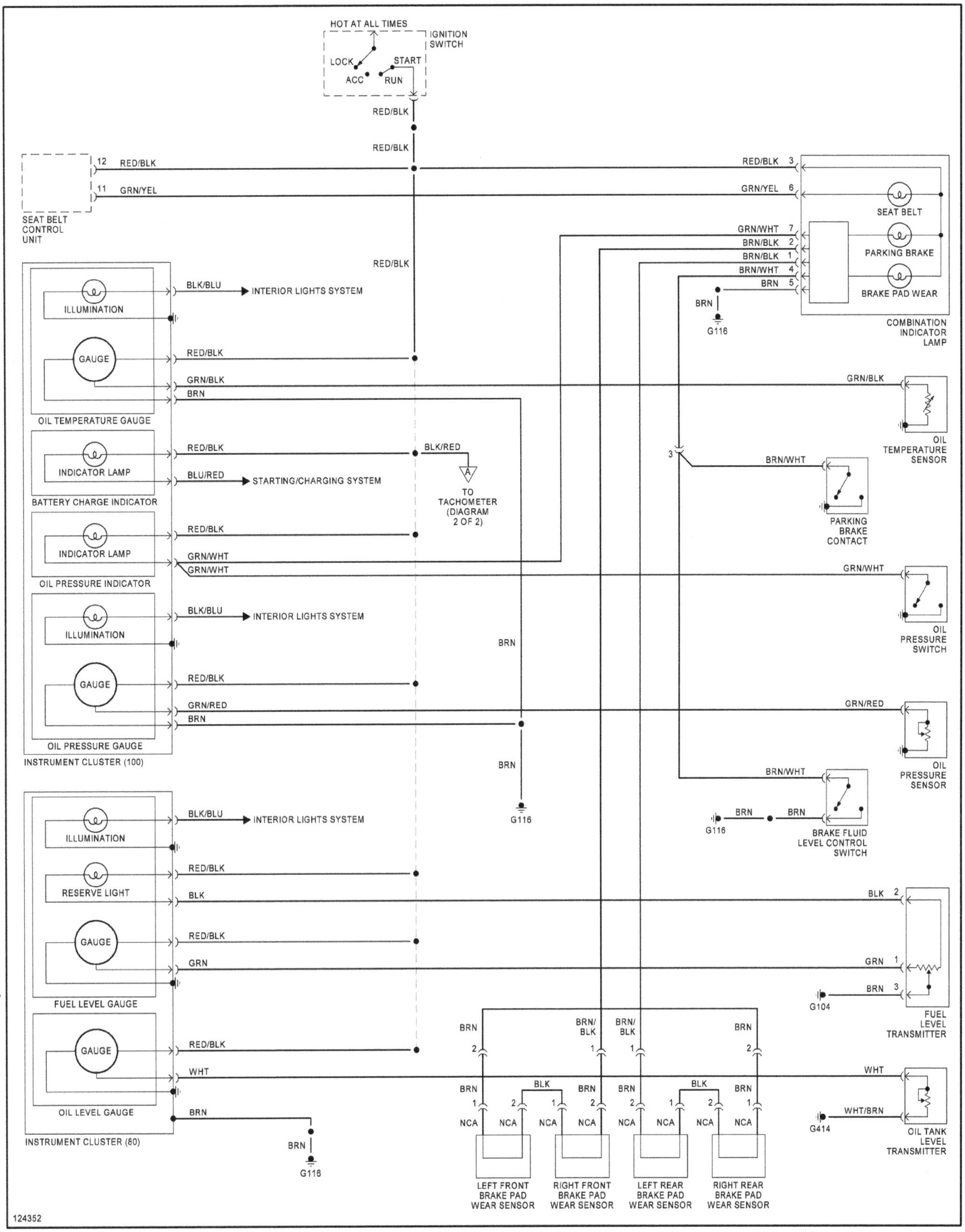

Instrument Panel

1986-1987 (2 of 2)

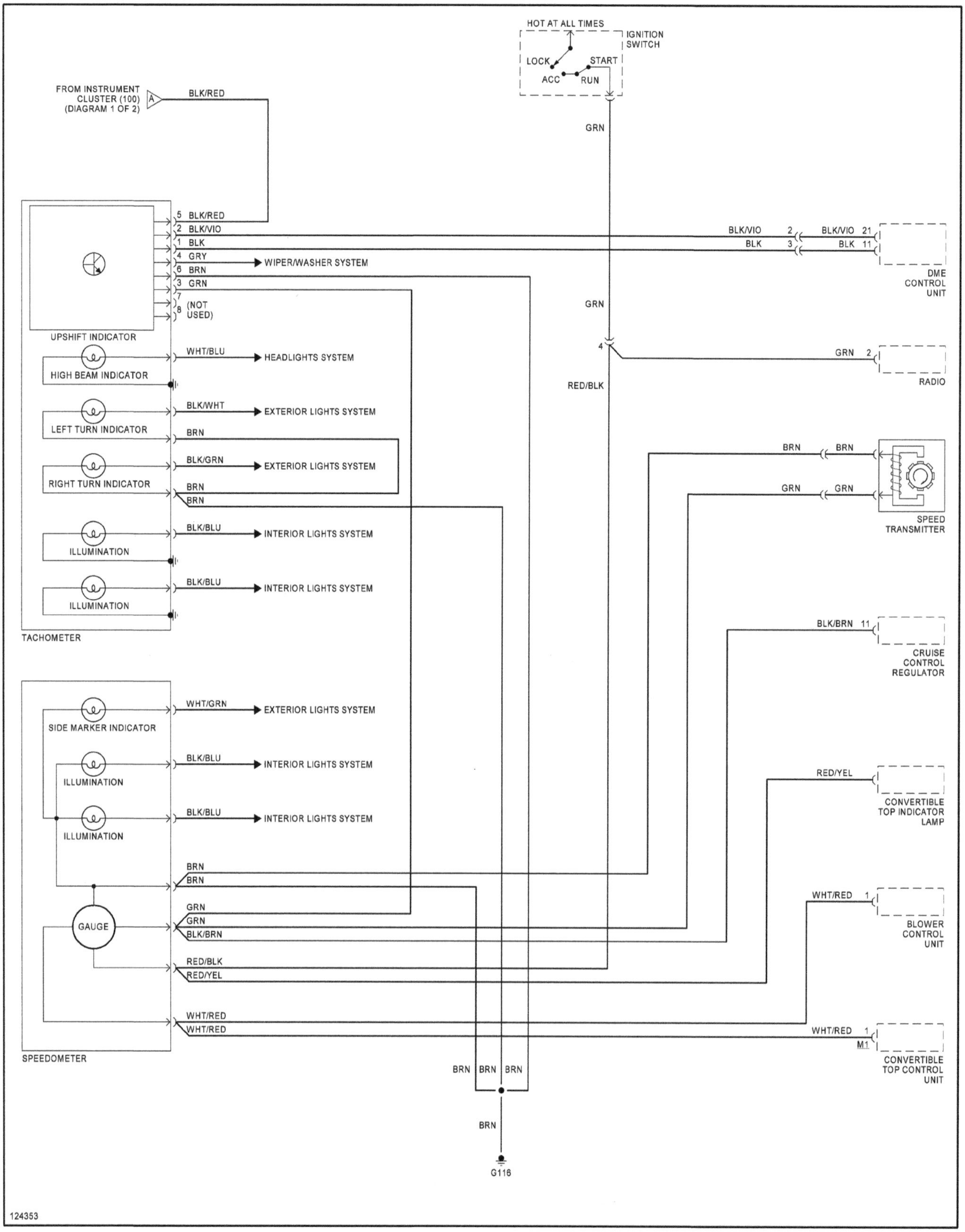

Instrument Panel

1988-1989 (1 of 2)

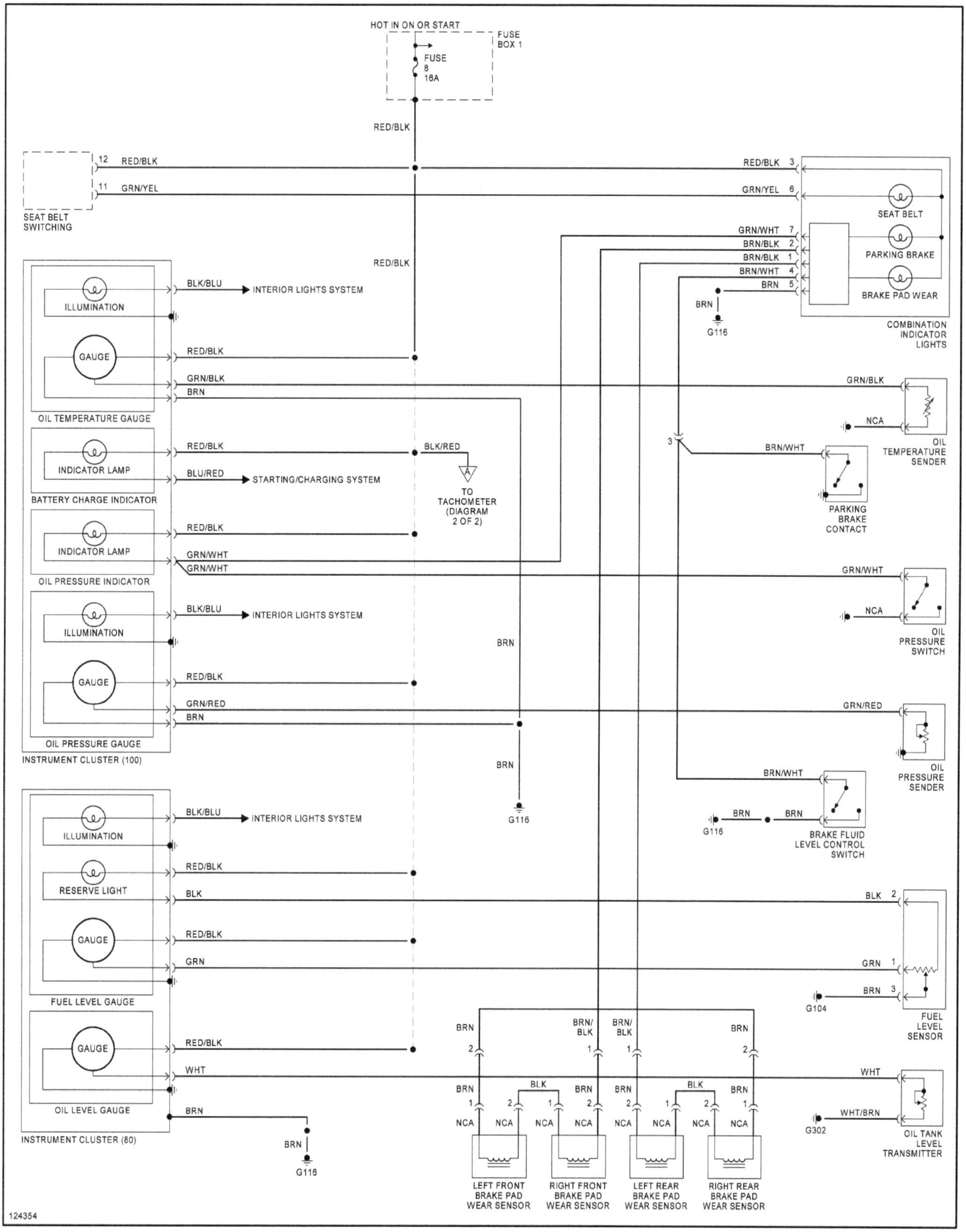

Instrument Panel
1988-1989 (2 of 2)

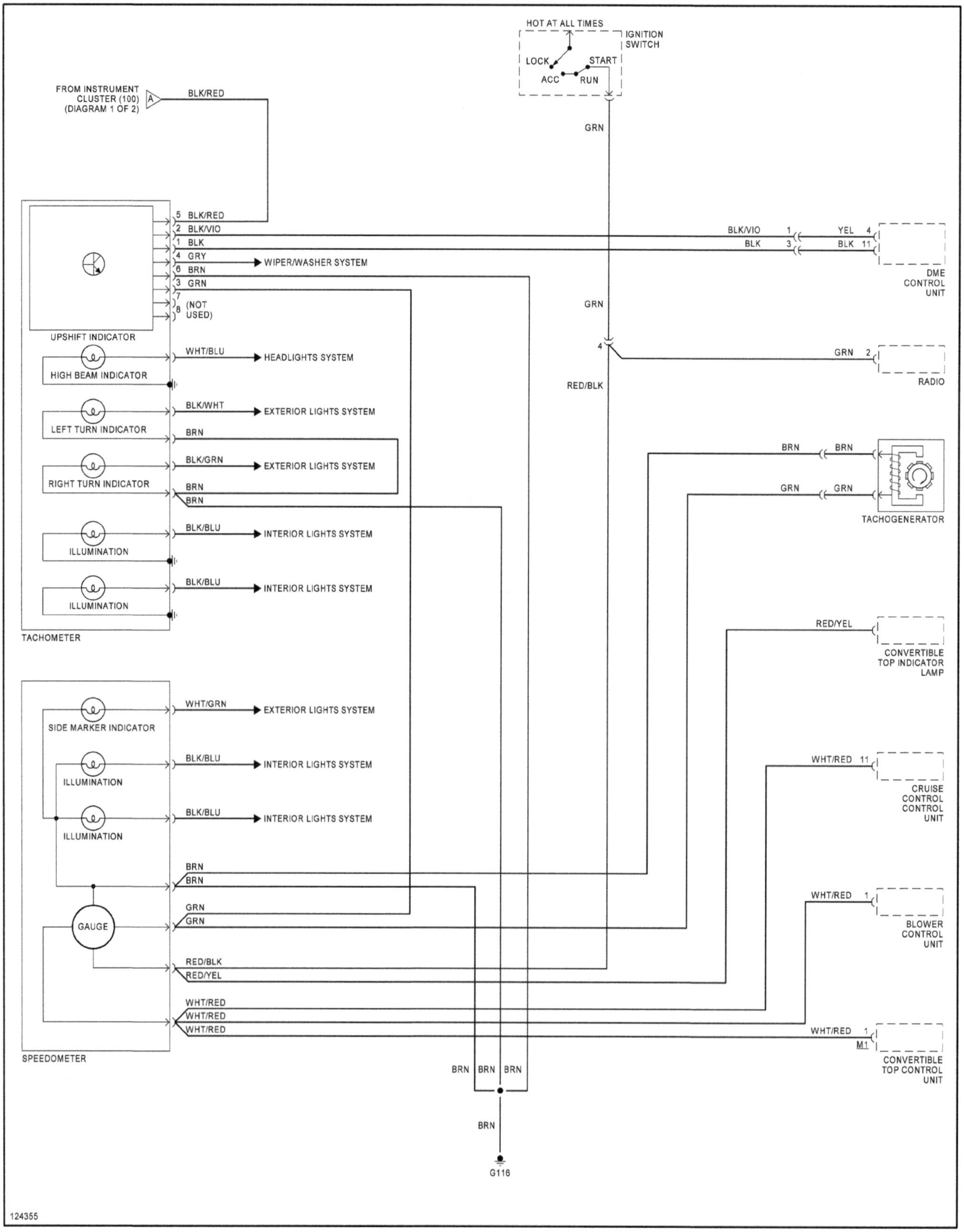

Sound Systems
1986-1989 without Audio Option Pack

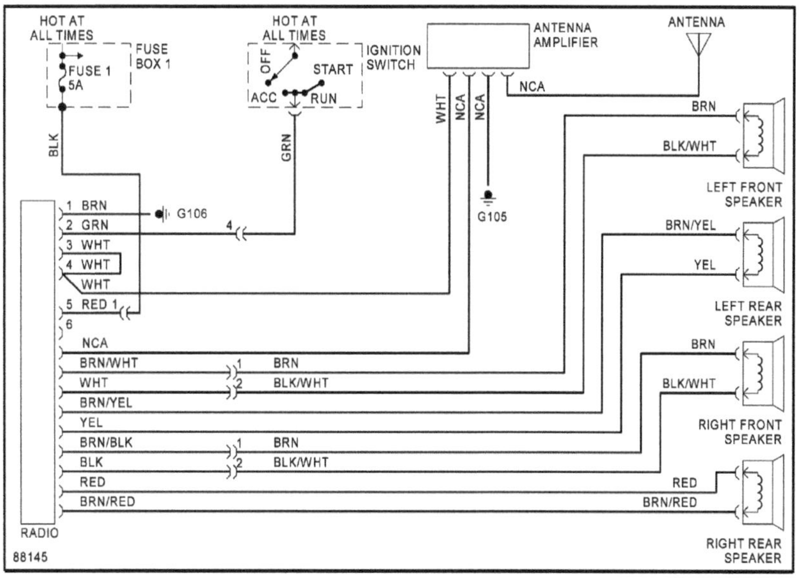

Sound Systems
with Audio Option Pack

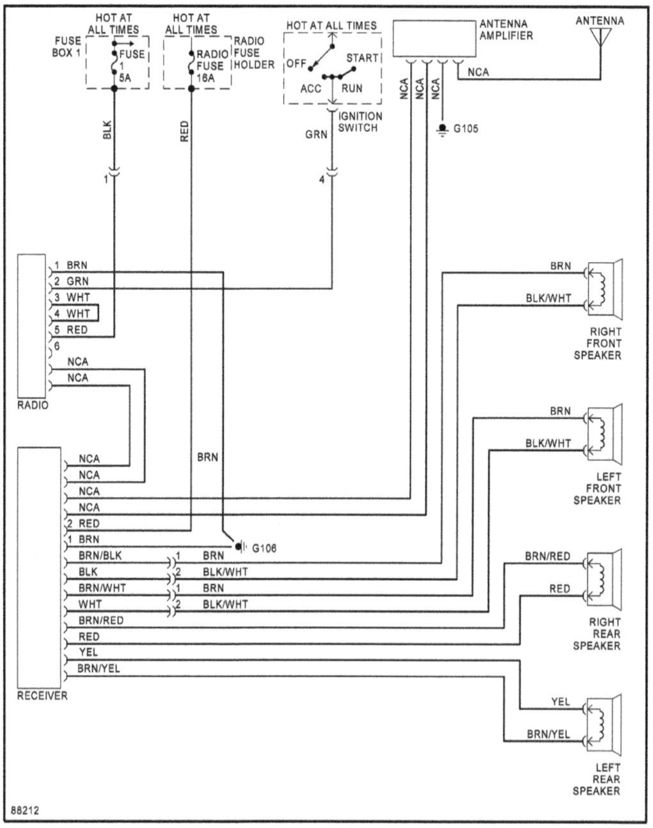

1986-1987 Sound Systems

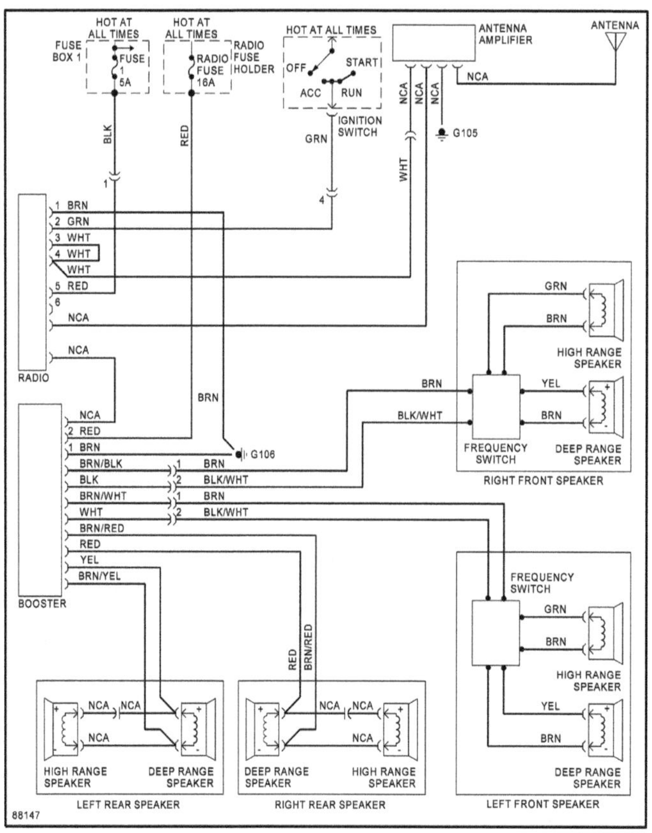

1988-1989 Sound Systems

Warning System

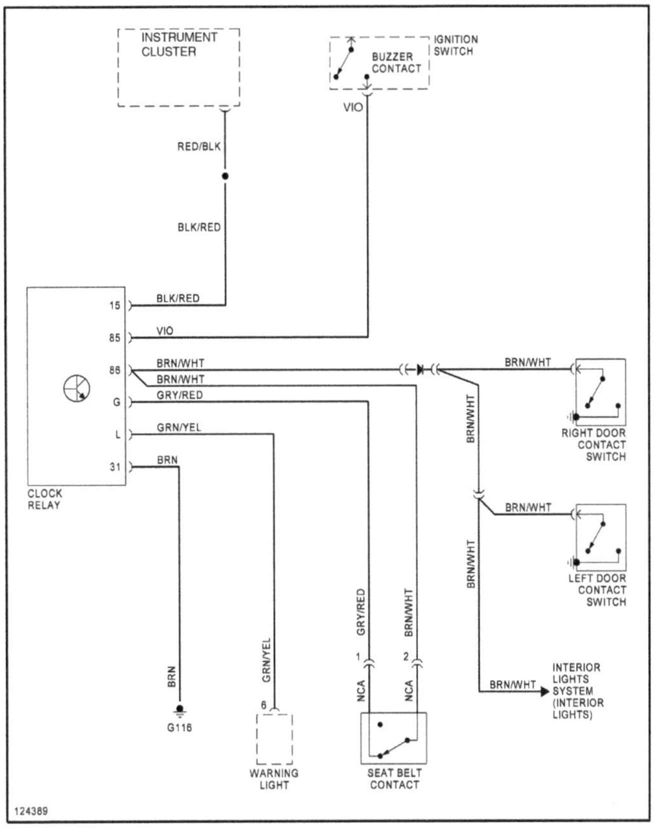

1984-1985

Warning System

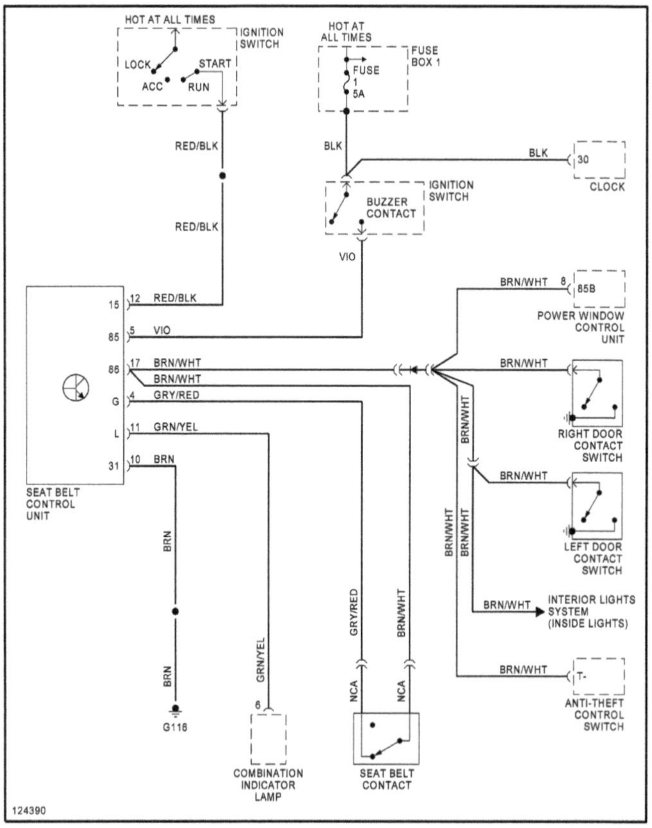

1986-1987 Warning System

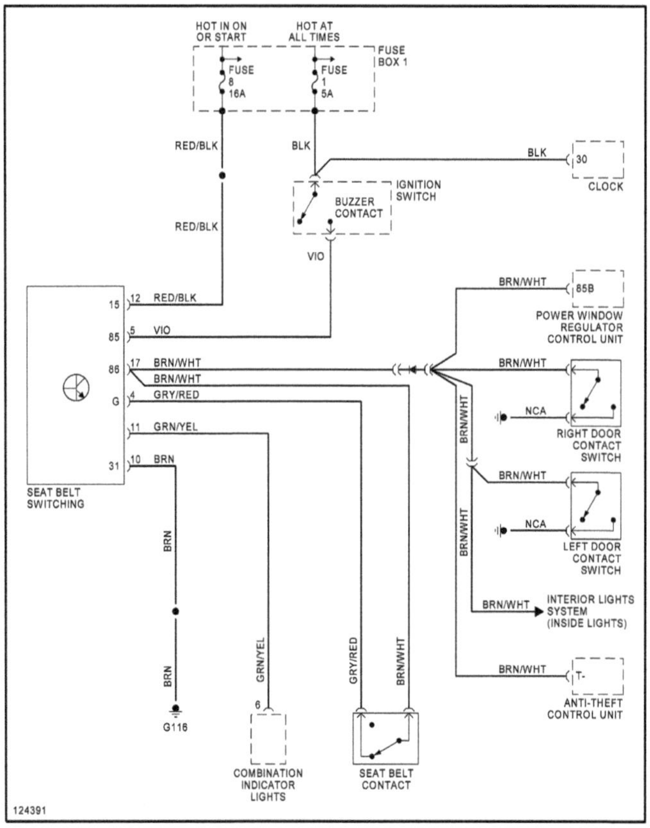

1988-1989 Warning System

Wiper/Washer and Headlight Washer

1984

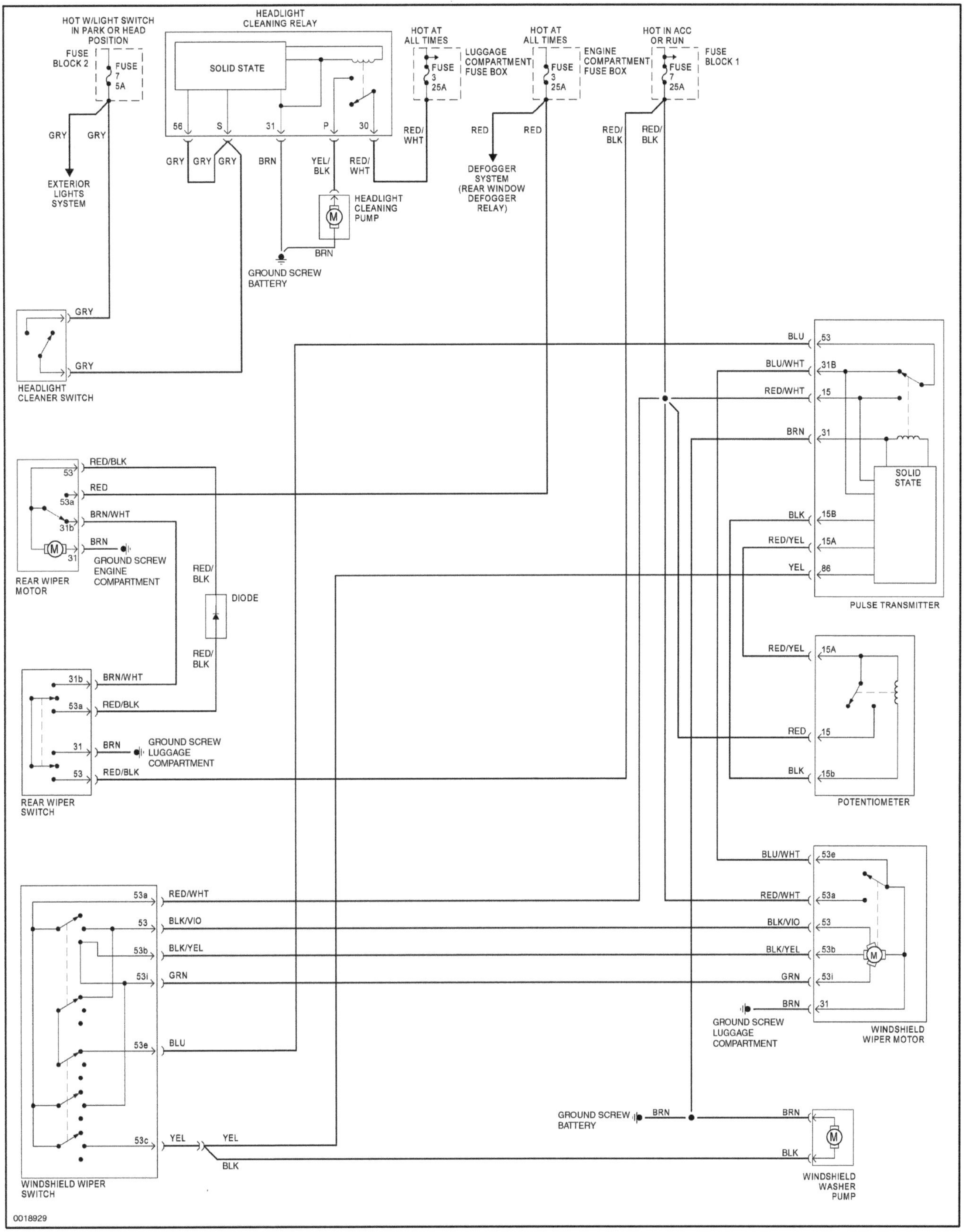

Wiper/Washer and Headlight Washer

1985

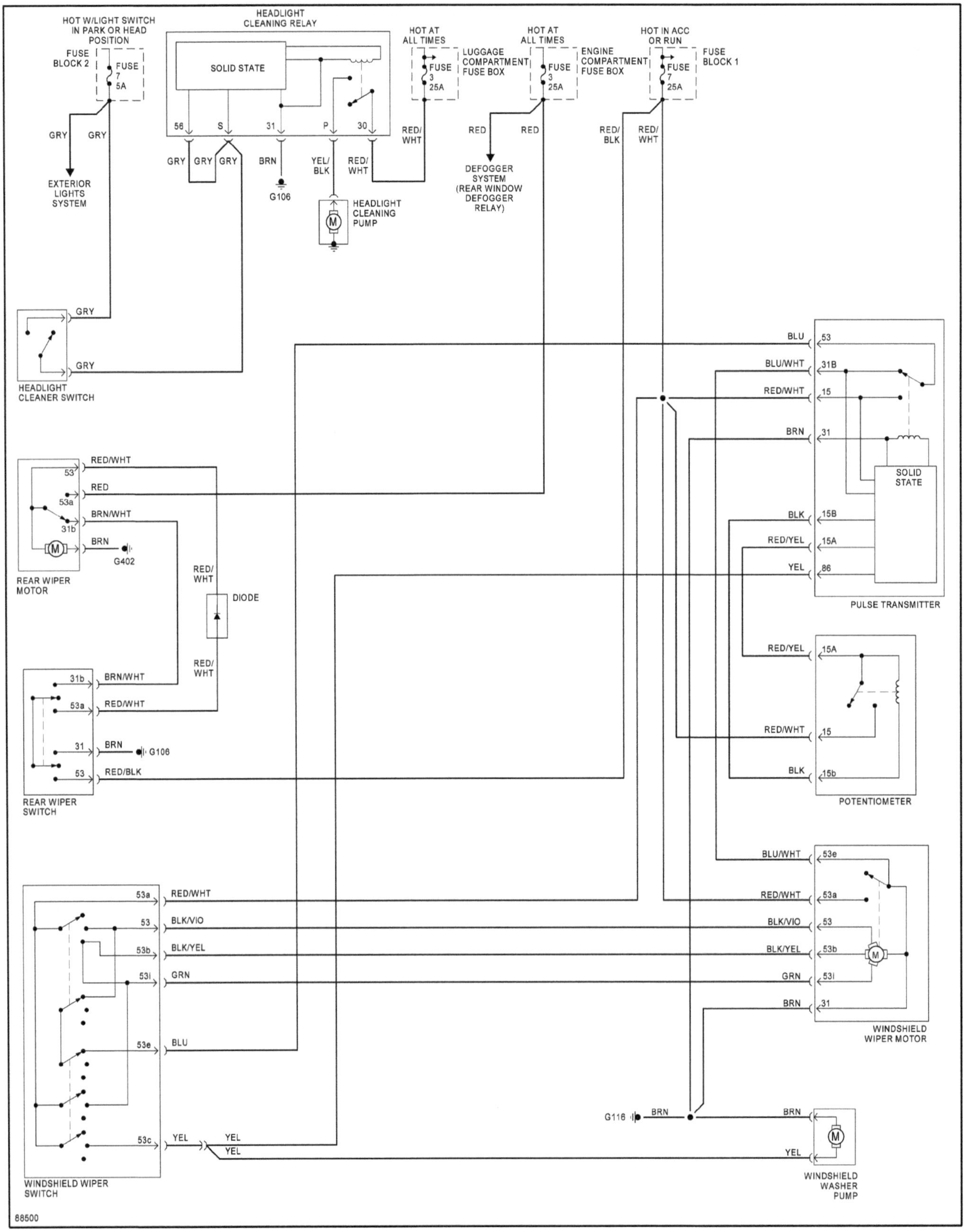

Wiper/Washer and Headlight Washer

1986-1989

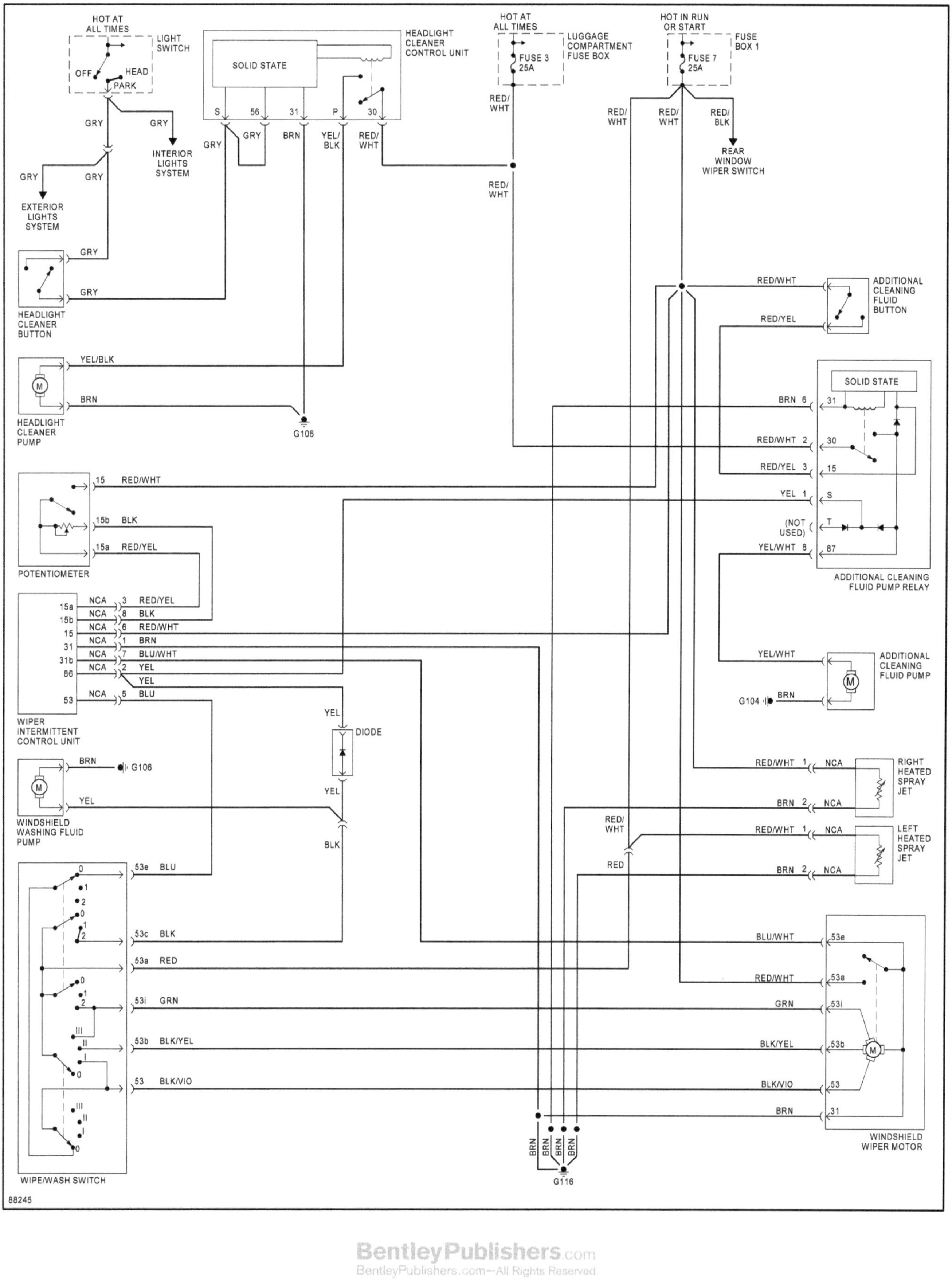

Rear Wiper/Washer
1986-1989

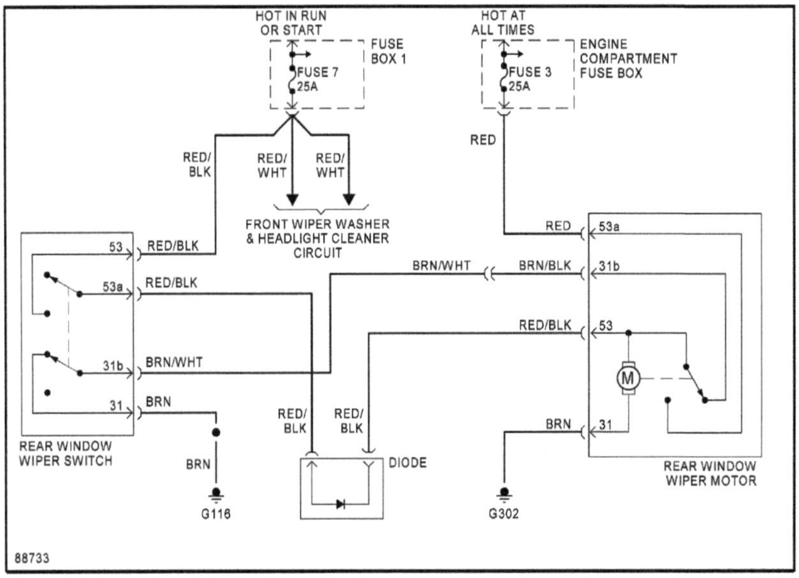

Mirrors

1984-1985

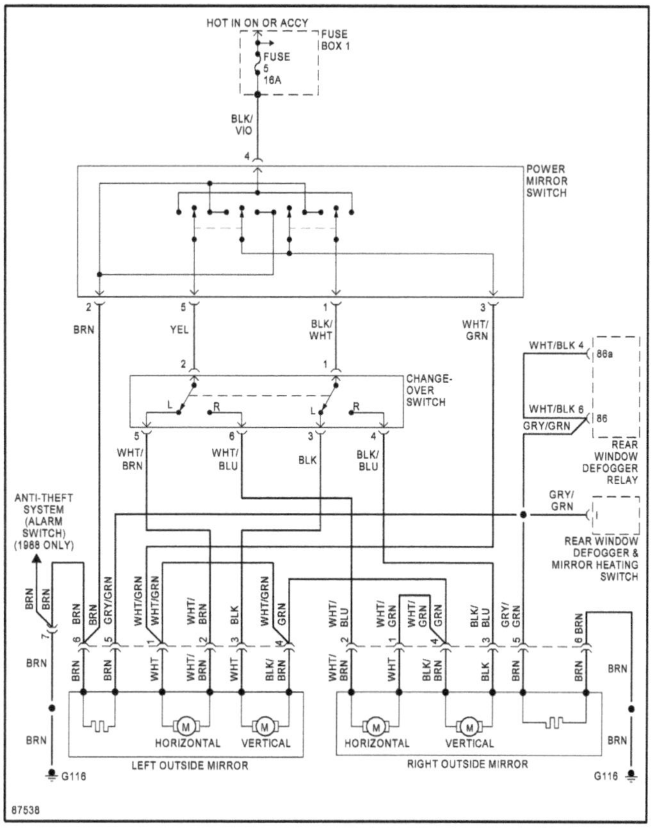

Mirrors

1986-1989

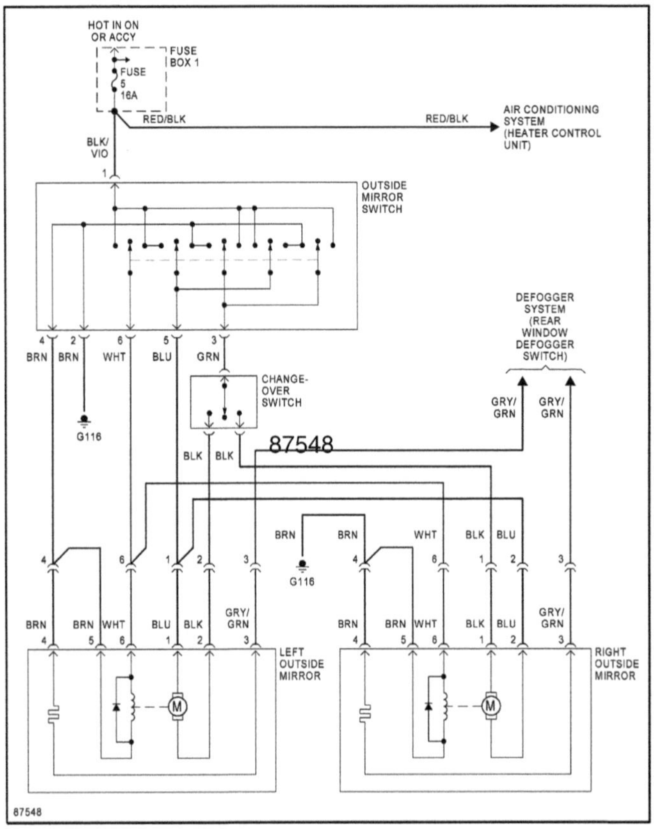

971 Electrical Component Locations

TABLES

GENERAL

This repair group covers fuse, relay, control module and ground location information. Electrical equipment and accessories installed varies depending on model year. To confirm that the proper electrical component has been identified, refer to **970 Electrical Wiring Diagrams**.

CAUTION —
- *Relay and fuse positions are subject to change and may vary from car to car. If questions arise, an authorized Porsche dealer is the best source for the most accurate and up-to-date information.*
- *A good way to verify a relay position is to compare the wiring colors at the relay socket to the colors indicated on the wiring diagrams located at the rear of this manual.*
- *Always switch the ignition off and remove the negative (–) battery cable before removing any electrical components.*
- *Prior to disconnecting the battery, read the battery disconnection cautions given at the front of this manual on page viii.*
- *Connect and disconnect ignition system wires, multiple connectors, and ignition test equipment leads only while the ignition is switched off.*
- *Only use a digital multimeter for electrical tests.*

Test equipment

Electrical and electronic automotive components require the use of low impedance test equipment.

Special Tools

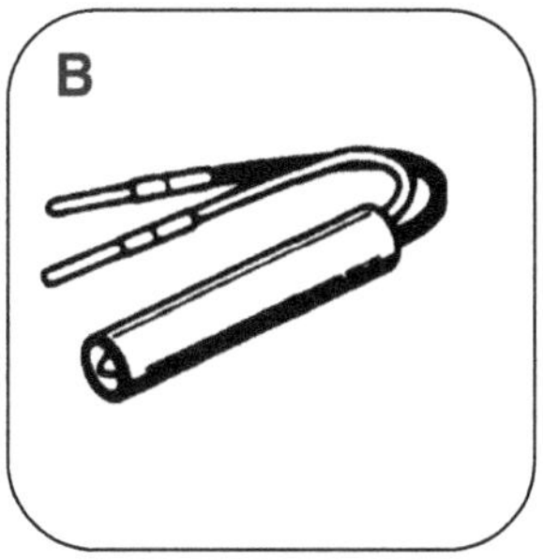

Special tool

A Fluke 87 . . . . Automotive digital multimeter
B 1115 . . . . LED tester with thin spade probes
(source: Baum Tools Unlimited)

FUSES, RELAYS AND CONTROL MODULES

The fuses, relays and control modules are located in various locations throughout the vehicle. Refer to the component location photos and tables in this repair group for electrical component locations.

Fuse positions

> **CAUTION —**
> - *Replace fuses with those of the same rating. Installing a fuse with higher rating can lead to circuit failure and may also start a fire.*
> - *Relay/fuse positions vary by model and equipment.*

Fuses no. 1-21 and are mounted in the front fuse/relay panel located in the luggage compartment. There are additional fuses in the engine compartment fuse/relay panel. The footwell blower fans behind the left and right kick panels are fused separately and the fuse holders are installed behind the kick panels as well.

Comprehensive lists of fuse positions for all years and models are given in **Table c** through **Table g**.

Relays and control modules

The luggage compartment and the engine compartment fuse/relay panel are illustrated in Fig. 971-1, Fig. 971-2 and Fig. 971-3.

Relay and module positions are included in the comprehensive list of electrical components in **Table b**. The last column of the table refers to photos or line art in this section which illustrates the component.

Front fuse and relay panel

The front fuse and relay panel is mounted in the left side of the luggage compartment.

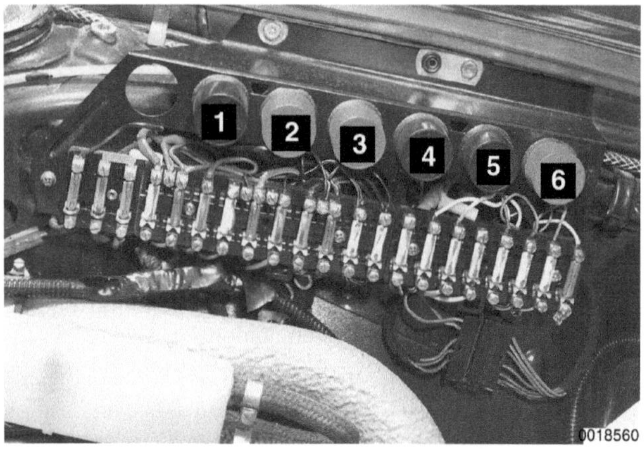

971-1 Front fuse and relay panel 1984-1986

Relay positions

1. Not used
2. Power windows
3. Cruise control
4. Horn
5. Fog lights
6. A/C condenser blower

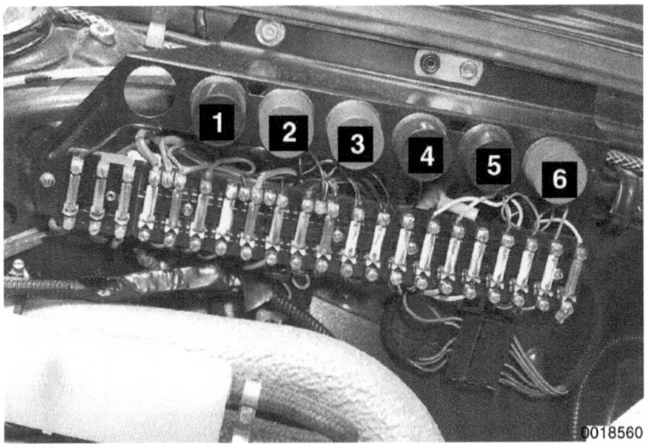

971-2 **Front fuse and relay panel 1987-1989**

Relay positions

1. Footwell blower motors
2. Oil cooler fan
3. Cruise control
4. Horn
5. Fog lights
6. A/C condenser blower

Engine compartment fuse and relay panel

The rear fuse and relay panel is located to the left of the engine.

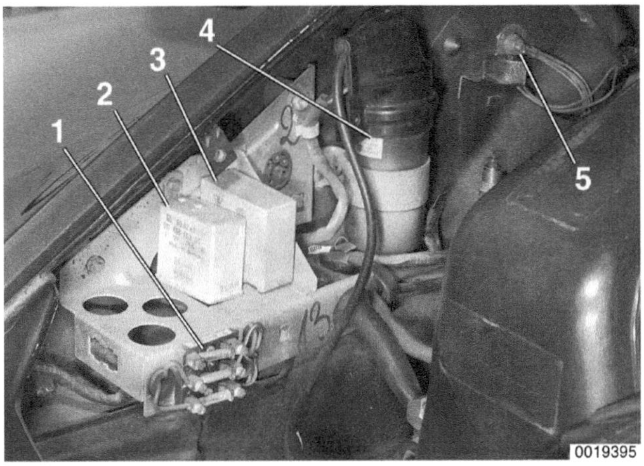

971-3 **Engine compartment fuse and relay panel**

1. Fuses
2. Heater blower relay
3. Rear window defogger relay
4. Ignition coil
5. Ground point G402

GROUND AND COMPONENT LOCATIONS

Ground locations

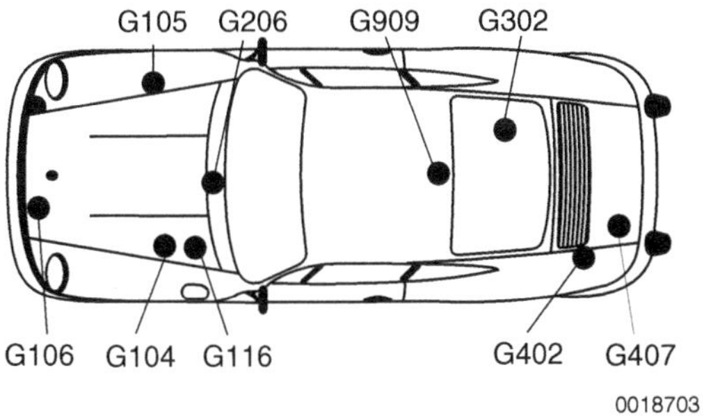

The grounds used in Porsche 911 Carrera models from 1985 to 1989 are shown in **Table a**.

Table a. Ground Locations (Quick Reference)

Ground	Location	Reference figure
G104	Luggage compartment, left (on inner fender panel at rear of fuse box)	971-6
G105	Electric antenna (right front fender inner panel)	
G106	Battery ground strap, luggage compartment (left front beside battery)	
G116	Trunk floor plate, left (below fuel filler tube)	971-7
G206	Passenger compartment (behind heater/defroster controls)	
G302	Ground strap, engine/transmission/body (in front right transmission mount)	
G402	Engine compartment, left (behind fuel filter)	971-3
G407	Engine compartment, left (on #1 intake runner)	971-18
G909	Roof, center, interior light and raised stop light ground (convertible only)	

Table b lists electrical components and their locations. Numbers in the **Reference figure** column refer to photos or illustrations in this group.

Table b. Component Locations

Component	Model Year	Location	Reference figure
A/C blower resistor	1984 - 1989	Under right front foot rest	971-15
A/C blower switch	1984 - 1989	Center console	
A/C clutch	1984 - 1989	Right side of engine mounted on A/C compressor	
A/C condenser blower	1984 - 1989	Luggage compartment, left front	971-4
A/C condenser blower relay	1984 - 1989	Luggage compartment fuse/relay panel, relay 6	971-1, 971-2
A/C evaporator blower	1984 - 1989	Luggage compartment, center, under access panel	971-5
A/C evaporator blower relay	1984 - 1989	Luggage compartment, center, under access panel	971-5
A/C high pressure switch	1984 - 1989	Not used	
A/C low pressure switch	1984 - 1998	Not used	
A/C temperature switch	1984 - 1989	Center console	
Additional blower left, right		See Footwell blower, left, right	
Additional blower motor resistor		See Footwell blower resistor	
Additional blower relay		See Footwell blower relay	
Air conditioner		See A/C	
Air flow meter		See Volume air flow sensor	
Alarm horn	1985 -1989	Underneath car, behind fuel pump	
Alarm system relay	1985-1989	Luggage compartment, left rear	
Alarm system control module	1985 - 1989	Luggage compartment, under fresh air blower	971-9
Alarm system control switch	1985 -1987	Driver door, above latch	
Alarm system control switch	1988 - 1989	Driver or passenger door handle	
Alternator	1984 - 1989	Rear of engine behind cooling fan	
Altitude sensor	1984 - 1989	Under driver seat	971-17
Antenna amplifier	1984 - 1989	Luggage compartment, below right defroster vent	971-12
Anti-theft control module		See Alarm system control module	
Ashtray light	1984 - 1989	Above ashtray	
Automatic heating control		See Heater blower switch	
Auxiliary fuse/relay panel	1984 - 1989	Engine compartment left side	
Back-up light switch	1984 - 1986	Next to gear selector rod at front of transmission	
Back-up light switch	1987 - 1989	Left side transmission housing	
Back-up lights	1984 - 1989	Taillights	
Battery	1984 - 1989	Left front luggage compartment	
Bimetallic switch		See A/C blower motor resistor	
Brake fluid level switch	1984 - 1989	In brake master cylinder cap	
Brake light switches	1984 - 1989	In luggage compartment mounted on brake master cylinder	971-8
Brake pad sensor	1984 - 1989	Inner brake pad, each wheel	
Buzzer contact	1984 -1989	In ignition switch	
Central locking button	1985 - 1989	Center console	
Central locking control module	1985 - 1989	Luggage compartment, near right defroster vent	971-12
Charge indicator light	1984 -1989	Tachometer	
Chime relay		See Warning relay	
Cigarette lighter	1984 - 1989	Next to glove box latch	

Table b. Component Locations

Component	Model Year	Location	Reference figure
Climate control		See A/C See Heater	
Clock	1984 - 1989	Dashboard, right side	
Clock relay		See Warning relay	
Coil		See Ignition coil	
Condenser blower motor		See A/C condenser blower	
Convertible top control module	1986 - 1989	Behind right side foot board	
Convertible top indicator light	1986 - 1989	In gauge cluster, between oil temperature and oil level gauges	
Convertible top micro switch control	1986 - 1989	Bolted to back side of passenger footrest	
Convertible top micro switch TDC	1986 - 1989	Behind right rear seat side panel, upper switch on top drive	
Convertible top motor front left	1986 - 1989	In front roof frame	
Convertible top motor front right	1986 - 1988	In front roof frame	
Convertible top motor left micro switch	1986 - 1989	In front roof frame, between guide pin and latch	
Convertible top motor left rear	1986 - 1989	Behind rear seat back rest	
Convertible top motor right micro switch	1986 - 1989	In front roof frame, between guide pin and latch	
Convertible top motor right rear	1986 - 1989	Behind rear seat back rest	
Convertible top switch	1986 - 1989	Above ignition switch	
Cruise control actuator	1984 - 1989	Engine compartment, to the left of left rear shock mount	
Cruise control module	1984 - 1989	Below radio, under dash	971-14
Cruise control switch	1984 - 1989	Right side steering column	
Cylinder head temperature sensor	1984 - 1989	Cylinder head 3	
Defogger		See Rear window defogger	
Defroster		See Windshield defroster	
Distributor	1984 - 1989	Left side of crankcase behind crankshaft pulley	
DME control module	1984 - 1989	Under driver seat	971-17
DME relay		See Main/fuel pump relay	
Door contact switch, left	1985 - 1989	Door hinge post	
Door contact switch, right	1985 - 1989	Door hinge post	
Door lock microswitch, left	1985 - 1989	On door latch	
Door lock microswitch, right	1985 - 1989	On door latch	
Door lock motor, left	1985 - 1989	Bottom rear of door	
Door lock motor, right	1985 - 1989	Bottom rear of door	
Drivers seat delaying relay		See Seat heater relay	
Emergency flasher		See Turn signal flasher	
Emergency flasher switch	1984 - 1989	Center console	
Engine compartment fan		See Heater blower motor	
Engine compartment fan relay		See Heater blower relay	
Fan temperature switch	1984 - 1989	Center of engine, flywheel end	
Flasher		See Turn signal flasher	
Footwell blower fuse, 10A (left, right)	1984 - 1989	Kickpanel (left, right)	971-16
Footwell blower motor (left, right)	1984 - 1989	Kickpanel (left, right)	971-16
Footwell blower relay	1987 - 1989	Luggage compartment fuse/relay, relay 1	971-2
Footwell blower resistor	1984 - 1989	Luggage compartment rear, below right defroster vent	971-11
Fresh air blower		See Windshield defroster blower	

Table b. Component Locations

Component	Model Year	Location	Reference figure
Front fog light relay	1984 - 1989	Luggage compartment fuse/relay panel, relay 5	971-1, 971-2
Front fog light switch	1984 - 1989	Above radio	
Front fog lights	1984 - 1989	Below front bumper	
Fuel injection		See DME	
Fuel injectors	1984 - 1989	Right and left sides of engine	
Fuel level gauge	1984 - 1989	Instrument cluster II	
Fuel level sender	1984 - 1989	Top of fuel tank in luggage compartment	
Fuel pump	1984 - 1989	Mounted to front suspension sub-frame	
Fuel pump relay		See Main/fuel pump relay	
Fuse holder 10 A (left, right)	1984 - 1989	With footwell blower, in kickpanel (left, right)	971-16
Fuse/relay panel, front	1984 - 1989	Luggage compartment, left front fender inner panel	971-1, 971-2
Fuse/relay panel, rear	1984 - 1989	Engine compartment, left	971-3
Glove compartment light	1984 - 1989	In dash, above glove compartment door	
Hazard warning switch	1984 - 1989	See Emergency flasher switch	
Headlight cleaning control module	1984 - 1989	Relay panel behind instrument gauges	971-13
Headlight cleaning pump	1984 - 1989	Luggage compartment, in front of battery	
Headlight cleaning switch	1984 - 1989	Dashboard, between gauges	
Headlight dimmer switch	1984 - 1989	Right side steering column	
Headlight switch	1984 - 1989	left side of dash next to ignition switch	
Headlight washer		See Headlight cleaning	
Heater blower motor	1984 - 1989	Engine compartment, left	971-18
Heater blower relay	1984 - 1989	Engine compartment fuse/relay panel	971-3
Heater blower switch	1984 - 1989	Between front seats	
Heater control	1984 - 1989	In dash on right side of steering column	
Heater lever contacts	1986 - 1989	Between front seats	
Heater temperature sensor	1984 - 1989	Left heat exchanger	
High beam indicator light	1984 - 1989	Tachometer gauge	
Horn contacts	1984 - 1989	Steering wheel pad	
Horn relay	1984 - 1989	Luggage compartment fuse/relay panel, relay 4	971-1, 971-2
Horns, high and low	1984 - 1989	Behind right headlight	
Idle control switch	1984 - 1989	Left side throttle body	
Idle control valve	1984 - 1989	Between intake runners	
Idle stabilizer		See Idle control valve	
Ignition coil	1984 - 1989	Left side engine compartment mounted to inner panel	971-3
Ignition switch	1984 - 1989	Dashboard, left side of steering column	
Inside temperature sensor blower	1984 - 1989	Dashboard, right of steering column	
Instrument lamps	1984 - 1989	Dashboard gauges	
Intensive cleaning fluid pump	1988 - 1989	Luggage compartment, left	
Interior light, left, right	1984 - 1989	Above doors	
Intermittent wiper control module		See Windshield wiper intermittent relay	
Intermittent wiper switch		See Windshield wiper intermittent switch	
Light switch	1984 - 1989	Dashboard, next to ignition switch	
Luggage compartment light	1984 - 1989	Luggage compartment hood, right rear	
Main/fuel pump relay	1984 - 1989	Under driver seat	

Table b. Component Locations

Component	Model Year	Location	Reference figure
Mirror control	1984 - 1989	Interior door panel below vent window	
Mirror heating switch		See Rear defogger switch	
Oil cooler fan	1987 - 1989	Behind right headlight	
Oil cooler fan relay	1987 - 1989	Luggage compartment fuse/relay panel, relay 2	971-2
Oil cooler thermoswitch	1986 - 1989	Top of oil cooler	
Oil level transmitter	1984 - 1989	In oil tank behind right rear wheel	
Oil pressure sending module	1984 - 1989	Right side of crankshaft pulley	
Oil pressure switch	1984 - 1989	Top of crankcase on flywheel end	
Oil temperature sensor	1984 - 1989	Right side crankshaft pulley	
Outside mirror heater	1985 - 1989	Outside mirror glass	
Oxygen sensor	1984 - 1989	Rear of catalytic converter	
Parking brake contacts	1984 - 1989	Hand brake bracket, rear	
Passenger seat delaying relay		See Seat heater relay	
Power door locks		See Central locking	
Power window motor	1986 - 1989	Inside door	
Power window relay	1987 - 1989	Relay panel, behind gauges	971-13
Power window relay	1984 - 1986	Luggage compartment fuse/relay panel, relay 2	971-1
Power window switch	1984 - 1989	Interior door panel	
Pulse transmitter	1984 - 1985	See Windshield wiper intermittent relay	
Radio	1984 - 1989	Dashboard, center	
Radio speaker, front left, front right	1984 - 1989	Interior door panel, left, right	
Radio speakers rear	1984 - 1989	Utility shelf, below rear window	
Rear window defogger	1984 - 1989	In rear window	
Rear window defogger and mirror heating switch	1984 - 1989	Above radio	
Rear window defogger relay	1984 - 1989	Engine compartment fuse/relay panel	971-3
Rear window wiper motor	1984 - 1989	Inside rear engine lid	
Rear window wiper switch	1984 - 1989	Dashboard, between gauges	
Reference sensor	1984 - 1989	Top left side of clutch bellhousing	
Reverse lights		See Back-up lights	
Seat adjusting motors	1985 - 1989	Seat bottom	
Seat adjusting switches	1985 - 1989	Seat bottom	
Seat belt contact	1984 - 1989	Seat belt latch	
Seat belt control module		See Clock relay	
Seat belt switch		See Seat belt contact	
Seat belt warning light	!984 - 1989	Dashboard	
Seat heater element, backrest	1985 - 1989	Seat back	
Seat heater element, cushion	1985 - 1989	Seat bottom	
Seat heater relay	1985 - 1989	Seat bottom	
Seat heater switch left	1985 - 1989	Seat bottom, left, with seat adjusting switches	
Seat heater switch right	1985 - 1989	Seat bottom, right, with seat adjusting switches	
Speed sensor	1984 - 1989	Top left side of clutch bellhousing	
Speed transmitter	1984 - 1989	See Speedometer pick-up	
Speedometer	1984 - 1989	Instrument cluster	

Table b. Component Locations

Component	Model Year	Location	Reference figure
Speedometer pick-up	1984 - 1986	Right side of transmission final drive above axle	
Speedometer pick-up	1987 - 1989	Left side transmission final drive	
Starter	1984 - 1989	Right side transmission bell housing	
Stop light switch		See Brake light switches	
Sunroof motor	1984 - 1989	Inside roof headliner between sunroof and rear window	
Sunroof switch	1984 - 1989	Above ignition switch	
Tachometer	1984 - 1989	Center gauge in instrument cluster	
TDC sensor	1984 - 1989	Top of crankcase on top of flywheel	
Temperature sensor, cylinder head	1984 - 1989	Cylinder head 3	
Tempostat		See Cruise control	
Thermo bimetal switch		See A/C blower resistor	
Throttle switch	1984 - 1989	Right side of throttle body	
Turn signal flasher	1984 - 1989	In luggage compartment left rear, in front of power window relay	971-13
Turn signal switch	1984 - 1989	Left side steering column	
Voltage regulator	1984 - 1989	Mounted to rear of alternator	
Volume air flow sensor	1984 - 1989	Top of engine mounted between intake runners	
Warning relay	1984 - 1989	Relay panel behind gas gauge	971-13
Window motor	1984 - 1989	See Power window motor	
Window switch	1984 - 1989	See Power window switch	
Windshield defroster blower	1984 - 1989	Luggage compartment, rear	
Windshield defroster switch	1984 - 1989	Beside radio	
Windshield washer nozzles, heated	1986 - 1989	Front cowl, right and left sides	
Windshield washer pump	1984 - 1989	Luggage compartment, in front of battery	971-4
Windshield washer switch	1984 - 1989	See Windshield wiper switch	
Windshield wiper diode	1984 - 1989	Luggage compartment, under fresh air blower	
Windshield wiper intermittent relay	1984 - 1989	Luggage compartment, behind fresh air blower	971-9
Windshield wiper intermittent switch	1986 - 1989	Dashboard, between speedometer and clock	
Windshield wiper motor	1984 - 1989	Behind speedometer and clock	
Windshield wiper pulse transmitter	1984 - 1989	See Windshield wiper intermittent relay	
Windshield wiper switch	1984 - 1989	Right side steering column	

971-4 Luggage compartment, left front

1. Windshield washer pump
2. A/C condenser blower

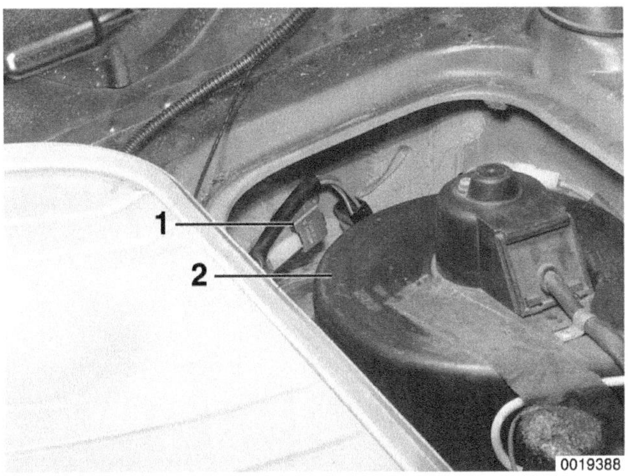

971-5 Luggage compartment, center, under hinged access panel

1. A/C evaporator blower relay
2. A/C evaporator blower

971-6 Luggage compartment, left, behind fuse/relay panel

1. Ground point G104

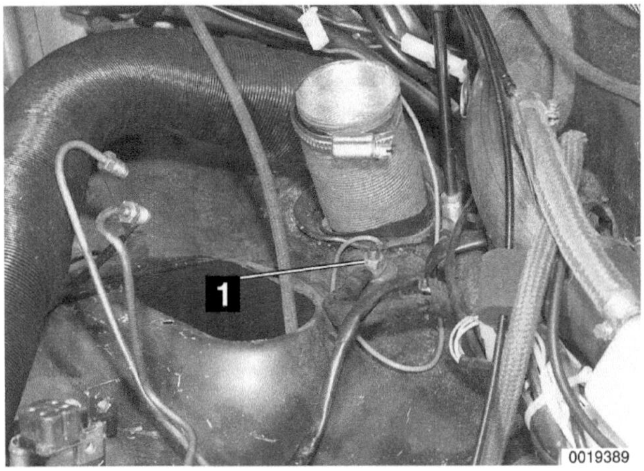

971-7 Luggage compartment, left rear (brake booster/master cylinder assembly removed)

1.	Ground point G116

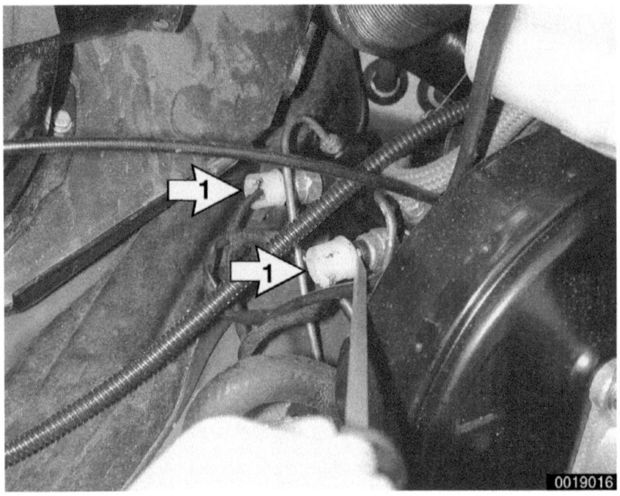

971-8 Luggage compartment, left rear, at brake master cylinder

1.	Brake light switch

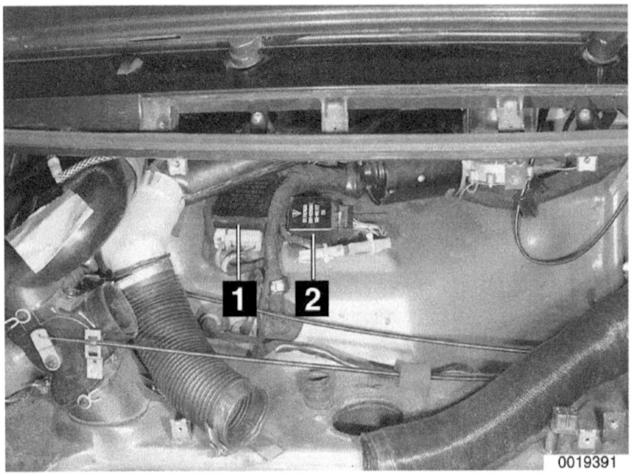

971-9 Luggage compartment, center rear (fresh air plenum removed)

1.	Alarm system control module
2.	Wiper motor intermittent relay

971-10 Luggage compartment, center rear

1. Windshield defroster blower

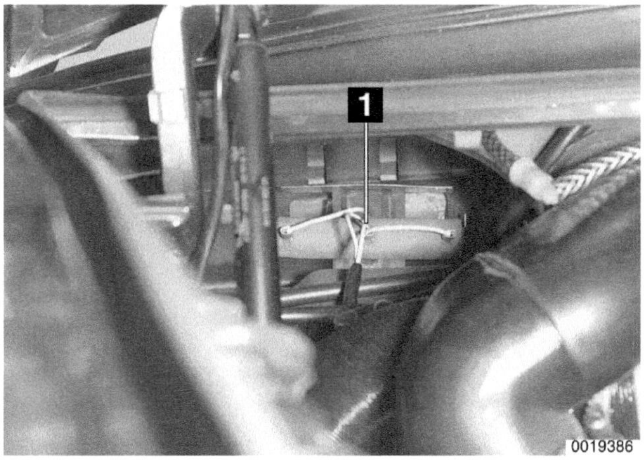

971-11 Luggage compartment, right rear

1. Footwell blower motor resistor (blowers in right and left kickpanels)

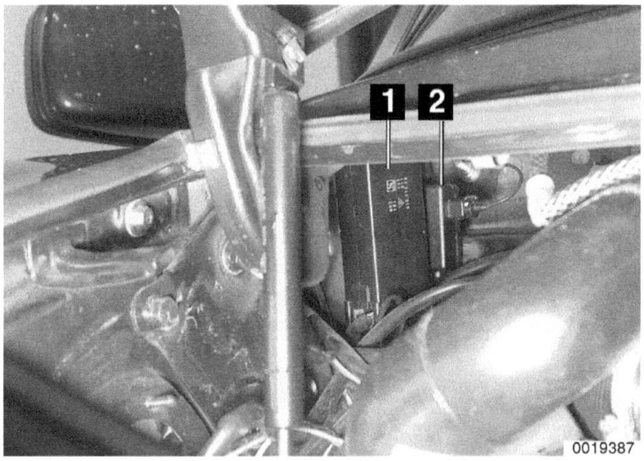

971-12 Luggage compartment, right rear

1. Central locking module
2. Antenna amplifier

971-13 Dashboard, left (left side instruments removed)

1. Power window relay
2. Warning chime relay
3. Windshield washer relay
4. Turn signal/emergency flasher relay (not shown)

971-14 Dashboard, center

1. Cruise control module

NOTE —

See also page 960-13 for alternate location of cruise control module on early models.

971-15 Front bulkhead, ahead of passenger seat

1. A/C blower resistor (bi-metallic switch)

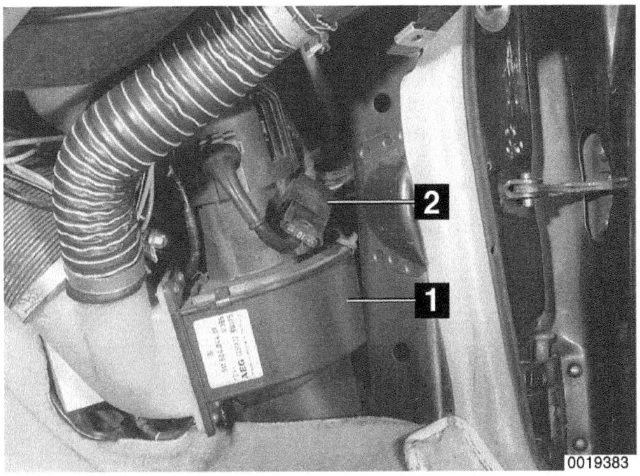

971-16 Behind kickpanel, right (left side is similar)

1. 10 A fuse holder for footwell blower
2. Footwell blower

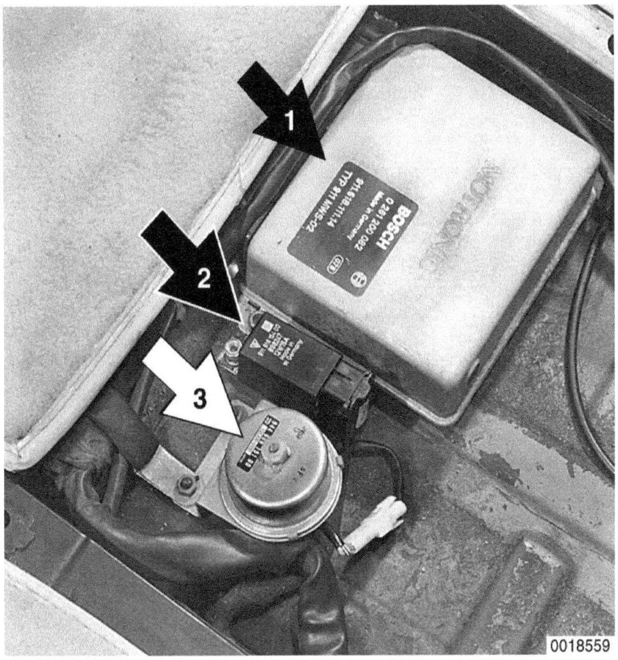

971-17 Under driver seat

1. DME control module
2. Main/fuel pump relay
3. Altitude sensor

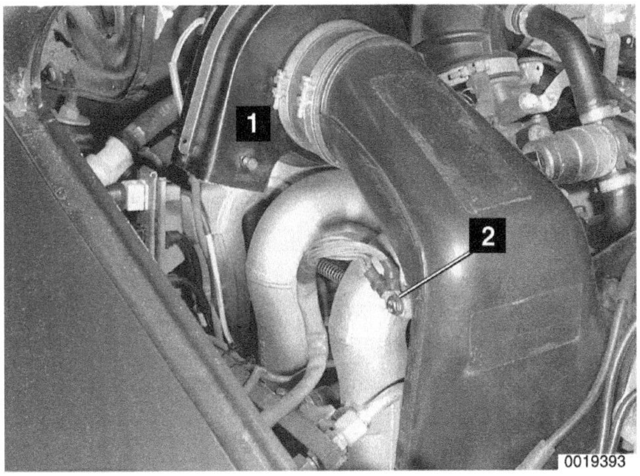

971-18 Left intake manifold, runner no.1

1. Heater blower motor
2. Ground point (G407) for DME components

Luggage compartment fuse/relay panel

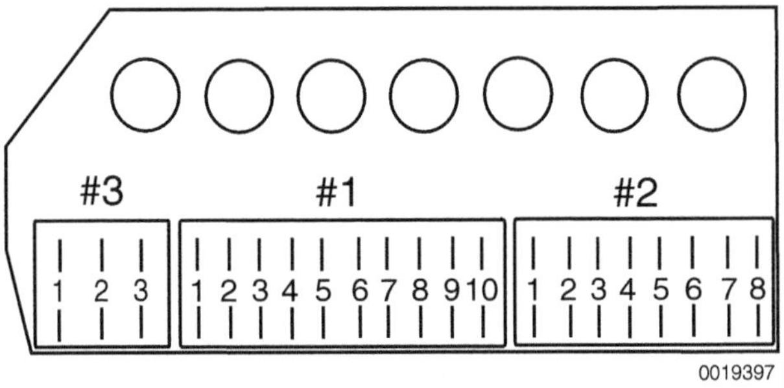

FUSE POSITION TABLES

Table c through **Table g** list fuse positions in the front fuse/relay panel located in the left side of the luggage compartment and the rear fuse/relay panel located in the engine compartment.

Table c. 1984 Fuse Positions

Fuse	Rating	Protected circuit
Luggage compartment fuse/relay panel #3		
1	25	Power windows Sunroof
2	25	Air conditioning system Seat adjustment
3	25	Headlight washer
Luggage compartment fuse/relay panel #1		
1	5	Clock Glove compartment light Interior light Luggage compartment light
2	16	Windshield defogger Emergency flasher
3	25	Fuel pump
4	8	Brake lights Cruise control
5	16	Outside mirror Heating/A/C control module
6	25	Cigarette lighter Windshield washer Rear window wiper
7	25	Fresh air blower Rear defogger switch light Rear window defogger relay
8	16	Back-up lights Emergency flasher relay Rear turn signal lights
9	5	Turn signal lights front (left)
10	5	Turn signal lights front (right)
Luggage compartment fuse/relay panel #2		
1	8	High beam (left) High beam indicator
2	8	High beam (right)
3	8	Low beam (left) Fog light switch indicator
4	8	Low beam (right)
5	5	Parking lights (left) Engine compartment light
6	5	Parking lights (right)
7	5	License plate lights
8	16	Fog light relay
Engine compartment fuse/relay panel		
1	16	Heater blower relay
2	25	Heater blower
3	25	Defogger Rear window wiper

> ***CAUTION —***
> *Fuse designations and locations are subject to change and may vary from car to car, depending on options.*

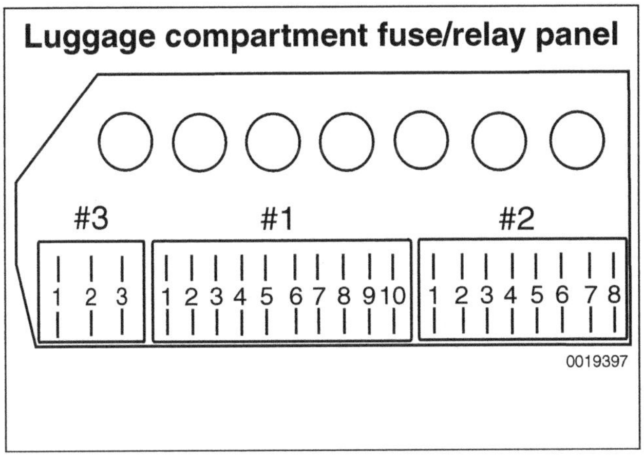

Table d. 1985 Fuse Positions

Fuse	Rating	Protected circuit
Luggage compartment fuse/relay panel #3		
1	25	Power windows Sunroof
2	25	Air conditioning system Seat Adjustment
3	25	Headlight washer
Luggage compartment fuse/relay panel #1		
1	5	Clock Glove compartment light Interior light Luggage compartment light
2	16	Windshield defogger Emergency flasher
3	25	Fuel pump
4	8	Brake lights Cruise control
5	16	Outside mirrors Heating/A/C control module
6	25	Cigarette lighter Fresh air blower Rear window defogger
7	25	Windshield washer Rear window wiper
8	16	Back-up light Emergency flasher relay Rear turn signal lights
9	5	Turn signal lights front (left)
10	5	Turn signal lights front (right)
Luggage compartment fuse/relay panel #2		
1	8	High beam (left) High beam indicator
2	8	High beam (right)
3	8	Low beam (left)
4	8	Low beam (right)
5	5	Parking lights (left) Engine compartment light
6	5	Parking lights (right)
7	5	License plate lights Headlight switch illumination Fog light switch illumination
8	16	Fog lights
Engine compartment fuse/relay panel		
1	16	Heater blower relay
2	25	Heater blower
3	25	Defogger Rear window wiper

> ***CAUTION —***
> *Fuse designations and locations are subject to change and may vary from car to car, depending on options.*

Luggage compartment fuse/relay panel

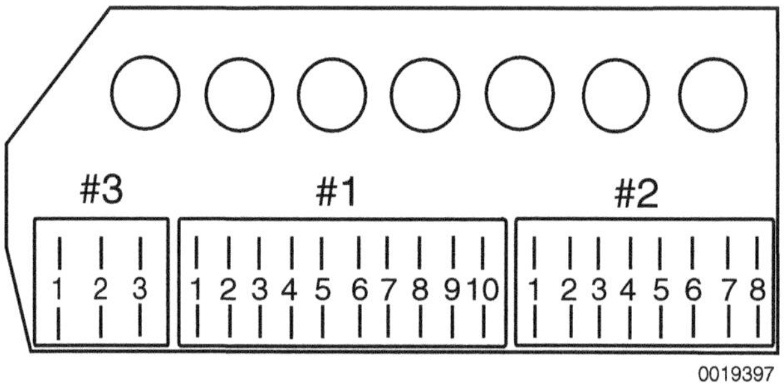

Table e. 1986 Fuse Positions

Fuse	Rating	Protected circuit
Luggage compartment fuse/relay panel #3		
1	25	Power windows Seat heating Sunroof
2	25	Air conditioning system Seat Adjustment
3	25	Convertible top Headlight washer
Luggage compartment fuse/relay panel #1		
1	5	Clock Glove compartment light Interior light Luggage compartment light Radio
2	16	Central locking system Emergency flasher
3	25	Fuel pump
4	8	Brake lights Cruise control
5	16	Outside mirrors Heating/A/C control module
6	25	Cigarette lighter Fresh air blower Rear window defogger
7	25	Heated washer nozzles Rear window wiper Windshield wiper switch Wiper delay relay
8	16	Back-up lights Emergency flasher relay Rear turn signal lights
9	5	Turn signal lights front (left)
10	5	Turn signal lights front (right)
Luggage compartment fuse/relay panel #2		
1	8	High beam (left) High beam indicator
2	8	High beam (right)
3	8	Low beam (left)
4	8	Low beam (right)
5	5	Parking lights (left) Engine compartment light
6	5	Parking lights (right)
7	5	License plate lights Headlight switch illumination Fog light switch illumination Rear fog light
8	16	Fog lights
Engine compartment fuse/relay panel		
1	16	Heater blower relay
2	25	Heater blower
3	25	Defogger Rear window wiper
Additional fuses		
	2.5	License plate lights (in luggage compt.)
	2.5	Convertible top control (in luggage compt.)

CAUTION —

Fuse designations and locations are subject to change and may vary from car to car, depending on options.

Luggage compartment fuse/relay panel

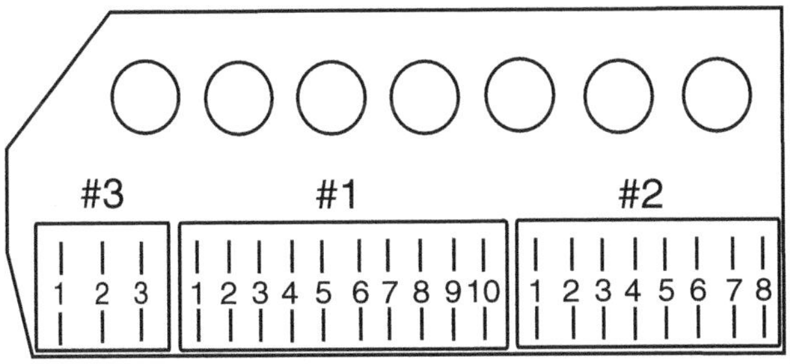

0019397

Table f. 1987 Fuse Positions

Fuse	Rating	Protected circuit
Luggage compartment fuse/relay panel #3		
1	25	Power windows Seat heating Sunroof
2	25	Air conditioning system Seat Adjustment
3	25	Convertible top Headlight washer
Luggage compartment fuse/relay panel #1		
1	5	Clock Glove compartment light Interior light Luggage compartment light Radio
2	16	Central locking system Emergency flasher
3	25	Fuel pump
4	8	Brake lights Cruise control
5	16	Outside mirrors Heating/A/C control module
6	25	Cigarette lighter Fresh air blower Rear window defogger
7	25	Heated washer nozzles Rear window wiper Windshield wiper switch Wiper delay relay
8	16	Back-up lights Emergency flasher relay Rear turn signal lights
9	5	Turn signal lights front (left)
10	5	Turn signal lights front (right)
Luggage compartment fuse/relay panel #2		
1	8	High beam (left) High beam indicator
2	8	High beam (right)
3	8	Low beam (left)
4	8	Low beam (right)
5	5	Parking lights (left) Engine compartment light
6	5	Parking lights (right)
7	5	License plate lights Headlight switch illumination Fog light switch illumination
8	16	Fog lights Rear fog light
Engine compartment fuse/relay panel		
1	16	Heater blower relay
2	25	Heater blower
3	25	Defogger Rear window wiper
Additional fuses		
	16	Radio (in luggage compt.)
	2.5	Convertible top control (in luggage compt.)

CAUTION —
Fuse designations and locations are subject to change and may vary from car to car, depending on options.

Luggage compartment fuse/relay panel

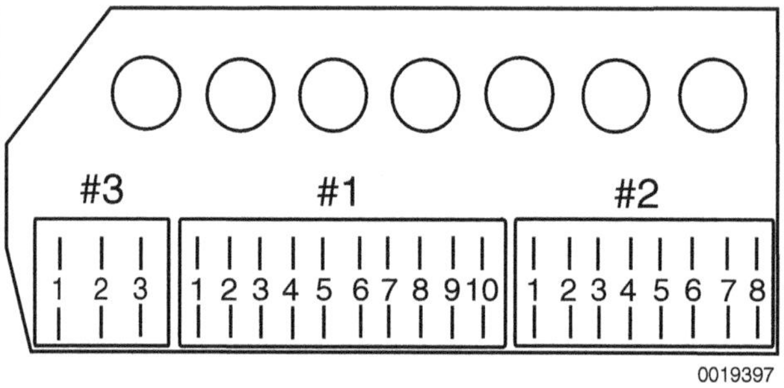

Table g. 1988-89 Fuse Positions

Fuse	Rating	Protected circuit
Luggage compartment fuse/relay panel #3		
1	25	Power windows Seat heating Sunroof
2	25	Additional blower Air conditioning system Seat Adjustment
3	25	Convertible top Headlight washer
Luggage compartment fuse/relay panel #1		
1	5	Clock Glove compartment light, Interior light Luggage compartment light Radio
2	16	Central locking system Emergency flasher
3	25	Fuel pump
4	8	Brake lights Cruise control
5	16	Outside mirrors Heating/A/C control module
6	25	Cigar lighter Fresh air blower Rear window defogger
7	25	Heated washer nozzles Rear window wiper switch Rear window washer Windshield wiper
8	16	Back-up lights Emergency flasher relay Instruments Rear turn signal lights
9	5	Turn signal lights front (left)
10	5	Turn signal lights front (right)
Luggage compartment fuse/relay panel #2		
1	8	High beam (left) High beam indicator
2	8	High beam (right)
3	8	Low beam (left)
4	8	Low beam (right)
5	5	Parking lights (left) Engine compartment light
6	5	Parking lights (right)
7	5	License plate lights
8	16	Fog lights Rear fog light
Engine compartment fuse/relay panel		
1	16	Heater blower relay
2	25	Heater blower
3	25	Defogger, Rear window wiper
Additional fuses		
	16	Radio (in luggage compt.)
	2.5	Convertible top control (in luggage compt.)

CAUTION —

Fuse designations and locations are subject to change and may vary from car to car, depending on options.

> **WARNING**
>
> *Your common sense, good judgement, and general alertness are crucial to safe and successful service work. Before attempting any work on your Porsche, read the warnings and cautions on page vii and the copyright page at the front of the manual. Review these warnings and cautions each time you prepare to work on your Porsche. Please also read any warnings and cautions that accompany the procedures in the manual.*

A

B

WARNING

Your common sense, good judgement, and general alertness are crucial to safe and successful service work. Before attempting any work on your Porsche, read the warnings and cautions on page vii and the copyright page at the front of the manual. Review these warnings and cautions each time you prepare to work on your Porsche. Please also read any warnings and cautions that accompany the procedures in the manual.

> ***WARNING***
>
> *Your common sense, good judgement, and general alertness are crucial to safe and successful service work. Before attempting any work on your Porsche, read the warnings and cautions on page vii and the copyright page at the front of the manual. Review these warnings and cautions each time you prepare to work on your Porsche. Please also read any warnings and cautions that accompany the procedures in the manual.*

F

WARNING

Your common sense, good judgement, and general alertness are crucial to safe and successful service work. Before attempting any work on your Porsche, read the warnings and cautions on page vii and the copyright page at the front of the manual. Review these warnings and cautions each time you prepare to work on your Porsche. Please also read any warnings and cautions that accompany the procedures in the manual.

I

J

K

L

> ***WARNING***
>
> *Your common sense, good judgement, and general alertness are crucial to safe and successful service work. Before attempting any work on your Porsche, read the warnings and cautions on page vii and the copyright page at the front of the manual. Review these warnings and cautions each time you prepare to work on your Porsche. Please also read any warnings and cautions that accompany the procedures in the manual.*

WARNING

Your common sense, good judgement, and general alertness are crucial to safe and successful service work. Before attempting any work on your Porsche, read the warnings and cautions on page vii and the copyright page at the front of the manual. Review these warnings and cautions each time you prepare to work on your Porsche. Please also read any warnings and cautions that accompany the procedures in the manual.

W

QUICK REFERENCE DATA

WARNING —

Your common sense, good judgement, and general alertness are crucial to safe and successful service work. Before attempting any work on your Porsche, read the warnings and cautions on page vii and the copyright page at the front of the manual. Review these warnings and cautions each time you prepare to work on your Porsche. Please also read any warnings and cautions that accompany the procedures in the manual.

Engine

Engine codes
- 1984-86 .... 930/21 (3.2L)
- 1987-89 .... 930/25 (3.2L)

Cylinder compression pressures (new engine)
- All models .... 8.9 bar (130 psi)

Engine oil capacity (with filter change)
- All SC models .... 10.0 liters (10.5 qt.)

Tightening torques
- Oil drain plug to crankcase .... 70 Nm (41 ft-lb)
- Oil drain plug to oil tank.......... 42 Nm (31 ft-lb)

Fuel Injection

Fuel injection system
- 930/21 - 930/25 .... DME (Digital Motor Electrics)
- Fuel requirements
- All models .... unleaded gasoline 92 AKI (96 RON) (min)
- Fuel tank capacity (approx.) .... 22.5 gal. (85 liters)

Ignition System

Spark plugs
- Type.......... WR7DC (copper) or WR7DP (platinum)
- Electrode gap .... 0.7-0.8 mm (0.028-0.032 in.)
- Firing order 1-6-2-4-3-5

Tightening torque
- Spark plug to cylinder head.......... 30 Nm (22 ft-lb)

Manual Transmission

1984-1986 with 915 transmission
- Oil type .... 75W/90(API-GL5)

Oil capacity
- Without oil cooler.......... 3.0 liters (3.2 qt.)
- With oil cooler .... 3.1 liters (3.3 qt.)

Tightening torque
- Drain plug to transmission.......... 24 Nm (18 ft-lb)

Manual Transmission

1987-1989 with G50 transmission
- Oil type .... 75W/90(API-GL5)

Oil capacity
- Without oil cooler .... 3.0 liters (3.2 qt.)

Tightening torque
- Drain plug to transmission .... 24 Nm (18 ft-lb)

Suspension

Standard wheel and tire specifications
- Front 6 J x 15 .... 185/70 VR 15
- Rear 7 J x 15 .... 215/60 VR 15

Optional wheel specifications
- Front 6 J x 16 .... 205/55 VR 16
- Rear 7 J x 16 .... 225/50 VR 16

Tightening torque
- Wheel nut to hub.......... 130 Nm (94 ft-lb)

Brake System

Brake fluid type.......... SAE DOT 4

Brake pads
- wear limit, front and rear.......... 2 mm (0.08 in.)

Brake rotors
- Wear limit (min. permissible thickness)
  - Front .... 22.0 mm (0.866 in.)
  - Rear .... 22.0 mm (0.866 in.)
- Minimum thickness after machining
  - Front .... 22.6 mm (0.890 in.)
  - Rear .... 22.6 mm (0.890 in.)

Air Conditioning System

Refrigerant
- Type R-12
- Quantity .... 1350 g (47 oz.)
- Oil quantity (total) .... 120 cc (4.0 oz.)

Wipers and Washer

Windshield wipers
- Wiper blade replacements .... 13 in.

Windshield washer
- Reservoir capacity .... 8.5 liters (5.8 qt.)

B® BENTLEY PUBLISHERS® | Automotive Reference™

Bentley Publishers, a division of Robert Bentley, Inc.
1734 Massachusetts Avenue
Cambridge, MA 02138 USA
800-423-4595 / 617-547-4170

Information that makes the difference®

BentleyPublishers®.com